Handbook of Neuroimaging Data Analysis

Chapman & Hall/CRC
Handbooks of Modern Statistical Methods

Series Editor

Garrett Fitzmaurice

Department of Biostatistics
Harvard School of Public Health
Boston, MA, U.S.A.

Aims and Scope

The objective of the series is to provide high-quality volumes covering the state-of-the-art in the theory and applications of statistical methodology. The books in the series are thoroughly edited and present comprehensive, coherent, and unified summaries of specific methodological topics from statistics. The chapters are written by the leading researchers in the field, and present a good balance of theory and application through a synthesis of the key methodological developments and examples and case studies using real data.

The scope of the series is wide, covering topics of statistical methodology that are well developed and find application in a range of scientific disciplines. The volumes are primarily of interest to researchers and graduate students from statistics and biostatistics, but also appeal to scientists from fields where the methodology is applied to real problems, including medical research, epidemiology and public health, engineering, biological science, environmental science, and the social sciences.

Published Titles

Handbook of Mixed Membership Models and Their Applications
Edited by Edoardo M. Airoldi, David M. Blei,
Elena A. Erosheva, and Stephen E. Fienberg

Handbook of Markov Chain Monte Carlo
Edited by Steve Brooks, Andrew Gelman,
Galin L. Jones, and Xiao-Li Meng

Handbook of Big Data
Edited by Peter Bühlmann, Petros Drineas,
Michael Kane, and Mark van der Laan

Handbook of Discrete-Valued Time Series
Edited by Richard A. Davis, Scott H. Holan,
Robert Lund, and Nalini Ravishanker

Published Titles Continued

Handbook of Design and Analysis of Experiments
Edited by Angela Dean, Max Morris,
John Stufken, and Derek Bingham

Longitudinal Data Analysis
Edited by Garrett Fitzmaurice, Marie Davidian,
Geert Verbeke, and Geert Molenberghs

Handbook of Spatial Statistics
Edited by Alan E. Gelfand, Peter J. Diggle,
Montserrat Fuentes, and Peter Guttorp

Handbook of Cluster Analysis
Edited by Christian Hennig, Marina Meila,
Fionn Murtagh, and Roberto Rocci

Handbook of Survival Analysis
Edited by John P. Klein, Hans C. van Houwelingen,
Joseph G. Ibrahim, and Thomas H. Scheike

Handbook of Spatial Epidemiology
Edited by Andrew B. Lawson, Sudipto Banerjee,
Robert P. Haining, and María Dolores Ugarte

Handbook of Missing Data Methodology
Edited by Geert Molenberghs, Garrett Fitzmaurice,
Michael G. Kenward, Anastasios Tsiatis, and Geert Verbeke

Handbook of Neuroimaging Data Analysis
Edited by Hernando Ombao, Martin Lindquist,
Wesley Thompson, and John Aston

Chapman & Hall/CRC
Handbooks of Modern Statistical Methods

Handbook of Neuroimaging Data Analysis

Edited by

Hernando Ombao
University of California
Irvine, USA

Martin Lindquist
Johns Hopkins University
Baltimore, Maryland, USA

Wesley Thompson
Institute of Biological Psychiatry, Denmark
University of California, San Diego, La Jolla, USA

John Aston
University of Cambridge, UK

CRC Press
Taylor & Francis Group
Boca Raton London New York

CRC Press is an imprint of the
Taylor & Francis Group, an **informa** business
A CHAPMAN & HALL BOOK

CRC Press
Taylor & Francis Group
6000 Broken Sound Parkway NW, Suite 300
Boca Raton, FL 33487-2742

First issued in paperback 2019

© 2017 by Taylor & Francis Group, LLC
CRC Press is an imprint of Taylor & Francis Group, an Informa business

No claim to original U.S. Government works

ISBN-13: 978-1-4822-2097-1 (hbk)
ISBN-13: 978-0-367-33069-9 (pbk)

Visit the Taylor & Francis Web site at
http://www.taylorandfrancis.com

and the CRC Press Web site at
http://www.crcpress.com

Contents

List of Figures — xix

List of Tables — xxxiii

Preface — xxxv

Contributors — xxxvii

I Overview — 1

1 Overview of the Handbook — 3

1.1 Introduction — 3
1.2 A Brief History of Neuroimaging — 4
1.3 Modalities — 4
1.4 Statistical Methods — 7
    1.4.1 Preprocessing — 8
    1.4.2 Methods in Structural Neuroimaging — 8
    1.4.3 Localizing Areas of Activation — 9
    1.4.4 Brain Connectivity — 9
    1.4.5 Analysis of Electroencephalograms — 10
    1.4.6 Multi-Modal Analysis — 10
1.5 Additional Topics — 11
    1.5.1 Meta-Analysis — 11
    1.5.2 Experimental Design in fMRI — 11
    1.5.3 Experimental Design in EEG — 11
    1.5.4 Imaging Genetics — 12
1.6 Conclusions — 12
Bibliography — 13

II Imaging Modalities — 15

2 Positron Emission Tomography: Some Analysis Methods — 17
John Aston

2.1 Introduction — 17
2.2 Background — 19
2.3 Tracer Kinetic Modelling: Compartmental Approaches — 20
    2.3.1 Plasma Input Functions Models — 20
    2.3.2 Reference Tissue Models — 23
2.4 Estimation and Statistical Methods — 24
    2.4.1 Non-Linear Least Squares — 24
    2.4.2 Basis Function Methods — 25
    2.4.3 Model Selection — 26

2.5 Other Modelling Approaches 27
    2.5.1 Graphical Methods 27
    2.5.2 Bayesian Approaches 28
    2.5.3 Non-Parametric Approaches 28
2.6 Further Modelling Considerations 29
Bibliography 30

3 Structural Magnetic Resonance Imaging   35

Wesley K. Thompson, Hauke Bartsch, and Martin A. Lindquist

3.1 Introduction 36
3.2 Image Acquisition 37
    3.2.1 MRI Physics 38
    3.2.2 Image Reconstruction 38
    3.2.3 MRI Sequences 38
    3.2.4 MRI Artifacts 39
3.3 Multi-Subject MRI 40
    3.3.1 Registration 40
        3.3.1.1 Volumetric Registration 41
        3.3.1.2 Surface-Based Registration 41
    3.3.2 Segmentation 43
        3.3.2.1 Foreground from Background Segmentation 43
        3.3.2.2 Brain Tissue Segmentation 44
    3.3.3 Templates and Atlases 45
    3.3.4 Morphometry 46
        3.3.4.1 Subcortical Volumes 46
        3.3.4.2 Voxel-Based Morphometry 47
        3.3.4.3 Deformation- and Tensor-Based Morphometry 47
        3.3.4.4 Surface-Based Measures 48
        3.3.4.5 Other Morphometric Measures 49
    3.3.5 Statistical Analyses 50
        3.3.5.1 Statistical Parametric Maps 51
3.4 Miscellaneous Topics 51
    3.4.1 Structural Integrity and Tumor Detection 51
    3.4.2 Anatomical References for Functional Imaging 52
    3.4.3 Multi-Center Studies 52
    3.4.4 Imaging Genetics 53
3.5 Glossary of MRI Terms 53
Bibliography 54

4 Diffusion Magnetic Resonance Imaging (dMRI)   65

Jian Cheng and Hongtu Zhu

4.1 Introduction to Diffusion MRI 66
    4.1.1 Diffusion Weighted Imaging (DWI) 66
        4.1.1.1 Diffusion Gradient Sequence 66
        4.1.1.2 Free Diffusion 67
        4.1.1.3 Restricted Diffusion 67
    4.1.2 Diffusion Tensor Imaging (DTI) 69
        4.1.2.1 Scalar Indices and Eigenvectors of Diffusion Tensor 69
4.2 High Angular Resolution Diffusion Imaging (HARDI) 70
    4.2.1 Generalization of Diffusion Tensor Imaging 70
        4.2.1.1 Mixture of Tensor Model 70

4.2.1.2 Generalized DTI (GDTI) 71
4.2.1.3 High-Order Tensor Model, ADC-Based Model 72
4.2.2 Diffusion Spectrum Imaging (DSI) 73
4.2.3 Hybrid Diffusion Imaging (HYDI) 74
4.2.4 Q-Ball Imaging (QBI) 75
4.2.4.1 Original Q-Ball Imaging 75
4.2.4.2 Exact Q-Ball Imaging 76
4.2.5 Diffusion Orientation Transform (DOT) 77
4.2.6 Spherical Deconvolution (SD) 77
4.2.7 Diffusion Propagator Imaging (DPI) 77
4.2.8 Simple Harmonic Oscillator Reconstruction and Estimation (SHORE) 78
4.2.9 Spherical Polar Fourier Imaging (SPFI) 78
4.3 Reconstruction 79
4.3.1 Noise Components and Voxelwise Estimation Methods 79
4.3.2 Spatial-Adaptive Estimation Methods 80
4.4 Tractography Algorithms 83
4.5 Uncertainty in Estimated Diffusion Quantities 86
4.6 Sampling Mechanisms 87
4.7 Registration 89
4.8 Group Analysis 90
4.9 Public Resources 93
4.9.1 Datasets 93
4.9.2 Software 94
4.10 Glossary 95
Bibliography 96

5 A Tutorial for Multisequence Clinical Structural Brain MRI 109

*Ciprian Crainiceanu, Elizabeth M. Sweeney, Ani Eloyan, and
Russell T. Shinohara*

5.1 Introduction 110
5.1.1 What Are These Images? 112
5.1.2 How Can We Handle sMRI? 113
5.1.3 What Are Some Major Pitfalls When Starting Working on sMRI? . 114
5.2 Data Structure and Intuitive Description of Associated Problems 114
5.3 Acquisition and Reconstruction 115
5.4 Preprocessing 116
5.4.1 Inhomogeneity Correction 117
5.4.1.1 Concepts 118
5.4.1.2 Practical Approaches, Software, and Application to Data . 118
5.4.2 Skull Stripping 119
5.4.2.1 Concepts 119
5.4.2.2 Practical Approaches, Software, and Application to Data . 119
5.4.3 Interpolation 119
5.4.3.1 Concepts 120
5.4.3.2 Practical Approaches, Software, and Application to Data . 121
5.4.4 Spatial Registration 121
5.4.4.1 Concepts 121
5.4.4.2 Practical Approaches, Software, and Application to Data . 123
5.4.5 Intensity Normalization 126
5.4.5.1 Concepts 127
5.4.5.2 Practical Approaches, Software, and Application to Data . 127

  5.5   Analysis 129
        5.5.1   Lesion Segmentation 129
        5.5.2   Lesion Mapping 130
        5.5.3   Longitudinal and Cross Sectional Intensity Analysis 131
  5.6   Conclusions 133
  Bibliography 133

6 Principles of Functional Magnetic Resonance Imaging   139
  Martin A. Lindquist and Tor D. Wager
  6.1   Introduction 139
  6.2   The Basics of fMRI Data 141
        6.2.1   Principles of Magnetic Resonance Signal Generation 141
                6.2.1.1   The MRI Scanner 141
                6.2.1.2   Basic MR Physics 142
                6.2.1.3   Image Contrast 143
        6.2.2   Image Formation 144
        6.2.3   From MRI to fMRI 146
  6.3   BOLD fMRI 148
        6.3.1   Understanding BOLD fMRI 148
        6.3.2   Spatial Limitations 150
        6.3.3   Temporal Limitations 151
        6.3.4   Acquisition Artifacts 152
  6.4   Modeling Signal and Noise in fMRI 153
        6.4.1   BOLD Signal 153
        6.4.2   Noise and Nuisance Signal 155
  6.5   Experimental Design 157
  6.6   Preprocessing 158
  6.7   Data Analysis 160
        6.7.1   Localization 160
        6.7.2   Connectivity 162
        6.7.3   Prediction 164
  6.8   Resting-State fMRI 165
  6.9   Data Format, Databases, and Software 166
  6.10 Future Developments 168
  Bibliography 169

7 Electroencephalography (EEG): Neurophysics, Experimental Methods,
**   and Signal Processing   175**
  Michael D. Nunez, Paul L. Nunez, and Ramesh Srinivasan
  7.1   Introduction 175
  7.2   The Neurophysics of EEG 177
  7.3   Synchronization and EEG 180
  7.4   Recording EEG 182
  7.5   Preprocessing EEG 183
  7.6   Artifact Removal 184
  7.7   Stationary Data Analysis 187
  7.8   Nonstationary Data Analysis 194
  7.9   Summary 197
  Bibliography 197

III STATISTICAL METHODS AND MODELS 203

8 Image Reconstruction in Functional MRI 205

Daniel B. Rowe

8.1 Introduction 205
8.2 The Fourier Transform 206
   8.2.1 One-Dimensional Fourier Transform 206
   8.2.2 Two-Dimensional Fourier Transform 210
8.3 FMRI Acquisition and Reconstruction 214
   8.3.1 The Signal Equation and k-Space Coverage 214
   8.3.2 Nyquist Ghost k-Space Correction 217
8.4 Image Processing 220
   8.4.1 Reconstruction Isomorphism Representation 220
   8.4.2 Image Processing Implications 222
8.5 Additional Topics and Discussion 228
   8.5.1 Complex-Valued fMRI Activation 229
   8.5.2 Discussion 230
Acknowledgments 230
Bibliography 230

9 Statistical Analysis on Brain Surfaces 233

Moo K. Chung, Seth D. Pollak, and Jamie L. Hanson

9.1 Introduction 234
9.2 Surface Parameterization 235
   9.2.1 Local Parameterization by Quadratic Polynomial 236
   9.2.2 Surface Flattening 236
   9.2.3 Spherical Harmonic Representation 237
9.3 Surface Registration 239
   9.3.1 Affine Registration 239
   9.3.2 SPHARM Correspondence 239
   9.3.3 Diffeomorphic Registration 241
9.4 Cortical Surface Features 241
   9.4.1 Cortical Thickness 241
   9.4.2 Surface Area and Curvatures 242
   9.4.3 Gray Matter Volume 242
9.5 Surface Data Smoothing 243
   9.5.1 Diffusion Smoothing 243
   9.5.2 Iterated Kernel Smoothing 245
   9.5.3 Heat Kernel Smoothing 246
9.6 Statistical Inference on Surfaces 248
   9.6.1 General Linear Models 248
   9.6.2 Multivariate General Linear Models 249
   9.6.3 Small-n Large-p Problems 250
   9.6.4 Longitudinal Models 251
   9.6.5 Random Field Theory 252
Bibliography 254

10 Neuroimage Preprocessing   **263**

Stephen C. Strother and Nathan Churchill

10.1 Introduction 264
10.2 Principles for Studying and Optimizing Preprocessing Pipelines 267
     10.2.1 The Utility of Simulated Datasets 268
     10.2.2 Quantifying the Impact of Preprocessing Changes 268
     10.2.3 The Neuroscientific Importance of Preprocessing Choices 269
10.3 Metrics for Evaluating Neuroimaging Pipelines 270
     10.3.1 Pseudo-ROC Curves 271
     10.3.2 Cluster Overlap Metrics 271
     10.3.3 Intra-Class Correlation Coefficient 272
     10.3.4 Spatial Pattern Reproducibility Using Correlations 275
     10.3.5 Similarity Metric Ranking Approaches 275
     10.3.6 Prediction Metrics 276
     10.3.7 Combined Prediction versus Spatial Reproducibility Metrics 276
10.4 Preprocessing Pipeline Testing in the Literature 277
     10.4.1 Between-Subject, MRI Brain Registration 277
     10.4.2 Preprocessing for fMRI Resting-State Analysis 278
     10.4.3 Preprocessing for fMRI Task-Based Analysis 280
10.5 A Case Study: Optimizing fMRI Task Preprocessing 280
     10.5.1 Optimization with Prediction and Reproducibility Metrics 280
     10.5.2 An Individual Subject's (P, R) Curves 284
     10.5.3 fMRI Preprocessing Pipeline Optimization with (P, R) Curves . . . 285
         10.5.3.1 Some fMRI Datasets 285
         10.5.3.2 Selecting Optimal Preprocessing Pipeline Steps 286
     10.5.4 Fixed and Individually Optimized Preprocessing Pipelines 289
     10.5.5 Independent Tests of Pipeline Optimization Results 292
10.6 Discussion of Open Problems and Pitfalls 293
Bibliography 295

11 Linear and Nonlinear Models for fMRI Time Series Analysis   **309**

Tingting Zhang, Haipeng Shen, and Fan Li

11.1 Introduction 309
11.2 The GLM: Single-Level Analysis 310
11.3 Modeling the Hemodynamic Response Function in the Time Domain . . . 312
     11.3.1 Parametric Models 313
     11.3.2 Nonparametric Models 317
     11.3.3 Comparison of Different HRF Estimation Methods 320
11.4 Hemodynamic Response Estimation in the Frequency Domain 321
11.5 Multi-Subject Analysis 322
     11.5.1 Semi-Parametric Approaches 323
11.6 Nonlinear Models 323
     11.6.1 The Balloon Model 324
     11.6.2 Volterra Series Model 325
     11.6.3 Bi-Exponential Nonlinear Model 326
     11.6.4 Volterra Series Models for Multi-Subject Data 326
11.7 Summary and Future Directions 327
Bibliography 328

12 Functional Neuroimaging Group Studies   **335**

Bertrand Thirion

12.1 Introduction 335
12.2 Variability of Brain Shape and Function 337
12.3 Mixed-Effects and Fixed-Effects Analyses 339
12.4 Group Analysis for Functional Neuroimaging 339
    12.4.1 Problem Setting and Notations 340
    12.4.2 Estimation 341
    12.4.3 Statistical Inference 342
    12.4.4 The Random Effects t-Test 343
12.5 Taking into Account the Spatial Context in Statistical Inference 344
12.6 Type I Error Control with Permutation Testing 346
12.7 Illustration of Various Inference Strategies on an Example 347
12.8 Conclusion 350
Bibliography 350

13 Corrections for Multiplicity in Functional Neuroimaging Data   **355**

Nicole A. Lazar

13.1 Introduction 355
13.2 Control of Familywise Error Rate 356
13.3 Control of False Discovery Rate 360
13.4 Accounting for Spatial Dependence 362
13.5 Summary 364
Bibliography 365

14 Functional Connectivity Analyses for fMRI Data   **369**

Ivor Cribben and Mark Fiecas

14.1 Introduction 369
14.2 Methods and Measures for FC 371
    14.2.1 Setup 371
    14.2.2 Cross-Correlation and Partial Cross-Correlation 371
    14.2.3 Stability Selection 372
    14.2.4 Cross-Coherence and Partial Cross-Coherence 373
    14.2.5 Mutual Information 374
    14.2.6 Principal and Independent Components Analyses 374
    14.2.7 Time-Varying Connectivity 375
14.3 Simulation Study 376
    14.3.1 Results 378
14.4 Functional Connectivity Analysis of Resting-State fMRI Data 381
    14.4.1 Data Description and Preprocessing 381
    14.4.2 Overview of the Estimation Procedure 382
    14.4.3 Results 382
14.5 Future Directions and Open Problems 384
Bibliography 389

15 Multivariate Decompositions in Brain Imaging   **399**

Ani Eloyan, Vadim Zipunnikov, Juemin Yang, and Brian Caffo

15.1 Introduction 399
15.2 Principal Component Analysis and Singular Value Decomposition 400
    15.2.1 Singular Value Decomposition 401
    15.2.2 Principal Components Analysis 401

    15.2.3 PCA in Brain Imaging 402
  15.3 Structured PCA Models 403
    15.3.1 Calculation of High-Dimensional PCA 404
  15.4 Independent Component Analysis 405
    15.4.1 ICA in Brain Imaging 406
    15.4.2 Homotopic Group ICA 407
    15.4.3 Computation of High-Dimensional ICA 407
  15.5 Discussion of Other Methods 409
  15.6 Acknowledgements 411
  Bibliography 411

16 Effective Connectivity and Causal Inference in Neuroimaging   **419**
Martin A. Lindquist and Michael E. Sobel
  16.1 Introduction 419
  16.2 Effective Connectivity 420
  16.3 Models of Effective Connectivity 422
    16.3.1 Structural Equation Models 422
    16.3.2 Dynamic Causal Models 427
    16.3.3 Granger Causality 429
  16.4 Effective Connectivity and Causation 432
  16.5 Conclusions 435
  Bibliography 436

17 Network Analysis   **441**
Cedric E. Ginestet, Mark Kramer, and Eric D. Kolaczyk
  17.1 Introduction 441
  17.2 Network Construction 443
    17.2.1 Notation 443
    17.2.2 Vertex Set 444
    17.2.3 Edge Set 446
    17.2.4 Thresholding Networks 446
  17.3 Descriptive Measures of Network Topology 447
    17.3.1 Characteristic Path Length 447
    17.3.2 Clustering Coefficient 448
    17.3.3 Degree Distribution 449
  17.4 Network Models 450
    17.4.1 Erdős–Rényi Random Graphs 451
    17.4.2 Small-World Networks 452
    17.4.3 Preferential Attachment 453
    17.4.4 Exponential Random Graph Models 453
    17.4.5 Stochastic Block Models 454
  17.5 Estimation and Comparison of Networks 455
    17.5.1 Statistical Parametric Networks 456
    17.5.2 Density-Integrated Topology 457
    17.5.3 Comparison of Weighted Networks 458
  17.6 Conclusion 460
  Bibliography 460

18 Modeling Change in the Brain: Methods for Cross-Sectional and Longitudinal Data   **467**

Philip T. Reiss, Ciprian M. Crainiceanu, Wesley K. Thompson, and Lan Huo

18.1 Introduction 468
18.2 Notation and Road Map 468
18.3 Cross-Sectional and Longitudinal Designs 469
    18.3.1 Cross-Sectional Designs 470
    18.3.2 Single-Cohort Longitudinal Designs 470
    18.3.3 Multi-Cohort Longitudinal Designs 471
18.4 Region-Wise Linear Models for the Mean 472
    18.4.1 What Is v, and What Is t? 472
    18.4.2 Cross-Sectional Data 473
    18.4.3 Longitudinal Data 473
        18.4.3.1 Mixed-Effects Models 473
        18.4.3.2 Marginal Models 474
    18.4.4 Relative Efficiency for Estimating the Mean Function 474
    18.4.5 Complications Due to Misalignment 475
    18.4.6 Borrowing Information "Spatially" 477
18.5 Nonlinear Models for the Mean 477
    18.5.1 Polynomial Models 477
    18.5.2 Nonparametric and Semiparametric Models 477
    18.5.3 Analyses with Repeated Cross-Sectional Subsamples 480
18.6 Beyond Modeling the Mean 482
    18.6.1 Individual-Specific Curves 482
    18.6.2 Modeling Components of Change: Longitudinal Functional Principal Component Analysis 483
    18.6.3 Modeling the Entire Age-Specific Distribution 485
    18.6.4 Modeling Local Rates of Change 485
18.7 Discussion 486
Bibliography 487

19 Joint fMRI and DTI Models for Brain Connectivity   **495**

F. DuBois Bowman, Sean Simpson, and Daniel Drake

19.1 Brain Connectivity 495
    19.1.1 Structural Connectivity 496
    19.1.2 Functional Connectivity 496
19.2 Single Modality Methods 497
    19.2.1 Methods for Functional Connectivity 497
        19.2.1.1 Defining the Spatial Scale for Connectivity Analysis 497
        19.2.1.2 Measures of Association 498
        19.2.1.3 Modeling Approaches 499
        19.2.1.4 Partitioning Methods 500
        19.2.1.5 Network Methods 500
    19.2.2 Methods for Effective Connectivity 502
    19.2.3 Determining Structural Connectivity 502
        19.2.3.1 Diffusion Weighted Imaging and DTI 503
        19.2.3.2 Tractography 504
19.3 Multimodal Approaches 505
    19.3.1 Sequential Procedures 506
    19.3.2 Functional Connectivity with Anatomical Weighting 508

19.3.3 Modeling Joint Activation and Structural Connectivity 509
    19.3.3.1 Functional Coherence 510
    19.3.3.2 Ascendancy 511
    19.3.3.3 Likelihood Function 511
19.3.4 Joint ICA 512
19.3.5 Multimodal Prediction Methods 512
19.4 Conclusion 513
Bibliography 515

20 Statistical Analysis of Electroencephalograms   **523**
Yuxiao Wang, Lechuan Hu, and Hernando Ombao
20.1 Introduction 524
20.2 Spectral Analysis of a Single-Channel EEG 524
  20.2.1 Brief Description of the Data 525
  20.2.2 Fourier-Domain Approach 525
    20.2.2.1 The Fourier Regression Model and Variance Decomposition 525
    20.2.2.2 The Spectrum of a Single-Channel Time Series 528
    20.2.2.3 Estimating the Spectrum via Periodograms 528
    20.2.2.4 Other Periodogram-Based Estimation Methods 531
    20.2.2.5 Examples of Smoothing Periodograms 531
    20.2.2.6 Multitaper Method (MTM) 531
  20.2.3 Time-Domain Approach 532
    20.2.3.1 Moving Average (MA) Model 532
    20.2.3.2 Autoregressive (AR) Model 533
    20.2.3.3 Autoregressive Moving Average (ARMA) Model 534
    20.2.3.4 The Spectra of MA, AR and ARMA Processes 535
    20.2.3.5 Second-Order Autoregressive [AR(2)] Model 537
    20.2.3.6 Estimating the Spectrum 538
  20.2.4 Estimating the Spectrum Using Multiple EEG Traces 539
    20.2.4.1 Other Averaged Estimators 539
    20.2.4.2 Estimating Power in Specific Frequency Bands 540
    20.2.4.3 Detecting Outliers 540
  20.2.5 Confidence Intervals 540
20.3 Spectral Analysis of Multichannel EEG 541
  20.3.1 Fourier-Domain Approach 542
    20.3.1.1 The Fourier–Cramér Representation 542
    20.3.1.2 The Spectral Matrix of an EEG 542
    20.3.1.3 Non-Parametric Estimator of the Spectral Matrix 544
  20.3.2 Time-Domain Approach 544
  20.3.3 Estimating Partial Coherence 545
  20.3.4 Estimating the Spectral Matrix Using Multiple EEG Traces 548
    20.3.4.1 Estimating the Spectral Matrix in Specific Frequency Bands 549
  20.3.5 Modeling and Inference on Connectivity 549
    20.3.5.1 Granger Causality 550
    20.3.5.2 Partial Directed Coherence (PDC) 550
    20.3.5.3 Summary of Metrics for Connectivity 551
20.4 Spectral Analysis for High-Dimensional Data 551
  20.4.1 Methods for Fitting VAR Model on Multivariate Time Series 552
    20.4.1.1 Least Squares Estimation 552
    20.4.1.2 LASSO 552
    20.4.1.3 LASSLS 553

20.4.2 EEG Data Analysis via LASSLS Methods 554
    20.4.2.1 VAR Modeling on High-Dimensional Multichannel EEG . . 554
    20.4.2.2 Inference on Directed Connectivity 555
20.5 Source Localization and Estimation 555
  20.5.1 Overview of Source Models for EEG Data 555
    20.5.1.1 Dipole Source Model 556
    20.5.1.2 Independent Source Model 558
    20.5.1.3 A Generalized Model of EEG Signals 559
  20.5.2 Inverse Source Reconstruction 559
    20.5.2.1 Parametric Methods 559
    20.5.2.2 Imaging Methods 560
    20.5.2.3 Summary 562
Bibliography 562

21 Advanced Topics for Modeling Electroencephalograms   567
Hernando Ombao, Anna Louise Schröder, Carolina Euán,
Chee-Ming Ting, and Balqis Samdin
21.1 Introduction 568
21.2 Clustering of EEGs 572
  21.2.1 Proposal: The Spectral Merger Clustering Method 572
    21.2.1.1 Total Variation Distance 572
    21.2.1.2 Hierarchical Spectral Merger Algorithm 573
  21.2.2 Analysis of Epileptic Seizure EEG Data 574
21.3 Change-Point Detection 580
  21.3.1 Existing Methods and Challenges 580
  21.3.2 The FreSpeD Method 582
    21.3.2.1 Comparison to the Other Approaches 584
  21.3.3 Analysis of the Multichannel Seizure EEG Data 585
    21.3.3.1 Seizure Localization 585
    21.3.3.2 Seizure Onset Estimation and Potential Precursors 585
21.4 Modeling Time-Varying Connectivity Using Switching Vector Autoregressive
  Models 588
  21.4.1 Background on Vector Autoregressive (VAR) Models 589
    21.4.1.1 Stationary VAR Model 589
    21.4.1.2 Time-Varying VAR Model 590
    21.4.1.3 Switching VAR (SVAR) Model 591
  21.4.2 Parameter Estimation 592
  21.4.3 Estimating Dynamic Connectivity States in Epileptic EEG 593
21.5 Best Signal Representation for Non-Stationary EEGs 599
  21.5.1 Overview of Signal Representations 599
  21.5.2 Overview of SLEX Analysis 600
  21.5.3 Selecting the Best SLEX Signal Representation 603
  21.5.4 SLEX Analysis of Multichannel Seizure EEG 606
21.6 Dual-Frequency Coherence Analysis 608
  21.6.1 Overview and Historical Development 610
  21.6.2 The Local Dual-Frequency Cross-Periodogram 612
  21.6.3 Formalizing the Concept of Evolutionary Dual-Frequency Spectra . 612
    21.6.3.1 Harmonizable Process: Discretized Frequencies 612
    21.6.3.2 A New Model: The Time-Dependent Harmonizable Process 613
    21.6.3.3 Dual-Frequency Coherence between Bands 613
  21.6.4 Inference on Local Dual Frequency Coherence 614

        21.6.5 Local Dual Frequency Coherence Analysis of EEG Data 615

              21.6.5.1 Description of the Data and Experiment 615

              21.6.5.2 Implementation Details 615

              21.6.5.3 Results and Discussion 616

      21.6.6 Conclusion 617

   21.7 Summary 618

Bibliography 621

Index   **627**

List of Figures

2.1   Images of [^{18}F]Fallypride, which is a high-affinity dopamine (D2/D3) receptor antagonist. Top image is binding potential in healthy human; bottom image is binding potential in control rat. Both are overlaid on their respective structural MRIs. Images come from studies as described in (8) and (9).   18

2.2   [^{11}C]PK11195 imaging of inflammation 10 days after stroke and coregistered MR scan. Images from study as described in (38).   21

3.1   Unfolding of a left hemisphere cortical triangulated surface from the top left (pial surface) clock-wise to the bottom left (spherical representation). The insets show the arrangement of four triangles across the unfolding steps. Gray-scale illustrates the curvature information that is computed in the folded state and is carried over to the spherical representation.   42

3.2   Sample locations for gene expressions in MNI space (spheres) relative to the FreeSurfer average surface transparent (left). Surface vertices are colored according to the identifier of the closest sample regions (right).   47

3.3   A source surface warped to a subset of 774 target surfaces representing the left and right human hippocampus. Calculating the cross-covariance matrix of the surface point coordinates and subsequently the eigenvalue decomposition of the matrix, we decompose the shape variability into decorrelated modes sorted by variance. Here the first 10 modes are displayed by varying the weight for the particular mode symmetrically around zero. The resulting two extreme surfaces are overlaid using transparency and highlight directions in which the particular mode varies.   50

4.1   Pulsed Gradient Spin-Echo (PGSE) Sequence.   67

4.2   DWI images for different b-values and gradients.   68

4.3   3D $\mathbf{x}$-space and 3D $\mathbf{R}$-space. EAPs in different regions in the brain reflect different micro-structures with isotropic diffusion, single fiber, and crossing fibers.   69

4.4   Diffusion tensor representation.   70

4.5   Tensor field and the scalar maps estimated from the monkey data with $b = 1500 s/mm^2$.   71

4.6   Fiber directions and ADC profiles with different b values, two kinds of ODFs, and EAP profiles with different R.   73

4.7   EAP in 3D $\mathbf{R}$-space and its two features, i.e. EAP profile (or called iso-surface of EAP) and ODF.   74

4.8   Simulation Results   84

4.9   Simulation Results   85

4.10   Several kinds of sampling in $\mathbf{q}$-space.   88

4.11   Real Data Results based on FADTTS   93

5.1   Multi-sequence MRI data for one subject. Three axial slices are shown on
       each row (letters A, B, C indicate a different slice going from the inferior
       (A) to superior (C) of the brain) indicating FLAIR (A1, B1, C1), T2 (A2,
       B2, C2), T1 (A3, B3, C3), and PD (A4, B4, C4). A small MS white matter
       lesion is visible in the A-slice images. Some larger MS lesions are visible
       closer to the ventricle in the B-slice images.   111
5.2   Dynamic Contrast Enhancing (DCE) volume after gadolinium injection.   112
5.3   A. Axial slice from a T1 -weighted volume obtained from a 7T scanner.
       Volumes from scanners with a higher magnetic field strength often con-
       tain more intensity inhomogeneity artifacts, as seen in this image. B. The
       inhomogeneity field for this slice as modeled by the N4 ITK algorithm.   .   117
5.4   An axial slice of from a 3T T1-weighted imaging of a patient with MS before
       (A, showing $Y_{ijm}(v_{ijm})$) and after (B, showing $Y_{ijm}(v_{ijm})S_{ijm}(v_{ijm})$) skull
       stripping using BET.   120
5.5   Steps of registration: a toy example.   125
5.6   Application of two software methods (the function flirt on the bottom left
       and ANTs affine on the bottom right) to register a real brain image (top
       left) to a template (top right).   126
5.7   Intensity Normalization Methods. First column: region of interest from
       patient with MCI shown before (A) and after (C) histogram matching.
       The red square indicates a region of gray matter on the unnormalized im-
       age that disappears after histogram matching. Second column: histograms
       (shades of gray indicate different study visits) of the gray matter be-
       fore (B) and after (D) histogram matching and (E) white stripe normal-
       ization for subjects in the Alzheimer's Disease Neuroimaging Initiative
       http://adni.loni.usc.edu/. Note the large proportion of gray matter mis-
       matched to background (zero intensity) after histogram matching.   128
5.8   A. An axial slice from the FLAIR volume from a patient with multiple
       sclerosis. B. Manual expert segmentation of the lesions from this slice.   .   129
5.9   Two slices from a histogram of lesions constructed using non-linear regis-
       tration of observed images in ANTs.   131
5.10   Voxel intensities for FLAIR (top panels), T1 (middle panels), and T2 (bot-
       tom panels) images in a lesion (labeled "lesion 14") for one subject over 8
       years at 40 visits.   132

6.1   The growth of fMRI   140
6.2   Differences in image contrast   144
6.3   Image recontruction   145
6.4   Structural and functional data   146
6.5   The structure of a brain volume   147
6.6   The hemodynamic response function and the evoked BOLD response   . .   149
6.7   Examples of fMRI artifacts.   153
6.8   An example of a statistical image.   162
6.9   Connectivity analysis in the brain   164
6.10   Multivoxel pattern analysis example.   165

7.1 Typical power spectra of 124 EEG channels of 66 seconds (epoch length $T = 2$ sec with $K = 33$ epochs) of data from a subject (male, 25 yrs) who fixated on a computer monitor with his eyes open. While EEG spectral band definitions vary from lab to lab and across different fields, bands are typically defined as follows: delta 1–4 Hz, theta 4–8 Hz, alpha 8–13 Hz, beta 13–20 Hz, and gamma > 20 Hz. Some groups also identify the mu rhythm which exists as a peak either in the alpha or beta bands and typically has high power over the motor cortex. In the eyes-open resting data, with some artifact power removed using Independent Component Analysis (ICA), we see peaks in the delta, theta, and alpha bands and some power in the beta band. However the dominant peaks in the spontaneous EEG are in the theta and alpha bands, which have different spatial distributions over the electrodes and are associated with different cognitive functions. Topographic scalp maps were generated by summing power across frequencies in the theta (left) and alpha (right) frequency bands and interpolating between electrodes, such that brighter values correspond to higher power. Alpha, which is empirically associated with the resting state, has maximum power over parietal channels as indicated by the right topographic scalp map. 176

7.2 Volume conduction models for EEG. (*a*) A dipole is shown in the inner sphere of a *4-concentric spheres head model* consisting of the inner sphere (brain) and three spherical shells representing CSF (cerebral spinal fluid), skull and scalp. The parameters of the model are the radii (r_1, r_2, r_3, r_4) of each shell and the conductivity ratios ($\sigma_x v \sigma_y sg \sigma_x v \sigma_z sg \sigma_x v \sigma_4$). Typical values are: radii (8, 8.1, 8.6, 9.2 cm) and conductivity ratios (0.2, 40, 1). This model is used in the simulations in this chapter. (*b*) A realistic shaped boundary element model (BEM) of the head. The brain and scalp boundaries were found by segmenting the images with a threshold, and opening and closing operations, respectively, while the outer skull boundary was obtained by dilation of the brain boundary (ASA, Netherlands). Although geometrically more accurate than the spherical model, the (geometrically) realistic BEM may be no more accurate than a concentric spheres model because tissue resistivities are poorly known. (*c*) A realistic finite element model (FEM) obtained from MRI. This model has potentially better accuracy than the BEM model because the skull is subdivided into three layers corresponding to hard (compact) and spongy (cancellous) bone layers. . . 179

7.3   (a) Time series of a dipole meso-source $\mathbf{P}(\mathbf{r},t)$ composed of a 6-Hz, 15-μV sine wave added to Gaussian random noise with a standard deviation of σ = 150 μV. The Gaussian random noise was low-pass filtered at 100 Hz. The sine wave has variance (power) equal 1% of the noise. (b) Power spectrum of the time series shown in Part (a). The power spectrum has substantial power at many frequencies other than 6 Hz. (c) Time series recorded by an electrode on the outer sphere (scalp) of a four-concentric-spheres model above the center of a dipole layer of diameter 3 cm. The dipole layer is composed of 32 dipole sources $\mathbf{P}(\mathbf{r},t)$ with time series constructed similar to Part (a) with independent Gaussian noise (uncorrelated) at each dipole source. Scalp potential was calculated for a dipole layer at a radius $r_z = 7.8$ cm in a four-concentric-spheres model. The model parameters were radii $(r_1, r_2, r_3, r_4) = (8, 8.1, 8.6, 9.2)$ and conductivity ratios $(\sigma_1/\sigma_2, \sigma_1/\sigma_3, \sigma_1/\sigma_4) = (0.2, 40, 1)$. Notice that the time series is smoother than in the case of the individual dipole source. (d) Power spectrum of the time series shown in Part (c). Note the peak at 6 Hz. (e) Time series similar to Part (c), but due to a dipole layer of diameter of 4 cm composed of 68 dipole sources. (f) Power spectrum of the time series shown in Part (e). (g) Similar time series to Part (c), but with a dipole layer of diameter 5 cm composed of 112 dipole sources. The presence of the 6-Hz sinusoid is obvious from the time series. (h) Power spectrum of the time series shown in Part (g). A large spectral peak at 6-Hz is evident. 181

7.4   (From top-left, clockwise) Power spectra, time courses, and spatial loading topographies of the first twelve independent components (ICs) from an Independent Component Analysis (ICA) of an EEG recording while a subject (male, 25 yrs) was fixating on a computer monitor. The ICs are ordered by their contribution to the total variance in the raw data. ICs that are likely to reflect artifact contribution can be removed from the raw EEG data. IC1 is indicative of an eye blink. IC6 and IC7 are indicative of temporary electrical discontinuities. IC12 is indicative of muscle artifact.   186

7.5   Example power spectra from a single subject (female, 22 yrs). The subject is at rest with eyes closed. (a) Power spectrum of a midline occipital channel with epoch length $T = 60$ sec and $K = 1$ epochs. The power spectrum appears to have two distinct peaks, one below 10 Hz and one above 10 Hz. (b) Power spectrum at a midline frontal channel with epoch length $T = 60$ sec and $K = 1$ epochs. Here only the peak below 10 Hz is visible. (c) Power spectra of a midline occipital channel calculated with two different choices of epoch length T and number of epochs K. The grey circles indicate the power spectrum with $T = 1$ sec and $K = 60$ epochs. The black circles indicate the power spectrum with $T = 2$ sec and $K = 30$ epochs. (d) Power spectra of a midline frontal channel calculated as in Part (c). . . . 189

7.6   (a) Plots of 30 (individual epoch) power spectra for the occipital channel shown in Figures 7.5a and 7.5c. (b) Plots of the same 30 individual epoch spectra for the frontal channel shown in Figures 7.5b and 7.5d. (c) Peak power histograms show the distribution of peak frequencies for the 30 epochs shown in Part (a). (d) Peak power histograms for the 30 epochs shown in Part (b). 192

7.7   Scalp potential coherence spectra from a subject (female, 22 yrs) at rest with eyes closed in order to maximize alpha coherence. Coherence was estimated with $T = 2$ sec ($\Delta f = 0.5$ Hz) in a 60-sec record. The head plot shows the location of 9 electrodes, labeled x and 1 through 8. Coherence spectra between electrode x and each of the other electrodes 1–8 are shown, with increasing separations along the scalp. Note that very close electrodes have higher coherence independent of frequency as predicted by the theoretical volume conductor model. Alpha band coherence is high for large electrode separations, apparently reflecting the large cortical source coherence. 193

7.8   (*a*) A typical Visual Evoked Response (VEP; also known as an Event-Related Potential; ERP) to a large, high-contrast sinusoidal grating stimulus recorded at 124 electrodes of a high-density 128 Electrical Geodesics, Inc. (EGI) cap. The VEP was calculated by averaging low-pass Butterworth filtered data (with a 20-Hz passband) across all trials in one subject (male, 23 yrs) and by subtracting each trial by the time average of 200 milliseconds before that trial's stimulus onset (to *baseline* the VEP). Topographies of traditional local peaks (Luck et al., 2000) are labeled with $P1$, $N1$, $P2$, and $N2$ indicating the first and second positive and negative peaks over posterior electrodes. The N1 and P2 components evoke typical bilateral responses over parietal electrodes. The P1 and N2 reflect other network behavior related to processing of the visual stimulus. (*b*) A time-varying, phase-locked power spectrum of the same average data at an electrode over the left parietal cortex calculated with a Morlet wavelet transform. As shown by the wavelet, the VEP can also be thought of as a phase-locked alpha response to the visual stimulus. 195

7.9   A *non-phase-locked* time-varying power spectrum from the same subject and visual task as in Figure 7.8 such that power was calculated using all $K = 157$ epochs time-locked to the motor response (button press given by the right or left hand). Data from a left-central electrode C3 in the 10–20 electrode placement system is presented. Mu power *"desynchronizes"* (i.e., decreases) approximately 200 ms before the button press, likely reflecting cognitive control over the motor response. Similar magnitude of mu power before the desynchronization is observed after the motor response. 196

8.1   Discrete constituent parts and their sum for the one-dimensional function. 208

8.2   Fourier transform of the time series. 209

8.3   Matrix representation of one-dimensional discrete Fourier transform. . . . 209

8.4   Matrix representation of one-dimensional discrete inverse Fourier transform. 210

8.5   Discrete constituent parts and their sum for the two-dimensional image. . 212

8.6   Matrix representation of two-dimensional discrete Fourier transform. . . . 213

8.7   Matrix representation of two-dimensional discrete inverse Fourier transform. 214

8.8   Standard GRE-EPI pulse sequence and k-space coverage. 215

8.9   Matrix representation of fMRI image reconstruction. 216

8.10   Magnitude and phase of real and imaginary reconstructed image. 217

8.11   Reconstructed raw k-space data with Nyquist ghost. 218

8.12   Magnitude and phase of reconstructed raw k-space data with Nyquist ghost. 219

8.13   Odd and even line phase discrepancy and navigator echoes. 219

8.14 Image and spatial frequencies for illustrative isomorphism example. . . . 221
8.15 Image reconstruction via isomorphism representation. 222
8.16 AMMUST processing operators, O. 224
8.17 Modified mean images from processing. 225
8.18 Modified correlations from image processing. 227
8.19 Complex-valued signal changes for activation. 229

9.1   Left: The outer cortical brain surface mesh consisting of 81,920 triangles.
      Measurements are defined at mesh vertices. Right: The part of the mesh
      is enlarged to show the convoluted nature of the surface. 235
9.2   The displacement vector fields of registering from the hippocampus tem-
      plate to two individual surfaces. The displacement vector field is obtained
      from the diffeomorphic image registration (6). The variability in the dis-
      placement vector field can be analyzed using a multivariate general linear
      model (24). 240
9.3   A typical triangulation in the neighborhood of $\mathbf{p} = \mathbf{p}_0$ in a surface mesh. 244
9.4   Schematic of hat kernel smoothing on a hippocampal surface. Given noisy
      functional on the surface, the Laplace–Beltrami eigenfunctions ψ_j are com-
      puted and their exponentially weighted Fourier coefficients $\exp -\lambda_j \sigma$ are
      multiplied as a form a regression. This process smoothes out the noisy
      functional signal with bandwidth σ. 247
9.5 F-statistics maps on testing the significant hippocampus shape difference
      on the income level while controlling for age and gender. The arrows
      show the deformation differences between the groups (high income—low
      income). The fixed effects result (left) is obtained by treating the repeat
      scans as independent. The mixed effects result (right) is obtained by explic-
      itly modeling the covariance structure of the repeat scans with a subject.
      Both results are not statistically significant under multiple comparisons
      even at 0.1 level. 249
9.6 F-statistics maps testing the interaction between the income level and age
      while controlling for gender in a linear mixed-effects model. The arrows
      show the deformation differences between the groups (high income—low
      income). Significant regions are only found in the tail and midbody regions
      of the right hippocampus. 252
9.7   The plots showing income level-dependent growth differences in the poste-
      rior (left) and midbody (right) regions of the right hippocampus. The red
      lines are the linear regression lines. Scans within a subject are identified
      by dotted lines. 253

10.1 The multi-stage experimental pipeline for neuroimaging studies (see text
      for details). 265
10.2 Examples of multi-modal preprocessing steps for fMRI (left) and simulta-
      neous EEG-fMRI (right) experiments (see text for details). 266
10.3 NPAIRS subsampling procedure producing training and test, split-half
      datasets from which are produced (A) spatial reproducibility metric, R,
      between split-half statistical parametric maps (SPM), and (B) a predic-
      tion metric, P, of the accuracy with which the training set SPM values are
      able to correctly predict the test set's scan condition/class labels. 281
10.4 Prediction versus reproducibility curves as a function of regularization and
      spatial smoothing for a SVM classifier applied to two conditions for one
      subject from the dataset of (65) (see text for details). 285

10.5   Steps in identifying the optimal set of fixed pipelines across all subjects (see text for details). Reproduced with modifications from Figure 2 of (30). 288

10.6   For 27 subjects performing the trails making task (TMT) analyzed with the CVA model we display the preprocessing steps selected ON (white) and OFF (black) for (left panel) 97 fixed group pipelines that cannot be statistically distinguished based on (prediction, reproducibility) metrics, and (right panel) 27 individually optimized pipelines (see text for details). 290

10.7   Plots of prediction (P) versus global SNR ($gSNR$) for 5 different preprocessing pipelines: (1) conservative ($CONS$, Red), (2) optimal fixed across all subjects (FIX, Green), (3) individually optimized (IND) for each subject's run and scanning session using optimization metrics, (3a. light blue) maximum reproducibility (IND–R), (3b. medium blue) minimum distance from (P=1, R=1) (IND–D), and (3c. dark blue) maximum prediction (IND–P). The left column shows the Recognition task results (REC), and the right column the Trail-Making Task results (TMT). The upper row illustrates results for a univariate, Gaussian Naive Bayes (GNB) predictive model, and the lower row for a multivariate, PCA-regularized canonical variates analysis (CVA) predictive model. See text for further details. . . 291

10.8   Plots of Jaccard activation overlap for individually optimized preprocessing pipelines using the minimum distance from (P–1, R=1) (IND–D) versus the conservative pipeline defined in the text ($CONS$) for (a) within-subject test-retest runs, and (b) between-subject overlap of run1-to-run1 and run2-to-run2. Individual session, run, and subject data points are not shown, but are summarized with 1 Standard Deviation ellipses (enclosing $\approx$68% of data points). Results are shown for the Recognition (REC) and Trail-Making Task (TMT) for predictive models using Gaussian Naive Bayes (GNB) and PCA-regularized canonical variates analysis (CVA). See text for further details. 293

11.1   (a) Height (H), time-to-peak (T), and width (W) of the HRF. (b) The GLM for an event-related design, where the stimulus is evoked at 0, 5, and 10 s. 314

12.1   Illustration of the deviation from normality of functional neuroimaging datasets: The map shows the p-value of a test (5) rejecting the normality of the distribution of the z-transformed activation statistics across subjects using the dataset presented in Section 12.7 ($n = 573$). Note that the Gaussian hypothesis is significantly rejected in *all* cortical regions. 336

12.2   Illustration of the relative magnitude of within- and between-subject variability, through the average between/within variance ratio of the dataset presented in Section 12.7. While the ratio is close to 1 in many regions, it is larger in regions that display a non-zero mean effect across the population (compare with Fig. 12.5). 338

12.3   Illustration of the difference between fixed- and mixed-effects inference: Given $n = 10$ observations associated with a given level of uncertainty (left), one can perform a fixed-effects inference that ignores cross-subject variability of the observations and thus leads to population effects with tight uncertainty (middle), or consider this variance and then obtain wider uncertainty estimates (right). Only the mixed effects model yields a valid inference on the population from which the observations were sampled. . 339

12.4 Likelihood of the observations displayed in Fig. 12.3 as a function of the parameters (β, γ^2) (which are 2 scalars in that case). It can be seen that the likelihood function has a unique maximum. 343

12.5 Difference between a mixed-effects model and a random-effects model for statistical analysis. (a) One-sample test that aims at detecting the mean effect on a computation task (from which a simple reading effect has been subtracted (24)), based on a sample of $n_{subjects} = 30$ subjects; top: mixed-effects model; bottom: random-effects model. Both maps are thresholded at the $p < 0.05$ level, corrected for multiple comparisons, using permutation testing. (b) Illustration of the difference on a two-sample test, where the differential effect of two MRI scanners on the activation obtained in the same task as (a). The mixed- (top) and random- (bottom) effects maps, both thresholded at $p < 0.05$, FWER-corrected by permutation, show the same effects, but again the mixed-effects inference is more sensitive. . . . 348

12.6 Impact of including the spatial context in the neuroimaging statistical inference procedure: the four maps above represent the activation related to the one-sample test presented in Fig. 12.5(a), thresholded at a significance level of $p < .05$, corrected for multiple comparison through an F-max permutation scheme. Voxel-level, cluster-level, TFCE and RPBI present increasing amounts of activations. 349

13.1 Simulations from the null. 357

13.2 Simulations from a mixture of alternative and null. 358

14.1 The true correlation structure for the multivariate normal dataset with 10 and 90 brain signals, respectively. 377

14.2 The distribution of CCor and CCoh for 1000 simulations of a multivariate normal dataset with 10 brain signals and a sparse FC structure. 378

14.3 The distribution of PCCor, PCCorG, PCCorGR, PCCorGRS, PCCoh and PCCohS for 1000 simulations of a multivariate normal dataset with 10 brain signals and a sparse FC structure. 379

14.4 The distribution of CCor and CCoh for 100 simulations of a multivariate normal dataset with 90 brain signals and an FC network structure obtained from the resting-state fMRI data in Section 14.4. 379

14.5 The distribution of PCCor, PCCorG, PCCorGR, PCCorGRS and PCCohS for 100 simulations of a multivariate normal dataset with 90 brain signals and an FC network structure obtained from the resting-state fMRI data in Section 14.4. 380

14.6 The distribution of the 4005 FC estimates using CCor and CCoh on the resting-state fMRI data. 383

14.7 CCoh plotted against the absolute value of CCor for the resting-state fMRI data. The identity line is the dashed blue line. 383

14.8 The distribution of the 4005 FC estimates of the conditional dependency methods (PCCor, PCCorG, PCCorGR, PCCorGRS, PCCoh, and PCCohS) on the resting-state fMRI data. 384

14.9 The unregularized estimates of the conditional dependencies plotted against each of their regularized counterparts in the resting-state fMRI data. The correlation coefficient r between each FC metric pair is computed between those values whose regularized estimate is non-null. 385

14.10 The marginal dependencies plotted against their corresponding conditional dependencies for the resting-state fMRI data. The zero line is the dashed blue line. 386

14.11 Voxel-wise FC analysis of the resting-state fMRI data, with the seed (represented by a purple dot) in the PCC, using CCor. 387

14.12 Voxel-wise FC analysis of the resting-state fMRI data, with the seed (represented by a purple dot) in the PCC, using CCoh. 387

14.13 CCoh plotted against the absolute value of CCor from the resting-state fMRI data. The identity line is the dashed blue line. 388

15.1 An example of the longitudinal design for the study presented by Zipunnikov et al. (106). Fractional Anisotropy (FA) in the Corpus Callosum of subjects diagnosed with MS. Darker color means lower FA. 404

15.2 Four networks (Auditory, Default Mode (DM), Motor, and Visual) computed for 20 subjects using the H-gICA algorithm. The color bar on the right shows the values corresponding to each color, namely the highest intensities are colored in red followed by the yellow and blue for voxels with lower intensities. 408

16.1 A simple recursive threee-variable SEM. 424

16.2 A simple two-variable DCM. 428

17.1 Meta-analytic functional coactivation network, based on data combining in excess of 1,600 neuroimaging studies published between 1985 and 2010. (A) A minimum spanning tree is used to locate nodes in relation to their topological proximity to each other. Different modules are coded by color, with the size of all nodes proportional to their weighted degree (strength). (B) Nodes in anatomical space, colored according to their number of activations and deactivations. (C) Nodes arranged in the same layout as A, and colored as in B. Note that the rich club concentrates most of the activations, whereas the periphery and particularly the default-mode network concentrates the deactivations. See (19), and Section 17.3.3 in this chapter, for a discussion of hubs. 445

17.2 Thresholded correlation matrices for four levels of a working memory experimental manipulation, called the N-back task, over a vertex set based on the AAL template with 90 cortical and subcortical ROIs. Adjacency matrices become sparser with increasing working memory load. These adjacency matrices were obtained by constructing group mean networks, using a mass univariate approach based on z-tests with FDR correction (base rate $\alpha_0 = .05$). Zero entries are denoted in black and edges in white. (See (35) for details.) 447

17.3 Graphical representations of mean SPNs, over four levels of a cognitive task. The mean SPNs for an N-back task in the coronal planes are here presented, after FDR correction (base rate $\alpha_0 = .05$). The locations of the nodes represent the stereotaxic centroids of the corresponding cortical regions. The orientation axis is indicated in italics: inferior–superior for the coronal section. The size of each node is proportional to its degree. (See (35) for a full description.) 456

17.4   Visualization of a differential SPN, summarizing the effect of a cognitive factor. Left and right sagittal section of a negative differential SPN, representing the significantly "lost" edges, due to the N-back experimental factor. The presence of an edge is here determined by the thresholding of p-values at .01, uncorrected. (See (35) for further details.)   457

17.5   Topological randomness and number of edges predict number of modules. (A) Relationship between the number of random rewirings of a regular lattice and the number of modules in such a network. Here, the number of edges is kept constant throughout all rewirings. (B) Relationship between the number of edges in a network and its number of modules for both regular (i.e., lattice) and random graphs. This shows that the number of modules tends to decrease as more edges are added to both types of networks. (C-D) Modular structures of regular (C) and random (D) networks for different number of edges, N_E. These networks are represented using the algorithm of (40), with different colors representing different modules. In all simulations, the number of vertices is $N_V = 112$, as in (5) and (33).   459

18.1   Forty simulated curves following a decline trajectory of form (18.2), with normally distributed horizontal shifts τ and vertical shifts Δ. The mean of the curves, shown in red, declines less steeply than any of the individual curves.   471

18.2   At left, the raw cortical thickness values at one vertex for a longitudinal sample (the curves are discussed below in §18.5.3). Upon fitting model (18.7), the estimated values with random subject effects removed, i.e. $\hat{\mu}(t_{ij}) + \hat{e}_{ij}$, are as shown at right; the red curves represent $\hat{\mu}(t) \pm 2$ standard errors.   475

18.3   False discovery rate-corrected p-values for restricted likelihood ratio tests of the null hypothesis H_{0v}: μ_v is linear, for each vertex v.   479

18.4   Left: Division of the brain into 3 clusters of similar cortical thickness trajectories. Right: Examples of mean function estimates for randomly chosen vertices within each cluster, with cluster mean function shown in black and number of vertices in the cluster shown above each subfigure.   479

18.5   Median 2nd-derivative penalty (over 20 random cross-sectional subsamples) plotted against median 3rd-derivative penalty, for cortical thickness trajectory estimates based on 2nd- and 3rd-derivative penalties respectively; each point corresponds to one vertex. Wilcoxon signed-rank tests were applied to the differences between the 2nd- and 3rd-derivative fits' GCV scores for the 20 subsamples (see the text). Blue and red dots represent vertices for which the 2nd- and 3rd-derivative penalty, respectively, performed better according to these tests.   481

18.6   Two vertices for which 2nd- and 3rd-derivative-penalized fits yield conflicting results regarding age of peak cortical thickness. Specifically, 2nd-derivative fits for most of the 20 random cross-sectional subsamples suggest mean cortical thickness is highest at the lower end of the age range, whereas most of the 3rd-derivative fits attain a peak at a later age (rug plots display the peak ages for each of the curve estimates). GCV scores indicate that the 3rd-derivative penalty yields superior curve estimates for the example at left, while the 2nd-derivative penalty performs better for the example at right.   482

18.7 Longitudinal functional PCA estimates for the corpus callosum FA data. (a) Estimated mean FA curve $\mu(v,t)$, for t set to visits 1, 2, 3, 4. (b) First eigenfunction $\phi_1^u(v)$ of the visit-specific process. (c) First eigenfunction of the RIRS process, represented as $\phi_1^0(v) + t\phi_1^1(v)$ for t set to visits 1, 2, 3, 4. (d) Second eigenfunction of the RIRS process, represented as $\phi_2^0(v) + t\phi_2^1(v)$. Shown in parentheses are the percent of total variation accounted for by each component. 484

18.8 Left: The line segment indicates an edge connecting two of the regions considered by Di Martino et al. (24). Center: Resting-state functional connectivity for this pair of regions plotted against age (blue dots represent controls; red, ASD), along with estimated 5th, 25th, 50th, 75th, and 95th percentile curves for the control data. Right: Histogram of quantile ranks for the ASD subjects, with respect to the estimated age-varying distribution for the controls. Each bar is divided into three age groups. 486

19.1 Functional connectivity map 497
19.2 Functional connectivity matrix 498
19.3 Generating brain networks from fMRI data 501
19.4 Functional Connectivity with Anatomical Weighting 509
19.5 Cluster Time Series 510
19.6 Multimodal prediction inputs 514

20.1 EEG topography. 526
20.2 Plot of EEG signals for 8 channels (1 epoch). 526
20.3 Plot of EEG signals for one channel (in the SMA region) for 160 epochs. 527
20.4 The Fourier coefficients are interpreted as the inner product (cross-covariance) between the EEG $\mathbf{X}$ and each of the Fourier waveforms ψ_k. . 529
20.5 Plot of EEG signal at channel 23 (in SMA region). 532
20.6 Power spectrum density estimate via periodogram, smoothed periodogram, multitaper method, and ARMA method. 533
20.7 The five leading Slepian sequences (taper functions) for T = 10000 and half-bandwidth = 2.5. 534
20.8 Left: Spectra for the AR(2) process with different peak frequency; $\Omega_s\xi = 2, 6, 10, 21, 40$ which correspond to delta, theta, alpha, beta and gamma frequency bands. Right: Simulated signals from the corresponding AR(2) process. 538
20.9 95% confidence interval for the power spectrum of signals at channel 23, computed using 160 epochs. 541
20.10 Decomposition of EEG signals into oscillations of different frequency bands. 543
20.11 EEG plot for multiple channels at one epoch, channels include 23 (SMA), 51 (left M1), 164 (right M1), 172 (right antPr), and 77 (left antPr). . . . 545
20.12 Periodogram (log scale) and smoothed periodogram (m = 1, 2, 5) of EEG channels 23 (SMA), 51 (left M1), 164 (right M1), 172 (right antPr), and 77 (left antPr), averaged over the delta band. 546
20.13 Periodogram (log scale) and smoothed periodogram (m = 1, 2, 5) of EEG channels 23 (SMA), 51 (left M1), 164 (right M1), 172 (right antPr), and 77 (left antPr), averaged over the theta band. 546
20.14 Periodogram (log scale) and smoothed periodogram (m = 1, 2, 5) of EEG channels 23 (SMA), 51 (left M1), 164 (right M1), 172 (right antPr), and 77 (left antPr), averaged over the alpha band. 547

20.15   Periodogram (log scale) and smoothed periodogram (m = 1, 2, 5) of EEG channels 23 (SMA), 51 (left M1), 164 (right M1), 172 (right antPr), and 77 (left antPr), averaged over the beta band. 547

20.16   Periodogram (log scale) and smoothed periodogram (m = 1, 2, 5) of EEG channels 23 (SMA), 51 (left M1), 164 (right M1), 172 (right antPr), and 77 (left antPr), averaged over the gamma band. 548

20.17   Estimate of VAR coefficients. 555

20.18   Connectivity at the theta band. 556

20.19   Connectivity at the beta band. 557

20.20   Connectivity at the gamma band. 558

20.21   Graphical representation of the latent source model. The directions of the arrows represent a dependence relationship. 559

21.1   Top: EEG recordings at the left temporal channel (T3). Bottom: right temporal channel (T4). Sampling rate 100 Hertz. Duration of recording is 500 seconds. 569

21.2   EEG scalp topography. 570

21.3   Left: EEG scalp topography. Right: traces of the EEG signals for one trial for leftward and rightward movements. 570

21.4   The TV distance measures the similarity between the two densities. The blue (pink) area is the value of the TV distance. 573

21.5   Dynamic of the hierarchical spectral merger algorithm. (a), (b), (c), and (g) show the clustering process for the spectra. (d), (e), (f) and (h) show the evolution of the estimated spectra, which improves when we merge the series on the same cluster. 574

21.6   EEG signal from channel T3. (a) Complete signal. (b) Segments involved in the analysis to compare EEG signals before (gray), during (red) and after (blue, green and yellow) the epileptic seizure. 575

21.7   Initial values of the TVD between each channel for time segments (a) 33001–34000, (b) 34201–35200, (c) 36201–37200, (d) 38001–39000 and (e) 44001–45000. 576

21.8   Trajectories of the minimum value of the TVD resulting using the SMC method. 577

21.9   Location of each cluster at the cortical surface for subsets before, during and after the epileptic seizure. Time segments (a) 33001–34000, (b) 34201–35200, (c) 36201–37200, (d) 38001–39000 and (e) 44001–45000. Notice that $SP1$ and $SP2$ are located inside the brain, as a visualization tool, we plotted these channels on the right upper corner. 578

21.10   Spectrum estimated for each channel by cluster for subsets before, during and after the epileptic seizure. Time segments (a) 33001–34000, (b) 34201–35200, (c) 36201–37200, (d) 38001–39000 and (e) 44001–45000. 579

21.11   Change points in the autospectra of channels T3 and T5. The figure shows time (x-axis) versus the number of frequency bands, where energy changes are detected (y-axis). 586

21.12   Cumulative sum of detected change points over all channels and/or channel pairs. 586

21.13   A framework for estimating quasi-stationary dynamic effective connectivity states in multi-channel EEG signals, based on TV-VAR and SVAR processes. 595

21.14 Estimation of state-related dynamics of directed connectivity between the channels P3, T3 and T4. (a) Estimated TV-VAR coefficient matrices (each vectorized with dimension $P^2 \times L = 3^2 \times 3 = 27$) using Kalman filtering at different time lags $\ell = 1, 2, 3$. (b) Inferred states (blue:normal; red:seizure) at each time point, for different stages of our general framework combining K-means clustering of the TV-VAR coefficients, $\widehat{S}_t^{\mathrm{KM}}$ (top), Switching KF, $\widehat{S}_t^{\mathrm{SKF}}$ (middle) and switching KS, $\widehat{S}_t^{\mathrm{SKS}}$ (bottom) based on a EM-estimated two-state SVAR(3) model. (c) EEG data overlaid by final estimated states. 597

21.15 Estimated VAR connectivity matrices (at three lags) between the three epileptic EEG channels, $\widehat{\Phi}_{[j],\ell}^{\mathrm{EM}}, \ell = 1, 2, 3$, for the non-ictal and the ictal brain state, $j = 1, 2$. 598

21.16 Estimates of directed coherence between EEGs for the normal (blue) and the epileptic (red) brain state, computed from the estimated VAR parameters in Figure 21.15. 598

21.17 Smooth window pairs $\Psi_{+,B}(u)$ and $\Psi_{-,\omega,B}(u)$. These windows can be stretched or compressed. In the top picture, B is approximately the rescaled interval $(\frac{500}{1000}, \frac{900}{1000})$; in the bottom picture, B is approximately $(\frac{500}{1000}, \frac{750}{1000})$. 600

21.18 Examples of the SLEX waveforms at different scales and locations. The SLEX waveforms can be dilated or compressed as well as shifted. 601

21.19 A SLEX library with level $J = 2$. The shaded blocks represent one basis from the SLEX library. 601

21.20 Time-varying spectra of the first and second SLEX principal components. 607

21.21 Left: SLEX coherence estimates between $T3$ and channels on the left side of the brain namely $F3, C3, P3$. Right: SLEX coherence estimates between $T3$ and the channels on the right side of the brain namely $F4, C4, P4, T4$. The color index here is actually coherence $\times$ 100 so that the values range from 0 to 100. 608

21.22 Left: SLEX coherence estimates between $T4$ and channels on the left side of the brain namely $F3, C3, P3, T3$. Right: SLEX coherence estimates between $T4$ and the channels on the right side of the brain namely $F4, C4, P4$. The color index here is actually coherence $\times$ 100 so that the values range from 0 to 100. 609

21.23 Decomposition of the electroencephalograms at the right frontal channel (FC4) and the left parietal channel (P3) into the delta (0–4 Hertz), alpha (8–12 Hertz) and beta (12–30 Hertz) oscillations. The goal is to estimate dependence between the alpha band oscillations at the FC4 channel with the beta band oscillations at the P3 channel. 610

21.24 Local evolutionary dual coherence estimate between alpha activity at the frontal-central channels and beta activity in the rest of the channels. Each plot has estimates for both the left and right conditions. Vertical dashed lines at time 0.5 sec for the left and right conditions represent time when the visual cue was presented. The color indicates the magnitude of the coherence estimates. White indicates insignificant coherence at FDR level 0.05. The black dots denote statistically significant differences between coherence values in the left vs. right conditions. 617

21.25 Local evolutionary dual coherence estimate between alpha activity at the central channels and beta activity in the rest of the channels. Each plot has estimates for both the left and right conditions. Vertical dashed lines at time 0.5 sec for the left and right conditions represent time when the visual cue was presented. The color indicates the magnitude of the coherence estimates. White indicates insignificant coherence at FDR level 0.05. The black dots denote statistically significant differences between coherence values in left vs. right conditions. 618

21.26 Local evolutionary dual coherence estimate between alpha activity at the parietal channels and beta activity in the rest of the channels. Each plot has estimates for both the left and right conditions. Vertical dashed lines at time 0.5 sec for the left and right conditions represent the time when the visual cue was presented. The color indicates the magnitude of the coherence estimates. White indicates insignificant coherence at FDR level 0.05. The black dots denote statistically significant differences between coherence values in left vs. right conditions. 619

21.27 Significant differences between the left vs. right directions were observed over the time interval $(600, 640)$ milliseconds which is also equivalent to 100–140 milliseconds post visual cue presentation. This figure refers to differences in coherence of alpha activity in *right frontal-central channels* with beta activity in the rest of the channels and differences in coherence of alpha activity in left frontal-central channels with beta activity in the rest of the channels. 620

List of Tables

4.1   The condition numbers and gradient sampling indices (GSI) of thirteen acquisition schemes in (123). 89

8.1   Some k-space, reconstruction, and image processing operations. 223

11.1   Summaries of notations in Section 11.2. 312

A.1   Example of statistical models used in neuroimaging. The first column describes the model, the second column describes how the data are ordered in the outcome vector, the third column shows the design matrix, and the last column illustrates the hypothesis tests and corresponding contrasts. Note, in the ANOVA example F-tests are used for all contrasts, whereas t-tests are used for the other examples. We use the notations from Eq. (12.13). This table is adapted from (25). 354

14.1   A table of abbreviations for the FC metrics. 371

19.1   Joint activation probabilities for regions i and j. 511

20.1   Summary of metrics for connectivity 551
20.2   EEG channel grouping. 554
20.3   Summary of methods for source reconstruction. 562

21.1   Frequency-specific proportion of change points and change magnitude, in percent. Change magnitude is measured as the sum over thresholded CUSUM statistics, over time, frequency and EEG channels and channel pairs. 588

Preface

With this handbook, we summarize the state-of-the-art in the statistical modeling and analysis of neuroimaging data and point out the current challenges. Our goal is that this will provide statisticians and other quantitative researchers with an introduction to important active research areas in neuroimaging data analysis, and provide them with a path to contribute to this rapidly emerging area covering both theoretical and applied statistics.

There are challenges to the statistical analysis of neuroimaging data. It is a massive data problem; in most cases the signal of interest is relatively weak; the data exhibit a complicated spatio-temporal correlation structure. Today statistics plays an integral role in neuroimaging data analysis. It is truly an exciting time to be involved in neuroimaging research as increasingly ambitious experiments are being performed daily. This creates a significant new demand, and an unmatched opportunity, for quantitative researchers working in the neurosciences.

Understanding human brain structure and function is arguably one of the most complex, important and challenging issues in science today. For this endeavor to be successful, researchers are needed to make sense of the massive amounts of data being generated. Our sincere hope is that this book can in some way aid in this development, by providing researchers with an introduction to the fascinating world of neuroimaging data analysis.

Contributors

John Aston
Statistical Laboratory
University of Cambridge
UK

Hauke Bartsch
Multi-Modal Imaging Laboratory
University of California, San Diego
USA

F. DuBois Bowman
Department of Biostatistics
Columbia University
USA

Brian Caffo
Department of Biostatistics
Johns Hopkins University
USA

Jian Cheng
Department of Radiology
University of North Carolina, Chapel Hill
USA

Moo K. Chung
Department of Biostatistics
University of Wisconsin, Madison
USA

Nathan Churchill
St. Michael's Hospital, Toronto
Canada

Ciprian Crainiceanu
Department of Biostatistics
Johns Hopkins University
USA

Ivor Cribben
Department of Finance and Statistical
    Analysis
University of Alberta School of Business
Canada

Daniel Drake
Department of Biostatistics
Columbia University
USA

Ani Eloyan
Department of Biostatistics
Brown University
USA

Carolina Euán
Departamento de Estadistica
Centro de Investigaciones en Matematicas
Mexico

Mark Fiecas
Department of Statistics
University of Warwick
UK

Cedric Ginestet
Department of Biostatistics
Institute of Psychiatry, Psychology and
    Neuroscience King's College London
UK

Jamie L. Hanson
Center for Developmental Science
University of Pittsburgh
USA

Lechuan Hu
Department of Statistics
University of California, Irvine
USA

Lan Huo
Department of Child and Adolescent
    Psychiatry
New York University
USA

Eric D. Kolaczyk
Department of Mathematics and Statistics
Boston University
USA

Mark Kramer
Department of Mathematics and Statistics
Boston University
USA

Nicole A. Lazar
Department of Statistics
University of Georgia
USA

Fan Li
Department of Statistical Sciences
Duke University
USA

Martin Lindquist
Department of Biostatistics
Johns Hopkins University
USA

Michael D. Nunez
Department of Cognitive Sciences
University of California, Irvine
USA

Paul L. Nunez
Cognitive Dissonance, LLC
USA

Seth D. Pollack
Department of Psychology
University of Wisconsin, Madison
USA

Hernando Ombao
Department of Statistics
University of California, Irvine
USA

Phil Reiss
Department of Child and Adolescent
   Psychiatry
New York University
USA
and

Department of Statistics
University of Haifa
Israel

Daniel B. Rowe
Department of Mathematics, Statistics and
   Computer Science
Marquette University, Wisconsin
USA

Balqis Samdin
Center for Biomedical Engineering
Universiti Teknologi Malaysia
Malaysia

Anna Louise Schröder
Department of Statistics
London School of Economics
UK

Haipeng Shen
Department of Innovation and Information
   Management
University of Hong Kong
Hong Kong

Russell T. Shinohara
Department of Biostatistics and
   Epidemiology
University of Pennsylvania
USA

Sean Simpson
Department of Biostatistical Sciences
Wake Forest University Health Sciences
USA

Michael E. Sobel
Department of Statistics
Columbia University
USA

Ramesh Srinivasan
Department of Biomedical Engineering
University of California, Irvine
USA

Stephen C. Strother
Rotman Research Institute, Baycrest
Canada

Elizabeth M. Sweeney
Department of Biostatistics
Johns Hopkins University
USA

Bertrand Thirion
Parietal team, Inria, France
Neurospin, France
Université Paris Saclay, France

Wesley Thompson
Department of Psychiatry
University of California, San Diego
USA
Institute of Biological Psychiatry
Mental Health Centre Sct. Hans, Mental
    Health Services
Denmark

Chee-Ming Ting
Center for Biomedical Engineering
Universiti Teknologi Malaysia
Malaysia

Tor Wager
Department of Psychology and
    Neuroscience
University of Colorado, Boulder
USA

Yuxiao Wang
Department of Statistics
University of California, Irvine
USA

Juemin Yang
Department of Biostatistics
Johns Hopkins University
USA

Vadim Zipunnikov
Department of Biostatistics
Johns Hopkins University
USA

Tingting Zhang
Department of Statistics
University of Virginia, Charlottesville
USA

Hongtu Zhu
Department of Biostatistics
University of North Carolina, Chapel Hill
USA

Part I

Overview

1

Overview of the Handbook

CONTENTS

1.1	Introduction	3
1.2	A Brief History of Neuroimaging	4
1.3	Modalities	4
1.4	Statistical Methods	7
	1.4.1 Preprocessing	8
	1.4.2 Methods in Structural Neuroimaging	8
	1.4.3 Localizing Areas of Activation	9
	1.4.4 Brain Connectivity	9
	1.4.5 Analysis of Electroencephalograms	10
	1.4.6 Multi-Modal Analysis	10
1.5	Additional Topics	11
	1.5.1 Meta-Analysis	11
	1.5.2 Experimental Design in fMRI	11
	1.5.3 Experimental Design in EEG	11
	1.5.4 Imaging Genetics	12
1.6	Conclusions	12
	Bibliography	13

1.1 Introduction

Neuroimaging has become an umbrella term for the ever-increasing number of minimally invasive medical imaging techniques used to study the brain. These include a variety of rapidly evolving technologies for measuring brain properties related to structure, function, development and disease pathophysiology. These technologies are currently being applied in a vast number of medical and scientific areas of inquiry, and have found applications in a wide variety of fields, including neuroscience, cognitive science, psychology, neurology, economics and political science. Today, understanding the brain has become arguably one of the most complex, important and challenging issues in science, and imaging has become an invaluable tool in this endeavor.

With this book we seek to summarize the state-of-the-art in the statistical analysis of neuroimaging data and point out the current challenges. Our hope is that this will provide statisticians and other quantitative researchers with an introduction to important active research areas in neuroimaging data analysis, and provide them with a path to contribute to this rapidly emerging area covering both theoretical and applied statistics. This brief chapter sets the stage for the rest of the book, and introduces key concepts that will be discussed further in subsequent chapters.

1.2   A Brief History of Neuroimaging

In the late 1800s William Rontgen revolutionized medicine, when he discovered the X-ray. This new discovery provided physicians with the ability for the first time to study the inside of a living subject without having to resort to actually cutting them open. Images were, and still are to this day, acquired by passing a beam of X-rays, through an object, and onto a photographic film on the opposite side. A fraction of the X-rays are either absorbed or scattered before reaching the other side and the resulting decrease in intensity (increased attenuation), is measured by its effect on the film. As increased attenuation gives rise to less darkening of the film, dense structure such as bone clearly appear in the resulting images.

Today, X-rays are omnipresent in hospitals throughout the world. These are primarily used to study damage to bones, teeth and lungs. However, they have inherent limitations in their ability to study basic anatomical structure. A fundamental problem is that they simply provide two-dimensional (2D) projections of an inherently three-dimensional (3D) object. While in many cases this is sufficient to provide useful information (such as identifying a bone fracture), it can also have serious consequences. For example, if one were to place a pair of scissors *on* top of the chest of a patient, the resulting X-ray would make it appear exactly as if they had been located inside of the patient!

These shortcomings led to the advent of more advanced imaging modalities, including computed tomography (CT), which provides improved anatomical information, positron emission tomography (PET), which additionally provides detailed metabolic and neuro-chemical information, and magnetic resonance imaging (MRI), which gives yet more detailed anatomical information than CT. These new techniques further revolutionized medicine, as they not only allowed for the cross-sectional representation of a 3D structure, but also in some cases provided the ability for the first time to measure physiological activity within a certain tissue or organ. Techniques such as functional magnetic resonance imaging (fMRI) inform about brain activity in response to outside stimuli by measuring quantities like the blood oxygen level. Together, imaging modalities such as PET, fMRI, electroencephalography (EEG) and magnetoencephalography (MEG) give researchers unprecedented access to the brain in action and allowed for numerous insights into human brain function.

These advances in medical imaging were made possible because of critical breakthroughs in diverse scientific disciplines such as physics, chemistry and engineering. However, their ultimate success has been aided by employing statistics, mathematics and computer science to process the increasingly complex spatio-temporal measurements provided by these new imaging modalities. While X-ray uses photographic film that can be developed by a technician, a common theme of these newer modalities is that the underlying measurements are not directly interpretable, but rather require the use of computers and reconstruction algorithms to transform the measured signal into images. Once images are acquired there is a need for sophisticated statistical methods to make sense of the data. This book attempts to highlight some of these methods.

1.3   Modalities

It is useful to separate neuroimaging into structural imaging, which deals with the study of brain structure and the diagnosis of disease and injury, and functional imaging, which can be used to study both cognitive and affective processes. There exist a number of different structural and functional imaging modalities, each of which has their relative pros and cons. Here we briefly introduce some of these techniques, with more detailed descriptions coming in later chapters, and additionally seek to compare and contrast them to one another.

Commonly used modalities for structural imaging include CT, MRI, and PET. The images obtained using each of these modalities differ fundamentally in what they seek to measure, and the amount of spatial information they provide. Ultimately, their usage depends upon the question one is interested in addressing, as well as their availability.

CT, also known as CAT (Computed Axial Tomography), was developed in the early 1970s, although the basic concept for a single slice was introduced in the 1900s by a radiologist named Vallebona. The term "tomography" has its root in the Greek word for "slice" and has generally been used to describe a class of applied and theoretical inverse problems related to medical imaging, as multiple lower-dimensional projections (slices) are transformed into full 3D scans. In the medical imaging context, CT is intended to denote a modality for performing transmission tomography. A CT scanner works by combining a series of X-rays acquired at many different angles, and uses computerized technology to combine these measurements to produce cross-sectional images of an object. To illustrate, assume that an X-ray is sent through the body along a line L. By measuring the attenuation of the X-ray one can compute the line integral $P_f(L)$ of the density $f(x, y)$ along L. Modern CT scanners use so-called fan-beam geometry (12), where there are n different arrays consisting of m lines. Each array has a common point on a fixed circle surrounding the object, and its lines are uniformly spaced in their angle over a range covering the object of interest. Reconstruction can be performed by solving an inverse problem using the so-called filtered back-projection algorithm. For excellent overviews see (3) and (18).

In CT, as well as in other transmission tomography modalities, the source of radiation is external to the patient and its positioning is known prior to attempting to reconstruct the image, which consists of a map of the attenuation density within the patient. In contrast, PET is an imaging modality for performing emission tomography. Here, the source of radiation is a compound that has been introduced into a subject and is distributed across the organ of interest. This is done by chemically attaching a radioactive label to a pharmacological agent, which is injected into the bloodstream. The agent is transported to the brain, where it binds to a specific class of receptors, depending on its biochemical nature. The PET scanner consists of a number of detectors positioned on a circular array. When a radioisotope decays, it emits a positron, which finds a nearby electron and annihilates, producing two photons propagating from the point of annihilation in opposite directions along a line of random orientation. Each photon is detected by one of the detectors surrounding the subject, allowing one to determine the line along which the decay occurred. The radioactive decay is typically modeled as a Poisson process, and the emission distribution estimated using the expectation-maximization (EM) algorithm. For more a more detailed description, see Chapter 2.

MRI is a procedure that uses a combination of a strong external magnetic field and radio frequency pulses to excite nuclei in the brain. Once the pulse is removed, the system seeks to return to equilibrium and the nuclei emit the absorbed energy as they relax. This creates a signal that the MR scanner is able to measure and use to obtain in vivo images of the human body. A system of gradient coils are used to control the spatial inhomogeneity of the magnetic field, so that each measurement can be approximated as the Fourier transform of the proton density at a single point in the frequency domain. The image can be reconstructed using the inverse Fourier transform. MRI is able to provide detailed anatomical scans of gray and white matter with a spatial resolution well below $1mm^3$ and has become an increasingly popular choice for obtaining high spatial resolution anatomical images of the brain. This topic is covered in Chapters 3 and 5.

PET imaging is complementary to MRI, as it permits estimation of the density of a variety of neurochemical receptors across the brain. PET provides a 3-D map of the distribution of labeled substances across the brain. CT is better suited for bone injuries, lung or chest imaging, and detecting certain cancers. In contrast, MRI is effective for studying soft tissue damage. An additional advantage of MRI is that it does not use radiation, in contrast to CT and PET. Since radiation can be harmful in repeated exposure, it may be particularly beneficial to use MRIs to study longitudinal changes in brain structure.

Diffusion magnetic resonance imaging (dMRI) is a class of techniques for measuring directional diffusion and reconstructing fiber tracts of the brain. Since water diffuses more quickly along (than across) axons, this can be used to map the microstructure and organization of white matter pathways. This provides a basis for studying the anatomical connectivity between different brain regions. It is used in the study of normal brain maturation and aging, as well as in clinical applications such as multiple sclerosis, epilepsy, metabolic disorders, and brain tumors. The theoretical underpinning of various mathematical and statistical methods associated with dMRI is described in greater detail in Chapter 4.

Recently, there has been an explosion of interest in using functional brain imaging to study both cognitive and affective processes. Commonly used modalities in this setting include PET, fMRI, EEG, and MEG. As for structural imaging, each functional imaging modality provides a different type of brain measurement. Each modality also has its own strengths and weaknesses with regards to spatial resolution (the ability to distinguish changes in an image across different spatial locations), temporal resolution (the ability to separate brain events in time), and invasiveness.

Both EEG and MEG reflect brain electrical activity with millisecond temporal resolution, and provide the most direct measurement of brain activity that can be obtained non-invasively. EEG, covered in Chapter 6, is based on the fact that electrical activity of active neurons (mostly cortical) produce currents that spread throughout the head. Once these currents reach the scalp they are recorded using EEG. An electroencephalogram is the recording of electrical activity from a number of electrodes (usually 32, 64, or 256) placed on the scalp. It measures voltage fluctuations resulting from ionic current flows within populations of neurons in the brain. Each channel measures the net contribution of electrical activity from the entire cortex, and is believed to capture mostly the net local activity of many neurons. The same currents produce magnetic fields, which can be measured on the scalp using MEG.

An important problem is to take the measurements from the electrodes and make inferences about the underlying sources in the brain that gave rise to them. This inverse problem is a difficult one and even with an infinite amount of EEG and MEG electrodes around the head, non-ambiguous localization of the activity inside the brain is impossible (the problem is not well-posed). However, by imposing reasonable modeling constraints, useful inferences about the source activity of interest can still be made. EEG in its own right has been analyzed to investigate the oscillatory activity of populations of neurons. Many studies have used the spectral properties of the EEG, i.e., decomposition of the total variance of the EEG into different frequency bands and cross-coherence between channels, to find associations between electrophysiology and behavior. Moreover, EEG has been used to monitor changes in brain activity arising from an epileptic seizure or following the administration of anesthetic drugs. Statistical methods and models for analyzing EEG are discussed in Chapters 20 and 21.

Functional MRI, covered in Chapter 6, is based on taking a sequence of particularly focused MRIs over time. The most common approach towards fMRI uses the Blood Oxygenation Level Dependent (BOLD) contrast, which allows for the measurement of the ratio of oxygenated to deoxygenated hemoglobin in the blood. Hence, it is important to note that fMRI does not measure neuronal activity directly, but rather the metabolic demands (oxygen consumption) of active neurons. When an area of the brain is in use, blood flow to that region also increases and these changes can be measured. The data acquired in an fMRI study consists of a sequence of 3D MRIs, each consisting of roughly 100,000 voxels, where the image intensity value corresponding to each voxel represents the spatial distribution of the nuclear spin density, which relates to blood oxygenation and flow, in the local area. During a standard fMRI experiment, between 100 and 1000 images of the whole brain are acquired. Image reconstruction is discussed in Chapter 8.

Each of the functional imaging modalities mentioned above have their own strengths and limitations, and provide a unique window into how the brain processes information and responds to stimuli. The temporal resolution of EEG and MEG is excellent, on the order of milliseconds, while fMRI and PET have time resolutions between seconds and minutes. In addition, activation in both PET and fMRI reflect changes in neural activity only indirectly, and they measure different biological processes related to brain activity, which may be broadly defined as the energy consumption of neurons. In contrast, the signals recorded by EEG/MEG directly reflect current generated by neurons within the brain. However, the spatial resolution of EEG and MEG is limited (on the order of 6 cm^3). The relatively good spatial resolution of PET and fMRI complement the precise timing information provided by EEG and MEG. The spatial resolution of PET is on the order of 200 mm^3. For fMRI, the resolution can be less than 1 mm^3 when performing high-field imaging in animals, but is typically on the order of 27–36 mm^3 for human studies. Thus, features such as cortical columns and even major sub-nuclei typically cannot be identified. The main advantage of MEG over EEG is that it has a relatively better spatial resolution, as magnetic fields are less distorted than electric fields by the skull and scalp. EEG tends to be sensitive to activity in more brain areas, but activity visible in MEG can also be localized with more accuracy.

In practice, PET and fMRI are the most widely used and provide the most anatomically specific information across the entire brain. The relatively good spatial resolution of PET and fMRI complement the precise timing information provided by EEG and MEG, and efforts have been made to combine the information obtained by each in isolation.

1.4 Statistical Methods

The statistical analysis of neuroimaging data is challenging for a number of reasons. First, it is a massive data problem akin to those faced in other modern statistical applications, such as genetics. Second, in most cases the signal of interest is relatively weak. Third, the data exhibit a complicated temporal and spatial noise structure.

Much of the early work on neuroimaging statistics involved solving so-called tomographic problems related to image reconstruction. Examples of important early work include Shepp and Logan's CT reconstruction algorithm (19), and Shepp and Vardi's PET reconstruction algorithm (21). In particular, Shepp and Vardi's algorithm provided the basis for many subsequent extensions in the statistics literature, including the suggested use of ordered Subsets Expectation Maximization (OSEM) (13), the addition of penalty terms to the likelihood function (4), and the use of Bayesian methods (1, 9, 11).

In the early 1990s there was a shift toward functional neuroimaging. A statistical advance that had a large influence on the field was Keith Worsley's seminal work on random field theory (RFT). To this day, multiple testing corrections in neuroimaging are often performed using RFT. RFT models a statistical image as a lattice representation of a continuous random field. Regions above a certain threshold are defined to be an excursion set. The

Euler Characteristic (EC) is a topological measure of an excursion set, and the expected EC is a good approximation of the family-wise error rate. Random field methods are able to approximate the upper tail of the maximum distribution, which is the part needed to find appropriate thresholds, and account for the spatial dependence in the data.

Because of these earlier advances, today, statistics plays an integral role in neuroimaging data analysis, and the number of statisticians involved in the field is increasing. This book seeks to explore some cutting-edge research topics in neuroimaging statistics. Below we provide an overview of methods covered in this book.

1.4.1   Preprocessing

Prior to any statistical analysis, all neuroimaging data undergoes a series of Preprocessing steps. The goals of preprocessing are (i) to minimize the influence of data acquisition and physiological artifacts; (ii) to check statistical assumptions and transform the data to meet these assumptions; and (iii) to standardize the locations of brain regions (across time or subjects) to achieve validity and sensitivity in group analyses.

While preprocessing is essential for the standard model assumptions required for statistical analysis to hold, there still needs to be a clear understanding of the effects they have on the data. For example, it is critical to study the interactions among the individual preprocessing steps. The neuroimaging modalities covered in this book require preprocessing pipelines of varying levels of complexity, with the most complex experimental and preprocessing pipelines required for modalities that have relatively high sampling rates in both space and time together with within-session, time-dependent experimental designs. Chapter 10 frames neuroimaging studies as an experimental pipeline with decisions made at multiple stages, from subject selection through data analysis, including the evaluation of pipeline efficacy using performance metrics.

1.4.2   Methods in Structural Neuroimaging

High-resolution structural imaging is used extensively in clinical settings, as it provides detailed anatomical information of the brain, is sensitive to many pathologies, and hence assists in the diagnosis of disease. Chapter 3 provides a brief overview of structural MRI, which has become the dominant modality in the field due to its spatial resolution, non-invasive nature, and wide availability. This is followed up by Chapter 5, which provides a detailed tutorial for analyzing structural MRI data with a particular focus on clinical settings.

Next, we review the widely used statistical analysis framework for data defined along the brain cortical and subcortical surfaces. The cerebral cortex has the topology of a 2D highly convoluted sheet. For data measured along curved non-Euclidean surfaces, traditional statistical analysis and data smoothing techniques based on the Euclidean metric structure are inefficient. Chapter 9 discusses basic concepts involved with differential geometry for surfaces and introduces various surface-based smoothing techniques used in the field. The smoothed surface data is then treated as smooth random fields and statistical inference will be naturally done within Worsley's RFT discussed above.

Changes in the human brain take many forms, which are studied in various fields including neurology, psychiatry, developmental psychology and radiology. Imaging studies, in particular using MRI, have helped to characterize the typical course of brain development in early life (7), and of brain aging in later life (2). With regard to abnormalities of cognition, behavior, and brain health, the focus has depended on the age group being studied. For example, research on disorders of children and adolescents has sought to pinpoint anomalies in neurodevelopmental trajectories (20), whereas studies of adult disorders have emphasized modeling disease-related brain changes (14). Chapter 18 aims to present a set of approaches with broad applicability to the study of change in the brain, and to highlight some areas where new statistical methods may help to drive scientific advances. A recurring theme will be the question of what can be inferred from cross-sectional versus longitudinal data.

1.4.3 Localizing Areas of Activation

One goal of functional neuroimaging is to determine which regions of the brain are active during a specific task. The use of linear regression to model activation in brain images and their modulation by various factors has enabled reliance on relatively simple estimation and statistical testing procedures. Specifically, the analysis of functional neuroimaging signals is typically carried out on a per-voxel basis, in the so-called mass univariate framework.

In this context, the voxel-wise general linear model (GLM) (25) has arguably become the dominant approach towards analyzing fMRI data. It tests whether variability in a voxel's time course can be explained by a set of a priori defined regressors that model predicted responses to psychological events of interest. GLMs treat the measured response as a linear combination of predictors plus error. Statistical inference of brain activation is performed using linear regression inference techniques such as the Student t-tests and analysis of variance. Chapter 11 introduces the GLM framework in the context of the analysis of fMRI data.

Most neuroimaging experiments are performed on groups of subjects. Multi-subject statistical analyses play an essential role in these experiments, making it possible to draw conclusions that hold with a prescribed confidence level for the population under study. Chapter 12 covers the general framework for group inference, mixed-effects model design and its simplifications, together with the various solutions that have been considered to improve the standard mass-univariate testing framework.

When performing mass-univariate analyses, the question of multiple-testing corrections cannot be ignored if valid statistical inference is to be obtained. Historically, the family-wise error rate has been a relevant quantity to control. However, for massive datasets this criterion is generally too conservative. Therefore, more recently, methods for the control of the false discovery rate have gained in popularity in the functional neuroimaging literature, as they are better able to accommodate a large number of tests carried out simultaneously. This topic is discussed in depth in Chapter 13.

Some studies collect neuroimaging data longitudinally over a span of many months or many years. The goal of these studies is to track changes in brain function or structure over a span of time. For example, in Alzheimer studies, clinicians study changes in neuroimaging biomarkers and their association with cognitive decline. In developmental studies, scientists investigate changes in brain structure and function and how these relate to behavior from pre-pubescent to young adulthood. Methods for analyzing longitudinal neuroimaging studies are discussed in Chapter 18.

1.4.4 Brain Connectivity

Researchers are becoming increasingly interested in determining how different brain regions are connected with one another. The idea is that in order to thoroughly understand brain function, researchers must study the interaction of brain regions, as a great deal of neural processing is performed by an integrated network of several brain regions. In the neuroimaging literature it is common to distinguish between anatomical, functional and effective connectivity. Anatomical connectivity describes the physical connections between different brain regions and is typically studied using techniques such as diffusion MRI (see Chapter 4). Functional connectivity is defined as the undirected association between two or more time series of measurements from different regions (see Chapter 14). Finally, effective connectivity is defined as the directed influence of one brain region on others (see Chapter 15). Each of these approaches has different goals and is performed using different methods.

Given the size and complexity of brain imaging datasets, dimension reduction and decomposition techniques are routinely used in analyses. In particular, multivariate decomposition methods, such as principal components analysis (PCA) and independent components analysis (ICA) are often used to analyze neuroimaging data. They also provide a means to assess functional connectivity, and have been critical in detecting so-called resting-state networks. Dimension reduction methods are discussed in detail in Chapter 16.

Finally, a recent trend is to view the brain as a network of interacting biological elements. This has motivated the use of network analytical methods in neuroimaging. Methodologically, the analysis of such networks falls under the emerging statistical field of complex network analysis. This is covered in Chapter 17.

1.4.5 Analysis of Electroencephalograms

The analysis of EEGs require, for the most part, methods and models that are different from fMRI. In Chapter 20, we cover time and spectral domain approaches to analyzing epochs of EEGs. The primary features of interest covered in this chapter are the spectrum and coherence. The former gives the relative contribution of different oscillatory activity to the total variance; the latter gives a measure of dependence (or correlation) between oscillations at different channels. In addition to spectral and coherence estimation, we also cover source estimation and penalized regression approaches when the number of channels is large. Chapter 21 gives a summary of some of the advanced approaches for modeling EEG. The key emphasis of the methods in this chapter is how to deal with non-stationarity in EEG data (i.e., when the statistical properties of the EEG such as correlation, variance, spectrum and so on evolve over time). This chapter covers methods for detecting change-points, representations using localized basis, evolution of network or clustering, and finally estimating changes in connectivity via switching vector auto-regressive processes.

1.4.6 Multi-Modal Analysis

In neuroimaging there has been a general trend toward using multiple modalities together in order to overcome the inherent limitations involved with using each approach for isolation. For example, EEG has an extremely high temporal resolution but a rather poor spatial resolution, while fMRI has high spatial resolution with lower temporal resolution. By combining these two modalities, it may be possible to harness the strengths of each technique and obtain both spatial and temporal information at a high resolution. In another example, diffusion tensor imaging (DTI) can be combined with fMRI to determine appropriate regions of the brain to include in connectivity models. This is discussed in Chapter 10. Finally, combining imaging and genetics has recently been seen as a way to study how a particular subset of polymorphisms may affect functional brain activity. The hope is that quantitative indicators of brain function could help facilitate the identification of the genetic determinants of complex brain-related disorders such as autism, dementia and schizophrenia (8).

Multi-modal imaging techniques promise to be important topics for future research in neuroimaging data analysis, and to fully realize their promise, novel statistical techniques will be needed. The problem of combining information from different modalities will be challenging to statisticians, if for no other reason than the sheer volume of data created. In addition, since the different modalities provide fundamentally different information, it may not always be immediately clear how to best combine this information.

1.5 Additional Topics

Due to space constraints, we are unfortunately unable to cover all the interesting and important topics involved in the analysis of neuroimaging data. Some additional topics that deserved their own chapters include meta-analysis, prediction, experimental design in fMRI and EEG, prediction and imaging genetics. Here we briefly touch on some of these topics.

1.5.1 Meta-Analysis

In the past decade, there has been tremendous growth in both the number and variety of neuroimaging studies being performed. For these reasons, it has thus become increasingly important to find ways to integrate and synthesize research findings in a meaningful way. Meta-analyses have become the primary research tool for accomplishing this goal ((23); (22)), as they allow for the pooling of multiple similar studies and can be used to evaluate the consistency of findings across labs, scanning procedures, and tasks. They also provide information about the specificity of findings in different brain regions or networks for particular task types. Meta-analyses have already been used to study many types of psychological processes, including cognitive control, working memory, decision making, language, pain, and emotion, and have also been used to summarize structural and functional brain correlates of disorders such as attention deficit disorder, schizophrenia, depression, chronic pain, anxiety disorders, and obsessive-compulsive disorder. See (16) for further examples and a discussion of commonly used techniques for performing meta-analysis.

1.5.2 Experimental Design in fMRI

Experimental design is a key component of any neuroimaging study, and there is a growing literature on how to construct an optimal design. What constitutes an optimal experimental design depends on the task, the ability to track changes introduced by the task over time, and what types of comparisons are of interest. In addition, in fMRI research, the delay and shape of the BOLD response, scanner drift and physiological noise all conspire to complicate experimental design. Not all designs with the same number of trials of a given set of conditions are equal, and the spacing and ordering of events is critical.

Several methods have been introduced for determining the optimal design parameters, and the sequencing of events that should be used in an experiment (e.g., (24); (17)). These methods define metrics for evaluating the estimation efficiency, detection power and randomness of the experiment, and apply search algorithms to optimize the design according to these criteria. As research hypotheses ultimately become more complicated, the need for more advanced experimental designs will only increase further, making this an interesting area of research.

1.5.3 Experimental Design in EEG

During the course of a neuroscience experiment (e.g., learning experiment), the neuronal response to stimuli (even to identical stimuli) may also evolve. Reference (5) develops the evolutionary locally stationary process model for studying changes in neuronal activity between regions engaged in the associate learning and reward. Many statistical models already take into account non-stationarity within a single trial of the experiment, but the evolution of brain dynamics across trials is often ignored. The model in (5) captures both sources of nonstationarity. Under the proposed model, the spectral density matrix evolves over time within a trial and also across trials. Their model produces estimates of coherence and the auto-spectrum that changes within a trial and also evolves across trials during the course of the experiment. An open question is how to model the evolution of the spectrum or coherence when there is correlation between trials.

A related approach, given in (15), developed a mixed effects Cramer spectral representation that models the effects of covariates on the power spectrum while accounting for potential correlations among the epochs collected from the same unit. The resulting log-spectrum has a functional mixed-effects (both deterministic and random) representation and is estimated using splines. Reference (6) has a similar modeling framework but the log spectrum (which also has a functional mixed-effects representation) is estimated using tree-structured wavelets. Both of these related approaches are developed only for single-channel or univariate time series. Models for multi-channel EEG that take into account between-trial correlation in an experimental design have yet to be developed.

It is known that differential brain response to sensory stimuli is very small compared to the overall magnitude of spontaneous EEG, stimuli are applied repeatedly and the ERP signals arising from the individual trials are averaged at the subject level. Reference (10) develops a model, based on a moving average of ERP across sliding trial windows, to capture such longitudinal trends across an experiment.

1.5.4   Imaging Genetics

Rapid advances are taking place in the study of human brain function. Concurrently, there have been significant advances in molecular genetics research. Integrating genetics with brain imaging is an important problem that has the potential to fundamentally alter our understanding of how the brain functions in diseased populations. For instance, it could provide a way to study how a particular subset of polymorphisms affects brain morphology, functional activity, and quantitative indicators of brain function could facilitate the identification of the genetic determinants of complex brain-related disorders such as autism, dementia and schizophrenia. These indicators may also aid in gene discovery and help researchers understand the consequences of specific genes or gene pathways at the level of systems neuroscience.

1.6   Conclusions

It is truly an exciting time to be involved in neuroimaging research. More and more increasingly ambitious experiments are being performed daily. This is creating a significant new demand, and an unmatched opportunity for quantitative researchers working in the neurosciences. Understanding the human brain is arguably one of the most complex, important and challenging issues in science today. For this endeavor to be successful, new researchers are needed to make sense of the massive amounts of data being generated. Our sincere hope is that this book can in some way aid in this development, by providing researchers with an introduction to the fascinating world of neuroimaging data analysis.

Bibliography

[1] Bouman, C. and Sauer, K. (1993), 'A generalized Gaussian image model for edge-preserving map estimation', *Image Processing, IEEE Transactions on* **2**(3), 296–310.

[2] Davatzikos, C. and Resnick, S. M. (2002), 'Degenerative age changes in white matter connectivity visualized in vivo using magnetic resonance imaging', *Cerebral cortex* **12**(7), 767–771.

[3] Epstein, C. L. (2008), *Introduction to the mathematics of medical imaging*, Siam.

[4] Fessler, J. et al. (1994), 'Penalized weighted least-squares image reconstruction for positron emission tomography', *Medical Imaging, IEEE Transactions on* **13**(2), 290–300.

[5] Fiecas, M. and Ombao, H. (2015), 'Modeling the evolution of dynamic brain processes during an associative learning experiment', *Journal of the American Statistical Association* p. Under revision.

[6] Freyermuth, J.-M., Ombao, H. and von Sachs, R. (2010), 'Tree-structured wavelet estimation in a mixed effects model for spectra of replicated time series', *Journal of the American Statistical Association*, **105**(49)) 634–646.

[7] Giedd, J. N., Lalonde, F. M., Celano, M. J., White, S. L., Wallace, G. L., Lee, N. R. and Lenroot, R. K. (2009), 'Anatomical brain magnetic resonance imaging of typically developing children and adolescents', *Journal of the American Academy of Child and Adolescent Psychiatry* **48**(5), 465.

[8] Glahn, D. C., Bearden, C. E., Barguil, M., Barrett, J., Reichenberg, A., Bowden, C. L., Soares, J. C. and Velligan, D. I. (2007), 'The neurocognitive signature of psychotic bipolar disorder', *Biological psychiatry* **62**(8), 910–916.

[9] Green, P. J. (1990), 'Bayesian reconstructions from emission tomography data using a modified em algorithm', *Medical Imaging, IEEE Transactions on* **9**(1), 84–93.

[10] Hasenstab, K., Sugar, C., Telesca, D., McEvoy, K., Jeste, S. and Senturk, D. (2015), 'Identifying longitudinal trends within EEG experiments', *Biometrics* p. In Press.

[11] Hebert, T. and Leahy, R. (1989), 'A generalized EM algorithm for 3-D Bayesian reconstruction from Poisson data using Gibbs priors', *Medical Imaging, IEEE Transactions on* **8**(2), 194–202.

[12] Horn, B. K. (1979), 'Fan-beam reconstruction methods', *Proceedings of the IEEE* **67**(12), 1616–1623.

[13] Hudson, H. M. and Larkin, R. S. (1994), 'Accelerated image reconstruction using ordered subsets of projection data', *Medical Imaging, IEEE Transactions on* **13**(4), 601–609.

[14] Jones, T. A., Allred, R. P., Jefferson, S. C., Kerr, A. L., Woodie, D. A., Cheng, S.-Y. and Adkins, D. L. (2013), 'Motor system plasticity in stroke models intrinsically use-dependent, unreliably useful', *Stroke* **44**(6 suppl 1), S104–S106.

[15] Krafty, R., Hall, M. and Guo, M. (2011), 'Functional mixed effects spectral analysis', *Biometrika*, **98**(3) 583–598.

[16] Lindquist, M. A. and Wager, T. D. (2015), *Principles of fMRI*, Leanpub.

[17] Liu, T. T. and Frank, L. R. (2004), 'Efficiency, power, and entropy in event-related fMRI with multiple trial types: Part i: Theory', *NeuroImage* **21**(1), 387–400.

[18] Natterer, F. (1986), *The mathematics of computerized tomography*, Vol. 32, Siam.

[19] Shepp, L. A. and Logan, B. F. (1974), 'The Fourier reconstruction of a head section', *Nuclear Science, IEEE Transactions on* **21**(3), 21–43.

[20] Treit, S., Lebel, C., Baugh, L., Rasmussen, C., Andrew, G. and Beaulieu, C. (2013), 'Longitudinal MRI reveals altered trajectory of brain development during childhood and adolescence in fetal alcohol spectrum disorders', *The Journal of Neuroscience* **33**(24), 10098–10109.

[21] Vardi, Y., Shepp, L. and Kaufman, L. (1985), 'A statistical model for positron emission tomography', *Journal of the American Statistical Association* **80**(389), 8–20.

[22] Wager, T. D., Lindquist, M. A., Nichols, T. E., Kober, H. and Van Snellenberg, J. X. (2009), 'Evaluating the consistency and specificity of neuroimaging data using meta-analysis', *NeuroImage* **45**(1), S210–S221.

[23] Wager, T. D., Lindquist, M. and Kaplan, L. (2007), 'Meta-analysis of functional neuroimaging data: Current and future directions', *Social cognitive and affective neuroscience* **2**(2), 150–158.

[24] Wager, T. D. and Nichols, T. E. (2003), 'Optimization of experimental design in fMRI: A general framework using a genetic algorithm', *Neuroimage* **18**(2), 293–309.

[25] Worsley, K. J. and Friston, K. J. (1995), 'Analysis of fMRI time-series revisited again', *Neuroimage* **2**(3), 173–181.

Part II

Imaging Modalities

2

Positron Emission Tomography: Some Analysis Methods

John Aston

University of Cambridge

CONTENTS

2.1	Introduction	....	17
2.2	Background	....	19
2.3	Tracer Kinetic Modelling: Compartmental Approaches	....	20
	2.3.1	Plasma Input Functions Models	....	20
	2.3.2	Reference Tissue Models	....	23
2.4	Estimation and Statistical Methods	....	24
	2.4.1	Non-Linear Least Squares	....	24
	2.4.2	Basis Function Methods	....	25
	2.4.3	Model Selection	....	26
2.5	Other Modelling Approaches	....	27
	2.5.1	Graphical Methods	....	27
	2.5.2	Bayesian Approaches	....	28
	2.5.3	Non-Parametric Approaches	....	28
2.6	Further Modelling Considerations	....	29
	Acknowledgements	....	30
	Bibliography	....	30

2.1 Introduction

Positron Emission Tomography (PET) is one of the older neuroimaging techniques, having been first realised in the early 1950s with roots even further back (42) and with significant development in the early 1980s. PET is a non-invasive imaging technique, but works on the principle of using injected radiotracers (radioactive compounds present in tracer amounts) with specific biochemical properties to mimic systems in the body or actions of other known non-radio-labelled chemical compounds. The emission of positrons is observed using a PET tomograph scanner, which can produce three-dimensional images over time of the concentrations of the radio-labelled compounds.

For neuroimaging, the primary initial role of PET was in tracking cerebral blood flow (12). While fMRI has now almost exclusively replaced PET for such usage, PET does have an intrinsic advantage over fMRI, in that injected radioactivity can be associated with measured radioactivity from the scanner, leading to absolute quantification in PET as opposed to the relative measurements delivered by fMRI. This is increasingly important in cancer studies (not limited to brain cancers, but to cancers anywhere in the body), as

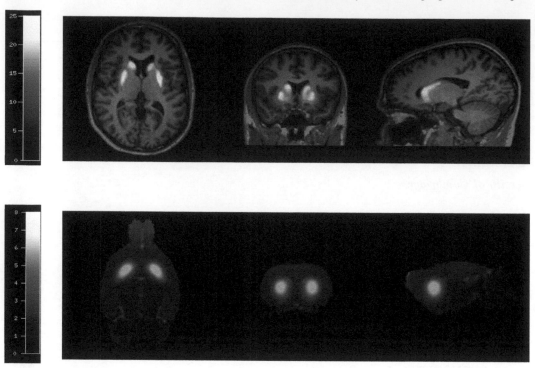

FIGURE 2.1
Images of $[^{18}F]$Fallypride, which is a high-affinity dopamine (D2/D3) receptor antagonist.
Top image is binding potential in healthy human; bottom image is binding potential in
control rat. Both are overlaid on their respective structural MRIs. Images come from studies
as described in (8) and (9).

radio-labelled Fluorodeoxyglucose or $[^{18}F]$FDG can be used for the measurement of
metabolism, with cancers naturally having high metabolic rates. This has led to a surge in
the use of PET in clinical oncology imaging studies (13).

The current primary use of PET in research is in the design and analysis of novel
radiotracers to interrogate different neurochemical systems. As can be seen in other chapters,
fMRI is primarily associated with blood flow because the magnetic differences in oxy and
deoxy hemoglobin can be used to understand blood flow changes. However, as PET tracers
can be designed to target any number of brain chemical systems, this makes PET ideal
for the study of a wide variety of neuroreceptor systems, such as the dopamine system,
the serotonergic system and the glutamate systems. In addition, specific pharmaceutical
products can be labelled, allowing quantitative *in vivo* investigation of their efficacy or
biodistribution (26). PET also allows targeted investigation of systems that may be involved
in neurological disease such as dementia (18). Indeed, in many settings, it is the only reliable
way to investigate these systems or compounds *in vivo*.

This chapter begins with a short overview of the PET measurement system, and then
proceeds to the mathematical and statistical models often used for the analysis of PET
data. PET is often used in both academic and pharmaceutical research, both on animal
and human subjects (see Figure 2.1). However, discussion in this chapter will concentrate
on the recording and analysis of human subject data. While different PET tracers will be
investigated, the study of compartmental and other modelling systems for the study of
general neuroreceptor tracers will be the main focus.

2.2   Background

Here we will give a very short introduction to PET, and some of its early uses. However, a comprehensive review of the background to PET is given in (32, 37).

PET, as an in vivo imaging procedure, works on the principle that compounds, which can be associated with various biochemical processes at a molecular level, can be synthesised to contain radioactively decaying atoms that produce positrons as part of their decay. Common examples of such atoms are ^{11}C, ^{15}O, and ^{18}F, and these can be produced using a cyclotron. These elements typically have short half-lives, for example approximately 20 minutes for ^{11}C, and therefore only remain in the body for relatively short periods of time after injection.

PET begins by taking the radio-labelled compound and injecting it into the subject. From there, the patient is placed into the PET scanner, which primarily consists of rings of detectors, crystals of bismuth germanium oxide or similar materials which are coupled to photomultiplier tubes. Events are detected when positrons collide with electrons (antimatter meeting matter) resulting in the generation of energy in the form of photons. In particular, in the case of positron annihilation, two gamma rays are produced almost exactly 180 degrees to each other (by conservation of momentum), and simultaneous detection on either side of the ring of detectors is considered an "event", where the position of the event is known up to a line. Reconstruction methods then take these line events and solve the resulting inverse problem (often using techniques such as filter back projection or order subset expectation maximization algorithms). While there are considerable mathematical and statistical aspects to solving these underlying inverse problems, these are not the focus of this chapter, as we will concentrate on working with the reconstructed images. However, issues such as signal attenuation, scatter correction as well as numerous other issues related to the physics of the underlying collection mechanism need to be included in the reconstruction (see (32) for an overview of physics and reconstruction in PET). However, anatomical information can be very useful in constraining such reconstructions and modern PET cameras are usually multimodal in that they combine PET and computed tomography (CT) or PET and magnetic resonance (MR) cameras in one machine.

Initially, the mainstay of PET was [^{15}O]Water PET, where a moderately automated process of producing radioactive water could be used. This was necessary as ^{15}O only has a half-life of a little over two minutes (122.24 secs). However, due to water's ready combination into blood, ^{15}O-Water PET could be used to track blood flow in the body, especially cerebral blood flow. This provided a quantitative measurement of blood flow into the brain and produced the first *in vivo* "brain-mapping" results. However, with the advent of fMRI (30), which does not require injection of radio-labelled compounds, [^{15}O]Water PET has fallen out of usage, despite being quantitative, in the sense that a known amount of radioactive material is injected, so detected radiation can be used to directly quantify blood flow, where most fMRI is purely relative within the scan.

Currently the most popular usage by far of PET is the use of [^{18}F]FDG to detect various forms of cancer. [^{18}F]FDG is a glucose analogue and so is especially present in areas where there is considerable metabolic activity, such as in cancerous cells. Therefore, "hot-spots" on clinical images can identify areas of current tumour activity and indeed, delineate such activity from surrounding areas of necrosis. This can therefore make PET a ready tool for the diagnosis of presence and even progression stage of tumours. [^{18}F]FDG can also be relatively easily produced and distributed even to fairly remote sites from the original production centre, particularly thanks to the long half-life of [^{18}F] (109.8 minutes). In addition, as will be mentioned later, the chemical kinetics of [^{18}F]FDG are relatively simple and easy to model, to the extent that a single scan taken around an hour after injection is often enough to produce useful quantification, without the need for more complex blood modelling that is needed with more complex neurochemical radioligands. This makes the resulting analysis of the $[^{18}F]FDG$ PET scans relatively simple for clinical usage (13).

However, modern PET analysis can yield considerably more specific insights than simple metabolism or blood flow. These are of most interest (with exceptions) in the study of brain neurochemical systems, where PET radiotracers can be designed to target specific neurochemical systems, for example, the dopamine system (50), the serotonergic system (10), the opioid receptor system (22), amyloid plaques and tau-proteins (18) and numerous others. While the production of such ligands often requires considerable complex radiochemistry, compared with say $[^{18}F]FDG$ or $[^{15}0]Water$, the ability to produce ligands that cross the blood–brain barrier allows insights into specific neurochemical systems, these insights not currently being routinely possible with any other imaging modalities. However, neurochemical differences and changes have been linked to almost every major neurological and psychiatric disorder (23). For example, neuroinflammation after stroke can clearly be seen using the PET tracer $[^{11}C]PK11195$; see Figure 2.2.

While the radiochemistry is complex, after reconstruction, the analysis of the resulting images is also not necessarily straightforward. Considerable mathematical and statistical modelling of the process needs to be performed in order to understand the underlying information that has been captured by the scans. This will be the subject of the rest of this chapter.

2.3   Tracer Kinetic Modelling: Compartmental Approaches

The major principle behind almost all PET (neuroreceptor) modelling is that the time activity curve data can be well modelled by a set of linear first order ordinary differential equations (although we will also examine methods which go beyond these assumptions later in the chapter). This assumption arises from the pharmacokinetic nature of the PET radiotracers. It is assumed that first-order kinetics govern the major phases that the PET tracer undergoes. Firstly, it has to cross the blood–brain barrier, and then it either binds specifically to the receptor of interest or non-specifically to other receptor systems or indeed some combination of the two. This binding can occur either reversibly or irreversibly (at least on the time scale of the scan), and it is also assumed that these processes are governed by first-order kinetics. In addition, all modelling assumes that the PET radioligand is only present in tracer quantities, and as such has no material effect on the system that it is measuring (i.e., there is little pharmacodynamics).

For a comprehensive reference concerning the mathematical models in many general situations for PET compartmental models, the reader is referred to (15). However, given the ubiquitous nature of the methods, they are summarised in the next subsections.

2.3.1   Plasma Input Functions Models

Depending on the availability of the data, methods involving plasma input functions are usually considered to be the gold standard. Plasma input functions can be derived from on-line arterial blood sampling that is carried out concurrently with the PET measurements. This allows the concentration of PET tracer in the blood to be calibrated to the activity recorded by the PET camera, allowing a true quantification to be given, as a known amount of radiotracer is injected into the patient. However, arterial blood sampling is both difficult

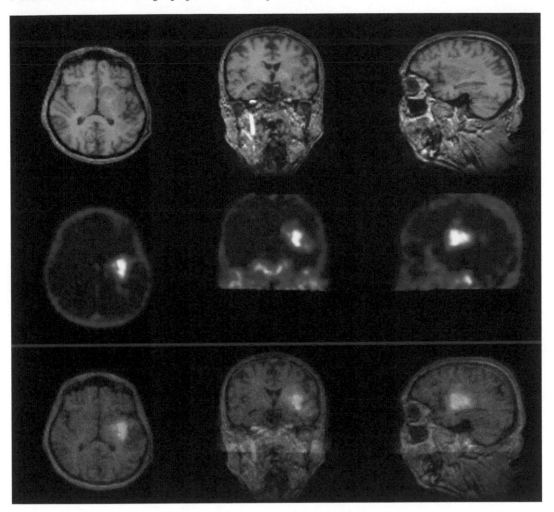

FIGURE 2.2
[^{11}C]PK11195 imaging of inflammation 10 days after stroke and coregistered MR scan. Images from study as described in (38).

to manage in practice and not something that is particularly enjoyable for the patient, hence the use of alternative, reference tissue models, as described in the next subsection.

Using the notation of (15), if plasma input data is available, then the linear system under investigation can be formulated as a state space model. For a given plasma input, $C_P(t)$

$$\dot{\mathbf{C}}_{\mathbf{T}}(t) = A\mathbf{C}_{\mathbf{T}}(t) + K_1 e_1 C_P(t)$$
$$C_T(t) = \mathbf{1}^T \mathbf{C}_{\mathbf{T}}(t)$$

with $\mathbf{C}_{\mathbf{T}}(0) = \mathbf{0}$. Here $\mathbf{C}_{\mathbf{T}}(t)$ represents the radiotracer concentration in each of the underlying compartments, while $C_T(t)$ is the summed total of all compartments. e_1 and $\mathbf{1}$ represent the unit vector with one in the first row and the vector of all ones, respectively. The two parameters of interest are the matrix A and the input rate from plasma to brain K_1. The matrix A is negative semi-definite (although not necessarily symmetric), with a negative diagonal and non-negative entries elsewhere. For simplicity, we have here ignored the blood volume component, but this can be added relatively easily into the above system, see (15) for details. The solution to this ODE system can easily then be framed in terms of the parameters of interest:

$$\mathbf{C_T}(t) = \int_0^t e^{A(t-s)} K_1 e_1 C_P(s) ds,$$

where the matrix exponential $e^{At} = \sum_{k=0}^{\infty} \frac{(At)^k}{k!}$.

It is often more useful to consider this result in terms of the impulse response function (IRF) $H(t)$ as this is often the quantity of interest as we want to compare estimates without the effect of differing inputs on the system. These IRFs can be determined in terms of microparameters ϕ, θ which due to the exponential nature of the solution, mean the overall solution can be written as

$$C_T = H(t) \bigotimes C_P(t) \tag{2.1}$$

where

$$H(t) = \sum_{i=1}^n \phi_i e^{\theta_i t}. \tag{2.2}$$

In this system, we allow two possible model configurations. Firstly, if all compartments are reversible (i.e., there is both inflow and outflow from a compartment), then the above $H(t)$ requires $\theta_i > 0$ for all i. If one compartment is irreversible (i.e., the tracer becomes trapped), then w.l.o.g $\theta_n = 0$. It is somewhat straightforward to see that by definition $\sum_{i=1}^n \phi_i = K_1$.

In many cases, mainly due to microparameter (in)stability when estimated from noisy data, macroparameters are usually considered instead. These are functions of $H(t)$, and therefore implicitly functions of the microparameters. In the case of reversible models, the volume of distribution in the tissue, V_T, is defined to be integral to infinity of the IRF, which given the exponential nature of the solution, is simply

$$V_T \doteq \int_0^{\infty} H(t) = \sum_{i=1}^n \frac{\phi_i}{\theta_i}, \tag{2.3}$$

while in the case of irreversible models, the K_I of the system is simply the long-run kinetic behaviour

$$K_I \doteq \lim_{t \to \infty} H(t) = \phi_n.$$

As a concrete example of a plasma input model, the two-tissue compartmental model is now considered. In this model, there are assumed to be two compartments in the brain along with the associated blood input. This yields the following compartmental setup:

$$\frac{dC_1(t)}{dt} = K_1 C_P(t) - (k_2 + k_3) C_1(t) + k_4 C_2(t)$$

$$\frac{dC_2(t)}{dt} = k_3 C_1(t) - k_4 C_2(t)$$

where k_2 indicate the rate constants from compartment 1 back to plasma, k_3 from compartment 1 to 2, and k_4 the reverse direction. While the exact form of the microparameters in terms of the rate constants is somewhat complex (see (15)), the V_T of this model turns out simply to be

$$V_T = \frac{K_1}{k_2} \left(1 + \frac{k_3}{k_4} \right)$$

where it can easily be seen that V_T is infinite if the model is irreversible (i.e., $k_4 = 0$). In this case, the long-run behaviour, K_I, is of greater interest and is given by

$$K_I = \frac{K_1 k_3}{k_2 + k_3}.$$

2.3.2 Reference Tissue Models

In many practical situations, it is preferable for both the experimenter and the patient to forego the collection of arterial blood samples. However, this leaves the issue of how to evaluate the models without the measurement of $C_P(t)$. One solution for this is to consider whether a region of the measured PET scan can be used as a surrogate for the input, a reference region. If it is known that a region is devoid of the neuroreceptor of interest (from post-mortem neuroanatomical studies, for example), then the relative binding, under certain assumptions, can be found between the target region and the reference region.

For simplicity, we will restrict discussion here to reversible reference tissue models, where the reference tissue model is assumed to be reversible and no blood volume component is present, but for a complete discussion again see (15). By assuming that the reference and target tissues are linked purely through the plasma driving input, it is possible to again derive a solution to the compartmental models. These can then be reformulated to express the target tissue response in terms of the reference response.

$$C_T(t) = H_R(t) \bigotimes C_R(t)$$

where the notation $H_R(t)$ is used explicitly to show that this impulse response is defined via an impulse with respect to the reference tissue. The corresponding definition is then

$$H_R(t) = \phi_0 \delta(t) + \sum_{i=1}^{m+n-1} \phi_i e^{\theta_i} \qquad (2.4)$$

where $\phi_0 = K_1/K_1^1$ and there are n compartments in the target tissue and m compartments in the reference tissue, and $\delta(t)$ is the usual Dirac delta function. Similar restrictions as before are placed on θ_i depending on the reversibility or not of the target model. In addition, the macroparameters V_T and K_1 are now normalised by the V_T^1 of the reference tissue.

In many situations, the assumption is made that the V_T between the target and reference tissues only differ through the specific binding in the target to the neuroreceptor of interest. In this situation, the binding potential is defined to be

$$BP_{ND} \doteq V_T/V_T^{(R)} - 1$$

in a slight abuse of notation with respect to the current nomenclature in PET (19).

A simple example of a reference tissue model is the full reference tissue model where a single-tissue compartmental model is assumed as the reference tissue and a two-compartmental model for the target. This essentially means that the full model can be expressed as

$$dC_{T_1}(t) = K_1 C_P(t) - (k_2 + k_3)C_{T_1}(t) + k_4 C_{T_2}(t)$$
$$dC_{T_2}(t) = k_3 C_{T_1}(t) - k_4 C_{T_2}(t)$$
$$dC_R(t) = K_1' C_P(t) - k_2' C_R(t)$$

where K_1' and k_2' are the parameters associated with the reference corresponding to the parameters in the target tissue, and the other rate constants are defined analogously to the two-tissue plasma input function model above. Again, while the actual IRF has a somewhat complicated relationship to the parameters, the macroparameter estimate of BP_{ND} can be calculated as

$$BP_{ND} = \frac{K_1 k_2' \left(1 + \frac{k_3}{k_4}\right)}{K_1' k_2} - 1$$

which equates exactly to the ratio of two V_T parameters, one from a two-tissue compartmental model and the other from a one-tissue compartmental model (used as the model for the reference region). In addition, identifiability of the model is a consideration at this point, and the more complex the model, the more concerns over identifiability exist, both from a theoretical perspective and in practice, as only noisy and limited data are observed to fit the model.

2.4   Estimation and Statistical Methods

In practice, in most situations, of primary interest in many neuroreceptor studies is the estimation of macroparameters from the compartmental fit. These are then used in a variety of situations, in similar ways to regression parameters from fMRI models. However, given the more complex non-linear nature of these compartmental models, priority in the PET literature has often been given to parameter estimation rather than statistical properties of the resulting estimators. In this section, the most common methods of parameter estimation will be reviewed, and where possible, linked to related statistical properties.

2.4.1   Non-Linear Least Squares

Non-linear least squares (NLLS) are generic algorithms to solve least squares problems where the data is a non-linear function of the parameters. Given the least squares nature of the optimization, it is often convenient for statistical evaluation to make a Gaussian assumption with regards to the observation errors, which are usually the only source of error to be considered in the analysis. As an aside, it is possible to determine compartmental models from stochastic ODEs but given this is rarely used in PET, it will be assumed that the only source of stochasticity in the data comes from observation error.

Given that the reversible and irreversible models above basically define non-linear regression problems in the parameters of interest, these can typically be solved using standard NLLS algorithms such as the Levenberg–Marquardt algorithm (27), or in some cases the Nelder–Mead simplex algorithm (29). However, NLLS algorithms in general tend to require certain amounts of monitoring, particularly with regards to their convergence and instability. While this is feasible if only one or two regions are being assessed, if a voxel-by-voxel analysis of the data is to be carried out, it is impossible to manually tune all the NLLS algorithm evaluations. For this reason, there have been a number of simplifications made in the PET literature to allow fast and convenient assessment of the models without the inherent instability. However, these methods (which will be considered in further detail below) aim to solve the NLLS optimisation problem, and as such, the statistical analysis can be considered somewhat independent of the optimization method chosen.

The statistical analysis of PET data routinely takes the form of computing the parameters across a group of scans (maybe patient versus controls, or test–retest variability) and then assessing whether there is a statistical difference based on a t-test between the two groups. However, this ignores the considerable information associated with the parameter estimates that can be obtained from the dynamic information in the scan. Given the non-linear nature of the model, much can be made of the association with standard statistical practice in non-linear regression settings (40).

Given a set of measurements, y_i at times t_i, $i = 1, \ldots, T$, it is assumed that the observed data follow the following model

$$y_i = C_T(t_i, \beta) + \epsilon_i, \quad \epsilon_i \sim N(0, \sigma^2)$$

which given our use of non-linear least squares, implies that solving the following objective function is of interest:

$$\widehat{\beta} = \mathrm{argmin}_\beta \{ (y_i - C_T(t_i, \beta))' \, (y_i - C_T(t_i, \beta)) \},$$

which is the classic NLLS problem and β is the vector of unknown parameters in the model. As can be shown, see for example (40), this allows a subsequent variance estimate to be found for the parameters

$$V(\widehat{\beta}) = \left[\left(\frac{\partial C_T(t_i, \widehat{\beta})}{\partial \beta} \right)' \left(\frac{\partial C_T(t_i, \widehat{\beta})}{\partial \beta} \right) \right]^{-1} \widehat{\sigma}^2$$

where $\widehat{\sigma}^2$ is a suitable estimator for the variance, usually based on the residuals to the fit, and the notation implying that the partial derivative is evaluated at the least squares estimate of the parameters.

This approach was utilised in (3) to provide improved estimates for the variance, where the analysis is explicitly shown for a simplified reference tissue model. In addition, it can be combined with the approach by (51) for fMRI to yield t-tests based on random effects with improved degrees of freedom, taking advantage of the extra local spatial information that is available if a voxel-by-voxel approach is used (2). This is achieved by spatially smoothing ratios of fixed- to random-effect variances and borrowing strength of the local spatial information over the highly precise (large degrees of freedom) fixed-effects estimates relative to the random-effects variances (low degrees of freedom), and using a Satterthwaite approximation to establish the final degrees of freedom of the smoothed estimates.

2.4.2   Basis Function Methods

Given the nature of the IRF, as given in (2.2) and (2.4), it is somewhat natural to think about using a basis function strategy to solve the NLLS problem. This is the fairly well-known statistical concept of separable least squares (40). It can be noted in (2.2) and (2.4) that the non-linear parameters do not depend on the linear parameters and vise versa, and indeed, the function can be optimised by considering a range of solutions for the non-linear exponents, and then solving the associated linear problems to obtain the linear parameters (which can be done in a very fast and stable manner, as it is now a linear problem).

The authors of (17) proposed this approach for PET modelling of the simplified reference tissue model, although the idea goes back much further to the paper of (7) if not further. Indeed, given the physiological nature of PET, there are natural bounds for the non-linear parameters to be physiologically meaningful. By careful setup of the optimisation problem, it is possible to exhaustively search across a set of basis functions (where the basis is defined by taking a discretised interval of the non-linear parameter space), and choose the optimal combination in terms of least squares error of a predefined subset of basis functions. For example, if a two-tissue compartmental model is to be fit to the data, then the optimal combination of two basis functions and their associated linear parameters can be found by selecting the best basis from a range of say 1000 functions over the physiologically plausible range. In this way, a stable least squares solution can be found, both computationally fast as well as stably (as only linear fits are being considered).

This approach also offers a multitude of extensions that can be related to model selection as will be discussed next.

2.4.3   Model Selection

Model selection is an important problem in any discipline, and PET is no exception. In most situations, the primary question of interest is how many compartments to choose. This problem is both statistical as well as practical, because in many situations a large number of compartments cannot be reliably estimated from the data regardless of the choice of statistical analysis (due to problems of identifiability), and thus this question in PET usually reduces to whether a one-, two-, or three-tissue compartmental model is most appropriate.

Standard model selection techniques, particularly Akaike's celebrated information criterion (AIC), are frequently used. These simply compare the likelihood of the model, as assessed through the least squares fit to the data, with a penalty of twice the number of additional parameters in the model. However, it has been noted (48) that in PET, it is often the case that the parameters or even the model is not the primary outcome of interest; it is often macroparameters such as V_T and BP_{ND}. These parameters can be commonly defined across a number of models, and thus, in the end, a weighted combination of models may well be more informative and reliable than any one individual model—a model combination approach rather than a model selection approach. In (48), the approach was taken to use the AIC weights to define a probabilistic weighting of the models, and then associate the V_T as a weighted combination of the individual model's V_T's. This can be further formalised into a Bayesian approach as will be discussed later in the chapter.

A further approach is very much related to the basis function approach used to solve the initial NLLS problem. The problem can be reformulated as

$$y_i = \Phi \omega + \epsilon_i$$

where Φ is the range of basis functions derived from the underlying model class (in this case the non-linear exponents in (2.2), for example), and ω are the linear parameters associated with them. In most cases the number of observations of y_i (defined as T above) is usually on the order of 20–40, while for reasonable solutions to the NLLS problem, 100–1000 basis functions are required. This is then an overcomplete system, which has no unique solution.

However, this type of problem has probably been the most studied issue in statistics in the last 15–20 years, thanks to the seminal papers on the LASSO (43) and basis pursuit (6). By reformulating the problem above as a joint L_1 and L_2 minimization problem

$$\widehat{\omega} = \mathrm{argmin}_\omega \{|y - \Phi\omega|_2 + \lambda |\omega|_1\}$$

where λ is the regularisation parameters helping to enforce sparsity, which in PET is usually calculated using cross-validation (16). In this way, combined parameter estimation and model selection can be performed. The number of non-zero elements in ω indicates the number of compartments selected in the model.

However, possibly the most widely used approach in PET, when it is of interest to consider an unknown compartmental structure, is PET spectral analysis (7). This basis function approach works in a similar fashion to the LASSO approaches although it predates these approaches and is based on the idea that a spectrum of peaks in the basis range corresponds to the underlying compartmental structure. The idea behind PET spectral analysis (not to be confused with spectral analysis in time series, which is a fundamentally different concept) is that given the non-negative constraints on the parameters (due to their arising from chemical kinetics, which enforce non-negativity on the parameters an a necessary condition of identifiability), the function minimization can be turned from a NLLS problem into a non-negative least squares problem (NNLS)

$$\widehat{\omega} = \mathrm{argmin}_{\omega \geq \mathbf{0}}\{|y - \Phi\omega|_2\},$$

which has a well-known algorithmic solution (under certain conditions) (24). It turns out that the exponential basis with a small number of components (less than half the number of time points) can be shown to still yield a unique solution to the minimisation even if the basis representation is overcomplete (see (14)).

However, as can be imagined from the resulting minimization problem, the parameter estimates are positively biased (due to the non-negativity constraint), and thus it is hard to give a rigorous statistical characterisation of the resulting estimates. However, (44) used a bootstrap approach to obtain approximate confidence intervals for the parameter estimates, particularly of V_T as obtained from PET spectral analysis.

2.5   Other Modelling Approaches

While direct tracer kinetic compartmental modelling is the mainstay of dynamic modelling in PET, other approaches, particularly graphical approaches, are also routinely used. While these approaches may or may not be explicitly based on compartmental assumptions, they abstract or even challenge the notion that compartmental models are necessarily the only possible process for modelling PET data.

2.5.1   Graphical Methods

The development of graphical models is intrinsically linked to the development of $[^{18}\mathrm{F}]$FDG. The Patlak plot (35) is based on the idea that when examining irreversible tracer uptake, after a certain amount of time, denoted t^*, the main features of the analysis will be that there is a linear relationship between the integrated plasma input and the measured tissue activity,

$$\frac{C_T(t)|_{t>t^*}}{C_P(t)} \approx \mathrm{Const} + K_I \frac{\int_0^t C_P(s)ds}{C_P(t)},$$

where the constant is related to the volume of distribution of the reversible part of the system. Whilst the formal derivation is given in (35), a heuristic derivation can be easily seen by considering the time-limiting behaviour of (2.2) for irreversible tracers.

The equivalent analysis for reversible regions was extended from the Patlak analysis by (25), and is known as a Logan plot. This uses a slightly modified form of the equation

$$\frac{C_T(t)|_{t>t^*}}{C_T(t)} \approx \mathrm{Const} + V_T \frac{\int_0^t C_P(s)ds}{C_P(t)}$$

where we have assumed there is vascular contribution to simplify exposition here. This would initially seem like an excellent way to characterise the volume of distribution, but on closer inspection, it is clear that there are statistical issues with such analysis. While it is often (fairly reasonably) assumed that C_P can be measured without noise, this is not the case for C_T, yet it is being used to normalise the regression. This is not the case for the Patlak plot where C_P is being used for the normalisation. It has indeed been shown by (41) that this is a major issue for the Logan plot, while updated estimators to account for the noise have been considered in (31).

Graphical analysis can also be extended, similarly to reference tissue models, but the caveats with respect to noise bias are more intrinsic in this case.

2.5.2 Bayesian Approaches

In parallel to much of the work on tracer kinetic modelling in PET from a more classical biomathematical point of view, there has also been work on explicitly Bayesian formulations of the process. This can take several forms. Some of the early work was on using Bayesian prior specification from parts of the image to help inform other parts of the analysis (45). This was carried out using wavelet analysis to place spatial prior information on the parameter estimates (see Section 2.6 for further examples of wavelet models in PET).

However, intrinsically computational considerations have dominated the ability to perform full Bayesian analysis in PET, particularly on a voxel-by-voxel basis, and it is only recently that this has even been a possibility. An early Bayesian formulation was that of Relevance Vector Regression for PET analysis, which took a sparse Bayesian Learning Approach to model selection and parameter estimation (36). The authors of (1, 11) used previous parametric images as prior information for parametric image estimation. By carefully choosing the spatial representation of the prior information, this was shown to be an effective way of estimating parameters in situations where there was a low signal-to-noise ratio, such as in parametric images.

The idea of using prior information was taken in a different direction by (53), which showed that model selection could be carried out using biologically informed priors derived from mathematical spectral analysis (in the sense of eigenanalysis). This was a computationally expensive analysis, but by efficient implementation, a full MCMC-based approach could be used, even in a parametric imaging situation, yielding not only point estimates, but also posterior distributions associated with parameters and model choices. This shows that modern Bayesian computational approaches are becoming a realistic alternative to classical approaches for PET imaging, and indeed across neuroimaging as a whole (5).

2.5.3 Non-Parametric Approaches

While compartmental analysis of PET time courses is by far the most prevalent approach in PET, there are, quite understandably, some concerns as to whether compartmental models are reasonable models for data from all regions of the brain. This was most eloquently argued in (34), which proposed using non-parametric methods to determine outcomes in PET. This requires a somewhat fundamental shift in the notions of the outcomes of interest. V_T and K_I are, by definition, associated with the models time-limiting behaviour. Non-parametric approaches can realistically only be defined on the data range itself, and extrapolation becomes increasingly problematic over longer ranges. For this reason, (34) proposed using measures related to $H(t)$ (as in (2.1)) but only evaluated on the time scales of the scan, rather than over the infinite horizon. This does not preclude a comprehensive biological setup for the resulting PET analysis, but rather places more emphasis on the interpretation of the function rather than model parameters.

Non-parametric approaches also have been used as preprocessing steps for subsequent compartmental analysis. This is due to the widely recognised problems of low signal-to-noise ratios in non-linear modelling (as mentioned in several contexts above). The authors of (21) gave a functional principal component preprocessing procedure that was both computationally efficient but also subsequently allowed better modelling with a compartmental approach due to the noise reduction achieved by the non-parametric smoothing via functional principal component analysis.

Very recent work, (20), has combined these two approaches to produce a functional data analysis approach to the entire deconvolution process. The main idea is that information can be borrowed across the brain (spatially) to get good estimates of the underlying functions in (2.1) that are represented in the data (without the assumption of an explicit model or class of models), and then surrogates of common parameters such as V_T (as in (2.3)) can be defined from these non-parametric estimates (similarly to the approaches suggested by (34, 33)), for example by replacing in (2.3) the infinite integral with a finite one, based on the length of the scan. The deconvolution can be somewhat easily handled due to the assumptions made as part of the functional data approach and particularly the positive nature of the input function, allowing a principal component analysis approach to deconvolution to be taken. These non-parametric methods have all recently been investigated on a number of different radiotracers and the simple approach taken by (20) has been shown to perform remarkably well and robustly in a number of settings (52).

2.6 Further Modelling Considerations

Although the above represent the majority of the considerations in modelling for PET, particularly at the time course level, there are numerous other issues that have statistical implications in the analysis. As mentioned earlier, we have ignored any blood volume component in the modelling above, although as shown in (15), this is fairly easily incorporated into the modelling and explicitly accounted for.

Another source of considerable heterogeneity in the data is that fact that many radiotracers are metabolised in the body, producing radio-metabolites. When arterial inputs are used, if the metabolites do not cross the blood–brain barrier, then these need to be corrected for within the input function. Statistical approaches, using non-linear mixed-effect models, have been suggested for such analysis (49). If they do cross the blood–brain barrier, then these need to be included in the model explicitly, depending on their affinity for the receptor system and their differing properties from the parent (or otherwise), and cause the standard models to no longer be appropriate in many cases.

Further considerations include issues such as partial volume correction. This is where, due to the limited resolution of the PET camera, the underlying time courses are mixtures of multiple sources. There is considerable literature on how to account for this using mathematical and statistical methods which combined additional multimodal information such as segmentations of structural MRIs (28, 39, 4). Indeed, this is one example, among many, of where it is of great importance to realise that voxel or ROI time courses do not arise independently from one another, and that a full statistical analysis really should take advantage of the considerable spatial data available. One successful way of achieving this in PET has been through the use of wavelets (47), where different spatial resolutions are examined simultaneously. This allows information to be borrowed across space, which can greatly enhance detection. Both theoretical considerations (46) and empirical studies (47) have shown this to be of significant use, reducing both bias and variance of the results (due to the non-linear models being considered). However, with the advent of modern big data techniques and the rapid growth in computational power, particularly taking advantage of parallel architectures for these kinds of problems, it is likely more complex full spatio-temporal analyses of PET data will soon be routine.

This chapter has only covered a fraction of the modelling techniques that are routinely used in PET, and indeed not covered some of the methods for techniques such as bolus-infusion experimental designs, for example. New techniques and software are routinely being developed to overcome many of the inherent difficulties mentioned above. This is because PET remains at the forefront of neuroreceptor mapping, and indeed is one area of neuroimaging where the analysis truly is at the interface of biomathematical and statistical modelling.

Acknowledgements

I am very grateful to my many collaborators, in particular, Federico Turkheimer and Roger Gunn for all their insights into PET modelling. The images were very kindly provided by Franklin Aigbirhio of the Wolfson Brain Imaging Centre in Cambridge. All errors are, of course, solely the responsibility of the author.

Bibliography

[1] Nathaniel M. Alpert and Fang Yuan. A general method of Bayesian estimation for parametric imaging of the brain. *NeuroImage*, 45(4):1183–1189, 2009.

[2] J.A.D. Aston. Statistical methods for functional neuroimaging data. PhD thesis, Imperial College, University of London, 2002.

[3] J.A.D. Aston, R.N. Gunn, K.J. Worsley, Y.Ma, A.C. Evans, and A. Dagher. A statistical method for the analysis of positron emission tomography neuroreceptor ligand data. *Neuroimage*, 12:245–256, 2000.

[4] John A.D. Aston, Vincent J. Cunningham, Marie-Claude Asselin, Alexander Hammers, Alan C. Evans, and Roger N. Gunn. Positron emission tomography partial volume correction: Estimation and algorithms. *Journal of Cerebral Blood Flow & Metabolism*, 22(8):1019–1034, 2002.

[5] John A.D. Aston and Adam M. Johansen. Bayesian inference on the brain: Bayesian solutions to selected problems in neuroimaging. *Current Trends in Bayesian Methodology with Applications*, page 1, CRC Press, 2015.

[6] Scott Shaobing Chen, David L. Donoho, and Michael A. Saunders. Atomic decomposition by basis pursuit. *SIAM Journal on Scientific Computing*, 20(1):33–61, 1998.

[7] V.J. Cunningham and T. Jones. Spectral analysis of dynamic PET studies. *Journal of Cerebral Blood Flow and Metabolism*, 13:15–23, 1993.

[8] Jeffrey W. Dalley, Tim D. Fryer, Laurent Brichard, Emma S.J. Robinson, David E.H. Theobald, Kristjan Lääne, Yolanda Peña, Emily R. Murphy, Yasmene Shah, Katrin Probst, et al. Nucleus accumbens d2/3 receptors predict trait impulsivity and cocaine reinforcement. *Science*, 315(5816):1267–1270, 2007.

[9] Natalia del Campo, Tim D. Fryer, Young T. Hong, Rob Smith, Laurent Brichard, Julio Acosta-Cabronero, Samuel R. Chamberlain, Roger Tait, David Izquierdo, Ralf Regenthal, et al. A positron emission tomography study of nigro-striatal dopaminergic mechanisms underlying attention: Implications for ADHD and its treatment. *Brain*, 136(11):3252–3270, 2013.

[10] Wayne C. Drevets, Ellen Frank, Julie C. Price, David J. Kupfer, Daniel Holt, Phil J. Greer, Yiyun Huang, Clara Gautier, and Chester Mathis. PET imaging of serotonin 1A receptor binding in depression. *Biological Psychiatry*, 46:1375–1387, 1999.

[11] Yu-Hua Dean Fang, Georges El Fakhri, John A. Becker, and Nathaniel M. Alpert. Parametric imaging with Bayesian priors: A validation study with 11 c-altropane PET. *NeuroImage*, 61(1):131–138, 2012.

[12] Richard S.J. Frackowiak, Gian-Luigi Lenzi, Terry Jones, and Jon D. Heather. Quantitative measurement of regional cerebral blood flow and oxygen metabolism in man using 150 and positron emission tomography: Theory, procedure, and normal values. *Journal of Computer Assisted Tomography*, 4(6):727–736, 1980.

[13] Sanjiv S. Gambhir. Molecular imaging of cancer with positron emission tomography. *Nature Reviews Cancer*, 2:683–693, 2002.

[14] R.N. Gunn. Mathematical modelling and identifiability applied to positron emission tomography. PhD thesis, University of Warwick, 1996.

[15] R.N. Gunn, S.R. Gunn, and V.J. Cunningham. Positron emission tomography compartmental models. *Journal of Cerebral Blood Flow and Metabolism*, 21(6):635–52, 2001.

[16] R.N. Gunn, S.R. Gunn, F.E. Turkheimer, J.A.D. Aston, and V.J. Cunningham. Positron emission tomography compartmental models: A basis pursuit strategy for kinetic modeling. *Journal of Cerebral Blood Flow and Metabolism*, 22:1425–1439, 2002.

[17] R.N. Gunn, A.A. Lammertsma, S.P. Hume, and V.J. Cunningham. Parametric imaging of ligand-receptor binding in PET using a simplified reference region model. *Neuroimage*, 6(4):279–287, 1997.

[18] Karl Herholz, S.F Carter, and M. Jones. Positron emission tomography imaging in dementia. *The British Journal of Radiology*, 2014.

[19] Robert B. Innis, Vincent J. Cunningham, Jacques Delforge, Masahiro Fujita, Roger N Gunn, James Holden, Sylvain Houle, Sung-Cheng Huang, Masanori Ichise, Hidehiro Iida, Hiroshi Ito, Yuichi Kimura, Robert A. Koeppe, Gitte Moos Knudsen, Juhani Knuuti, Adriaan A. Lammertsma, Marc Laruelle, Ralph Paul Maguire, Mark Mintun, Evan D. Morris, Ramin Parsey, Julie Price, Mark Slifstein, Vesna Sossi, Tetsuya Suhara, John Votaw, Dean F. Wong, and Richard E. Carson. Consensus nomenclature for in vivo imaging of reversibly binding radioligands. *Journal of Cerebral Blood Flow and Metabolism*, 27:1533–1539, 2007.

[20] C.R. Jiang, J.A.D. Aston, and J.L. Wang. A functional approach to deconvolve dynamic neuroimaging data. *Journal of the American Statistical Association*, page in press, 2015.

[21] Ci-Ren Jiang, John A.D. Aston, and Jane-Ling Wang. Smoothing dynamic positron emission tomography time courses using functional principal components. *NeuroImage*, 47:184–193, 2009.

[22] Anthony K.P. Jones, Hiroshi Watabe, Vin J. Cunningham, and Terry Jones. Cerebral decreases in opioid receptor binding in patients with central neuropathic pain measured by [11C] diprenorphine binding and PET. *European Journal of Pain*, 8(5):479–485, 2004.

[23] Terry Jones and Eugenii A. Rabiner. The development, past achievements, and future directions of brain pet. *Journal of Cerebral Blood Flow & Metabolism*, 32(7):1426–1454, 2012.

[24] C.L. Lawson and R.J. Hanson. *Solving Least Squares Problems*. Prentice-Hall, New York, 1974.

[25] J.Logan, J.S. Fowler, N.D. Volkow, A.P. Wolf, S.L. Dewey, D.J. Schlyer, R.R. MacGregor, R.Hitzemann, B.Bendriem, S.J. Gatley, and D.R. Christman. Graphical analysis of reversible radioligand binding from time-activity measurements applied to [N-11C-methyl]-(-)-cocaine PET studies in human subjects. *Journal of Cerebral Blood Flow and Metabolism*, 10(5):740–747, 1990.

[26] Paul M. Matthews, Eugenii A. Rabiner, Jan Passchier, and Roger N. Gunn. Positron emission tomography molecular imaging for drug development. *British Journal of Clinical Pharmacology*, 73(2):175–186, 2012.

[27] Jorge J. Moré. The Levenberg-Marquardt algorithm: Implementation and theory. In *Numerical Analysis*, pages 105–116. Springer, 1978.

[28] Hans W. Muller-Gartner, Jonathan M. Links, Jerry L. Prince, R. Nick Bryan, Elliot McVeigh, Jeffrey P. Leal, Christos Davatzikos, and J. James Frost. Measurement of radiotracer concentration in brain gray matter using positron emission tomography: Mri-based correction for partial volume effects. *J Cereb Blood Flow Metab*, 12(4):571–83, 1992.

[29] John A. Nelder and Roger Mead. A simplex method for function minimization. *The Computer Journal*, 7(4):308–313, 1965.

[30] Seiji Ogawa, Tso-Ming Lee, Alan R. Kay, and David W. Tank. Brain magnetic resonance imaging with contrast dependent on blood oxygenation. *Proceedings of the National Academy of Sciences*, 87(24):9868–9872, 1990.

[31] R. Todd Ogden. Estimation of kinetic parameters in graphical analysis of PET imaging data. *Statistics in Medicine*, 22(22):3557–3568, 2003.

[32] John M. Ollinger and Jeffrey A. Fessler. Positron-emission tomography. *Signal Processing Magazine*, 14:43–55, 1997.

[33] Finbarr O'Sullivan, Mark Muzi, David M. Mankoff, Janet F. Eary, Alexander M. Spence, and Kenneth A. Krohn. Voxel-level mapping of tracer kinetics in PET studies: A statistical approach emphasizing tissue life-tables. *Annals of Applied Statistics*, 8:1065–1094, 2014.

[34] Finbarr O'Sullivan, Mark Muzi, Alexander M. Spence, David M. Mankoff, Janet N. O'Sullivan, Niall Fitzgerald, George C. Newman, and Kenneth A. Krohn. Nonparametric residue analysis of dynamic PET data with application to cerebral FDG studies in normals. *Journal of the American Statistical Association*, 104(486):556–571, 2009.

[35] C.S. Patlak, R.G. Blasberg, and J. D. Fenstermacher. Graphical evaluation of blood-to-brain transfer constants from multiple-time uptake data. *Journal of Cerebral Blood Flow and Metabolism*, 3:1–7, 1983.

[36] Jyh-Ying Peng, John A.D. Aston, Roger N. Gunn, Cheng-Yuan Liou, and John Ashburner. Dynamic positron emission tomography data-driven analysis using sparse Bayesian learning. *Medical Imaging, IEEE Transactions on*, 27(9):1356–1369, 2008.

[37] Peter E. Valk, Dale L. Bailey, David W. Townsend, and Michael N. Maisey. *Positron Emission Tomography: Basic Science and Clinical Practice*. Springer, 2004.

[38] Christopher J.S. Price, Dechao Wang, David K. Menon, Joe V. Guadagno, Marcel Cleij, Tim Fryer, Franklin Aigbirhio, Jean-Claude Baron, and Elizabeth A. Warburton. Intrinsic activated microglia map to the peri-infarct zone in the subacute phase of ischemic stroke. *Stroke*, 37(7):1749–1753, 2006.

[39] Olivier G Rousset, Yilong Ma, Alan C Evans, et al. Correction for partial volume effects in PET: Principle and validation. *Journal of Nuclear Medicine*, 39(5):904–911, 1998.

[40] G.A.F. Seber and C.J. Wild. *Nonlinear Regression*. Wiley, 1989.

[41] Mark Slifstein and Marc Laruelle. Effects of statistical noise on graphic analysis of PET neuroreceptor studies. *Journal of Nuclear Medicine*, 41(12):2083–2088, 2000.

[42] Michel M. Ter-Pogossian. The origins of positron emission tomography. *Seminars in Nuclear Medicine*, 22(3):140–149, 1992. Positron Emission Tomography: Part I.

[43] Robert Tibshirani. Regression shrinkage and selection via the lasso. *Journal of the Royal Statistical Society. Series B (Methodological)*, **58**(1): 267–288, 1996.

[44] F. Turkheimer, L. Sokoloff, A. Bertoldo, G. Lucignani, M. Reivich, J. L. Jaggi, and K. Schmidt. Estimation of component and parameter distributions in spectral analysis. *Journal of Cerebral Blood Flow and Metabolism*, 18:1211–1222, 1998.

[45] Federico E. Turkheimer, J.A.D. Aston, M-C. Asselin, and R. Hinz. Multi-resolution Bayesian regression in pet dynamic studies using wavelets. *Neuroimage*, 32(1):111–121, 2006.

[46] Federico E. Turkheimer, John A.D. Aston, Richard B. Banati, Cyril Riddell, and Vincent J. Cunningham. A linear wavelet filter for parametric imaging with dynamic pet. *Medical Imaging, IEEE Transactions on*, 22(3):289–301, 2003.

[47] Federico E. Turkheimer, Matthew Brett, Dimitris Visvikis, and Vincent J. Cunningham. Multiresolution analysis of emission tomography images in the wavelet domain. *Journal of Cerebral Blood Flow & Metabolism*, 19(11):1189–1208, 1999.

[48] Federico E. Turkheimer, Rainer Hinz, and Vincent J. Cunningham. On the undecidability among kinetic models: From model selection to model averaging. *Journal of Cerebral Blood Flow & Metabolism*, 23(4):490–498, 2003.

[49] Mattia Veronese, Roger N. Gunn, Stefano Zamuner, and Alessandra Bertoldo. A nonlinear mixed effect modelling approach for metabolite correction of the arterial input function in pet studies. *NeuroImage*, 66:611–622, 2013.

[50] Henry N. Wagner, H. Donald Burns, Robert F. Dannals, Dean F. Wong, Bengt Langstrom, Timothy Duelfer, J. James Frost, Hayden T. Ravert, Jonathan M. Links, Shelley B. Rosenbloom, Scott E. Lukas, Alfred V. Kramer, and Michael J. Kuhar. Imaging dopamine receptors in the human brain by positron tomography. *Science*, 221:1264–1266, 1983.

[51] K.J. Worsley, C. Liao, J.A.D. Aston, V. Petre, G.H. Duncan, and A.C. Evans. A general statistical analysis for fMRI data. *Neuroimage*, 15(1):1–15, 2002.

[52] Francesca Zanderigo, Ramin V. Parsey, and R. Todd Ogden. Model-free quantification of dynamic PET data using nonparametric deconvolution. *Journal of Cerebral Blood Flow & Metabolism*, bf 35(8):1368–1379, 2015.

[53] Yan Zhou, John A.D. Aston, and Adam M. Johansen. Bayesian model comparison for compartmental models with applications in positron emission tomography. *Journal of Applied Statistics*, 40(5):993–1016, 2013.

3

Structural Magnetic Resonance Imaging

Wesley K. Thompson

University of California, San Diego, Department of Psychiatry

Hauke Bartsch

University of California, San Diego, Multi-Modal Imaging Laboratory

Martin A. Lindquist

John Hopkins University, Department of Biostatistics

CONTENTS

3.1	Introduction		36
3.2	Image Acquisition		37
	3.2.1	MRI Physics	38
	3.2.2	Image Reconstruction	38
	3.2.3	MRI Sequences	38
	3.2.4	MRI Artifacts	39
3.3	Multi-Subject MRI		40
	3.3.1	Registration	40
		3.3.1.1 Volumetric Registration	41
		3.3.1.2 Surface-Based Registration	41
	3.3.2	Segmentation	43
		3.3.2.1 Foreground from Background Segmentation	43
		3.3.2.2 Brain Tissue Segmentation	44
	3.3.3	Templates and Atlases	45
	3.3.4	Morphometry	46
		3.3.4.1 Subcortical Volumes	46
		3.3.4.2 Voxel-Based Morphometry	47
		3.3.4.3 Deformation- and Tensor-Based Morphometry	47
		3.3.4.4 Surface-Based Measures	48
		3.3.4.5 Other Morphometric Measures	49
	3.3.5	Statistical Analyses	50
		3.3.5.1 Statistical Parametric Maps	51
3.4	Miscellaneous Topics		51
	3.4.1	Structural Integrity and Tumor Detection	51
	3.4.2	Anatomical References for Functional Imaging	52
	3.4.3	Multi-Center Studies	52
	3.4.4	Imaging Genetics	53
3.5	Glossary of MRI Terms		53
	Bibliography		54

This chapter summarizes some basic concepts related to structural magnetic resonance imaging (MRI). The focus is primarily on T_1-weighted structural MRI, and covers image acquisition, processing, extraction of morphometric measures, and statistical analyses for multi-subject cross-sectional studies.

3.1　Introduction

This chapter briefly summarizes some basic concepts related to *structural magnetic resonance imaging* (structural MRI). We will focus primarily on so-called T_1-weighted structural MRI, and cover image acquisition, processing, and morphometric analysis for multi-subject cross-sectional studies. Other chapters cover a diverse range of MR imaging modalities, including diffusion tensor imaging (DTI) and functional MRI (fMRI), as well as applications to longitudinal imaging studies. Hereafter, in this chapter, we generally refer to "structural MRI" as simply "MRI" for the sake of brevity. Good overviews of MRI are given in (90, 56, 66). These sources also contain more technical references regarding MRI physics and image acquisition, if the reader wishes to delve further into this topic.

The adult human brain is a complex-structured organ weighing around 1.5 kilograms and consisting of roughly 1,500 cubic centimeters of volume. Brain size varies in the population, and is correlated with age, overall body size, and gender (67, 52, 53). The two principal cell types within the brain are *neurons*, which process information, and *glial* cells, which play a variety of supporting roles. In total there are 100 billion or more neurons, and perhaps 10 times as many glial cells in the adult human brain (95). *Gray matter* (GM) refers to brain tissue wherein cell bodies and dendrites predominate, whereas *white matter* (WM) refers to brain tissue where there is a higher proportion of myelinated axons. In addition, the vasculature of the human brain is extensive and detectable via MRI (34, 96).

Macroscopically, the human brain is composed of the brain stem, the cerebellum, and the cerebrum. The cerebrum consists of two cerebral hemispheres, connected to each other by the corpus callosum. Each cerebral hemisphere is covered by a six-layered sheet of GM (the *cerebral neocortex*), roughly 1 to 4 mm thick, with WM in the interior. The cerebral cortices are deeply folded, and folding patterns exhibit broad-scale similarities across adult humans (29, 31). A cortical ridge is called a *gyrus*, while a depression is called a *sulcus*. It has been shown that roughly two thirds of the cerebral neocortex is hidden within sulci (36).

Each cerebral hemisphere is divided into frontal, parietal, temporal, and occipital lobes. Each lobe can be further subdivided into cortical regions, based on, for example, cytoarchitecture (17), geometrical landmarks (31), or genetic differences (17). Additionally, there are a number of GM subcortical structures lying underneath the cerebral neocortex, including structures belonging to the limbic system and the basal ganglia. Cerebral hemispheres are roughly bilaterally symmetric, though there are some systematic asymmetries in shape and function (123). In particular, most regions of the cerebrum have homologous left and right versions.

MRI is a flexible imaging modality, and different types of images can be generated to emphasize contrast related to different tissue characteristics. For example, T_1-weighted MRI provides good contrast between gray- and white- matter tissues, with GM appearing as dark gray and WM as lighter gray. *Cerebrospinal fluid* (CSF), a clear fluid contained within the ventricles and the subarachnoid spaces, appears as dark regions in T_1-weighted MRI. Typically, a T_1-weighted image is *segmented* into these three tissue types, as will be described below. They can be further subdivided into regions using expert manual tracing, or by automatic parcellation algorithms. The resulting data can then be utilized to characterize the spatial extent and distribution of different tissue types in an individual's brain, including volumes and shapes of subcortical structures, and the volume, thickness, and surface area of cerebral cortex. Other tissue properties and aspects of brain morphometry can also be extracted from MRI signals.

Morphometric analyses of MR images have been very widely applied in the biomedical research literature over the past 20 years. The broad range of applications include the assessment of normative brain structure, e.g., development in children and adolescents (48, 49), atrophy and cognitive impairments in later life (86), the connection between brain structure and intellectual capabilities (57), personality traits (45), and so forth. Other applications focus on the impact of illness and disease on brain structure. Significant differences have been found in the morphometric properties of cases and controls in studies of psychiatric illnesses such as autism (14), schizophrenia (64), and depression (82), as well as in studies of disorders directly affecting brain structure, such as Alzheimer's (69) and Parkinson's disease (18). More recently, *imaging genetics* has focused on the relationship between genetic information and structural MRI via familial or twin (17) and genome-wide association studies (115). Another use of MRI includes the production of high-resolution anatomical references for co-localizing functional activations obtained, for example, from fMRI (109, 35). A very important clinical use for MRI, which we briefly describe below, is the detection and localization of focal and space-filling brain tumors, edema, and necrotic tissues for use in surgical planning (11).

There are several freely available software packages for preprocessing MRI data and performing morphometric analysis of the human brain. Some of the more popular packages include the Statistical Parametric Mapping package (http://www.fil.ion.ucl.ac.uk/spm/), FreeSurfer (https://surfer. nmr.mgh.harvard.edu/), the FMRIB Software Library (FSL, http://fsl.fmrib. ox.ac.uk/fsl/fslwiki/), and AFNI (http://afni.nimh.nih.gov/). While the image processing pipelines contain broadly similar elements across each of these packages, there are some major differences. Additionally, many imaging labs create their own individualized pipelines either by extracting and adapting elements of these packages or by creating new processing scripts, depending on the particular needs of the research being performed. Statisticians involved in the analysis of MRI data are generally well served by gaining a substantial degree of familiarity about the processing pipeline and how these impact the types of possible inferences.

Note, while it is not the goal of this chapter to endorse or promote any particular processing and analysis package, we will to some extent focus on the processing and morphometric analyses originally developed within the FreeSurfer, and to a lesser extent, the SPM frameworks.

The remainder of this chapter is organized as follows. In Section 3.2 we describe the basics of structural image acquisition and preprocessing. Section 3.3 describes the steps involved in multi-subject MRI research studies, including registration and segmentation, various techniques for extracting morphometric features from structural images, and statistical analyses of the resulting data. We conclude the chapter with a brief discussion of miscellaneous issues in Section 3.4.

3.2 Image Acquisition

Nuclear magnetic resonance (NMR) imaging is one of the workhorses for non-invasive, clinical interrogation of the soft tissue structure in the brain. Since its humble origins in the 1980s, and a rebranding[1] to MR imaging (MRI), it has evolved into a multi-billion dollar

---

[1]Rebranding to prevent association with ionizing radiation.

industry. MRIs can be used to differentiate between brain tissue types, including normal and abnormal tissue, and therefore provides information about morphology and a diagnostic tool to detect disease.

3.2.1   MRI Physics

In order to understand the terminology used in MRI, we first take a brief look at the physics at the core of this imaging technique. MRI creates series of stacked two-dimensional (2d) images based on interactions between radiofrequency (RF) electromagnetic fields and atomic nuclei after the tissue has been placed in a strong magnetic field maintained inside an MRI scanner. Only atomic nuclei with an odd number of protons and neutrons have an angular momentum and they act as tiny magnetic dipoles that spin around their axis of rotation. The scanner is able to detect concerted changes in precession from a large number of spins. The focus is typically on the single protons found in the nuclei of hydrogen atoms within water and fat, as they are the most frequent dipoles found in the brain.

Without an external magnetic field, the spin axes are oriented randomly. Inside the scanner's strong magnetic field, the spins align over time and eventually precess around the field's direction. In order to measure tissue properties, MRI measures macroscopic tissue magnetization caused by the imbalance of nuclei that are orientated parallel or anti-parallel to the scanner's magnetic field. The scanner uses an RF pulse to flip the spin orientation between the two orientations. The amplitude of the pulse determines the flip angle, and after switching off the RF pulse, the nuclei will briefly precess synchronously, resulting in measurable electromagnetic radiation picked up by the scanner's receiver coil as an echo.[2] Due to interactions with other molecules, the spin of nuclei will quickly fall out of phase with a transversal relaxation time constant T_2. The spin axes will realign with the scanner magnetic field axis again after the longitudinal relaxation time T_1. This value depends on the rate of energy transfer of the spins with their neighbors in the tissue lattice. For example, the T_1 time is shorter in fat than in pure water, since carbon molecules in fatty acids provide more efficient energy transfer due to spin-lattice interactions. Note basic MR physics is discussed further in Chapter 6. More detailed information can be found in (56) and (66).

3.2.2   Image Reconstruction

The signal picked up by the scanner's receiver coil contains a large signal component due to spin precession. One method for image reconstruction removes the precession by multiplication with a sine and a cosine function oscillating at, or near, the frequency of precession followed by low-pass filtering. After multiplication, both signals are combined and result in a demodulated complex signal that is interpreted as the Fourier transform of the tissue transverse magnetization. An inverse complex Fourier transform is used to assemble the series of 2d MR magnitude images, which is exported for image viewing. Phase information is usually discarded but special applications such as flow imaging use this information (83).

3.2.3   MRI Sequences

The tissue-dependent time scales for magnetic relaxation, given by T_1 and T_2, as well as the density of protons (PD) are measured by applying precisely timed combinations of RF pulses and secondary magnetic gradients. The variables that are changed between these

---

[2]If the echo was caused by spins getting into alignment, the sequence is also called spin-echo sequence. They can also be caused by applying secondary gradients in faster gradient-echo sequences.

sequences are the amplitude of the RF signal (flip angle), the time after which the echo is measured (echo time, TE) and the time interval between consecutive RF pulses (repetition time, TR). A short repetition time relative to the tissues T_1 relaxation time and a short echo time relative to the tissues T_2 time will result in predominantly T_1-weighted images. In T_1-weighted images, the fat present in WM appears bright, GM appears dark gray, and CSF appears black. In a similar manner, MRI can be used to produce predominantly T_2- and PD-weighted images. For T_2-weighted images (long TR and long TE), water present in CSF and GM appears bright, while air appears dark. PD-weighted images (long TR and short TE) provide good contrast between GM (bright) and WM (darker gray), but little contrast between brain and CSF.

Sequence development is an active field of research and can be used to probe tissue in many different ways. For example, in the time between when spins are initially aligned and measured, some might travel out of the inspected region resulting in a signal drop in T_2-weighted images, indicative of molecular motion caused by diffusion. Probing this motion in different orientations is the basis for diffusion-weighted imaging; see Chapter 4. Other sequences try to suppress the diffusion signal to better probe disease processes such as inflammation and demyelination, by increased contrast for associated changes in local water and lipid content. Similarly, T_2^* is the combined effect of T_2 and local inhomogeneities in the magnetic field. While certain sequences attempt to eliminate the effects of these inhomogeneities, others try to emphasize them. The latter types of procedures form the basis of so-called blood-oxygen-level-dependent (BOLD) functional MRI (fMRI); see Chapter 6.

3.2.4   MRI Artifacts

Discrete sampling and filtering in the Fourier domain during image reconstruction can produce a variety of different image artifacts. Artifacts can cause serious changes in image intensity and deformations in the image structure. It is therefore useful to understand some of these artifacts to guide image interpretation.

Because the signal is acquired in the Fourier domain (or k-space) with a given sampling interval, it is periodic in nature and the image repeats itself. The replication interval depends on the inverse of the sampling interval and the frequency bandwidth of the signal. If the field of view (FOV) of the image is set up improperly, this can result in overlap between adjacent replicates. Visually, this results in copies of the imaged structure appearing above and below the image. As an example, the nose of the subject might intersect the back of the brain. Another potential artifact related to the repeated structure is a ghosting artifact caused by non-matching phase shifts between the two demodulating sine and cosine signals. In this case, the copy of the object has low signal amplitude and appears shifted and superimposed on the image.

Another common image artifact is caused by the finite sampling interval in Fourier space. Any sufficiently abrupt change in image intensity will produce a ringing artifact in the reconstructed image, as an infinite number of frequencies would be required to represent an instantaneous jump in intensity. Further artifacts, which distort the image regionally, include chemical shift and magnetic susceptibility artifacts caused by non-matching RF pulse frequencies used to address spins in space. Artifacts can also be caused by patient motion during image acquisition, appearing as repeated bright features across the image. A source of intensity variation that is smoothly varying over space is a multiplicative bias field caused by interference of the RF coil signal with the imaged brain. Dependent on the scanner, the center of the brain can then appear brighter or darker compared to the periphery.

Most image artifacts caused by signal reconstruction can be controlled with appropriate acquisition sequences and filtering in Fourier space. However, most image analysis is performed long after image acquisition, and without access to the original k-space data. Manual inspection of the reconstructed images is therefore advised to identify inferior image quality, and their removal from further analysis may be necessary. Bias field correction and magnetic susceptibility artifacts are removed as a post-processing step after reconstruction. Whereas the bias field can be estimated and removed from the image due to its slowly varying characteristics, magnetic susceptibility artifacts require specialized imaging sequences in which two scans are obtained with orthogonal distortions. Using elastic registration of the two images, the mean-corrected image can be obtained.

3.3 Multi-Subject MRI

After artifact correction and transformation to Euclidean space, multi-subject MRI analyses typically require a spatial *registration* (or normalization) step. This entails estimating a mapping between each individual's image and a stereotactic template (6), thus allowing the different individuals brains to be compared. The normalized image is then *segmented* into tissue types, including GM, WM, and CSF. Smooth intensity non-uniformity is corrected for concomitantly with registration and tissue segmentation. Atlas-based methods can further segment the images into cortical and subcortical regions of interest (ROIs) (20). The key to successful registration and segmentation is the incorporation of prior knowledge regarding the spatial topology of the brain, as well as information (e.g., a forward model) regarding MR image formation (6, 58). Morphometric measures can then be computed from individual MR images (e.g., (3)) and used as outcomes in statistical analyses, for example demonstrating association with diagnostic status (e.g., (53)). If there are many such measures, as is the case with whole-brain (voxelwise) analyses, significance levels of statistical tests need to be adjusted to account for inflation of Type-I error rates due to multiple comparisons (94).

3.3.1 Registration

The objective of registration is to map intensity images $J_n : \mathbb{R}^3 \to \mathbb{R}^+$, $n = 1, \ldots, N$, in native space onto a template image T, where N is the number of subjects (39). Registration requires the estimation of transforms $f_n : \mathbb{R}^3 \to \mathbb{R}^3$, which take coordinates $r = (r_1, r_2, r_3)$ in native space and map them to stereotactic coordinates $f_n(r)$ in the template space. Ideally, the transforms are (at least approximately) diffeomorphic, and are therefore smooth, continuous mappings with derivatives that are invertible while preserving the topology of the brain (23, 2, 124). The resolution of the registration should also be high enough to ensure that *partial volume effects* (MRI signal from voxels containing multiple tissue types) are minimized. This typically entails having roughly 1-mm isotropic voxels (3). Moreover, transforms have to leave relevant subject-level differences in the volume and shape of brain substructures as identifiable, even after correcting for global features such as the position in the scanner and overall head volume, of no, or secondary, interest. This information can be preserved by enforcing smoothness constraints that correct only for global size and shape variables (3), or by recording differences in the transformations f_n across subjects (5, 2).

Note, in some registration schemes the template image is instead mapped onto each individual image (2). Ultimately, there are technical considerations that might influence the direction of the mapping. If we want to display an image in a template, or atlas space, after registration, the opposite transformation from atlas onto image space is convenient to use. This is because the sampling of the image in atlas space is done using the inverse transformation (from atlas into image space). Using the inverse transformation sampling scheme guarantees that every voxel in the atlas space has a corresponding intensity from image space using trilinear interpolation. This technique removes the requirement of computing the inverse of the image to atlas space transformation, which might be ill-defined if the transformation is not volume preserving.

3.3.1.1 Volumetric Registration

The majority of *volumetric registration* approaches proposed in the literature first perform a global 6- or 12-parameter transform, which can be expressed as 4×4 matrices in Euclidean space. The global transform is rigid (6 parameters: translation, rotation) or affine (12 parameters: rigid plus anisotropic scale and shear). These global transforms are often followed by local, non-linear transforms to match the image with the template on fine-scale structure (6, 20).

Estimation of transform parameters typically involves minimization of a cost function subject to biologically motivated constraints. Special care has to be taken if the template and mapped images are derived from different image modalities. In general, images can be aligned if their two-way histograms are in a low-entropy state. This is usually the case if brain tissue shows arbitrary, but uniform, intensities in both the template and image. If both template and image are from the same modality, a simple Euclidian distance, or correlation measure, can be used at the core of the cost function. Multi-modal registration, such as between MRI and computer tomography (CT) images, instead requires cost functions based on mutual information (104).

Several authors have proposed incorporating biologically motivated constraints using Bayesian models, and obtaining *maximum a posteriori* (MAP) estimates (6, 39, 40). Let $\boldsymbol{\theta}$ denote the model parameters, and $\mathcal{D}$ the data. By Bayes' rule we have $p(\boldsymbol{\theta}|\mathcal{D}) \propto p(\mathcal{D}|\boldsymbol{\theta})p(\boldsymbol{\theta})$. The MAP estimate $\widehat{\boldsymbol{\theta}}_{\mathrm{MAP}}$ of $\boldsymbol{\theta}$ is

$$\widehat{\boldsymbol{\theta}}_{\mathrm{MAP}} = \mathrm{argmin}_{\boldsymbol{\theta}} \left\{ -\log p(\mathcal{D}|\boldsymbol{\theta}) - \log p(\boldsymbol{\theta}) \right\}. \tag{3.1}$$

For example, (2) proposes a local deformation model with a velocity field consisting of a linear combination of first-degree B-splines. The coefficients $\boldsymbol{\theta}$ of the splines are given a prior distribution $p(\boldsymbol{\theta})$ consisting of a zero-mean Gaussian with a covariance structure based on membrane, bending, or linear elastic energy. The likelihood $p(\mathcal{D}|\boldsymbol{\theta})$ is a normal approximation based on squared differences between the transformed image and template intensities, summed across voxel mid-points. MAP estimates are obtained by minimizing the expression on the right side of Eq. 3.1 and the resulting velocity field is numerically integrated to obtain a diffeomorphic mapping.

3.3.1.2 Surface-Based Registration

Surface-based registration attempts to incorporate the topology of the cerebral cortex to construct transforms onto a template space (29, 41, 38, 76, 101). Unlike subcortical volumes, the cerebral cortex has the topology of a highly convoluted 2d sheet (29). Volumetric registration methods do not preserve this topology, as two voxels can be close neighbors in Eulidean space but lie far apart with respect to distance across the cortical surface. Moreover, cortical geography and function are best mapped following the local orientation of the cortex (121).

An exemplar surface-based registration algorithm is given by (29). After an initial voxel-based segmentation of WM, the cerebral hemispheres are cut along the corpus callosum and the pons, removing subcortical GM structures and resulting in a single mass of connected WM voxels for each hemisphere. The gray–white cortical boundary of each hemisphere is tessellated and smoothed. These tessellated surfaces are then repositioned by minimizing the intensity differences between the image and target intensity values, subject to energy

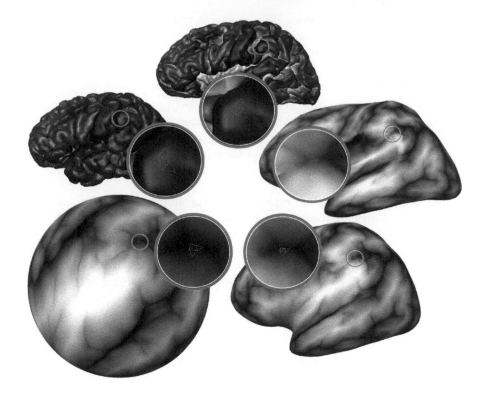

FIGURE 3.1

Unfolding of a left hemisphere cortical triangulated surface from the top left (pial surface) clock-wise to the bottom left (spherical representation). The insets show the arrangement of four triangles across the unfolding steps. Gray-scale illustrates the curvature information that is computed in the folded state and is carried over to the spherical representation.

functionals tangential and normal to the surface, which smooth the surface and encourage uniform spacing of vertices on the inflated cortices. These priors are built into the model via a Bayesian formulation, and the MAP estimate is obtained as described in Equation (3.1). The tessellated surfaces are given a spherical topology by "capping" the midbrain. Cortical thickness and surface area can then be computed at each vertex of the tessellation, as described below.

The initial steps of the mapping from subject surface to atlas surface are depicted in Figure 3.1. Correspondence between image and atlas is calculated by using the similarity between gyri and sulci patterns, measured by local Gaussian curvature. Figure 3.1 (top left) shows the folded pial surface of the left hemisphere (eyes pointing to the left) and a set of six vertices that are connected by four triangles as a close-up. The top to top-right figures show an operation that unfolds the cortex using a forced geometric approach. Gray values indicate the curvature information obtained from the initial pial surface, which is carried over (dark positive curvature in gyri, light negative curvature in sulci). The transformation is space preserving. The unfolded shape is used for visualization because the general shape is still recognizable and features in the suli can be visualized. From this unfolded state (top right) a spherical reconstruction (bottom left) creates a representation that makes the tangential surface displacement explicit (Gauss map). It can be seen in the series of insets that the surface vertices undergo a series of deformations but keep their pattern of connections. Locations in the spherical reconstruction can therefore be mapped back to the corresponding location on the unfolded pial surface.

A similar spherical representation of the atlas is used as a target for registration. Vertices of the subject brain are moved while minimizing the summed distanced between corresponding subject and atlas curvature points. This procedure stretches and shifts the pattern of surface vertices in the tangential direction of the surface to best fit the curvature observed in the atlas. Surface area expansion for each vertex can now be calculated as the change in area between the spherical representation and the original pial surface representation the procedure started with. Furthermore, after minimization, subject surface vertices get assigned the region of interest of the closest atlas vertex. Region of interest measures for thickness and surface area are computed by summing up vertex measures for all vertices that share the same label.

Automated methods have also been proposed that incorporate features of both surface-based and volumetric registration (71, 105). The goal of these registration methods is to preserve correspondences in cortical folding patterns while simultaneously aligning sub-cortical volumes, thereby providing more accurate comparisons of cortical and subcortical morphology across subjects, or more accurate co-registration of structural and functional images.

3.3.2   Segmentation

Image segmentation consists of classifying voxels into categories based on their intensities, location, and prior knowledge regarding neurobiology and MR image formation. Accurate classification depends on the fact that non-brain and brain voxels have different intensity distributions, as do voxels containing different brain tissue types. Note that MR image intensities are non-negative, because after Fourier reconstruction they are converted into magnitude images, thus placing limitations on appropriate probability distributions for voxel intensities. In particular, the distribution of noise from MR images is Rician, not Gaussian (54). Gaussian approximations can be adequate at the signal-to-noise ratio observed in MR images for gray and white matter brain tissue. However, as CSF and air can have intensities very close to zero, and distribution noise will be increasingly skewed and approach a Rayleigh distribution.

3.3.2.1   Foreground from Background Segmentation

The most basic use of classification in MRI is the separation of foreground and background based on a global intensity threshold. It is important that the classifier used for this separation is insensitive to the contrast and brightness differences common to biomedical images. A simple algorithm used for separation is based on the assumption that the general shape of the imaged object is known beforehand, so that the proportion of voxels belonging to foreground (head) and background (air) can be assumed to be approximately known. The intensity threshold for classification is then based on the cumulative distribution function (cdf) of voxel intensities. The performance of this algorithm is sufficient to provide robust and automatic brightness and contrast adjustments for image viewing on medical workstations where MRI measurements are mapped to voxel brightness using linear transfer functions. More complex histogram equalization procedures are sometimes used in research settings (15).

Intensity thresholds can also be defined if foreground and background intensities appear as a bi-modal distribution (119). This is usually the case for MRI, since brain tissue appears bright in front of the more dark-appearing air. A commonly used unsupervised clustering algorithm is Otsu's method (97), in which the optimal threshold is defined as the one that minimizes the intra-class variance. A histogram of the image intensities is computed followed by an exhaustive search through all possible thresholds. The threshold that yields the minimal intra-class variance is then used for classification. Another unsupervised classification algorithm based on computing thresholds is the IsoData (108) algorithm, which uses an iterative procedure to compute a threshold. From an initial starting threshold the mean of both parts of the intensity histogram is computed. The threshold is then moved so it is centered between the two means and the procedure is repeated until convergence.

All global threshold procedures rely on prior image corrections for intensity inhomogeneities and sufficient signal to noise. They are useful during the initial stages of image processing because they are fast and unsupervised. They can also be applied in a hierarchical manner to compute multiple thresholds but most often they are used to provide informed initialization to more complex, model-based image segmentation algorithms. For example, in the case of segmenting objects with blurred edges in front of a high background signal, *hysteresis thresholding* uses an initial high threshold to define cores of regions that belong to objects of interest. Region growing from these initial seed regions using a secondary threshold results in the final object classification. Non-local threshold algorithms have also been proposed and cope with inhomogeneous intensity variations in images (93, 87). Other algorithms use derived image features such as edges for segmentation. One of the most widely used edge- or gradient-based algorithms is *watershed segmentation* (13). High gradient edges are interpreted as rims separating uniform low gradient areas. A flooding procedure is used to region grow, starting from the low gradient regions. As the high gradient image edges separate the different basins from each other, they are interpreted as segmentation borders separating brain from background (37).

3.3.2.2 Brain Tissue Segmentation

In addition to separating foreground and background, MR image segmentation seeks to classify voxels contained within the brain into different tissue types (GM, WM, and CSF) (3, 4, 8). Class labels can also include partial volume categories (28). Atlas-based methods further partition GM voxels into ROIs (29, 40, 20). In the simplest case, each voxel has a single label assignment. Binary labels are then used to create annotated images that share resolution and voxel size with the original image data. Both files are merged to overlay the information appropriately for visualization. Non-binary label assignments can also be used to store posterior probabilities for labels to be assigned to a particular volume. These probabilities can then be converted to binary labels by thresholding (28).

A commonly used probabilistic classifier is the Finite Gaussian Mixture Model (FGMM) (130, 40, 4, 28). Let $\delta_{n,i} \in C$ denote the indicator for class membership for the ith voxel, $i = 1, \ldots, I$, in the nth image, where $C = \{1, \ldots, C\}$ is the set of possible tissue classes. The probability of the entire image $\boldsymbol{J}_n$ can be derived by assuming that all of the voxels are independent, and is given by

$$
\begin{aligned}
p(\boldsymbol{J}_n) &= \prod_{i=1}^{I} \sum_{c=1}^{C} p(\boldsymbol{J}_n(\boldsymbol{r}_i)|\delta_{n,i} = c)p(\delta_{n,i} = c) \\
&= \prod_{i=1}^{I} \sum_{c=1}^{C} \frac{1}{\sqrt{2\pi\sigma_c^2}} \exp\left\{-\frac{(J_n(\boldsymbol{r}_i) - \mu_c)^2}{2\sigma_c^2}\right\} p(\delta_{n,i} = c)
\end{aligned}
\tag{3.2}
$$

where $p(\delta_{i,n} = c)$ is the prior probability that the ith voxel is in tissue class c. In this model, prior probabilities are spatially stationary. Estimates can be obtained by pre-identifying (using prior anatomical knowledge) voxels that are highly likely to be from a given tissue type, and estimating class means and variances (μ_c, σ_c^2), $c = 1, \ldots, C$ based on these voxels.

Equation (3.2) makes a number of unrealistic assumptions. For example, neighboring voxels are likely correlated with one another. Hence, a number of authors have incorporated a Markov Random Field (MRF) formulation (130, 28, 39), that allows the prior probabilities in Equation (3.2) to be expressed as $p(\delta_{n,i} = c|\boldsymbol{\delta}_{\mathcal{N}_i})$, where $\mathcal{N}_i$ is the set of neighbors of the ith voxel and $\boldsymbol{\delta}_{\mathcal{N}_i}$ are their class labels. This spatial prior can be expressed as a Gibbs distribution according to the Hammersley–Clifford theorem (12). Additionally, the distributions of partial volume voxel intensities can be modeled as a function of the proportion of each tissue class contained within the voxel (110). Prior distributions can also incorporate information about local differences in MR image intensities and spatial distribution of structures (39). Non-parametric, information-theoretic approaches have also been proposed that do not rely on Gaussianity (51, 28).

The steps involved in registration and segmentation are sometimes performed sequentially (6, 3). However, several authors have proposed methods that perform registration and segmentation in a simultaneous manner (40, 4). This is beneficial because if the segmentation of each subject's brain image were known *a priori*, the optimal transform to register each image to a common template would be straightforward to compute. Conversely, if the optimal transform were known for each image, segmentation would be much easier to preform (39). Other refinements to registration and segmentation of multi-subject MRI data, often tailored to specific scenarios such as longitudinal change or detection of disease states, is an ongoing and active area of research (26, 74, 77, 102, 127, 27).

3.3.3 Templates and Atlases

Registration and segmentation of multi-subject MRI depends on the use of a common *template*, or an intensity image to serve as a common target for registration across subjects. Templates provide a 3d stereotactic coordinate system that can be used to report results that are comparable across studies. An *atlas* consists of a pair of images, a template and a corresponding segmentation (labeled image) (20). Atlas-based image segmentation is essentially a registration problem, since registering an MR image to the template automatically gives a segmentation from the corresponding labeled image (39, 31, 20). This procedure works as long as the atlas is appropriate, i.e., labels represented in the atlas have a representation in the MR image. Brain tumors and surgical interventions can invalidate this assumption, as well as brain atlases that are inappropriate for the imaged subject. For example, in young children the area of the ventricles can appear partially collapsed, resulting in poor registration if one uses an atlas built using adult subjects.

An atlas may be *single-subject*, i.e., based on a high-resolution image of one individual. The advantage of single-subject atlases is that they allow resolution at a fine scale. The disadvantage is that a single subject is not necessarily representative of the population of interest. One of the first human brain atlases was created by Talairach and Tournoux to give a stereotactic coordinate system and labeling for brain surgeries (117, 118, 36). This is a single-subject atlas, made from drawings of slices from the brain of a 60-year-old woman. Using two landmarks easily visible on MRI images (the anterior and posterior commissure), this atlas uses a proportional grid of labeled regions. Brodmann areas, based on cytoarchitectural differences in the human cortex (17), are used for labeling of cortical structures. Talairach coordinates were first used for structural MRI in the early 1990s (36) and are still often used for reporting areas of activation in functional imaging studies (80). Another commonly used single-subject atlas is the *Colin27* atlas, created from 27 high-resolution images from a single subject (63).

Alternatively, an atlas may be *population-based*, i.e., averaged across a number of representative individuals. These atlases retain only features that are common across the majority of subjects, at the cost of a loss of potentially informative resolution (39). Population-based atlases include those of the Montreal Neurological Institute (MNI). The first MNI atlas was based on MR images of 305 young healthy subjects, with stereotactic coordinates approximating those of Talairach and Tournoux (35). The MNI atlas has been updated several times over the years (36), including the International Consortium for Brain Mapping (ICBM152) atlas, based on 152 high-resolution MR images from a young adult population (88). Averaged templates are also often constructed using (a subset of) the subjects from the current study itself. For example, (65) used a random sample of the control subjects to construct a *minimal deformation target* template in an MRI study of Alzheimer's disease and mild cognitive impairment.

In addition to volumetric atlases, there are several atlases of the human cortical surface (41, 29, 31, 122, 32). For example, (31) collected T_1 MRI data from 40 subjects: 10 young adults, 10 middle-aged adults, 10 elderly adults, and 10 patients with Alzheimer's disease. Cerebral hemispheres were manually divided into 34 regions, based primarily on cortical (sulcal) geography. Spherical representations of the cortical surfaces were created for each of the 40 images, and were then registered together (42). Each point on the surface was then probabilistically assigned to one of the 34 regions (43).

Dozens of other atlases have been published in the literature (see (20) for a review). These atlases are often tailored to specific populations, e.g., pediatric subjects, elderly subjects, or diseased populations. Atlas labels can also be derived from biological sources other than direct anatomical or imaging information. For example, the Allen Human Brain Atlas (113) is based on gene expression data from two post-mortem adult male brains, whereas the atlas of (17) is based on genetic correlations of cortical surface area computed from an MRI study of 406 twins.

Figure 3.2 depicts an example of MRI atlas construction using the Allen Brain Atlas data. The Allen Brain Atlas project derived gene expression from small volumetric ROIs that had been cut from the original brain and analyzed using microarrays. Expression on several thousand genes is available at the center of mass of these volumetric ROIs. Coordinates in the Allen Brain Atlas are stored in MNI space, which is also used in the FreeSurfer atlas. Therefore, no explicit volume-based registration is required. The sampling density in the Allen Brain Atlas changes between cortical and sub-cortical regions. As a first approximation we can therefore assume that a vertex in the surface atlas can be linked to the closest sample region of the Allen Brain Atlas given a Euclidean distance measure. In order to improve the mapping, a subset of points can be used that represents cortical sample regions of a single hemisphere only. Figure 3.2 shows the mapping of gene expression sample regions as a color overlay on the FreeSurfer average surface. Sample regions have been generated from slabs of tissue that are visible in the bands of similar color that are an artifact of the brain preparation procedure. Using this preprocessing workflow gene expression pattern for each micro-array well can now be assigned to cortical surface vertices for subsequent statistical analysis.

3.3.4   Morphometry

3.3.4.1   Subcortical Volumes

A number of methods have been introduced to estimate volumes of subcortical GM structures. A simple approach is to run a segmentation algorithm to partition brain tissue into GM and other categories, followed by expert manual tracing of gray-matter subcortical structures based on neuroanatomy (e.g., as in (55)). Volumes are computed by counting the number of voxels falling within the tracing and multiplying by the voxel dimensions. To ensure that the manual tracings are reliable, multiple raters typically perform them, and an intra-class correlation coefficient (ICC), or some other measure of reliability, is reported.

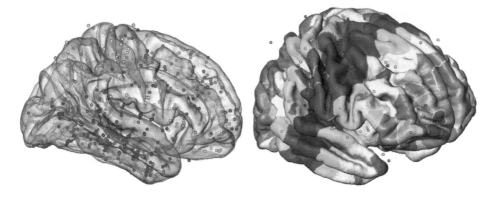

FIGURE 3.2
Sample locations for gene expressions in MNI space (spheres) relative to the FreeSurfer average surface transparent (left). Surface vertices are colored according to the identifier of the closest sample regions (right).

Moreover, if subcortical volumes are used as dependent measures in a statistical analysis, the manual tracing is done blinded to the independent variables of interest, such as diagnostic status. Many automated methods have also been developed for assessment of subcortical volumes (68, 9, 106, 7). These typically use atlas-based, or shape- and appearance-based, algorithms to segment subcortical structures and volumes are computed as described above.

3.3.4.2 Voxel-Based Morphometry

While volumetric analysis of pre-defined ROIs is relatively straightforward, there may be more general morphometric measures that are more highly correlated with independent variables of interest. If there are no strong *a priori* hypotheses regarding specific subcortical volumes, a whole-brain voxel-wise analysis may be able to discover such relationships. One such approach is termed *voxel-based morphometry* (VBM) (3, 53, 89). Briefly, VBM proceeds by first registering each subject MR image to a common stereotactic space. Registered images are segmented into brain tissue types and smoothed. Each voxel i of the smoothed image n is associated with a number $0 \leq p_{ni} \leq 1$, which represents the local concentration of GM tissue. After a logistic transformation, voxel GM intensities are entered into a general linear model (GLM) to study their relationship with independent variables, while controlling for covariates of no interest. This is a massively univariate approach, i.e., a separate GLM is fit for each voxel $i = 1, \ldots, I$, where I may be in the hundreds of thousands, and it is critical that multiple testing adjustments be performed to guard against inflated Type-I errors.

3.3.4.3 Deformation- and Tensor-Based Morphometry

VBM is a local measure of GM tissue intensities across individuals. In contrast, *deformation-based morphometry* (DBM) and *tensor-based morphometry* (TBM) yield more global measures of structural differences (5, 3, 24, 26, 65, 75). Both DBM and TBM utilize information from the registration-to-template transforms $\boldsymbol{f}_n, 1 \leq n \leq N$, to summarize morphometric differences across individuals. The advantages of these approaches is that they do not depend on strong *a priori* hypotheses about which ROIs are associated with the independent variables of interest, and therefore they allow for more subtle characterization of global or regional morphometric differences than simply volume or GM concentration.

The DBM approach, proposed by (5), performs an affine registration followed by a nonlinear deformation consisting of discrete cosine basis functions. The cosine basis function coefficients $\boldsymbol{\theta}_n$ are estimated for each image $\boldsymbol{J}_n$, resulting in K-dimensional vectors $\widehat{\boldsymbol{\theta}}_n$, where K is large (over a thousand). Removing the effects of brain size and position, these parameters are placed in an $N \times K$ matrix $\boldsymbol{A}$, where each row summarizes the deformation field encoding shape differences for the nth subject. A principal components analysis is then performed to reduce the dimensionality to a relatively small number of parameters per image (around 20) that capture most of the subject-to-subject variation. Finally, a Hotelling T^2 or MANCOVA can be performed on the resulting low-dimensional characterization of the deformation fields to assess associations with independent variables, possibly controlling for covariates of no interest. Note, DBM can also be used to focus on positional differences in specific ROIs (5).

TBM is similar to DBM, in that properties of the non-linear deformations $\boldsymbol{f}_n$ are used to characterize shape differences among individual images. Each mapping $\boldsymbol{f}_n$ can be represented as a discrete displacement vector field. A Jacobian matrix can computed by taking the gradient of the deformation at each voxel of the template image, giving rise to a tensor field that characterizes local displacements for each MR image (3). Taking the determinant of the Jacobian of $\boldsymbol{f}_n$ at each voxel gives local volumetric differences across subjects relative to the template image (24, 65). Subcortical volumes can be obtained by integrating the Jacobian determinants over voxels contained within segmented substructures. As an alternative, surface-based TBM has also been proposed (26).

3.3.4.4 Surface-Based Measures

Surface-based morphometric measures include cortical surface area, thickness, and volume. In human prenatal and perinatal data, surface area is primarily related to cortical column counts whereas thickness is more closely related to neuron counts within columns (107). Surface area and cortical thickness appear to have independent genetic determinants (100). In later life, changes in both surface area and thickness may be driven by changes at the level of synapses, dendrites, and spines (116). Subject-level variation in cortical volume appears to be more closely related to surface area than to thickness, though thinning of cortices is highly prevalent in later life and probably accounts for a substantial amount of normative change with age, as well as volumetric losses caused by disorders such as Alzheimer's disease (98, 67, 33, 116).

As described above, the FreeSurfer software implements a semi-automated approach to surface reconstruction (30, 29, 41), resulting in over 160,000 polygonal tessellation vertices for each hemisphere. Each subject's cortical surface is mapped to a spherical atlas space using a diffeomorphic registration procedure based on folding patterns. The surface alignment method uses the entire pattern of surface curvature at every vertex across the cortex to register individual subjects to atlas space (42). Using the surface between white and gray matter (white matter surface) and the surface between gray matter and cerebral spinal fluid (pial surface), it is straightforward to calculate the cortical thickness for each pial surface vertex (38). The computation uses the assumption already implicit in the surface generation that, although convoluted due to cortical folding, both surfaces run parallel to each other with a known minimal and maximal distance. In the (inverse) direction of the pial surface normal at each vertex, a ray is calculated until it hits the white matter surface. The distance traversed by the ray is used as the cortical thickness measured at the pial surface point. Notice that this procedure is not symmetric with respect to the surface chosen for the ray-to-surface intersection and becomes noisier in regions with high surface curvature. Several other algorithms for measuring cortical thickness have been proposed in the literature (85, 76, 25).

Surface area is also computed as a vertex-based measure that reflects the size of the area of adjacent triangles (37). This, of course, depends on the number of vertices that are generated during surface tessellation and on the uniformity of the tessellation. These are both properties of the surface-generating algorithms. In order to remove this ambiguity, surfaces are matched against an atlas surface. As each surface is matched to the same atlas, changes in surface area between subjects can be compared with one another. Because this vertex measure depends on the chosen atlas, the area is referred to as the *surface area expansion factor*. The atlas mapping also allows for atlas-based regions of interest to be carried over to each subject's surface. These regions are defined as collections of triangles and correspond to functional regions of the brain.

3.3.4.5 Other Morphometric Measures

There are many other ways to summarize important aspects of brain morphometry. For example, one can extract the location of the regional center of mass relative to the subject's location or the number of disconnected regions. However, selection of appropriate measures for a particular analysis and set of subjects should be done carefully to prevent an overwhelming number of multiple comparisons.

One set of measures assesses the degree of cortical folding, or *gyrification*, of the cerebral cortex (84, 112, 111). For example, in (111) the outermost surface of the brain, without folds, is defined with morphological closing operations. Next, hundreds of circular and overlapping regions are defined on that outermost surface, and each region is matched with the outline of a corresponding region on the pial surface, which may be buried and/or folded underneath the outermost defined surface. Then, gyrification is calculated quantitatively as the ratio between overall buried/folded cortex (on the pial surface) to the visible cortex on the outermost surface. This measure has been studied mainly in relation to neuropsychiatric disorders (81, 99, 126, 92).

Another set of measures can be created using image analysis methods such as marching cubes to compute surface representations from segmented ROIs. These surfaces allow us to further extend the list of measures that can be calculated for a given labelled region, with the hope that some might relate to behavior, disease, or function. We can calculate the area enclosing the region or regions, and use aspect ratios as measures of elongation and sphericity, surface curvature, tortuosity, and so forth. As a specific example, we introduce a procedure that learns optimal shape measures from the given set of input shapes.

A surface can be represented as a collection of points in 3d space. Connecting each group of three points with a triangle, the surface depicts the region's border. If the same region is segmented in another subject, a new surface can be obtained. Repeating the process for a selection of subjects, we can ask what the shape variability of that particular ROI is and how it might relate to other phenotypic measures. In particular, we can compute the mean shape and variation around the mean shape that are indicative of the shape space spanned by the input surfaces. This type of analysis provides a convenient data-driven decomposition of the observed shape variability.

In order to perform this analysis we need to represent the different shapes using the same triangulated surface topology. A point or vertex identified in one surface by a landmark has to correspond to the same landmark on all the other surfaces. Solving this problem is non-trivial, since each surface is obtained from a separate run of the marching cubes algorithm without guarantee of vertex correspondence between separate runs. We solve this problem by introducing a preprocessing stage. First, a single surface is identified as the source surface. For each subject, the source surface is now aligned and warped until it approximates the shape of the subject's target surface. The different instantiations of the source surface can now be used as stand-ins during the subsequent analysis steps since they all share the

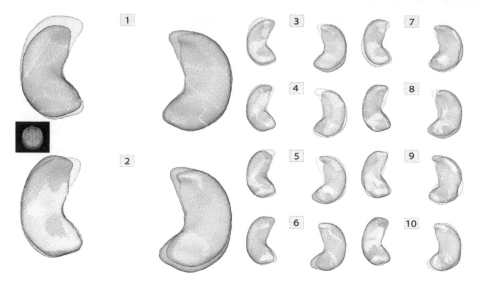

FIGURE 3.3

A source surface warped to a subset of 774 target surfaces representing the left and right human hippocampus. Calculating the cross-covariance matrix of the surface point coordinates and subsequently the eigenvalue decomposition of the matrix, we decompose the shape variability into de-correlated modes sorted by variance. Here the first 10 modes are displayed by varying the weight for the particular mode symmetrically around zero. The resulting two extreme surfaces are overlaid using transparency and highlight directions in which the particular mode varies.

topology of the source surface. Figure 3.3 shows a source surface warped to a subset of 774 target surfaces representing the left and right human hippocampus, a region essential for memory formation. The registration and surface warping was performed independently for the left and right homologues, but for subsequent steps, both left and right hemisphere hippocampus surfaces are combined. Calculating the cross-covariance matrix of the surface point coordinates and the eigenvalue decomposition of the matrix (principal component analysis), we decompose the shape variability into de-correlated modes sorted by variance. In Figure 3.3 the first 10 modes are displayed by varying the weight for the particular mode symmetrically around zero. The resulting two extreme surfaces are overlaid using transparency and highlight directions in which the particular mode varies.

Analyzing the modes of variation is instructive in itself, but we can also derive from them a compact representation of the shape of each target surface. Each target surface can be expressed in the space of shape modes by a weight vector that, when multiplied with the modes matrix and added together, approximates the target surface. Because modes are sorted by variance, we can limit the weight vector to a few entries for the modes that explain the most variance and remove weights that explain little in terms of variance. The resulting non-zero weights can be used as compact morphological measures for shape in subsequent analysis.

3.3.5 Statistical Analyses

Once a particular set of morphometric measures are produced from a collection of multi-subject MR images, these are typically entered into a statistical analysis to determine if individual differences are related to independent variables of interest. Most often this consists of performing a t-test or F-test for each morphometric measure if the independent variable is categorical, or a correlation coefficient if it is continuous. If one wants to include covariates of no interest, a general linear model (GLM) is often employed; see Chapter 9. Within this framework it is possible to perform group comparisons, and find brain correlates of various covariates such as age or disease progression.

The parametric statistical tests applied in the analysis of MRI data generally requires that the noise distribution does not depart too strongly from a Gaussian distribution. As noted above, the MR image intensities are non-negative since after Fourier reconstruction they are converted into magnitude images. This results in noise governed by a Rician distribution, which is substantially different from Gaussian only if the signal to noise of the data is small (61). Many factors affect SNR in MR images, including magnet strength and image acquisition parameters such as slice thickness, field of view, TR, TE, and flip angle (60). A correction for the intensity bias caused in regions with low signal-to-noise ratio (SNR) has been proposed in (54). In general, structural images are typically spatially smoothed, which both increases SNR and promotes a more Gaussian error distribution (3). Indeed, some authors have argued that modeling noise distributions as Rician adds considerable complexity to analyses with few tangible benefits (1).

3.3.5.1 Statistical Parametric Maps

In many instances, each MR image is summarized by a single, or at most a few, morphometric measures. Reporting the results of statistical analyses in this scenario differs little from other types of biomedical research studies. If multiple statistical tests are performed, one can control for inflation of Type I error rates via Bonferroni adjustments to p-values or by resampling-based algorithms such as those of Westfall–Young (128, 46).

In contrast, whole-brain voxel-wise approaches such as DBM, result in tens or hundreds of thousands of statistical models, one for each voxel (3). The resulting collection of statistical tests (e.g., t, chi-square, F, or Hotelling T^2 tests), after thresholding at a given critical value, is called a *statistical parametric map* (SPM) (30, 103). The appropriate threshold can be chosen via a number of criteria. Typically, one differentiates between methods that control the family-wise error rate (the probability of at least one false positive) or the false detection rate (the expected proportion of incorrectly rejected null hypothesis). Standard Bonferroni adjustments tend to be far too conservative, especially since the effective number of tests is typically much smaller than the number of voxels due to spatially correlated noise (16). Common multiple testing adjustments include random field theory (129, 21), permutation tests (59), and false discovery rate (47, 114). Chapter 9 covers random field theory; for additional review see (79). For additional review see (79). An alternative approach towards controlling for multiple testing in whole-brain analyses is to use multivariate statistical methods. This is what is done in the TBM analysis outlined above.

3.4 Miscellaneous Topics

3.4.1 Structural Integrity and Tumor Detection

MRI has proven to be extremely useful in detecting abnormalities in the brain. This could entail atrophy changes in the brain related to normal aging and disease progression (70), or the appearance of tumors. In fact, MRI is often able to detect many types of tumors at earlier stages of development than many other medical imaging modalities. The use of automated methods for MRI brain tumor segmentation is critical, as it provides important information for both medical diagnosis and surgical planning (73). It also offers the possibility of freeing doctors from the burdens involved in manual labeling. However, automated brain segmentation is a difficult problem due to inherent differences in the characteristics of different tumors, including their size, shape, location and intensity. For these reasons, manual tracing and delineation of tumors remains the gold standard.

More generally, MRI can be used to provide the information needed to both qualitatively and quantitatively describe the integrity of gray and white matter structures in the brain. For example, changes in the structural integrity of myelin can be measured with MRI as myelin breakdown increases the water content in white matter (10). In addition, MRI can together with diffusion weighted MRI, be used to provide a picture of white matter integrity. Also, since brain function may depend on the integrity of brain structure, it can be used together with fMRI to examine the impact of tissue loss or damage on functional responses.

3.4.2   Anatomical References for Functional Imaging

In any given functional MRI study, a number of high-resolution structural scans are collected and used for both preprocessing and presentation purposes. Typically, fMRI data is of relatively low spatial resolution (compared to structural MRI), and therefore provides relatively little anatomical detail. It is therefore useful to map the results obtained from an fMRI analysis onto a higher-resolution structural MR image for presentation purposes. This process, referred to as *co-registration*, is typically performed using either a rigid body or affine transformation. Because the functional and structural images are collected using different sequences and focus on different tissue properties with differences in average intensities, it is generally recommended that transformation parameters be estimated by maximizing the mutual information between the two images. Typically, a single structural image is co-registered to the first or mean functional image. Co-registration is also a necessary step for subsequent normalization of the functional data onto a template. Here the high-resolution structural image is transformed to a standard template and the same transformations are thereafter applied to the functional images that were previously co-registered to the structural scan; see Chapter 6 for more detail.

Finally, we are often interested in focusing our analysis on certain targeted regions of interest (ROI). In certain experiments with a targeted hypothesis of interest, one can pre-specify a set of anatomical regions *a priori* and perform statistical analysis solely on data associated with these regions. This can help minimize problems with multiple comparisons by limiting the required number of statistical tests. The structural ROIs can be defined using automated anatomical labeling of subject-specific data, allowing one to define ROIs for each subject based on their anatomy, or by using standard brain atlases.

3.4.3   Multi-Center Studies

Multi-center MRI studies are becoming an increasingly common method to answer scientific questions that would be difficult or impossible to address with a single site study (50). Multi-center studies can be especially advantageous with rare conditions, where recruiting subjects in sufficient numbers in a single geographic area would be difficult. However, the benefit from the increase in power from access to more subjects is potentially offset by differences in scanner properties that introduce substantial site effects to MRI data. If multiple sites acquire the image data, each scanner might introduce variations in the data that can hinder the detection of weak correlations or that may introduce spurious correlations. Documenting auxiliary measures such as the identity of the imaging scanner (i.e., device serial number) and the version of the software performing image reconstruction are therefore essential additional information that can lead to increased power to reject or confirm hypothesis.

A few MRI reliability studies have been performed. For example, (72) found that T_1-weighted MRI volumes produced from automated segmentation algorithms are fairly reliable as long as they are produced on the same platform and field strength. In light of recent articles critiquing the reliability of neuroscience research (19, 91), this promises to be an important area for future research.

3.4.4 Imaging Genetics

Imaging genetics is a relatively new field that attempts to associate genetic variation in a population to measures derived from structural or functional imaging (62, 120). If there are only one, or a few, genetic loci of interest, standard MRI analyses suffice. However, as with many complex phenotypes, morphometric properties of the human brain are most likely the product of many genetic loci, each with small effect (17). A genome-wide association study (GWAS) may involve millions of separate tests of association for each MR-derived outcome (120). This is a type of analysis that will clearly lead to enormous multiple testing problems.

Some authors have proposed statistical methods promoting sparse solutions for both domains simultaneously (e.g., (125)). Chen et al. (17) takes a different tack, using voxel-wise cortical surface area measures from an MRI twin study. The genetic correlation of surface area is computed for each pair of voxels, which is used to form a similarity matrix. A fuzzy clustering algorithm is applied to this similarity matrix, with the number of clusters selected via a silhouette coefficient. The end result is a probabilistic atlas of the human cortex based on genetically informed cluster assignment. Note, this approach reduces dimensionality dramatically, and reflects the fact that genetic effects are unlikely to be sparse.

3.5 Glossary of MRI Terms

Axon: A long threadlike part of a neuron, along which impulses are conducted from the cell body.

Brain Atlas: Reference brains where relevant brain structures are placed in a standardized coordinate system.

Cerebrospinal Fluid (CSF): A watery fluid that flows in the ventricles and around the surface of the brain and spinal cord.

Dendrite: A branched extension of a neuron, along which impulses received from other cells are transmitted to the cell body.

Echo Time (TE): The time between an excitation pulse and the start of data acquisition.

Field of View (FOV): The extent of an image in space.

Field Strength: The strength of the static magnetic field of the MR scanner. It is typically measured in units of Tesla.

Flip Angle: The change in angle of the net magnetization immediately following excitation.

Glial Cell: A cell that surround neurons and provides them with support and nutrition.

Gray Matter (GM): Tissue of the brain and spinal cord that consists primarily of nerve cell bodies and branching dendrites.

Image Registration: The process of transforming images into a standard coordinate system.

Image Segmentation: The process of partitioning the image according to different tissue types.

MPRAGE: Magnetization Prepared Rapid Acquisition Gradient Echo (MPRAGE) is a specialized pulse sequence defined for rapid acquisition.

Neuron: A nerve cell for processing and transmitting electrical signals.

Proton Density-Weighted Image: Images providing information about the number of protons contained within each voxel.

Pulse Sequence: The set of magnetic field gradients and oscillating magnetic fields used to create MR images.

Region of Interest (ROI): An anatomical area of the brain of particular importance.

Repetition Time (TR): The time between two successive excitation pulses.

Slice: The cross-section of the brain being imaged.

SPGR: Spoiled Gradient-Echo (SPGR) is a pulse sequence that destroy and remaining transverse magnetization at the end of each excitation.

Statistical Parametric Map (SPM): An image where brain voxels are labeled according to the results of a statistical test.

T_1-Weighted Image: Images providing information about different tissues' T_1 values.

T_2-Weighted Image: Images providing information about different tissues' T_2 values.

Template Image: A standardized image used as the target in the process of image registration.

Voxel: A volume element. The basic unit of measurement of an MRI.

White Matter (WM): Tissue of the brain and spinal cord that consists primarily of nerve fibers and their myelin sheaths.

Bibliography

[1] Daniel W Adrian, Ranjan Maitra, and Daniel B Rowe. *Stat*, 2(1):303–316, 2013.

[2] J. Ashburner. A fast diffeomorphic image registration algorithm. *NeuroImage*, 38(1):95–113, 2007.

[3] John Ashburner and Karl J Friston. *Neuroimage*, 11(6):805–821, 2000.

[4] John Ashburner and Karl J Friston. Unified segmentation. *Neuroimage*, 26(3):839–851, 2005.

[5] John Ashburner, Chloe Hutton, Richard Frackowiak, Ingrid Johnsrude, Cathy Price, Karl Friston, et al. Identifying global anatomical differences: Deformation-based morphometry. *Human Brain Mapping*, 6(5-6):348–357, 1998.

[6] John Ashburner, P Neelin, DL Collins, A Evans, and K Friston. Incorporating prior knowledge into image registration. *Neuroimage*, 6(4):344–352, 1997.

[7] Kolawole Oluwole Babalola, Brian Patenaude, Paul Aljabar, Julia Schnabel, David Kennedy, William Crum, Stephen Smith, Tim Cootes, Mark Jenkinson, and Daniel Rueckert. An evaluation of four automatic methods of segmenting the subcortical structures in the brain. *Neuroimage*, 47(4):1435–1447, 2009.

[8] Mohd Ali Balafar, Abdul Rahman Ramli, M Iqbal Saripan, and Syamsiah Mashohor. Review of brain MRI image segmentation methods. *Artificial Intelligence Review*, 33(3):261–274, 2010.

[9] Vincent Barra and J-Y Boire. Automatic segmentation of subcortical brain structures in mr images using information fusion. *Medical Imaging, IEEE Transactions on*, 20(7):549–558, 2001.

[10] George Bartzokis, Jeffrey L Cummings, David Sultzer, Victor W Henderson, Keith H Nuechterlein, and Jim Mintz. White matter structural integrity in healthy aging adults and patients with Alzheimer disease: A magnetic resonance imaging study. *Archives of Neurology*, 60(3):393–398, 2003.

[11] Stefan Bauer, Roland Wiest, Lutz-P Nolte, and Mauricio Reyes. A survey of MRI-based medical image analysis for brain tumor studies. *Physics in Medicine and Biology*, 58(13):R97, 2013.

[12] Julian Besag. Spatial interaction and the statistical analysis of lattice systems. *Journal of the Royal Statistical Society. Series B (Methodological)*, pages 192–236, 1974.

[13] S Beucher and C Lantuejoul. International workshop on image processing (real-time edge and motion detection/estimation). 1979.

[14] Nathalie Boddaert, Nadia Chabane, H Gervais, CD Good, M Bourgeois, MH Plumet, C Barthelemy, MC Mouren, E Artiges, Y Samson, et al. Superior temporal sulcus anatomical abnormalities in childhood autism: A voxel-based morphometry MRI study. *Neuroimage*, 23(1):364–369, 2004.

[15] Alberto Boschetti, Nicola Adami, Riccardo Leonardi, and Masahiro Okuda. High dynamic range image tone mapping based on local histogram equalization. In *Multimedia and Expo (ICME), 2010 IEEE International Conference on*, pages 1130–1135. IEEE, 2010.

[16] Matthew Brett, Will Penny, and Stefan Kiebel. Introduction to random field theory. *Human Brain Function*, 2, 2004.

[17] Korbinian Brodmann. *Vergleichende Lokalisationslehre der Grosshirnrinde in ihren Prinzipien dargestellt auf Grund des Zellenbaues*. Barth, 1909.

[18] Emma J Burton, Ian G McKeith, David J Burn, E David Williams, and John T O'Brien. Cerebral atrophy in Parkinson's disease with and without dementia: A comparison with Alzheimer's disease, dementia with Lewy bodies and controls. *Brain*, 127(4):791–800, 2004.

[19] Katherine S Button, John PA Ioannidis, Claire Mokrysz, Brian A Nosek, Jonathan Flint, Emma SJ Robinson, and Marcus R Munafò. Power failure: Why small sample size undermines the reliability of neuroscience. *Nature Reviews Neuroscience*, 14(5):365–376, 2013.

[20] Mariano Cabezas, Arnau Oliver, Xavier Lladó, Jordi Freixenet, and Meritxell Bach Cuadra. A review of atlas-based segmentation for magnetic resonance brain images. *Computer Methods and Programs in Biomedicine*, 104(3):e158–e177, 2011.

[21] Jin Cao, Keith J Worsley, et al. The detection of local shape changes via the geometry of Hotelling's t^2 fields. *The Annals of Statistics*, 27(3):925–942, 1999.

[22] Chi-Hua Chen, ED Gutierrez, Wesley K Thompson, Matthew S Panizzon, Terry L Jernigan, Lisa T Eyler, Christine Fennema-Notestine, Amy J Jak, Michael C Neale, Carol E Franz, et al. Hierarchical genetic organization of human cortical surface area. *Science*, 335(6076):1634–1636, 2012.

[23] Gary E Christensen, Richard D Rabbitt, Michael I Miller, Sarang C Joshi, Ulf Grenander, and Thomas A Coogan. Topological properties of smooth anatomic. In *Information Processing in Medical Imaging*, volume 3, page 101, Springer, 1995.

[24] MK Chung, KJ Worsley, T Paus, C Cherif, DL Collins, JN Giedd, JL Rapoport, and AC Evans. A unified statistical approach to deformation-based morphometry. *NeuroImage*, 14(3):595–606, 2001.

[25] Moo K Chung, Steven M Robbins, Kim M Dalton, Richard J Davidson, Andrew L Alexander, and Alan C Evans. Cortical thickness analysis in autism with heat kernel smoothing. *NeuroImage*, 25(4):1256–1265, 2005.

[26] Moo K Chung, Keith J Worsley, Steve Robbins, Tomáš Paus, Jonathan Taylor, Jay N Giedd, Judith L Rapoport, and Alan C Evans. Deformation-based surface morphometry applied to gray matter deformation. *NeuroImage*, 18(2):198–213, 2003.

[27] WR Crum, T Hartkens, and DLG Hill. Non-rigid image registration: Theory and practice. 2014.

[28] Meritxell Bach Cuadra, Leila Cammoun, Torsten Butz, Olivier Cuisenaire, and JP Thiran. Comparison and validation of tissue modelization and statistical classification methods in t_1-weighted MR brain images. *Medical Imaging, IEEE Transactions on*, 24(12):1548–1565, 2005.

[29] Anders M Dale, Bruce Fischl, and Martin I Sereno. Cortical surface-based analysis: I. segmentation and surface reconstruction. *Neuroimage*, 9(2):179–194, 1999.

[30] Anders M Dale and Martin I Sereno. Improved localizadon of cortical activity by combining EEG and MEG with MRI cortical surface reconstruction: A linear approach. *Journal of Cognitive Neuroscience*, 5(2):162–176, 1993.

[31] Rahul S Desikan, Florent Ségonne, Bruce Fischl, Brian T Quinn, Bradford C Dickerson, Deborah Blacker, Randy L Buckner, Anders M Dale, R Paul Maguire, Bradley T Hyman, et al. An automated labeling system for subdividing the human cerebral cortex on MRI scans into gyral based regions of interest. *Neuroimage*, 31(3):968–980, 2006.

[32] Christophe Destrieux, Bruce Fischl, Anders Dale, and Eric Halgren. Automatic parcellation of human cortical gyri and sulci using standard anatomical nomenclature. *Neuroimage*, 53(1):1–15, 2010.

[33] Bradford C Dickerson, Eric Feczko, Jean C Augustinack, Jenni Pacheco, John C Morris, Bruce Fischl, and Randy L Buckner. Differential effects of aging and Alzheimer's disease on medial temporal lobe cortical thickness and surface area. *Neurobiology of Aging*, 30(3):432–440, 2009.

[34] Henri M Duvernoy, S Delon, and JL Vannson. Cortical blood vessels of the human brain. *Brain Research Bulletin*, 7(5):519–579, 1981.

[35] Alan C Evans, D Louis Collins, SR Mills, ED Brown, RL Kelly, and Terry M Peters. 3d statistical neuroanatomical models from 305 MRI volumes. In *Nuclear Science Symposium and Medical Imaging Conference, 1993., 1993 IEEE Conference Record.*, pages 1813–1817. IEEE, 1993.

[36] Alan C Evans, Andrew L Janke, D Louis Collins, and Sylvain Baillet. Brain templates and atlases. *Neuroimage*, 62(2):911–922, 2012.

[37] Bruce Fischl. FreeSurfer. *Neuroimage*, 62(2):774–781, 2012.

[38] Bruce Fischl and Anders M Dale. Measuring the thickness of the human cerebral cortex from magnetic resonance images. *Proceedings of the National Academy of Sciences*, 97(20):11050–11055, 2000.

[39] Bruce Fischl, David H Salat, Evelina Busa, Marilyn Albert, Megan Dieterich, Christian Haselgrove, Andre van der Kouwe, Ron Killiany, David Kennedy, Shuna Klaveness, et al. Whole brain segmentation: Automated labeling of neuroanatomical structures in the human brain. *Neuron*, 33(3):341–355, 2002.

[40] Bruce Fischl, David H Salat, André JW van der Kouwe, Nikos Makris, Florent Ségonne, Brian T Quinn, and Anders M Dale. Sequence-independent segmentation of magnetic resonance images. *Neuroimage*, 23:S69–S84, 2004.

[41] Bruce Fischl, Martin I Sereno, and Anders M Dale. Cortical surface-based analysis: II: Inflation, flattening, and a surface-based coordinate system. *Neuroimage*, 9(2):195–207, 1999.

[42] Bruce Fischl, Martin I Sereno, Roger BH Tootell, Anders M Dale, et al. High-resolution intersubject averaging and a coordinate system for the cortical surface. *Human Brain Mapping*, 8(4):272–284, 1999.

[43] Bruce Fischl, André van der Kouwe, Christophe Destrieux, Eric Halgren, Florent Ségonne, David H Salat, Evelina Busa, Larry J Seidman, Jill Goldstein, David Kennedy, et al. Automatically parcellating the human cerebral cortex. *Cerebral cortex*, 14(1):11–22, 2004.

[44] Karl J Friston. Statistical parametric mapping. In *Neuroscience Databases*, pages 237–250. Springer, 2003.

[45] Simona Gardini, C Robert Cloninger, and Annalena Venneri. Individual differences in personality traits reflect structural variance in specific brain regions. *Brain Research Bulletin*, 79(5):265–270, 2009.

[46] Youngchao Ge, Sandrine Dudoit, and Terence P Speed. Resampling-based multiple testing for microarray data analysis. *Test*, 12(1):1–77, 2003.

[47] Christopher R Genovese, Nicole A Lazar, and Thomas Nichols. Thresholding of statistical maps in functional neuroimaging using the false discovery rate. *Neuroimage*, 15(4):870–878, 2002.

[48] Jay N Giedd. Structural magnetic resonance imaging of the adolescent brain. *Annals of the New York Academy of Sciences*, 1021(1):77–85, 2004.

[49] Jay N Giedd and Judith L Rapoport. Structural MRI of pediatric brain development: What have we learned and where are we going? *Neuron*, 67(5):728–734, 2010.

[50] Gary H Glover, Bryon A Mueller, Jessica A Turner, Theo GM van Erp, Thomas T Liu, Douglas N Greve, James T Voyvodic, Jerod Rasmussen, Gregory G Brown, David B Keator, et al. Function biomedical informatics research network recommendations for prospective multicenter functional MRI studies. *Journal of Magnetic Resonance Imaging*, 36(1):39–54, 2012.

[51] Erhan Gokcay and Jose C. Principe. Information theoretic clustering. *Pattern Analysis and Machine Intelligence, IEEE Transactions on*, 24(2):158–171, 2002.

[52] Jill M Goldstein, Larry J Seidman, Nicholas J Horton, Nikos Makris, David N Kennedy, Verne S Caviness, Stephen V Faraone, and Ming T Tsuang. Normal sexual dimorphism of the adult human brain assessed by in vivo magnetic resonance imaging. *Cerebral Cortex*, 11(6):490–497, 2001.

[53] Catriona D Good, Ingrid S Johnsrude, John Ashburner, Richard NA Henson, KJ Fristen, and Richard SJ Frackowiak. A voxel-based morphometric study of ageing in 465 normal adult human brains. In *Biomedical Imaging, 2002. 5th IEEE EMBS International Summer School on*, pages 16–pp. IEEE, 2002.

[54] Hákon Gudbjartsson and Samuel Patz. The Rician distribution of noisy MRI data. *Magnetic Resonance in Medicine*, 34(6):910–914, 1995.

[55] Raquel E Gur, Veda Maany, P David Mozley, Charlie Swanson, Warren Bilker, and Ruben C Gur. Subcortical MRI volumes in neuroleptic-naive and treated patients with schizophrenia. *American Journal of Psychiatry*, 155(12):1711–1717, 1998.

[56] E Mark Haacke, Robert W Brown, Michael R Thompson, and Ramesh Venkatesan. Magnetic resonance imaging. *Physical Principles and Sequence Design*, 1999.

[57] Richard J Haier, Rex E Jung, Ronald A Yeo, Kevin Head, and Michael T Alkire. The neuroanatomy of general intelligence: sex matters. *NeuroImage*, 25(1):320–327, 2005.

[58] Xiao Han and Bruce Fischl. Atlas renormalization for improved brain MR image segmentation across scanner platforms. *Medical Imaging, IEEE Transactions on*, 26(4):479–486, 2007.

[59] Satoru Hayasaka and Thomas E Nichols. Validating cluster size inference: Random field and permutation methods. *Neuroimage*, 20(4):2343–2356, 2003.

[60] R Edward Hendrick, J Bruce Kneeland, and David D Stark. Maximizing signal-to-noise and contrast-to-noise ratios in flash imaging. *Magnetic Resonance Imaging*, 5(2):117–127, 1987.

[61] R Mark Henkelman. Measurement of signal intensities in the presence of noise in MR images. *Medical Physics*, 12(2):232–233, 1985.

[62] Derrek P Hibar, Jason L Stein, Omid Kohannim, Neda Jahanshad, Andrew J Saykin, Li Shen, Sungeun Kim, Nathan Pankratz, Tatiana Foroud, Matthew J Huentelman, et al. Voxelwise gene-wide association study (GENEWAS): Multivariate gene-based association testing in 731 elderly subjects. *Neuroimage*, 56(4):1875–1891, 2011.

[63] Colin J Holmes, Rick Hoge, Louis Collins, Roger Woods, Arthur W Toga, and Alan C Evans. Enhancement of MR images using registration for signal averaging. *Journal of Computer Assisted Tomography*, 22(2):324–333, 1998.

[64] Robyn Honea, Tim J Crow, Dick Passingham, and Clare E Mackay. Regional deficits in brain volume in schizophrenia: A meta-analysis of voxel-based morphometry studies. *American Journal of Psychiatry*, 162(12):2233–2245, 2005.

[65] Xue Hua, Alex D Leow, Neelroop Parikshak, Suh Lee, Ming-Chang Chiang, Arthur W Toga, Clifford R Jack Jr, Michael W Weiner, and Paul M Thompson. Tensor-based morphometry as a neuroimaging biomarker for Alzheimer's disease: An MRI study of 676 AD, MCI, and normal subjects. *Neuroimage*, 43(3):458–469, 2008.

[66] Scott A Huettel, Allen W Song, and Gregory McCarthy. *Functional Magnetic Resonance Imaging*, volume 1. Sinauer Associates Sunderland, MA, 2004.

[67] Kiho Im, Jong-Min Lee, Oliver Lyttelton, Sun Hyung Kim, Alan C Evans, and Sun I Kim. Brain size and cortical structure in the adult human brain. *Cerebral Cortex*, 18(9):2181–2191, 2008.

[68] Dan V Iosifescu, Martha E Shenton, Simon K Warfield, Ron Kikinis, Joachim Dengler, Ferenc A Jolesz, and Robert W McCarley. An automated registration algorithm for measuring MRI subcortical brain structures. *Neuroimage*, 6(1):13–25, 1997.

[69] Clifford R Jack, Matt A Bernstein, Nick C Fox, Paul Thompson, Gene Alexander, Danielle Harvey, Bret Borowski, Paula J Britson, Jennifer L Whitwell, Chadwick Ward, et al. The Alzheimer's disease neuroimaging initiative (ADNI): MRI methods. *Journal of Magnetic Resonance Imaging*, 27(4):685–691, 2008.

[70] Clifford R Jack, Ronald C Petersen, Yue Cheng Xu, Stephen C Waring, Peter C O'Brien, Eric G Tangalos, Glenn E Smith, Robert J Ivnik, and Emre Kokmen. Medial temporal atrophy on MRI in normal aging and very mild Alzheimer's disease. *Neurology*, 49(3):786–794, 1997.

[71] Anand A Joshi, David W Shattuck, Paul M Thompson, and Richard M Leahy. Surface-constrained volumetric brain registration using harmonic mappings. *Medical Imaging, IEEE Transactions on*, 26(12):1657–1669, 2007.

[72] Jorge Jovicich, Silvester Czanner, Xiao Han, David Salat, Andre van der Kouwe, Brian Quinn, Jenni Pacheco, Marilyn Albert, Ronald Killiany, Deborah Blacker, et al. MRI-derived measurements of human subcortical, ventricular and intracranial brain volumes: Reliability effects of scan sessions, acquisition sequences, data analyses, scanner upgrade, scanner vendors and field strengths. *Neuroimage*, 46(1):177–192, 2009.

[73] Michael R Kaus, Simon K Warfield, Arya Nabavi, Peter M Black, Ferenc A Jolesz, and Ron Kikinis. Automated segmentation of MR images of brain tumors 1. *Radiology*, 218(2):586–591, 2001.

[74] Ali R Khan, Moo K Chung, and Mirza Faisal Beg. Robust atlas-based brain segmentation using multi-structure confidence-weighted registration. In *Medical Image Computing and Computer-Assisted Intervention–MICCAI 2009*, pages 549–557. Springer, 2009.

[75] F Lepore, Caroline Brun, Yi-Yu Chou, Ming-Chang Chiang, Rebecca A Dutton, Kiralee M Hayashi, Eileen Luders, Oscar L Lopez, Howard J Aizenstein, Arthur W Toga, et al. Generalized tensor-based morphometry of HIV/AIDS using multivariate statistics on deformation tensors. *Medical Imaging, IEEE Transactions on*, 27(1):129–141, 2008.

[76] Jason P Lerch and Alan C Evans. Cortical thickness analysis examined through power analysis and a population simulation. *Neuroimage*, 24(1):163–173, 2005.

[77] Kelvin K Leung, Josephine Barnes, Gerard R Ridgway, Jonathan W Bartlett, Matthew J Clarkson, Kate Macdonald, Norbert Schuff, Nick C Fox, and Sebastien Ourselin. Automated cross-sectional and longitudinal hippocampal volume measurement in mild cognitive impairment and Alzheimer's disease. *Neuroimage*, 51(4):1345–1359, 2010.

[78] Martin A Lindquist et al. The statistical analysis of fMRI data. *Statistical Science*, 23(4):439–464, 2008.

[79] Martin A Lindquist and Amanda Mejia. Zen and the art of multiple comparisons. *Psychosomatic Medicine*, 77(2):114–125, 2015.

[80] Martin A Lindquist and Tor D Wager. Meta-analyses in functional neuroimaging. In Arthur W Toga, editor, *Brain Mapping: An Encyclopedic Reference*, pages 661–665. Academic Press, Elsevier, 2015.

[81] Tao Liu, Darren M Lipnicki, Wanlin Zhu, Dacheng Tao, Chengqi Zhang, Yue Cui, Jesse S Jin, Perminder S Sachdev, and Wei Wen. Cortical gyrification and sulcal spans in early stage Alzheimer's disease. *PloS One*, 7(2):e31083, 2012.

[82] Valentina Lorenzetti, Nicholas B Allen, Alex Fornito, and Murat Yücel. Structural brain abnormalities in major depressive disorder: A selective review of recent MRI studies. *Journal of affective disorders*, 117(1):1–17, 2009.

[83] Joachim Lotz, Christian Meier, Andreas Leppert, and Michael Galanski. Cardiovascular flow measurement with phase-contrast MR imaging: Basic facts and implementation 1. *Radiographics*, 22(3):651–671, 2002.

[84] E Luders, PM Thompson, KL Narr, AW Toga, L Jancke, and C Gaser. A curvature-based approach to estimate local gyrification on the cortical surface. *Neuroimage*, 29(4):1224–1230, 2006.

[85] David MacDonald, Noor Kabani, David Avis, and Alan C Evans. Automated 3-d extraction of inner and outer surfaces of cerebral cortex from MRI. *NeuroImage*, 12(3):340–356, 2000.

[86] Daniel S Marcus, Tracy H Wang, Jamie Parker, John G Csernansky, John C Morris, and Randy L Buckner. Open access series of imaging studies (OASIS): Cross-sectional MRI data in young, middle aged, nondemented, and demented older adults. *Journal of Cognitive Neuroscience*, 19(9):1498–1507, 2007.

[87] Kanti V Mardia and TJ Hainsworth. A spatial thresholding method for image segmentation. *Pattern Analysis and Machine Intelligence, IEEE Transactions on*, 10(6):919–927, 1988.

[88] John Mazziotta, Arthur Toga, Alan Evans, Peter Fox, Jack Lancaster, Karl Zilles, Roger Woods, Tomas Paus, Gregory Simpson, Bruce Pike, et al. A probabilistic atlas and reference system for the human brain: International Consortium for Brain Mapping (ICBM). *Philosophical Transactions of the Royal Society of London. Series B: Biological Sciences*, 356(1412):1293–1322, 2001.

[89] Andrea Mechelli, Cathy J Price, Karl J Friston, and John Ashburner. Voxel-based morphometry of the human brain: Methods and applications. *Current Medical Imaging Reviews*, 1(2):105–113, 2005.

[90] Donald G Mitchell and Mark Cohen. *MRI Principles*. Saunders Philadelphia, 1999.

[91] Marcus Munafò, Simon Noble, William J Browne, Dani Brunner, Katherine Button, Joaquim Ferreira, Peter Holmans, Douglas Langbehn, Glyn Lewis, Martin Lindquist, et al. Scientific rigor and the art of motorcycle maintenance. *Nature Biotechnology*, 32(9):871–873, 2014.

[92] Pranav Nanda, Neeraj Tandon, Ian T Mathew, Christoforos I Giakoumatos, Hulegar A Abhishekh, Brett A Clementz, Godfrey D Pearlson, John Sweeney, Carol A Tamminga, and Matcheri S Keshavan. Local gyrification index in probands with psychotic disorders and their first-degree relatives. *Biological Psychiatry*, 2013.

[93] Wayne Niblack. *An Introduction to Digital Image Processing*. Strandberg Publishing Company, 1985.

[94] Thomas E Nichols. Multiple testing corrections, nonparametric methods, and random field theory. *Neuroimage*, 62(2):811–815, 2012.

[95] John Nolte. *The Human Brain in Photographs and Diagrams*. Elsevier Health Sciences, 2013.

[96] Seiji Ogawa and Tso-Ming Lee. Magnetic resonance imaging of blood vessels at high fields: In vivo and in vitro measurements and image simulation. *Magnetic Resonance in Medicine*, 16(1):9–18, 1990.

[97] Nobuyuki Otsu. A threshold selection method from gray-level histograms. *Automatica*, 11(285-296):23–27, 1975.

[98] Bente Pakkenberg and Hans Jørgen G Gundersen. Neocortical neuron number in humans: Effect of sex and age. *Journal of Comparative Neurology*, 384(2):312–320, 1997.

[99] Lena Palaniyappan and Peter F Liddle. Aberrant cortical gyrification in schizophrenia: A surface-based morphometry study. *Journal of Psychiatry & Neuroscience: JPN*, 37(6):399, 2012.

[100] Matthew S Panizzon, Christine Fennema-Notestine, Lisa T Eyler, Terry L Jernigan, Elizabeth Prom-Wormley, Michael Neale, Kristen Jacobson, Michael J Lyons, Michael D Grant, Carol E Franz, et al. Distinct genetic influences on cortical surface area and cortical thickness. *Cerebral Cortex*, 19(11):2728–33, 2009.

[101] Dimitrios Pantazis, Anand Joshi, Jintao Jiang, David W Shattuck, Lynne E Bernstein, Hanna Damasio, and Richard M Leahy. Comparison of landmark-based and automatic methods for cortical surface registration. *Neuroimage*, 49(3):2479–2493, 2010.

[102] Brian Patenaude, Stephen M Smith, David N Kennedy, and Mark Jenkinson. A Bayesian model of shape and appearance for subcortical brain segmentation. *Neuroimage*, 56(3):907–922, 2011.

[103] William D Penny, Karl J Friston, John T Ashburner, Stefan J Kiebel, and Thomas E Nichols. *Statistical Parametric Mapping: The Analysis of Functional Brain Images*. Academic Press, 2011.

[104] Josien PW Pluim, JB Antoine Maintz, and Max A Viergever. Mutual-information-based registration of medical images: a survey. *Medical Imaging, IEEE Transactions on*, 22(8):986–1004, 2003.

[105] Gheorghe Postelnicu, Lilla Zollei, and Bruce Fischl. Combined volumetric and surface registration. *Medical Imaging, IEEE Transactions on*, 28(4):508–522, 2009.

[106] Stephanie Powell, Vincent A Magnotta, Hans Johnson, Vamsi K Jammalamadaka, Ronald Pierson, and Nancy C Andreasen. Registration and machine learning-based automated segmentation of subcortical and cerebellar brain structures. *Neuroimage*, 39(1):238–247, 2008.

[107] Pasko Rakic. Specification of cerebral cortical areas. *Science*, 241(4862):170–176, 1988.

[108] T W Ridler and S Calvard. Picture thresholding using an iterative selection method. *IEEE Transactions on Systems, Man and Cybernetics*, 8(8):630–632, 1978.

[109] Ziad S Saad, Daniel R Glen, Gang Chen, Michael S Beauchamp, Rutvik Desai, and Robert W Cox. A new method for improving functional-to-structural MRI alignment using local Pearson correlation. *Neuroimage*, 44(3):839–848, 2009.

[110] Peter Santago and H Donald Gage. Quantification of MR brain images by mixture density and partial volume modeling. *Medical Imaging, IEEE Transactions on*, 12(3):566–574, 1993.

[111] Marie Schaer, Meritxell Bach Cuadra, Nick Schmansky, Bruce Fischl, Jean-Philippe Thiran, and Stephan Eliez. How to measure cortical folding from MR images: A step-by-step tutorial to compute local gyrification index. *Journal of Visualized Experiments: JoVE*, (59), 2012.

[112] Marie Schaer, Meritxell Bach Cuadra, Lucas Tamarit, François Lazeyras, Stephan Eliez, and J Thiran. A surface-based approach to quantify local cortical gyrification. *Medical Imaging, IEEE Transactions on*, 27(2):161–170, 2008.

[113] Elaine H Shen, Caroline C Overly, and Allan R Jones. The Allen human brain atlas: Comprehensive gene expression mapping of the human brain. *Trends in Neurosciences*, 35(12):711–714, 2012.

[114] Hai Shu, Bin Nan, Robert Koeppe, et al. Multiple testing for neuroimaging via hidden Markov random field. *arXiv preprint arXiv:1404.1371*, 2014.

[115] Jason L Stein, Xue Hua, Jonathan H Morra, Suh Lee, Derrek P Hibar, April J Ho, Alex D Leow, Arthur W Toga, Jae Hoon Sul, Hyun Min Kang, et al. Genome-wide analysis reveals novel genes influencing temporal lobe structure with relevance to neurodegeneration in Alzheimer's disease. *Neuroimage*, 51(2):542–554, 2010.

[116] Andreas B Storsve, Anders M Fjell, Christian K Tamnes, Lars T Westlye, Knut Overbye, Hilde W Aasland, and Kristine B Walhovd. Differential longitudinal changes in cortical thickness, surface area and volume across the adult life span: Regions of accelerating and decelerating change. *The Journal of Neuroscience*, 34(25):8488–8498, 2014.

[117] Jean Talairach. *Atlas d'anatomie stéréotaxique du télencéphale: études anatomo-radiologiques. Atlas of stereo-taxic anatomy of the telencephalon.* Masson, 1967.

[118] Jean Talairach and Pierre Tournoux. Co-planar stereotaxic atlas of the human brain. 3-dimensional proportional system: An approach to cerebral imaging. 1988.

[119] RW Thatcher, M Camacho, A Salazar, C Linden, C Biver, and L Clarke. Quantitative MRI of the gray–white matter distribution in traumatic brain injury. *Journal of Neurotrauma*, 14(1):1–14, 1997.

[120] Paul M Thompson, Jason L Stein, Sarah E Medland, Derrek P Hibar, Alejandro Arias Vasquez, Miguel E Renteria, Roberto Toro, Neda Jahanshad, Gunter Schumann, Barbara Franke, et al. The enigma consortium: Large-scale collaborative analyses of neuroimaging and genetic data. *Brain Imaging and Behavior*, pages 1–30, 2014.

[121] Alan Tucholka, Virgile Fritsch, Jean-Baptiste Poline, and Bertrand Thirion. An empirical comparison of surface-based and volume-based group studies in neuroimaging. *Neuroimage*, 63(3):1443–1453, 2012.

[122] David C Van Essen and Donna L Dierker. Surface-based and probabilistic atlases of primate cerebral cortex. *Neuron*, 56(2):209–225, 2007.

[123] David C Van Essen, Matthew F Glasser, Donna L Dierker, John Harwell, and Timothy Coalson. Parcellations and hemispheric asymmetries of human cerebral cortex analyzed on surface-based atlases. *Cerebral Cortex*, 22(10):2241–2262, 2012.

[124] Tom Vercauteren, Xavier Pennec, Aymeric Perchant, and Nicholas Ayache. Diffeomorphic demons: Efficient non-parametric image registration. *NeuroImage*, 45(1):S61–S72, 2009.

[125] Maria Vounou, Thomas E Nichols, Giovanni Montana, Alzheimer's Disease Neuroimaging Initiative, et al. Discovering genetic associations with high-dimensional neuroimaging phenotypes: A sparse reduced-rank regression approach. *Neuroimage*, 53(3):1147–1159, 2010.

[126] Gregory L Wallace, Briana Robustelli, Nathan Dankner, Lauren Kenworthy, Jay N Giedd, and Alex Martin. Increased gyrification, but comparable surface area in adolescents with autism spectrum disorders. *Brain*, ePub May 28, 2013.

[127] Hongzhi Wang, Sandhitsu R Das, Jung Wook Suh, Murat Altinay, John Pluta, Caryne Craige, Brian Avants, and Paul A Yushkevich. A learning-based wrapper method to correct systematic errors in automatic image segmentation: Consistently improved performance in hippocampus, cortex and brain segmentation. *NeuroImage*, 55(3):968–985, 2011.

[128] Peter H Westfall. *Resampling-Based Multiple Testing: Examples and Methods for p-Value Adjustment*, volume 279. John Wiley & Sons, 1993.

[129] Keith J Worsley. Local maxima and the expected Euler characteristic of excursion sets of χ 2, f and t fields. *Advances in Applied Probability*, pages 13–42, 1994.

[130] Yongyue Zhang, Michael Brady, and Stephen Smith. Segmentation of brain MR images through a hidden Markov random field model and the expectation-maximization algorithm. *Medical Imaging, IEEE Transactions on*, 20(1):45–57, 2001.

4

Diffusion Magnetic Resonance Imaging (dMRI)

Jian Cheng

Department of Radiology and Biomedical Imaging Research Institute, University of North Carolina at Chapel Hill

Hongtu Zhu

Department of Biostatistics and Biomedical Imaging Research Institute, University of North Carolina at Chapel Hill

CONTENTS

4.1	Introduction to Diffusion MRI			66
	4.1.1	Diffusion Weighted Imaging (DWI)		66
		4.1.1.1	Diffusion Gradient Sequence	66
		4.1.1.2	Free Diffusion	67
		4.1.1.3	Restricted Diffusion	67
	4.1.2	Diffusion Tensor Imaging (DTI)		69
		4.1.2.1	Scalar Indices and Eigenvectors of Diffusion Tensor	69
4.2	High Angular Resolution Diffusion Imaging (HARDI)			70
	4.2.1	Generalization of Diffusion Tensor Imaging		70
		4.2.1.1	Mixture of Tensor Model	70
		4.2.1.2	Generalized DTI (GDTI)	71
		4.2.1.3	High-Order Tensor Model, ADC-Based Model	72
	4.2.2	Diffusion Spectrum Imaging (DSI)		73
	4.2.3	Hybrid Diffusion Imaging (HYDI)		74
	4.2.4	Q-Ball Imaging (QBI)		75
		4.2.4.1	Original Q-Ball Imaging	75
		4.2.4.2	Exact Q-Ball Imaging	76
	4.2.5	Diffusion Orientation Transform (DOT)		77
	4.2.6	Spherical Deconvolution (SD)		77
	4.2.7	Diffusion Propagator Imaging (DPI)		77
	4.2.8	Simple Harmonic Oscillator Reconstruction and Estimation (SHORE)		78
	4.2.9	Spherical Polar Fourier Imaging (SPFI)		78
4.3	Reconstruction			79
	4.3.1	Noise Components and Voxelwise Estimation Methods		79
	4.3.2	Spatial-Adaptive Estimation Methods		80
4.4	Tractography Algorithms			83
4.5	Uncertainty in Estimated Diffusion Quantities			86
4.6	Sampling Mechanisms			87
4.7	Registration			89
4.8	Group Analysis			90
4.9	Public Resources			93
	4.9.1	Datasets		93
	4.9.2	Software		94

4.10   Glossary 95
       Bibliography 96

4.1   Introduction to Diffusion MRI

Since the 1980s, diffusion magnetic resonance imaging (dMRI) as a magnetic resonance imaging (MRI) technique has been widely used to track the effective diffusion of water molecules, which is hindered by many obstacles (e.g., fibers or membranes), in the human brain *in vivo*. Because water molecules tend to diffuse slowly across white matter fibers and diffuse fast along such fibers, the use of dMRI to track water diffusion allows one to map the microstructure and organization of those white matter pathways (17). Quantitatively measuring the diffusion process is critical for a quantitative assessment of the integrity of anatomical connectivity in white matter and its association with brain functional connectivity. Its clinical applications include normal brain maturation and aging, cerebral ischemia, multiple sclerosis, epilepsy, metabolic disorders, and brain tumors, among many others. Although there are several nice review papers and monographies on dMRI (15, 5, 101, 81), this chapter was written for the readers who are interested in the theoretical underpinning of various mathematical and statistical methods associated with dMRI. Due to limitations of space, we are unable to cite all important papers in the dMRI literature.

4.1.1   Diffusion Weighted Imaging (DWI)

4.1.1.1   Diffusion Gradient Sequence

In diffusion weighted imaging (DWI), imaging signals can be made sensitive to diffusion through some diffusion gradient sequences. A standard diffusion gradient sequence used in dMRI is the Pulsed Gradient Spin-Echo (PGSE) sequence proposed by Stejskal and Tanner (126). See Figure 4.1 for the sketch map of the PGSE sequence. The PGSE sequence consists of two gradient pulses $\mathbf{G}(t)$ with duration time δ. Although it is common to use rectangular gradient lobes in the PGSE sequence, there are other kinds of gradient lobes commonly used in dMRI (24, chap. 9).

(i) The first 90° radio-frequency (RF) pulse translates the spins into the transverse plane, i.e. the x-y plane, considering the $\mathbf{B}_0$ is along the z-axis. Then the spins precess around $\mathbf{B}_0$ with RF ω_0. Due to local magnetic field inhomogeneities, some spins slow down and some spins speed up.[1]

(ii) After time Δ between two pulses, the second 180° RF pulse refocuses the phase of spins so that slower spins lead ahead and the fast ones trail behind. The spin echo process occurs when the spins recover their net magnetization.

(iii) The scanner coils receive the diffusion signal at echo time $t = TE$ after the two pulses.

Mathematically, the diffusion gradient $\mathbf{G}(\cdot)$ sequence can be written as

$$\mathbf{G}(t) = \{H(t_1) - H(t_1 - \delta) + H(t_2) - H(t_2 - \delta)\}\mathbf{G}, \qquad (4.1)$$

where $t_2 = t_1 + \Delta$, $H(\cdot)$ is the heaviside step function, and $\mathbf{G} = \|\mathbf{G}\|\mathbf{u}$, in which $\mathbf{u} \in S^2$ represents the gradient direction.

---

[1] http://en.wikipedia.org/wiki/Spin_echo

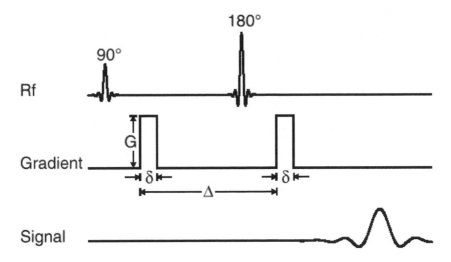

FIGURE 4.1
Pulsed Gradient Spin-Echo (PGSE) sequence introduced by Stejskal and Tanner (126). δ is the duration of the diffusion gradient pulses and Δ is the time between two diffusion gradient pulses.

4.1.1.2 Free Diffusion

Let $S(b)$ and $S(0)$ be, respectively, DWI signal at the diffusion weighting factor b and at $b = 0$, where b is defined below. Given the PGSE sequence with the gradient sequence in (4.1), the diffusion weighted **signal attenuation** $E(b) = S(b)/S(0)$ is given by the Stejskal–Tanner equation (126)

$$E(b) = \exp(-bD), \tag{4.2}$$

where D is known as the Apparent Diffusion Coefficient (ADC), which reflects the property of surrounding tissues. In general, ADC D depends on $\mathbf{G}$ in a complex way, but *free diffusion* assumes that D is only dependent on the direction of $\mathbf{G}$, i.e. $\mathbf{u} = \mathbf{G}/\|\mathbf{G}\|$. The early works in dMRI reported that the ADC D is a scalar and independent of gradient direction $\mathbf{u}$ (102). Then, Dr. Basser introduced the diffusion tensor (14) to represent ADC as $D(\mathbf{u}) = \mathbf{u}^T \mathbf{D} \mathbf{u}$, where $\mathbf{D}$ is a 3×3 symmetric positive definite matrix, called the **diffusion tensor**. This method is the well an known Diffusion Tensor Imaging (DTI). See Section 4.1.2 for more materials DTI.

The b factor is given by $b = \gamma^2 \delta^2 \tau \|\mathbf{G}\|^2$, where γ is the proton gyromagnetic ratio, and $\tau = \Delta - \delta/3$ is used to describe the effective diffusion time (25, 14). The b value is dependent on the sequence, and it is different in different kinds of lobes in the diffusion sequence (24, Chap. 9). The signal intensity at each voxel in DWI is dependent on both surrounding structures and given a weighted magnetic gradient (25). See Figure 4.2 for the DWI images $S(b)$ with different b values and different gradient directions $\mathbf{u}$. It can be seen that the DWI images are very noisy, especially for large b values.

4.1.1.3 Restricted Diffusion

Although we have obtained (4.2) for the measured diffusion signal attenuation for free diffusion, the diffusion of water molecules is hindered by surrounding tissues, especially in white matter. We derive the diffusion signal attenuation $E(\mathbf{G}, \Delta, \delta) = S(\mathbf{G}, \Delta, \delta)/S(0)$ under such restricted diffusion as follows, where $S(\mathbf{G}, \Delta, \delta)$ is the DWI signal associated with imaging parameters $(\mathbf{G}, \Delta, \delta)$. For each voxel in $\mathbf{x}$-space, let $\rho(\mathbf{R}_0)$ denote the spin

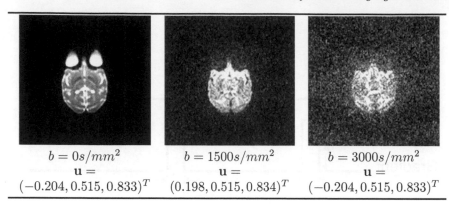

$b = 0s/mm^2$	$b = 1500s/mm^2$	$b = 3000s/mm^2$
$\mathbf{u} =$	$\mathbf{u} =$	$\mathbf{u} =$
$(-0.204, 0.515, 0.833)^T$	$(0.198, 0.515, 0.834)^T$	$(-0.204, 0.515, 0.833)^T$

FIGURE 4.2
DWI images for different b values and gradients. The data is from one of the subjects in a real monkey dataset.

density at initial time $t = 0$ and $P(\mathbf{R}_\Delta|\mathbf{R}_0)$ denote the probability that the spin moves from $\mathbf{R}_0$ at $t = 0$ to $\mathbf{R}_\Delta$ at $t = \Delta$. Then $E(\mathbf{G}, \Delta, \delta)$ can be represented as (126)

$$\int_{\mathbb{R}^3} \rho(\mathbf{R}_0) \int_{\mathbb{R}^3} P(\mathbf{R}_\Delta|\mathbf{R}_0) \exp(i\gamma(\mathbf{R}_\Delta - \mathbf{R}_0)^T(\int_0^\delta \mathbf{G}(t)dt))d\mathbf{R}_\Delta d\mathbf{R}_0. \qquad (4.3)$$

We simplify $E(\mathbf{G}, \Delta, \delta)$ under the narrow pulse condition, that is, $0 \approx \delta << \Delta$. We first define the Ensemble Average Propagator (EAP) as

$$P(\mathbf{R}) = \int_{\mathbb{R}^3} \rho(\mathbf{R}_0)P(\mathbf{R}_0 + \mathbf{R}|\mathbf{R}_0)d\mathbf{R}_0, \quad \mathbf{R} = \mathbf{R}_\Delta - \mathbf{R}_0 \qquad (4.4)$$

where $\mathbf{R} = ||\mathbf{R}||\mathbf{r}$ is the displacement vector in $\mathbf{R}$-space and $\mathbf{r}$ is a unit vector. EAPs in different regions in the brain reflect the different micro-structures and reveal fiber directions. See Figure 4.3 for diffusion data in a six-dimensional space including i.e. 3D $\mathbf{k}$-space and 3D $\mathbf{q}$-space. Since $\mathbf{G}(t)$ is a constant during δ, we may introduce a $\mathbf{q}$ vector in $\mathbf{q}$-space as

$$\mathbf{q} = q\mathbf{u} = (2\pi)^{-1}\gamma \int_0^\delta \mathbf{G}(t)dt = (2\pi)^{-1}\gamma\delta\mathbf{G}. \qquad (4.5)$$

Thus, $E(\mathbf{G}, \Delta, \delta)$ can be written as

$$E(\mathbf{q}) = \int_{\mathbb{R}^3} P(\mathbf{R}) \exp\left(2\pi i \mathbf{q}^T \mathbf{R}\right) d\mathbf{R} = \mathscr{F}_{3D}^{-1}\{P(\mathbf{R})\}(\mathbf{q}), \qquad (4.6)$$

where $\mathscr{F}_{3D}$ and $\mathscr{F}_{3D}^{-1}$, respectively, denote the Fourier transformation and its inverse transformation. Without confusion, we call both $E(\mathbf{q})$ and $E(b)$ diffusion signals. Since $P(\mathbf{R}) = P(-\mathbf{R})$ holds due to the principle of microscopic detailed balance, we have $E(\mathbf{q}) = E(-\mathbf{q}) = \mathscr{F}\{P(\mathbf{R})\}(\mathbf{q})$ leading to

$$\int_{\mathbb{R}^3} P(\mathbf{R}) \exp(-2\pi i \mathbf{q}^T \mathbf{R})d\mathbf{R} = \int_{\mathbb{R}^3} P(\mathbf{R}) \cos(2\pi \mathbf{q}^T \mathbf{R})d\mathbf{R}. \qquad (4.7)$$

Since the DWI data $S(\mathbf{q})$ is the Fourier transform of the $\mathbf{k}$-space signal and the EAP is another Fourier transform of $E(\mathbf{q})$, we can represent $P(\mathbf{R})$ as

$$P(\mathbf{R}) = \mathscr{F}_{3D}\{E(\mathbf{q})\}(\mathbf{R}) = \mathscr{F}_{3D}\{\frac{S(\mathbf{G}, \Delta, \delta)}{S(0)}\}(\mathbf{R}). \qquad (4.8)$$

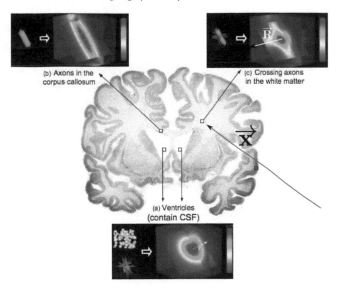

FIGURE 4.3
3D **x**-space and 3D **R**-space. EAPs in different regions in the brain reflect different micro-structures with isotropic diffusion, single fiber, and crossing fibers. The image is taken from (46) with the original figures adapted from (72) and the brain museum (www.brainmuseum.org/Specimens).

Due to $E(\mathbf{q}) = E(-\mathbf{q})$, $P(\mathbf{R})$ can be further written as

$$\int_{\mathbb{R}^3} E(\mathbf{q}) \exp(-2\pi i \mathbf{q}^T \mathbf{R}) d\mathbf{q} = \int_{\mathbb{R}^3} E(\mathbf{q}) \cos(2\pi \mathbf{q}^T \mathbf{R}) d\mathbf{q}. \tag{4.9}$$

4.1.2 Diffusion Tensor Imaging (DTI)

In (14), Dr. Basser proposed to model the ADC as a quadratic form parameterized by the diffusion tensor $\mathbf{D}$. Then the Stejskal–Tanner equation becomes

$$E(b) = \exp(-b\mathbf{u}^T \mathbf{D} \mathbf{u}). \tag{4.10}$$

The diffusion tensor $\mathbf{D} = \sum_{k=1}^{3} \lambda_k \mathbf{e}_k \mathbf{e}_k^T \in Sym_3^+$ is independent of the b value and gradient direction $\mathbf{u}$, where $\{(\lambda_k, \mathbf{e}_k)\}_{k \leq 3}$ are eigenvalue–eigenvector pairs such that $\lambda_1 \geq \lambda_2 \geq \lambda_3$ and Sym_3^+ is the space of 3×3 symmetric positive definite matrices. In free diffusion, $P(\mathbf{R})$ is given by

$$P(\mathbf{R}) = \mathscr{F}_{3D}\{\exp(-4\pi^2 \tau \mathbf{q}^T \mathbf{D} \mathbf{q})\} = \phi(\mathbf{R}|\mathbf{0}, 2\tau\mathbf{D}), \tag{4.11}$$

where $\phi(\mathbf{R}|\mathbf{0}, 2\tau\mathbf{D})$ denotes the Gaussian density with mean $\mathbf{0}$ and covariance $2\tau\mathbf{D}$. See Figure 4.4 for the sketch map of the tensor representation and free diffusion along fibers.

4.1.2.1 Scalar Indices and Eigenvectors of Diffusion Tensor

Several scalar indices based on $\mathbf{D}$ have been widely used in various biomedical studies. The two most important indices include Fractional Anisotropy (FA) and Mean Diffusivity

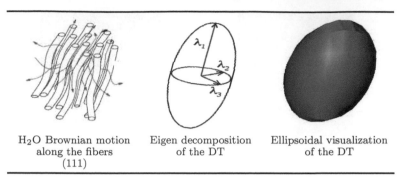

H$_2$O Brownian motion   Eigen decomposition   Ellipsoidal visualization
  along the fibers   of the DT   of the DT
       (111)

FIGURE 4.4
Diffusion tensor representation from (46).

(MD) (109) given by

$$\text{FA} = \sqrt{\frac{3}{2}} \sqrt{\frac{(\lambda_1 - \bar{\lambda})^2 + (\lambda_2 - \bar{\lambda})^2 + (\lambda_3 - \bar{\lambda})^2}{\lambda_1^2 + \lambda_2^2 + \lambda_3^2}}, \qquad \text{MD} = \bar{\lambda}, \qquad (4.12)$$

where $\bar{\lambda} = (\lambda_1 + \lambda_2 + \lambda_3)/3$. In (143), linear, planar, and spherical measures are introduced as

$$\text{LA} = (\lambda_1 - \lambda_2)/(3\bar{\lambda}), \quad \text{PA} = 2(\lambda_2 - \lambda_3)/(3\bar{\lambda}), \quad \text{SA} = \lambda_3/\bar{\lambda}. \qquad (4.13)$$

The eigenvectors of $\mathbf{D}$ are also very useful. When $\lambda_1 > \lambda_2$, the eigenvector corresponding to the largest eigenvalue is expected to be parallel to the local fiber orientation. In practice, the red-blue-green (RGB) map is used to describe the fiber directions. The tensor $\mathbf{D}$ itself can be visualized by a ellipsoid, then the tensor field becomes the ellipsoid field. See Figure 4.5 for the tensor field and various scalar maps estimated from an monkey data with $b = 1500s/mm^2$, 30 directions, where the Geodesic Anisotropy (GA) is introduced in (108).

4.2   High Angular Resolution Diffusion Imaging (HARDI)

The term High Angular Resolution Diffusion Imaging (HARDI) was first proposed by Tuch (137, 136) in order to have a more precise angular characterization of the diffusion signal. In this chapter, **HARDI methods** include all diffusion modeling methods beyond DTI. The HARDI methods for single shell (only one b value) data are called sHARDI methods. The HARDI methods for multiple shell (multiple b values) data are called mHARDI methods.

4.2.1   Generalization of Diffusion Tensor Imaging

4.2.1.1   Mixture of Tensor Model

The mixture of tensor model assumes that the signal is a mixture of signals generated from multiple tensors $\{\mathbf{D}_k\}_{k=1}^{K}$ given by

$$E(b) = \sum_{k}^{K} w_k \exp(-b\mathbf{u}^T \mathbf{D}_k \mathbf{u}) \quad \text{with} \quad \sum_{k=1}^{K} w_k = 1 \quad \text{and} \quad w_k \geq 0. \qquad (4.14)$$

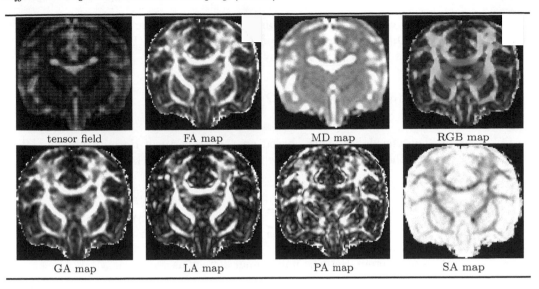

FIGURE 4.5
Tensor field and the scalar maps estimated from the monkey data with $b = 1500s/mm^2$ and 30 dircctions.

Due to some biological priors and computational complexity, the number of tensors is normally less than 3, typically $K = 2$. Computationally, although various optimization algorithms may be developed to estimate $\mathbf{D}_k$s (136), they may be computationally unstable and sensitive to the initial point. Moreover, for single shell data, the isotropic part of $\mathbf{D}_i$, i.e. Trace($\mathbf{D}_i$)/3, and w_i are undistinguishable (87). Thus, some constraints on tensors are normally imposed in model (4.14). For example, the two minimal eigenvalues λ_2 and λ_3 in $\mathbf{D}_i$ can be chosen as the same value. Tensors $\{\mathbf{D}_i\}$ can be chosen as one isotropic tensor and other anisotropic tensor with $\lambda_2 = \lambda_3 = 0$, which is called the **ball and stick model** (74). CHARMED model (10) considers $\{\mathbf{D}_i\}$ as a hindered diffusion part that is close to Gaussian diffusion, and a restricted diffusion part that is non-Gaussian diffusion. The mixture of tensor model has been widely used to generate synthetic data for evaluation because many quantities have closed forms in this model.

Model (4.14) has three major limitations. (i) Selecting the number of tensors is an open problem. (ii) The optimization process strongly depends on the initial point and is computationally inefficient. (iii) The radial decay of the mixture of tensor model is close to, but is not, the Gaussian function. Consider the number of tensors is $K = 2$ in Figure 4.6; along a given direction, one component decays fast and the other one decays slowly. For large b values, the component with slow decay dominates the signal.

4.2.1.2 Generalized DTI (GDTI)

In the GDTI model (92, 91), the signal is represented as

$$E(\mathbf{q}) = \exp(\sum_{l=2}^{L}(2\pi i)^l(\Delta - \frac{l-1}{l+1}\delta)D^{(l)}_{i_1 i_2 \ldots i_l}q_{i_1}q_{i_2}\cdots q_{i_l}), \qquad (4.15)$$

where we use the Einstein summation convention, i.e., $D^{(l)}_{i_1 i_2 \ldots i_l}q_{i_1}q_{i_2}\cdots q_{i_l} = \sum_{i_1=1}^{3}\sum_{i_2=1}^{3}\cdots\sum_{i_l=1}^{3}D^{(l)}_{i_1 i_2 \ldots i_l}q_{i_1}q_{i_2}\cdots q_{i_l}$. When $L = 2$, the GDTI becomes the DTI model in Eq. (4.10).

The generalized diffusion coefficients $D_{i_1 i_2 \ldots i_l}^{(l)}$ can be estimated by the least square fitting the samples of $\ln E(\mathbf{q})$. Due to the relationship between $E(\mathbf{q})$ and $P(\mathbf{R})$ in (4.6), $E(\mathbf{q})$ is the characteristic function of $P(\mathbf{R})$ (91) given by

$$E(\mathbf{q}) = \exp(\sum_{l=0}^{L} (-2\pi i)^l Q_{i_1 i_2 \ldots i_l}^{(l)} q_{i_1} \cdots q_{i_l} / l!), \qquad (4.16)$$

where $Q_{i_1 i_2 \ldots i_l}^{(l)}$s are the cumulants. Then, based on the property of Hermite polynomial $He_n(x)$, we can obtain the closed form for the EAP by using the Gram–Charlier A series (91), which leads to

$$P(\mathbf{R}) = N(\mathbf{R}|\mathbf{0}, Q_{i_1 i_2}^{(2)})(1 + \sum_{l=3}^{L} \frac{Q_{i_1 i_2 \ldots i_l}^{(l)}}{l!} He_{i_1 i_2 \ldots i_l}^{(l)}(\mathbf{R})), \qquad (4.17)$$

where $He_{i_1 i_2 \ldots i_l}^{(l)}(\mathbf{R})$ is the l-order Hermite polynomial defined as

$$(-1)^l \exp\left(-0.5 \mathbf{R}^T \mathbf{R}\right) \left(\frac{\partial}{\partial R_{i_1}} \frac{\partial}{\partial R_{i_2}} \cdots \frac{\partial}{\partial R_{i_l}}\right) \exp\left(0.5 \mathbf{R}^T \mathbf{R}\right).$$

There are several major limitations associated with the GDTI. It models the ADC using the polynomial basis, which is not orthogonal. Although theoretically the ADC can be modeled as infinite terms, in practice a truncated order L is needed in Eq. (4.15). However, it was proved in (95) that the Gaussian distribution is the only distribution that has a finite number of non-zero cumulants. Thus a truncation order L only results in a reasonable PDF if the EAP is Gaussian and $L = 2$ in this case. For other cases, the estimated EAP and cumulants are theoretically problematic. Moreover, estimation of the PDF from its cumulants is known to be very problematic.

4.2.1.3 High-Order Tensor Model, ADC-Based Model

The High-Order Tensor (HOT) model (106) assumes

$$E(\mathbf{q}) = \exp(-4\pi^2 \tau q^2 D(\mathbf{u})). \qquad (4.18)$$

The ADC is independent of radial part q, and can be represented as

$$D(\mathbf{u}) = \sum_{n_1 + n_2 + n_3 = L} D_{n_1 n_2 n_3} u_1^{n_1} u_2^{n_2} u_3^{n_3}, \qquad (4.19)$$

where $\mathbf{u} = (u_1, u_2, u_3)^T \in \mathbb{S}^2$, and L is even due to $D(\mathbf{u}) = -D(\mathbf{u})$. Moreover, $\{u_1^{n_1} u_2^{n_2} u_3^{n_3}\}_{n_1 + n_2 + n_3 = L}$ is the **homogeneous polynomial basis** restricted in $\mathbb{S}^2$, which is also called the **High-Order Tensor (HOT) basis** in dMRI domain. When $L = 2$, HOT model is just the DTI model in Eq. (4.10).

In the HOT, the diffusion signal decays as a mono-exponential function, which is called **mono-exponential decay assumption**, given by

$$E(q\mathbf{u}) = E(q_0 \mathbf{u})^{q^2/q_0^2}, \qquad (4.20)$$

where q_0 and q are any two different radii. Compared with the GDTI, which is model-free method, HOT is model-based. The mono-exponential decay assumption is not satisfied in real signal decay (88), but it can be a good approximation of the signal, especially when the b value is around $1500 s/mm^2$ (105).

$$\text{ADC } b = 1500s/mm^2 \quad \text{ADC } b = 3000s/mm^2 \quad \text{ODF by Tuch } \Phi_t(\mathbf{r}) \quad \text{ODF by Wedeen } \Phi_w(\mathbf{r})$$

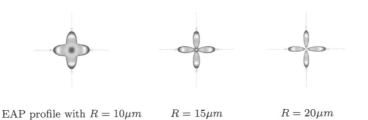

$$\text{EAP profile with } R = 10\mu m \qquad R = 15\mu m \qquad R = 20\mu m$$

FIGURE 4.6
Fiber directions and ADC profiles with different b values, two kinds of ODFs, and EAP profiles with different radius R. The data was generated from the mixture of tensor model with two tensors that have the eigenvalues $[1.7, 0.3, 0.3] \times 10^{-3}\text{mm}^2/s$ and a crossing angle of $90°$. We set $\tau = \frac{1}{4\pi^2}$ such that $b = q^2$. The long sticks with blue color along the x- and y-axes are the fiber directions. The short sticks with yellow color are the detected maxima of the spherical functions.

The HOT model often uses single shell data in Figure 4.10(c), which is a kind of sHARDI method. Historically, people used both the High-Order Tensor basis (106) and Spherical Harmonic (SH) basis (62) to estimate ADC from a measured signal. Theoretically, these two bases are equivalent to each other. In HARDI literatures, the maximal order of the SH basis or the order of the HOT basis must be higher than 4, because the order 2 of the SH basis and HOT basis are equivalent to the tensor model. Normally 4 or 6 is used in practice.

ADC modeling, like the HOT method, has its intrinsic and fatal limitation, i.e. both the maxima and the minima of ADC profile $D(\mathbf{u})$ are inconsistent with the fiber directions when $L > 2$ (139). Figure 4.6 demonstrates the ADC $D(\mathbf{u})$ for the synthetic data generated from the mixture of tensor model with a crossing angle of $90°$. It shows that the maxima of ADC do not agree with the fiber directions. Even in this simple mixture of tensor model, the ADC D is actually dependent on the b value, and the mono-exponential decay assumption is violated. For the data with different b values, the ADC is determined by $D = -b^{-1} \ln E(\mathbf{q})$, which means D is dependent on b if $E(\mathbf{q}) = \sum_{i=1}^{K} \exp(-b\mathbf{u}^T \mathbf{D}_i \mathbf{u})$. Although there is a coincidence that the minima of ADC agree with the fiber directions in this specific case of the mixture of tensor model with $90°$, the minima and maxima of the ADC generally have a complex relation with fiber directions.

4.2.2 Diffusion Spectrum Imaging (DSI)

Due to Eq. (4.9) for the narrow pulse assumption $\delta \ll \Delta$, a straightforward idea is to estimate $P(\mathbf{R})$ using fast Fourier transform from exhaustive signal samples (142). For instance, (142) used 515 DWI images in a Cartesian sampling lattice in $\mathbf{q}$-space and the signal in

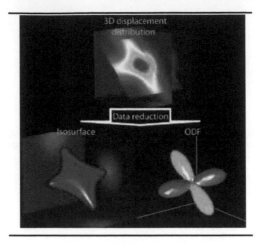

FIGURE 4.7
EAP in 3D **R**-space and its two features, i.e. EAP profile (or called iso-surface of EAP) and ODF. The figure is from (72).

q-space was premultiplied by a Hanning window to obtain smooth attenuation of the signal at high **q** values. Thus interpolation and extrapolation are normally performed on given signal samples $\{E(\mathbf{q}_i)\}$ before numerical Fourier transform.

In (142), the **EAP profile** is EAP at a given radius R_0 as

$$P(R_0\mathbf{r}) = P(R\mathbf{r})|_{R=R_0}. \tag{4.21}$$

The maxima of the EAP profile were used to describe fiber directions later in many HARDI works (105, 11, 48). See Figure 4.6 for the EAP profile with different radius R. The larger the radius R, the sharper the EAP profile is. However, the EAP profile with large R has more estimation error. Thus normally $R = 15\,\mu m$ is used in the EAP profile to detect the fiber directions (105, 48).

In (142), the ODF is defined as

$$\Phi_w(\mathbf{r}) \overset{\text{def}}{=} \int_0^\infty P(\mathbf{R})R^2 \mathrm{d}R. \tag{4.22}$$

Since $\Phi_w(\mathbf{r})$ is the marginal distribution of EAP $P(\mathbf{R})$, the integration of $\Phi_w(\mathbf{r})$ over $\mathbb{S}^2$ is naturally 1. Reference (142) proposed to first estimate the EAP via numerical Fourier transform and then estimate the ODF in Eq. (4.22) by numerical integration. Historically there are several kinds of ODFs which can be seen in the following of this section. Like the EAP profile, the maxima of ODFs are also normally assumed to be the directions of underlying fibers. Please see Figure 4.6 for the EAP in 3D space and its two features, i.e. the EAP profile and ODF.

4.2.3   Hybrid Diffusion Imaging (HYDI)

Hybrid Diffusion Imaging (HYDI) uses multiple shell sampling to measures the diffusion signal in **q**-space (147, 148). See Figure 4.10(d). The HYDI data in the shell with low b values can be modeled by DTI, whereas those data with high b values can be modeled by using Q-Ball Imaging and other sHARDI methods. The whole HYDI dataset can be used in DSI after re-griding data from multiple shells to the Cartesian lattice.

Two useful scalar features of EAP in HYDI include the Return-to-Origin probability (RTO) and the Mean Squared Displacement (MSD). The RTO denoted by P_o is the EAP value when $R = 0$, that is, $P_o = P(\mathbf{R})|_{R=0} = P(0) = \int_{\mathbb{R}^3} E(\mathbf{q})\mathrm{d}\mathbf{q}$. The RTO is the probability of water molecules that minimally diffuse within the diffusion time Δ. The RTO map can be used in tissue segmentation and some other applications (147). The MSD is the variance of the EAP, that is, $MSD = \int_{\mathbb{R}^3} P(\mathbf{R})\mathbf{R}^T\mathbf{R}\mathrm{d}\mathbf{R}$. In (148), the ODF by Tuch $\Phi_t(\mathbf{r})$ in Eq. (4.24) is proportional to the integration of $E(\mathbf{q})$ in the orthogonal plane

$$\Pi_{\mathbf{r}} = \{q\mathbf{u} :\ \mathbf{u}^T\mathbf{r} = 0\}. \tag{4.23}$$

It is an important relation between the ODF by Tuch in **R**-space and the signal $E(\mathbf{q})$ in **q**-space, and it is used in exact QBI to estimate ODFs analytically (1, 133). See Section 4.2.4 for more details on QBI and ODFs.

4.2.4   Q-Ball Imaging (QBI)

Q-Ball Imaging (QBI) is currently the most widely used HARDI method. QBI was proposed to estimate the several kinds of ODFs, not the EAP, from single shell sampling demonstrated in Figure 4.10(c), rather than Cartesian sampling inside a given ball used in DSI in 4.10(b).

4.2.4.1   Original Q-Ball Imaging

QBI was first proposed by Dr. Tuch in (135, 134) in a numerical way and then was improved in an analytical way based on the Spherical Harmonic basis in (7, 47). Dr. Tuch originally proposed to estimate a kind of ODF defined as

$$\Phi_t(\mathbf{r}) \overset{\text{def}}{=} \frac{1}{Z} \int_0^\infty P(R\mathbf{r})\mathrm{d}R, \tag{4.24}$$

where Z is the normalization factor that makes $\int_{\mathbb{S}^2} \Phi_t(\mathbf{r})\mathrm{d}\mathbf{r} = 1$. This $\Phi_t(\mathbf{r})$ is different from the ODF $\Phi_w(\mathbf{r})$ defined in Eq. (4.22).

In QBI, Dr. Tuch proposed to estimate $\Phi_t(\mathbf{r})$ directly from samples of $E(\mathbf{q})$ in single-shell data based on the Funk–Radon transform (FRT). See Figure 4.10(c) for the sketch map of single shell sampling. For single-shell data with $b = 4\pi^2\tau q_0^2$, the FRT of $E(\mathbf{q})$ (134) in direction $\mathbf{r}$, denoted as $\text{FRT}\{E(q_0\mathbf{u})\}(\mathbf{r})$, is the **circle integration** in the orthogonal plane, i.e.

$$\int_{\Pi_{\mathbf{r}}} E(q\mathbf{u})\delta(q - q_0)q\mathrm{d}q\mathrm{d}\mathbf{u} = q_0 \int_{\mathbf{u}\in\mathbb{S}^2} E(q_0\mathbf{u})\delta(\mathbf{u}^T\mathbf{r})\mathrm{d}\mathbf{u}, \tag{4.25}$$

where $\Pi_{\mathbf{r}}$ is defined in Eq. (4.23). The ODF $\Phi_t(\mathbf{r}')$ can be written as

$$(2Z)^{-1} \int_{-\infty}^{\infty} \int_0^{2\pi} \int_0^\infty P(r_{\mathbf{R}}, \phi_{\mathbf{R}}, z_{\mathbf{R}})\delta(r_{\mathbf{R}})\delta(\phi_{\mathbf{R}})r_{\mathbf{R}}\mathrm{d}r_{\mathbf{R}}\mathrm{d}\theta_{\mathbf{R}}\mathrm{d}z_{\mathbf{R}} \approx$$

$$2\pi q_0 Z^{-1} \int_{-\infty}^{\infty} \int_0^{2\pi} \int_0^\infty P(r_{\mathbf{R}}, \phi_{\mathbf{R}}, z_{\mathbf{R}})J_0(2\pi q_0 r_{\mathbf{R}})r_{\mathbf{R}}\mathrm{d}r_{\mathbf{R}}\mathrm{d}\theta_{\mathbf{R}}\mathrm{d}z_{\mathbf{R}}, \tag{4.26}$$

which equals $Z^{-1}\text{FRT}\{E(q_0\mathbf{u})\}(\mathbf{r}')$, where $J_0(\cdot)$ is the Bessel function of the first kind with order 0. The key idea of Eq. (4.26) is to approximate the delta function using Bessel function $0.5aJ_0(ax)$. As q_0 increases, $2\pi q_0 J_0(2\pi q_0 r_{\mathbf{R}})$ will be very close to the delta function, since the lobes of J_0 become more concentrated around the origin point. However, the signal has smaller values for larger q_0, which results in a low signal-to-noise ratio (SNR). Thus there is a tradeoff for q_0 between approximation accuracy and SNR. Normally QBI works suggest data with b values around $3000s/mm^2$ (134, 47).

Dr. Tuch proposed to estimate the circle integration in FRT using numerical integration (134). The numerical QBI was later replaced by analytical QBI based on $E(q_0\mathbf{u}) = \sum_{l=0}^{L} \sum_{m=-l}^{l} c_{lm} Y_l^m(\mathbf{u})$, where $Y_l^m(\mathbf{u})$ is the symmetric real spherical harmonic with order l and degree m (7). We have

$$\widetilde{\Phi_t}(\mathbf{r}) = Z^{-1}\text{FRT}\{E(q_0\mathbf{u})\}(\mathbf{r}) = Z^{-1}\sum_{l=0}^{L}\sum_{m=-l}^{l} 2\pi P_l(0) c_{lm} Y_l^m(\mathbf{u}) \qquad (4.27)$$

where $P_l(0)$ is the Legendre polynomial of order l evaluated at 0. In practice the ODF by Tuch $\Phi_t(\mathbf{r})$ in Eq. (4.24) is very smooth. The peaks of the ODF are only a little higher than the baseline values. Dr. Tuch proposed a min-max normalization method for visualization of $\Phi_t(\mathbf{r})$ to enhance the peaks of ODFs. However, the min-max normalization also enhances the peaks of the ODFs in the area with isotropic diffusion. Compared with ADC-based modeling like the HOT method, the maxima of ODFs agree with the fiber directions. Dr. Tuch also proposed a useful scalar index, named Generalized Fractional Anisotropy (GFA), to describe the anisotropy of the ODFs given by

$$\text{GFA}\{\Phi_t(\mathbf{r})\} \overset{\text{def}}{=} \sqrt{\frac{N\sum_{i=1}^{N}(\Phi_t(\mathbf{r}_i) - \langle\Phi_t(\mathbf{r})\rangle)^2}{(n-1)\sum_{i=1}^{N}\Phi_t(\mathbf{r}_i)^2}}, \qquad (4.28)$$

where $\langle\Phi_t(\mathbf{r})\rangle$ is the mean of $\Phi_t(\mathbf{r})$.

QBI has several major limitations. (i) The ODF by Wedeen $\Phi_w(\mathbf{r})$ defined in Eq. (4.22) is theoretically sharper than $\Phi_t(\mathbf{r})$. (ii) The estimation via FRT has an intrinsic blurring effect, which leads to smoothed ODFs. (iii) QBI actually assumes the radial part of $E(\mathbf{q})$ as a delta function, which is unrealistic. The burring effect from FRT is the direct consequence from this assumption of radial decay. (iv) QBI cannot be used in multiple-shell data, because the data from different b values leads to different ODFs from FRT.

4.2.4.2 Exact Q-Ball Imaging

Exact QBI was proposed by several groups independently (148, 133) to estimate ODFs through the famous **projection-slice theorem** in Fourier transform[2]: the projection of $P(R\mathbf{r})$ along direction $\mathbf{r}$, i.e. the radial integration, equals the integration of $E(\mathbf{q})$ in the orthogonal plane $\Pi_\mathbf{r}$. Thus we have the following corollary, which is a straightforward result of the above proposition and has been used to estimate both ODF by Tuch and ODF by Wedeen (148, 1, 131, 133, 39). The $\Phi_t(\mathbf{r})$ and $\Phi_w(\mathbf{r})$ can be written as

$$\Phi_t(\mathbf{r}) = Z^{-1}\int_{\Pi_\mathbf{r}} E(\mathbf{q})\mathrm{d}\mathbf{q}, \quad \Phi_w(\mathbf{r}) = (4\pi)^{-1} - (8\pi^2)^{-1}\int_{\Pi_\mathbf{r}} q^{-1}\Delta_b E(\mathbf{q})\mathrm{d}\mathbf{q},$$

where Δ_b is the Laplace–Beltrami operator in $\mathbb{S}^2$. If $E(\mathbf{q})$ follows the mono-exponential decay assumption (1), then we have

$$\Phi_w(\mathbf{r}) = \frac{1}{4\pi} + \frac{1}{16\pi^2}\int_{\mathbb{S}^2} \Delta_b \ln\left(-\ln E(q_0\mathbf{u})\right)\delta(\mathbf{u}^T\mathbf{r})\mathrm{d}\mathbf{u}. \qquad (4.29)$$

By representing $\ln\left(-\ln E(q_0\mathbf{u})\right) = \sum_{l=0}^{L}\sum_{m=-l}^{l} c_{lm} Y_l^m(\mathbf{u})$, and considering $\Delta_b Y_l^m(\mathbf{u}) = -l(l+1)Y_l^m(\mathbf{u})$, we have $\Phi_w(\mathbf{r}) = \frac{1}{4\pi} - \frac{1}{8\pi}l(l+1)P_l(0)c_{lm}Y_l^m(\mathbf{u})$. Although the mono-exponential decay assumption is better than the delta function assumption, it is still a strong and unrealistic assumption of radial decay; even the signal generalized by a simple mixture of tensor model does not follow this assumption as shown in Figure 4.6. Exact QBI is a kind of sHARDI method and leads to different results for the data from different shells.

---

[2]http://en.wikipedia.org/wiki/Projection-slice_theorem

4.2.5 Diffusion Orientation Transform (DOT)

Diffusion Orientation Transform (DOT) was proposed by Dr. Özarslan in (105) to estimate the EAP profile $P(R\mathbf{r})$ from single-shell data under the mono-exponential decay assumption in Eq. (4.18). It can be regarded as an estimator of EAP in exact QBI methods. Consider the plane wave equation as

$$\cos(2\pi\mathbf{q}^T\mathbf{R}) = 4\pi \sum_{l=0}^{\infty} \sum_{m=-l}^{l} (-1)^{l/2} j_l(2\pi qR) Y_l^m(\mathbf{u}) Y_l^m(\mathbf{r}), \qquad (4.30)$$

where $j_l(\cdot)$ is the l-order spherical Bessel function. Thus, we have

$$P(R\mathbf{r}) = \sum_{l=0}^{\infty} \sum_{m=-l}^{l} \left(\int_{\mathbb{S}^2} Y_l^m(\mathbf{u}) I_l(R,\mathbf{u}) \mathrm{d}\mathbf{u} \right) Y_l^m(\mathbf{r}), \qquad (4.31)$$

where $I_l(R,\mathbf{u}) = 4\pi(-1)^{l/2} \int_0^{\infty} E(\mathbf{q}) j_l(2\pi qR) q^2 dq$ can be calculated analytically based on the samples of ADC $\{D(\mathbf{u}_j)\}$. Then a least square fitting can be used to obtain the coefficients of $P(R_0\mathbf{r})$ under the SH basis from $\{I_l(R_0,\mathbf{u}_j)\}$.

Reference (105) validated the mono-exponential decay assumption through synthetic data generated from the cylinder model. It showed that signal decay can be approximated well as a mono-exponential function around $b = 1500s/mm^2$. For the b value larger than $3000s/mm^2$, the mono-exponential decay assumption is not well satisfied, and the data with large b value has low SNR. Thus $1500s/mm^2$ seems to be the optimal b value for DOT. Note like original QBI and exact QBI, DOT cannot handle multiple-shell data, since the data in different shells leads to different EAP profiles.

4.2.6 Spherical Deconvolution (SD)

Spherical Deconvolution (SD) methods (130, 129) generalize the mixture model from the discrete case to the continuous case by assuming

$$E(q\mathbf{u}) = \int_{\mathbb{S}^2} \Phi_f(\mathbf{r}) R(\mathbf{r}^T\mathbf{u}) \mathrm{d}\mathbf{r}, \qquad (4.32)$$

where $\Phi_f(\mathbf{r})$ is the fiber ODF (fODF) and $R(\mathbf{r}^T\mathbf{u})$ is the fiber response signal generated from one fiber. The spherical deconvolution is a model-based method because it assumes the typical signal $R(\mathbf{r}^T\mathbf{u})$ and linear combination in the convolution. The mixture of tensor model suffers from the model selection of the number of tensors and local minima of cost function. In contrast, SD can be solved analytically by considering the closed-form expression of spherical integration using the SH basis representation of the $E(q\mathbf{u})$ and $R(\mathbf{r}^T\mathbf{u})$ using SHs (129, 49). SD can be used for DWI signal $E(q\mathbf{u})$ or ODFs/EAPs calculated from DWI signal (49). However the direct usage of SD for the DWI signal is theoretically better because the estimation of ODFs/EAPs from DWI signal always suffers from noise and insufficient samples. SD methods normally obtain many false positive fODF peaks when the signal is less anisotropic. The false positive peaks can be largely reduced by considering the non-negativity constraint of the estimated fODFs (129, 38). Conventional SD is only for single-shell data. However by introducing a 3-dimensional fiber response function, SD methods can be used in multi-shell data (38).

4.2.7 Diffusion Propagator Imaging (DPI)

Diffusion Propagator Imaging (DPI) was proposed to model the signal $E(\mathbf{q})$ as the solution of Laplace's equation (48). DPI can be seen as a generalization of the QBI method to handle multiple-shell data, although this generalization has many problems. In DPI, the signal is assumed to be

$$E(q\mathbf{u}) = \sum_{l=0}^{L} \sum_{m=-l}^{l} \left(\frac{c_{lm}}{q^{l+1}} + d_{lm}q^l\right) Y_l^m(\mathbf{u}).\tag{4.33}$$

Then, the EAP is estimated from incomplete 3D integration inside the ball with a given radius q_{max}, which is the maximum q value used in DPI acquisition, since the complete integration in $\mathbb{R}^3$ does not converge. After obtaining the coefficients $\{c_{lm}\}$ and $\{d_{lm}\}$, several EAP features can be calculated analytically based on incomplete radial integration. However, DPI has some limitations. (i) The estimated ODFs suffer from two incomplete integrations including one for EAP estimation and the other for ODF estimation. (ii) It is unclear how to choose q_{max}. (iii) The DPI model does not satisfy some priors of signal $E(\mathbf{q})$, which brings intrinsic modeling errors. It also cannot represent an isotropic Gaussian signal.

4.2.8 Simple Harmonic Oscillator Reconstruction and Estimation (SHORE)

SHORE was proposed by Dr. Özarslan in (104) for 3D signals. In SHORE, $E(\mathbf{q})$ in 3D (41) is represented by

$$E(\mathbf{q}) = \sum_{n=0}^{N} \sum_{l=0}^{2n} \sum_{m=-l}^{l} a_{nlm} B_{nlm}^{\mathrm{SHO3}}(\mathbf{q}|\zeta),\tag{4.34}$$

where $B_{nlm}^{\mathrm{SHO3}}(\mathbf{q}|\zeta) = G_{nl}(q|\zeta)Y_l^m(\mathbf{u})$ and $G_{nl}(q|\zeta)$ is a given function and depends on l. SHORE is model-free, since the SHO basis is a complete basis in $\mathbb{R}^3$. The linear analytical solutions are very fast and avoid numerical integration. However, the representation in (4.34) is not non-negative in nature. After estimating $\{a_{nlm}\}$, the EAP can be analytically calculated as

$$P(\mathbf{R}) = \sum_{n=0}^{N} \sum_{l=0}^{2n} \sum_{m=-l}^{l} a_{nlm}(-1)^n G_{nl}(R|\frac{1}{4\pi^2\zeta})Y_l^m(\mathbf{r}).\tag{4.35}$$

The two kinds of ODFs can also be analytically derived from the estimated coefficients.

4.2.9 Spherical Polar Fourier Imaging (SPFI)

Spherical Polar Fourier Imaging (SPFI) was first proposed by Dr. Assemlal in (11) in a numerical way and then improved by (40, 39). SPFI represents the diffusion signal $E(\mathbf{q})$ as a linear combination of the SPF basis, i.e.,

$$E(\mathbf{q}) = \sum_{n=0}^{N} \sum_{l=0}^{L} \sum_{m=-l}^{l} a_{nlm} B_{nlm}^{\mathrm{SPF}}(\mathbf{q}|\zeta),\tag{4.36}$$

where $B_{nlm}^{\mathrm{SPF}}(\mathbf{q}|\zeta) = G_n(q|\zeta)Y_l^m(\mathbf{u})$, in which $\kappa_n(\zeta)^2 = 2n!/[\zeta^{3/2}\Gamma(n+3/2)]$ and $G_n(q|\zeta) = \kappa_n(\zeta)\exp\left(-q^2/(2\zeta)\right) L_n^{1/2}(q^2/\zeta)$. The SPF basis is a 3D orthonormal basis with SHs in the spherical part and Gaussian Laguerre functions in the radial part. SPFI is model-free, since the SPF basis is complete in their domains. The linear analytical solutions are very fast and avoid numerical integration. However, the representation Eq. (4.36) is not non-negative in nature. After we estimate the coefficients of the diffusion signal under SPF basis, the EAP and its various features, e.g., ODFs, RTO, can be obtained in an analytical way (40, 39).

4.3 Reconstruction

DMRI data consists of n DWIs with n measurements $\{(S(\mathbf{q}_i; \mathbf{v}), \mathbf{r}_i, b_i) : i = 1, \cdots, n\}$ at voxel $\mathbf{v}$ in a common space $\mathcal{V}$. A reconstruction step in dMRI is to estimate ODF $\Phi_w(\mathbf{r}; \mathbf{v})$, and EAP $p(\mathbf{R}; \mathbf{v})$ at each voxel $\mathbf{v} \in \mathcal{V}$. To design an efficient method to accurately reconstruct the ODF and EAP, one must address three key components of dMRI data including (i) the model for the diffusion signal attenuation; (ii) the noise components in dMRI data; and (iii) the spatial/functional nature of dMRI data. Since various models for $E(\mathbf{q}; \mathbf{v})$ have been extensively reviewed above, we focus on the last two key components here. If such models for $E(\mathbf{q}; \mathbf{v})$ were misspecified, one would not expect to accurately reconstruct the ODF and/or EAP. We will discuss why the last two components of dMRI data are critical for dMRI reconstruction.

4.3.1 Noise Components and Voxelwise Estimation Methods

DWIs inherently contain varying amounts of noise that must be modeled or corrected appropriately if ODFs and EAPs are to be estimated accurately; failure to do so may lead to a biased estimate of the ODF (or EAP) and to incorrect estimates of their associated invariant measures (e.g., GFA). The measured diffusion weighted signals, however, can contain varying amounts of noise of diverse origins, including noise from stochastic variation, numerous physiological processes, eddy currents, artifacts from the differing magnetic field susceptibilities of neighboring tissues, rigid body motion, nonrigid motion, and many others (82, 83). Some noise components, such as bulk motion from cardiac pulsation and head or body movement, generate unusual observations, or statistical outliers. Previous studies have shown that those noise components can introduce substantial bias into measurements and estimations made from those images, such as invariant measures and fiber tracts in diffusion tensor images (152, 96, 35). Identifying and reducing these noise components in DWIs is essential to improving the validity and accuracy of DTI studies designed to map brain structure and function.

Two types of approaches, including various robust statistical methods and diagnostic methods, have been proposed to address the 'non-random' noise components in DWIs. Robust statistical methods produce the estimators that are insensitive to significant deviations from the model assumption, while incorporating the properties of classic statistics (76, 43). Specifically, in DTI, several robust approaches have been used to exclude outliers from the diffusion signal attenuation in order to improve the accuracy of tensor estimation (34, 35, 96). These proposed algorithms, however, only work properly for a small number of outliers in the case of high SNR. Diagnostic methods based on some influence measures (e.g., Cook's distance) can isolate outliers caused by certain noise components, including motion artifacts (167). An adaptive estimation procedure can be used to refit to dMRI data in order to obtain refined estimators by downweighting outliers.

In the presence of random noise only, the signal intensity in DWIs acquired from a single coil follows a Rician distribution, denoted by $S(\mathbf{q}_i; \mathbf{v}) \sim R(\mu_i(\mathbf{q}_i, \mathbf{v}), \sigma^2(v))$ (122). As shown (167), a general Rician regression model was introduced and an expectation-maximization (EM) algorithm was first proposed to calculate the maximum likelihood estimate of unknown parameters. Moreover, the Rician distribution can be well approximated by a normal distribution $N(\sqrt{\mu_i(\mathbf{q}_i, \mathbf{v})^2 + \sigma(\mathbf{v})^2}, \sigma(\mathbf{v})^2)$ when SNR≥ 2 and $N(\mu_i(\mathbf{q}_i, \mathbf{v}), \sigma(\mathbf{v})^2)$ when SNR≥ 5. The log-transformed signal intensity $\log S(\mathbf{q}_i; \mathbf{v})$ approximately follows a weighted Gaussian distribution $N(\log(\mu_i(\mathbf{q}_i, \mathbf{v})), \sigma^2(v)/\mu_i(\mathbf{q}_i, \mathbf{v})^2)$ (13, 6, 165). For DTI in (4.10), we have $\log(\mu_i(\mathbf{q}_i, \mathbf{v})) = \log S(0) - b\mathbf{u}^T \mathbf{D} \mathbf{u}$. An efficient weighted least-square algorithm was developed to reconstruct $\mathbf{D}$ (165). If DWIs are acquired from multiple coils, $S(\mathbf{q}_i; \mathbf{v})$ is non-central Chi (nc-χ) distributed, provided that the k space is fully sampled and no correlations between the coil data exists (2). Recently, the estimation method under the nc-χ noise has been developed to estimate $\mathbf{D}$ in (23).

Raw HARDI images, as a result of elevated b factor and decreased voxel size, suffer from depressed SNRs, which make the problem of reconstructing HARDI data of practical importance and challenging. Most HARDI reconstruction algorithms directly assume that

$$f(E(\mathbf{q}_i; \mathbf{v})) = \mathbf{x}_i^T \beta(\mathbf{v}) + \epsilon_i(\mathbf{v}), \qquad (4.37)$$

where $f(\cdot)$ is a given transformation function (e.g., $f(s) = s$ or $f(s) = \log(s)$), $\mathbf{x}_i$ is a $p \times 1$ vector of covariates, which depends on $\mathbf{q}_i$ (or $(\mathbf{b}_i, r_i)$), $\beta(\mathbf{v})$ is a $p \times 1$ vector of regression coefficients, and $\epsilon_i(\mathbf{v})$ is an error term with mean zero and variance $\sigma_i^2(\mathbf{v})$. Model (4.37) is general enough to cover many existing HARDIs. In the literature, for GDTI and HOT, it is common to set $f(E(\mathbf{q}_i; \mathbf{v})) = \log(E(\mathbf{q}_i; \mathbf{v}))$ and represent $\log(E(\mathbf{q}_i; \mathbf{v}))$ as a polynomial function of $\mathbf{q}_i$, whereas for most other HARDIs, such as QBI or DOT, it is common to set $f(E(\mathbf{q}_i; \mathbf{v})) = E(\mathbf{q}_i; \mathbf{v})$ and approximate $E(\mathbf{q}_i; \mathbf{v})$ by a linear combination of some basis functions, such as the spherical polar Fourier basis.

Most HARDI methods focus on the reconstruction of $\beta(\mathbf{v})$ by solving a regularized linear least-squares optimization problem

$$\widehat{\beta}(\mathbf{v}) = \operatorname{argmin}_{\beta(\mathbf{v})} ||\mathbf{y}(\mathbf{v}) - \mathbf{X}\beta(\mathbf{v})||^2 + \rho(\beta(\mathbf{v}); \lambda(\mathbf{v})), \qquad (4.38)$$

where $\mathbf{y}(\mathbf{v}) = (f(E(\mathbf{q}_1; \mathbf{v})), \cdots, f(E(\mathbf{q}_n; \mathbf{v})))^T$, $\mathbf{X}$ is an $n \times p$ matrix with the i-th row being $\mathbf{x}_i$, and $\rho(\beta(\mathbf{v}); \lambda(\mathbf{v}))$ is a penalty function with $\lambda(\mathbf{v})$ being a tuning parameter. Different penalty functions, such as LASSO and the Laplacian–Beltrami, can be used in (4.38) (128, 47, 98). Recently, as discussed in (98) and references therein, there is a high interest in developing the compression sensing technique for dMRI applications. In Bayesian statistics, $-\log(\rho(\beta(\mathbf{v}); \lambda(\mathbf{v})))$ can be regarded as the prior of $\beta(\mathbf{v})$.

Existing methods based on (4.37) and (4.38) have at least three major limitations. First, these methods largely ignore the stochastic noise components of the DW signal that we discussed above. Therefore, it may lead to bias and loss of efficiency in the estimation of the ODF and EAP. Second, these methods perform reconstruction independently at each voxel, which essentially ignores the functional nature of the DWI data at different voxels in space. Third, most HARDI reconstruction algorithms often use some heuristic methods to determine a single value of $\lambda(\mathbf{v})$ for all voxels. However, both theoretically and numerically, the selection of the tuning parameter across voxels plays a critical role in ensuring the nice properties of the regularized estimators (28).

4.3.2 Spatial-Adaptive Estimation Methods

Recently, there is a growing interest in developing spatial-adaptive estimation methods for the HARDI/DTI reconstruction in order to characterize the spatial/functional nature of DWIs. Until recently, a number of different approaches have been developed starting from smoothing raw DWIs (50, 19, 18, 115, 20, 21), smoothing procedures in tensor space (99, 60), smoothing procedures in ODF space (85, 68), spatial DTI (127, 155, 154, 93), to spatial HARDI, which reconstructs and denoises all ODFs simultaneously (113). The key idea of these methods is to explicitly incorporate spatial smoothness constraints into various HARDI reconstruction algorithms. The key assumption of this type of approach is that the orientation and anisotropy of any single fiber population are expected to vary smoothly along the dominant fiber orientation, except at the boundaries between tracts and interfaces with gray matter structures and cerebrospinal fluid spaces. Mathematically, DWI can be characterized as a convolution of a piecewise smooth function with various MRI noises.

Most spatial-adaptive methods can be classified into three categories including (i) the denoising of raw DWI data, (ii) the denoising of the estimated tensor/EAP/ODF field, and (iii) simultaneous smoothing and estimation of DWI data. In the first category, most approaches to DWI denoising are designed to incorporate the stochastic components of raw DWI data with their spatial smoothness by using either regularization methods or nonparametric statistical methods. The DWI denoising method is to denoise the observed $\{(S(\mathbf{q}_i; \mathbf{v}) : \mathbf{v} \in \mathcal{V})_{i \geq 1}\}$ in order to calculate the denoised DWIs, denoted by $\{(S_*(\mathbf{q}_i; \mathbf{v}) : \mathbf{v} \in \mathcal{V})_{i \geq 1}\}$. The regularization-based denoising methods estimate $S_*(\mathbf{q}_i; \mathbf{v})$ by solving a regularized optimization problem

$$\operatorname{argmin}_{S_*(\mathbf{q}_i; \mathbf{v})} \int_{\mathbf{v} \in \mathcal{V}} \{\ell(S(\mathbf{q}_i; \mathbf{v}), S_*(\mathbf{q}_i; \mathbf{v})) + \rho(S_*(\mathbf{q}_i; \mathbf{v}), \lambda(\mathbf{v}))\} dL(\mathbf{v}), \qquad (4.39)$$

where $-\ell(S(\mathbf{q}_i; \mathbf{v}), S_*(\mathbf{q}_i; \mathbf{v}))$ is usually chosen to be the log-likelihood function of the DW signal at voxel $\mathbf{v}$, $\rho(\cdot, \cdot)$ is a pre-specified penalty function, such as total variation, and $L(\mathbf{v})$ is a measure defined on $\mathcal{V}$. Various penalty functions include those associated with total variation schemes, Markov random fields, and Perona–Malik-like smoothing. In Bayesian statistics, $\log(-\rho(S_*(\mathbf{q}_i; \mathbf{v}), \lambda(\mathbf{v})))$ is the prior of $S_*(\mathbf{q}_i; \mathbf{v})$ in the $(\mathbf{q}, \mathbf{v})$−space.

Nonparametric statistical methods incorporate both spatial proximity and similarity measure to calculate weighted averages of 'similar' DW signals in order to explicitly account for the piecewisely smooth nature of imaging data with jumps and edges. These similar signals can be incorporated in denoising from both the spatial $\mathbf{v}$-space and the diffusion $\mathbf{q}$-space. Some well-known methods include non-local means (NLM) and unbiased NLM algorithms, propagation-separation methods, anisotropic Wiener filtering, the bilateral filter, and the Sigma filter, among others (110, 8, 151, 27, 84). For instance, NLM uses small sub-images, called patches, to denoise the image by accounting for the redundancy in natural images, especially in textured parts (8, 27, 84). Based on the Rician noise, a NLM for DWI data can be formulated as

$$\text{NLM}(S(\mathbf{q}, \mathbf{v})) = \sqrt{\sum_{(\mathbf{q}', \mathbf{v}') \in V_{(\mathbf{q}, \mathbf{v})}} w((\mathbf{q}, \mathbf{v}), (\mathbf{q}', \mathbf{v}'))S(\mathbf{q}', \mathbf{v}')^2 - 2\sigma^2}, \qquad (4.40)$$

where $w((\mathbf{q}, \mathbf{v}), (\mathbf{q}', \mathbf{v}'))$ is defined by the distance of the patches centered in $(\mathbf{q}, \mathbf{v})$ and $(\mathbf{q}', \mathbf{v}')$, and σ is a global noise variance. The NLM can be performed separately for different $\mathbf{q}$ or jointly for all signals in $\mathbf{q}$ space by defining $w(\cdot, \cdot)$ based on vector-valued patches (51, 145). When the noise variance is unknown, it can be estimated jointly from all signals in $\mathbf{q}$ space via a linear minimum mean square error (LMMSE) estimator (3, 132). However, NLM has some limitations for piecewise smooth images when the noise is not small (8). This is exactly the case for DWI. In contrast, the propagation-separation (PS) method is very efficient at smoothing noisy piecewise smooth images and dealing with edges (18), even though such methods cannot proceed efficiently in textured regions. The key idea of the PS method is to construct a sequence of nested local neighborhoods (or patches) adapted to DW signals in its neighboring voxels and then adaptively estimate $S_*(\mathbf{q}_i; \mathbf{v})$ at each voxel. Although the PS method is computationally extensive, it is robust to the selection of kernel window sizes and patch shapes at different locations.

In the second category, most methods perform denoising on the estimation tensor/ODF/EAP results by using either regularization methods or nonparametric statistical methods (97, 108, 99, 68, 162, 157, 61, 121, 119, 54, 37). For DTI, various regularization-based denoising methods estimate a tensor field $\{\mathbf{D}(\mathbf{v}) : \mathbf{v} \in \mathcal{V}\}$ based on the estimated tensor field $\{\widehat{\mathbf{D}}(\mathbf{v}) : \mathbf{v} \in \mathcal{V}\}$ (36, 45). Mathematically, it can be formulated as

$$\text{argmin}_{\{\mathbf{D}(\mathbf{v}):\mathbf{v}\in\mathcal{V}\}} \int_{\mathbf{v}\in\mathcal{V}} d(\widehat{\mathbf{D}}(\mathbf{v}), \mathbf{D}(\mathbf{v})) + \rho(\mathbf{D}(\mathbf{v}), \lambda(\mathbf{v}))\}dL(\mathbf{v}), \qquad (4.41)$$

where $d(\widehat{\mathbf{D}}(\mathbf{v}), \mathbf{D}(\mathbf{v}))$ is usually chosen to be a pre-specified distance between $\widehat{\mathbf{D}}(\mathbf{v})$ and $\mathbf{D}(\mathbf{v})$ and $\rho(\cdot, \cdot)$ is a pre-specified penalty function on the tensor field. Similar to (4.39), various penalty functions can be developed for the tensor field based on total variation schemes, Markov random fields, and Perona–Malik-like smoothing. Since $\mathbf{D}$ lies in a curved space, one has to face additional theoretical and computational challenges.

For DTI, nonparametric statistical methods have been developed to estimate an intrinsic 'expectation' (or 'median') of a symmetric positive definite matrix response $\mathbf{D}$, given a voxel location $\mathbf{v}$ from a set of estimated diffusion tensors $(\mathbf{v}_1, \widehat{\mathbf{D}}(\mathbf{v}_1)), \ldots, (\mathbf{v}_m, \widehat{\mathbf{D}}(\mathbf{v}_m))$, in which $\mathbf{v}$ may belong to the set of $\{\mathbf{v}_1, \cdots, \mathbf{v}_m\}$. Mathematically, it can be formulated by solving a weighted estimator of $\mathbf{D}(\mathbf{v})$ that is defined by

$$\tilde{\mathbf{D}}^\alpha(\mathbf{v}) = \text{argmin}_{\mathbf{D}(\mathbf{v})} \sum_{i=1}^m w(\mathbf{v}_m, \mathbf{v}) g(\widehat{\mathbf{D}}_m(\mathbf{v}_m), \mathbf{D}(\mathbf{v}))^\alpha, \qquad (4.42)$$

where $\alpha \geq 1$, $w(\mathbf{v}, \mathbf{v}')$ is defined by the weighted 'distance' of voxels $\mathbf{v}$ and $\mathbf{v}'$, and $g(\mathbf{D}(\mathbf{v}), \mathbf{D}(\mathbf{v}'))$ is the geodesic distance between $\mathbf{D}(\mathbf{v})$ and $\mathbf{D}(\mathbf{v}')$. Moreover, $\tilde{\mathbf{D}}^1(\mathbf{v})$ and $\tilde{\mathbf{D}}^2(\mathbf{v})$ are, respectively, an intrinsic median estimator and an intrinsic least square estimator of $\mathbf{D}(\mathbf{v})$ (61, 60, 157, 30). Two commonly used metrics, including the trace metric and the log-Euclidean metric, are usually chosen for the geodesic distance on the space of symmetric positive definite matrices (9, 157, 54). Directly solving (4.42) is equivalent to the calculation of an intrinsic local constant estimator of $\mathbf{D}(\mathbf{v})$. In (157), the authors propose a general intrinsic local polynomial regression estimate for the analysis of symmetric positive definite matrices as responses. For each metric, they develop a cross-validation bandwidth selection method, derive the asymptotic bias, variance, and normality of the intrinsic local constant and local linear estimators, and compare their asymptotic mean-square errors. For the ODF, (55, 68) develop an intrinsic local constant estimator of the ODF in order to smooth ODF imaging, but its related statistical theory has not been established yet.

In the third category, a few methods have been developed to perform simultaneous smoothing and estimation of DTI by using either regularization methods or nonparametric statistical methods (127, 154, 93). Specifically, (93) proposed to solve a regularized optimization problem

$$\text{argmin}_{\{\mathbf{D}(\mathbf{v}):\mathbf{v}\in\mathcal{V}\}} \int_{\mathbf{v}\in\mathcal{V}} \ell_n(\{S(\mathbf{q}_i; \mathbf{v})\}_{i\geq 1}; \mathbf{D}(\mathbf{v}))dL(\mathbf{v}) \qquad (4.43)$$

$$+\lambda_1 \int_{\mathbf{v}\in\mathcal{V}} \int_{W(\mathbf{v})} \omega(\mathbf{v}, \mathbf{v}')g(\mathbf{D}(\mathbf{v}), \mathbf{D}(\mathbf{v}'))d\mathbf{v}d\mathbf{v}',$$

where λ_1 is a tuning parameter, $\ell_n(\{S(\mathbf{q}_i; \mathbf{v})\}_{i\geq 1}; \mathbf{D}(\mathbf{v}))$ is usually chosen to be the log-likelihood function of the observed DW signals $S(\mathbf{q}_i; \mathbf{v})\}_{i\geq 1}$, $\omega(\mathbf{v}, \mathbf{v}')$ are the regularization weights, and $W(\mathbf{v})$ is the search window at voxel $\mathbf{v}$. Reference (93) proposed to use the weighting function of NLM to construct $\omega(\mathbf{v}, \mathbf{v}')$ and use the total Bregman divergence to design $g(\mathbf{D}(\mathbf{v}), \mathbf{D}(\mathbf{v}'))$.

A multiscale adaptive regression modelling (MARM) framework based on the PS method can be used to carry out simultaneous smoothing and estimation of DTI/ODF/EAP (90, 127). Specifically, let $B(\mathbf{v}, h_s) = \{\mathbf{v}' : ||\mathbf{v}' - \mathbf{v}||_2 \leq h_s\}$ be a sequence of balls centered at $\mathbf{v}$ with increasing radii $\{h_s\}$ such that $h_0 = 0 < h_1 < \cdots < h_S = r_0$. At each voxel $\mathbf{v}$, MARM iteratively maximizes a weighted objective function as

$$\hat{\mathbf{D}}(\mathbf{v}; h_s) = \operatorname{argmax}_{\mathbf{D}(\mathbf{v})} \sum_{\mathbf{v}' \in B(\mathbf{v}, h_s)} \omega(\mathbf{v}, \mathbf{v}'; h_s) \ell_n(\{S(\mathbf{q}_i; \mathbf{v}')\}_{i \geq 1}; \mathbf{D}(\mathbf{v})),$$

where $\omega(\mathbf{v}, \mathbf{v}'; h_s)$s are adaptive weights calculated at each radius h_s and allow us to incorporate data from neighboring voxels $\mathbf{v}' \in B(\mathbf{v}, h_s)$. At each voxel $\mathbf{v}$, we will obtain a sequence of estimators of $\mathbf{D}(\mathbf{v})$ as follows:

$$\hat{\mathbf{D}}(\mathbf{v}; h_0) \to \{\omega(\mathbf{v}, \mathbf{v}'; h_1)\} \to \cdots \to \{\omega(\mathbf{v}, \mathbf{v}'; h_S)\} \to \hat{\mathbf{D}}(\mathbf{v}; h_S). \tag{4.44}$$

When $s = 0$, $\hat{\mathbf{D}}(\mathbf{v}; h_0)$ reduces to the estimator of $\hat{\mathbf{D}}(\mathbf{v})$ for the voxel-wise method. Compared with the regularization method in (4.43), MARM should be more robust to higher noise levels and the selection of kernel window sizes and patch shapes at different locations. Finally, the adaptive weights in MARM can be extended to include the weighting function of NLM and/or existing biological information (e.g., fiber tracks). See Figure 4.8 (a)–(c) for an illustration of the use of MARM for ODF reconstruction based on QBI.

However, methods for each category have some advantages and disadvantages. (i) For the first category, these methods have been criticized for ignoring the fact that raw diffusion weighted signals acquired at different $\mathbf{q}$-values are highly associated with each other in each voxel. Moreover, in white matter regions, the SNRs vary dramatically across different $\mathbf{q}$ values. Since these methods primarily use different weights to smooth the raw diffusion weighted images independently, such methods are prone to accumulate biases from all DWIs, which can lead to large biases in the estimated tensor/ODF/EAP images.

(ii) For the second category, these methods have been criticized for ignoring the stochastic components of the raw DWI data and directly smoothing the estimated tensor/ODF/EAP based on a specific metric of the tensor/ODF/EAP space. Since each estimated DTI is estimated by using all diffusion signals in each voxel, the estimated tensors can be shown to be asymptotically normal distributed with zero mean by using the central limit theorem. The estimated tensors are asymptotically unbiased and thus it is not critical to model the distribution of the stochastic components of the raw DWI data. Moreover, a key advantage of these methods is to use the same set of weights to smooth the raw DWI data across all $\mathbf{q}$-values. However, if the originally estimated tensor/ODF/EAP field is biased, then these methods may not be able to reduce the biases in the smoothed tensor/ODF/EAP field.

(iii) For the third category, these methods are computationally more expensive, but a key advantage of these methods is to adaptively determine the weights at each voxel and then apply them to the raw DW signals. They avoid the potential biases introduced by those methods for the first category. Moreover, since MARM refits the raw DWI data at each bandwidth, it avoids the potential biases introduced by the voxelwise method.

4.4 Tractography Algorithms

Many tractography algorithms have been proposed to map fibers through the entire brain based on the estimated principal direction/ODF field (59, 116, 77, 89, 100). The algorithms can be categorized into two main groups: local and global methods. Local methods use local ODF information to independently construct fibers path by path. Local methods can be grouped into two classes: deterministic and probabilistic. Deterministic algorithms usually start at seed voxels and follow the local principal directions/ODFs estimated by the diffusion

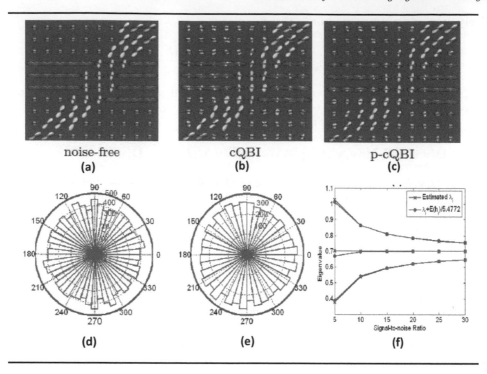

FIGURE 4.8
Simulation Results. The first row shows ODF reconstruction results based on simulated data with twisted crossing: true ODF field (a), estimated ODF fields based on cQBI (b), and MARM-cQBI (c). The second row shows results from a simulation study of the isotropic tensor $\mathbf{D} = \mathrm{diag}(0.7, 0.7, 0.7)$ (units: $10^{-3}\mathrm{mm}^2/s$): the angle histogram plots of θ based on 10,000 simulated DW datasets at SNR $= 20$ (d) and 10,000 eigenvectors (e) simulated from the theoretical distribution given in (165), respectively, where $\theta \in [0, 2\pi]$ is the subcomponent of $(1, \theta, \phi)$, the spherical coordinate of the eigenvector corresponding to the largest eigenvalue. (f) shows the theoretical means of the estimated three eigenvalues and the mean value of estimated eigenvalues as a function of SNR from 5 to 30 based on 10,000 simulated datasets.

model in order to generate sequences of points that are considered on major fibers. Several deterministic tractography algorithms include streamline algorithms and more elaborated tensor deflection algorithms, among others. Probabilistic algorithms repeatedly use Monte Carlo simulations (e.g., Markov chain Monte Carlo) to statistically generate the principal directions and then apply some deterministic methods for tracking fiber bundles. Such methods produce maps of 'probability' for each voxel to be crossed by a random track and the probabilistic maps of connectivity between any two ROIs. See (78, 141, 89) for a nice review of various tractography algorithms and references therein. An advantage of local methods is their computational efficiency. However, the local methods can be very sensitive to noise components in DWIs, which can significantly affect the final tracking result.

Figure 4.9 showed fiber tracts across several ROIs by the deterministic local tractography method in MRtrix (See Section 4.9.2). The subject is from the Q3 dataset in the Human Connectome Project (HCP), where $b = 1000, 2000, 3000 s/mm^2$, 90 directions per shell. Constrained SD (129) was performed for all $90x3$ volumes using a 3D fiber response function (38) to estimate fODFs. Then MRtrix is used for deterministic local tractography.

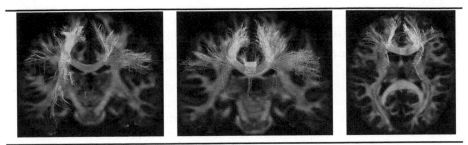

FIGURE 4.9
Fiber tracts of a subject in HCP Q3 dataset by a deterministic local tractography used in MRtrix.

We consider a stochastic differential equation model with measurement errors for local tractography methods (86, 118). Specifically, let $\mathbf{v}(t)$ be the true fiber trajectory in R^3. The stochastic differential equation model assumes

$$\frac{d\mathbf{v}(t)}{dt} = \mathbf{e}(\mathbf{v}), \quad t \geq 0 \;\; \text{with} \;\; \mathbf{v}(0) = \mathbf{v}_0, \tag{4.45}$$

where $\mathbf{e}(\mathbf{v})$ is the true fiber direction at location $\mathbf{v}$ and $\mathbf{v}_0$ is the position of the seed location. Based on dMRI data, one is able to obtain an estimate of the true fiber direction field, denoted by $\{\hat{\mathbf{e}}(\mathbf{v}) : \mathbf{v} \in \mathcal{V}\}$, such that

$$\hat{\mathbf{e}}(\mathbf{v}) = \mathbf{e}(\mathbf{v}) + \epsilon(\mathbf{v}), \tag{4.46}$$

where $\epsilon(\mathbf{v})$ is a zero-mean stochastic process. Numerically, let $\delta > 0$ be a fixed approximation step and a sequence of points $t_k = k\delta$ for $k = 0, 1, \ldots, [T/\delta]$. By using Euler's approximation, one can solve (4.45) by iteratively updating

$$\mathbf{v}(t_k) = \mathbf{v}(t_{k-1}) + \hat{\mathbf{e}}(\mathbf{v}(t_{k-1})) \quad \text{for} \;\; k = 1, \cdots, [T/\delta], \;\; \mathbf{v}(t_0) = \mathbf{v}_0. \tag{4.47}$$

Global methods reconstruct all detectable fibers of the brain simultaneously. It reconstruct fibers by finding a configuration that best describes the whole set of measured data (94). The reconstructed fibers are built by small line elements, each of them reflecting a part of the whole diffusion anisotropy. Elements being connected in lines eventually form reconstructed fibers. An advantage of global methods is stable with respect to noise and imaging artifacts. However, the global methods are often computationally time-consuming.

We consider a Bayesian approach for the global methods as follows. Let $\mathcal{M}$ be the assumed fiber model in $\mathcal{V} \subset R^3$ and $\mathcal{S} = \{(S(\mathbf{q}_i; \mathbf{v}), \mathbf{q}_i) : i = 1, \cdots, n\}$ be observed DWI data. One needs to specify a sampling distribution of $\mathcal{S}$ given $\mathcal{M}$, denoted by $p(\mathcal{S}|\mathcal{M})$, and a prior distribution of $\mathcal{M}$, $p(\mathcal{M})$. For the sampling distribution, one can use the dMRI models discussed above. The key idea and challenge of the global tracking methods lies in how to specify the prior of the fiber model, $p(\mathcal{M})$. In (116), the authors used small line (fiber) segments $L_S(\mathbf{v}) = (\mathbf{v}, \mathbf{r}(\mathbf{v}))$ consisting of a continuous spatial position $\mathbf{v} \in \mathcal{V}$ and an orientation $\mathbf{r}(\mathbf{v})$ that can form chains to represent the individual fibers. A mixture model of the product of a stick model in orientation space and an isotropic Gaussian in the spatial domain is assumed for $p(\mathcal{S}|\mathcal{M})$. A simple interaction model is assumed for all connected segments, which leads to $p(\mathcal{M})$. The simulated annealing algorithm is used to calculate the posterior mode of $\hat{\mathcal{M}} = \text{argmax}_{\mathcal{M}} p(\mathcal{M}|\mathcal{S})$, where $p(\mathcal{M}|\mathcal{S}) \propto p(\mathcal{S}|\mathcal{M})p(\mathcal{M})$ is the posterior distribution of $\mathcal{M}$ given $\mathcal{S}$.

Despite the increasing availability of different tractography algorithms, there are many open questions in the quantification of these fiber-tracking methods.

- **It is critical to develop a reliable evaluation and validation system for tractography algorithms** (44). The diffusion community has developed several evaluation measures (e.g., connectivity analysis) and two well-known phantoms including the FiberCup phantom dataset and the HARDI reconstruction challenge phantom to evaluate various diffusion models and tractography algorithms (59). An important finding is that probabilistic tractography algorithms lead to many false positives and should be used with caution (44). Much more research should be done on the design of evaluation measures and more realistic phantoms that are close to human brain in various settings.

- **Development of computationally efficient global tractography algorithms needs more attention.** It is critical to develop a more reasonable fiber model $p(\mathcal{M})$ in order to better estimate the true fiber tracts and quantify their uncertainties. Moreover, optimizing $p(\mathcal{M}|\mathcal{S})$ is computationally challenging due to the presence of a large number of parameters and their non-convexity.

4.5   Uncertainty in Estimated Diffusion Quantities

Because of the noise that is inherent in DWI data, calculated tensors/ODFs/EAPs and their associated quantities (e.g., eigenvalues and principal directions) generally differ from the true ones, producing uncertainty in their estimation. To establish dMRI as a reliable and widely accepted technique, it is critical to quantify such uncertainty in various estimated diffusion quantities. Such quantification is very important for addressing many scientific questions in neuroscience and for designing and carrying out large DWI-related clinical studies.

Two classes of methods, including Monte Carlo and theoretical methods, have been developed to quantify estimated diffusion tensors and their eigenspace components. The Monte Carlo methods consist of (i) a statistical model for diffusion weighted signals; (ii) the choice of an estimation method; and (iii) the quantification of uncertainty in estimated diffusion quantities based on Monte Carlo simulations. In contrast, besides (i) and (ii), the theoretical methods use some mathematical and statistical techniques to directly approximate the uncertainty of estimated diffusion quantities instead of using Monte Carlo simulations.

Recent theoretical calculations based on perturbation theory and asymptotic theory have accurately approximated the uncertainty of the estimated eigenvalues and eigenvectors, as well as the bias that is introduced by sorting by their magnitudes eigenvalues in both degenerate and nondegenerate tensors (6, 165). Those calculations have shown in particular that the uncertainty in identifying a tensor's principal direction is determined primarily by whether the overall morphology of the tensor is degenerate or not (165). The results in the asymptotic theory allow us to delineate the stochastic behavior of estimated eigenvalues and eigenvectors for degenerate tensors, whereas those in the perturbation theory cannot. See Figure 4.8 (d)–(f) for an illustration of theoretical results for the isotropic tensor.

The Monte Carlo methods include both simulation studies and bootstrapping methods. Based on (4.10) and the Rician noise model, previous simulation studies have shown that estimated eigenvalues are always distinct and that their estimated FA is always greater than zero, regardless of whether the tensor is degenerate (i.e., oblate, prolate, or isotropic) or nondegenerate (109, 16, 22). Thus, one always incorrectly identifies the principal directions of tensors within regions that contain isotropic or oblate tensors in real DWI.

Bootstrapping methods, including repetition and the wild bootstrap, have been widely used to numerically quantify the uncertainty of eigenvalues, eigenvectors, and diffusion properties (144). Repetition bootstrap in DTI requires repeated measurements in each gradient direction, because it resamples with replacement the raw DW images in each of those directions. The accuracy of the repetition bootstrap depends on the number of repeated measurements in each direction. The wild bootstrap is a model-based method that resamples the residuals of the linear regression model used to estimate the tensor at each voxel. In particular, it is applicable to most DTI acquisition schemes, including the standard acquisition of one measurement per direction, unlike the repetition bootstrap (144).

One has to use the wild and repetition bootstrap methods with extra caution, since these methods have been used in the dMRI literature without any theoretical justification. However, such justification is necessary for producing any scientifically meaningful measure of diffusion uncertainty in which we are interested. In (158), the authors examine several fundamental issues associated with the two bootstrap methods by using both theoretical arguments and extensive Monte Carlo simulations. The two bootstrap methods are invalid for quantifying the uncertainty of the parameters for some tensors, such as the principal direction of an isotropic or a degenerate tensor. The validity of the wild bootstrap strongly depends on the correct specification of the fitted model used to estimate a tensor. Because the wild bootstrap resamples the residuals of the fitted tensor model, resampled tensors may not reflect the true characteristics of DTs in real DWIs.

There are many open questions in the quantification of the uncertainty in various estimated diffusion quantities.

- **For HARDI, little has been done to quantify the uncertainty of estimated ODFs and EAPs and their associated quantities based on the voxelwise estimation methods.** Moreover, if one uses more complex spatial-adaptive estimation methods, such as PS, to estimate the ODF and EAP, such quantification becomes more difficult due to spatial smoothness and the presence of spatial correlation. According to the best of our knowledge, nothing has ever been done on such quantification.

- **How to quantify the uncertainty of estimated fiber tracks is largely unknown.** Although there are a few attempts at quantifying of uncertainty in estimated tractography from both numerical and theoretical perspectives (63, 22, 86), several critical issues remain open and need further theoretical investigation. Theoretically, (86) first proved some asymptotic/stochastic properties of the estimated tractography based on models (4.45)–(4.46). More research should be done along this direction. In contrast, although some existing DTI packages produce some uncertainty measures in the tractography results, it is unclear whether such measures are valid from a methodological perspective. For instance, one approach is to calculate the probability that two regions are connected based on local tractography algorithms and Monte Carlo methods, such as bootstrap. However, such probability may be positively correlated with the true probability that the two regions are connected, but they are not the same. Such probability should be used with great caution.

4.6 Sampling Mechanisms

An important design issue is how to select a set of gradients and b values $\{(\mathbf{r}_i, b_i) : i = 1, \ldots, n\}$ or q values $\{\mathbf{q}_i : i = 1, \ldots, n\}$ in order to accurately estimate tensor/ODF/EAP across all voxels. Statistically, this is an optimal design problem (12). Different sampling schemes in **q**-space have been developed in the literature (Figure 4.10). See (32) for an extensive review of various acquisition strategies in q−space. There are three principles for comparing different acquisition strategies including antipodal symmetry, being isotropic, and reconstruction. When there is no prior in the underlying tensor/ODF/EAP field, the first two principles have motivated people to uniformly arrange points on the sphere with central symmetry. Based on the third principle, various acquisition schemes have been developed to optimize tensor reconstruction (64, 107, 123).

To select an efficient set of q values, one must address three key components of dMRI data including (i) a statistical model for diffusion weighted signals; (ii) the choice of an appropriate optimality criterion, denoted by $L(\{\mathbf{q}_i\}_{i \leq n})$; and (iiii) optimizing the optimality criterion with respect to $\{\mathbf{q}_i : i = 1, \ldots, n\}$. For (i), various models for dMRI signals have been developed above. For (ii), the optimality criterion is usually developed to quantify the uncertainty of the objective of interest, such as tensors and fiber tracts. The existing optimality criteria largely depend on estimated tensor/ODF/EAP and their invariant measures. In (iii), one needs to use some optimization algorithms to solve $\{\hat{\mathbf{q}}_i\}_{i \leq n} = \text{argmin } L(\{\mathbf{q}_i\}_{i \leq n})$. Since the optimality criterion may not be convex, calculating $\{\hat{\mathbf{q}}_i\}_{i \leq n}$ is not a trivial problem at all.

As an illustration, we consider the reconstruction of a diffusion tensor field, denoted by $\{\beta(\mathbf{v}) : \mathbf{v} \in \mathcal{V}\}$, based on model (4.37). We consider the covariance matrix of $\hat{\beta}(\mathbf{v})$ at voxel $\mathbf{v}$, denoted by $C(\hat{\beta}(\mathbf{v}))$. As shown in (165), $C(\hat{\beta}(\mathbf{v}))$ depends on the SNR, the b value, the number of of baseline acquisitions, denoted by m, the diffusion tensor matrix, and the gradient encoding scheme. Let $p(\beta(\mathbf{v}))$ be the prior distribution of $\beta(\mathbf{v})$, which may represent prior knowledge of the underlying fiber orientations of the tissue being imaged. A Bayesian criterion function can be written as

$$\text{GSI}(m, b, \text{SNR}, \mathbf{x}) = \int \Psi\{C(\hat{\beta}(\mathbf{v}))\}p(\beta(\mathbf{v}))d\beta(\mathbf{v}), \qquad (4.48)$$

where $\Psi\{\cdot\}$ is a pre-specified function, such as the trace. We can use GSI as an index to compare different DT acquisition schemes.

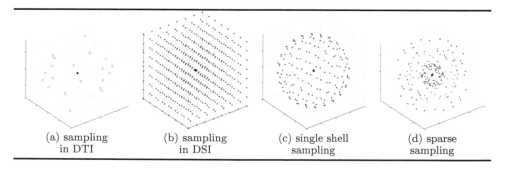

(a) sampling in DTI   (b) sampling in DSI   (c) single shell sampling   (d) sparse sampling

FIGURE 4.10
Several kinds of sampling in **q**-space. The black dot in $\mathbf{q} = (0, 0, 0)^T$ is the baseline image without diffusion gradient. Note that although we showed sampling in $\mathbb{R}^3$, normally only samples in a half space are used, e.g., $(0, 0, 1)\mathbf{q} = q_z \geq 0$. (a) Sampling used in DTI, normally less than 20 DWI images are used; (b) dense Cartesian sampling used in DSI. Note in practice the Cartesian samples inside a given Ball are used. (c) Single shell sampling used in sHARDI methods, e.g., QBI, DOT etc. (d) Sparse sampling used in mHARDI methods, e.g., DPI, SHORE, SPFI. Note although normally multiple-shell sampling is used, any sampling scheme can be used in mHARDI methods.

TABLE 4.1
The condition numbers and gradient sampling indices (GSI) of thirteen acquisition schemes in (123).

Scheme name	Condition number	Number of directions	m	Repetition	GSI$(m, 900, 15, \mathbf{x})$ $\times 10^7$
Tetrahedral	9.148	6	1	20	2.944
Cond6	5.989	6	1	20	1.398
Decahedral	2.749	10	2	12	0.8080
Jones noniso	2.562	7	1	18	0.8083
Dual-gradient	2.000	6	1	20	0.7052
Jones10	1.624	10	2	12	0.5865
Jones20	1.615	20	3	6	0.6177
Jones30	1.595	30	5	4	0.5886
Papadakis	1.587	12	2	10	0.5269
Jones6	1.583	6	1	20	0.5954
Muthupallai	1.581	6	1	20	0.5829
Tetraortho	1.528	7	1	18	0.5127
DSM	1.323	6	1	20	0.6014

To compare the adequacy of differing image acquisition strategies, we calculated the values of GSI for thirteen data acquisition schemes that were used previously to demonstrate the importance of the condition number in the determination of noise characteristics for particular acquisition schemes (see Tables 1 and 2 of (123)). Table 4.6 presents the GSI values for each of the thirteen acquisition schemes at $b = 900s/mm^2$. Although the DSM scheme in (123) has the smallest condition number, its GSI was larger than those of the Papadakis and Tetraortho schemes. In terms of both the number of images and GSI, it seems that the Papadakis scheme is the best among these thirteen strategies.

There are many open questions in the design of sampling mechanisms for various HARDI models.

- **Little has been done on the design of effective sampling mechanisms for various HARDI models.** Several key difficulties include the choice of $L(\{\mathbf{q}_i\}_{i\leq n})$ and its optimization, particularly for a multiple q-shell acquisition.

- **Nothing has been done on the use of the uncertainty of estimated fiber tracks to design sampling mechanism.** Since our primary objective of interest is to reconstruct fiber tracks, it is important to develop some design criterion based on the uncertainty of estimated fiber tracks. As discussed above, since there is a lack of theoretical results on the uncertainty of estimated fiber tracks, it is impossible to develop an optimal sampling mechanism based on such results.

4.7 Registration

The above sections focus on estimation of the diffusion signal, EAP and ODF for an individual subject. Spatial alignment, also called image registration, is an important issue for group analysis. Image registration has been studied for decades in the medical image analysis domain, and in the last ten years some methods have been proposed to align DWI data, or the estimated tensor/ODF/EAP data. Although image registration techniques in dMRI originate from vector-valued image registration methods, these are not applicable to directly apply vector-valued image registration methods to diffusion data (4). An image registration method for diffusion data normally includes two aspects. One is the spatial alignment of 3D anatomical structures, and the other one is the re-orientation of local diffusion profiles. Some methods propose to perform these two steps separately for diffusion data (4, 73, 33, 112, 149), i.e. perform re-orientation after spatial alignment. Some other methods propose to consider the re-orientation issue inside the cost function of the registration and perform these two steps iteratively (29, 42, 153, 67, 150, 56, 161).

Reference (4) first addressed the re-orientation issue in dMRI and proposed "finite strain" (FS) and "preservation of principal direction" (PPD) to orient the tensor image, where PPD gives the best performance. The idea of PPD is to keep the tensor shape while rotating the principal direction of the tensor using the Jacobian matrix of the deformation field, and the idea of FS is to rotate the tensor using a rotate matrix extracted from the Jacobian matrix of the deformation field. It was shown that FS cannot consider shearing or scaling effect of transforms (4). PPD is widely used in tensor registration (29, 42, 33). Re-orientation of the DWI signal, ODF and EAP data is more complicated compared with tensor data. Since the FS method is much simpler than PPD, some HARDI methods use FS to orient the ODF or EAP represented by the SH basis due to the closed form of rotation of the SH basis (26, 67). Some other methods separate diffusion signals and ODFs using some kind of basis functions, then re-orientate the basis function separately and combine the re-orientated functions together (73, 112, 161). The basis functions in (73) are delta functions, and those in (52, 112, 161) are fiber response functions.

There are some open questions in the registration of diffusion data.

- **Does the registration need to be performed in diffusion signals or estimation results (tensors, ODFs, or EAPs)?** Most methods perform registration on tensors, ODFs, or EAPs, because the estimation results are more spatially meaningful than DWI signals. However these methods are largely affected by the reconstruction methods used. Some registration methods were performed directly in DWI signals. However, these registration methods still need to consider a model for re-orientation DWI signals.

- **How can re-orientation be done?** It is well accepted that registration of diffusion data needs re-orientation. However, there is no consistent and well-accepted way to do re-orientation. PPD is well accepted for tensor data, which assumes that the shape of the local diffusion profile does not change. However, the current state-of-the-art registration methods in HARDI changes the shape of local diffusion profiles for re-orientation (73, 52, 112, 161).

- **It is critical to develop a reliable evaluation and validation system for registration algorithms.** Unfortunately, little has ever been done on such development due to many fundamental difficulties, such as the two open questions discussed above.

4.8 Group Analysis

In the current literature, there exist three major approaches to the group analysis of diffusion imaging data: region-of-interest (ROI) analysis, voxel-based analysis, and fiber tract- based analysis (124, 103, 125).

The ROI analysis can be performed by registering individual subject DWI images to an atlas and then averaging diffusion properties in some manually drawn ROIs of the atlas (125). Subsequently, a group analysis can be carried out to correlate all statistics at each ROI or across multiple ROIs with covariates of interest. An advantage of ROI analysis is that processing is relatively simple and robust against imperfect registration. The three drawbacks of ROI analysis include the difficulty in identifying meaningful ROIs, particularly the long curved structures common in fiber tracts, the instability of statistical results obtained from ROI analysis, and the partial volume effect in relatively large ROIs (70, 164). A stringent assumption of ROI analysis is that diffusion properties in all voxels of the same ROI are essentially homogeneous, which is largely false for dMRI data. Moreover, this form of analysis leads to limited localization of findings.

Voxel-based analysis has been widely used in neuroimaging studies. It involves registering each subject into a study-specific reference space and fitting a statistical model to the smoothed and registered diffusion property imaging data from multiple subjects at each voxel to generate a parametric map of test statistics (or p-values). Subsequently, a multiple-comparison procedure such as false discovery rate is applied to correct for multiple comparisons across the many voxels of the imaging volume (146). The major drawbacks of voxel-based analysis include poor alignment quality and the arbitrary choice of smoothing extent (124, 80). Moreover, in practice, one has to interpret the findings based on voxelwise comparison of the eigenvector and/or tensor images with great caution (120, 162), since they are very sensitive to alignment inaccuracies compared with FA images.

Fiber tract–based analysis has received growing interest, since it may be more robust to alignment inaccuracies, while directly incorporating fiber tract information (124, 103, 160, 70, 164, 71). There are three major fiber tract–based analysis methods including tract-based spatial statistics (TBSS), medial sheet based analysis, and fiber tract analysis. A *tract based spatial statistics* (TBSS) framework was developed to construct local diffusion properties along the white matter skeleton and then perform pointwise hypothesis tests on the skeleton (124). However, TBSS does not have an explicit tract representation that can be uniquely linked to individual fibers throughout the brain, while the use of maximal FA values renders TBSS sensitive to DWI artifacts.

A medial model–based framework was developed for the statistical analysis of diffusion properties on the medial manifolds of fiber tracts followed by testing pointwise hypotheses on the medial manifolds (160). The framework consists of effectively modeling six sheet-like fasciculi by using deformable medial representations, averaging and combining tensor-based features along directions locally perpendicular to the tracts, and pointwise statistical analysis. However, it is limited to the sheet-like white matter tracts and relies on expert-driven segmentation of the fasciculi.

A fiber tract analysis framework was developed for the statistical analysis of diffusion properties along major fiber tracts followed by using functional data analysis (70, 164, 138, 66, 69, 104, 53). See (138) for an overview of the fiber tract analysis framework developed at UNC-Chapel Hill. The fiber tract analysis framework consists of using anatomically informed curvilinear regions to analyze diffusion at specific locations all along fiber tracts, taking weighted averages at each step along the fiber bundles, an unbiased atlas-building step, and a Functional Analysis of Diffusion Tensor Tract Statistics (FADTTS) pipeline. This form of analysis results in highly localized statistics that can be visualized back on the individual fiber bundles. Moreover, there is great interest in developing new fiber registration methods for group analysis (65, 57, 169, 140, 79).

A set of FADTTS has been developed for delineating the structure of the variability of multiple diffusion properties or tensors along major white matter fiber bundles and their association with a set of covariates for both cross-sectional and longitudinal studies (164, 156, 163, 159, 75, 166). The advantages of FADTTS are that they are capable of modelling the structured inter-subject variability by a functional principal component analysis method, testing the joint effects by a global test statistic and local test statistics, and constructing simultaneous confidence bands of the interested effects through a resampling method. Statistically, as shown in various simulations and real data analysis, these statistical methods in FADTTS are much more powerful than the standard voxel-wise methods.

As an illustration, we applied FADTTS to study the spatial-temporal dynamics of white-matter fiber tracts in a clinical study of neurodevelopment. There are 298 high-quality scans available for 137 children with 83 males and 54 females. As a graphical illustration, FA measures were plotted along the genu and splenium of the corpus callosum for each subjects within each age group (Figure 4.11 (a)). FA measures were also plotted for 35 subjects along the genu tract (Figure 4.11 (b) and (c)). An obvious increasing trend for the values of FA were observed at nearly all grid points, especially from neonate to the first year.

For the genu tract, we fitted a functional mixed-effects model (FMEM) in (156) to the FA curves, denoted by $y_{ij}(s)$, from all 137 subjects. Specifically, FMEM is given by

$$y_{ij}(s) = \mathbf{x}_{ij}^T B(s) + \mathbf{z}_{ij}^T \xi_i(s) + \eta_{ij}(s) + \epsilon_{ij}(s), \tag{4.49}$$

where $\mathbf{x}_{ij} = (1, \mathrm{Dir}_{ij}, \mathrm{G}_i, \mathrm{Age}_{ij1}, \mathrm{Age}_{ij2})^T$, $\mathbf{z}_{ij} = (1, \mathrm{Age}_{ij1}, \mathrm{Age}_{ij2})^T$ and Age_{ij1} (Age_{ij2}) is an indicator variable indicating whether a subject belongs to the first (second) year age group. The coefficient functions related to Age_{ij1} and Age_{ij2} can be used to investigate whether there is some change from neonate to the first year of life, from the first year to the second year, and from neonate to the second year. Moreover, in model (4.49), $\eta_{ij}(s)$ primarily characterizes within-curve spatial correlation structure, while $\xi_i(s)$ primarily characterizes the subject-level variations and within-subject spatial-temporal correlation. Then we estimated the functional coefficients. For hypothesis testing, we constructed the global test statistic to test the gender, number of gradient directions and age effects on FA values. We approximated the p value of the global test using the resampling method with 5,000 replications. Finally, we constructed the 95% simultaneous confidence bands for the functional coefficients.

The hypothesis testing results show that there are significant age and number of gradient direction effects on FA, RD and AD values. The FA are significantly different between neonate versus the first year, and between the first year versus the second year with p value $< .0001$, far smaller than a 0.05 significance level. It is observed from Figure 4.11 (b) that mean FA values increase from neonate to the first year and then from the first year to the second year. Moreover, the change from the neonate to the first year is larger than that from the first year to the second year. No gender difference in FA was found for the genu tract.

There are many open questions in the joint analysis of diffusion imaging data and other data.

- **All fiber tract–based methods including FADTTS are only applicable to these major white matter tracts in which one can establish common localization across subjects.** However, the centroid of the localization of white matter lesion could vary across time and subjects. In some heterogeneous populations, it is possible that tract-specific changes occur in only a subset of subjects. In these scenarios, none of group analysis methods discussed above would be appropriate.

- **There is an urgent demand for the development of functional regression methods** for the analysis of repeated functional data and clinical data obtained from longitudinal and familial studies. Although there is a handful of papers on the development of statistical models and their estimation methods for repeated functional data, the methodology for dealing with such data is still in its infancy, and further computational and theoretical development is greatly needed.

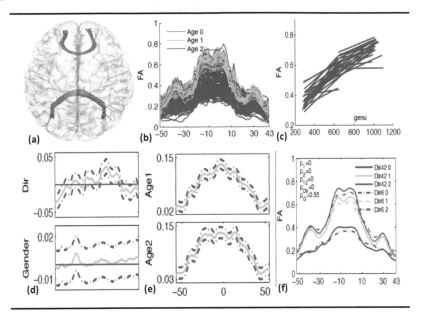

FIGURE 4.11

(a) The commissural bundles of the genu and splenium of the corpus callosum. (b) FA values along the genu tract for all 137 subjects in each age group. (c) FA values varying over age at a selected location along the genu tract. (d) and (e) 95% simultaneous confidence bands for varying coefficient functions for FA along the genu of the corpus callosum tract. The solid curves are the estimated coefficient functions, and the dashed curves are the 95% confidence bands. The thin horizontal line is the line crossing the origin. (f) p_1 is the p value for the difference in the diffusion measure between neonate and the first year, p_2 is the p value for the difference in the diffusion measure between neonate and the second year, p_{12} is the p value for the difference between the first year and the second year, p_G is the p value for the gender effect, p_{Dir} is the p value for the effect of the number of gradient directions.

- **There is an urgent demand for the development of** high-dimensional risk prediction models by integrating and identifying important white matter tracts, functional images, and biological markers for risk prediction. These models can have a great impact in public health from disease prevention, to detection, to treatment selection. For instance, it is interesting to consider generalized functional linear models, in which a scalar outcome (e.g., diagnostic group) is used as the response and fiber bundle diffusion properties are used as varying covariate functions (or functional predictor) (114, 117).

4.9   Public Resources

4.9.1   Datasets

Public datasets are important for reproducible research. For synthetic data simulation, people normally use a mixture of tensor models or cylinder models to generate the ground truth of signals (105), then corrupt the ground truth signal using Rician noise. Here are some real public datasets for simulation and evaluation.

- Q1, Q2, Q3 data from HCP: http://www.humanconnectome.org/data/, three shells, staggered 90 direction per shell (31), $b = 1000, 2000, 3000s/mm^2$, 1.25 mm isotropic voxels.

- HARDI data for Stanford: http://purl.stanford.edu/yx282xq2090, single shell HARDI, 150 directions, $b = 2000s/mm^2$.

- Phantom data for fiber cup 2009 (58): http://lnao.lixium.fr/spip.php?rubrique79, three shells, the same 64 directions with twice the scans per shell, $b = 650, 1500, 2000s/mm^2$.

- HARDI reconstruction challenge: http://hardi.epfl.ch/static/events/2013ISBI, 3 predetemined sampling schemes including DTI ($b = 1200s/mm^2$, 32 directions), HARDI ($b = 3,000s/mm^2$, 64 directions), and Heavyweight ($b < 12,000s/mm^2$, 515 acquisitions).

- DWI datasets from the Alzheimer's Disease Neuroimaging Initiative (ADNI): http://www. adni-info.org/, 264 subjects with a total of 799 DWI datasets, $b = 1,000s/mm^2$, and 41 directions.

4.9.2 Software

There is a branch of open source codes and software for diffusion MRI data processing. Here are a recommendation list.

- FSL: http://fsl.fmrib.ox.ac.uk/fsl/fslwiki/FDT/UserGuide. Eddy current correction, tensor estimation, multi-tensor estimation, registration of scalar images, deterministic tracking, probabilistic tracking, TBSS, QBI, SD.

- 3D Slicer: http://www.slicer.org/. DWI denoise, DTI, fiber tracking, visualization of tensors/fibers.

- CAMINO: http://www.cs.ucl.ac.uk/research/medic/camino. DTI, multi-tensor estimation, QBI, SD, PASMRI, data simulation, tensor registration, uniform sampling scheme based on minimization of electrostatic energy, peak detection, visualization of tensors/ODFs/fibers/peaks.

- MITK: http://www.mitk.org/. DTI, QBI, global fiber tracking, deterministic fiber tracking, peak detection, TBSS, data simulation, visualization of tensors/ODFs/fibers.

- MRtrix: http://www.brain.org.au/software/mrtrix/. DTI, SD, deterministic/probabilistic fiber tracking, visualization of tensors/ODFs/fibers/peaks.

- MRI Studio: https://www.mristudio.org. DTI, deterministic fiber tracking, visualization of tensors/fibers.

- Trackvis: http://www.trackvis.org, DTI, QBI, DSI, fiber tracking, visualization of fibers.

- MEDINRIA: http://med.inria.fr/. DTI, deterministic fiber tracking, visualization of tensors/fibers.

- DTI-TK: http://dti-tk.sourceforge.net/pmwiki/pmwiki.php. Tensor estimation, tensor registration, image format conversion, visualization of tensors/fibers.

- Fibernavigator: http://scilus.github.io/fibernavigator/. Visualization of tensors/ODFs/ fibers/peaks.

- DIPY: https://github.com/nipy/dipy. DTI, SD, QBI, DSI, fiber tracking, visualization of tensors/ODFs/fibers/peaks.

- DSI-studio : http://dsi-studio.labsolver.org/. DTI, QBI, DSI, fiber tracking, visualization of ODFs/fibers.

- SSPM: http://www.nitrc.org/projects/sspm/. Group analysis toolbox for carrying out voxel-based analysis and fiber tract–based analysis.

- NAMICDTIFIBER: https://www.nitrc.org/projects/namicdtifiber/. UNC/Utah NAMIC DTI Fiber Analysis Framework for carrying out fiber tract analysis.

4.10 Glossary

ADC: Apparent Diffusion Coefficient

DOT: Diffusion Orientation Transform

DPI: Diffusion Propagator Imaging

DSI: Diffusion Spectrum Imaging

DTI: Diffusion Tensor Imaging

DWI: Diffusion Weighted Imaging

EAP: Ensemble Average Propagator

FA: Fractional Anisotropy

GDTI: Generalized Diffusion Tensor Imaging

GFA: Generalized Fractional Anisotropy

HARDI: High Angular Resolution Diffusion Imaging

HOT: High-Order Tensor

MD: Mean Diffusivity

ODF: Orientation Distribution Function

PGSE: Pulsed Gradient Spin-Echo

QBI: Q-Ball Imaging

SH: Spherical Harmonic

SHORE: Simple Harmonic Oscillator Reconstruction and Estimation

SNR: Signal-To-Noise ratio

SPFI: Spherical Polar Fourier Imaging

Bibliography

[1] I. Aganj, C. Lenglet, G. Sapiro, E. Yacoub, K. Ugurbil, and N. Harel. Reconstruction of the orientation distribution function in single and multiple shell q-ball imaging within constant solid angle. *Magnetic Resonance in Medicine*, 2:554–566, 2010.

[2] S. Aja-Fernández, A. Tristán-Vega, and W.S. Hoge. Statistical noise analysis in GRAPPA using a parametrized noncentral Chi approximation model. *Magn Reson Med*, 65:1195–1206, 2011.

[3] Santiago Aja-Fernández, Carlos Alberola-López, and C-F Westin. Noise and signal estimation in magnitude MRI and Rician distributed images: A LMMSE approach. *Image Processing, IEEE Transactions on*, 17(8):1383–1398, 2008.

[4] D. C. Alexander, C. Pierpaoli, P. J. Basser, and J. C. Gee. Spatial transformations of diffusion tensor magnetic resonance images. *IEEE Transactions on Medical Imaging*, 20:1131–1139, 2001.

[5] Daniel C Alexander. Multiple-fiber reconstruction algorithms for diffusion MRI. *Annals of the New York Academy of Sciences*, 1064(1):113–133, 2005.

[6] A. W. Anderson. Theoretical analysis of the effects of noise on diffusion tensor imaging. *Magnetic Resonance in Medicine*, 46:1174–1188, 2001.

[7] A.W. Anderson. Measurement of fiber orientation distributions using high angular resolution diffusion imaging. *Magnetic Resonance in Medicine*, 54(5):1194–1206, 2005.

[8] E. Arias-Castro, J. Salmon, and R. Willett. Oracle inequalities and minimax rates for nonlocal means and related adaptive kernel-based methods. *SIAM J. Imaging Sciences*, 5:944–992, 2012.

[9] V. Arsigny, Fillard P, Pennec X, and Ayache N. Geometric means in a novel vector space structure on symmetric positive definite matrices. *SIAM. J. Matrix Anal. Appl.*, 29:328–347, 2007.

[10] Y. Assaf, R.Z. Freidlin, G.K. Rohde, and P.J. Basser. New modeling and experimental framework to characterize hindered and restricted water diffusion in brain white matter. *Magnetic Resonance in Medicine*, 52(5):965–978, 2004.

[11] Haz-Edine Assemlal, David Tschumperlé, and Luc Brun. Efficient and robust computation of PDF features from diffusion MR signal. *Medical Image Analysis*, 13:715–729, 2009.

[12] A. C. Atkinson and A. N. Donev. *Optimum Experimental Designs*. Clarendon Press, Oxford, 1992.

[13] P. J. Basser, J. Mattiello, and D. LeBihan. Estimation of the effective self-diffusion tensor from the NMR spin echo. *Journal of Magnetic Resonance Series B*, 103:247–254, 1994.

[14] P. J. Basser, J. Mattiello, and D. LeBihan. MR diffusion tensor spectroscopy and imaging. *Biophysical Journal*, 66:259–267, 1994.

[15] Peter J. Basser and Derek K. Jones. Diffusion-tensor MRI: Theory, experimental design and data analysis—A technical review. *NMR in Biomedicine*, 15(7-8):456–467, 2002.

[16] Peter J. Basser and Sinisa Pajevic. A normal distribution for tensor-valued random variables applications to diffusion tensor MRI. *IEEE Transactions on Medical Imaging*, 22:785–794, 2003.

[17] Peter J. Basser and Carlo Pierpaoli. Microstructural and physiological features of tissues elucidated by quantitative-diffusion-tensor MRI. *Journal of Magnetic Resonance*, 111:209–219, 1996.

[18] S. Becker, K. Tabelow, S. Mohammadi, N. Weiskopf, and J. Polzehl. Adaptive smoothing of multi-shell diffusion weighted magnetic resonance data by msPOAS. Technical report, Weierstrass-Institute, 2012.

[19] S. Becker, K. Tabelow, H.U. Voss, A. Anwander, R.M. Heidemann, and J. Polzehl. Position-orientation adaptive smoothing of diffusion weighted magnetic resonance data (POAS). *Med. Image Anal.*, 16:1142–1155, 2012.

[20] SMA Becker, K. Tabelow, S. Mohammadi, N. Weiskopf, and J. Polzehl. Adaptive smoothing of multi-shell diffusion weighted magnetic resonance data by msPOAS. *NeuroImage*, 95:90–105, 2014.

[21] SMA Becker, Karsten Tabelow, Henning U. Voss, Alfred Anwander, Robin M. Heidemann, and Jörg Polzehl. Position-orientation adaptive smoothing of diffusion weighted magnetic resonance data (POAS). *Medical Image Analysis*, 16(6):1142–1155, 2012.

[22] T.E. Behrens, M.W. Woolrich, M. Jenkinson, H. Johansen-Berg, R.G. Nunes, S. Clare, P.M. Matthews, J.M. Brady, and S.M. Smith. Characterization and propagation of uncertainty in diffusion-weighted mr imaging. *Magn Reson Med.*, 50:1077–1088, 2003.

[23] L. Beltrachini, N. von Ellenrieder, and C.H. Muravchik. Error bounds in diffusion tensor estimation using multiple-coil acquisition systems. *Magn Reson Imaging*, 31:1372–1383, 2013.

[24] M. A. Bernstein, K. F. King, and X. Zhou, editors. *Handbook of MRI Pulse Sequences*. Elsevier Academic Press, 2004.

[25] D. Le Bihan, E. Breton, D. Lallemand, P. Grenier, E. Cabanis, and M. Laval-Jeantet. MR imaging of intravoxel incoherent motions: application to diffusion and perfusion in neurologic disorders. *Radiology*, 161:401–407, 1986.

[26] Luke Bloy and Ragini Verma. Demons registration of high angular resolution diffusion images. In *Biomedical Imaging: From Nano to Macro, 2010 IEEE International Symposium on*, pages 1013–1016. IEEE, 2010.

[27] A. Buades, B. Coll, and J. M. Morel. Imaging denoising methods: A new nonlocal principle. *SIAM Review*, 52:113–147, 2010.

[28] P. Buhlmann and S. van de Geer. *Statistics for High-Dimensional Data: Methods, Theory and Applications*. Springer, 2011.

[29] Y. Cao, M.I. Miller, S. Mori, R.L. Winslow, and L. Younes. Diffeomorphic matching of diffusion tensor images. In *Computer Vision and Pattern Recognition Workshop, 2006. CVPRW'06. Conference on*, pages 67–67. IEEE, 2006.

[30] O. Carmichael, J. Chen, D. Paul, and J. Peng. Diffusion tensor smoothing through weighted Karcher means. *Electron. J. Stat.*, 7:1913–1956, 2013.

[31] E. Caruyer, J. Cheng, C. Lenglet, G. Sapiro, T. Jiang, R. Deriche, et al. Optimal design of multiple Q-shells experiments for Diffusion MRI. In *Computational Diffusion MRI - MICCAI Workshop*, 2011.

[32] Emmanuel Caruyer. Q-space diffusion MRI: Acquisition and signal processing. PhD thesis, University College London, UK, 2012.

[33] Can Ceritoglu, Kenichi Oishi, Xin Li, Ming-Chung Chou, Laurent Younes, Marilyn Albert, Constantine Lyketsos, Peter van Zijl, Michael I Miller, and Susumu Mori. Multi-contrast large deformation diffeomorphic metric mapping for diffusion tensor imaging. *Neuroimage*, 47(2):618–627, 2009.

[34] L.C. Chang, D.K. Jones, and C. Pierpaoli. Restore: Robust estimation of tensors by outlier rejection. *Magn. Reson. Med.*, 53:1088–1095, 2005.

[35] L.C. Chang, L. Walker, and C. Pierpaoli. Informed restore: A method for robust estimation of diffusion tensor from low redundancy datasets in the presence of physiological noise artifacts. *Magn. Reson. Med.*, 68:1654–1663, 2012.

[36] C. Chefd'hotel, D. Tschumperlé, R. Deriche, and O. Faugeras. Regularizing flows for constrained matrix-valued images. *Journal of Mathematical Imaging and Vision*, 20(1-2):147–162, 2004.

[37] Jian Cheng, Rachid Deriche, Tianzi Jiang, Dinggang Shen, and Pew-Thian Yap. Non-negative spherical deconvolution (NNSD) for estimation of fiber orientation distribution function in single-/multi-shell diffusion MRI. *NeuroImage*, 101:750–764, 2014.

[38] Jian Cheng, Rachid Deriche, Tianzi Jiang, Dinggang Shen, and Pew-Thian Yap. Non-Negative Spherical Deconvolution (NNSD) for Fiber orientation distribution function estimation. In *Computational Diffusion MRI and Brain Connectivity*, pages 81–93, 2014.

[39] Jian Cheng, Aurobrata Ghosh, Rachid Deriche, and Tianzi Jiang. Model-free, regularized, fast, and robust analytical orientation distribution function estimation. In *Medical Image Computing and Computer-Assisted Intervention — MICCAI*, 13(Pt 1):648–656, 2010.

[40] Jian Cheng, Aurobrata Ghosh, Tianzi Jiang, and Rachid Deriche. Model-free and analytical EAP reconstruction via spherical polar fourier diffusion MRI. In *Medical Image Computing and Computer-Assisted Intervention — MICCAI*, 13(Pt 1):590–597, 2010.

[41] Jian Cheng, Tianzi Jiang, and Rachid Deriche. Theoretical analysis and practical insights on EAP estimation via a unified HARDI framework. In *Computational Diffusion MRI — MICCAI Workshop*, 2011.

[42] Ming-Chang Chiang, Alex D. Leow, Andrea D. Klunder, Rebecca A. Dutton, Marina Barysheva, Stephen E. Rose, Katie L. McMahon, Greig I. De Zubicaray, Arthur W. Toga, and Paul M. Thompson. Fluid registration of diffusion tensor images using information theory. *Medical Imaging, IEEE Transactions on*, 27(4):442–456, 2008.

[43] R. D. Cook and S. Weisberg. *Residuals and Influence in Regression*. London: Chapman and Hall, 1982.

[44] Marc-Alexandre Côté, Gabriel Girard, Arnaud Boré, Eleftherios Garyfallidis, Jean-Christophe Houde, and Maxime Descoteaux. Tractometer: Towards validation of tractography pipelines. *Medical Image Analysis*, 17(7):844–857, 2013.

[45] O. Coulon, D. C. Alexander, and S. Arridge. Diffusion tensor magnetic resonance image regularization. *Medical Image Analysis*, 8:47–67, 2004.

[46] M. Descoteaux. High angular resolution diffusion MRI: From local estimation to segmentation and tractograpghy. PhD thesis, INRIA Sophia Antipolis, 2008.

[47] M. Descoteaux, E. Angelino, S. Fitzgibbons, and R. Deriche. Regularized, fast and robust analytical Q-ball imaging. *Magnetic Resonance in Medicine*, 58:497–510, 2007.

[48] M. Descoteaux, R. Deriche, D.L. Bihan, J.F. Mangin, and C. Poupon. Multiple q-shell diffusion propagator imaging. *Medical Image Analysis*, 2010.

[49] M. Descoteaux, R. Deriche, T. R. Knösche, and A. Anwander. Deterministic and probabilistic tractography based on complex fiber orientation distributions. *IEEE Transactions in Medical Imaging*, 28:269–286, 2008.

[50] M. Descoteaux, N. Wiest-Daesslé, S. Prima, C. Barillot, and R. Deriche. Impact of Rician adapted non-local means filtering on HARDI. *Medical Image Computing and Computer-Assisted Intervention*, 5242:122–130, 2008.

[51] M. Descoteaux, N. Wiest-Daessle, S. Prima, C. Barillot, and R. Deriche. Impact of Rician Adapted Non-local Means Filtering on HARDI. In *MICCAI*, 2008.

[52] T. Dhollander, W. Van Hecke, F. Maes, S. Sunaert, and P. Suetens. Spatial transformations of high angular resolution diffusion imaging data in q-space. In *Computational Diffusion MRI — MICCAI Workshop*, pages 73–83, 2010.

[53] C.Z. Di, C. M. Crainiceanu, B. S. Caffo, and Punjabi N. M. Multilevel functional principal component analysis. *Annals of Applied Statistics*, 3:458–488, 2009.

[54] I. L. Dryden, A. Koloydenko, and D. Zhou. Non-Euclidean statistics for covariance matrices, with applications to diffusion tensor imaging. *Annals of Applied Statistics*, 3:1102–1123., 2009.

[55] J. Du, A. Goh, S. Kushnarev, and A. Qiu. Geodesic regression on orientation distribution functions with its application to an aging study. *NeuroImage*, 87:416–426, 2014.

[56] Jia Du, Alvina Goh, and Anqi Qiu. Diffeomorphic metric mapping of high angular resolution diffusion imaging based on Riemannian structure of orientation distribution functions. *Medical Imaging, IEEE Transactions on*, 31(5):1021–1033, 2012.

[57] S. Durrleman, P. Fillard, X. Pennec, A. Trouvé, and N. Ayache. Registration, atlas estimation and variability analysis of white matter fiber bundles modeled as currents. *NeuroImage*, 55(3):1073–1090, 2011.

[58] P. Fillard, M. Descoteaux, A. Goh, S. Gouttard, B. Jeurissen, J. Malcolm, A. Ramirez-Manzanares, M. Reisert, K. Sakaie, F. Tensaouti, et al. Quantitative evaluation of 10 tractography algorithms on a realistic diffusion MR phantom. *NeuroImage*, 2011.

[59] P. Fillard, M. Descoteaux, A. Goh, S. Gouttard, B. Jeurissen, J. Malcolm, A. Ramirez-Manzanares, M. Reisert, K. Sakaie, F. Tensaouti, T. Yo, J. F. Mangin, and C. Poupon. Quantitative evaluation of 10 tractography algorithms on a realistic diffusion MR phantom. *NeuroImage*, 56:220–234, 2011.

[60] P.T. Fletcher and S. Joshi. Riemannian geometry for the statistical analysis of diffusion tensor data. *Signal Processing*, 87:250–262, 2007.

[61] P.T. Fletcher, S. Venkatasubramanian, and S. Joshi. The geometric median on Riemannian manifolds with application to robust atlas estimation. *NeuroImage*, 45:S143–S152, 2009.

[62] L. R. Frank. Characterization of anisotropy in high angular resolution dffusion-weighted MRI. *Magnetic Resonance in Medicine*, 47:1083–1099, 2002.

[63] O. Friman, G. Farneback, and C. F. Westin. A Bayesian approach for stochastic white matter tractography. *IEEE Trans Med Imaging*, 25:965–978, 2006.

[64] W. Gao, H.T. Zhu, and W. Lin. A unified optimization approach for diffusion tensor imaging technique. *Neuroimage*, 44(3):729–741, 2009.

[65] E. Garyfallidis, O. Ocegueda, D. Wassermann, and M. Descoteaux. Robust and efficient linear registration of white-matter fascicles in the space of streamlines. *NeuroImage*, 117, 2015.

[66] X. Geng, S. Gouttard, A. Sharma, H. Gu, M. Styner, and W. Lin. Quantitative tract-based white matter development from birth to age 2 years. *NeuroImage*, 61:542–557, 2012.

[67] X.J. Geng, T.J. Ross, H. Gu, W. Shin, W. Zhan, Y.P. Chao, C.P. Lin, N. Schuff, and Y. Yang. Diffeomorphic image registration of diffusion MRI using spherical harmonics. *Medical Imaging, IEEE Transactions on*, 30(3):747–758, 2011.

[68] I. Goh, C. Lenglet, P. M. Thompson, and R. Vidald. A nonparametric Riemannian framework for processing high angular resolution diffusion images and its applications to ODF-based morphometry. *Neuroimage*, 56:1181–1201, 2011.

[69] A.J. Goldsmith, C.M. Crainiceanu, B.S. Caffo, and D. Reich. Penalized functional regression analysis of white-matter tract profiles in multiple sclerosis. *NeuroImage*, 57:431–439, 2011.

[70] C. B. Goodlett, P. T. Fletcher, J. H. Gilmore, and G. Gerig. Group analysis of DTI fiber tract statistics with application to neurodevelopment. *NeuroImage*, 45:S133–S142, 2009.

[71] S. Greven, C. Crainiceanu, B. Caffo, and D. Reich. Longitudinal functional principal components analysis. *Electronic Journal of Statistics*, 4:1022–1054, 2010.

[72] P. Hagmann, L. Jonasson, P. Maeder, J.P. Thiran, Van J. Wedeen, and R. Meuli. Understanding Diffusion MR Imaging Techniques: From scalar diffusion-weighted imaging to diffusion tensor imaging and beyond. *RadioGraphics*, 26:S205–S223, 2006.

[73] Xin Hong, Lori R. Arlinghaus, and Adam W. Anderson. Spatial normalization of the fiber orientation distribution based on high angular resolution diffusion imaging data. *Magnetic Resonance in Medicine*, 61(6):1520–1527, 2009.

[74] T. Hosey, G. Williams, and R. Ansorge. Inference of multiple fiber orientations in high angular resolution diffusion imaging. *Magnetic Resonance in Medicine*, 54(6):1480–1489, 2005.

[75] Z. W. Hua, D. B. Dunson, J. H. Gilmore, M. Styner, and H. T. Zhu. Semiparametric Bayesian local functional models for diffusion tensor tract statistics. *NeuroImage*, 63:460–674, 2012.

[76] P. J. Huber. *Robust Statistics*. Wiley Series in Probability and Statistics, 1981.

[77] S. Jbabdi and T. E Behrens. Long-range connectomics. *Annals of the New York Academy of Sciences*, 1305(1):83–93, 2013.

[78] S. Jbabdi and H. Johansen-Berg. Tractography: Where do we go from here? *Brain Connect.*, 1:169–183, 2011.

[79] Y. Jin, Y. Shi, L. Zhan, B. A. Gutman, G. I. de Zubicaray, K. L. McMahon, M. J. Wright, A. W. Toga, and P. M. Thompson. Automatic clustering of white matter fibers in brain diffusion MRI with an application to genetics. *NeuroImage*, 100:75–90, 2014.

[80] D. K. Jones, M. R. Symms, M. Cercignani, and R. J. Howard. The effect of filter size on VBM analyses of DT-MRI data. *NeuroImage*, 26:546–554, 2005.

[81] D.K. Jones. *Diffusion MRI: Theory, Methods, and Applications*. Oxford University Press, Oxford, New York, 2011.

[82] D.K. Jones and M. Cercignani. Twenty-five pitfalls in the analysis of diffusion MRI data. *NMR Biomed*, 23:803–820, 2010.

[83] D.K. Jones, T.R. Knösche, and R. Turner. White matter integrity, fiber count, and other fallacies: The do's and don'ts of diffusion MRI. *Neuroimage*, 73:239–254, 2013.

[84] V. Katkovnik, A. Foi, K. Egiazarian, and J. Astola. From local kernel to nonlocal multiple-model image denoising. *International Journal of Computer Vision*, 86:1–32, 2010.

[85] Y. Kim, P.M. Thompson, A.W. Toga, L. Vese, and L. Zhan. HARDI denoising: Variational regularization of the spherical apparent diffusion coefficient SADC. *Information Processing in Medical Imaging*, pages 515–527, 2009.

[86] V. Koltchinskii, L. Sakhanenko, and S. Cai. Integral curves of noisy vector fields and statistical problems in diffusion tensor imaging: Nonparametric kernel estimation and hypotheses testing. *The Annals of Statistics*, pages 1576–1607, 2007.

[87] B.W. Kreher, J.F. Schneider, I. Mader, E. Martin, J. Hennig, and K.A. Il'yasov. Multitensor approach for analysis and tracking of complex fiber configurations. *Magnetic Resonance in Medicine*, 54(5):1216–1225, 2005.

[88] P.W. Kuchel, A. Coy, and P. Stilbs. NMR "diffusion-diffraction" of water revealing alignment of erythrocytes in a magnetic field and their dimensions and membrane transport characteristics. *Magnetic Resonance in Medicine*, 37(5):637–643, 1997.

[89] Mariana Lazar. Mapping brain anatomical connectivity using white matter tractography. *NMR in Biomedicine*, 23(7):821–835, 2010.

[90] Y. Li, H. Zhu, D. Shen, W. Lin, J. H. Gilmore, and J. G. Ibrahim. Multiscale adaptive regression models for neuroimaging data. *Journal of the Royal Statistical Society: Series B*, 73:559–578, 2011.

[91] C. Liu, R. Bammer, B. Acar, and M. E. Moseley. Characterizing non-Gaussian diffusion by using generalized diffusion tensors. *Magnetic Resonance in Medicine*, 51:925–937, 2004.

[92] C. Liu, R. Bammer, and M. E. Moseley. Generalized diffusion tensor imaging (GDTI): A method for characterizing and imaging diffusion anisotropy caused by non-Gaussian diffusion. *Israel Journal of Chemistry*, 43:145–154, 2003.

[93] M. Liu, B.C. Vemuri, and R. Deriche. A robust variational approach for simultaneous smoothing and estimation of DTI. *NeuroImage*, 67:33–41, 2013.

[94] J-F. Mangin, P. Fillard, Y. Cointepas, D. Le Bihan, V. Frouin, and C. Poupon. Toward global tractography. *NeuroImage*, 80:290–296, 2013.

[95] J. Marcinkiewicz. Sur une propriété de la loi de gauss. *Mathematische Zeitschrift*, 44(1):612–618, 1939.

[96] II. Maximov, F. Grinberg, and N.J. Shah. Robust tensor estimation in diffusion tensor imaging. *J. Magn. Reson.*, 213:136–144, 2011.

[97] T. McGraw, B.C. Vemuri, Y Chen, M. Rao, and T. Mareci. DT-MRI denoising and neuronal fiber tracking. *Medical Image Analysis*, 8(2):95–111, 2004.

[98] S. Merlet. Compressive sensing in diffusion MRI. PhD thesis, University of Nice-Sophia Antipolis, 2013.

[99] C.C. Moraga, C. Lenglet, R. Deriche, and J. Ruiz-Alzola. A Riemannian approach to anisotropic filtering of tensor fields. *Signal Process*, 87:217–352, 2007.

[100] S. Mori and P. C. M. van Zijl. Fiber tracking: principles and strategies: A technical review. *NMR in Biomedicine*, 15:468–480, 2002.

[101] S. Mori, S. Wakana, P.C.M. van Zijl, and L.M. Nagae-Poetscher. *MRI Atlas of Human White Matter*. Am Soc Neuroradiology, 2005.

[102] M.E. Moseley, Y. Cohen, J. Mintorovitch, J. Kucharczyk, J. Tsuruda, P. Weinstein, and D. Norman. Evidence of anisotropic self-diffusion. *Radiology*, 176:439–445, 1990.

[103] L.J. O'Donnell, C.-F. Westin, and A.J. Golby. Tract-based morphometry for white matter group analysis. *Neuroimage*, 45:832–844, 2009.

[104] E. Özarslan, CG Koay, T. Shepherd, S. Blackband, and PJ Basser. Simple harmonic oscillator based reconstruction and estimation for three-dimensional q-space MRI. In *International Society for Magnetic Resonance in Medicine (ISMRM)*, page 1396, 2009.

[105] E. Özarslan, T. M. Shepherd, B. C. Vemuri, S. J. Blackband, and T. H. Mareci. Resolution of complex tissue microarchitecture using the diffusion orientation transform (DOT). *NeuroImage*, 31:1086–1103, 2006.

[106] E. Özarslan, B.C. Vemuri, and T.H. Mareci. Generalized scalar measures for diffusion MRI using trace, variance, and entropy. *Magnetic Resonance in Medicine*, 53(4):866–876, 2005.

[107] N. G. Papadakis, D. Xing, C.L.H. Huang, L.D. Hall, and T. A. Carpenter. A comparative study of acquisition schemes for diffusion tensor imaging using MRI. *Journal of Magnetic Resonance*, 137(1):67–82, 1999.

[108] X. Pennec, P. Fillard, and N. Ayache. A Riemannian framework for tensor computing. *International Journal of Computer Vision*, 66:41–66, 2006.

[109] C. Pierpaoli and P. Basser. Toward a quantitative assessment of diffusion anisotropy. *Magnetic Resonance in Medicine*, 36:893–906, 1996.

[110] J. Polzehl and V. G. Spokoiny. Adaptive weights smoothing with applications to image restoration. *J. R. Statist. Soc. B*, 62:335–354, 2000.

[111] C. Poupon. Détection des faisceaux de fibres de la substance blanche pour l'étude de la connectivité anatomique cérébrale. PhD thesis, Ecole Nationale Supérieure des Télécommunications, December 1999.

[112] David Raffelt, J. Tournier, Jurgen Fripp, Stuart Crozier, Alan Connelly, Olivier Salvado, et al. Symmetric diffeomorphic registration of fibre orientation distributions. *NeuroImage*, 56(3):1171–1180, 2011.

[113] A. Raj, C. Hess, and P. Mukherjee. Spatial HARDI: Improved visualization of complex white matter architecture with Bayesian spatial regularization. *Neuroimage*, 54:396–409, 2011.

[114] J. O. Ramsay and B. W. Silverman. *Functional Data Analysis*. Springer-Verlag, New York, 2005.

[115] S. Rao, J. G. Ibrahim, G. Cheng, P.T. Yap, and H.T. Zhu. SR-HARDI: Spatially regularizing high angular resolution diffusion imaging. *Journal of Computational and Graphical Statistics*, page in press, 2016.

[116] M. Reisert, I. Mader, C. Anastasopoulos, M. Weigel, S. Schnell, and V. Kiselev. Global fiber reconstruction becomes practical. *NeuroImage*, 54:955–962, 2011.

[117] P.T. Reiss and R.T. Ogden. Functional generalized linear models with images as predictors. *Biometrics*, 66:61–69, 2010.

[118] L. Sakhanenko. Numerical issues in estimation of integral curves from noisy diffusion tensor data. *Statistics & Probability Letters*, 82(6):1136–1144, 2012.

[119] A. Schwartzman. Random ellipsoids and false discovery rates: Statistics for diffusion tensor imaging data. Ph.D. thesis, Stanford University, 2006.

[120] A. Schwartzman, W. Mascarenhas, and J. E. Taylor. Inference for eigenvalues and eigenvectors of Gaussian symmetric matrices. *Ann. Statist.*, 36:1423–1431, 2008.

[121] X. Shi, M. Styner, J. Lieberman, J. G. Ibrahim, W. Lin, and H. T. Zhu. Intrinsic regression models for manifold-valued data. *Medical Image Computing and Computer-Assisted Intervention–MICCAI*, pages 192–199, 2009.

[122] J. Sijbers, A.J. den Dekker, P. Scheunders, and D. Van Dyck. Maximum-likelihood estimation of Rician distribution parameters. *IEEE Transactions on Image Processing*, 17:357–361, 1998.

[123] S. Skare, M. Hedehus, M. E Moseley, and T. Q. Li. Condition number as a measure of noise performance of diffusion tensor data acquisition schemes with MRI. *Journal of Magnetic Resonance*, 147(2):340–352, 2000.

[124] S. M. Smith, M. Jenkinson, H. Johansen-Berg, D. Rueckert, T. E. Nichols, C. E. Mackay, K. E. Watkins, O. Ciccarelli, M.Z. Cader, P.M. Matthews, and T. E. Behrens. Tract-based spatial statistics: Voxelwise analysis of multi-subject diffusion data. *NeuroImage*, 31:1487–1505, 2006.

[125] L. Snook, C. Plewes, and C. Beaulieu. Voxel based versus region of interest analysis in diffusion tensor imaging of neurodevelopment. *NeuroImage*, 34:243–252, 2007.

[126] E.O. Stejskal and J.E. Tanner. Spin diffusion measurements: Spin echoes in the presence of a time-dependent field gradient. *The Journal of Chemical Physics*, 42:288–292, 1965.

[127] Karsten Tabelow, Jörg Polzehl, Vladimir Spokoiny, and Henning U. Voss. Diffusion tensor imaging: Structural adaptive smoothing. *NeuroImage*, 39(4):1763–1773, 2008.

[128] Robert Tibshirani. Regression shrinkage and selection via the lasso. *J. Roy. Statist. Soc. Ser. B*, 58(1):267–288, 1996.

[129] J. Tournier, F. Calamante, and A. Connelly. Robust determination of the fibre orientation distribution in diffusion MRI: Non-negativity constrained super-resolved spherical deconvolution. *NeuroImage*, 35(4):1459–1472, 2007.

[130] J.D. Tournier, F. Calamante, D. Gadian, and A. Connelly. Direct estimation of the fiber orientation density function from diffusion-weighted MRI data using spherical deconvolution. *NeuroImage*, 23:1176–1185, 2004.

[131] A. Tristán-Vega, C.F. Westin, and S. Aja-Fernández. Estimation of fiber orientation probability density functions in high angular resolution diffusion imaging. *NeuroImage*, 47(2):638–650, 2009.

[132] Antonio Tristán-Vega and Santiago Aja-Fernández. DWI filtering using joint information for DTI and HARDI. *Medical Image Analysis*, 14:205–218, 2010.

[133] Antonio Tristán-Vega, Carl-Fredrik Westin, and Santiago Aja-Fernández. A new methodology for the estimation of fiber populations in the white matter of the brain with the Funk-Radon transform. *NeuroImage*, 49:1301–1315, 2010.

[134] D. S. Tuch. Q-ball imaging. *Magnetic Resonance in Medicine*, 52:1358–1372, 2004.

[135] D.S. Tuch. *Diffusion MRI of Complex Tissue Structure*. PhD thesis, MIT, 2002.

[136] D.S. Tuch, T.G. Reese, M. R. Wiegell, N. Makris, J.W. Belliveau, and Van J. Wedeen. High angular resolution diffusion imaging reveals intravoxel white matter fiber heterogeneity. *Magnetic Resonance in Medicine*, 48:577–582, 2002.

[137] D.S. Tuch, R.M. Weisskoff, J.W. Belliveau, and V.J. Wedeen. High angular resolution diffusion imaging of the human brain. In *Proceedings of the 7th Annual Meeting of ISMRM*, 1999.

[138] Audrey R Verde, Francois Budin, Jean-Baptiste Berger, Aditya Gupta, Mahshid Farzinfar, Adrien Kaiser, Mihye Ahn, Hans Johnson, Joy Matsui, Heather C Hazlett, et al. UNC-Utah NA-MIC framework for DTI fiber tract analysis. *Frontiers in Neuroinformatics*, 7, 2013.

[139] E.A.H. von dem Hagen and R. M. Henkelman. Orientational diffusion reflects fiber structure within a voxel. *Magnetic Resonance in Medicine*, 48:454–459, 2002.

[140] D. Wassermann, Y. Rathi, S. Bouix, M. Kubicki, R. Kikinis, M. Shenton, and C. Westin. White matter bundle registration and population analysis based on Gaussian processes. In *Information Processing in Medical Imaging*, pages 320–332. Springer, 2011.

[141] V. J. Wedeen, D.L. Rosene, R. Wang, G. Dai, F. Mortazavi, P. Hagmann, J. H. Kaas, and W.Y.I. Tseng. The geometric structure of the brain fiber pathways. *Science*, 335:1628–1634, 2012.

[142] Van J. Wedeen, P. Hagmann, W.Y. I. Tseng, T. G. Reese, and R. M. Weisskoff. Mapping complex tissue architecture with diffusion spectrum magnetic resonance imaging. *Magnetic Resonance In Medicine*, 54:1377–1386, 2005.

[143] C. F. Westin, S.E. Maier, H. Mamata, A. Nabavi, F.A. Jolesz, and R. Kikinis. Processing and visualization for diffusion tensor MRI. *Medical Image Analysis*, 6:93–108, 2002.

[144] B. Whitcher, J. J. Wisco, N. Hadjikhani, and D. S. Tuch. Statistical group comparison of diffusion tensors via multivariate hypothesis testing. *Magnetic Resonance in Medicine*, 57:1065–1074, 2007.

[145] N. Wiest-Daesslé, S. Prima, P. Coupé, S. P. Morrissey, and C. Barillot. Non-local means variants for denoising of diffusion-weighted and diffusion tensor MRI. In *Medical Image Computing and Computer-Assisted Intervention–MICCAI 2007*, pages 344–351. Springer, 2007.

[146] K. J. Worsley, J. E. Taylor, F. Tomaiuolo, and J. Lerch. Unified univariate and multivariate random field theory. *NeuroImage*, 23:189–195, 2004.

[147] Y.C. Wu and A. L. Alexander. Hybrid diffusion imaging. *NeuroImage*, 36:617–629, 2007.

[148] Yu-Chien Wu, Aaron S. Field, and Andrew L. Alexander. Computation of diffusion function measures in q-space using magnetic resonance hybrid diffusion imaging. *IEEE Transactions On Medical Imaging*, 27:858–865, 2008.

[149] P. Yap, G. Wu, H.T. Zhu, W. Lin, and D. Shen. Timer: Tensor image morphing for elastic registration. *NeuroImage*, 47:549–563, 2009.

[150] Pew-Thian Yap, Yasheng Chen, Hongyu An, Yang Yang, John H Gilmore, Weili Lin, and Dinggang Shen. Sphere: Spherical harmonic elastic registration of HARDI data. *NeuroImage*, 55(2):545–556, 2011.

[151] L. P. Yaroslavsky. *Digital Picture Processing*, volume 9 of Springer Ser. Inform. Sci. Springer, Berlin, 1985.

[152] A. Yendikia, K. Koldewynb, S. Kakunooria, N. Kanwisherb, and B. Fischla. Spurious group differences due to head motion in a diffusion MRI study. *NeuroImage*, 88:79–90, 2014.

[153] B.T.T. Yeo, T. Vercauteren, P. Fillard, J.M. Peyrat, X. Pennec, P. Golland, N. Ayache, and O. Clatz. DT-REFinD: Diffusion tensor registration with exact finite-strain differential. *Medical Imaging, IEEE Transactions on*, 28(12):1914–1928, 2009.

[154] T. Yu and P. Li. Spatial shrinkage estimation of diffusion tensors on diffusion weighted imaging data. *Journal of the American Statistical Association*, 108:864–875, 2013.

[155] T. Yu, C. Zhang, A. L. Alexander, and R. J. Davidson. Local tests for identifying anisotropic diffusion areas in human brain with DTI. *Annals of Applied Statistics*, 7:201–225, 2013.

[156] Y. Yuan, J. H. Gilmore, X. Geng, M. Styner, K. Chen, J. L. Wang, and H. T. Zhu. FMEM: Functional mixed effects modeling for the analysis of longitudinal white matter tract data. *NeuroImage*, 84:753–764, 2014.

[157] Y. Yuan, H. Zhu, W. Lin, and J. S. Marron. Local polynomial regression for symmetric positive definite matrices. *Journal of Royal Statistical Society B*, 74:697–719, 2012.

[158] Y. Yuan, H. T. Zhu, J. Ibrahim, W.L. Lin, and B.G. Peterson. A note on bootstrapping uncertainty of diffusion tensor parameters. *IEEE Transactions on Medical Imaging*, 27:1506–1514, 2008.

[159] Y. Yuan, H.T. Zhu, M. Styner, J. H. Gilmore, and J. S. Marron. Varying coefficient model for modeling diffusion tensors along white matter bundles. *Annals of Applied Statistics*, 7:102–125, 2013.

[160] P. A. Yushkevich, H. Zhang, T.J. Simon, and J. C. Gee. Structure-specific statistical mapping of white matter tracts. *Neuroimage*, 41:448–461, 2008.

[161] Pei Zhang, Marc Niethammer, Dinggang Shen, and Pew-Thian Yap. Large deformation diffeomorphic registration of diffusion-weighted images with explicit orientation optimization. In *Medical Image Computing and Computer-Assisted Intervention–MICCAI 2013*, pages 27–34. Springer, 2013.

[162] H. T. Zhu, Y. S. Cheng, J. G. Ibrahim, Y. M. Li, C. Hall, and W. L. Lin. Intrinsic regression models for positive definitive matrices with applications in diffusion tensor images. *Journal of the American Statistical Association*, 104:1203–1212, 2009.

[163] H. T. Zhu, R. Z. Li, and L. L. Kong. Multivariate varying coefficient model for functional responses. *Annals of Statistics*, 40:2634–2666, 2012.

[164] H. T. Zhu, M. Styner, N. S. Tang, Z. X. Liu, W. L. Lin, and J.H. Gilmore. FRATS: Functional regression analysis of DTI tract statistics. *IEEE Transactions on Medical Imaging*, 29:1039–1049, 2010.

[165] H. T. Zhu, H. P. Zhang, J. G. Ibrahim, and B. G. Peterson. Statistical analysis of diffusion tensors in diffusion-weighted magnetic resonance image data (with discussion). *Journal of the American Statistical Association*, 102:1085–1102, 2007.

[166] H.T. Zhu, L. Kong, R. Li, M. Styner, G. Gerig, W. Lin, and J. H. Gilmore. Fadtts: functional analysis of diffusion tensor tract statistics. *NeuroImage*, 56:1412–1425, 2011.

[167] H.T. Zhu, Y.M. Li, J. G. Ibrahim, X.Y. Shi, and B.S. Peterson. Rician regression models for magnetic resonance images. *Journal of the American Statistical Association*, 104:623–637, 2009.

[168] V. Zipunnikov, B.S. Caffo, D.M. Yousem, C. Davatzikos, B.S. Schwartz, and C. Crainiceanu. Functional principal components model for high-dimensional brain imaging. *NeuroImage*, 58:772–784, 2011.

[169] O. Zvitia, A. Mayer, R. Shadmi, S. Miron, and H. K. Greenspan. Co-registration of white matter tractographies by adaptive-mean-shift and Gaussian mixture modeling. *IEEE Transactions on Medical Imaging*, 29(1):132–145, 2010.

5

A Tutorial for Multisequence Clinical Structural Brain MRI

Ciprian Crainiceanu and Elizabeth M. Sweeney

Johns Hopkins University

Ani Eloyan

Brown University

Russell T. Shinohara

University of Pennsylvania

CONTENTS

5.1	Introduction	....	110
	5.1.1	What Are These Images?	112
	5.1.2	How Can We Handle sMRI?	113
	5.1.3	What Are Some Major Pitfalls When Starting Working on sMRI? ..	114
5.2	Data Structure and Intuitive Description of Associated Problems		114
5.3	Acquisition and Reconstruction		115
5.4	Preprocessing		116
	5.4.1	Inhomogeneity Correction	117
		5.4.1.1 Concepts	118
		5.4.1.2 Practical Approaches, Software, and Application to Data .	118
	5.4.2	Skull Stripping	119
		5.4.2.1 Concepts	119
		5.4.2.2 Practical Approaches, Software, and Application to Data .	119
	5.4.3	Interpolation	119
		5.4.3.1 Concepts	120
		5.4.3.2 Practical Approaches, Software, and Application to Data .	121
	5.4.4	Spatial Registration	121
		5.4.4.1 Concepts	121
		5.4.4.2 Practical Approaches, Software, and Application to Data .	123
	5.4.5	Intensity Normalization	126
		5.4.5.1 Concepts	127
		5.4.5.2 Practical Approaches, Software, and Application to Data .	127
5.5	Analysis		129
	5.5.1	Lesion Segmentation	129
	5.5.2	Lesion Mapping	130
	5.5.3	Longitudinal and Cross Sectional Intensity Analysis	131
5.6	Conclusions		133
	Bibliography		133

5.1   Introduction

High-resolution structural magnetic resonance imaging (sMRI) is used extensively in clinical practice, as it provides detailed anatomical information of the living organism, is sensitive to many pathologies, and assists in the diagnosis of disease (1). Applications of sMRI cover essentially every part of the human body from toes to brain and a wide variety of diseases from stroke, cancer, and multiple sclerosis (MS), to internal bleeding and torn ligaments. Since the introduction of MRI in the 1980s, the noninvasive nature of the technique, the continuously improving resolution of images, and the wide availability of MRI scanners have made sMRI instantly recognizable in the popular literature (30). Indeed, when one is asked to have an MRI in a clinical context it is almost certainly an sMRI. These images are fundamentally different from functional MRI (fMRI) in size, complexity, measurement target, type of measurement, and intended use. While fMRI aims to study brain activity, sMRI reveals anatomical information. This distinction is important as the scientific problems and statistical techniques for fMRI and sMRI analysis differ greatly (35), yet confusion between the two continues to exist in the statistical literature and among reviewers. Despite the enormous practical importance of sMRI, few biostatisticians have made research contributions in this field. This may be due to the subtle aspects of sMRI, the relatively steep learning curve, and the lack of contact between biostatisticians and the scientists working in clinical neuroimaging. Our goal is reduce the price of entry, accelerate learning, and provide the information required to progress from novice to initiated sMRI researcher.

This chapter is designed to provide a tutorial for sMRI research, introduce some major unsolved scientific problems in brain imaging of patients with neurological disorders and diseases, and describe the important technical problems associated with data analysis. Image acquisition and preprocessing, usually addressed through preprocessing pipelines, will also be discussed, as in our experience, it has been impossible to separate the image preprocessing pipeline from later analysis. The paper is accompanied by sMRI for two subjects with multiple sclerosis at two visits together with the associated R code that can be used to open, visualize, and conduct small statistical analyses. These studies have been pre-processed using the preprocessing steps outlined in this chapter. Data and software relevant to this chapter can be found at http://biostat.jhsph.edu/ ccrainic/software.html.

sMRI plays a fundamental role in the diagnosis, management, and study of MS, stroke, cancer, traumatic brain injury, and Alzheimer's disease (AD). There are fundamental differences between identifying affected tissues in different neurological disorders, but our experience has centered on the development of translational methods for sMRI in MS. Thus, in this chapter, we focus mainly on MS as an example to demonstrate preprocessing and analysis techniques and problems that extend across pathologies.

An sMRI study of the brain typically consists of several different sMRI sequences, often including T1-weighted (T1), T2-weighted (T2), T2-weighted Fluid Attenuated Inversion Recovery (FLAIR), and Proton Density-weighted (PD) imaging. Other sequences are continuously being researched and may become standard in future sMRI studies. Moreover, the type of magnet (1.5T, 3T, or 7T), the brand of MRI scanner, and the choice of scanning parameters can induce major differences between images, even if they are of the same sequence type. We refer to the sMRI collection of two or more sequences as multi-sequence sMRI. We will distinguish multi-sequence sMRI from multi-modality imaging, which refers to the combination of at least two different types of imaging, for example sMRI and Computed Tomography (CT), or CT and Positron Emitted Tomography (PET).

From a data perspective, every sequence yields a three-dimensional array, with each entry representing a voxel (three-dimensional pixel). The size of the voxels depends on the

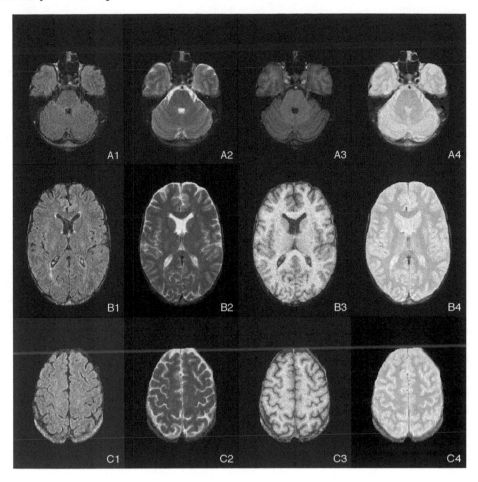

FIGURE 5.1
Multi-sequence MRI data for one subject. Three axial slices are shown on each row (letters A, B, C indicate a different slice going from the inferior (A) to superior (C) of the brain) indicating FLAIR (A1, B1, C1), T2 (A2, B2, C2), T1 (A3, B3, C3), and PD (A4, B4, C4). A small MS white matter lesion is visible in the A-slice images. Some larger MS lesions are visible closer to the ventricle in the B-slice images.

acquisition parameters and provides the resolution of the image. Figure 5.1 displays data from a standard sMRI sequence protocol for three slices shown in the three rows. The voxel size for these images has been interpolated to $1 \times 1 \times 1$ mm (interpolation of sMRI is discussed in detail in Section 5.4.3). Slices are displayed, moving from the inferior to the superior brain and are labeled A, B, and C, respectively. Each column corresponds to a different sequence: FLAIR (A1, B1, C1), T2 (A2, B2, C2), T1 (A3, B3, C3), and PD (A4, B4, C4). An intuitive way to think about the different sequences is that they are slices through the brain seen through different filters. Making such plots in R (44) is relatively easy using the `oro.nifti` R package (74). After setting the working directory to the location of the compressed FLAIR volume, the following lines of code will load the volume and plot one axial slice of the FLAIR image:

```
library(oro.nifti)
flair <- readNIfTI('FLAIRnorm.nii.gz', reorient=TRUE)
image(flair[,,50])
```

A plot of the sagittal, coronal and axial view from the slices [111, 132, 102] ([sagittal slice, coronal slice, axial slice]) can be obtained using the command `orthographic`:

```
orthographic(flair, xyz = c(111, 132, 102))
```

While R packages may change, improve, or become obsolete, we currently like `oro.nifti`, because it is relatively easy to use and allows us to work directly with compressed files. This is a big advantage when working on large studies and/or transferring files. Once the magic of staring of the pictures is gone, some important technical questions remain. Most importantly: 1) What are these images? 2) How can we model sMRI? and 3) What are some major pitfalls when starting working on sMRI? We now address these questions.

5.1.1   What Are These Images?

At the most basic level, every sMRI volume is a 3-dimensional (3D) array, with dimensions determined according to the acquisition parameters. For example, the FLAIR volume shown in Figure 5.1 is stored as a 3D array and the 50th axial slice (moving from the inferior to the superior of the brain) is stored in `flair[,,50]`. This FLAIR image is interpolated to a voxel size of $1 \times 1 \times 1$ mm and contains $182 \times 218 \times 182$ voxels, or about 7 million voxels. As shown in Figure 5.1, this MRI study contains 4 sequences, for a total of around 30 million voxels for the entire study. In contrast, fMRI are 4-dimensional (4D) matrices, where time is the fourth dimension. Similarly, Dynamic Contrast Enhanced MRI (DCE-MRI) (55, 56) is also 4D, though here we focus on 3D sMRI.

Figure 5.2 displays an axial slice of the T1 image obtained post-gadolinium injection. A gadolinium chelate is a paramagnetic substance that can be injected in the blood stream and makes blood appear hyperintense in the T1 image. When a sequence of such images is taken before and after injection, for the purpose of observing and quantifying the blood dynamics into the brain, the sequence is referred to as DCE-MRI. Figure 5.2 shows one time point from a DCE-MRI. Alternatively, only one post-contrast injection image may be acquired and this is referred to as a post-gad T1 image. The image contains an MS lesion surrounded by a hyperintense ring, which indicates blood with a higher concentration of

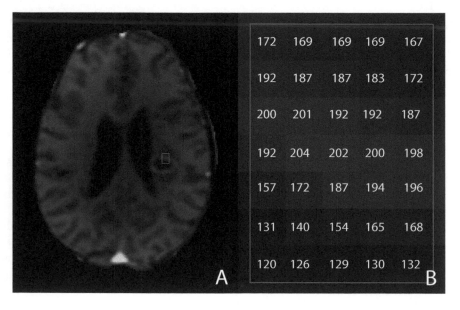

FIGURE 5.2
Dynamic Contrast Enhancing (DCE) volume after gadolinium injection.

gadolinium. A small red box in Figure 5.2A is magnified in Figure 5.2B. Each voxel in the magnification contains both the intensity and the associated numerical data. For example, the largest value in this rectangle is 204, and corresponds to the most hyperintense shade. Images are just representations of data using a particular mapping from real numbers to a gray (or color) scale. Simple manipulations of this mapping can lead to dramatic changes in contrasts, at least in the way they appear to the human eye. The representation appears to be reasonable as the correspondence between known and represented anatomy and pathology is remarkable. Surprisingly, even if one tried to cluster intensities of voxels across the entire brain, there is overlap between various tissue classes, simply because the same intensity can easily appear in two different parts of the brain. For example, there are many areas in the normally appearing white matter that have roughly the same intensity with the ring around the lesion.

A natural question then becomes: what are the data units and how comparable are these units across subjects, visits, and studies? Unfortunately, standard sMRI are unitless. Thus, the size of the units is comparable within the same sequence, but taking the difference between two sequences of the same type is meaningless. Thus, before conducting any sort of analysis on these images, image intensity normalization is a crucial step. We will discuss methods for intensity normalization in Section 5.4.4.

5.1.2 How Can We Handle sMRI?

An important characteristic of brain imaging data is that it is big. In most computing environments, loading into memory even a small sample of sMRI studies is not recommended. This raises questions about data storage and handling for conducting analyses where data can be accessed one or a few images at a time. We recommend storing data using a folder structure of the type:

`D:/study_type/subject_id/visit_k/sMRIsequence_name.nii.gz`

A separate file containing subject identifiers, visit information, covariates and health outcomes can be stored as a master file. Some researchers prefer to have the visit identifier and the subject identifier in the file name to avoid confusion. Regardless of preferences, careful naming and organization of the data is a crucial step towards more sophisticated analyses. As a basic rule, for population-level analyses the naming system and directory structure must be consistent, script-friendly, intuitive, and documented.

The compressed files are quite small (around 3Mb), though loading and de-compressing hundreds of such files in the computer memory can slow down and even crash computers. We have found that the most robust approach is to upload the minimum number of images necessary for performing the analysis. For example, if one is interested in calculating the mean FLAIR image of spatially registered images, then one can simply open one image at a time and use an iterative formula for calculating the mean. If $\widehat{\mu}_n$ is an estimator of the mean using the first n observations Y_i, $i \geq 1$, then $\widehat{\mu}_n = \frac{n-1}{n}\widehat{\mu}_{n-1} + \frac{1}{n}Y_n$. Similar formulas exist for more complex operations, such as sequential updating of covariance operators.

The R computational environment is familiar to biostatisticians and the R environment has many packages designed for neuroimaging. However, neuroimaging has been developed primarily outside of statistics, with a distinctly different software and analytic culture. Indeed, in neuroimaging MATLAB®, Python (69), and C are used extensively. Learning these languages is especially useful for direct collaboration. There is an extensive collection of useful neuroimaging software; in this chapter we will cover those which we have found to work particularly well. For example, Medical Image Processing, Analysis, and Visualization (MIPAV) is powerful for data visualization, exploratory analysis, spatial inhomogeneity corrections, segmentation, spatial registration, and many other tasks.

5.1.3   What Are Some Major Pitfalls When Starting Working on sMRI?

The biggest mistake in neuroimaging analysis is to look for an application that illustrates a particular biostatistical modeling idea. A "method backwards" approach is problematic in any discipline, but it is especially dangerous in imaging. A reasonably deep understanding of imaging, image preprocessing, and imaging literature can save time, avoid "wheel re-invention", and maintain focus on scientifically relevant and important problems. Thus, we advocate a "problem forward" approach, where biostatisticians work directly with collaborators, learn about the details of data acquisition, and identify the most important problems. Like every technology-intensive field, imaging requires developing a basic set of skills that allows us to understand, formulate, and help solve the most important problems. While biostatisticians "get to play in everyone else's backyard" (John Tukey, Bell Labs, Princeton University), there must be rules about "playing". We have found the neuroscientific community to be incredibly welcoming and open to informed biostatistical ideas and approaches, when biostatisticians are open to learning the necessary background for working with neuroimaging data.

Another pitfall is to not understand the dangers that lurk in neuroimaging. Here, we warn of a few. First, there is much biological variation between brains and, in addition, neurological disease can deform the brain quite dramatically. Therefore, methods that are reasonably well developed for healthy brains tend to fail badly on brains that exhibit various pathologies. Second, magnetic coils create spatial inhomogeneities that could be quite large and vary with the subjects and time of the scan. Spatial inhomogeneity corrections, such as N3 (60) or N4ITK (67), work quite well and are standard in most imaging processing platforms; however, subtle bias fields remain and can strongly affect quantitative analyses. We discuss in detail the inhomogeneity corrections in Section 5.4.1. Third, for many preprocessing steps, a number of different methods exist and little work has been done to evaluate and compare these methods. For example, it is common to hear statements of the type "my registration method to a template is better" or "this segmentation approach works well". Often there is little evidence supporting such statements, and these judgements are based solely on the qualitative inspection of images. There is a need for validation and replication work as well as understanding human qualitative assessment of images. A fourth major pitfall is to assume that problems in neuroimaging have been solved. The range, complexity, and diversity of unsolved problems is astonishing. Indeed, although much progress has been made, registration, intensity normalization, longitudinal co-registration, spatial inhomogeneity correction, segmentation, and population-level analyses are all open areas. Fifth, quantifying associations between imaging and health is a hard problem that needs to be well understood and addressed. Indeed, brain characteristics are extremely heterogeneous across individuals, while longitudinal changes tend to be much smaller. For example, in a study of fractional anisotropy, a measure derived from diffusion weighted imaging, of the corpus callosum the longitudinal variability over 4–5 years, only accounted for 2 to 3% of the total observed variability (22). This raises important problems for biostatisticians in many neuroimaging studies where the signal, if it exists, sits under a pile of noise.

5.2   Data Structure and Intuitive Description of Associated Problems

It is useful to describe the data structure and discuss sMRI from a notation perspective. We denote by $Y_{ijm}(v_{ijm})$ the intensity of the mth, $m = 1, \ldots, M$, sequence of the sMRI data at the jth study visit, $j = 1, \ldots, J_i$, of the ith subject, $i = 1, \ldots, I$, at the voxel v_{ijm}. For the data accompanying this chapter, $I = J = 2$, and $M = 4$ resulting in a total of 16 images. For those cases when there is only one sMRI per subject (e.g. cross-sectional imaging studies), the index j could be omitted. As the indexes i, j, and k in v_{ijk} indicate, images are typically not registered, in the sense that voxels do not have the same interpretation between the same sMRI sequences, visits, or subjects. A transformation of images that ensures that the voxel depends only on the subject, that is $v_{ijm} = v_i$, is called co-registration. A transformation of the image to a template, $X(v)$, where the voxel does not depend on the subject is called registration to a template or simply registration. While co-registration is less controversial and current software seems to handle it well, registration to a template raises multiple problems, especially in brains affected by disease. We will discuss registration and co-registration in Section 5.4.4.

A major problem in imaging is that images may have spatial inhomogeneities. More precisely, this means that the intensity of the image in various tissues (e.g. fat, white matter) varies by the location in the brain. This can be quite obvious when, for example, the inferior part of the brain is brighter than the superior. This can lead to serious problems, as gray matter in the inferior part of the brain may actually be "whiter" than the white matter in the superior part. Spatial inhomogeneities vary in severity, and can often be very subtle. Such subtle distortions would be discarded by a human observer, but may create serious problems when one tries to analyze data. For example, they have been shown to have a large negative effect on MS lesion segmentation (65). Thus, an image with spatial inhomogeneities will have the local intensity distributions in the same tissue vary across locations in the brain. The problem of inhomogeneity correction depends on the definition of "tissue" and requires distribution matching across various tissue types and brain locations. This is a tough problem with imperfect, but reasonable solutions. This is discussed in detail in Section 5.4.1. A quick way to diagnose spatial inhomogeneities is to visually identify white matter, fat, gray matter, and bone regions from various parts of the brain and plot the histograms of intensities for each such region separately. A less effective, but faster alternative is to compare the histograms of axial, sagittal, and coronal distributions. Of course, tissue type proportions should vary by slice, but reasonable approximations can be obtained. Another alternative is to use and visualize an aggressive smoother that would hide biological information, but would highlight unusual spatial patterns of image intensity.

Whenever one is interested in analyzing more than one sequence, it is useful for the units in which $Y_{ijm}(v_{ijm})$ is expressed to have the same interpretation and be on the same scale. As we mentioned earlier, this is not the case in sMRI, which can raise fundamental questions related to quantification of population-level effects. Indeed, if data are not on the same scale, even taking the differences between two images does not make sense. A transformation of image intensities from the raw image to an interpretable scale is called image intensity normalization. This should not be mistaken for image registration, which is also often referred to in practice as "image normalization". In Section 5.4.5 we will discuss the statistical principles of image normalization and we will discuss various ways of conducting image intensity normalization.

5.3 Acquisition and Reconstruction

The contrast of an sMRI volume is the relative difference of signal intensities within the volume. When an MRI scan is acquired, changing the scanning parameters changes the contrast of the volume to produce the different sequences, such as FLAIR, T1, T2, and

PD. The scanning parameters that contribute to the contrast of an image are the flip angle (FA), the repetition time (TR), the inversion time (TI), and the echo time (TE). A more detailed description of image scanning parameters can be found in (6). Small changes in the scanning parameters can result in different contrasts. For example, two volumes may both be a "FLAIR" volume, but if acquired with different scanning parameters, can have different image contrasts. MRI physicists are continually working to develop new imaging techniques in the form of different combinations of these parameters to produce higher-quality volumes. It is therefore desirable, but quite difficult, to develop algorithms that are robust to changes in the scanning parameter. Variability in the contrasts can also arise from the strength of the magnet used for imaging. The magnet strength is measured in teslas and the unit is abbreviated as T. Currently, common field strengths for sMRI are 1.5T and 3T, and some research institutes also use 7T magnets for research purposes (6). Slice thickness and the in-plane resolution of the original volumes are also important, as the volumes may be interpolated during image preprocessing. Information about the scanning parameters, slice thickness, and field strength of the magnet can often be found in a section of the image file called the header of the sMRI volume.

During acquisitions, imaging artifacts can arise due to the imaging hardware or from subject motion. It is well established that the introduction of artifacts associated with patient motion and the variability associated with scanners can significantly degrade the accuracy of results from further analysis (21). Therefore, volumes from the scanner typically undergo either a manual or automatic quality control to assure that volumes with artifacts are removed before analysis. References (21) and (28) both propose automated methods for assessing the quality of an image.

5.4   Preprocessing

After the sMRI is acquired and reconstructed, data are preprocessed for analysis. It is often hard to define exactly what "preprocessing" means, as it will vary by study, scientist, or even analysis. Indeed, preprocessing and image analysis are closely linked, with preprocessing often having a dramatic impact on the analytic results. Thus, it is important for the biostatisticians working with sMRI data to have knowledge of the preprocessing steps and their potential impacts on the downstream analyses. For the purpose of this paper, we divide image preprocessing into four main steps: 1) inhomogeneity correction; 2) spatial interpolation; 3) skull stripping; 4) spatial registration; and 5) intensity normalization. A detailed description of each step with software and data applications is provided in this section. These steps are typically executed in this order in an image preprocessing pipeline and depend on various choices and optimality criteria. A pipeline is a choice of a particular set of image preprocessing steps that can be applied to many images. While we simplify here for understanding, an additional complication is that the order, steps, and algorithm for each step are not agreed upon in the community. Part of the reason for the plurality of pipelines is that it is difficult to quantify the difference in quality between preprocessing pipelines. Developing improved algorithms for image preprocessing and methods for quantifying the quality of pipelines is an area filled with opportunities.

There are many tools for creating image preprocessing pipelines. The choice of tools should be based upon the tool's availability, results quality, and computational feasibility for large collections of images. While we do believe that there is no universally best pipeline, a non-exhaustive list of popular sMRI pipelines includes the Analysis for Functional NeuroImaging (AFNI) (11), the Advanced Normalization Tools packages

(ANTs) (2), the FMRIB Software Library (FSL) (24), the LONI Pipeline Processing Environment (45), the Java Image Science Toolkit (JIST) (37) implemented in Medical Image Processing Analysis and Visualization (MIPAV) (38), and the Statistical Parametric Mapping (SPM) (41) implemented in Matlab. An exciting new tool for R users is ANTsR : Advanced Normalization Tools with R http://stnava.github.io/ ANTsR/index.html, a preprocessing pipeline that can be run through R. Image pipelines often fail on a subset of images, which, left uncorrected, can seriously impact downstream analyses. Therefore, another quality control step must be performed, which often consists of a qualitative visual inspection of the preprocessed images by an expert.

5.4.1 Inhomogeneity Correction

MRI intensity inhomogeneity is the slow variation of intensities within a tissue class in an image. In the literature, intensity inhomogeneity is also referred to as intensity nonuniformity, shading, the bias field or the gain field. Spatial inhomogeneity can be caused by the MRI scanner or by the properties of the object that is being imaged. The latter cause is hard to control and account for, but is relatively small in lower magnetic field intensity scanners. Inhomogeneity may raise analytical challenges, because basic assumptions of various models and techniques may be violated. A consequence can be that methods developed for images with no, small, or known inhomogeneity field patterns may fail in heterogeneous imaging studies where inhomogeneity fields can be quite large or have unexpected distributions. For example, many segmentation algorithms use image intensity thresholding on one or more images that are known to discriminate well between specific tissue classes. However, if the same tissue has different intensity distributions at different locations in the brain then segmentation algorithms can be seriously affected. For example, Figure 5.3A displays a T1 volume from a 7T scanner, while Figure 5.3B displays the estimated inhomogeneity field,

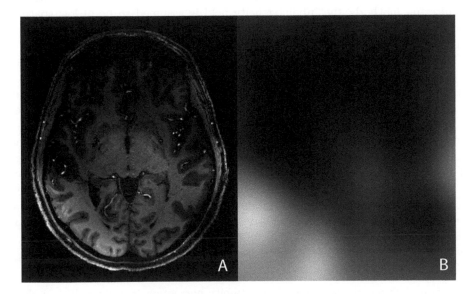

FIGURE 5.3
A. Axial slice from a T1 -weighted volume obtained from a 7T scanner. Volumes from scanners with a higher magnetic field strength often contain more intensity inhomogeneity artifacts, as seen in this image. B. The inhomogeneity field for this slice as modeled by the N4 ITK algorithm.

indicating higher-intensities in the left-bottom corner. Thus, gray matter in this area has higher intensities than white matter areas in other areas of the brain (gray matter looks "whiter" than white matter). While resolution and biological details are sharper in higher intensity scanners, spatial inhomogeneity is known to increase with the field strength of the magnet.

5.4.1.1   Concepts

The inhomogeneity field of an image is commonly modeled multiplicatively. For a voxel v, the observed intensity in an image is $Y_m(v)$, where, for simplicity, we have dropped the subject and visit indexes used in Section 5.2. Conceptually, the observed image is modeled as $Y_m(v) = \alpha(v)X_m(v) + e_v$, where $X_m(v)$ is the true voxel intensity, $\alpha(v)$ is the multiplicative inhomogeneity field, and e_v is an additive error assumed to follow a Gaussian distribution. The additive error is often ignored and data are modeled additively on the log-intensity scale:

$$\log\{Y_m(v)\} = \log\{X_m(v)\} + \log\{\alpha(v)\}. \tag{5.1}$$

References (26) and (71) provide comprehensive reviews of methods to correct for image inhomogeneities. The inhomogeneity field can be corrected for prospectively through phantom scans, the use of multiple coils, or special sequences. We focus on retrospective methods to estimate the field from the data. Clearly, the deconvolution model 5.1 requires strong assumptions to ensure identifiability. While each method used for estimation makes slightly different technical assumptions, intuitively they all make assumptions about the degree of variation in the $\log\{X_m(v)\}$ and $\log\{\alpha(v)\}$ processes. Typically, though often not explicitly, one assumes that $\log\{\alpha(v)\}$ varies spatially much slower than $\log\{X_m(v)\}$. Under this assumption, an aggressive smoother (e.g. 3D kernel smoother with a large window) of the image could be viewed as an estimator of $\log\{\alpha(v)\}$. The majority of inhomogeneity correction methods can be grouped as (1) filtering, (2) surface fitting, and (3) statistical models. In filtering methods, the inhomogeneity field is assumed to be of low spatial frequency and the signal of the anatomical structures in the image of high frequency. The inhomogeneity field can then be removed using a low-pass filter (a.k.a. aggressive or over smoothing). Surface fitting methods use a tissue segmentation first, which is then followed by smoothing within tissue classes. Statistical models assume that the inhomogeneity field follows a particular random process distribution. We will take exception to this nomenclature, as all these approaches are based on statistical models. However, we provide the accepted categorization to help with communication.

When the true inhomogeneity field is not available, criteria used to assess the performance of inhomogeneity correction methods include: 1) variance over the entire image or segmented portions of the image; 2) coefficient of variation over the image; and 3) joint coefficient of variation between two tissue classes. When the true inhomogeneity field is available, the mean square error between the derived and true inhomogeneity field is calculated. Other important considerations for assessing these methods are stability, computer requirements, and CPU time (26).

5.4.1.2   Practical Approaches, Software, and Application to Data

The most commonly used method for inhomogeneity correction is a statistical model, the nonparametric nonuniform intensity normalization (N3) correction (60). The method assumes that $f(v) = \log\{\alpha(v)\}$ and $u(v) = \log\{X_m(v)\}$ are two independent random variables with distributions F and U, respectively. The distribution of the sums of these two random variables is the convolution of F and U. N3 searches for the inhomogeneity field to maximize the frequency content of the image intensity distribution and constrains the inhomogeneity field to be modeled as a Gaussian distribution with small variance. More recently, an improvement and extension of the N3 method has been proposed, the N4ITK (67). Code for N3 and N4ITK is publicly available and has already been implemented in many pipelines. Figure 5.3B shows the inhomogeneity field as modeled by the N4 ITK algorithm (67).

5.4.2 Skull Stripping

Skull stripping is the process of extracting the brain from an image by removing the background and all other tissue. More specifically, the problem is to estimate $S_{ijm}(v_{ijm}) \in \{0,1\}$, the indicator variable of brain tissue being contained in voxel v_{ijm} from the images of each subject at each visit, $Y_{ijm}(v_{ijm})$. This process, which may be considered a segmentation task (see also Section 5.5.1), is necessary for the identification of tissue to be studied. Errors in skull stripping can produce both fictitious effects and reduce power if key regions of the brain are erroneously removed.

5.4.2.1 Concepts

While dozens of techniques have been proposed for this task over the past two decades, the most common method remains the brain extraction tool (BET) (61). BET is a simple technique that aims to iteratively fit a mesh around the surface of the brain, and has been shown to have performance superior to competing methods, although it has documented limitations including a propensity to include extracerebral tissue in the brain mask (17, 31). While several hybrid methods (7, 27) have been proposed by integrating generative and classification techniques to produce methods that are robust to differences between scanners and protocols, no solution has satisfactorily solved the problem. Thus, most image analysis groups still resort to manual correction after automatic skull stripping. Recently, multi-atlas label fusion techniques (25, 72) have shown great promise for skull-stripping (33, 14); these methods use several deformable registrations to compare the subject under study with a library of other images for which manual skull-stripped images (called atlases) exist. Labels are then obtained from each atlas and averaged (or fused) across atlases. Patch-based techniques (16) have also shown great promise with significantly lower computational burden. As new methods are developed, many authors submit results for active comparison to a validation resource, and comparisons are publicly available (53).

5.4.2.2 Practical Approaches, Software, and Application to Data

As BET (61) is so commonly used, we demonstrate its application as an easy-to-use and computationally practical approach. BET was first implemented in FSL (62), but now is available in other image processing packages including MIPAV and JIST (37). Using MIPAV, skull-stripping can be achieved using BET on a T1-weighted image in less than a minute on a standard personal computer and an example of the results is shown in Figure 5.4.

5.4.3 Interpolation

Interpolation transforms a discrete array of numbers into a continuous image. As we saw in Figure 5.2, sMRI are arrays of intensity values that have been sampled on a grid. When performing operations such as image registration, magnification, image reslicing and resampling, and surface rendering, it is desirable to have a continuous image and to know the approximate values of the image at points other than those on the original grid. Interpolation is also key in multi-sequence MRI studies as the different sequences are often acquired at differing resolutions.

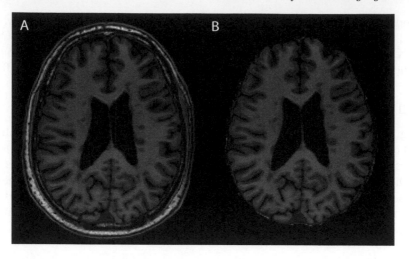

FIGURE 5.4
An axial slice of from a 3T T1-weighted imaging of a patient with MS before (A, showing $Y_{ijm}(v_{ijm})$) and after (B, showing $Y_{ijm}(v_{ijm})S_{ijm}(v_{ijm})$) skull stripping using BET.

5.4.3.1   Concepts

An extensive review and comparison of interpolation methods in medical image analysis can be found in (32) and (41). The most common interpolation methods in sMRI analysis are truncated and windowed sinc, nearest neighbors, linear, quadratic, cubic b-splines, cubic, Lagrange, and Gaussian interpolation. Consider the voxel $v = (x_v, y_v, z_v)$ with coordinates (x_v, y_v, z_v) that were not necessarily among the coordinates where data were sampled. Then the image can be interpolated at v as

$$Y_m(v) = Y_m(x_v, y_v, z_v) = \sum_{p,r,s} Y_m(p, r, s)h(x_v - p, y_v - r, z_v - s)$$

where the function $h(\cdot, \cdot, \cdot)$ is the interpolation kernel and the summation is done over all p, r, s where data are observed. To provide some intuition, we describe the interpolation kernels for "one nearest neighbor", "linear" and "windowed sinc". For one nearest neighbor, the interpolation kernel is

$$h(x) = \begin{cases} 1 & 0 \leq |x| \leq 0.5 \\ 0 & \text{elsewhere.} \end{cases}$$

For linear interpolation the interpolation kernel is

$$h(x) = \begin{cases} 1 - |x| & 0 \leq |x| \leq 0.5 \\ 0 & \text{elsewhere.} \end{cases}$$

The sinc function is $\text{sinc}(x) = \frac{\sin(\pi x)}{\pi x}$. If N denotes the number of supporting points used for interpolation, then the windowed sinc interpolation kernel is

$$h(x) = \begin{cases} \frac{\sin(\pi x)}{\pi x} & 0 \leq |x| \leq \frac{N}{2} \\ 0 & \text{elsewhere.} \end{cases}$$

5.4.3.2 Practical Approaches, Software, and Application to Data

In sMRI preprocessing, interpolation is linked closely to spatial registration. As an image is spatially aligned to another image or a template, the image being registered must be interpolated to determine the values of the registered image in the new coordinate space. Interpolation methods are typically a tuning parameter for registration and are important as they can impact the clarity of the image after registration. Windowed sinc has been shown to produce good results in accordance with the number of supporting points used in the interpolation, but can become quite computationally intensive as the number of supporting points increases (32).

5.4.4 Spatial Registration

Registration is the process of determining the spatial alignment and correspondence between images. Consider the case when one is interested in registering image $Y(p, r, s)$ to $\widetilde{Y}(p, r, s)$, where $p = 1, \ldots, P$, $r = 1, \ldots, R$, and $s = 1, \ldots, S$ are the indexes of the three-dimensional arrays. The dimension of both arrays are $P \times R \times S$; when the arrays have different dimensions, interpolation, as described in Section 5.4.3, can be applied to make the dimensions of the array equal. If we identify a voxel with its array index $v = (p, r, s)$, then the product of registration is a bijective transformation map, $v \to T(v)$, from one image reference system to another. The registered image in the reference system of $\widetilde{Y}(\cdot, \cdot, \cdot)$ is then $Y\{T(p, r, s)\}$ whereas results or images can be obtained in the "native space" by using the back transformation $v \to T^{-1}(v)$. It is important to note that registration is a transformation of space and does not affect image intensities; however, image intensities can be used to find optimal transformations in a specific class of transformations. In this section we focus on intra-subject registration, also referred to as spatial normalization. Inter-subject registration and registration to a group template image are discussed in the Analysis section. Reference (23) provides a detailed summary of registration methods and we use the notation introduced in this text in the following description.

5.4.4.1 Concepts

There is an infinite number of transformations $v \to T(v)$ and they range from useless to useful. For example, given two images expressed on the same 3D grid one can perfectly transform each individual point from the first image to each individual point in the second image. Such a transformation may or may not respect some order and is characterized by the degree of smoothness (number of degrees of freedom). A random assignment of indexes would have $P \times R \times S!$ degrees of freedom, with most transformations being useless and uninformative. Here we will describe a few useful transformations, including rigid and affine registrations, and we will provide the necessary tools for non-linear and diffeomorphic approaches.

Rigid registration is the simplest type of registration and consists of one translation and one rotation. Thus, 3D rigid registrations have six degrees of freedom, 3 associated with the translation vector, $t = (t_x, t_y, t_z)$ in the x, y and z directions, and 3 associated with the rotation parameters $\theta = (\alpha, \beta, \gamma)$. For a voxel $v = (i, j, k)$, the rigid transformation can be written as

$$T_{\text{rigid}}(v) = Rv + t$$

where

$$R = \begin{bmatrix} \cos\beta\cos\gamma & \cos\alpha\sin\gamma + \sin\alpha\sin\beta\cos\gamma & \sin\alpha\sin\gamma - \cos\alpha\sin\beta\cos\gamma \\ -\cos\beta\sin\gamma & \cos\alpha\cos\gamma - \sin\alpha\sin\beta\sin\gamma & \sin\alpha\cos\gamma + \cos\alpha\sin\beta\sin\gamma \\ \sin\beta & -\sin\alpha\cos\beta & \cos\alpha\cos\beta \end{bmatrix}.$$

An affine registration has the same form as the rigid and can be written as $T_{\text{affine}}(v) = Av+t$, with the difference that the matrix A is not constrained to be a rotation matrix. Thus, the total number of degrees of freedom of 3D affine transformation is 12 with 9 degrees of freedom corresponding to the 9 entries of the matrix A and 3 corresponding to the translation vector t.

Choosing the registration class (e.g. rigid or affine) is a crucial step, though one still needs to estimate the parameters of registration in the induced spaces. This is done through the minimization of a particular utility function. There are three main ways of constructing a utility function using: 1) landmarks; 2) surface fitting; and 3) voxel-similarity metrics. In landmark-based registration, fiducial markers or landmark points identified by hand in the images are used as points of reference. A fiducial marker is an object that is placed in the field of view in an image. These landmarks replace the original frame of reference and transformations are applied to reduce a particular distance between them, which could include minimizing the geometric distance or a combination between the geometric distance and the intensities in the image. Either the Root Mean Square Error (RMSE) or Fiducial Registration Error (FRE) can be optimized to select the registration parameters. The error of the registration can be assessed by reporting the Target Registration Error (TRE) for the non-landmark areas in the image. Landmark-based registration requires the identification of landmarks, which can be time consuming and may be prone to observer error. Therefore, finding landmarks automatically and reliably is an active area of research. An excellent comprehensive description of shape analysis and landmark-based registration can be found in (15).

Surface-based registration ((13), (12), (18)) takes into account the different geometric structures of the brain to improve the intra-subject variability of brain shapes after registration. Several methods exist for surface-based registration including methods available in the widely used software FreeSurfer (http://surfer.nmr.mgh.harvard.edu). In surface registration, the idea is to transform the cerebral cortex into a new space so that the gyri and sulci on the cortex are matched.

Voxel-similarity-based registration methods are popular, as they do not require the identification of landmarks or segmentation of the image. Here the registration T is optimized by a function of the voxel values in the two images. For images of the same modality, the sum of shared differences (SSD) can be used

$$SSD = \frac{1}{n} \sum_{v \in \Omega} |Y\{T(v)\} - \widetilde{Y}(v)|,$$

where Ω is the image domain of the two images, $Y\{T(v)\}$ is image after the transformation T is applied, and $\widetilde{Y}(v)$ is the voxel intensity of the target image. Another popular similarity function is correlation coefficient (CC) between the intensity values in the two images and is defined as

$$CC = \frac{\sum_{v \in \Omega}[Y\{T(v)\} - \bar{Y}^T][\widetilde{Y}(v) - \bar{\widetilde{Y}}]}{\sqrt{\sum_{v \in \Omega}[Y\{T(v)\} - \bar{Y}^T]^2 \cdot \sum_{v \in \Omega}[\widetilde{Y}(v) - \bar{\widetilde{Y}}]^2}},$$

where $\bar{Y}^T = \sum_{v \in \Omega} Y\{T(v)\}/n$ and $\bar{\widetilde{Y}} = \sum_{v \in \Omega} Y(v)/n$. For images that are obtained as different sequences or even different modalities, the intensities of the images can differ quite substantially. Thus, joint entropy is often used as an alternative registration optimization method. For a vector of probabilities $p = (p_1,\ldots,p_K)$ the entropy is $H(p) = -\sum_k p_k \log(p_k)$. Entropy can be thought of as a measure of information contained in the image. Maximum entropy, $\log(K)$, is obtained when $p_1 = \ldots = p_K = 1/K$. Minimum entropy, 0, is obtained when $p_1 = 1$ and $p_2 = \ldots = p_K = 0$. Maximum entropy corresponds to a perfectly chaotic environment (e.g. random assignment of shades of gray to an image), whereas minimum entropy corresponds to a perfectly organized system (e.g. assigning the same shade of gray to the entire image.) Registration is often done through minimizing the joint entropy

$$H(A, B) = \sum_a \sum_b p_{AB}(a, b) \log p_{AB}(a, b),$$

where $p_{AB}^T(a, b)$ is the joint probability of the pair of image values a in image A and b in image B being observed at the same voxel. As image intensities can be completely different, minimizing $H(A, B)$ directly does not typically work. Instead, the histogram of each image's intensities can be partitioned into quantiles and one can assess that the two images have the same intensity at voxel v if the image intensities fall within the same inter-quantile interval of the image-specific intensity distribution. Thus, the summation in the joint entropy formula refers to summation according to the two inter-quantile ranges corresponding to the two images. In a scatter plot of image intensities at the corresponding voxels, $H(A, B)$ is a symmetric measure of how far the points are from a line. A major problem in this context is that there are many background (non-tissue) voxels, which could dominate the joint entropy. To mitigate the effect of the background voxels, joint entropy can be replaced by mutual information

$$I(A, B) = H(A) + H(B) - H(A, B).$$

The mutual information is a measure of the mutual dependence of the two images.

In practice, the effect of these measures is often not completely understood and minor assumptions can have serious effects on the results of registration. Note, for example, that the quantile transformation that we have introduced above is, essentially, a histogram matching approach for signal intensities. Such approaches can be seriously affected if the relative intensities in images have different distributions. For example, a brain with larger ventricles will have a larger number of voxels in the cerebro-spinal fluid (CSF) than a standard template. Similarly, a brain with a large lesion with specific intensity properties will have a histogram with a fundamentally altered shape. Ignoring the effects of pathology and between-subject variability can have large effects on the results of registration and is one of the dirty, unspoken of, secrets of registration.

5.4.4.2 Practical Approaches, Software, and Application to Data

We now discuss and visualize some simple examples showing the process of registration. Registration can be thought of as a collection of steps that transform the image into the template space. Recall that registration is a transformation on the voxel location and not of image intensities; image intensities can be used to optimize the transformation using, for example, differences between the transformed image and a template. Suppose that we observe the 2-dimensional (2D) image depicted in Figure 5.5 (top left panel) that needs to be transformed to the template space (top right panel). In this example, the template image is a clockwise rotation of the observed image by a $\pi/2$ degree angle and a shift. Hence, we can use the following rotation matrix to transform the observed data into the template space

$$R = \begin{bmatrix} \cos(\pi/2) & -\sin(\pi/2) \\ \sin(\pi/2) & \cos(\pi/2) \end{bmatrix}.$$

Thus, for each pixel with coordinates $x = (x_1, x_2)^T$ we obtain the coordinates in the new space $y = (y_1, y_2)^T$ as $y = Rx$. The resulting image is shown in the left middle panel of Figure 5.5. With a shift in the X coordinate, we may completely register the observed image into the template space. The shift can be incorporated in the transformation as follows:

$$
\begin{bmatrix} y_1 \\ y_2 \\ 1 \end{bmatrix} = \begin{bmatrix} \cos(\pi/2) & -\sin(\pi/2) & -101 \\ \sin(\pi/2) & \cos(\pi/2) & 0 \\ 0 & 0 & 1 \end{bmatrix} \begin{bmatrix} x_1 \\ x_2 \\ 1 \end{bmatrix}.
$$

A comparison between the top-right and top-middle panels in Figure 5.5 indicates that the resulting image is very similar to the image in the template space; in fact, because this is a toy example, they are identical.

A rigid transformation is useful in cases when several images for one subject are acquired over a relatively short period of time, as we expect the images to be similar except that the subject may have changed positions between the image acquisitions. If the acquired image has different voxel dimensions than the template image, we may want to use an affine transformation. In a second example, the observed data is simply a $\pi/4$ degree rotation of the template image while the pixel size is half that of the pixel size of the template image. We may use the following transformation matrix to transform the observed image into the template space.

$$
\begin{bmatrix} y_1 \\ y_2 \\ 1 \end{bmatrix} = \frac{1}{2} \begin{bmatrix} \cos(-\pi/4) & -\sin(-\pi/4) & 0 \\ \sin(-\pi/4) & \cos(-\pi/4) & 0 \\ 0 & 0 & 1 \end{bmatrix} \begin{bmatrix} x_1 \\ x_2 \\ 1 \end{bmatrix}
$$

The image in Figure 5.5 is based on rounding the noninteger coordinates. Thus, as described in Section 5.4.3, interpolation of the intensities is needed to obtain the complete image. As discussed above, once the user chooses the parameterization (e.g. rigid, affine, etc.) and the objective function to be minimized, the transformation matrix can be estimated.

Several software packages exist to estimate the transformation matrix for 3D image data, including the Advanced Normalization Tools (ANTS) (2), FMRIB Software Library (FSL) (24), Medical Image Processing Analysis and Visualization (MIPAV) (http://mipav.cit.nih.gov) and Statistical Parametric Mapping (SPM) http://www.fil.ion.ucl.ac.uk/spm/.

The top left image in Figure 5.6 shows one slice of a 3-dimensional T1 image. Suppose that data from multiple subjects need to be registered into the template space shown in the top right panel; this is the Montreal Neurological Institute (MNI) template. The following code in FSL can be used to obtain the affine transformation matrix from the image into the template space

```
flirt -in Brain.nii.gz -ref Template.nii.gz
-out Brain_affine.nii.gz      -omat affine.mat
```

where Brain.nii.gz contains the image and Template.nii.gz is the template image. The resulting Brain_affine.nii.gz will be the image transformed into the template space, finally, the affine.mat will show the transformation matrix.

$$
\begin{bmatrix} y_1 \\ y_2 \\ y_3 \\ 1 \end{bmatrix} = \begin{bmatrix} -1.089 & 0.001 & 0.025 & 186.19 \\ 0.006 & -1.104 & 0.054 & 224.17 \\ -0.006 & -0.005 & 1.173 & -13.196 \\ 0 & 0 & 0 & 1 \end{bmatrix} \begin{bmatrix} x_1 \\ x_2 \\ x_3 \\ 1 \end{bmatrix}
$$

A similar transformation can be obtained using the ANTs software. However, the function ANTS provides the transformation matrix as an outcome and a second function has to be used to transform the image into the new space based on the estimated matrix from ANTS.

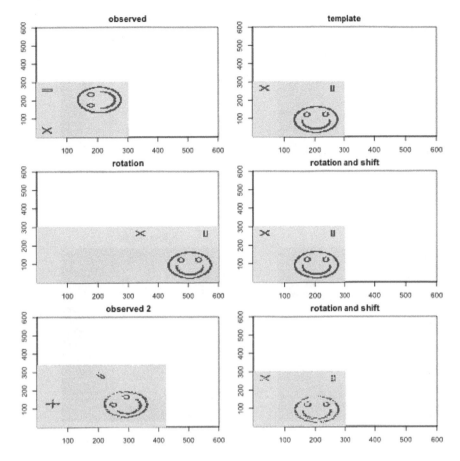

FIGURE 5.5
Steps of registration: a toy example.

```
ANTS 3 -m MI[Brain.nii.gz, Template.nii.gz, 1, 4] -o img.nii.gz
     -r Gauss[3,0] -i 0
```

```
WarpImageMultiTransform 3 Brain.nii.gz BrainWarp.nii.gz  -R
     Template.nii.gz  -i imgAffine.txt
```

If the brain structures are similar up to affine differences, then the methods described above will likely produce reasonable results. However, in some cases, one area of the brain may need to be transformed more than surrounding tissue or tissue from a different part of the brain. Approaches that go beyond affine transformations are typically referred to as non-linear and diffeomorphic approaches. Here is one parameterization in ANTs to obtain a non-linear transformation:

```
ANTS 3 -m CC[Brain.nii.gz, Template.nii.gz, 1, 8] -o img.nii.gz
     -r Gauss[1,1] -i 30 * 20 * 5 -t SyN[0.25]
```

```
WarpImageMultiTransform 3 Brain.nii.gz BrainWarp.nii.gz  -R
     Template.nii.gz  -i imgAffine.txt BrainInverseWarp.nii.gz
```

(2) provides an overview of how one can use ANTs to obtain non-linear transformations of

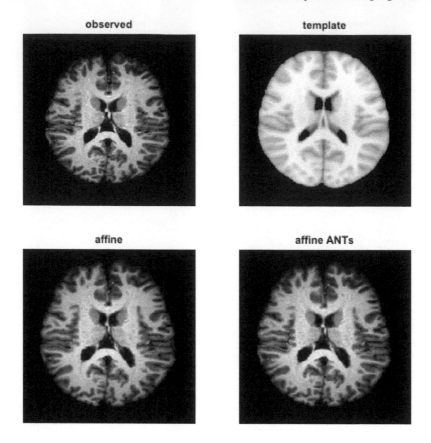

FIGURE 5.6
Application of two software methods (the function flirt on the bottom left and ANTs affine on the bottom right) to register a real brain image (top left) to a template (top right).

the images from patients with disease, by focusing the optimization function on the area containing the healthy tissue.

5.4.5   Intensity Normalization

Conducting any type of population-level analysis or inference on data usually requires that the units of measurement have the same meaning for every subject and visit. Indeed, when units are different, even calculating a simple average is not possible. This is a well-known problem in sMRI image analysis, as these modalities are measured in arbitrary units that depend on many factors including the scanner, protocol, and manual adjustments made by the radiology technician acquiring the images. The importance of intensity normalization has been emphasized by numerous publications in the imaging literature (39, 52, 73). The normalization process should produce units that: 1) have a common interpretation at the tissue-type level; 2) are replicable; 3) preserve the rank of intensities; 4) have similar distributions for the same tissues of interest across and within patients; 5) are not influenced by biological abnormality or population heterogeneity; 6) are minimally sensitive to noise and artifacts; and 7) do not result in the loss of information either due to pathology or other phenomena. These principles, proposed in Shinohara et al. (57) and referred to as the statistical principles of image normalization (SPIN), guide the mathematical formulation described in the next section.

5.4.5.1 Concepts

Consider the image intensity $Y_{ij}(v)$ at each voxel v expressed in arbitrary units and measured for subject i at visit j. Normalization is any transformation of the type $Y_{ij}(v) \rightarrow N_{ij}\{Y_{ij}(v)\}$. It is useful to conceptualize the histogram of intensities $Y_{ij}(v)$ as a mixture of densities

$$h_{ij}(x) = \sum_{k=1}^{K} w_{ijk} f_{ijk}(x), \qquad (5.2)$$

where $f_{ijk}(x)$ are the subject/visit-specific intensity densities of empty space and known tissues, such as white matter, gray matter, cerebrospinal fluid, bone, skin, and lesions. The weights $w_{ijk} \geq 0$ sum to 1 and represent the relative weights of components $k = 1, \ldots, K$. This includes both cases with and without pathology, as the weight for lesions can be allowed to be 0. Careful inspection of SPIN 4 suggests that after normalization $f_{ijk}(\cdot)$ should be as close to one another as possible for all i and j and for any fixed k. Thus, a natural starting point would be to consider transformations that reduce the distance between the $f_{ijk}(x)$ for any fixed k. Together with SPIN 1, this suggests the existence of the following theoretical model in normalized space for all images: $g_{ij}^{N}(x) = \sum_{k=1}^{K} w_{ijk} g_k(x)$, where the densities $g_k(x)$ are independent of subjects and/or visits, though the weights assigned to these densities depend on subject and visit and may be the measure of interest in medical studies. The fundamental difficulty of normalization is to find a transformation from $h_{ij}(x)$ to $g_{ij}^{N}(x)$ that respects the ordering of distributions and their mutual distances in the normalized space.

5.4.5.2 Practical Approaches, Software, and Application to Data

The most widely used image normalization techniques have centered on histogram matching (39, 40, 52). First, a template histogram is constructed, usually by averaging across a group of subjects in a training set. Then, a nonlinear transformation $N_{ij}(\cdot)$ (often a linear spline) is estimated for each subject at each visit to minimize any deviations between the normalized histogram $N_{ij}\{Y_{ij}(v)\}$ and the template. Although histogram matching methods produce replicable results, they are based on assumptions that are often violated: 1) the tissue-type distribution is the same across subjects and visits; 2) the absence of abnormal pathology; and 3) the absence of technical artifacts. For example, Figure 5.7 shows how the assumption of common distributions of tissue throughout the head can cause mismatching of gray matter (GM) to cerebrospinal fluid (CSF); note how a normal-appearing part of the brain (raw data shown in the top panel) is induced to show erosion of GM by histogram normalization (histogram normalized data shown in the bottom left panel).

An alternative method (58, 64) is to match a particular subdistribution consisting of a reference tissue as well as possible. We call this tissue-specific histogram normalization. It is important to note that although the distribution of intensities in the reference tissue are matched, regions of normal tissue with intensities outside the range of the reference tissue are not necessarily comparable; the true normalization function $N_{ij}(\cdot)$ is often nonlinear (due to differences in protocol, etc.) and thus normalizing with respect to white matter may not result in normalized gray matter intensities. Tissue-specific histogram normalization does, however, maintain the natural variability in other tissue types, allowing for the study of pathology.

Assume for the moment that for every subject and visit we have an area of reference tissue (a sub-mask, usually of the white matter). Then we can accurately estimate $f_{ij1}(x)$ (say $k = 1$ for white matter) for each i and j and obtain a normalized estimator that has

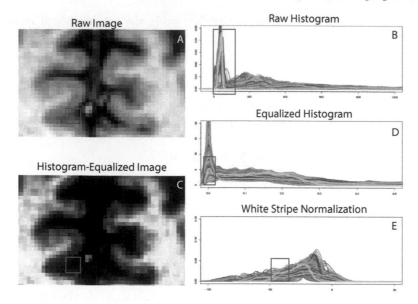

FIGURE 5.7

Intensity Normalization Methods. First column: region of interest from patient with MCI shown before (A) and after (C) histogram matching. The red square indicates a region of gray matter on the unnormalized image that disappears after histogram matching. Second column: histograms (shades of gray indicate different study visits) of the gray matter before (B) and after (D) histogram matching and (E) white stripe normalization for subjects in the Alzheimer's Disease Neuroimaging Initiative http://adni.loni.usc.edu/. Note the large proportion of gray matter mismatched to background (zero intensity) after histogram matching.

mean zero and variance one, $f_{ij1}^N(x) = \sigma_{ij1} f_{ijk}(\mu_{ij1} + \sigma_{ij1}x)$, where μ_{ij1} and σ_{ij1} are the mean and standard deviation of $f_{ij1}(x)$, respectively. An estimator of $g_1(x)$ is the average of $f_{ij1}^N(x)$ and linear normalization with respect to the white-matter distribution is

$$h_{ij}^N(x) = \sum_{k=1}^{K} w_{ijk}[\sigma_{ij1} f_{ijk}(\mu_{ij1} + \sigma_{ij1}x)]. \tag{5.3}$$

All units are expressed in multiples of standard deviations, σ_{ij1}, of the white-matter intensities, and zero is the average intensity of white matter. This method relies on the availability of a reference tissue (usually white matter) mask indicating a region of the brain that should be comparable across subjects. This is often unavailable before intensity normalization, however, and (57) proposed and validated a fully automatic method that avoids this problem.

Consider a T1 sMRI, $Y_{ij}(v)$, for subject $i = 1, \ldots, n$. (57) use normal-appearing white matter (NAWM) as a reference color, since it is the largest and most contiguous brain tissue. To identify the distribution of NAWM intensities, (57) use a penalized spline smoother (48) to estimate the mode of the intensity histogram in white matter, μ_{ij1}^* (the largest non-background peak). To estimate the variability within NAWM on the raw image, they estimate the standard deviation σ_{ij1}^* of intensities in $\Omega_{i,\tau} = \{v : H_{ij}^{-1}[H(\mu_{ij1}^*) - \tau] < Y_{ij}(v) < H_{ij}^{-1}[H(\mu_{ij1}^*) + \tau]\}$, which is referred to as the white stripe (where $H_{ij}(x) = \int_{-\infty}^{x} h_{ij}(x)\, dx$). Here τ is a quantile tolerance in the original space of intensities. The estimation of μ_{ij1}^* and σ_{ij1}^* was found to be remarkably robust across several thousand images (failure rate < 1%). If the family of densities $f_{ij1}(v)$ can be parameterized by two parameters then $\mu_{ij1} =$

$\psi_1(\mu_{ij1}^*, \sigma_{ij1}^*)$ and $\sigma_{ij1} = \psi_2(\mu_{ij1}^*, \sigma_{ij1}^*)$ (proof follows from the method of moments). Thus, matching μ_{ij1}^* and σ_{ij1}^* (estimable directly from the white stripe without prior segmentation) results in matching μ_{ij1} and σ_{ij1}, and this tissue-specific histogram normalization method has been shown to perform well in large multicenter studies.

5.5 Analysis

5.5.1 Lesion Segmentation

Segmentation is the labeling of voxels in an image according to particular properties of the voxel (e.g. the type of tissue the voxel contains). Examples include segmenting and comparing the cingulate gyrus of subjects with schizophrenia and healthy controls (43) and segmenting the hippocampus to assess volume loss due to chronic heavy drinking (5). A review of methods for segmentation of brain sMRI can be found in (42) and (3).

Here we describe the problem of segmentation of brain lesions in multiple sclerosis (MS) from a single sMRI study. MS is an inflammatory disease of the brain and spinal cord characterized by demyelinating lesions that can be observed with sMRI (49). In MS quantitative analyses of sMRI, the number and volume of lesions in an image are essential for diagnosing and monitoring the disease (47). In practice, MS lesions are manually segmented by experts from sMRI. Figure 5.8 shows an example of a manual segmentation of lesion voxels. As manual or semi-automated segmentations of images are time consuming, costly, and prone to large inter- and intra-observer variability (59), development of automated methods of lesion segmentation is an active field of research (19). Reviews of current lesion segmentation methods can be found in (19), (35), and (36). LesionTOADs is a readily available software for lesion segmentation that runs in the JIST pipeline environment (54). An excellent resource for lesion segmentation data is the MS lesion segmentation challenge 2008 (63). This database includes sMRI volumes acquired at the Children's Hospital Boston and University of North Carolina along with expert manual segmentations.

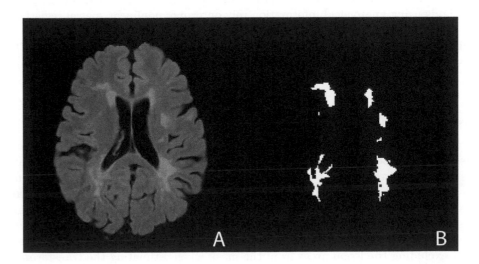

FIGURE 5.8
A. An axial slice from the FLAIR volume from a patient with multiple sclerosis. B. Manual expert segmentation of the lesions from this slice.

Lesion segmentation is a classification problem. In the literature, supervised classifiers are trained on expert manual segmentations of lesion voxels or unsupervised classifiers use clustering methods to identify lesion voxels. The covariates or features in the model are derived from the different imaging sequences Y_m. From these images anatomical information derived from atlases and/or the voxel-level intensity information from an imaging modality $Y_m(v)$ can be used for classification (36). To illustrate the problem, we describe a lesion segmentation method (65) that is particularly fast and accurate. Let $L_i(v) = 1$ be the lesion indicator for voxel v of subject i obtained from manual segmentation; $L_i(v) = 0$ indicates that the voxel v is not part of a lesion. We concatenate the manual segmentations from each voxel for all subjects into a single vector L. Similarly, let X denote the design matrix of features derived from different imaging modalities for all voxels and subjects. We can then model the probability that a lesion contains a voxel using, for example, logistic regression:

$$\text{logit} P\{L(v) = 1\} = X(v)\beta.$$

Recently, it has been found that for lesion segmentation the particular classification algorithm is less important than the development of the features (66). Indeed, the classification method (e.g. random forest, support vector machine, or logistic regression) offers basically no information on what actually improves prediction. In spite of this, methods continue to be labeled according to the classification method and not to the feature space. In a seminal paper, (24, p.1) notes "the extra performance to be achieved by more sophisticated classification rules, beyond that attained by simple methods, is small. It follows that if aspects of the classification problem are not accurately described (e.g., if incorrect distributions have been used, incorrect class definitions have been adopted, inappropriate performance comparison criteria have been applied, etc.), then the reported advantage of the more sophisticated methods may be incorrect." We subscribe to this view, though we describe it more plastically as "there are very few reasonable ways of cutting the potato." Here, the potato is, of course, the cloud of points in the feature space.

A major difficulty in automated segmentation of MRI is due to variable imaging acquisition parameters (36, 19). Most segmentation methods have tuning parameters that are adjusted to a particular dataset and may not generalize to a new dataset with different acquisition parameters. Lesion segmentation is also closely intertwined with image preprocessing. Methods using anatomical information rely on proper nonlinear spatial registration to a template image. The use of image intensities requires high-quality inhomogeneity correction and intensity normalization, otherwise population-level modeling is hopeless. There are even lesion segmentation methods in the literature that iterate between lesion segmentation and inhomogeneity correction (68).

5.5.2   Lesion Mapping

In Section 5.4.4 we discussed spatial registration methods for structural images. There are two main issues in the context of registration in the presence of lesions: 1) finding registration approaches that work reasonably well for many subjects with heterogeneous brain image distortions due to disease; 2) deciding whether lesion segmentation should be performed before or after registration. If lesions are hand-segmented in the native or in the rigidly registered space, then one can use the lesion masks to extract the areas that are within the lesion and use only the healthy brain for registration. Both ANTs and FSL provide the capability of excluding the lesion areas from the mask when estimating the transformation matrices. For instance, in ANTs, this can be performed by adding an option in the function call.

```
ANTS 3 -m MI[Brain.nii.gz, Template.nii.gz, 1, 4] -o img.nii.gz
```

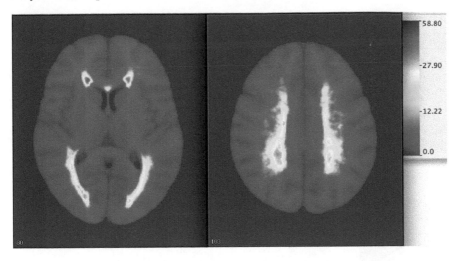

FIGURE 5.9
Two slices from a histogram of lesions constructed using non-linear registration of observed images in ANTs.

```
-r Gauss[3,0] -i 0  -x lesion_mask.nii.gz
```

Another approach to alleviate the effect of lesions on registration is to fill in the lesion areas by the average of the normally appearing white matter intensities and then estimate the transformation matrix for the resulting images (8, 50). After obtaining the transformation matrices for each subject, their corresponding lesion masks can be transformed into the template space by applying the transformation.

```
WarpImageMultiTransform 3 lesion_mask.nii.gz lesion_maskWarp.
nii.gz  -R       Template.nii.gz  -i imgAffine.txt
```

After registering all the mask maps into the template space, we can plot a lesion histogram (also known as a statistical atlas) that shows the spatial prevalence of lesions in the population; Figure 5.9 shows such a lesion histogram based on 98 subjects with MS. These maps can be compared between treatment groups, disease subtype groups, and for other investigations of the spatial distribution of pathology.

Once lesion localization is determined on a template, one could study the association between lesion localization and health outcomes. For example, (10) showed that there is an association between lesion load and disability, whereas (9) showed that cortical atrophy occurs even in MS patients with low disability scores. Lesion localization was found to be associated with gray matter volume reduction (51, 4), disease severity scores (70, 20), and cognitive impairment (46). Given the sensitivity to registration algorithms, it seems necessary to further investigate the sensitivity of these results to registration, multiplicity, and confounders, such as disease duration.

5.5.3 Longitudinal and Cross Sectional Intensity Analysis

Another very important type of analysis is to directly analyze the intensities in the image after intensity normalization. Indeed, after intensity normalization, we can compare and quantify the histograms of intensities within each sequence or combined across sequences. This could be done at the brain level, which requires only intensity normalization, or at the

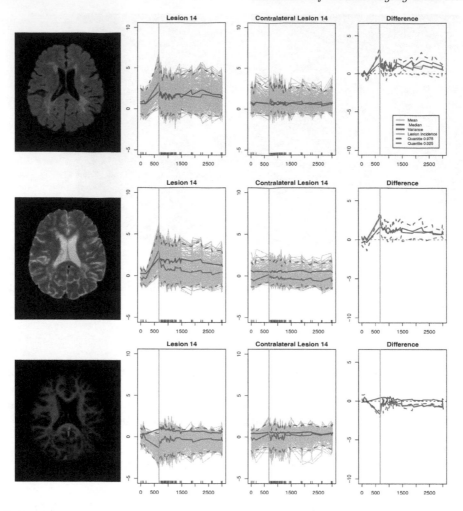

FIGURE 5.10

Voxel intensities for FLAIR (top panels), T1 (middle panels), and T2 (bottom panels) images in a lesion (labeled "lesion 14") for one subject over 8 years at 40 visits.

tissue level, which would also require a segmentation algorithm. For example, at the population level one may be interested in the intensity distribution in white matter. This could be done by obtaining the histogram intensities, stacking them in a matrix and conducting a PCA or other dimensionality reduction approach. A similar analysis could be conducted at the lesion level if one is interested in cross-sectional differences in lesion intensity distributions. Another possibility is to study the association between these distributions and health outcomes.

To provide a view of what could be achieved, consider a study of natural history of MS conducted by Daniel Reich at the National Institutes of Health (NIH). Figure 5.10 displays the voxel-specific intensities of the FLAIR (top panels), T1 (middle panels), and T2 (bottom panels) images in a lesion (labeled "lesion 14") for one subject over 8 years at 40 visits. The x-axis represents time in days and the y-axis represents image intensities as standard deviations of white matter intensities; see Section 5.4.5 for more details. The first column displays the three sequences along an axial slice, with MS lesions being visible around the ventricles, especially in the FLAIR and T1 images. The second column displays the corresponding voxel intensities, while the third column displays the voxel intensities in the contralateral part of the brain corresponding to the lesion. The contralateral area of the brain is used here as the control. The fourth column provides the difference in voxel intensities between each voxel and its contralateral correspondent voxel.

The orange vertical line indicates the first time when lesion 14 was identified using the SuBLIME algorithm, (64), an algorithm designed to segment a new and enlarging MS lesion; due to the nature of the observational study, the lesion may have occurred earlier. Indeed, before the lesion is observed there is approximately a 1.5-year gap between the MRI visits. Visits are indicated as a rug on the x-axis; notice that there are few visits in the beginning followed by many monthly visits after lesion 14 was detected. This is just one example of the data, and similar plots can now be obtained for all lesions. This raises important scientific questions, such as: 1) How many patterns of intensities are there before and after the lesion is detected? 2) Are there subtle changes in image intensities before the lesion is detected that could predict the lesion localization or timing? 3) Are certain white matter areas with specific intensity patterns more prone to lesion formation?

5.6   Conclusions

We could have written a book. Perhaps we should have written a book. After finishing this chapter, we realize that we barely scratched the surface of one of the most exciting areas of research in biostatistics. Far from being exhaustive, this chapter introduces fundamental analytic concepts in neuroimaging that are pertinent to healthy subjects, but especially to those who suffer from brain diseases. While the number of biostatisticians who work in this area is incredibly small given the importance of the problem, those who "get hooked" become passionate about looking in the most quantitative way possible into the brain and dealing with complex analytic problems. Far from sending the message that this area of research is closed, our own perception is that research is just now starting. Important concepts have already been introduced and progress has been achieved. However, without knowing what has already been achieved, the research community will simply reinvent the wheel.

We would like to thank the people who donated their brain data to science in the hope of better scientific understanding of devastating diseases such as MS, cancer, and stroke. We hope that they get and feel better. We would also like to thank our collaborators Daniel Reich from the National Institute for Neurological Diseases and Stroke of the NIH and Peter Calabresi from Johns Hopkins University for their continuous scientific guidance and for allowing the data to be shared.

Bibliography

[1] Atlas, S. (2009). *Magnetic Resonance Imaging of the Brain and Spine*, volume 1. Lippincott Williams & Wilkins.

[2] Avants, B., Tustison, N., and Song, G. (2009). Advanced normalization tools (ANTS). *Insight J.*

[3] Balafar, M. A., Ramli, A. R., Saripan, M. I., and Mashohor, S. (2010). Review of brain MRI image segmentation methods. *Artificial Intelligence Review*, 33(3):261–274.

[4] Bendfeldt, K., Blumhagen, J., Egger, H., Loetscher, P., Denier, N., Kuster, P., Traud, S., Mueller-Lenke, N., Naegelin, Y., Gass, A., Hirsch, J., Kappos, L., Nichols, T., Radue, E., and Borgwardt, S. (2010). Spatiotemporal distribution pattern of white matter lesion volumes and their association with regional grey matter volume reductions in relapsing-remitting multiple sclerosis. *Human Brain Mapping*, 31(10):1542–1555.

[5] Beresford, T. P., Arciniegas, D. B., Alfers, J., Clapp, L., Martin, B., Du, Y., Liu, D., Shen, D., and Davatzikos, C. (2006). Hippocampus volume loss due to chronic heavy drinking. *Alcoholism: Clinical and Experimental Research*, 30(11):1866–1870.

[6] Bernstein, M. A., King, K. F., and Zhou, X. J. (2004). *Handbook of MRI pulse sequences*. Elsevier.

[7] Carass, A., Cuzzocreo, J., Wheeler, M. B., Bazin, P.-L., Resnick, S. M., and Prince, J. L. (2011). Simple paradigm for extra-cerebral tissue removal: algorithm and analysis. *NeuroImage*, 56(4):1982–1992.

[8] Ceccarelli, A., Jackson, J., Tauhid, S., Arora, A., Gorky, J., Dell'Oglio, E., Bakshi, A., Chitnis, T., Khoury, S., Weiner, H., et al. (2012). The impact of lesion in-painting and registration methods on voxel-based morphometry in detecting regional cerebral gray matter atrophy in multiple sclerosis. *American Journal of Neuroradiology*, 33(8):1579–1585.

[9] Charil, A., Dagher, A., Lerch, J., Zijdenbos, A., Worsley, K., and Evans, A. (2007). Focal cortical atrophy in multiple sclerosis: relation to lesion load and disability. *Neuroimage*, 34(2):509–517.

[10] Charil, A., Zijdenbos, A., Taylor, J., Boelman, C., Worsley, K., Evans, A., and Dagher, A. (2003). Statistical mapping analysis of lesion location and neurological disability in multiple sclerosis: Application to 452 patient datasets. *Neuroimage*, 19(3):532–544.

[11] Cox, R. W. (1996). AFNI: Software for analysis and visualization of functional magnetic resonance neuroimages. *Computers and Biomedical Research*, 29(3):162–173.

[12] Dale, A., Fischl, B., and Sereno, M. (1999). Cortical surface-based analysis. I. Segmentation and surface reconstruction. *Neuroimage*, 9(2):179–94.

[13] Davatzikos, C. (1997). Spatial transformation and registration of brain images using elastically deformable models. *Comput Vision and Image Understanding*, 66(2):207–222.

[14] Doshi, J., Erus, G., Ou, Y., Gaonkar, B., and Davatzikos, C. (2013). Multi-atlas skull-stripping. *Academic Radiology*, 20(12):1566–1576.

[15] Dryden, I. and Mardia, K. (1998). *Statistical shape analysis*. John Wiley & Sons, Chichester.

[16] Eskildsen, S. F., Coupé, P., Fonov, V., Manjón, J. V., Leung, K. K., Guizard, N., Wassef, S. N., Østergaard, L. R., and Collins, D. L. (2012). Beast: Brain extraction based on nonlocal segmentation technique. *NeuroImage*, 59(3):2362–2373.

[17] Fennema-Notestine, C., Ozyurt, I. B., Clark, C. P., Morris, S., Bischoff-Grethe, A., Bondi, M. W., Jernigan, T. L., Fischl, B., Segonne, F., Shattuck, D. W., et al. (2006). Quantitative evaluation of automated skull-stripping methods applied to contemporary and legacy images: Effects of diagnosis, bias correction, and slice location. *Human Brain Mapping*, 27(2):99–113.

[18] Fischl, B., Sereno, M., and Dale, A. (1999). Cortical surface-based analysis. II: Inflation, flattening, and a surface-based coordinate system. *Neuroimage*, 9(2):195–207.

[19] García-Lorenzo, D., Francis, S., Narayanan, S., Arnold, D. L., and Collins, D. L. (2013). Review of automatic segmentation methods of multiple sclerosis white matter lesions on conventional magnetic resonance imaging. *Medical Image Analysis*, 17(1):1–18.

[20] Ge, T., Muller-Lenke, N., Bendfeldt, K., Nichols, T., and Johnson, T. (2012). Analysis of multiple sclerosis lesions via spatially varying coefficients. *Annals of Applied Statistics*.

[21] Gedamu, E. L., Collins, D., and Arnold, D. L. (2008). Automated quality control of brain MR images. *Journal of Magnetic Resonance Imaging*, 28(2):308–319.

[22] Greven, S., Crainiceanu, C. M., Caffo, B., and Reich, D. (2010). Longitudinal functional principal component analysis. *Electronic Journal of Statistics*, 4:1022–1054.

[23] Hajnal, J. V. and Hill, D. L. (2010). *Medical image registration*. CRC Press.

[24] Hand, D. (2006). Classifier technology and the illusion of progress. *Statistical Technology*, 21(1):1–15.

[25] Heckemann, R. A., Hajnal, J. V., Aljabar, P., Rueckert, D., and Hammers, A. (2006). Automatic anatomical brain MRI segmentation combining label propagation and decision fusion. *NeuroImage*, 33(1):115–126.

[26] Hou, Z. (2006). A review on MR image intensity inhomogeneity correction. *International Journal of Biomedical Imaging*, 2006.

[27] Iglesias, J. E., Liu, C.-Y., Thompson, P. M., and Tu, Z. (2011). Robust brain extraction across datasets and comparison with publicly available methods. *Medical Imaging, IEEE Transactions on*, 30(9):1617–1634.

[28] Ihalainen, T., Sipilä, O., and Savolainen, S. (2004). MRI quality control: Six images studied using eleven unified image quality parameters. *European Radiology*, 14(10):1859–1865.

[24] Jenkinson, M., Beckmann, C. F., Behrens, T. E., Woolrich, M. W., and Smith, S. M. (2012). FSL. *Neuroimage*, 62(2):782–790.

[30] Kevles, B. (1997). *Naked to the bone: Medical imaging in the twentieth century*. Rutgers University Press.

[31] Lee, J.-M., Yoon, U., Nam, S. H., Kim, J.-H., Kim, I.-Y., and Kim, S. I. (2003). Evaluation of automated and semi-automated skull-stripping algorithms using similarity index and segmentation error. *Computers in Biology and Medicine*, 33(6):495–507.

[32] Lehmann, T. M., Gonner, C., and Spitzer, K. (1999). Survey: Interpolation methods in medical image processing. *Medical Imaging, IEEE Transactions on*, 18(11):1049–1075.

[33] Leung, K. K., Barnes, J., Modat, M., Ridgway, G. R., Bartlett, J. W., Fox, N. C., and Ourselin, S. (2011). Brain maps: An automated, accurate and robust brain extraction technique using a template library. *Neuroimage*, 55(3):1091–1108.

[35] Lindquist, M. A. et al. (2008). The statistical analysis of fMRI data. *Statistical Science*, 23(4):439–464.

[35] Lladó, X., Ganiler, O., Oliver, A., Martí, R., Freixenet, J., Valls, L., Vilanova, J. C., Ramió-Torrentà, L., and Rovira, À. (2012a). Automated detection of multiple sclerosis lesions in serial brain MRI. *Neuroradiology*, 54(8):787–807.

[36] Lladó, X., Oliver, A., Cabezas, M., Freixenet, J., Vilanova, J. C., Quiles, A., Valls, L., Ramió-Torrentà, L., and Rovira, À. (2012b). Segmentation of multiple sclerosis lesions in brain mri: A review of automated approaches. *Information Sciences*, 186(1):164–185.

[37] Lucas, B. C., Bogovic, J. A., Carass, A., Bazin, P.-L., Prince, J. L., Pham, D. L., and Landman, B. A. (2010). The Java Image Science Toolkit (JIST) for rapid prototyping and publishing of neuroimaging software. *Neuroinformatics*, 8(1):5–17.

[38] McAuliffe, M. J., Lalonde, F. M., McGarry, D., Gandler, W., Csaky, K., and Trus, B. L. (2001). Medical image processing, analysis and visualization in clinical research. In *Computer-Based Medical Systems, 2001. CBMS 2001. Proceedings. 14th IEEE Symposium on*, pages 381–386. IEEE.

[39] Nyul, L. G. and Udupa, J. K. (1999). On standardizing the MR image intensity scale. *Magnetic Resonance in Medicine*, 42(6):1072–1081.

[40] Nyul, L. G., Udupa, J. K., and Zhang, X. (2000). New variants of a method of MRI scale standardization. *IEEE Transactions on Medical Imaging*, 19(2):143–150.

[41] Penny, W., Friston, K., Ashburner, J., Kiebel, S., and Nichols, T. (2007). *Statistical Parametric Mapping: The Analysis of Functional Brain Images*. Academic Press.

[42] Pham, D. L., Xu, C., and Prince, J. L. (2000). Current methods in medical image segmentation. *Annual review of biomedical engineering*, 2(1):315–337.

[43] Priebe, C. E., Miller, M. I., and Tilak Ratnanather, J. (2006). Segmenting magnetic resonance images via hierarchical mixture modelling. *Computational Statistics & Data Analysis*, 50(2):551–567.

[44] R Core Team (2014). *R: A Language and Environment for Statistical Computing*. R Foundation for Statistical Computing, Vienna, Austria.

[45] Rex, D. E., Ma, J. Q., and Toga, A. W. (2003). The LONI pipeline processing environment. *Neuroimage*, 19(3):1033–1048.

[46] Rossi, F., Giorgio, A., Battaglini, M., Stromillo, M., Portaccio, E., Goretti, B., Federico, A., Hakiki, B., Amato, M., and De Stefano, N. (2012). Relevance of brain lesion location to cognition in relapsing multiple sclerosis. *PLoS One*, 7(11).

[47] Rovira, À. and León, A. (2008). MR in the diagnosis and monitoring of multiple sclerosis: An overview. *European Journal of Radiology*, 67(3):409–414.

[48] Ruppert, D., Wand, M. P., and Carroll, R. J. (2003). *Semiparametric Regression*. Cambridge University Press.

[49] Sahraian, M. A., Radue, E.-W., and Gass, A. (2008). *MRI Atlas of MS Lesions*. Springer.

[50] Sdika, M. and Pelletier, D. (2009). Nonrigid registration of multiple sclerosis brain images using lesion inpainting for morphometry or lesion mapping. *Human Brain Mapping*, 30(4):1060–1067.

[51] Sepulcre, J., Goni, J., Masdeu, J., Bejarano, B., Vlez de Mendizbal, N., Toledo, J., and Villoslada, P. (2009). Contribution of white matter lesions to gray matter atrophy in multiple sclerosis: Evidence from voxel-based analysis of T1 lesions in the visual pathway. *Archives of Neurology*, 66(2):173–179.

[52] Shah, M., Xiao, Y., Subbanna, N., Francis, S., Arnold, D. L., Collins, D. L., and Arbel, T. (2011). Evaluating intensity normalization on MRIs of human brain with multiple sclerosis. *Medical Image Analysis*, 15(2):267–282.

[53] Shattuck, D. W., Prasad, G., Mirza, M., Narr, K. L., and Toga, A. W. (2009). Online resource for validation of brain segmentation methods. *NeuroImage*, 45(2):431–439.

[54] Shiee, N., Bazin, P.-L., Ozturk, A., Reich, D. S., Calabresi, P. A., and Pham, D. L. (2010). A topology-preserving approach to the segmentation of brain images with multiple sclerosis lesions. *NeuroImage*, 49(2):1524–1535.

[55] Shinohara, R., Crainiceanu, C., Caffo, B., Gaitn, M., and Reich, D. (2011a). Population-wide principal component-based quantification of blood-brain-barrier dynamics in multiple sclerosis. *Neuroimage*, 57(4):1430–1446.

[56] Shinohara, R., Goldsmith, A., Mateen, F., Crainiceanu, C., and Reich, D. (2012). Predicting breakdown of the blood-brain barrier in multiple sclerosis without contrast agents. *American Journal of Neuroradiology*, 33(8):1586–1590.

[57] Shinohara, R., Sweeney, E., Goldsmith, J., Shiee, N., Mateen, F., Calabresi, P., Jarso, S., Pham, D., Reich, D., and Crainiceanu, C. (2013). Normalization techniques for statistical inference from magnetic resonance imaging. In *UPenn Biostatistics Working Papers*, page Working Paper 36. http://biostats.bepress.com/upennbiostat/art36.

[58] Shinohara, R. T., Crainiceanu, C. M., Caffo, B. S., Gaitán, M. I., and Reich, D. S. (2011b). Population-wide principal component-based quantification of blood-brain-barrier dynamics in multiple sclerosis. *NeuroImage*, 57(4):1430–1446.

[59] Simon, J., Li, D., Traboulsee, A., Coyle, P., Arnold, D., Barkhof, F., Frank, J., Grossman, R., Paty, D., Radue, E., et al. (2006). Standardized MR imaging protocol for multiple sclerosis: Consortium of MS centers consensus guidelines. *American Journal of Neuroradiology*, 27(2):455–461.

[60] Sled, J. G., Zijdenbos, A. P., and Evans, A. C. (1998). A nonparametric method for automatic correction of intensity nonuniformity in MRI data. *Medical Imaging, IEEE Transactions on*, 17(1):87–97.

[61] Smith, S. M. (2002). Fast robust automated brain extraction. *Human Brain Mapping*, 17(3):143–155.

[62] Smith, S. M., Jenkinson, M., Woolrich, M. W., Beckmann, C. F., Behrens, T. E., Johansen-Berg, H., Bannister, P. R., De Luca, M., Drobnjak, I., Flitney, D. E., et al. (2004). Advances in functional and structural MR image analysis and implementation as FSL. *Neuroimage*, 23:S208–S219.

[63] Styner, M., Lee, J., Chin, B., Chin, M., Commowick, O., Tran, H., Markovic-Plese, S., Jewells, V., and Warfield, S. (2008). 3D segmentation in the clinic: A grand challenge II: MS lesion segmentation. *MIDAS Journal*, 2008:1–6.

[64] Sweeney, E., Shinohara, R., Shea, C., Reich, D., and Crainiceanu, C. (2013a). Automatic lesion incidence estimation and detection in multiple sclerosis using multisequence longitudinal MRIs. *American Journal of Neuroradiology*, 34(1):68–73.

[65] Sweeney, E., Shinohara, R., Shie, N., Mateen, F., Chudgar, A., Cuzzocreo, J., Calabresi, P., Pham, D., Reich, D., and Crainiceanu, C. (2013b). OASIS is automated statistical inference for segmentation, with applications to multiple sclerosis lesion segmentation in MRI. *NeuroImage Clinical*, 2:402–413.

[66] Sweeney, E. M., Vogelstein, J. T., Cuzzocreo, J. L., Calabresi, P. A., Reich, D. S., Crainiceanu, C. M., and Shinohara, R. T. (2014). A comparison of supervised machine learning algorithms and feature vectors for MS lesion segmentation using multimodal structural MRI. *PLOS ONE*, 9(4):e95753.

[67] Tustison, N. J., Avants, B. B., Cook, P. A., Zheng, Y., Egan, A., Yushkevich, P. A., and Gee, J. C. (2010). N4ITK: Improved N3 bias correction. *Medical Imaging, IEEE Transactions on*, 29(6):1310–1320.

[68] Van Leemput, K., Maes, F., Vandermeulen, D., Colchester, A., and Suetens, P. (2001). Automated segmentation of multiple sclerosis lesions by model outlier detection. *Medical Imaging, IEEE Transactions on*, 20(8):677–688.

[69] van Rossum, G. and de Boer, J. (1991). Interactively testing remote servers using the Python programming language. *CWI Quarterly*, 4(4):283–303.

[70] Vellinga, M., Geurts, J., Rostrup, E., Uitdehaag, B., Polman, C., Barkhof, F., and Vrenken, H. (2009). Clinical correlations of brain lesion distribution in multiple sclerosis. *Journal of Magnetic Resonance Imaging*, 29(4):768–773.

[71] Vovk, U., Pernus, F., and Likar, B. (2007). A review of methods for correction of intensity inhomogeneity in MRI. *Medical Imaging, IEEE Transactions on*, 26(3):405–421.

[72] Warfield, S. K., Zou, K. H., and Wells, W. M. (2004). Simultaneous truth and performance level estimation (staple): An algorithm for the validation of image segmentation. *Medical Imaging, IEEE Transactions on*, 23(7):903–921.

[73] Weisenfeld, N. and Warfield, S. (2004). Normalization of joint image-intensity statistics in MRI using the Kullback-Leibler divergence. In *Biomedical Imaging: Nano to Macro, 2004. IEEE International Symposium on*, pages 101–104. IEEE.

[74] Whitcher, B., Schmid, V. J., and Thornton, A. (2011). Working with the DICOM and NIfTI data standards in R. *Journal of Statistical Software*, 44(6):1–28.

6

Principles of Functional Magnetic Resonance Imaging

Martin A. Lindquist

Department of Biostatistics; Johns Hopkins University

Tor D. Wager

Department of Psychology & Neuroscience; University of Colorado at Boulder

CONTENTS

6.1	Introduction		139
6.2	The Basics of fMRI Data		141
	6.2.1	Principles of Magnetic Resonance Signal Generation	141
		6.2.1.1 The MRI Scanner	141
		6.2.1.2 Basic MR Physics	142
		6.2.1.3 Image Contrast	143
	6.2.2	Image Formation	144
	6.2.3	From MRI to fMRI	146
6.3	BOLD fMRI		148
	6.3.1	Understanding BOLD fMRI	148
	6.3.2	Spatial Limitations	150
	6.3.3	Temporal Limitations	151
	6.3.4	Acquisition Artifacts	152
6.4	Modeling Signal and Noise in fMRI		153
	6.4.1	BOLD Signal	153
	6.4.2	Noise and Nuisance Signal	155
6.5	Experimental Design		157
6.6	Preprocessing		158
6.7	Data Analysis		160
	6.7.1	Localization	160
	6.7.2	Connectivity	162
	6.7.3	Prediction	164
6.8	Resting-State fMRI		165
6.9	Data Format, Databases, and Software		166
6.10	Future Developments		168
	Bibliography		169

6.1 Introduction

Functional Magnetic Resonance Imaging (fMRI) is a non-invasive technique for studying brain activation. It measures changes in blood oxygenation and blood flow related to neuronal activity, providing researchers with the means to study human brain function *in vivo*,

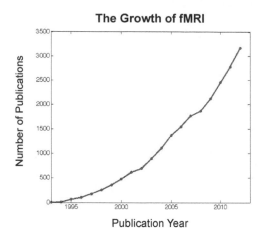

FIGURE 6.1

The yearly number of publications in PubMed that mention the term fMRI either in its title or abstract between 1993 to 2012.

either in response to a certain task or when at rest. During the past two decades fMRI has provided researchers with an unprecedented access to the inner workings of the human brain, which in turn has led to new insights into how the brain processes information.

The data acquired in an fMRI study consists of a sequence of 3-D magnetic resonance images (MRIs), each made up of a number of uniformly spaced volume elements, or voxels. The voxels partition the brain into a large number of equally sized cubes. A typical image may consist of roughly 100,000 voxels, where the image intensity value corresponding to each voxel represents the spatial distribution of the nuclear spin density, which relates to blood oxygenation and flow, in the local area. During an fMRI experiment, 100–1,000 such 3-D images of the whole brain are acquired. In addition, a standard fMRI experiment consists of multiple subjects (e.g, 10–50), potentially brought in for multiple scanning sessions, each consisting of a number of replications of a certain experimental task.

Clearly, the amount of available data from a single experiment is extremely large, and the analysis of fMRI data is an example of the type of modern big-data problem that is fundamentally changing the quantitative sciences. In addition, the data exhibit a complicated temporal and spatial noise structure with a relatively weak signal (though, with appropriate methods, these signals across the brain can be highly predictive of psychological and clinical states). Hence, the available data is not only massive in scale but also complex, making the statistical analysis of fMRI data a difficult task.

The field that has grown around the acquisition and analysis of fMRI data has experienced rapid growth in the past several years and found applications in a wide variety of fields, including neuroscience, psychology, medicine, economics and political science. The use of fMRI data is also central to a number of emerging fields, such as cognitive neuroscience, affective neuroscience, social cognitive neuroscience, and neuroeconomics. In these areas fMRI data is being combined with data on performance and psychophysiology to yield new exciting models of human thought, emotion, and behavior.

This explosive growth is illustrated by the exponential increase, shown in Fig. 6.1, of the number of yearly publications in PubMed that mention the term fMRI in either its title or abstract. In the early 1990's only a handful of such papers were published yearly, while in more recent years this number has increased to over 3,000 papers/year. In addition, more and more methodological papers appear each year, and the field has become fertile ground for the development and application of cutting-edge statistical methods. Researchers entering the field of MRI methods development come from diverse backgrounds: statistics, computer science, engineering, mathematical psychology, mathematics, and physics.

The rapid pace of development, as well as the interdisciplinary nature of the diverse fields that use fMRI data, presents an enormous challenge to researchers. The ability to move the field forward requires strong collaborative teams with expertise in a number of disciplines, including psychology, neuroanatomy, neurophysiology, physics, biomedical engineering, signal processing, and statistics. Of course, true interdisciplinary collaboration is extremely challenging, as all members of the research team must know enough about the other disciplines to be able to talk intelligently with experts in each discipline. Hence, making an impact in this exciting new area requires some initial start-up costs.

The goal of this chapter is to review the basic principles involved in the acquisition and analysis of fMRI data in enough detail to highlight the most important issues and concerns. The hope is that this will provide quantitative researchers with a basic understanding about the relevant research questions and how to apply their knowledge to these questions in an appropriate manner. We will also attempt to provide an overall road map to the kinds of study design and analysis options that are available and highlight some of their limitations. This chapter will be more focused on breadth rather than depth, with more detailed descriptions of many of the topics found in the later chapters of the book.

6.2 The Basics of fMRI Data

Functional MRI uses a standard magnetic resonance imaging (MRI) scanner to acquire a series of brain volumes that can be used to study dynamic changes in brain activation. In order to understand the manner in which fMRI data is acquired, one must first focus on the acquisition of a single static 3-D image. For this reason, while the focus of this chapter is on functional imaging, we must necessarily begin by reviewing data acquisition and reconstruction techniques used to obtain a static MRI of the brain. This closely follows the description in Chapter 3. After this review, we will transition our focus towards the particular issues involved with acquiring data meant for use in an fMRI study.

Proper understanding of the data acquisition and reconstruction procedures associated with fMRI is complex, requiring background in both MR physics and signal processing. Thus, the description that follows is abbreviated. For a more in-depth discussion see, for example, excellent references such as (21) or (23).

6.2.1 Principles of Magnetic Resonance Signal Generation

In this section we outline the physical bases of fMRI. We begin by providing some basic background on the MR scanner and continue by illustrating how it can be used to generate signal, and in turn how this signal can be used to construct an image. While these topics are common with the acquisition of MRI images, we conclude the section with discussing particular issues associated with fMRI.

6.2.1.1 The MRI Scanner

An MRI scanner is a large and versatile piece of hardware. Its main component is a superconducting electromagnet with an extremely strong static magnetic field, typically varying from 1.5–7.0 Tesla in human brain research. To place this into context, the Earth's magnetic field is only 0.00005 Tesla. Thus, the field strength is strong enough to pull magnetic objects into its core. Because the static field is always active, it is critical to observe caution when bringing objects into the MR scanner room. However, it is important to note that there are no known long-term effects on biological tissue, making the technique attractive for scanning humans.

A second critical component of the scanner is the radio frequency coils, hardware coils close to the object being imaged (e.g., the head) that can be used to generate and receive energy at the resonance frequency of the volume being imaged. They are turned on and off during the course of data acquisition.

A third component is the gradient coils, which are electromagnetic coils that can be used to create spatial variation in the strength of the magnetic field in a controlled manner. As we will see, this is critical for the ability to encode spatial information into the signal that is necessary for the creation of images.

MR scanners are extremely versatile, as they can be used to study both brain structure and function in multiple ways. Different types of images can be generated to emphasize contrast related to different tissue characteristics. In addition, the scanner can be used to study the directional patterns of water diffusion — diffusion-weighted imaging (DWI) used to measure white-matter tracts — elastic properties of brain tissue, flow of cerebrospinal fluid, and other properties. Hence, the same scanner is used to acquire structural MRIs, functional MRIs, and perform diffusion tensor imaging (DTI) of white-matter tracts; see Chapters 3 and 4. This is extremely beneficial as it allows for the acquisition of several different types of images on a specific subject during a given scanning session. In particular, structural images are always acquired as part of a standard fMRI scanning session, as they play an integral part in subsequent preprocessing of the data; see Section 6.6 and Chapter 10.

6.2.1.2   Basic MR Physics

All magnetic resonance imaging techniques rely on a core set of physical principles. To properly understand these principles, one should begin by looking at a single atomic nucleus and illustrate its impact on the generated MR signal. In particular we focus on hydrogen atoms consisting of a single proton (^{1}H atoms), as they are the most commonly used nuclei in MRI due both to their magnetic properties and abundance in the human body.

Protons can be viewed as positively charged spheres that are always spinning about their axis. This gives rise to a net magnetic moment along the direction of the axis of the spins, which is the source of the signal we seek to measure. Unfortunately, it is not possible to measure the magnetization of a single proton using an MRI scanner. Instead, we must focus our attention on measuring the net magnetization of the ensemble of all nuclei within a chosen volume. The net magnetization, denoted M, can be represented as a vector with two components. The first is a longitudinal component, which is parallel to the magnetic field, and the second a transverse component perpendicular to the field.

In the absence of an external magnetic field, the individual nuclei are randomly oriented with respect to one another and therefore do not give rise to a net magnetization. However, when placed into a strong magnetic field, the nuclei align with the field, creating a net longitudinal magnetization in the direction of the field. While aligned, the nuclei precess about the field with an angular frequency determined by the Larmor frequency, but at a random phase with respect to one another.

In order to measure the net magnetization of the nuclei within a certain volume, one must perturb the equilibrium and observe the reaction. A radiofrequency (RF) electromagnetic field pulse causes the nuclei to absorb the energy at a particular frequency band, and become "excited." Conceptually, we can imagine this process as the RF pulse aligning the phase of the precessing nuclei and tipping them over into the transverse plane. This causes the longitudinal magnetization to decrease, and establishes a new transversal magnetization.

After the RF pulse is removed, the system seeks to return to equilibrium. Now the nuclei emit the absorbed energy as they "relax." This causes the transverse magnetization to disappear, in a process known as transversal relaxation, while the longitudinal magnetization grows back to its original size in a process referred to as longitudinal relaxation. During this time, a signal is created that can be measured using a receiver coil.

Longitudinal relaxation represents the restoration of net magnetization along the longitudinal direction as the nuclei return to their original aligned state. It is seen as an exponential recovery in magnetization described by a time constant T_1. Transverse relaxation is the loss of net magnetization in the transverse plane due to loss of phase coherence. Since the net magnetization depends upon the combined contribution of a large number of nuclei, its value is largest when all the nuclei are in phase. However, the removal of the RF pulse causes the nuclei to de-phase, causing an exponential decay in magnetization described by a time constant T_2. Both the T_1 and T_2 values depend upon tissue type, and it is this property that allows for the creation of structural MR images that can be used to differentiate between different tissue types.

The term T_2^* is similar to T_2, but also depends on local inhomogeneities in the magnetic field caused by changes in blood flow and oxygenation. These inhomogeneities cause the nuclei to de-phase quicker than they normally would. Certain pulse sequences are able to eliminate the effects of these inhomogeneities, while others seek to emphasize them. Thus it is possible to produce images sensitive primarily to T_1, T_2, or T_2^*. The T_2^* signal provides the basis for functional MRI, as it is sensitive to neurovascular changes that accompany psychological and behavioral function.

6.2.1.3 Image Contrast

One of the reasons for the versatility of the MR scanner is its ability to create images based on a variety of different contrasts that are sensitive to both the number and properties of the nuclei being imaged. To illustrate, assume that the initial value of the net magnetization prior to excitation is given by the value M_0. By altering how often we excite the nuclei (TR) and how soon after excitation we begin data collection (TE), we can control which characteristic of the tissue is emphasized. This relationship can be seen by noting that the measured signal is approximately equal to

$$M_0(1 - e^{-TR/T_1})e^{-TE/T_2}. \tag{6.1}$$

If one, for example, chooses a long TR and short TE value, the signal will be approximately equal to M_0, which in turn is proportional to the number of nuclei (or protons) in the tissue. Hence, these settings can be used to produce so-called proton-density images that provide maps over how hydrogen is distributed across the sample. When the TE is short ($\sim 20\ ms$), but the TR is of intermediate length, we instead get T_1-weighted images, which are typically used to reveal anatomical structure. Finally, for T_2-weighted images, another type of structural image, a long TR and an intermediate TE should be chosen. Because T_1 and T_2 vary with tissue type, T_1- and T_2-weighted images can be used to provide detailed representations of the boundaries between gray matter, white matter, and cerebrospinal fluid (CSF).

Because T_2^* is sensitive to flow and oxygenation, T_2^*-weighting can be used to create images of brain function. T_2^*-weighted images are obtained in a similar manner to T_2-weighted images. The difference lies in manner in which the pulse sequence uses the magnetic gradients. This is beyond the scope of this chapter, but interested readers are referred to (21)

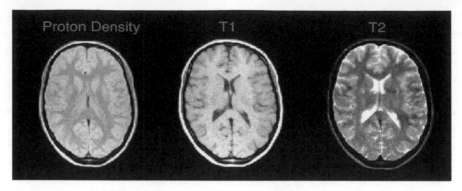

FIGURE 6.2
Examples of proton density, T_1, and T_2-weighted images.

or (23). See Fig. 6.2 for examples of the difference in proton density, T_1, and T_2-weighted images. In particular, note that the images highlight different anatomical properties of the underlying sample, and their usage depends upon the goals of the study.

6.2.2   Image Formation

The goal of exciting nuclei in the MRI scanner is ultimately to obtain enough information to be able to construct an image of the underlying sample. Any image is represented by a matrix of numbers that correspond to spatial locations. The images generally depict the spatial distribution of some property of the nuclei within the sample. This could be the density of nuclei, their mobility, or the relaxation time of the tissues in which they reside. Pulse sequences define particular manipulations of RF pulses and the shape of the magnetic field that allow us to reconstruct the acquired data into a map of the underlying signal sources, i.e., the hydrogen atoms, and obtain images of the brain.

While most MRIs are 3-D representations of the brain, they are almost always constructed through the acquisition of a series of 2-D slices. This acquisition can be performed in either a sequential or interleaved manner. To illustrate, consider that we are interested in acquiring a total of N_z slices of the brain. Using a sequential scheme, the slices are acquired in order, in either an ascending or descending manner. Using an interleaved scheme, one instead collects data in an alternating slice order. This can minimize the risk of signal bleeding from an adjacent slice that was previously excited. Both sequential and interleaved sequences have different pros and cons, though a detailed discussion is beyond the scope here.

In general, the process of exciting nuclei only provides information about the net magnetization within the slice. In order to construct a meaningful image of the brain we must find ways to extract information about the spatial contributions to the net magnetization, i.e., how much each individual voxel contributes to its value. This is done using a system of gradient coils that manipulate the magnetic field within the chosen slice and sequentially control its spatial inhomogeneity. In short, this process allows one to express each measurement of the signal as the Fourier transformation of the spin density at a single point in the frequency domain, or k-space, as it is commonly called in the field.

Conceptually, we can think of k-space being sampled at a number of discrete locations $(k_x(t_j), k_y(t_j))$ at time t_j. Here $t_j = j\Delta_t$ is the time of the jth measurement, where Δ_t depends on the sampling bandwidth of the scanner; typically it takes values in the range of $250 - 1000\ \mu s$. See Chapter 3 or (35) for more detail. Mathematically, the measurement of

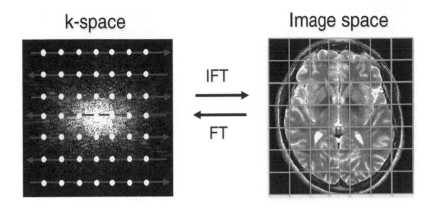

FIGURE 6.3
The raw data obtained from the MRI scanner is sampled in k-space, typically on a uniform grid. Using the inverse Fourier transform (IFT) the data can be transformed into image space where data analysis is performed.

the MR signal at the jth time point of a readout period can be written

$$S(t_j) \approx \int_x \int_y M(x,y)e^{-2\pi i(k_x(t_j)x+k_y(t_j)y)}dxdy, \tag{6.2}$$

where $M(x,y)$ is the spin density at the point (x,y). This is the entity that we seek to measure at each voxel of the brain.

To reconstruct a single MR image, one needs to sample a large number of individual k-space measurements. The exact number depends upon the desired image resolution. For example, to fully reconstruct a 64×64 image, a total of 4096 separate measurements are required, each sampled at a unique coordinate of k-space. There is a time cost involved in sampling each point, and therefore the time it takes to acquire an image is directly related to its spatial resolution.

There exist a variety of approaches towards sampling k-space. In echo-planar imaging (EPI) k-space is sampled uniformly around its origin (42). This allows for the quick and easy reconstruction of the image using the Fast Fourier Transform (FFT); see Fig. 6.3 for an illustration. Alternatively, one can use non-uniform trajectories, such as the Archimedean spiral (17). While these trajectories provide a number of benefits relating to speed and signal-to-noise ratio, the FFT algorithm cannot be directly applied to the non-uniformly sampled raw data. Instead, the raw data are typically interpolated onto a Cartesian grid in k-space and thereafter the FFT is applied to reconstruct the image. Image reconstruction is described in more detail in Chapter 8.

The k-space signal is measured over two channels, and therefore the raw k-space data will be complex valued. It is assumed that both the real and imaginary components are measured with independent normally distributed error. Since the Fourier transformation is a linear operation, the reconstructed data will also be complex valued in each voxel, with both parts following a normal distribution. In the final stage of the reconstruction, the complex valued measurements are separated into magnitude and phase components. In the vast majority of studies only the magnitude portion of the images are used in the data analysis, while the phase portion is discarded.

It is important to note that the magnitude images no longer follow a normal distribution, but rather a Rice distribution (20). The shape of this distribution depends on the signal-to-noise (SNR) ratio within the voxel. For the case when no signal is present, it behaves like a Rayleigh distribution. When the SNR is high, the distribution will be approximately Gaussian.

6.2.3 From MRI to fMRI

The data acquisition and reconstruction techniques discussed so far provide the means for obtaining a static 3-D image of the brain. However, changes in brain hemodynamics in response to neuronal activity impact the local intensity of the MR signal. Therefore, a sequence of properly acquired T_2^*-weighted brain images allow for the study of changes in brain function over time. This can be achieved by repeatedly measuring T_2^*-weighted images of the brain every few seconds. The time between successive 3-D brain volumes is referred to as the repetition time, or TR. In most fMRI experiments the TR is on the order of 2 seconds. However, depending on the research question, values can vary from anywhere between 0.1–6 seconds.

During the course of the scanning session the subject is either asked to perform a certain task, experience an induced psychological or behavioral state, or simply rest. The types of tasks that are performed depend upon the research question and can vary from the relatively mundane (e.g. tapping your fingers) to the complex. Fig. 6.4 shows an example of a T_1-weighted structural scan, as well as T_2^*-weighted functional images collected under two conditions. In recent years it has become increasingly common to perform so-called resting-state scans, where the subject lies still without actually performing an explicit task. These types of studies have been used to investigate synchronous activations in spatially distinct regions of the brain, which are thought to reflect functional systems supporting cognitive processes, and this will be discussed further in Section 14.4.

A standard fMRI experiment gives rise to massive amounts of data. It can consist of multiple subjects (usually 10–50, though some larger studies now include hundreds or thousands of participants) potentially brought in for multiple scanning sessions. Each session consists of a number of runs, or replications of a certain experimental task. Each run consists

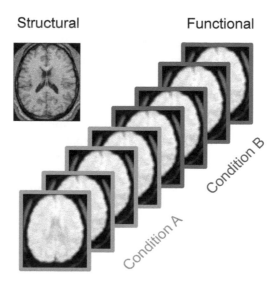

FIGURE 6.4
Example of a structural T_1-weighted image and corresponding T_2^*-weighted functional images measured during two conditions.

of a series of brain volumes, each volume is made up of multiple slices, and each slice contains many voxels that have an intensity values associated with it. On top of that, multiple high-resolution structural scans are collected for preprocessing and presentation purposes, and often diffusion tensor imaging (DTI) is also collected to inform subsequent network analyses. Hence, the amount of available data from a single experiment is enormous.

There are a number of critical determinations that go into designing the manner in which fMRI data is acquired. Certain issues directly related to the type of magnet and pulse sequences that are used, may ultimately depend on the particular lab where the data is being collected and could be outside of the researchers' control. However, there still remain many decisions that should be carefully discussed within the research team before beginning the process of acquiring the data.

One set of decisions concerns the desired spatial and temporal resolution of the study. The temporal resolution determines our ability to separate brain events in time. In fMRI its value depends upon how quickly each individual image is acquired, i.e., the TR. In contrast, the spatial resolution determines our ability to distinguish changes in an image across different spatial locations. The manner in which fMRI data is collected makes it impossible to simultaneously increase both, as increases in temporal resolution limit the number of k-space measurements that can be made in the allocated sampling window and thereby directly influence the spatial resolution of the image. Therefore there are inherent trade offs required when determining the appropriate spatial and temporal resolutions used in an fMRI experiment.

As previously mentioned, 3-D brain volumes are typically acquired in a series of axial slices (over the xy-plane with a fixed z value) of a certain slice thickness. Each slice is measured over a certain spatial extent, referred to as the field-of-view (FOV). The matrix size, or the number of voxels acquired in the xy-plane, determines the spatial resolution with higher values giving better resolution. Together the slice thickness, FOV and matrix size determine the size of the voxel. See Fig. 6.5 for an illustration. For example, consider that we choose to acquire a series of thirty 5-mm slices with a FOV of 192 mm, as this is typically sufficient to cover the entire brain. In addition, suppose the matrix size is 64×64, corresponding to 4096 voxels in the brain. Thus, the dimensions of each voxel will be $3 \times 3 \times 5$ mm.

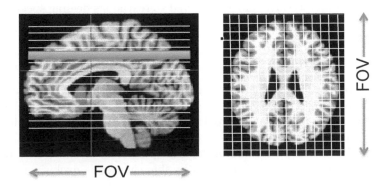

FIGURE 6.5
Each brain volume consists of multiple axial slices measured over a certain spatial extent, denoted the field-of-view (FOV). The matrix size, or the number of voxels acquired in the xy-plane, determines the spatial resolution with higher values giving better resolution. Together, the slice thickness, FOV and matrix size determine the size of the voxel.

Structural images tend to have high spatial resolution, but as they are static images, they lack any temporal resolution and generally do not reflect function at all. These are typically T_1-weighted images, as these are useful for distinguishing between different types of tissue. One of the benefits of MRI as an imaging technique is its ability to provide detailed anatomical scans of gray and white matter with a spatial resolution well below 1 mm^3. However, the time needed to acquire such an image is prohibitively high and currently not feasible for use in functional studies. Hence, by necessity, functional images have lower spatial resolution, but higher temporal resolution. As such they can be used to relate changes in signal to an experimental manipulation. The spatial resolution is typically on the order of $3 \times 3 \times 5$ mm, corresponding roughly to image dimensions on the order of $64 \times 64 \times 30$, which can readily be sampled with a resolution of approximately 2 seconds. However, with modern high-resolution imaging, combined with higher field strengths and new acquisition techniques, it is possible to achieve much higher resolution. For example, with simultaneous multi-slice acquisition at 7T, it is possible to acquire $2 \times 2 \times 2$ mm data across the brain in less than 1 second.

Regardless of these limitations, fMRI still provides relatively high spatial resolution compared with many other functional imaging techniques, including positron emission tomography (PET), electroencephalography (EEG), and magnetoencephalography (MEG). This balance of spatial and temporal resolution, together with its non-invasiveness, accounts for the popularity of the technique compared to other modalities. Advances in high-field imaging and parallel acquisition methods promise to further push the boundaries of both spatial and temporal resolution.

6.3   BOLD fMRI

The ability to connect measures of brain activation, obtained using fMRI, with the underlying neuronal activity that caused them, will greatly impact the choice of statistical procedure as well as the subsequent conclusions that can be made from the experiment. Therefore from a modeling perspective, it is critical to gain some basic understanding of brain physiology to better understand the data at hand. Again, the overview presented here is by necessity brief and interested readers are referred to textbooks dealing specifically with the subject (e.g., (23)).

6.3.1   Understanding BOLD fMRI

The most popular approach for performing fMRI uses the Blood-Oxygenation-Level-Dependent (BOLD) contrast (50, 28), measured using the difference in signal between a series of T_2^*-weighted images. Other methods for performing fMRI are available, but less widely used. These include several varieties of Arterial Spin Labeling (ASL), which use pulse sequences sensitive to blood volume or cerebral perfusion. Because it is by far the most common method currently used, our focus in this chapter is exclusively on BOLD physiology.

BOLD imaging takes advantage of differences in the magnetic properties of oxygenated and deoxygenated hemoglobin. As neural activity increases, so do the metabolic demands for oxygen and nutrients in affected regions of the brain. Neural firing signals the extraction of oxygen from hemoglobin in the blood. This extraction causes the hemoglobin to become paramagnetic as iron atoms are more exposed to the surrounding water. This creates small distortions in the magnetic field that cause a decrease in T_2^*, leading to a faster decay of

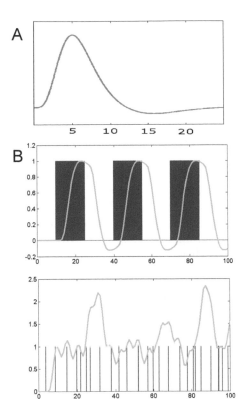

FIGURE 6.6

(A) The standard canonical HRF model used in fMRI data analysis. (B) The BOLD response modeled as the convolution of the stimulus function and the HRF. Here we see examples using both a block and event-related design.

the signal and a local decrease in BOLD signal. A subsequent over-compensation in blood flow increases the amount of oxygenated hemoglobin, leading to reduced signal loss and increased BOLD signal in the affected region.

Initially, fMRI was performed by injection of contrast agents with paramagnetic properties such as iron. However, the discovery that the T_2^* relaxation rate of oxygenated hemoglobin was longer than that of deoxygenated hemoglobin led to the advent of BOLD imaging, which has since come to dominate the field.

BOLD fMRI allows us to study the hemodynamic responses to neural firing. The change in the MR signal caused by a neural event is typically referred to as the hemodynamic response function (HRF); see Fig. 6.6A for an illustration. The increased metabolic demands due to neuronal activity lead to increases in the inflow of oxygenated blood to active regions of the brain. Since more oxygen is supplied than actually consumed, this leads to a decrease in the concentration of deoxygenated hemoglobin, which leads to an increase in signal. This positive rise in signal has an onset approximately 1–2 seconds after the onset of neural activity and peaks 5–8 seconds after peak neural activity. After reaching its peak level, the BOLD signal decreases to a below-baseline level, which is sustained for roughly 10 seconds. This effect, known as the post-stimulus undershoot, is due to the fact that blood flow decreases more rapidly than blood volume, thereby allowing for a greater concentration of deoxygenated hemoglobin in previously active brain regions.

A number of studies have shown evidence of a decrease in oxygenation levels in the time immediately following neural activity, giving rise to decreased BOLD signal in the first half second following activation. This is believed to be due to oxygen extraction taking place prior to the inflow of oxygenated blood, and is usually referred to as the initial dip (42, 41). The ratio of the amplitude of the dip compared to the positive BOLD signal depends on the strength of the magnet and has been reported to be roughly 20% at 3 Tesla. There is also evidence that the dip may be more localized to areas of neural activity (e.g., (70, 34) than the subsequent rise, which appears less spatially specific. Due in part to these reasons, the negative response has so far not been reliably observed and its existence remains controversial (37).

Clearly, the BOLD signal only provides an indirect measure of the quantity we actually seek to measure, which is the underlying neural activation. It is therefore important to understand how well the BOLD signal reflects actual increases in neuronal firing. The answer to this question is complex, and understanding the physiological basis of the BOLD response has long been a topic of intense research interest (e.g. (6)). In short, it has been shown that the BOLD signal corresponds closely to the local electrical field potential surrounding a group of cells, which is likely to reflect changes in post-synaptic activity under many conditions (38). However, neural activity and BOLD signal may, under other conditions, become decoupled. Thus, the BOLD signal is only likely to reflect a portion of the changes in neural activity in response to a task or psychological state. For this reason many regions may exhibit changes in neural activity that are missed because they do not change the net metabolic demand of the region.

6.3.2   Spatial Limitations

One of the primary benefits of fMRI is that it provides relatively high spatial resolution compared with many other functional imaging modalities, such as PET, MEG and EEG. The spatial resolution of fMRI can be made to be less than 1 mm^3 in high-field imaging of animals, but is typically on the order of 27–36 mm^3 for human studies. For these reasons, features such as cortical columns and sub-nuclei cannot easily be identified, and it may be difficult to study certain small-scale features using fMRI.

The limiting factors in fMRI include signal strength and the point-spread function of BOLD imaging, which typically extends beyond the actual neural activation sites into draining veins. The fact that only a fraction of the oxygen that flows to an active region is actually extracted, leads to oxygen-rich blood entering the venous system, thus increasing the BOLD signal in areas removed from the active neurons. This point-spread function decreases as the magnetic field strength increases, and interacts with head movement and physiological noise.

In addition, there are multiple analysis choices that ultimately limit the spatial resolution in most studies. First, it is common to spatially smooth fMRI data prior to analysis, which leads to a decrease in the effective resolution of the data. Second, making inferences about populations of subjects requires analyzing groups of individuals, each with inherent differences in brain structure. Usually, individual brains are aligned to one another through a registration or warping process, which introduces substantial blurring and noise in the group average.

One can potentially improve inferences in space by advances in both the manner data is acquired and preprocessed. For example, an important innovation in the area of acquisition has been the use of multiple coils with different spatial sensitivities to simultaneously measure k-space (61, 54). This approach, known as parallel imaging, allows for an increase in the amount of data that can be collected in a given time window. Hence, it can be used to either increase the spatial resolution of an image or decrease the amount of time required to sample an image with a certain specified spatial resolution. Parallel imaging techniques have already had a great influence on the way data is collected, and their role will only increase. The appropriate manner to deal with the acquisition and reconstruction of multi-coil data is a key direction for future research.

Similarly, in the area of preprocessing, the introduction of enhanced spatial inter-subject normalization techniques and improved smoothing techniques could help researchers avoid the most dramatic effects of blurring the data. In addition, adaptive smoothing techniques can be used to more efficiently retain boundaries between different tissue types.

6.3.3   Temporal Limitations

The temporal resolution of an fMRI study depends on the TR, which in most fMRI studies ranges from 0.5–4.0 seconds. Clearly, these values indicate a fundamental disconnect between the underlying neuronal activity, which takes place on the order of tens of milliseconds, and the temporal resolution of the study. However, the statistical analysis of fMRI data is primarily focused on using the positive rise in the BOLD response to study the underlying neural activity. Hence, the limiting factor in determining the appropriate temporal resolution is generally not considered the speed of data acquisition, but instead the speed of the underlying evoked hemodynamic response to a neuronal event. Since inference is based on oxygenation patterns taking place 5–8 seconds after activation, TR values in the range of 2 seconds have generally been considered adequate.

However, the currently used resolutions are not conducive to efficient modeling of physiological artifacts present in the fMRI signal. For example, heart rate and respiration give rise to periodic fluctuations that are difficult to model due to violations of the Nyquist criteria, which states that it is necessary to have a sampling rate at least twice as high as the frequency of the periodic function one seeks to model. At standard temporal resolutions this is clearly violated, and heart rate and respiration are often left un-modeled. Because of aliasing, these fluctuations tend to be distributed throughout the time course, giving rise to temporal autocorrelation in the signal. As fMRI signal generally suffers from low signal-to-noise ratio (SNR) and physiological artifacts potentially make up a large portion of the noise component, this may be a serious impediment. There has recently been active research in increasing the temporal resolution of fMRI studies, making TRs on the order of hundreds of milliseconds possible (34, 29, 48). These advances may ultimately allow us to circumvent many of these issues.

Because of the relatively low temporal resolution and the sluggish nature of the hemodynamic response, inference regarding when and where activation is taking place is based on oxygenation patterns outside of the immediate vicinity of the underlying neural activity we want to base our conclusions on. Since the time-to-peak positive BOLD response occurs in a larger time scale than the speed of brain operations, there is a risk of unknown confounding factors influencing the ordering of time-to-peak relative to the ordering of brain activation in different regions of interest. For these reasons it is difficult to determine the absolute timing of brain activity using fMRI. However, studies have shown (46, 47) that the relative timing within a voxel in response to different stimuli can be accurately captured in a well-designed experiment. There are also indications that focusing inference on features related to the initial dip can help alleviate concerns (33) regarding possible confounders. However, these types of studies require significant increases in temporal resolution and the ability to rapidly acquire data becomes increasingly important.

Another way of improving inferences in time is through appropriate experimental design. In principal it is possible to measure the HRF at a finer temporal resolution than the TR as long as the onsets of repeated stimuli are jittered in time relative to when the data is collected (12). For example, if the onset of a repeated stimulus is shifted by half a TR in a fraction of the stimuli, it may possible to estimate the HRF at a temporal resolution of $TR/2$, compared to a resolution of TR if jittering is not performed.

A series of recent technological developments referred to as "multiband" or "simultaneous multi-slice" MRI (29, 48), have sped up the temporal resolution of fMRI acquisitions by approximately an order of magnitude (i.e., from 2 s to 0.2 s, for whole-brain imaging), and appear likely to offer the possibility for even further acceleration. In contrast to standard acquisition techniques, multiband MRI excites multiple slices simultaneously, and the MR signals from these slices are then separated using multiple receiver coils and the aid of special encoding techniques. The introduction of multiband MRI promises to change the manner in which fMRI data is acquired and how it is analyzed.

6.3.4   Acquisition Artifacts

As with almost all types of physical measurements, fMRI data can be corrupted by acquisition artifacts. These artifacts arise from a variety of sources, including head movement, brain movement and vascular effects related to periodic physiological fluctuations, and reconstruction and interpolation processes. In particular, fMRI data often contain transient spike artifacts and slow drift over time related to a variety of sources, including magnetic gradient instability, RF interference, and movement-induced and physiologically induced inhomogeneities in the magnetic field. These artifacts will likely lead to violations of the assumptions of normally and identically distributed errors that are commonly made in subsequent statistical analysis. Unless they are properly dealt with, the consequences will include reduced power in group-level analysis and potentially increased false positives in single-subject inference. As always, it is critical to examine the data (preferably in as raw a form as possible), in order to diagnose problems. However, this can be challenging given the massive amount of fMRI data collected.

A significant source of signal variations includes the substantial slow drift of the signal across time. The presence of this low-frequency noise component in fMRI can obscure results related to a psychological process of interest and produce false positive results. Therefore, it is usually removed statistically prior to or during analysis. A consequence of slow drift is that it is often impractical to use fMRI for designs in which a process of interest only unfolds slowly over time or only happens once, such as the experience of strong emotions, though alternative techniques such as Arterial Spin Labeling or BOLD functional connectivity may be suitable for this purpose. The vast majority of fMRI designs use discrete events that can be repeated many times over the course of the experiment. For example, the most common method for studying emotional responses in fMRI is the repeated presentation of pictures with emotional content.

Susceptibility artifacts in fMRI occur because magnetic gradients near air and fluid sinuses and at the edges of the brain cause local inhomogeneities in the magnetic field that affects the signal, causing distortion in echo-planar sequences and blurring and dropout in spiral sequences. These problems increase at higher field strengths and provide a significant barrier for performing effective high-field fMRI studies. Not all scanner/sequence combinations can reliably detect BOLD activity near these sinuses. The regions most affected include the orbitofrontal cortex, inferior temporal cortex, hypothalamus, and amygdala. Some signal may be recovered by using optimized sequences or improved reconstruction algorithms.

Figure 6.7 shows examples of several types of artifacts, including susceptibility artifacts that are endemic to fMRI and several other types that can usually be controlled.

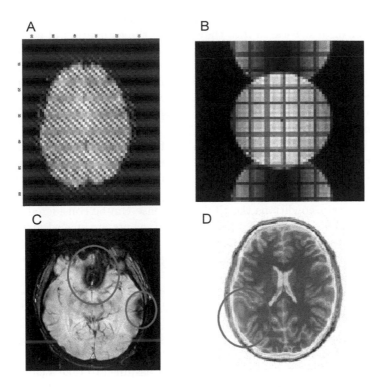

FIGURE 6.7

Examples of common fMRI artifacts: (A) k-space artifact; (B) ghosting in a phantom; (C) susceptibility artifact (dropout); and (D) normalization artifact.

6.4 Modeling Signal and Noise in fMRI

In order to appropriately model fMRI data, it is important to gain a better understanding of the components present in a BOLD fMRI time series. In general, it consists of the BOLD signal, which is the component of interest, a number of nuisance parameters and noise. Here we discuss each component in detail and discuss modeling strategies.

6.4.1 BOLD Signal

The evoked BOLD response is a complex, nonlinear function of the results of neuronal and vascular changes (66, 6), which complicates our ability to appropriately model its behavior. The shape of the response depends both upon the applied stimulus and the hemodynamic response to neuronal events. A number of methods for modeling the BOLD response, as well as the underlying HRF, exist in the literature. A major difference between the methods lies in how the relationship between the stimulus and BOLD response is modeled. In particular, we differentiate between non-linear physiological-based models, such as the Balloon model (6, 56), and models that assume a linear time invariant (LTI) system.

The Balloon model describes the dynamics of cerebral blood volume and de-oxygenation and their effects on the resulting BOLD signal. It consists of a set of ordinary differential equations that model changes in blood volume, blood inflow, deoxyhemoglobin and flow-inducing signal and describes how these changes impact the observed BOLD response. While models of this type tend to be more biophysically plausible than their linear counterparts, they have a number of drawbacks. First, they require the estimation of a large number of model parameters. Second, they do not always provide reliable estimates with noisy data, and third, they do not provide a direct framework for performing inference. In general, they are not yet considered feasible alternatives for performing whole-brain multi-subject analysis of fMRI data in cognitive neuroscience studies. However, they are a critical component in the study of brain connectivity using Dynamic Causal Modeling (DCM). This is discussed in Section 6.7.2 and in more detail in Chapter 16.

While the flexibility of nonlinear models is attractive, linear models allow for robust and interpretable characterizations in noisy systems. It is common to assume a linear relationship between neuronal activity and BOLD response, where linearity implies that the magnitude and shape of the evoked HRF does not depend on any of the preceding stimuli. Studies have shown that under certain conditions the BOLD response can be considered approximately linear with respect to the stimulus (6), particularly if events are spaced at least 5 seconds apart, though there are still some nonlinearities (10%) at 5-second spacing (47). However, other studies have found that nonlinear effects in rapid sequences (e.g., stimuli spaced less than 2 seconds apart) can be quite large (3, 66).

The ability to assume linearity is important, as it allows the relationship between stimuli and the BOLD response to be modeled using a linear time-invariant system, in which assumed neuronal activity (based on task manipulations) constitutes the input, or impulse, and the HRF is the impulse response function. In a linear system framework, the signal at time t, $x(t)$, is modeled as the convolution of a stimulus function $v(t)$ and the hemodynamic response $h(t)$, i.e., $x(t) = (v * h)(t)$. Here $h(t)$ is either assumed to take a canonical form, or alternatively modeled using a set of linear basis functions. See Fig. 6.6B for two examples.

An LTI system is characterized by its' scaling, superposition and time-invariance properties. Scaling implies that if the input is scaled by a factor b, then the BOLD response will be scaled by the same factor. This is important as it implies that the amplitude of the measured signal provides a measure of the amplitude of neuronal activity. Therefore the relative difference in amplitude between two conditions can be used to infer that the neuronal activity was similarly different. For this reason much of the activation analyses performed on fMRI data is focused on studying contrasts between the responses to stimuli at different levels. Superposition implies that the response to two different stimuli applied together is equal to the sum of the individual responses. Finally, time-invariance implies that if a stimulus is shifted by a time t, then the response is similarly shifted by t. These three properties allow us to differentiate between responses in various brain regions to multiple closely spaced stimuli.

When using an LTI system, as with any model, one makes a number of assumptions. First, it is assumed that the BOLD response is linear. Studies have shown that this is reasonable (6), though some departures from linearity have been observed. For example, there is some evidence of refractory effects, which are reductions in amplitude of a response as a function of inter-stimulus intervals. This may cause the amplitude of closely spaced stimuli to be overestimated. Second, it is assumed that the neural activity function is correctly modeled. As this is typically assumed to be equal to the experimental paradigm, one must assume this provides a reasonable proxy for the underlying neuronal activity. Third, it is assumed that the HRF is correctly modeled. Often researchers assume a canonical shape for the HRF. A popular choice is the linear combination of two gamma functions, i.e.

$$h(t) = \left(\frac{t^{\alpha_1-1}\beta_1^{\alpha_1}e^{-\beta_1 t}}{\Gamma(\alpha_1)} c \frac{t^{\alpha_2-1}\beta_2^{\alpha_2}e^{-\beta_2 t}}{\Gamma(\alpha_2)} \right), \tag{6.3}$$

where $\alpha_1 = 6$, $\alpha_2 = 16$, $\beta_1 = \beta_2 = 1$, $c = 1/6$ and Γ represents the gamma function. This particular shape, seen in Fig. 6.6A, is based on empirical findings from data extracted from the visual cortex.

However, it is critically important to note that the timing and shape of the HRF are known to vary across the brain, within an individual and across individuals (1, 59). Part of the variability is due to the underlying configuration of the vascular bed, which may cause differences in the HRF across brain regions in the same task for purely physiological reasons (64). Another source of variability is differences in the pattern of evoked neural activity in regions performing different functions related to the same task.

One of the major shortfalls when analyzing fMRI data is that users typically assume a canonical HRF (19), which may lead to mis-modeling of the signal in large portions of the brain (39). It is therefore important that these regional variations are accounted for when modeling the BOLD signal. This is often handled by modeling the HRF using multiple basis functions, and using a linear combination to better fit the evoked BOLD responses. To illustrate, suppose we model the HRF as a linear combination of temporal basis functions, $f_i(t)$, such that

$$h(t) = \sum \beta_i f_i(t). \tag{6.4}$$

Then the BOLD response can be rewritten:

$$x(t) = \sum \beta_i (s * f_i)(t) \tag{6.5}$$

where each corresponding β_i describes the weight of the ith component.

The ability to use basis sets to capture variations in hemodynamic responses depends both on the number and shape of the reference waveforms that are used in the model. For example, the finite impulse response (FIR) basis set consists of one free parameter for every time-point following stimulation in every cognitive event type that is modeled (16, 18). It can be used to estimate HRFs of arbitrary shape for each event type in every voxel of the brain. Another approach is to use the canonical HRF together with its temporal and dispersion derivatives to allow for small shifts in both the onset and width of the HRF. Other choices of basis sets include those composed of principal components (1, 68), cosine functions (71), radial basis functions (56), spectral basis sets (30) and inverse logit functions (32). For a critical evaluation of a number of commonly used basis sets, see (32) and (31).

6.4.2 Noise and Nuisance Signal

The measured fMRI signal is corrupted by random noise and various nuisance components that arise from hardware limitations and the subjects themselves. One source of variability are the fluctuations in the MR signal intensity caused by thermal motion of electrons within the subject and the scanner gives rise to noise that tends to be highly random and independent of the experimental task. The amount of thermal noise increases linearly as a function of the field strength of the scanner, with higher field strengths giving rise to more noise. However, it does not tend to exhibit spatial structure and averaging the signal over multiple voxels can minimize the effects.

Another source of variability in the signal is due to scanner drift, caused by scanner instabilities, which result in slow changes in voxel intensity over time, so-called low-frequency noise. The amount of drift varies across space, and it is important to include this source of variation in your models. Because of drift, most of the power in the time course lies in the low-frequency portion of the signal. To remove the effects of drift, it is common to remove fluctuations below a specified frequency cutoff using a high-pass filter. This can be performed either by applying a temporal filter as a preprocessing step, or by including covariates of no interest into the model. As an example of the latter, the drift can be modeled using a pth

order polynomial function or a series of low-frequency cosine functions. The most important issue when using a high-pass filter is to ensure that the fluctuations induced by the task design are not in the range of frequencies removed by the filter, as this may cause us to throw out the signal of interest. Hence, the ultimate choice in how to model the drift needs to be made with the experimental design in mind.

When subjects move their heads in the MRI scanner, the sequence of measurements corresponding to a given voxel in the resulting images may actually be composed of values originating from different brain locations. This necessitates motion correction, in which, prior to analysis, researchers estimate the between-scan movement using a rigid body transformation, and then realign the images. However, this procedure does not correct for so-called spin history artifacts, or changes in the magnetic field caused by head motion that lead to nonlinear, time-varying distortion of the resulting brain images, and there has been some debate on how to deal with these residual artifacts (25).

Some researchers (25, 40) suggest including motion regressors as nuisance covariates in the model for the BOLD response to adjust for this error, arguing that this yields estimates that seem more reasonable than those obtained when these covariates are omitted. But as head motion tends to be task related, there is concern that inclusion of these covariates can lead to underestimating the signal component due to genuine activation (53). Reference (10) argues that inclusion of motion regressors has a generally detrimental effect on activation for subjects with minimal head motion. In contrast, in recent work we have shown (57) that omitting signal components due to systematic error correlated with the task will generally lead to biased estimates of the effects of interest in a standard GLM analysis. Thus, we recommend that motion regressors be included in models of the BOLD response.

Physiological noise due to patient respiration and heartbeat can, as previously discussed, cause fluctuations in signal across both space and time. Physiological noise can in certain situations be directly estimated from the data (33). Some of it can be removed using a properly designed band-pass filter. However, in most studies, with TR values ranging from 2–4s, one cannot hope to estimate and remove the effects of heart rate and respiration based solely on the observed fMRI time series. According to the Nyquist theorem, it is necessary to have a sampling rate at least twice as high as the frequency of the periodic function one seeks to model. If the TR is too slow, which is true in most fMRI studies, there will be problems with aliasing. In this situation the periodic fluctuations will be distributed throughout the time course, giving rise to temporal autocorrelation. Some groups have therefore begun directly measuring heartbeat and respiration during scanning and using this information to remove signal related to physiological fluctuations from the data (9). This is done either as a preprocessing step, or by including these terms as covariates in the model.

In standard time series analysis, model identification techniques are used to determine the appropriate type and order of a noise process. In fMRI data analysis this approach is not feasible due to the large number of time series being analyzed, and noise models are typically specified *a priori*. Noise in fMRI is typically modeled using either an AR(p), with p set to either 1 or 2, or an ARMA(1,1) process (55). Here the autocorrelation is generally thought to be due to unmodeled nuisance signal. If these terms are properly removed, there is evidence that the resulting error term corresponds to white noise (40).

In our own work, we typically use an auto-regressive process of order 2. The reason we choose an AR model over an ARMA model is that it allows us to use method of moments rather than maximum likelihood procedures to estimate the noise parameters. This significantly speeds up computation time when repeatedly fitting the model to tens of thousands of time series. Choosing the order of the AR process to be 2 has been empirically determined to provide the most parsimonious model that is able to account for autocorrelation present in the signal due to aliased physiological artifacts.

6.5   Experimental Design

The experimental design of an fMRI study is complicated as it not only involves the standard issues relevant to all psychological experiments, but also issues related to data acquisition and stimulus presentation. Not all designs with the same number of trials of a given set of conditions are equal, and the spacing and ordering of events is critical. What constitutes an optimal experimental design depends on the psychological nature of the task, the ability of the fMRI signal to track changes introduced by the task over time, and the specific comparisons that one is interested in making. In addition, as the efficiency of the subsequent statistical analysis is directly related to the experimental design, it needs to be carefully considered during the design process.

A good experimental design attempts to maximize both statistical power and psychological validity. The statistical performance can be characterized by its estimation efficiency (the ability to estimate the shape of the HRF), or its detection power (the ability to detect which voxels are active). The psychological validity is often measured by the randomness of the stimulus presentation (e.g., balanced transitional probabilities across trial types), as this helps control for issues related to anticipation, habituation and boredom; however, whether a predictable vs. unpredictable design is psychologically undesirable depends heavily on the particular paradigm and task. Thus, when designing an experiment there are inherent trade-offs between estimation efficiency, detection power and randomness. The optimal balance between the three ultimately depends on the goals of the experiment and the combination of conditions that are of primary interest. For example, a design used to localize areas of brain activation may stress high detection power at the expense of estimation efficiency and randomness. Conversely, designs that attempt to link activity to particular events or time periods during processing may emphasize estimation efficiency at the expense of detection power.

There are two major classes of designs used in most fMRI experiments, namely blocked designs and event-related designs (though these can be intermixed or hybridized); see Fig. 6.6B for examples of each. In a blocked design, experimental conditions are separated into extended time intervals, or blocks. For example, one might repeat the process of interest during an experimental block (A) and have the subject rest during a control block (B). The A-B comparison can then be used to compare differences in signal between the two conditions. Increasing the length of each block will lead to a larger evoked response during the task. This increases the separation in signal between blocks, which in turn leads to higher detection power. However, it is still important to include multiple transitions between conditions as otherwise differences in signal due to low-frequency drift may be confused for differences in task conditions and to ensure that the same mental processes are evoked throughout each block. If block lengths are too long, this assumption may be violated due to the effects of fatigue and/or boredom.

Another benefit of blocked designs is that they are robust to uncertainties in the shape of the HRF. This holds because the predicted response depends upon the total activation caused by a series of stimuli, making it less sensitive to variations in the shape of responses to individual stimulus. However in contrast, block designs provide imprecise information about the particular processes that activated a brain region and cannot readily be used to directly estimate important features of the HRF, such as the onset, width or time-to-peak.

In an event-related design the stimulus consists of short discrete events, such as brief light flashes, whose timing can be randomized. These types of designs are flexible and allow for the estimation of key features of the HRF (e.g., onset and width) that can be used to make inference about the relative timing of activation across conditions and about sustained activity. Event-related designs allow one to discriminate the effects of different conditions as long as one either inter-mixes events of different types or varies the inter-stimulus interval between trials. Another advantage of event-related designs is that the effects of fatigue, boredom, and systematic patterns of thought unrelated to the task during long inter-trial intervals can be avoided. A drawback is that the power to detect activation is typically lower than for block designs, though the ability to obtain data over more trials per unit time can counter this loss of power.

What constitutes an optimal experimental design ultimately depends on the task, as well as on the ability of the fMRI signal to track changes introduced by the task over time. It also depends on what types of comparisons are of interest. The delay and shape of the BOLD response, scanner drift and physiological noise all conspire to complicate experimental design for fMRI. Several methods have been introduced that allow researchers to optimally select the design parameters, as well as the sequencing of events that should be used in an experiment (65, 36, 26, 27). These methods define fitness measures for the estimation efficiency, detection power and randomness of the experiment, and apply search algorithms (e.g., the genetic algorithm) to optimize the design according to certain specified criteria. When defining the fitness metrics it is typically assumed that the subsequent data analysis will be performed in the general linear model (GLM) framework described in Section 9.6.1.

The use of more complex nonlinear models requires different considerations when defining appropriate fitness metrics. An important consideration relates to assumptions made regarding the shape of the HRF and the noise structure. The inclusion of flexible basis functions and correlated noise in the model will modify the tradeoffs between estimation efficiency and detection power, and potentially alter what constitutes an optimal design. Hence, even seemingly minor changes to the model formulation can potentially have a large impact on the efficiency of the design. Together these issues complicate the design of experiments and significant work remains to find the appropriate balance between them. As research hypotheses ultimately become more complicated, the need for more advanced experimental designs will only increase further.

6.6   Preprocessing

Prior to statistical analysis, fMRI data typically undergoes a series of preprocessing steps aimed at removing artifacts and validating model assumptions. The main goals are to minimize the influence of data acquisition and physiological artifacts, to validate statistical assumptions and to standardize the locations of brain regions across subjects in order to achieve increased validity and sensitivity in group-level analysis.

When analyzing fMRI data it is typically assumed that all of the voxels in a particular 3-D brain volume were acquired simultaneously. Further, it is assumed that each data point in a specific voxel's time series only consists of signal from that particular voxel; an assumption that is invalid if the subject moves between scans. Finally, when performing group analysis and making population inference, all individual brains are assumed to be registered, so that each voxel is located in the same anatomical region for all subjects. Preprocessing is used to condition the data in ways that increase the plausibility of all of these assumptions. Without appropriate preprocessing of the data prior to analysis, none of these assumptions hold and the resulting statistical analysis would be invalid.

The major steps in the fMRI preprocessing pipeline are slice acquisition-timing correction ("slice-time" correction), realignment, co-registration of structural and functional images, normalization of brains to a group template, and smoothing. Below we briefly describe each step. For more detail see Chapter 10.

Slice-time correction: A typical assumption in the analysis of fMRI data is that all voxels within a 3-D image are acquired simultaneously. In reality, different slices from the same volume are acquired sequentially in time relative to one another. Thus, many researchers seek to estimate the signal intensity in all voxels at a common standardized time point in the acquisition period. This can be done by interpolating the signal intensity at the chosen time point from the same voxel in previous and subsequent acquisitions. Some researchers do not use slice timing, as it adds interpolation error to the data, and instead use (a) more flexible hemodynamic models to account for variations in acquisition time across the brain, or (b) more rapid acquisition sequences in which multiple slices are acquired simultaneously.

Realignment: A major source of error is subject movement during the course of the experiment. Excessive motion may cause the intensity in a given voxel to be contaminated by signal from neighboring voxels. For these reasons it is critical to realign each individual image to compensate for movement. This is typically done by first choosing a reference image, either the first or mean image, and then applying a rigid body transformation to all other images in the time series to match it. This allows the images to be translated (shifted in the x, y, and z directions) and rotated (altered roll, pitch, and yaw) to match the reference image. An iterative algorithm is used to search for the parameter estimates that provide the best match between a target image and the reference image. Matching is typically performed by minimizing the sums of the squared differences between the two images. Once the best match is found, the data is interpolated into the new space.

Realignment is critically important when analyzing fMRI data. It is able to correct for small movements of the head. However, it is important to note that it is unable to correct for the more complex spin-history artifacts created by the motion. For this reason the estimated motion parameters at each time point are saved for later inspection and are often included in subsequent analysis as covariates. It is not uncommon to have to exclude subjects that move their heads too much during the course of the scan. While there do not exist firm rules stating how much movement should be considered too much, more than 1.5 mm displacement within a scanning session is typically considered to be problematic.

Co-registration: Functional MRI data is typically of low spatial resolution and therefore provides relatively little anatomical detail. It is therefore common to map the results obtained from the analysis of functional data onto a high-resolution structural MR image for presentation purposes. This process is referred to as co-registration, and is typically performed using either a rigid body (6 parameters) or an affine (12 parameters) transformation. Because functional and structural images are collected with different sequences and different tissue classes have different average intensities, using a least squares difference method to match images is often not appropriate. Instead, it is preferable to estimate the parameters of the transformation by maximizing the mutual information between the two images. Typically, a single structural image is co-registered to the first or mean functional image. Co-registration is also a necessary step for subsequent normalization procedures (described below). Here high-resolution structural images (T_1 and/or T_2) are used for warping to a standard template and localization. The same transformations are thereafter applied to the functional images, which produce the activation statistics, so the results can be mapped onto a standard space.

Normalization: For group analysis, it is necessary for each voxel to lie within the same brain structure for all subjects. However, individual brains clearly have differences in both shapes and features. That said, there are certain regularities shared by every non-pathological brain, and normalization attempts to register each subject's anatomy with a standardized atlas space defined by a template brain, reported in the standard coordinate systems of the Montreal Neurologic Institute (MNI), or that of Talairach and Tournoux (59). Normalization can be linear, involving simple registration of the gross shape of the brain, or nonlinear, involving warping to match local features. This warping consists of shifting the locations of voxels by different amounts depending on their original location.

Inter-subject registration is perhaps the largest source of error in group-level analysis of fMRI data. For these reasons it is important to inspect each normalized brain as a quality control step. This can be done in a number of ways, and researchers should develop a set of standardized procedures to assess the results.

Smoothing: Many researchers apply a spatial smoothing kernel to the functional data, blurring the image intensities in space. One reason is to improve inter-subject registration, by eliminating intra-subject differences. A second reason is that Gaussian Random Field Theory, a popular multiple-comparisons correction procedure, assumes that the variations across space are continuous and normally distributed. Smoothing typically involves convolution with a Gaussian kernel, which is a 3-D normal probability density function often described by the full width of the kernel at half its maximum height (FWHM) in millimeters. One estimate of the amount of smoothing required to meet the assumption is a FWHM of 3 times the voxel size (e.g., 9 mm for 3 mm voxels).

6.7   Data Analysis

There are several common objectives in the analysis of fMRI data. These include localizing regions of the brain activated by a certain task, determining distributed networks that correspond to brain function and making predictions about psychological or disease states. Many of these objectives are related to understanding how induced or measured psychological states lead to changes in brain activity (a combination of neural and glial function), and others are related to the analysis of ongoing spontaneous fluctuations. All these objectives are intrinsically statistical in nature, and this area is the primary domain of statisticians currently involved in the field. For this reason, the material covered in this section is the focus of many subsequent chapters of this book. Here we simply provide a brief overview of relevant topics, but refer to future chapters for a more in-depth treatment.

6.7.1   Localization

To date, the most common use of fMRI has been to localize areas of the brain that activate in response to a certain task. These types of human brain mapping studies are instrumental for increasing our understanding of brain function. The general linear model (GLM) approach is the most common statistical method for assessing relationships between tasks and brain activity (69). Here the data is considered to be a linear combination of model functions plus noise. The model functions are assumed to have known shapes, but with unknown amplitudes that need to be estimated. The GLM can be used to estimate whether the brain responds to a single event type, to compare different types of events, and to assess correlations between brain activity and behavioral or other psychological variables.

In a typical fMRI experiment, the predictors are related to psychological events, and the outcome variable is the signal from a certain brain voxel or a region of interest. Analysis is typically massively univariate, meaning that a separate GLM analysis is performed at every voxel in the brain, and summary statistics are saved in maps of statistic values across the brain. This approach assumes an improbable independence between voxel pairs. Typically dependencies between voxels are dealt with later using random field theory, which makes assumptions about the spatial dependencies between voxels.

Using the GLM, the data at each voxel can be expressed as

$$\mathbf{Y} = \mathbf{X}\beta + \epsilon, \tag{6.6}$$

where $\epsilon \sim N(\mathbf{0}, \mathbf{V})$. Here $\mathbf{Y}$ represents the data (a vector of length T), $\mathbf{X}$ is a design matrix containing information about various signal components, and $\mathbf{V}$ is a covariance matrix that incorporates information about temporal autocorrelations in the data. The latter is typically modeled using either an autoregressive (AR) or auto-regressive moving-average (ARMA) model.

Though the model formulation is simple, difficulties arise when attempting to construct an appropriate design matrix $\mathbf{X}$. This process is complicated by a number of factors, including the fact that the BOLD response contains low-frequency drift and artifacts related to head movement and cardiopulmonary-induced brain movement, the neural response shape may not be known, and the hemodynamic response varies in shape across the brain. For these reasons the design matrix $\mathbf{X}$ usually consists of both nuisance parameters (corresponding to drift components and the estimated motion parameters) and signal of interest.

The simplest version of the GLM assumes that both the stimulus function and the HRF are known. As discussed in Section 6.4.1 the stimulus is assumed to be equivalent to the experimental paradigm, while the HRF is typically modeled using a canonical HRF; see Fig. 6.6A. If unwilling to assume a fixed canonical HRF, it is possible to model the shape by expressing the HRF as the linear combination of a number of basis functions. Here each basis function is convolved with the stimulus and entered as a separate column of the design. See Chapter 11 for more information about the GLM.

Often we are interested in combining the results for individual subjects in order to perform group-level inference. Because of the hierarchical structure of the data, an appropriate approach towards analyzing multi-subject fMRI data is to use either a multi-level or mixed-effects GLM model. For example, we can express the first-level model for subject i as $\mathbf{Y}_i = \mathbf{X}\beta_i + \epsilon_i$, where $\epsilon_i \sim N(\mathbf{0}, \mathbf{V}_i)$. The second level model can in turn be written, $\beta_i = \beta + \eta_i$, where $\eta_i \sim N(0, \sigma^2)$. This can easily be fit using most statistical software packages. However, in the neuroimaging community it is often approximated by performing a GLM on each subject, and thereafter using the resulting activation parameter estimates in a "second-level" group analysis; see Chapter 12.

Regardless of whether one performs single-subject or group-level analysis, the procedure follows the same general format. First, one fits a statistical model (e.g., the GLM) to data from a certain voxel in the brain. Next, the estimated model parameters are used to test for an effect of interest, for example $H_0 : \beta_1 - \beta_2 = 0$. This procedure is then repeated for each voxel across the brain, and the results are summarized in a statistical image; see the left panel of Fig. 6.8. The final step is to determine which voxels actually show a statistically significant effect. The results of neuroimaging studies are often summarized as a set of activated regions, such as those shown in the right panel of Fig. 6.8. These types of summaries describe brain activation by color-coding voxels whose test statistics exceed a certain statistical threshold for significance. The implication is that these voxels are activated by the experimental task.

Of course, a crucial decision is the choice of which threshold to use when deciding whether voxels should be deemed "active" or not. In many fields, test statistics whose p-values are below 0.05 are considered sufficient evidence to reject the null hypothesis, with an acceptable false positive rate of 0.05. However, in brain imaging on the order of 100,000 hypothesis tests (one for each voxel) are tested at a single time. Hence, using a voxelwise alpha of 0.05 implies that 5% of the voxels on average will show false positive results. Hence, we would actually expect on the order of 5,000 false positive results. Thus, even

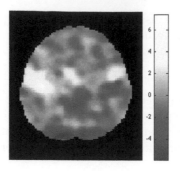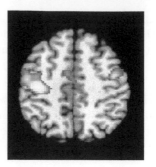

FIGURE 6.8
(Left) An example of a statistical image consisting of test statistic values at each voxel of the brain. (Right) The thresholded statistical map indicates active regions of the brain.

if an experiment produces no true activation, there is a good chance that without a more conservative correction for multiple comparisons, the activation map will show a number of activated regions, leading to erroneous conclusions.

The traditional way to deal with this multiplicity problem is to adjust the threshold so that the probability of obtaining a false positive is simultaneously controlled for every voxel (i.e., statistical test) in the brain. In neuroimaging, a variety of different approaches towards controlling the false positive rate are commonly used; see (22) and Chapter 13 for more detail. The fundamental difference between methods used is whether they control for the family-wise error rate (FWER) or the false discovery rate (FDR).

6.7.2 Connectivity

Previously, fMRI data was primarily used to construct maps representing regions of the brain activated by specific tasks. In recent years there has been increased interest in augmenting this type of analysis with connectivity studies that describe how various brain regions interact and how these interactions depend on experimental conditions. A number of methods have been suggested in the fMRI literature to quantify brain connectivity. Their appropriateness depends upon (i) what type of conclusions one is interested in making; (ii) what type of assumptions one is willing to make; (iii) the level of the analysis; (iv) the modality used to obtain the data; and (v) the number of brain regions that are included in the analysis.

The term connectivity is an umbrella term that has been used to refer to a number of related aspects of brain organization. In the neuroimaging literature it is common to distinguish between anatomical, functional and effective connectivity (17, 58). Anatomical connectivity deals with the description of how different brain regions are physically connected, and can be approached using techniques such as diffusion tensor imaging (DTI; see Chapter 4). Functional connectivity (see Chapter 14) and effective connectivity (see Chapter 16) study the functional relationships between different brain regions.

Functional connectivity is defined as the undirected association between two or more fMRI time series and/or performance and physiological variables. It makes statements about the structure of relationships among brain regions. However, methods used to access functional connectivity usually do not make any assumptions about the underlying biology and tend to be data-driven in nature.

A wide variety of methods have been proposed to study functional connectivity. The simplest approaches simply compare the bivariate correlations between regions of interest, or between a seed region and all other voxels across the brain. Recently, inverse covariance estimation methods have been applied, using the fact that for multivariate normal data, conditional independence between variables (regions) corresponds to zero entries in the inverse covariance matrix. This allows for the efficient investigation of a large number of regions simultaneously.

Other approaches towards investigating functional connectivity include using various multivariate decomposition methods to identify task-related patterns of brain activation without making a priori assumptions about its form. These methods include principal components analysis (PCA); (2)) and independent components analysis (ICA); (18, 44). These methods involve decomposing the time-by-voxel data matrix, $\mathbf{Y}$, into a set of spatial and temporal components according to some criteria (e.g., independence between components). They are discussed in detail in Chapter 15 and have been especially fruitful for analyzing so-called resting-state data, where the subjects do not perform an explicit task. This is an area of intense research where standard methods for localizing brain activation are not applicable due to the lack of stimulus.

Effective connectivity is defined as the directed influence of one brain region on the physiological activity recorded in other brain regions. It claims to make statements about causal effects among tasks and regions. Usually, methods that assess effective connectivity make anatomically motivated assumptions and restrict inference to networks comprising a number of pre-selected regions of interest.

Effective connectivity analyses are inherently model dependent. Typically, a small set of regions and a proposed set of connections are specified *a priori*, and tests of fit are used to compare a small number of alternative models and assess the statistical significance of individual connections. Because connections may be specified directionally (with hypothesized causal influences of one area on another), the models are typically thought to imply causal relationships. Because there are many possible models, the choice of regions and connections must be anatomically motivated. Thus, most effective connectivity depends upon a neuroanatomical model that describes which areas are connected, and a mathematical model that describes how areas are connected. Popular methods for assessing effective connectivity include structural equation modeling (SEM) (38), Granger causality (48), and dynamic causal modeling (DCM) (13).

Effective connectivity is popular because it is thought to provide powerful conclusions. However, the validity of these conclusions depends on certain assumptions being correct. They are often poorly specified and difficult to check, which is a major shortcoming of the field. The main problem is that results are discussed in terms of the applied estimation algorithm rather than carefully defined estimands of interest. This is discussed in more detail in Chapter 16.

In many situations we seek to create networks consisting of a large number of non-overlapping brain regions. Networks can be represented using graphs, which are mathematical structures that can be used to model pair-wise relationships between variables. They consist of sets of nodes (or vertices) V and corresponding links (or edges) E that connect pairs of vertices. A graph $G = (V, E)$ may be defined as being either undirected or directed with respect to how the edges connect one vertex to another; see Fig. 6.9 for an example.

As the number of regions included in the analysis approaches the hundreds, it can often be difficult to make sense of these vast amounts of data. Network analysis attempts to characterize these networks using a small number of meaningful summary measures. The hope is that comparisons of network topologies between groups of subjects may reveal connectivity abnormalities related to neurological or psychiatric disorders. As an example, there is growing interest in using graph theoretical approaches to investigate the organizational principles of large-scale brain networks. These studies have shown clear topological organization of the brain, including modularity, small-worldness, and the existence of highly connected network

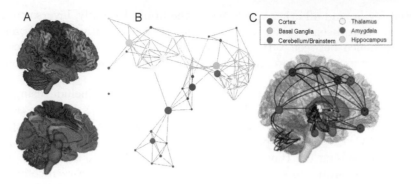

FIGURE 6.9

(A) A functional connectivity-based parcellation of the brain. (B) Co-activations in studies of disgust, from a meta-analysis of 148 neuroimaging studies. The nodes (circles) are regions or networks, color-coded by anatomical system. The edges (lines) reflect co-activation between pairs of regions or networks, assessed based on the joint distribution of activation intensity. The size of each circle reflects how strongly it connects disparate networks. C) The same connections in the anatomical space of the brain.

hubs. These network properties are thought to provide important implications for health and disease. See (58) or Chapter 17 for more detail.

6.7.3 Prediction

There is growing interest in using fMRI data for the classification of mental disorders and predicting the early onset of disease. In addition, there is interest in developing methods for predicting stimuli directly from functional data. The ability to do so opens the possibility of inferring information about subjective human experience directly from brain activation patterns.

Predicting brain states is challenging and requires the application of novel statistical and machine learning techniques (52). Various techniques have successfully been applied to fMRI data in which a classifier is trained to discriminate between different brain states and then used to predict the states in a new set of fMRI data. The application of machine learning methods to fMRI data is often referred to as multi-voxel pattern analysis (MVPA). As the name indicates, instead of focusing on single voxels, MVPA uses pattern-classification algorithms applied to multiple voxels to decode patterns of activity. In MVPA the goal is to determine the model parameters that allow for the most accurate prediction of new observations.

When applied to fMRI data, the result is often a pattern of weights across brain regions that can be applied to new brain activation maps in order to quantify the degree to which the pattern responds to a particular type of event. For example, consider the situation where we want to categorize subjects into two groups (A or B) based on their brain activation. Here the input features are measurements over all V voxels in the brain contained in the vector x. We now seek to find a weighting of these voxels, represented by the vector w of length V, so that if $xw > 0$ we categorize a subject as belonging to group A and if $xw < 0$ we categorize them as belonging to group B. Here the elements of the vector w consists of voxel-specific weights that can be mapped back onto 3D space and interpreted for scientific reasons; see Fig. 6.10.

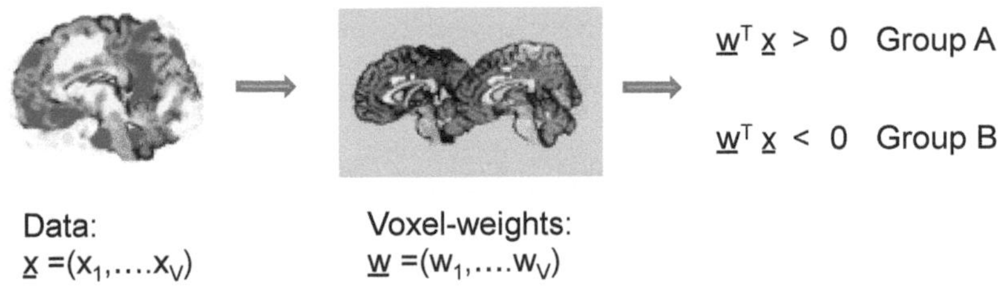

FIGURE 6.10

An MVPA example from (67). The input features are measurements over all V voxels in the brain contained in the vector **x**. These voxels are weighted by the vector **w** of length V, so that if $xw > 0$ we categorize a subject as belonging to group A and if $xw < 0$ we categorize them as belonging to group B.

When applying MVPA methods to fMRI data, it is particularly important to make analysis choices that balance interpretability with predictive power. Certain methods may give good predictions but the resulting voxel weights may be difficult to interpret and may not generalize well to new subjects.

Currently, one of the most exciting areas in neuroimaging is the work being done in decoding our thoughts based on brain activity. Here the idea is to study a person's activation and thereafter try to re-create what the subject is seeing or doing. In order to do this properly, researchers need a good mathematical model of how the brain functions and high-speed computing. Although there are a number of companies that are starting to pursue brain decoding for purposes ranging from market research to lie detection, most researchers are more excited about what this process can teach us about the way the brain functions.

6.8 Resting-State fMRI

Clearly the brain is always at work, even in the absence of an explicit task. In fact, according to certain estimates, task-related changes in neuronal metabolism only account for about 5% of the brain's total energy consumption. Resting state fMRI (rs-fMRI) is a relatively new approach to functional imaging that is used to identify synchronous BOLD changes in multiple brain regions while subjects lie in the scanner but do not perform a task (4).

Using rs-fMRI it has been shown that fluctuations in the low-frequency portion of the BOLD signal show strong correlations in spatially distant regions of the brain. While the exact mechanisms driving these correlations remain unclear, it is hypothesized that it may be due to fluctuations in spontaneous neural activity. Neuroscientists are increasingly interested in studying the correlation between spontaneous BOLD signals from different brain regions in order to learn more about the intrinsic functional connectivity of the brain.

Already rs-fMRI has revealed several large-scale spatial patterns of coherent signal in the brain during rest, corresponding to functionally relevant resting-state networks (RSNs). These networks are thought to reflect the baseline neuronal activity of the brain. A number of RSNs have been consistently observed both across groups of subjects and in repeated scanning sessions on the same subject. RSNs are localized to grey matter, and are thought to reflect functional systems supporting core perceptual and cognitive processes. Many regions that are co-activated during active tasks also show resting state connectivity, as brain regions with similar functionality tend to express similar patterns of spontaneous

BOLD activation. Sometimes subsets of RSNs appear to be either up- or down-regulated during specific cognitive tasks.

Resting-state fMRI is based on studying low-frequency BOLD fluctuations. Functionally relevant, spontaneous BOLD oscillations have been found in the lower frequency ranges (0.01–0.08 Hz). This is separable from frequencies corresponding to respiratory (0.1–0.5 Hz) and cardiovascular (0.6–1.2 Hz) signal. Typical resting experiments are of the order of 5–10 min, though the identification of an optimal duration of an rs-fMRI session and the possible need for multiple sessions remains an open issue. In addition, there is no consensus as to whether data should be collected while subjects are asleep or awake, and with eyes open or closed.

Preprocessing of rs-fMRI data typically follows the same pipeline applied to standard task-related BOLD FMRI. However, there are a few important differences. High-pass temporal filtering applied to task FMRI data may be overly aggressive with respect to removing some of the relevant frequency information. Often the data is band-pass filtered at (0.01–0.08 Hz). As it has been shown that non-neuronal physiological signals may interfere with resting-state BOLD data, the removal of confounding signals, such as respiratory or cardiovascular noise, has been shown to considerably improve the quality of data attributable to neural activity. It has therefore become common practice in rs-fMRI research to explicitly monitor these signals, and retrospectively correct for their confounding effects post-acquisition.

In addition, the global mean signal, at least six motion parameters estimated in the pre-processing stage, the cerebrospinal fluid (CSF), and the white matter signals are commonly removed prior to analysis in order to reduce the effects of head motion and non-neuronal BOLD fluctuations. However, the removal of the global signal has proven to be particularly controversial. In the past few years, there has been increased attention given to observed anti-correlations between RSNs. Anti-correlations between the components of the default-mode and attention networks have been consistently observed. However, recently there has been a lot of debate about these findings (e.g., (49)), as it is thought that global signal regression will induce a bias towards finding anti-correlations between RSNs.

Because of the lack of task, rs-fMRI is attractive as it removes the burden of experimental design, subject compliance, and training demands. It is particularly attractive for studies of development and clinical populations. In addition, it is easy to tack on a resting-state scan even when performing task-based experiments. For these reasons, the amount of available resting-state data has exploded and there is a growing subfield around the acquisition and analysis of rs-fMRI data.

As a final note, one of the primary benefits with rs-fMRI is the ability to compare data across research labs, as experiments do not need to be synchronized. This has led to a number of large-scale data sharing initiatives (e.g. 1000 Functional Connectomes Project).

6.9   Data Format, Databases, and Software

It is critical that new researchers interested in getting involved in fMRI research gain some basic understanding of the format of the data and have access to the tools needed to read, analyze and visualize them. In this section we discuss the formats that are typically used to store fMRI data, a variety of databases that will allow researchers to access fMRI data, and a number of freely available software packages that can be used to begin analyzing the data.

MRI data is usually stored in binary data files as either 8- or 16-bit integers. Additional information about the data, called meta-data, which includes image dimensions and type, are also stored. Structural MRI images are generally stored as 3-D data files, while functional MRI data can either be stored as a series of 3-D files, or as a single 4-D file.

Most MRI scanners save their reconstructed data to a file format called DICOM (Digital Imaging and Communications in Medicine). The data is generally stored slice-wise and contains a large amount of meta-data about image acquisition settings and the subject. Although DICOM is the standard format for outputting data from the MRI scanner, it is necessary to convert to other formats before performing the actual data analysis.

There are two main file formats typically used for data analysis, namely Analyze and NIFTI. The Analyze file format originates from the software package with the same name. It stores each dataset in two separate files. The first is the data file, which contains the binary data and has the extension .img. The second is the header file, which contains the header file and has the extension .hdr. NIFTI (Neuroimaging Informatics Technology Initiative) is another file format designed to promote compatibility among programs using fMRI datasets. It extends Analyze by storing additional meta-data, such as affine matrices, data arrangement, and slice order information. It comes in two different formats. The first format combines the header and binary data into a single file with the extension .nii. The second format uses the same extensions as Analyze, i.e., .hdr and .img, keeping the meta- and binary data separate from one another.

In recent years there has been an increased movement towards openly sharing fMRI data between researchers in the field. The goal is to emulate similar data sharing initiatives in other disciplines, such as genetics, that have led to important advances. The first such effort was the fMRI Data Center (fMRIDC) which has come to consist of 107 fMRI datasets. More recently, there has been a particular focus on sharing rs-fMRI data, as this type of data is particularly easy to compare across research labs, as experiments do not need to be synchronized. Perhaps, the most well-known repository of this kind is the 1000 Functional Connectomes Project (FCP), which to date consists of resting-state data on roughly 5,000 subjects from sites all around the world. The OpenFMRI Project (openfmri.org) is a relatively recent project that is particularly dedicated to the free and open sharing of data from task-based fMRI studies. Finally, on an institutional level the Human Connectome Project (HCT) is a project supported by the National Institutes of Health with the stated goal of mapping the human connectome. It will make available task-based and rs-fMRI on over 1,000 subjects, as well as data obtained using other modalities.

In addition to the above-mentioned efforts, there has been longstanding interest in sharing the activation coordinates reported in papers for use in meta-analysis. This data consists of the spatial locations of peaks of activation (peak coordinates), reported in the standard coordinate systems of the Montreal Neurologic Institute (MNI). This information can be used to pool multiple separate but similar studies, which can be used to evaluate the consistency of findings across labs, scanning procedures, and task variants, as well as the specificity of findings in different brain regions or networks to particular task types. There are several such databases containing peak coordinates that can be used for meta-analysis, including Brainmap (http://brainmap.org) and Neurosynth (http://neurosynth.org).

Finally, there are a number of free open source software packages used in the neuroimaging community that can downloaded freely from the web, that are relatively easy to use. The three most popular are SPM, FSL and AFNI. The most commonly used software is SPM (Statistical Parametric Mapping) (51), which consists of a set of MATLAB functions for preprocessing, analyzing, and displaying fMRI and PET data. It has a very active development community and there exist a large variety of add-ons to the core tools. FSL (FMRIB Software Library) (24) is written in C++, and has a number of unique tools at its disposal. It provides a comprehensive library of image analysis and statistical tools for fMRI, MRI and DTI brain imaging data. Finally, AFNI (Analysis of Functional NeuroImages) (11) is written in C. It consists of a series of programs for processing, analyzing, and displaying fMRI data. Of particular interest to statisticians might be its use of functions from the statistical software package R.

6.10   Future Developments

It is truly an exciting time to be involved in neuroimaging research. More and more increasingly ambitious experiments are being performed each day. This is creating a significant new demand, and an unmatched opportunity, for quantitative researchers working in the neurosciences. Understanding the human brain is arguably among the most complex, important and challenging issues in science today. For this endeavor to be successful, a legion of neuro-quants is needed to make sense of the massive amounts of data being generated.

With this rapid development, new research questions are opening up every day. Below we have made a decidedly non-inclusive list of exciting research areas that we feel will be increasingly important in coming years. However, by the time this book is published this list could no doubt be expanded even further!

Longitudinal Imaging Studies: Longitudinal neuroimaging studies have become increasingly common in recent years. Here the same subjects are repeatedly scanned over a prolonged period of time and changes in brain structure and function are assessed. These types of studies promise to play an important role as they have the potential to answer critical questions regarding brain development, aging, neuro-degeneration, and recovery from traumatic brain injuries or stroke.

However, with the increased amount of information, the resulting datasets will be even larger and more complex than their single-session counterparts. In addition to the standard problems of modeling within-session relationships, there is here the additional problem of correctly modeling the between-session variation and the appropriate inclusion of potential time-varying covariates. The analysis of longitudinal and repeated measures data has a long history in statistics, and it is time these ideas are moved into the high-dimensional analysis of longitudinal fMRI data.

High Temporal Resolution Multiband Data: Recently, a series of technological developments referred to as multiband MRI, have sped up the temporal resolution of fMRI acquisitions by approximately an order of magnitude (i.e., from 2 s to 0.2 s, for whole-brain imaging), and appear likely to offer the possibility for even further acceleration. As previously mentioned, most BOLD fMRI data are acquired by sequentially acquiring a series of two-dimensional slices. In contrast, multiband MRI excites multiple slices simultaneously, and the MR signals from these slices are then separated using multiple receiver coils and the aid of special encoding techniques. Thus, multiband MRI combines hardware and software innovations to significantly speed up fMRI acquisitions.

This unprecedented temporal resolution provides new challenges with regard to statistical analysis, but also offers new opportunities. Most analytic approaches to fMRI data are based on assumptions that may be inappropriate or suboptimal for the rapidly sampled data obtained using multiband approaches. Therefore, there is an opportunity to create new methods for the analysis of multiband data that will enhance the specificity and sensitivity of BOLD fMRI outcome measures and harness the increased information content. This is an exciting area where quantitative scientists promise to play an important role.

Harnessing Large-Scale Data Bases: As previously mentioned, there have been numerous efforts to construct large-scale imaging databases. These endeavors have been performed on both a grassroots (e.g., the 1000 Functional Connectomes Project) and institutional level (e.g., the Human Connectome Project), and databases consisting of more than 1,000 subjects are becoming increasingly available.

Many times, the largest imaging datasets tend to be collected for reasons other than a specific targeted scientific hypothesis. In these cases, data quality, sampling bias, missing data, data quality and the availability of covariates or outcomes tend to be problematic. This is in contrast to experiments consisting of a small number of subjects, where tight experimental control leads to direct measures for testing hypotheses of interest at the expense of lower power. Broadly speaking, large datasets used in isolation are useful for exploratory or predictive exercises, while small datasets, useful for knowledge creation and confirmatory analyses, are critically hampered by low power.

As the ultimate goal is to create new scientific knowledge, it would be ideal to use information from large datasets to inform the analysis of small sample data. Large datasets can be explored with the goal of establishing norms and priors for the use in small, more targeted datasets. To achieve this goal we believe that there is a need for more quantitative researchers to become involved in the field in order to make sense of the massive amounts of data being generated.

Multi-Modal Analysis: All methods used in the human neuro-behavioral sciences have limitations, and fMRI is of course no exception. Therefore the current trend is towards increasingly interdisciplinary approaches that use multiple methodologies to overcome some of the limitations of each method used in isolation. For example, fMRI data are increasingly combined with EEG and MEG data to improve temporal precision, among other benefits. Advances in engineering and signal processing, for example, allow EEG and fMRI data to be collected simultaneously. Neuroimaging data is also being combined with transcranial magnetic stimulation to integrate the powerful ability of neuroimaging to observe brain activity with the ability afforded by TMS to manipulate brain function and examine causal effects (5). Finally, integrating genetics with brain imaging is seen as a way to study how a particular subset of polymorphisms may affect functional brain activity. In addition, quantitative indicators of brain function could facilitate the identification of the genetic determinants of complex brain-related disorders such as autism, dementia and schizophrenia (15).

Each of these multi-modal approaches promises to be an important topic of future research, and to fully realize their promise, novel statistical techniques will be needed. Ultimately, combining information from different modalities will be challenging to data analysts, if for no other reason than that the amount of data will significantly increase. In addition, since different modalities are measuring fundamentally different quantities, it is not immediately clear how to best combine the information. However, clearly, this is an extremely important problem that has already started to become a major area of research.

Bibliography

[1] G.K. Aguirre, E.Zarahn, and M.D'Esposito. The variability of human, BOLD hemodynamic responses. *NeuroImage*, 8(4):360–369, 1998.

[2] A.H. Andersen, D.M. Gash, and Avison M.J. Principal component analysis of the dynamic response measured by fMRI: A generalized linear systems framework. *Magnetic Resonance in Medicine*, 17:785–815, 1999.

[3] R.M. Birn, Z.S. Saad, and P.A. Bandettini. Spatial heterogeneity of the nonlinear dynamics in the fMRI bold response. *NeuroImage*, 14:817–826, 2001.

[4] Bharat Biswal, F. Zerrin Yetkin, Victor M. Haughton, and James S. Hyde. Functional connectivity in the motor cortex of resting human brain using echo-planar MRI. *Magnetic Resonance in Medicine*, 34(4):537–541, 1995.

[5] D.E. Bohning, A.P. Pecheny, C.M. Epstein, A.M. Speer, D.J. Vincent, W. Dannels, and M.S. George. Mapping transcranial magnetic stimulation (TMS) fields in vivo with MRI. *Neuroreport*, 8:2535–2538, 1997.

[6] G.M. Boynton, S.A. Engel, G.H. Glover, and D.J. Heeger. Linear systems analysis of functional magnetic resonance imaging in human v1. *J. Neurosci*, 16:4207–4221, 1996.

[7] R.B. Buxton, E.C. Wong, and L.R. Frank. Dynamics of blood flow and oxygenation changes during brain activation: The balloon model. *Magnetic Resonance in Medicine*, 39:855–864, 1998.

[8] V.D. Calhoun, T. Adali, G.D. Pearlson, and J.J. Pekar. Spatial and temporal independent component analysis of functional MRI data containing a pair of task-related waveforms. *Human Brain Mapping*, 13:43–53, 2001.

[9] Catie Chang, John P. Cunningham, and Gary H. Glover. Influence of heart rate on the bold signal: The cardiac response function. *Neuroimage*, 44(3):857–869, 2009.

[10] Nathan W. Churchill, Anita Oder, Herve Abdi, Fred Tam, Wayne Lee, Christopher Thomas, Jon E. Ween, Simon J. Graham, and Stephen C. Strother. Optimizing preprocessing and analysis pipelines for single-subject fMRI. I. Standard temporal motion and physiological noise correction methods. *Human Brain Mapping*, 33(3):609–627, 2012.

[11] Robert W. Cox. AFNI: Software for analysis and visualization of functional magnetic resonance neuroimages. *Computers and Biomedical Research*, 29(3):162–173, 1996.

[12] A.M. Dale. Optimal experimental design for event-related fMRI. *Human Brain Mapping*, 8:109–114, 1999.

[13] K.J. Friston, L. Harrison, and W. Penny. Dynamic causal modelling. *NeuroImage*, 19:1273–1302, 2003.

[14] K.J. Friston. Functional and effective connectivity in neuroimaging: A synthesis. *Human Brain Mapping*, 2:56–78, 1994.

[15] D.C. Glahn, P.M. Thompson, and J. Blangero. Neuroimaging endophenotypes: Strategies for finding genes influencing brain structure and function. *Human Brain Mapping*, 28:488–501, 2007.

[16] G.H. Glover. Deconvolution of impulse response in event-related BOLD fMRI. *NeuroImage*, 9(4):416–429, 1999.

[17] G.H. Glover. Simple analytic spiral k-space algorithm. *Magnetic Resonance in Medicine*, 42(2):412–415, 1999.

[18] C. Goutte, F.A. Nielsen, and L.K. Hansen. Modeling the haemodynamic response in fMRI using smooth fir filters. *IEEE Trans Med Imaging*, 19:1188–1201, 2000.

[19] J. Grinband, T.D. Wager, M. Lindquist, V.P. Ferrera, and J. Hirsch. Detection of time-varying signals in event-related fMRI designs. *NeuroImage*, 43(3):509–520, 2008.

[20] H. Gudbjartsson and S. Patz. The Rician distribution of noisy MRI data. *Magnetic Resonance in Medicine*, 34:910–914, 1995.

[21] Mark E. Haacke, Robert W. Brown, Michael R. Thompson, and R. Venkatesan. *Magnetic Resonance Imaging: Physical Principles and Sequence Design*. Wiley-Liss, 1999.

[22] S. Hayasaka and T.E. Nichols. Combining voxel intensity and cluster extent with permutation test framework. *NeuroImage*, 23:54–63, 2004.

[23] Scott A. Huettel, Allen W. Song, and Gregory Mccarthy. *Functional Magnetic Resonance Imaging*. Sinauer Associates, 2004.

[24] Mark Jenkinson, Christian F. Beckmann, Timothy E.J. Behrens, Mark W. Woolrich, and Stephen M Smith. FSL. *Neuroimage*, 62(2):782–790, 2012.

[25] Tom Johnstone, Kathleen S. Ores Walsh, Larry L. Greischar, Andrew L. Alexander, Andrew S. Fox, Richard J. Davidson, and Terrence R. Oakes. Motion correction and the use of motion covariates in multiple-subject fMRI analysis. *Human Brain Mapping*, 27(10):779–788, 2006.

[26] Ming-Hung Kao, Abhyuday Mandal, Nicole Lazar, and John Stufken. Multi-objective optimal experimental designs for event-related fMRI studies. *NeuroImage*, 44(3):849–856, 2009.

[27] Ming-Hung Kao, Abhyuday Mandal, and John Stufken. Constrained multiobjective designs for functional magnetic resonance imaging experiments via a modified non-dominated sorting genetic algorithm. *Journal of the Royal Statistical Society: Series C (Applied Statistics)*, 61(4):515–534, 2012.

[28] Kenneth K. Kwong, John W. Belliveau, David A. Chesler, Inna E. Goldberg, Robert M. Weisskoff, Brigitte P. Poncelet, David N. Kennedy, Bernice E. Hoppel, Mark S. Cohen, and Robert Turner. Dynamic magnetic resonance imaging of human brain activity during primary sensory stimulation. *Proceedings of the National Academy of Sciences*, 89(12):5675–5679, 1992.

[29] David J. Larkman, Joseph V. Hajnal, Amy H. Herlihy, Glyn A. Coutts, Ian R. Young, and Gösta Ehnholm. Use of multicoil arrays for separation of signal from multiple slices simultaneously excited. *Journal of Magnetic Resonance Imaging*, 13(2):313–317, 2001.

[30] C. Liao, K.J. Worsley, J-B. Poline, G.H. Duncan, and A.C. Evans. Estimating the delay of the response in fMRI data. *NeuroImage*, 16:593–606, 2002.

[31] M.A. Lindquist, J.M. Loh, L. Atlas, and T.D. Wager. Modeling the hemodynamic response function in fMRI: Efficiency, bias and mis-modeling. *NeuroImage*, 2008.

[32] M.A. Lindquist and T.D. Wager. Validity and power in hemodynamic response modeling: A comparison study and a new approach. *Human Brain Mapping*, 28:764–784, 2007.

[33] M.A. Lindquist, C.H. Zhang, G. Glover, and L.A. Shepp. Acquisition and statistical analysis of rapid 3D fMRI data. *Statistica Sinica*, 2008.

[34] M.A. Lindquist, C.H. Zhang, G. Glover, and L.A. Shepp. Rapid three-dimensional functional magnetic resonance imaging of the negative bold response. *Journal of Magnetic Resonance*, 2008.

[35] Martin A. Lindquist et al. The statistical analysis of fMRI data. *Statistical Science*, 23(4):439–464, 2008.

[36] T.T. Liu and L.R. Frank. Efficiency, power, and entropy in event-related fMRI with multiple trial types: Part I: Theory. *NeuroImage*, 21:387–400, 2004.

[37] N.K. Logothetis. Can current fMRI techniques reveal the micro-architecture of cortex? *Nature Neuroscience*, 3:413, 2000.

[38] Nikos K. Logothetis, Jon Pauls, Mark Augath, Torsten Trinath, and Axel Oeltermann. Neurophysiological investigation of the basis of the fMRI signal. *Nature*, 412(6843):150–157, 2001.

[39] J.M. Loh, M.A. Lindquist, and T.D. Wager. Residual analysis for detecting mis-modeling in fMRI. *Statistica Sinica*, 2008.

[40] T.E. Lund, K.H. Madsen, K. Sidaros, W.L. Luo, and T.E. Nichols. Non-white noise in fMRI: Does modelling have an impact? *NeuroImage*, 29:54–66, 2006.

[41] D. Malonek and A. Grinvald. The imaging spectroscopy reveals the interaction between electrical activity and cortical microcirculation: Implication for optical, PET and MR functional brain imaging. *Science*, 272:551–554, 1996.

[42] P. Mansfield. Multi-planar image formation using NMR spin echoes. *Journal of Physics*, C10:L55–L58, 1977.

[43] A. McIntosh and F. Gonzalez-Lima. Structural equation modeling and its application to network analysis in functional brain imaging. *Human Brain Mapping*, 2:2–22, 1994.

[44] M.J. McKeown and S. Makeig. Analysis of fMRI data by blind separation into inde-pendant spatial components. *Human Brain Mapping*, 6:160–188, 1998.

[45] Andrea Mechelli, Will D. Penny, Cathy J. Price, Darren R. Gitelman, and Karl J. Fris-ton. Effective connectivity and intersubject variability: Using a multisubject network to test differences and commonalities. *Neuroimage*, 17(3):1459–1469, 2002.

[46] R.S. Menon, D.C. Luknowsky, and J.S Gati. Mental chronometry using latency resolved functional MRI. *Proc. Natl. Acad Sci. USA*, 95(18):10902–10907, 1998.

[47] F.M. Miezin, L. Maccotta, J.M. Ollinger, S.E. Petersen, and R.L. Buckner. Charac-terizing the hemodynamic response: Effects of presentation rate, sampling procedure, and the possibility of ordering brain activity based on relative timing. *NeuroImage*, 11, 2000.

[48] Steen Moeller, Essa Yacoub, Cheryl A. Olman, Edward Auerbach, John Strupp, Noam Harel, and Kâmil Uğurbil. Multiband multislice GE-EPI at 7 tesla, with 16-fold ac-celeration using partial parallel imaging with application to high spatial and temporal whole-brain fMRI. *Magnetic Resonance in Medicine*, 63(5):1144–1153, 2010.

[49] Kevin Murphy, Rasmus M. Birn, Daniel A. Handwerker, Tyler B. Jones, and Peter A. Bandettini. The impact of global signal regression on resting state correlations: Are anti-correlated networks introduced? *Neuroimage*, 44(3):893–905, 2009.

[50] S. Ogawa, D.W. Tank, R. Menon, J.M. Ellerman, S.G. Kim, H. Merkle, and K. Ugurbil. Intrinsic signal changes accompanying sensory simulation: Functional brain mapping and magnetic resonance imaging. *Proceedings of the National Academy of Sciences*, 89:5951–5955, 1992.

[51] William D. Penny, Karl J. Friston, John T. Ashburner, Stefan J. Kiebel, and Thomas E. Nichols. *Statistical Parametric Mapping: The Analysis of Functional Brain Images: The Analysis of Functional Brain Images.* Academic Press, 2011.

[52] Francisco Pereira, Tom Mitchell, and Matthew Botvinick. Machine learning classifiers and fMRI: A tutorial overview. *Neuroimage*, 45(1):S199–S209, 2009.

[53] Russell A. Poldrack, Jeanette A. Mumford, and Thomas E. Nichols. *Handbook of Functional MRI Data Analysis.* Cambridge University Press, 2011.

[54] K.P. Pruessmann, M. Weiger, M.B. Scheidegger, and P. Boesiger. Sense: Sensitivity encoding for fast MRI. *Magnetic Resonance in Medicine*, 42:952–956, 1999.

[55] P.L. Purdon, V. Solo, R.M. Weissko, and E. Brown. Locally regularized spatiotemporal modeling and model comparison for functional MRI. *NeuroImage*, 14:912–923, 2001.

[56] J.J. Riera, J. Watanabe, I. Kazuki, M. Naoki, E. Aubert, T. Ozaki, and R. Kawashima. A state-space model of the hemodynamic approach: Nonlinear filtering of bold signals. *NeuroImage*, 21:547–567, 2004.

[57] A. Roebroeck, E. Formisano, and R. Goebel. Mapping directed influence over the brain using Granger causality and fMRI. *NeuroImage*, 25:230–242, 2005.

[58] Mikail Rubinov and Olaf Sporns. Complex network measures of brain connectivity: Uses and interpretations. *Neuroimage*, 52(3):1059–1069, 2010.

[59] D.L. Schacter, R.L. Buckner, W. Koutstaal, A.M. Dale, and B.R. Rosen. Late onset of anterior prefrontal activity during true and false recognition: An event-related fMRI study, *NeuroImage*, 6(4):259–269, 1997.

[60] Michael E. Sobel and Martin A. Lindquist. Causal inference for fMRI time series data with systematic errors of measurement in a balanced on/off study of social evaluative threat. *Journal of the American Statistical Association*, (just-accepted):00–00, 2014.

[61] D.K. Sodickson and W.J. Manning. Simultaneous acquisition of spatial harmonics (smash): Fast imaging with radiofrequency coil arrays. *Magnetic Resonance in Medicine*, 38:591–603, 1997.

[62] Olaf Sporns. *Networks of the Brain.* MIT Press, 2011.

[63] Jean Talairach and Pierre Tournoux. Co-planar stereotaxic atlas of the human brain. 3-dimensional proportional system: An approach to cerebral imaging. 1988.

[64] A.L. Vazquez, E.R. Cohen, V. Gulani, L. Hernandez-Garcia, Y. Zheng, G.R. Lee, S.G. Kim, J.B. Grotberg, and D. C. Noll. Vascular dynamics and bold fMRI: CBF level effects and analysis considerations. *NeuroImage*, 32:1642–1655, 2006.

[65] T.D. Wager and T.E. Nichols. Optimization of experimental design in fMRI: A general framework using a genetic algorithm. *NeuroImage*, 18:293–309, 2003.

[66] T.D. Wager, A. Vazquez, L. Hernandez, and D.C. Noll. Accounting for nonlinear BOLD effects in fMRI: Parameter estimates and a model for prediction in rapid event-related studies. *NeuroImage*, 25(1):206–218, 2005.

[67] Tor D. Wager, Lauren Y. Atlas, Martin A. Lindquist, Mathieu Roy, Choong-Wan Woo, and Ethan Kross. An fMRI-based neurologic signature of physical pain. *New England Journal of Medicine*, 368(15):1388–1397, 2013.

[68] M.W. Woolrich, T.E. Behrens, and S.M. Smith. Constrained linear basis sets for HRF modelling using variational Bayes. *NeuroImage*, 21(4):1748–1761, 2004.

[69] K.J. Worsley and K.J. Friston. Analysis of fMRI time-series revisited-again. *NeuroImage*, 2:173–181, 1995.

[70] E. Yacoub, T.H. Le, and X. Hu. Detecting the early response at 1.5 tesla. *NeuroImage*, 7:S266, 1998.

[71] E. Zarahn. Using larger dimensional signal subspaces to increase sensitivity in fMRI time series analyses. *Hum Brain Mapp*, 17:13–16, 2002.

7

Electroencephalography (EEG): Neurophysics, Experimental Methods, and Signal Processing

Michael D. Nunez

Department of Cognitive Sciences, University of California, Irvine

Paul L. Nunez

Cognitive Dissonance, LLC

Ramesh Srinivasan

Department of Biomedical Engineering, University of California, Irvine

CONTENTS

7.1	Introduction	175
7.2	The Neurophysics of EEG	177
7.3	Synchronization and EEG	180
7.4	Recording EEG	182
7.5	Preprocessing EEG	183
7.6	Artifact Removal	184
7.7	Stationary Data Analysis	187
7.8	Nonstationary Data Analysis	194
7.9	Summary	197
	Bibliography	197

7.1   Introduction

Electroencephalography (EEG) is the measurement of the electric potentials on the scalp surface generated (in part) by neural activity originating from the brain. The sensitivity of EEG to changes in brain activity on such a millisecond time scale is the major advantage of EEG over other brain imaging modalities such as functional magnetic resonance imaging (fMRI) or near-infrared spectroscopy (NIRS) that operate on time scales in the seconds to minutes range. Over the past 100 years, neuroscientists and clinical neurologists have made use of EEG to obtain insight into cognitive or clinical disease state by applying a variety of signal processing and statistical analyses to EEG time series. More recently there has been growing interest in making use of statistical modeling of EEG signals to directly control physical devices in brain–computer interfaces. In this chapter we provide an introduction to EEG generation and measurement as well as the experimental designs that optimize information acquired by EEG.

EEG has statistical properties that vary over time and space. That is, we assume the data recorded are observations of a spatio-temporal stochastic process. The starting point

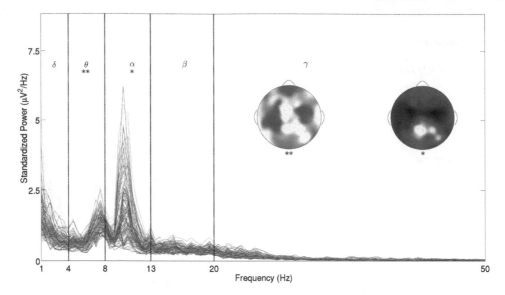

FIGURE 7.1

Typical power spectra of 124 EEG channels of 66 seconds (epoch length $T = 2$ sec with $K = 33$ epochs) of data from a subject (male, 25 yrs) who fixated on a computer monitor with his eyes open. While EEG spectral band definitions vary from lab to lab and across different fields, bands are typically defined as follows: delta 1–4 Hz, theta 4–8 Hz, alpha 8–13 Hz, beta 13–20 Hz, and gamma > 20 Hz. Some groups also identify the mu rhythm which exists as a peak either in the alpha or beta bands and typically has high power over the motor cortex. In the eyes-open resting data, with some artifact power removed using Independent Component Analysis (ICA), we see peaks in the delta, theta, and alpha bands and some power in the beta band. However the dominant peaks in the spontaneous EEG are in the theta and alpha bands, which have different spatial distributions over the electrodes and are associated with different cognitive functions. Topographic scalp maps were generated by summing power across frequencies in the theta (left) and alpha (right) frequency bands and interpolating between electrodes, such that brighter values correspond to higher power. Alpha, which is empirically associated with the resting state, has maximum power over parietal channels as indicated by the right topographic scalp map.

of most EEG analysis is to consider the properties of the time series at each electrode, such as spectral power, in relation to sensory stimulation, cognitive processing, or clinical disease state. EEG power is typically split up into bands which correspond to different spectral peaks that relate to behavior or cognitive state. These bands are typically defined as the delta (1–4 Hz), theta (4–8 Hz), alpha (8–13 Hz), beta (13–20 Hz), and gamma (> 20 Hz) bands and can have high-power over spatially distinct regions on the scalp, as shown by an EEG recording of a subject at rest in Figure 7.1. Because this EEG was recorded using a *high-density*, 128-electrode net, the topographic scalp maps could be found by interpolating values between electrodes. Modern EEG systems typically use a large number of electrodes (ranging from 64 to 256) to provide coverage over most of the scalp, enabling analyses of *spatial* properties of EEG, such as correlation or coherence between electrode sites.

In this chapter we first review physical properties of EEG recordings, in order to model the relationship between potentials on the scalp and current sources in the brain. Models of volume conduction of current passing through the head imply that EEG signals are strongly influenced by the synchronization of neural current sources; EEG is as much a measure of neural synchrony as neural activity. We then introduce issues in EEG recording and preprocessing, with particular emphasis on the problem of artifacts. The last two sections consider the types of EEG analyses appropriate for different experimental designs.

7.2   The Neurophysics of EEG

Scalp potentials are believed to be generated by *millisecond-scale modulations* of synaptic current sources at neuron surfaces (Lopes da Silva and Storm van Leeuwen 1978; Nunez 1981, 1995; 2000a,b; Lopes da Silva 1999), while single neuron firings of *action potentials* are mainly absent in scalp activity due to other, inactive neurons contributing to low-pass temporal filtering (Nunez and Srinivasan, 2006; Buzsáki, 2006). Action potential time scales are typically on the order of less than 1 ms, while synaptic potentials occur on the order of 10 ms or more—a time scale more consistent with the oscillations observed in EEG. Any patch of cortex 3–5 mm in diameter and traversing all cortical layers contains $\sim 10^6$ neurons and perhaps 10^{10} synapses (Nunez, 1995). Excitatory and inhibitory synapses inject current into the cell bodies and induce extracellular potentials (via return currents) of opposite polarity. These extracellular currents yield the electrical potentials recorded on the scalp with EEG.

When measurement of the potential is taken at "large" distance away from the source region, a complex current distribution in a small volume can be approximated by a dipole or more accurately, *a dipole moment per unit volume* (Nunez and Srinivasan, 2006). The (current) dipole moment per unit volume is an intermediate scale vector function based on the distribution of positive and negative micro-current sources in each local tissue mass, typically applied to cortical columns. The dipole approximation to cortical current sources provides a basis for realistic models of EEG signals. A "large" distance in this case is at least 3 or 4 times the distance between the effective poles of the dipole. In the context of EEG recording, the dipole approximation appears valid for potentials in superficial cortical tissue with a maximum extent in any dimension of roughly 0.5 cm or less. This is because superficial gyral surfaces are located at roughly 1.5–2 cm from scalp electrodes, separated by a thin layer of passive tissue, including cerebrospinal fluid (CSF), skull, and scalp.

Thus, the current sources in the brain that generate EEG can be modeled in terms of dipole moment per unit volume $\boldsymbol{P}(\boldsymbol{r'}, t)$ for any time t where the location vector $\boldsymbol{r'}$ spans the volume of the brain. For convenience of this discussion, the brain volume may be parceled into N small tissue masses of volume ΔV(e.g., 3 mm x 3 mm x 3 mm), each producing its vector dipole moment, such that, in an adult brain, N $\sim 10^5$ to 10^6. The strength and orientation of each vector depends on the distribution and synchrony of excitatory and inhibitory post-synaptic potentials within the tissue mass (Nunez and Srinivasan, 2006). The potential on the scalp surface can then be expressed as a weighted sum (or integral) of contributions from all these sources. In most models, each volume element $V(\boldsymbol{r'})$ is located only within the superficial cortex since current sources in deeper tissues, such as the thalamus or midbrain, typically contribute very little to scalp potentials (Nunez and Srinivasan, 2006). Thus, the volume integral may be reduced to a surface integral over the folded cortical surface

$$\Phi_S(\mathbf{r}, t) = \int_{B'} G_H(\mathbf{r}, \mathbf{r}')\mathbf{P}(\mathbf{r}', t) dV(\mathbf{r}'). \tag{7.1}$$

The weighting term G_H is the Green's function for volume conduction of current passing through the tissues of the head. It depends on both the location of source $\boldsymbol{r'}$ and the location of the scalp electrode $\boldsymbol{r}$. The Green's function can be thought of as the *impulse response function* between sources and surface locations and contains all geometric and conductive information about the head as a volume conductor. G_H will be larger for superficial sources in the visible gyral crowns of the cortex than for deeper sources, such as in the sulcal walls (folded surfaces) or sources on the mesial (underside) of the brain.

Any model of head volume conduction, i.e., any form of the function G_H, is only an approximation. Magnetic Resonance Imaging (MRI) can provide geometric information by imaging the boundaries between tissue compartments with different electrical conductivity (the inverse of resistivity). Numerical methods such as the Boundary Element Method (BEM) or Finite Element Methods (FEM) may then be used to estimate G_H, by employing MRI to determine tissue boundaries as shown in the examples in Figure 2 (b and c). However, the geometric model obtained from MRI is still only approximate due to limits in spatial resolution (typically 2–5 mm). And even if we were able to obtain perfect geometric information, the head model would still only be approximate due to the substantial uncertainty in our knowledge of tissue conductivities (Nunez and Srinivasan, 2006).

The poor conductivity of the skull is the feature that most strongly determines volume conduction in the head. Estimates of the conductivity of the skull vary widely depending on whether the estimate is in vivo or in vitro and differ between skull samples from different regions of the head. Skull itself is composed of three layers of different conductivity which vary in thickness across the head (Nunez and Srinivasan, 2006) and are not easily measured with MRI (Ding and Srinivasan, 2006).

Despite these uncertainties in the geometry and conductivity of the head, the gross features of volume conduction are captured by any model that includes a poorly conducting skull layer in between conductive soft tissue. A simple estimate of G_H can be provided by models which consist of three concentric spherical shells (brain, skull, and scalp) or four shells when including a CSF layer as shown in Figure 7.2a, such that these models have been useful in a number of simulation studies (Nunez et al., 1994; Srinivasan et al., 1998; Nunez and Srinivasan, 2006). These models also have the advantage of easy checking of computational accuracy as analytic solutions for these models have been obtained (Nunez and Srinivasan, 2006). The following gross features of head volume conduction are captured by this model: (12.1) the poor conductivity of the skull results in very little current entering the skull from the brain. (12.3) Current is expected to mostly flow radially through the skull into the scalp as current follows the path of least resistance. Exceptions are holes in the skull like the nasal passages. (21.12) All of the current is contained in the scalp, as no current can enter the surrounding air. (21.13) Very little current is expected to enter the body because of the high resistance of the neck, such that the head can be considered a closed object to first approximation. In all models that contain these essential features, the tangential spread of current within the scalp leads to the "smearing" of the scalp potential, i.e., low-pass spatial filtering, resulting in the low spatial resolution of EEG as compared to direct recordings on the brain surface. Thus this model captures the fact that EEG is a direct measurement of the current flowing in the scalp.

There is considerable interest in the EEG literature in developing methods to estimate the current source distribution in the brain $P(r',t)$ from the EEG recording and a volume conduction model of the head. However, this *inverse problem is ill-posed*; given the potential distribution on the surface of the scalp, it is not possible to estimate the source distribution without additional assumptions (Nunez and Srinivasan, 2006). That is, for any given G_H estimate and true scalp potential $\Phi_S(r, t)$, there are a large number of solutions for $P(r',t)$. Although in some cases, for example an epileptic focus in the cortex, it may be reasonable to assume a single isolated source $P(r',t)$ in order to find a solution. Another popular approach that is widely adopted is to use Tikhanov regularization to obtain a minimum L^2 norm estimate of $P(r',t)$ (Hauk, 2004). While this approach is mathematically tractable, there is no apparent theoretical reason why neuroscientists should seek solutions

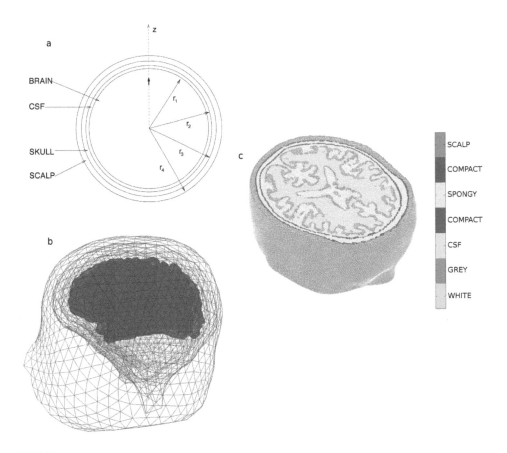

FIGURE 7.2

Volume conduction models for EEG. (*a*) A dipole is shown in the inner sphere of a *4-concentric spheres head model* consisting of the inner sphere (brain) and three spherical shells representing CSF (cerebral spinal fluid), skull and scalp. The parameters of the model are the radii (r_1, r_2, r_3, r_4) of each shell and the conductivity ratios ($\sigma_x v \sigma_y sg \sigma_x v \sigma_z sg \sigma_x v \sigma_4$). Typical values are: radii (8, 8.1, 8.6, 9.2 cm) and conductivity ratios (0.2, 40, 1). This model is used in the simulations in this chapter. (*b*) A realistic shaped boundary element model (BEM) of the head. The brain and scalp boundaries were found by segmenting the images with a threshold, and opening and closing operations, respectively, while the outer skull boundary was obtained by dilation of the brain boundary (ASA, Netherlands). Although geometrically more accurate than the spherical model, the (geometrically) realistic BEM may be no more accurate than a concentric spheres model because tissue resistivities are poorly known. (*c*) A realistic finite element model (FEM) obtained from MRI. This model has potentially better accuracy than the BEM model because the skull is subdivided into three layers corresponding to hard (compact) and spongy (cancellous) bone layers.

with a minimum L^2 norm. More recently, methods based on Bayesian inference have been developed which have the advantage of making assumptions explicit. These methods allow for the possibility of model validation and allow for comparisons between models based on different assumptions (Baillet and Garnero, 1997; Wipf and Nagarajan, 2009). They also allow for the use of prior information, for example by making use of fMRI information to influence the source solutions (Henson et al., 2010).

7.3   Synchronization and EEG

The magnitude of the scalp potential recorded with EEG can change for several reasons related to source synchronization. The large changes in scalp amplitude that occur when brain state changes are believed to be due mostly to distributed synchronization changes. That is, large-scale synchronization increases (or decreases) over cm scales in the tangential direction across the cortex will cause increases (or decreases) in scalp potential if there are no other changes. With this knowledge, EEG scientists and clinicians have adopted the label *desynchronization* to indicate large amplitude reductions (Pfurtscheller and Lopes da Silva 1999). Although at any one location r' in the brain, small-scale changes in synaptic source synchronization will *also* change the magnitude (or source strength) of $\boldsymbol{P}(r',t)$ (Nunez and Srinivasan, 2006), we will concentrate in this section on large-scale synchronization that does not change the magnitude of each individual dipole moment $\boldsymbol{P}(r',t)$. Instead, we will show that large-scale changes across different locations r' cause the scalp potential in Eq. (7.1) to decrease; this is because the integral approaches zero as more random positive and negative dipole moments $\boldsymbol{P}(r',t)$ at different cortical locations r' cancel. However if multiple locations r' have similar dipole moments per unit volume $\boldsymbol{P}(r',t)$ then synchronization occurs, which leads to larger observed scalp potentials.

We expect that the source of any EEG signal will never simply correspond to source activity in only one of the volume elements. Because neurons are highly interconnected, most EEG signals are generated by sources with spatial extent, i.e., patches of cortical tissue. Figure 7.3 shows examples of scalp potentials simulated in a concentric spheres volume conduction model due to a single dipole (a; corresponding to a patch of diameter < 3 mm) as well as dipole layers of diameter ranging from 3–5 cm (Figures 7.3c, 7.3e, and 7.3g). For simplicity, we only make use of radial dipole sources in this example as similar effects could be found with dipole layers of arbitrary orientation. Each dipole layer is composed of dipole sources with time series that are constructed by adding a 6-Hz sinusoid of fixed amplitude $A=15$ to a Gaussian random processes with mean $\mu=0$ and standard deviation $\sigma=150$. *The 6-Hz components are synchronized across the dipole layers, whereas all other frequencies will have random phases.* Each source signal is an independent random time series representing the potential across the cortical surface given by the dipole layer. The source time series of a single dipole source (i.e., the dipole layer of very small size) is plotted in Figure 7.3a. The magnitude of the 6 Hz sinusoid is only 1% of the total variance of each dipole source, and the sinusoid is not observable in the dipole time series. Figure 3b shows the estimated potential measured at an electrode on the scalp directly above the center of a dipole layer of diameter 3 cm, based on a four concentric spheres model of the head (see Figure 7.3 caption for details of the head model). The time series exhibits a smoother appearance compared to the source time series. And as the diameter of the dipole layer is increased from 3 to 4 to 5 cm (i.e., increasing synchrony), the calculated surface potential becomes more obviously sinusoidal (Figures 7.3c through 7.3h).

Clearly, spatial synchrony is a least as important as the strength of the source in the generation of scalp potentials. Most (99%) of the source activity in these examples is uncorrelated Gaussian noise across the dipoles in a patch of width 5 cm. Yet, the scalp potential will appear smooth and periodic reflecting mostly relatively small magnitude (1% of) source activity that is synchronous across all sources in the dipole layer. The effect of volume conduction is to sum the source activity at the scalp electrode, so the asynchronous source time series contributes minimally due to noise cancellation. The synchronized 6-Hz signal is emphasized and the scalp potential is remarkably sensitive to the size of the dipole layer.

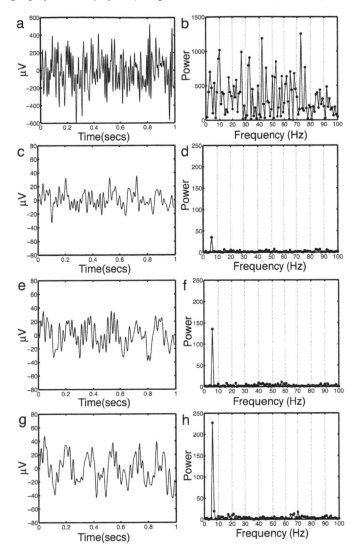

FIGURE 7.3

(*a*) Time series of a dipole meso-source $\mathbf{P}(\mathbf{r},t)$ composed of a 6-Hz, 15-μV sine wave added to Gaussian random noise with a standard deviation of $\sigma = 150$ μV. The Gaussian random noise was low-pass filtered at 100 Hz. The sine wave has variance (power) equal 1% of the noise. (*b*) Power spectrum of the time series shown in Part (*a*). The power spectrum has substantial power at many frequencies other than 6 Hz. (*c*) Time series recorded by an electrode on the outer sphere (scalp) of a four-concentric-spheres model above the center of a dipole layer of diameter 3 cm. The dipole layer is composed of 32 dipole sources $\mathbf{P}(\mathbf{r},t)$ with time series constructed similar to Part (*a*) with independent Gaussian noise (uncorrelated) at each dipole source. Scalp potential was calculated for a dipole layer at a radius $r_z = 7.8$ cm in a four-concentric-spheres model. The model parameters were radii $(r_1, r_2, r_3, r_4) = (8, 8.1, 8.6, 9.2)$ and conductivity ratios $(\sigma_1/\sigma_2, \sigma_1/\sigma_3, \sigma_1/\sigma_4) = (0.2, 40, 1)$. Notice that the time series is smoother than in the case of the individual dipole source. (*d*) Power spectrum of the time series shown in Part (*c*). Note the peak at 6 Hz. (*e*) Time series similar to Part (*c*), but due to a dipole layer of diameter of 4 cm composed of 68 dipole sources. (*f*) Power spectrum of the time series shown in Part (*e*). (g) Similar time series to Part (*c*), but with a dipole layer of diameter 5 cm composed of 112 dipole sources. The presence of the 6-Hz sinusoid is obvious from the time series. (*h*) Power spectrum of the time series shown in Part (*g*). A large spectral peak at 6-Hz is evident.

We have previously quantified this effect as spatial filtering by volume conduction (Srinivasan et al., 1998; Nunez and Srinivasan, 2006; Srinivasan et al., 2007). *One important implication is that spatial filtering by volume conduction can generally be expected to filter the temporal structure of source activity in the scalp EEG.* If sources with different time series take place in dipole layers of different sizes, EEG favors signals that are synchronized broadly over the cortical surface. The magnitude of any scalp EEG signal is determined not only by the source strength but also by spatial properties of the source such as its size and synchrony. Thus, we anticipate that EEG recorded within the brain (known as electrocorticography or ECoG) will have quite different properties than EEG recorded on the scalp. Neither signal is a more accurate representation of brain activity; instead they emphasize different spatial scales of synchronization in the brain.

7.4   Recording EEG

Every EEG recording involves at least 3 electrodes, two measurement electrodes and a ground electrode. Brain sources $P(r, t)$ (current dipole moments per unit volume) and biological artifacts generate the majority of scalp potential differences $V_2(t) - V_1(t)$. Environmental electric and magnetic fields also contribute to the measured scalp potential due mostly to capacitive coupling of body and electrode leads to power line fields. However, the amplifier ground electrode placed on the scalp, nose, or neck provides a reference voltage to the amplifier to prevent amplifier drift and facilitate better common mode rejection by serving as a reference for the differential amplifier (Nunez and Srinivasan, 2006).

Typically one electrode is singled out as the "reference electrode"; the remaining electrodes are characterized as "recording" electrodes. But electrode pairs are always required to measure scalp potentials because such recording depends on current passing through a measuring circuit (Nunez and Srinivasan, 2006). There are no monopolar recordings in EEG; all recordings are bipolar. Every EEG recording depends on the location of both recording and "reference" electrodes. Therefore any particular choice of reference placement offers possible advantages and disadvantages depending on actual source locations.

But, in general, we do not know the location of the sources prior to recording EEG, so no ideal reference location is likely to be found in advance. Reference strategies have often been adopted in EEG laboratories without a clear understanding of the attendant biases imposed on the recording. The linked-ears or linked-mastoids reference, a historically popular reference choice with cognitive scientists, is one such idea with minimal theoretical justification, but nevertheless persists in a number of laboratories. In EEG, *we generally measure potential differences between two locations on the head, and these differences depend on both electrode locations*, as well as on all brain generator configurations and locations.

In most EEG practice, the potentials at all the other electrode sites (typically 32-256) are recorded with respect to the reference electrode. The position of these electrodes varies considerably across laboratories. Standard electrode placement strategies make use of the 10-20, 10-10, and 10-5 electrode placement systems (Oostenveld and Praamstra, 2001). These systems are widely but not universally used. For larger numbers of channels (¿ 64), other electrode placement systems have been developed in order to obtain more uniform sampling of scalp potential, which is advantageous for source localization and high resolution EEG methods (Tucker, 1993). The reference point is largely arbitrary; it is special only because we choose to record potential differences with respect to one fixed location. But we do have the option of changing the effective reference to another recording site further down the processing chain by simple subtraction.

The average reference (also called the common average reference or global average reference) has become commonplace in EEG studies and has some theoretical justification (Bertrand et al. 1985). When recording from N electrodes located at scalp locations $r_n, n = 1, 2, \ldots, N$, the measured potentials $V(r_n)$ are related to the true scalp potential $\Phi(r_n)$ (measured with respect to "infinity") by

$$V(r_n) = \Phi(r_n) - \Phi(r_R) \tag{7.2}$$

where r_n is the position of the nth electrode and r_R is the reference electrode site. If we sum over all N electrodes, the potential with respect to infinity at the reference site can be written in terms of the scalp potentials as

$$\Phi(r_R) = \frac{1}{N} \left(\sum_{n=1}^{N} \Phi(r_n) - \sum_{n=1}^{N} V(r_n) \right). \tag{7.3}$$

The first term on the right side of Eq. 7.3 is the average of the scalp surface potential at all recording sites. Theoretically, this term vanishes if the mean of the potentials approximates a surface integral over a closed surface containing all current within the volume. Only minimal current flows from the head through the neck, even with reference electrode placed on the body, so a reasonable approximation considers the head to be a closed volume that confines all current. The surface integral of the potential over a volume conductor containing dipole sources must be zero as a consequence of current conservation (Bertrand et al., 1985). If we make this assumption, the reference potential can be estimated by the second term on the right side of Eq. 7.3; that is, by averaging the measured potentials at all electrodes and changing the sign of this average. This reference potential (i.e., the average across electrodes) can thus be added to each measurement $V(r_n)$, thereby estimating the reference-free potential $\Phi(r_n)$ (potential with respect to "infinity") at each location r_n.

However, since we cannot measure the potentials on a closed surface surrounding the brain, the first term on the right side of Eq. 7.3 will not generally vanish. The distribution of potential on the underside of the head (within the neck region) cannot be measured. Furthermore, the average potential for any group of electrode positions, given by the second term on the right side of Eq 7.3, is only an approximation of the surface integral. For example, the average potential is expected to be a very poor approximation if applied with the standard 10–20 electrode system with 21 electrodes. As the number of electrodes increases to 64 or more, the error in the approximation is expected to decrease. Thus, like any other choice of reference, the average reference provides biased estimates of reference-independent potentials. Nevertheless, when used in studies with large numbers of electrodes (say 128 or more), we have found that the average reference performs reasonably well as an estimate of reference-independent potentials (Srinivasan et al., 1998).

7.5 Preprocessing EEG

Measured EEG signals have been amplified and filtered by analog circuits to remove both low- and high-frequency noise as well as power at frequencies greater than the Nyquist limit, established by the sampling rate of the analog-to-digital converter (ADC). The discrete sampling of continuous signals is a well-characterized problem in time series acquisition and analysis (Bendat and Piersol, 2001). The central concept is the Nyquist criterion: $f_{dig} > 2f_{max}$ where f_{dig} is the digitization rate or sampling rate and f_{max} is the highest frequency present in the time series. For instance, if the highest frequency in a signal is 20 Hz (cycles/sec), a minimum sampling rate of 40 Hz (one sample every 25 ms) is required to record the signal discretely without aliasing. Aliasing is the misrepresentation of a high-frequency signal as a low-frequency signal because the sampling rate used during analog-to-digital conversion is lower than the Nyquist limit. If a time series has been aliased by under-sampling, no digital signal processing method can undo the aliasing because the necessary information for this procedure has been lost. In conventional EEG practice, a sampling rate is selected and the aliasing error is avoided by applying (in hardware) a low-pass filter to the analog signal that eliminates power at frequencies greater than the maximum frequency determined by the Nyquist limit. The low-pass filter is typically applied with a cut-off frequency 2.5 times smaller than the sampling rate. This more restrictive limit, known as the Engineer's Nyquist criterion, accounts for the possibility of phase-locking between the sampling and high-frequency components of the signal (Bendat and Piersol, 2001). The analog signal from each channel is sampled at perhaps 200 to 1000 times per second, assigned numbers proportional to instantaneous amplitude (digitized), and converted from ADC units to volts. These samples can then be stored digitally in conventional EEG practice or further processed online (e.g., using an FFT) in certain clinical or BCI applications.

The choice of filter settings requires some care. Clearly a low-pass filter must be set to insure removal of power at the very high frequencies determined by the Nyquist criterion. However, severe low-pass filtering runs the risk of removing obvious muscle artifact at high frequencies (which would indicate time segments of data that potentially needs to be discarded), while passing muscle artifact at frequencies overlapping with EEG that can be easily mistaken for EEG (Fisch et al., 1999). For example, imagine using an analog filter to remove most power at frequencies greater than 20 Hz, thereby obtaining a much cleaner looking signal. However, the remaining signal might well contain significant muscle artifact in roughly the 15–30 Hz range (beta band), which is much harder to identify without the information at higher frequencies. Such subtle artifact could substantially reduce the signal-to-noise ratio in the beta band. Some EEG systems have notch filters to remove power line interference (60 Hz in the Americas; 50 Hz in Europe, Australia, and Asia). However, the presence of power line noise in the recorded EEG signal is an easy way to detect electrodes that develop high contact impedances (or come off entirely) during the recording. If EEG processing and analysis is based on FFT or other spectral analysis methods, the presence of moderate 60-Hz noise will have no practical effect on results at lower frequencies, which contain most of the EEG information.

7.6 Artifact Removal

A substantial portion of the electrical signals recorded from EEG systems originate from outside the brain (Nunez and Srinivasan, 2006; Whitman et al., 2007). For example, in some areas on the head, close to the ears, eyes, and neck, we expect electrical signals originating in the cortex to have magnitude as much as 200 times lower than electrical signals from muscle activity (Fitzgibbon et al., 2015). Furthermore, movement of the head will generate artifacts over a large number of electrodes. Potentials generated from sources other than cortical activity are dubbed *artifact,* and a major challenge in EEG analysis is to detect these signals and remove them from EEG recordings.

Artifact can either be biological in nature, such as muscle activity, or due to environmental factors such as electric fields caused by the common AC standard and temporary potential shifts due to movement. Biological artifact is typically caused by electrical po-

tentials generated by muscle activity. The recording of muscle activity is known as electromyography (EMG) and typically originates from the eyes, face, and neck, but also from
muscles all over the body (Whitman et al., 2007). Another source of common artifact is the
rhythmic beating of arteries in the temples or neck and potentials from distant but large
muscles in the heart (electrocardiography or EKG). Transient muscle artifact can be due
to head movements, eye blinks, lateral eye movement, or jaw clenching; all of which may
display different spatio-temporal patterns of potentials on the scalp.

In order to better draw inference about brain activity, multiple procedures have been
developed to reduce the contribution of artifact in EEG recordings. Ocular artifact such as
eye blinks and lateral eye movements can be automatically removed using regression methods (Gratton et al., 1983). In a typical ocular regression method, electrodes are placed near
the eyes to record electrooculographic (EOG) signals, potentials generated by musculature
associated with the eyes. The effect of these EOG signals on the other EEG channels is
then estimated with linear regression. The total influence of the EOG signals on the EEG is
then removed by subtracting the product of the EOG signals and the regression coefficient
estimates (Schlögl et al., 2007).

Independent Component Analysis (ICA; Bell and Sejnowski, 1995) has become an important tool for identifying and removing artifact. Independent component analysis (ICA)
refers to a class of blind source-separation algorithms used to decompose linear mixtures
of data. For example, some ICA algorithms find linear mixtures of variables that are maximally non-Gaussian by searching for mixtures with either minimum mutual information or
maximum kurtosis (Makeig et al. 1996; Jung et al., 1998). In practice these methods often
yield non-normal mixtures that have distributions with outliers. The two most widely used
algorithms in the EEG literature are the FastICA (Hyvärinen & Oja, 1997) and InfoMax
ICA (Bell & Sejnowski, 1995; Delorme & Makeig, 2004).

ICA assumes that there is a linear mixture of the EEG data V (a channels c by time
t matrix) such that the independent components M (a component k by time t matrix)
are given by $M = W^{-1}V$ where W^{-1} is the matrix consisting of k by c weights. Some of
the resulting components have been shown to well represent some specific types of artifact
(Delorme et al. 2007). The components evaluated to reflect artifact can then be removed
from the data by inverting the equation using a reduced matrix W_L to remove the artifact
components.

There are two caveats with this approach. First, the identification of the artifact component is inherently a subjective judgement. Some artifact sources are easy to identify such
as eye blinks, eye movements, and temporary electrical discontinuities (perhaps due to a
reference electrode or ground electrode displacement during head movement). But artifacts
due to muscle are far more subtle. Second, the effect of reducing the number of sources in
M is to reduce the rank of the data matrix, which potentially influences further analysis by
reducing the amount of possible EEG mixtures.

To perform an ICA-based artifact removal procedure, the continuous recording is first
split into 1- to 3-second epochs, usually based on the trial structure of the experiment in
cognitive experiments. Epochs that obviously contain artifact rather than EEG, usually
due to gross movements by the subject, can then be rejected by visual inspection, or by
examining trials with high variance compared to other trials. Not removing this one-off data
hinders the ability of the ICA algorithm to isolate typical artifacts such as eye blinks. After
this "precleaning" step, an algorithm is run with the EEG data as input to obtain an ICA
decomposition.

Typical graphical representations of Independent Components (ICs) are topographic
maps of the inverse weights, component spectra, and component time series or average
component time series across epochs. Figure 7.4 provides typical graphical representations
of the 12 components that describe the most variance in a subject's EEG data using InfoMax

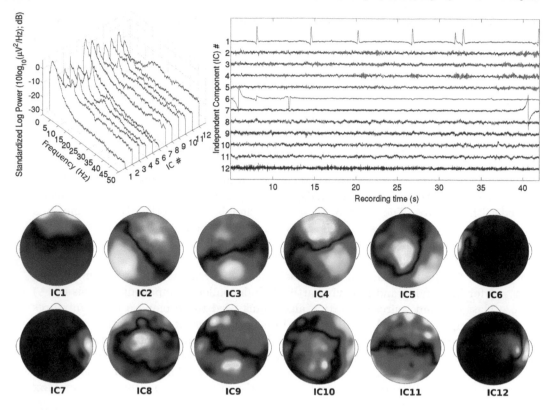

FIGURE 7.4
(From top-left, clockwise) Power spectra, time courses, and spatial loading topographies of
the first twelve independent components (ICs) from an Independent Component Analysis
(ICA) of an EEG recording while a subject (male, 25 yrs) was fixating on a computer
monitor. The ICs are ordered by their contribution to the total variance in the raw data. ICs
that are likely to reflect artifact contribution can be removed from the raw EEG data. IC1 is
indicative of an eye blink. IC6 and IC7 are indicative of temporary electrical discontinuities.
IC12 is indicative of muscle artifact.

ICA. The EEG was collected from a subject at rest who fixated on a cross on a monitor
for 42 seconds. Due to properties of the weight matrix W (columns represented as the
circular head plots in Figure 7.4 corresponding to each component), the component spectra
and the component time series, we identified 4 components that could be indicative of
artifact. Component 1 (IC1) most likely captures the electrical potentials due to eye blinks.
Indicative of eye blinks, the channel weights indicate that all the component information is
located near the eyes. The power spectrum has one peak in a low-frequency band because
the time series has high-amplitude waveforms located sparsely in time (which occur once
per blink). IC12 is probably muscle artifact, perhaps due to facial tension. It contributes
to the EEG recording mainly at peripheral electrodes, and its spectrum has high power at
high frequencies and low power at typical resting EEG frequencies (such as alpha rhythm,
around 10 Hz). Furthermore, the topography's spatial frequency is too high (i.e., too focal)
as this spatial frequency is near impossible for EEG to obtain due to the properties of head
volume conduction which acts as a low-pass spatial filter. Similarly, IC6 captures data that
cannot be due to brain activity because its weight is only at one electrode and the power
spectrum exhibits a $1/f$ frequency falloff (a property of electrical *"pink"* noise). As indicated

by its time course, this component is probably a mix of a temporary electrical discontinuity at about 6 seconds and a horizontal eye movement at about 12 seconds. IC7 is similar in its properties to IC6 and captures only a temporary electrical discontinuity at 41 seconds. The rest of the ICs most likely reflect cortical electrical activity or mixtures of cortical electrical activity and muscle artifact. These ICs contain peaks in alpha (8–13 Hz) and/or beta frequency bands (13–20 Hz) and have lower spatial frequency distributions typical of EEG.

We recommend keeping EEG and artifact mixtures in the data unless very specific properties of the EEG are of interest a priori. There is empirical evidence to suggest that ICA algorithms do not isolate many types of muscle artifact, especially task-related artifact, and thus rejecting ICs that do not clearly represent artifact becomes very subjective (Shackman et al., 2009). Furthermore, the efficacy of ICA to reduce *all* EMG artifact remains controversial at best (Olbrich et al., 2011; McMenamin et al., 2011). However if one must analyze a dataset that has a large quantity of muscle artifact, there may be a few indicators of EEG data that do not originate in the brain. For instance, an IC representing EEG or an EEG–EMG mixture may have a constant distribution of sample variances over all trials if the subject is in the same cognitive state and is doing the same task. In contrast, irregular EMG components will typically only have large variances on only a few trials.

In order to reduce subjectivity of the artifact independent component (IC) removal process and reduce the time demand on performing artifact removal, some progress has been made on automatic rejection of artifact components. ADJUST is an algorithm that uses properties of the components, such as spatial weight distributions on the scalp, variance, and kurtosis of the components' potentials, to automatically label components as eye blinks, vertical eye movements, horizontal eye movies, or generic potential discontinuities so that they can be subtracted from the recording (Mognon et al., 2011).

No known modern artifact correction technique is perfect for muscle artifact removal, and no EEG recording is completely immune to muscle artifact (Whitman et al., 2007). This is particularly the case for the neck and face muscle variety; thus, good recording and analysis practices are still the best approach for reducing artifact in EEG recordings. Subjects should be told to remain still and minimize jaw clenching, and the electrode cap or net should be positioned tightly (but comfortably) on the subject. Muscle artifact exhibits broadband frequency spectra with substantial relative power above 15 Hz; therefore analyses of the delta (1–4 Hz), theta (4–8 Hz), alpha (8–13 Hz) and mu (11–14 Hz) bands are typically more robust to muscle artifact contamination.

7.7 Stationary Data Analysis

The starting point of most EEG data analysis is *spectral analysis* to assess statistical properties of amplitude and phase of multiple EEG frequency bands. Even when the final goal of the analysis does not involve spectral analysis, examining the spectrum of the EEG is a useful starting point for evaluating data quality and for communication of more complex methods. The spectrum obtained by applying the Fourier transform to a single EEG epoch or time window provides information about its frequency content. Fourier transform algorithms yield estimates of Fourier coefficients that reflect both the amplitude and phase of the oscillations within one frequency band. Fast Fourier Transforms (FFT) are one class of algorithms that are particularly useful, and a number of important issues in practical FFT analysis are detailed in several texts (see Bendat and Piersol 2011, for example). Other Fourier analysis or spectra-like algorithms such as multi-taper analysis (Percival and

Walden, 1993), autoregressive models (Ding et al., 2000), wavelet analysis (Lachaux et al. 2002), and Hilbert transforms (Bendat and Piersol 2011; Le van Quyen et al. 2001; and Deng and Srinivasan, 2010) have potential applications in EEG, particularly in the analysis of short epochs characterizing EEG behavior after an experimental stimulus. Any of these algorithms can be used to carry out spectral analysis of time series, but an FFT-based analysis provides a quick and easy assessment of the spectrum.

The amplitude spectrum of one epoch of EEG is an exact representation of the frequency content of that particular time window, but only provides one observation about the random process generating the signal. The full ensemble of K epochs $\{V_k(t)\}$ can be used to estimate statistical properties of the random process generating the EEG under the assumption of weak stationarity (Bendat and Piersol, 2011). Weak stationarity is obtained if the mean and variance of the signal do not change with time. This can be verified by obtaining an estimate of mean and variance at each time point across epochs. Typically, the weak stationarity assumption is reasonable in the analysis of spontaneous EEG in resting-state experiments; it is not reasonable in any experiment where a sensory stimulus is presented and/or a motor response is obtained from the subject.

Estimating the *power spectrum* from an ensemble of epochs yields an estimate of the variance of the signal as a function of frequency. This is a particularly useful approach because EEG contains oscillatory activity in distinct frequency bands that are associated with different brain states. First, for each epoch $V_k(t)$, Fourier coefficients $F_k(f_n)$ are obtained by applying a Fourier transform, perhaps using the FFT. Then the power spectrum may be estimated from the ensemble of observations by summing over K epochs, given in Eq. (7.4).

$$P(f_n) = \frac{2}{K} \sum_{k=1}^{K} F_k(f_n) F_k^*(f_n) = \frac{2}{K} \sum_{k=1}^{K} |F_k(f_n)|^2 \quad n = 1, 2, ..., N/2 - 1 \qquad (7.4)$$

When applying the FFT, the frequency resolution $\Delta f = 1/T$ of the resulting power spectrum depends inversely on the length of each observation T as $f_n = n/T$ where n indexes the frequency band. The equation for the power spectrum is multiplied by a factor of two because the Fourier transform provides amplitudes split between positive and negative complementary phases and only amplitudes at positive phases are usually calculated. If the mean value of the signal is zero, the power spectrum summed over all frequencies is equal to the variance in the signal, a relationship known as *Parseval's theorem* (Bendat and Piersol 2011). The square root of the power spectrum, the *amplitude spectrum*, places more emphasis on non-dominant spectral peaks. Any algorithm used to obtain Fourier coefficients can be used to approximate Eq. (7.4); although the definition of frequency bands depends on the algorithm. Eq. (7.4) provides a definition of the EEG power spectrum in units that depend on the frequency resolution Δf. In order for the results to be compared across all choices of epoch length, the power spectrum is sometimes normalized by the frequency resolution Δf to express power in units of μV^2 per Hz.

Before the rise of widespread access to computational power and use of the fast Fourier transform, the power spectrum of a time series was typically calculated in a two-stage procedure. First the *autocorrelation function* was estimated and then the Fourier transform of the autocorrelation function was calculated. The result is equivalent to the power spectrum of the signal. The autocorrelation function is the covariance of the signal with itself as a function of lag:

$$R_{VV}(\tau) = E[V(t)V(t-\tau)]. \qquad (7.5)$$

Like the power spectrum, the true autocorrelation function is an unknown statistical property of the time series and can only be estimated. In Eq. (7.5) the lag variable τ is

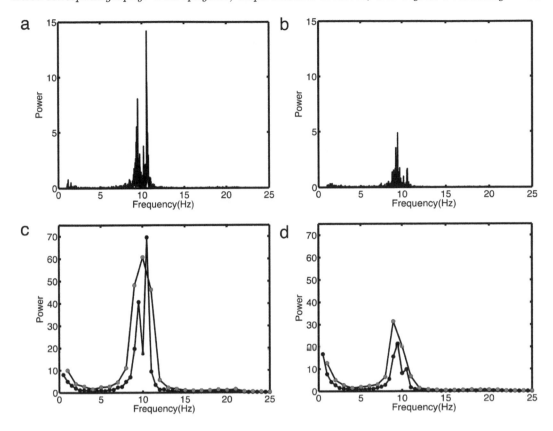

FIGURE 7.5

Example power spectra from a single subject (female, 22 yrs). The subject is at rest with eyes closed. (*a*) Power spectrum of a midline occipital channel with epoch length $T = 60$ sec and $K = 1$ epochs. The power spectrum appears to have two distinct peaks, one below 10 Hz and one above 10 Hz. (*b*) Power spectrum at a midline frontal channel with epoch length $T = 60$ sec and $K = 1$ epochs. Here only the peak below 10 Hz is visible. (*c*) Power spectra of a midline occipital channel calculated with two different choices of epoch length T and number of epochs K. The grey circles indicate the power spectrum with $T = 1$ sec and $K = 60$ epochs. The black circles indicate the power spectrum with $T = 2$ sec and $K = 30$ epochs. (*d*) Power spectra of a midline frontal channel calculated as in Part (c).

defined over positive and negative values. The autocorrelation function contains exactly the same spectral information as the time series of epoch length T if the domain of τ is $[-T/2, T/2]$. The Fourier transform of the autocorrelation function is then equal to the power spectrum of the signal (Bendat and Piersol, 2001). However in modern spectral analysis, the Fourier transform is usually directly calculated before calculating the power spectrum.

Estimation of the power spectrum involves tradeoffs in frequency resolution, statistical power, and weak stationarity. For example, consider the choices involved in analyzing a 60-second EEG record. Figure 7.5 demonstrates power spectra of two EEG channels, one occipital and one frontal, recorded with the subject's eyes closed and at rest. The power spectra were obtained using an epoch length T = 60 seconds ($\Delta f = 0$ .017 Hz) and no epoch averaging ($K = 1$ epochs). With this choice, the FFT of the entire record is obtained (exact spectra of the two EEG signals), but no information about the statistical properties of the underlying random process is gained. Note that the power spectrum of the occipital channel (Figure 7.5a) contains two peaks, one below 10 Hz and a larger peak above 10 Hz. The frontal channel (Figure 7.5b) shows a larger peak below 10 Hz. By examining the other channels it was found that the two peaks have distinct spatial distributions over the scalp, suggesting they have different source distributions. Each peak is surrounded by power in sidebands (adjacent frequency bins) of the two peak frequencies. *The signals are stochastic processes occupying relatively narrow bands in the frequency spectrum.*

To analyze the 60-sec signal properly, we must decide how to divide the record into epochs to implement Eq. (7.4). The choice is a compromise between the advantage of good frequency resolution yielded by long epochs (large T and small K) and the statistical power of our estimate gained by using a larger number of epochs (small T and large K). If a frequency resolution of $\Delta f = 0.5$ Hz is chosen, the record is segmented into $K = 30$ epochs of length $T = 2$ sec. If frequency resolution is reduced to $\Delta f = 1$ Hz, we divide the record into $K = 60$ epochs of length $T = 1$ sec. Figures 7.5c and 7.5d show the power spectra of the frontal and occipital channels with $\Delta f = 1$ Hz (grey circles) and $\Delta f = 0.5$ Hz (black circles). The power spectra at the occipital and frontal channels are both dominated by alpha rhythm oscillations. At the occipital electrode (Figure 7.5c) two separate peak frequencies are at 9.5 and 10.5 Hz are evident with $\Delta f = 0.5$ Hz, but this separation is not revealed with $\Delta f = 1$ Hz, where only a single peak frequency at 10 Hz is evident. Lowering frequency resolution has a similar effect at the frontal channel (Figure 7.5d), but since there is very little power at 10.5 Hz, the only clear peak appears at 9 Hz. Thus, by choosing a lower frequency resolution we observe different peak frequencies at the two sites, while choosing higher frequency resolution results in pairs of frequency peaks at both sites but with different magnitudes.

By examining the power spectra for the occipital (Figure 7.6a) and frontal sites (Figure 7.6b) for individual epochs with $\Delta f = 0.5$ Hz, evidence is found for two different oscillations within the alpha band. At the occipital channel, individual epochs display two distinct peaks at 9.5 Hz and 10.5 Hz. The first 15 epochs show a strong response at 10.5 Hz but the later epochs show a stronger response at 9.5 Hz. The dominant frequency in each epoch is summarized in the peak power histograms in Figure 7.5c showing that individual epochs displayed peak frequencies at both 9.5 Hz and 10.5 Hz. By contrast, very few epochs have a peak frequency of 10.5 Hz at the frontal site (Figure 7.5d); most epochs have peak frequencies either at 9.5 Hz or in the delta band (< 2 Hz). Note that during most epochs with strong delta activity in Figure 7.5d the alpha peaks are attenuated.

In the previous example, electrodes at different locations show different magnitudes of two distinct oscillations with center frequencies at 9.5 Hz and 10.5 Hz. The natural next step is to measure correlation between electrode sites to assess spatial statistics of the EEG. This is motivated by the idea that correlation of EEG signals should reflect functional connectivity of the brain. Neurons in distant (and nearby) cortex are connected by axons which form the white matter beneath the gray matter consisting more of cell bodies (Nunez, 1995). *Coherence* between two electrodes is a correlation coefficient (squared) that measures the consistency of relative phase between signals in a specific frequency band (and is weakly dependent on large amplitude changes). In EEG signals, coherence mainly reflects the consistency of phase differences across epochs, potentially reflecting axon transmission delays (Nunez and Srinivasan, 2006). Coherence ranges from 0 (indicating random phase differences) to 1 (indicating identical phase differences). Like the Pearson correlation coefficient, coherence is a unknown statistical property whose true value depends on the entire theoretical population of all similar epochs and can only be estimated using the given sample of epochs.

One way to estimate coherence between an electrode pair is to first calculate the *cross spectrum* (Bendat and Piersol, 2011). The cross spectrum, provided in Equation (7.6), is a measure of the joint spectral properties of two channels (i.e., a measure of the Fourier coefficient *covariance*). Fourier coefficients at a certain frequency f_n at each electrode are multiplied and averaged over K epochs to calculate the cross spectrum,

$$C_{uv}(f_n) = A_{uv}E^{j\phi_{uv}} = \frac{2}{K}\sum_{k=1}^{K} F_{uk}(f_n)F_{vk}^*(f_n) \quad n = 1, 2, ..., N/2 - 1, \qquad (7.6)$$

where F_{vk}^* refers to the complex conjugate of the Fourier coefficient at channel v. Note that the cross spectrum between a channel and itself ($u = v$) is equal to the power spectrum (Eq. (7.4)) and is real valued.

In general, the cross spectrum is complex valued and can be thought of in terms of its magnitude (or cross-power) A_{uv} and phase ϕ_{uv} information. The phase of the cross spectrum is the average phase difference between the two channels. This average phase difference is called the *relative phase*. The magnitude information A_{uv} is analogous to the ordinary covariance between two time series, i.e., the *covariance* across observations between two channels in one frequency band. Coherence at f_n is defined by the squared magnitude of the cross spectrum standardized by the product of the power spectra (which are variances) of each signal

$$\gamma_{uv}^2(f_n) = \frac{|C_{uv}(f_n)|^2}{P_u(f_n)P_v(f_n)} \quad n = 1, 2, ..., N/2 - 1. \qquad (7.7)$$

Coherence can be thought of as the fraction the variance at frequency f_n in channel u that can be explained by a constant amplitude and phase change (i.e., a *linear transformation* of Fourier coefficients) of the data at frequency f_n obtained at channel v.

Coherence between channels is most sensitive to phase differences. For instance, if the phase difference is constant over epochs between channels u and v, coherence is maximized and is equal to 1. If the relative phase between channels u and v varies across epochs, coherence will be less than one. Furthermore, if the phase difference is random across all epochs, the coherence estimate will approach zero as the number of epochs (K) increases. A coherence of 0.4 in frequency band f_n indicates that 40% of the variance at one channel can be explained by a linear transformation of the other channel. However, this does not imply that a linear relationship actually *exists*, only that the relationship (or part of the relationship) can be approximated by a linear transformation. If the activity at the two channels is truly related nonlinearly (e.g. have a quadratic relationship), the coherence would provide information as to how well the nonlinear relationship can be approximated by a linear relationship.

If the goal is to estimate phase synchronization independent of amplitude fluctuations, Fourier coefficients in Equation (7.7) could be normalized by amplitude to obtain *phase-only* coherence. Phase synchronization can also be evaluated with measures of the relative phase distribution across or within epochs (Tass et al., 1998). On the other hand, one reason to include amplitude as in Eqs. (7.6) and (7.7) is that coherence measures are weighted in favor of epochs with large amplitudes. Trials with large amplitudes are preferred if large amplitudes are indicative of high signal-to-noise ratios. If only epoch phase information were used, equal emphasis would be placed on low- and high- amplitude epochs in estimates of phase synchronization. This would potentially reduce the overall signal-to-noise ratio of the analysis.

Typically the goal of EEG coherence studies is to estimate the *functional connectivity* of the brain. Coherent activity between pairs of electrode sites may be either synchronous (zero phase difference) or asynchronous (constant phase difference). In other words, asynchronous coherent signals are essentially "synchronous with a time lag." Figure 7.7 shows the coherences between an electrode x and a ring of electrodes at progressively greater distances from x, labeled 1–9. The subject is at rest with eyes closed, a state in which coherence is

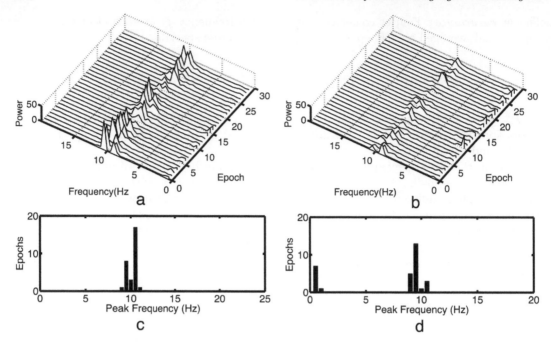

FIGURE 7.6

(*a*) Plots of 30 (individual epoch) power spectra for the occipital channel shown in Figures 7.5a and 7.5c. (*b*) Plots of the same 30 individual epoch spectra for the frontal channel shown in Figures 7.5b and 7.5d. (*c*) Peak power histograms show the distribution of peak frequencies for the 30 epochs shown in Part (a). (*d*) Peak power histograms for the 30 epochs shown in Part (b).

usually high in the alpha band (Nunez, 1995). The estimated coherence between electrode x and electrode n is labeled $x{:}n$. The electrode positions are shown and the distance between adjacent electrodes is about 2.7 cm. At the closest electrode pair $x{:}1$, coherence is very high (above 0.75) at all frequencies (i.e., coherence is generally independent of temporal frequency). This effect was predicted by the observed effect of volume conduction of current spreading through the head; the effect of volume conduction is to mix the brain signals at each electrode, which correlates them at all frequencies. This theoretical prediction of coherence due to volume conduction suggests that *the main effect of volume conduction is to artificially inflate coherences at short to moderate distances, and that this effect is independent of frequency.*

As the electrode separation is increased, the pair $x{:}2$ shows lower coherences, but at most frequencies coherence is still above 0.4, suggesting a strong component of coherence that is independent of frequency. A peak is visible in the alpha band at 9.5 Hz, but it is difficult to evaluate this peak, since there is also a broad elevation of coherence. As the sensor separation is further increased ($x{:}3$), the floor of the coherences reduces further to about 0.2, and a peak becomes more evident in the 18-Hz range. This electrode is about 10 cm away from electrode x, consistent with model predictions of volume conduction effects on EEG coherence for electrodes separated by less than 10 cm (Srinivasan et al., 1998). For pairs of electrodes involving sites over the temporal lobes, $x{:}4$ and $x{:}5$, the floor of the coherences approaches zero at most frequencies except for the alpha band where a second peak at 10.5 Hz is visible. At a very long distance ($x{:}8$) the coherences are again elevated

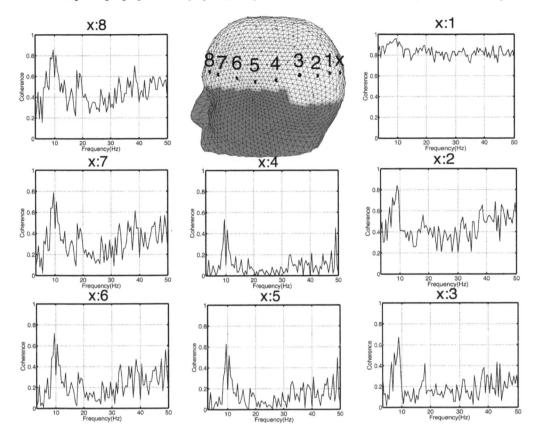

FIGURE 7.7

Scalp potential coherence spectra from a subject (female, 22 yrs) at rest with eyes closed in order to maximize alpha coherence. Coherence was estimated with $T = 2$ sec ($\Delta f = 0.5$ Hz) in a 60-sec record. The head plot shows the location of 9 electrodes, labeled x and 1 through 8. Coherence spectra between electrode x and each of the other electrodes 1–8 are shown, with increasing separations along the scalp. Note that very close electrodes have higher coherence independent of frequency as predicted by the theoretical volume conductor model. Alpha band coherence is high for large electrode separations, apparently reflecting the large cortical source coherence.

across all frequencies, suggesting a very small volume conduction effect at long distances as also observed in modeling studies (Srinivasan et al., 1998; Srinivasan et al., 2007).

Spatial statistics, like coherence, will always be influenced by volume conduction effects. In the case of coherence, a simple rule of requiring a separation distance of 10 cm or more can be used to minimize volume conduction effects. Another approach is to use the surface Laplacian to spatially high-pass filter the EEG signals, minimizing the effect of volume conduction (Nunez et al., 1994; Srinivasan et al., 1998). And as new methods are deployed to evaluate EEG spatial statistics (e.g. to make inferences about connectivity), more modeling studies are needed to evaluate the impact of volume conduction in order to interpret these new measures.

7.8   Nonstationary Data Analysis

A stationary time series is a random process whose statistical distribution is invariant over time. A *weakly* stationary time series is one in which the mean, variance, and autocorrelation of the random process are invariant to shifts in the time at which the sample records are obtained. Many EEG analyses assume weak stationarity in order to evaluate EEG in different brain states such as resting with eyes closed or eyes open, during mental calculations, or during different stages of sleep.

However, in many experiments, especially those related to cognitive functions, a stimulus is presented and/or a motor response is elicited and often both. This structure of events in an experimental trial implies that the EEG time series is nonstationary, as we can reasonably expect that the statistics of the signals will depend on time relative to the events. Clearly the EEG before a stimulus as presented has a different statistical structure than EEG following the stimulus. In fact we are interested in uncovering these differences!

Many experimenters analyze the time-varying mean of EEG data epochs following an experimental stimulus. The mean responses are called *evoked potentials* (EP) or *event related potentials* (ERP). These types of experimental designs are ubiquitous in the cognitive neuroscience literature. Because we may postulate that only the signals relevant to the stimulus remain after we average many trials of EEG response data, the EP or ERP is calculated as the ensemble mean of observations across epochs. If the subject views the same stimulus (or stimulus category) in all epochs (e.g. separate experimental trials) then we can expect the average EEG signal to approximate the response of ongoing EEG activity to the stimulus, such that the mean of the potential at each electrode varies as a function of time and is given by

$$\mu(t) = \frac{1}{K} \sum_{k=1}^{K} V_k(t). \tag{7.8}$$

Note that in order to calculate an evoked response, EEG is averaged across the dimension of epochs, not across the dimension of time.

Figure 7.8 shows an example of a visual evoked potential, or the time-varying mean at 124 electrodes of a high-density EEG cap, following presentation of a visual stimulus. Much of the ERP literature focuses on the peaks and troughs of this waveform (Figure 8a) and its relationship to perception and cognition. These are labeled P1, N1, P2, etc, reflecting the direction and order of the peaks (Luck et al., 2000). The inset topographic maps of the potential at the peaks indicate some spatial distribution, although naturally most of the response is over the back of the head, where visual cortex is located. Visual inspections suggests that this signal is an oscillation with a period of about 120 ms, which is confirmed by the wavelet spectrogram (Figure 7.8b) at one channel.

Spectral analysis methods as defined in Section 7.7 can be used to evaluate time-varying power and coherence of the EEG signals. The assumption in these analyses is that over narrow windows of time, e.g., 200 ms, the EEG can be considered stationary. This window is then moved over the epoch. A more convenient approach to calculate time-varying power and coherence is to use the complex Morlet wavelet transform to estimate time-varying Fourier coefficents (Lachaux et al., 2002). Two examples of time-varying power calculated using the Morlet wavelet transform are presented in Figures 7.8b and 7.9. Figure 7.8b shows the *phase-locked* change in alpha power *after* a visual stimulus is presented. In order to calculate phase-locked power, power of the time-average of epochs given in Eq. (7.8) can be calculated such that $K = 1$ in Eq. (7.4), which removes all signal that is not phase-locked. Figure 7.9 shows a non-phase-locked "desynchronization" (i.e., decrease) in alpha power in

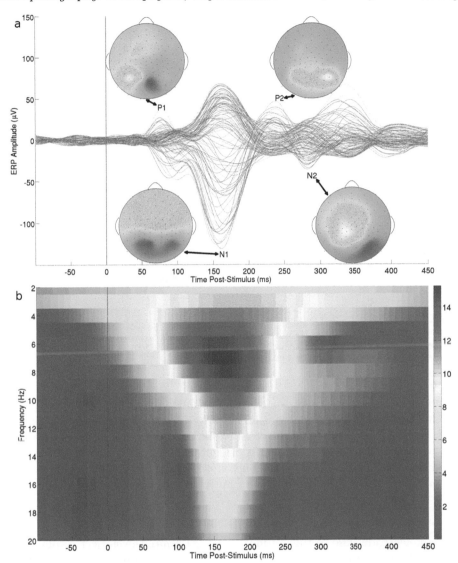

FIGURE 7.8

(*a*) A typical Visual Evoked Response (VEP; also known as an Event-Related Potential; ERP) to a large, high-contrast sinusoidal grating stimulus recorded at 124 electrodes of a high-density 128 Electrical Geodesics, Inc. (EGI) cap. The VEP was calculated by averaging low-pass Butterworth filtered data (with a 20-Hz passband) across all trials in one subject (male, 23 yrs) and by subtracting each trial by the time average of 200 milliseconds before that trial's stimulus onset (to *baseline* the VEP). Topographies of traditional local peaks (Luck et al., 2000) are labeled with $P1$, $N1$, $P2$, and $N2$ indicating the first and second positive and negative peaks over posterior electrodes. The N1 and P2 components evoke typical bilateral responses over parietal electrodes. The P1 and N2 reflect other network behavior related to processing of the visual stimulus. (*b*) A time-varying, phase-locked power spectrum of the same average data at an electrode over the left parietal cortex calculated with a Morlet wavelet transform. As shown by the wavelet, the VEP can also be thought of as a phase-locked alpha response to the visual stimulus.

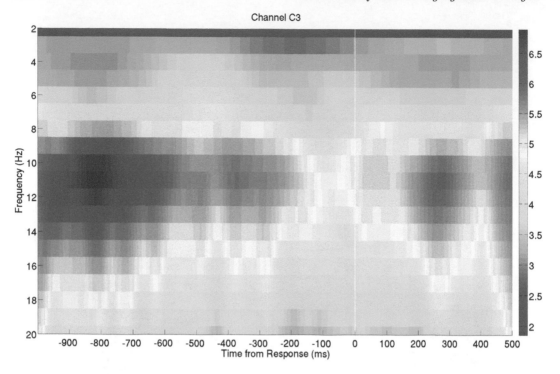

FIGURE 7.9

A *non-phase-locked* time-varying power spectrum from the same subject and visual task as in Figure 7.8 such that power was calculated using all $K = 157$ epochs time-locked to the motor response (button press given by the right or left hand). Data from a left-central electrode C3 in the 10–20 electrode placement system is presented. Mu power *"desynchronizes"* (i.e., decreases) approximately 200 ms before the button press, likely reflecting cognitive control over the motor response. Similar magnitude of mu power before the desynchronization is observed after the motor response.

electrode C3 (an electrode close to the left motor-cortex) *before* the subject responded with a button press. Power in the alpha band at electrodes over motor cortex are sometimes called the "mu rhythm" to differentiate it from spectral changes in the alpha band not associated with a motor response.

Steady-state evoked potentials (SSEPs) are another type of evoked potential, associated with experimental designs known as "frequency-tagging" (Ding et al., 2006; Deng and Srinivasan, 2010). SSEPs refer to a frequency-tagging paradigm where EEG responses are observed in narrow frequency bands corresponding to the frequencies and harmonics of a stimulus (Regan, 1977). Steady-state responses can be evoked with flickering visual stimuli (steady-state visual evoked potentials; SSVEPs), modulated auditory stimuli (steady-state auditory evoked potentials; SSAEPs), or rhythmic somatosensory stimuli (steady-state somatosensory evoked potentials, SSSEPs). SSEP paradigms are preferred by many neuroscientists due to their large signal-to-noise ratios. SSVEPs, SSAEPs, and SSSEPs have also been shown to track attention (Tiitinen et al., 1993, Morgan et al., 1996, and Giabbiconi et al., 2004).

SSEPs are narrow band signals with high signal-to-noise ratios minimizing the problem of artifacts, which are typically broadband. Our group has also shown that SSVEP data can be used to classify video game players versus non-gamers (Krishnan et al., 2013), differentiate individual differences in attention (Ding et al., 2006; Bridwell et al., 2013), and predict individual differences in perceptual decision making (Nunez et al., 2015).

7.9   Summary

In this chapter we introduce the physical basis of EEG recording and its implications for the type of brain processes observable with EEG. Practical EEG recording poses additional challenges, especially in relation to physiological artifacts such as muscle artifacts. We recommend that EEG analysis begin with spectral analysis methods, which provide a foundation from which more advanced statistical models can be developed and foster better communication with other EEG scientists. Spectral analysis is ubiquitous in the EEG literature in both stationary and non-stationary experimental designs. Finally, in research involving spatial statistics, the effect of volume conduction must be explicitly considered to make meaningful inferences about functional connectivity in the brain.

Bibliography

[1] Bell, A. J., and Sejnowski, T. J. An information-maximization approach to blind separation and blind deconvolution. *Neural Computation*, 7(21.16):1129–1159, 1995.

[2] Belouchrani A., Abed-Meriam K., Cardoso J.F., and Moulines E., A blind source separation technique using second-order statistics. *IEEE Transactions on Signal Processing.* 45:434–444, 1997.

[3] Bendat J.S. and Piersol A.G., *Random Data. Analysis and Measurement Procedures*, Fourth Edition. New York: Wiley, 2011.

[4] Bertrand, O., Perrin, F., and Pernier, J. A theoretical justification of the average reference in topographic evoked potential studies. *Electroencephalography and Clinical Neurophysiology/Evoked Potentials Section*, 62(21.16):462–464, 1985.

[5] Baillet, S., and Garnero, L. A Bayesian approach to introducing anatomo-functional priors in the EEG/MEG inverse problem. *Biomedical Engineering, IEEE Transactions on*, 44(21.15):374-385, 1997.

[6] Bridwell, D. A., Hecker, E. A., Serences, J. T., and Srinivasan, R. Individual differences in attention strategies during detection, fine discrimination, and coarse discrimination. *Journal of neurophysiology*, 110(21.12):784–794, 2013.

[7] Buzsáki, G, *Rhythms of the Brain*. New York: Oxford University Press, 2006.

[8] Cardoso JF Blind signal separation: Statistical principles. *Proc IEEE*, 9:2009–2025, 1998.

[9] Delorme, A. and Makeig, S. EEGLAB: An open source toolbox for analysis of single-trial EEG dynamics including independent component analysis. *Journal of Neuroscience Methods*, 134(12.1):9–21, 2004.

[10] Delorme, A, Sejnowski, T, Makeig, S. Enhanced detection of artifacts in EEG data using higher-order statistics and independent component analysis. *Neuroimage*, 34(21.13):1443–49, 2007.

[11] Deng, S., Srinivasan, R., Lappas, T. and D'Zmura, M. EEG classification of imagined syllable rhythm using Hilbert spectrum methods. *Journal of Neural Engineering*, 7(4):046006, 2010.

[12] Deng, S. and Srinivasan, R. Semantic and acoustic analysis of speech by functional networks with distinct time scales. *Brain Research*, 1346:132–144, 2010.

[13] Ding, J., Sperling, G., and Srinivasan, R. Attentional modulation of SSVEP power depends on the network tagged by the flicker frequency. *Cerebral cortex*, 16(7):1016–1029. 2006.

[14] Ding, M., Bressler, S.L., Yang, W., and Liang, H. Short-window spectral analysis of cortical event-related potentials by adaptive multivariate autoregressive modeling: data preprocessing, model validation, and variability assessment. *Biological cybernetics*, 83(1):35–45. 2000.

[15] Fisch, B.J. and Spehlmann, R., Fisch and Spehlmann's *EEG primer: basic principles of digital and analog EEG*. Elsevier Health Sciences, 1999.

[16] Fitzgibbon, S.P., DeLosAngeles, D., Lewis, T.W., Powers, D.M.W., Grummett, T.S., Whitham, E.M., Ward, L.M., Willoughby, J.O., and Pope, K.J. Automatic determination of EMG-contaminated components and validation of independent component analysis using EEG during pharmacologic paralysis. *Clinical Neurophysiology*, 2015.

[17] Gratton, G., Coles, M. G., and Donchin, E. A new method for off-line removal of ocular artifact. *Electroencephalography and Clinical Neurophysiology*, 55(21.13):468–484, 1983.

[18] Giabbiconi, C. M., Dancer, C., Zopf, R., Gruber, T., and Müller, M. M. Selective spatial attention to left or right hand flutter sensation modulates the steady-state somatosensory evoked potential. *Cognitive Brain Research*, 20(12.1):58–66, 2004.

[19] Hauk, O. Keep it simple: A case for using classical minimum norm estimation in the analysis of EEG and MEG data. *Neuroimage*, 21(21.13):1612–1621, 2004.

[20] Henson, R. N., Flandin, G., Friston, K. J., and Mattout, J. A parametric empirical Bayesian framework for fMRI-constrained MEG/EEG source reconstruction. *Human brain mapping*, 31(10):1512–1531, 2010.

[21] Hyvärinen, A., and Oja, E. A fast fixed-point algorithm for independent component analysis. *Neural Computation*, 9(21.17):1483–1492, 1997.

[22] Jung, T. P., Humphries, C., Lee, T. W., Makeig, S., McKeown, M. J., Iragui, V., and Sejnowski, T. J. Extended ICA removes artifacts from electroencephalographic recordings. *Advances in Neural Information Processing Systems*, 894–900, 1998.

[23] Krishnan, L., Kang, A., Sperling, G., and Srinivasan, R. Neural strategies for selective attention distinguish fast-action video game players. *Brain topography*, 26(12.1):83–97, 2013.

[24] Lachaux J. P., Lutz, A., Rudrauf, D., Cosmelli, D., Le Van Quyen, M., Martinerie, J., and Varela, F. Estimating the time-course of coherence between single-trial brain signals: An introduction to wavelet coherence. *Electroencephalography and Clinical Neurophysiology*, 32(21.12):157–74, 2002.

[25] Le Van Quyen, M., Foucher, J., Lachaux, J., Rodriguez, E., Lutz, A., Martinerie, J., and Varela, F. J. Comparison of Hilbert transform and wavelet methods for the analysis of neuronal synchrony. *J Neurosci Methods.* 111(12.3):83-98, 2001.

[26] Lopes da Silva FH and Storm van Leeuwen W, The cortical alpha rhythm in dog: The depth and surface profile of phase. In MAB Brazier MAB and H Petsche (Eds) *Architectonics of the Cerebral Cortex.* New York: Raven Press, pp. 319–333, 1978.

[27] Lopes da Silva, F. H. Dynamics of EEGs as signals of neuronal populations: models and theoretical considerations. In E Niedermeyer and FH Lopes da Silva (Eds) *Electroencephalography. Basic Principals, Clinical Applications, and Related Fields, 4th edition.* London: Williams and Wilkins, pp. 76–92, 1999.

[28] Luck, S. J., Woodman, G. F., and Vogel, E. K. Event-related potential studies of attention. *Trends in Cognitive Sciences,* 4(11):432–440, 2000.

[29] Makeig, S., Bell, A. J., Jung, T. P., and Sejnowski, T. J. Independent component analysis of electroencephalographic data. *Advances in Neural Information Processing Systems,* 145–151, MIT Press, 1996.

[30] Makeig S., Debener S., Onton J., and Delorme A., Mining event-related brain dynamics, *Trends in Cognitive Science,* 8(21.15):204–210, 2004.

[31] McMenamin, B.W., Shackman, A.J., Greischar, L.L., and Davidson, R.J. Electromyogenic artifacts and electroencephalographic inferences revisited. *Neuroimage,* 54(1):4–9, 2011.

[32] Morgan, S. T., Hansen, J. C., and Hillyard, S. A. Selective attention to stimulus location modulates the steady-state visual evoked potential. *Proceedings of the National Academy of Sciences,* 93(10):4770–4774, 1996.

[33] Mognon, A., Jovicich, J., Bruzzone, L., and Buiatti, M. ADJUST: An automatic EEG artifact detector based on the joint use of spatial and temporal features. *Psychophysiology,* 48(12.3):229–240, 2011.

[34] Murias, M., Swanson, J. M., and Srinivasan, R. Functional connectivity of frontal cortex in healthy and ADHD children reflected in EEG coherence. *Cerebral Cortex.* 17:1788-1799, 2007.

[35] Nunez, M. D., Srinivasan, R., and Vandekerckhove, J. Individual differences in attention influence perceptual decision making. *Frontiers in Psychology,* 8, 2015.

[36] Nunez, P.L. A study of origins of the time dependencies of scalp EEG: I-theoretical basis. *Biomedical Engineering, IEEE Transactions on,* (3):271–280, 1981.

[37] Nunez, P. L., Silberstein, R. B., Cadusch, P. J., Wijesinghe, R. S., Westdorp, A. F., and Srinivasan, R. A theoretical and experimental study of high resolution EEG based on surface Laplacians and cortical imaging. *Electroencephalography and Clinical Neurophysiology,* 90(12.1):40–57, 1994.

[38] Nunez, P. L. *Neocortical Dynamics and Human EEG Rhythms.* New York: Oxford University Press, 1995.

[39] Nunez, P. L. Toward a quantitative description of large scale neocortical dynamic function and EEG. *Behavioral and Brain Sciences* 23: 371–398 (target article), 2000a.

[40] Nunez, P. L. Neocortical dynamic theory should be as simple as possible, but not simpler. *Behavioral and Brain Sciences* 23: 415–437 (response to commentary by 18 neuroscientists), 2000b.

[41] Nunez, P. L. and Srinivasan, R. *Electric Fields of the Brain: The Neurophysics of EEG, 2nd edition.* New York: Oxford University Press, 2006.

[42] Olbrich, S., Jödicke, J., Sander, C., Himmerich, H., and Hegerl, U. ICA-based muscle artefact correction of EEG data: What is muscle and what is brain?: Comment on McMenamin et al. *Neuroimage*, 54(1):1–3, 2011.

[43] Oostenveld, R., and Praamstra, P. The five percent electrode system for high-resolution EEG and ERP measurements. *Clinical Neurophysiology*, 112(21.13):713–719, 2001.

[44] Percival, D. B. and Walden, A. T., *Spectral Analysis for Physical Applications: Multitaper and Conventional Univariate Techniques,* Cambridge U Press, Cambridge, UK, 1993.

[45] Petsche, H. and Etlinger, S. C. *EEG and Thinking. Power and Coherence Analysis of Cognitive Processes.* Vienna: Austrian Academy of Sciences, 1998.

[46] Pfurtscheller, G. and Lopes da Silva, F. H. Event-related EEG/MEG synchronization and desynchronization: Basic principles. *Electroencephalography and Clinical Neurophysiology* 110:1842–1857, 1999.

[47] Phillips, C., Rugg, M. D., and Friston, K. J. Anatomically informed basis functions for EEG source localization: Combining functional and anatomical constraints. *NeuroImage*, 16(21.12):678–695, 2002.

[48] Polich, J., and Kok, A. Cognitive and biological determinants of P300: an integrative review. *Biological psychology*, 41(12.3):103–146, 1995.

[49] Pope, K. J., Fitzgibbon, S. P., Lewis, T. W., Whitham, E. M., and Willoughby, J. O. Relation of gamma oscillations in scalp recordings to muscular activity. *Brain Topography*, 22(12.1):13–17, 2009.

[50] Regan, D. Steady-state evoked potentials. *JOSA*, 67(11):1475–1489, 1977

[51] Schlögl, A., Keinrath, C., Zimmermann, D., Scherer, R., Leeb, R., and Pfurtscheller, G. A fully automated correction method of EOG artifacts in EEG recordings. *Clinical Neurophysiology*, 118(12.1):98–104, 2007.

[52] Shackman, A. J., McMenamin, B. W., Slagter, H. A., Maxwell, J. S., Greischar, L. L., and Davidson, R. J. Electromyogenic artifacts and electroencephalographic inferences. *Brain Topography*, 22(12.1):7–12, 2009.

[53] Srinivasan, R., and Deng, S. Multivariate spectral analysis of the electroencephalogram: Power, coherence, and second-order blind identification. 2012.

[54] Srinivasan, R., Nunez, P.L., Tucker, D.M., Silberstein, R.B., and Cadusch, P.J. Spatial sampling and filtering of EEG with spline Laplacians to estimate cortical potentials. *Brain Topography*, 8:355–366, 1996.

[55] Srinivasan, R., Nunez, P.L., and Silberstein, R.B. Spatial filtering and neocortical dynamics: Estimates of EEG coherence. *IEEE Transactions on Biomedical Engineering*, 45:814–826, 1998.

[56] Srinivasan, R. Anatomical constraints on source models for high-resolution EEG and MEG derived from MRI. *TCRT* 5:389–399, 2006.

[57] Srinivasan, R., Winter, W. R., Ding, J., and Nunez, P.L., EEG and MEG coherence: Measures of functional connectivity at distinct spatial scales of neocortical dynamics. *Journal of Neuroscience Methods* 166:41–52, 2007.

[58] Tang, A. C., Sutherland, M. T., and McKinney, C. J., Validation of SOBI components from high-density EEG, *NeuroImage*, 25:539–553, 2004.

[59] Tass, P., Rosenblum, M. G., Weule, J., Kurths, J., Pikovsky, A., Volkmann, J., Schnitzler, A. and Freund, H. J. Detection of n: m phase locking from noisy data: Application to magnetoencephalography. *Physical Review Letters*, 81(15):3291, 1998.

[60] Tiitinen, H., Sinkkonen, J., Reinikainen, K., Alho, K., Lavikainen, J., and Näätänen, R. Selective attention enhances the auditory 40-Hz transient response in humans. *Nature*, 364(6432):59–60, 1993.

[61] Tucker, D. M. Spatial sampling of head electrical fields: The geodesic sensor net. *Electroencephalography and Clinical Neurophysiology*, 87(21.12):154–163, 1993.

[62] Wipf, D., and Nagarajan, S. A unified Bayesian framework for MEG/EEG source imaging. *Neuroimage*, 44(21.12):947–966, 2009.

[63] Whitham, E. M., Pope, K. J., Fitzgibbon, S.P., Lewis, T., Clark, C. R., Loveless, S., Broberg, M., Wallace, A., DeLosAngeles, D., Lillie, P., Hardy, A., Fronsko, R., Pulbrook, A. and Willoughby, J.O. Scalp electrical recording during paralysis: Quantitative evidence that EEG frequencies above 20 Hz are contaminated by EMG. *Clinical Neurophysiology*, 118(21.18):1877–1888, 2007.

[64] Whitham, E. M., Lewis, T., Pope, K. J., Fitzgibbon, S. P., Clark, C. R., Loveless, S., DeLosAngeles, D., Wallace, A. K., Broberg, M., and Willoughby, J. O. Thinking activates EMG in scalp electrical recordings. *Clinical Neurophysiology: Official Journal of the International Federation of Clinical Neurophysiology*, 119(21.15):1166, 2008.

[65] Wu, J., Quinlan, E. B., Dodakian, L., McKenzie, A., Kathuria, N., Zhou, R. J., Augsburger, R., See, J., Le, V. H., Srinivasan, R., and Cramer, S. C. Connectivity measures are robust biomarkers of cortical function and plasticity after stroke. *Brain*, 138(21.18):2359–2369, 2015.

[66] Yuval-Greenberg, S., Tomer, O., Keren, A. S., Nelken, I., and Deouell, L. Y. Transient induced gamma-band response in EEG as a manifestation of miniature saccades. *Neuron.* 58(21.12):429–441, 2008.

[67] Ziehe, A., Laskov, P., Muller, K.R., Nolte, G. A linear least-squares algorithm for joint diagonalization. *Proceedings ICA2003* 469–474, 2003.

Part III

STATISTICAL METHODS
AND MODELS

8

Image Reconstruction in Functional MRI

Daniel B. Rowe
Department of Mathematics, Statistics, and Computer Science
Marquette University

CONTENTS

8.1	Introduction	205
8.2	The Fourier Transform	206
	8.2.1 One-Dimensional Fourier Transform	206
	8.2.2 Two-Dimensional Fourier Transform	210
8.3	FMRI Acquisition and Reconstruction	214
	8.3.1 The Signal Equation and k-Space Coverage	214
	8.3.2 Nyquist Ghost k-Space Correction	217
8.4	Image Processing	220
	8.4.1 Reconstruction Isomorphism Representation	220
	8.4.2 Image Processing Implications	222
8.5	Additional Topics and Discussion	228
	8.5.1 Complex-Valued fMRI Activation	229
	8.5.2 Discussion	230
	Acknowledgments	230
	Bibliography	230

8.1 Introduction

In the world we live in with all of our technological advances, we still have a relatively poor understanding of how the healthy human brain works, let alone how it is disfunctioning due to disease or injury. Our brain is one of the most complicated systems in the universe, which is exactly what makes it one of the last but most exciting scientific frontiers. Magnetic resonance imaging (MRI), for which Drs. Paul Lauterbur and Peter Mansfield won the 2003 Nobel Prize, is an ideal noninvasive imaging technique to see inside the human brain.

Structural or anatomical MRI has been an invaluable tool for the diagnosis and monitoring of human neurological ailments. Functional MRI (fMRI) is a lesser known type of MRI that allows us to observe the cognitively active brain in action. In 1992 there were three fMRI publications in close succession: the first by Bandettini (1), the second by Kwong (2) and the third by Ogawa (3). These three papers established fMRI with the blood-oxygen-level-dependent (BOLD) signal, which is a neural correlate and does not require exogenous contrast.

In MRI and fMRI, the measurements taken by the scanner are not voxel values. The actual measurements taken by the scanner are to a good approximation, complex-valued (real and imaginary) spatial frequencies. Small magnetic field gradients are, changed in time that result in changing location in spatial frequency space. While changing spatial frequency location, complex-valued spatial frequency measurements are taken. This Chapter aims to provide the reader with the necessary technical abilities and conceptual understandings to perform and understand fMRI image reconstruction and processing. The outline of the chapter is as follows. Section 8.2 describes the discrete Fourier transform, which is absolutely invaluable for understanding fMRI measurements. Section 8.3 describes Fourier encoding and complex-valued fMRI measurements that are discrete inverse Fourier transform reconstructed into a complex-valued image. Section 8.4 presents the discrete inverse Fourier image reconstruction process with an isomorphism representation so that signal and image processing steps can be described and their effects quantified. Section 8.5 summarizes the additional related topic of complex-valued time series activation models and a discussion reviewing the covered topics and future areas. In this chapter, the magnetic resonance (MR) physics will not be described, but the basic measurement and image reconstruction processes will be discussed. The discussion here will be limited to Cartesian k-space spatial frequency sampling.

8.2   The Fourier Transform

As previously noted, the measurements taken by the MRI scanner are, to a good approximation, complex-valued spatial frequencies. That is, the measurements taken by the scanner are ideally the Fourier transform (4) of the object being imaged. In this section, the one-dimensional discrete Fourier transform and discrete inverse Fourier transform will be described then extended to the two-dimensional discrete Fourier transform and inverse Fourier transform. What we will see is that regardless of whether we have a one-dimensional or two-dimensional signal, the Fourier transform selects out the constituent cosine and sine frequencies that make up the signal. In Section 8.3 we will see that the MRI scanner measures (to a good approximation) the constituent cosine and sine frequencies in an image and we inverse Fourier transform these frequencies in order to reconstruct an image.

8.2.1   One-Dimensional Fourier Transform

The mathematics for the one-dimensional discrete Fourier transform will be described then an example involving a time series of measurements will be presented. The one-dimensional discrete Fourier transform $F(q\Delta\nu)$ of a time series $y(t)$ sampled at N times Δt apart is defined as in Equation 8.1 below

$$f(q\Delta\nu) \;=\; \sum_{p=1}^{N} y(p\Delta t)e^{-i2\pi(p-1)(q-1)/N} \tag{8.1}$$

for $p, q = 1, \ldots, N$. The difference in temporal frequency $\Delta\nu$ between $f(q\Delta\nu)$ and $f((q + 1)\Delta\nu)$ is $\Delta\nu = 1/(N\Delta t)$ with $i = \sqrt{-1}$ being the imaginary unit. The one-dimensional discrete Fourier transform $f(q\Delta\nu)$ at a given frequency $q\Delta\nu$ in general consists of both a real and an imaginary part. In order to satisfy the very important Shannon–Nyquist sampling criteria (5, 6), which allows the resolution of temporal frequencies, the time series must be sampled at twice its highest constituent temporal frequency. The highest temporal frequency that can be resolved and hence the bounds on the horizontal frequency axis is $\nu_{max} = 1/(2\Delta t)$. If the time series being discrete Fourier transformed is completely real-valued, then the discrete temporal frequencies have Hermitian symmetry where the bottom half is the conjugate transpose of the top half.

With the zero frequency of the frequency spectrum centered, the one-dimensional discrete inverse Fourier transform of $f(q\Delta\nu)$ is defined as in Equation 8.2 below

$$y(p\Delta t) \;=\; \frac{1}{N} \sum_{q=-N/2}^{N/2-1} f(q\Delta\nu)e^{+i2\pi pq/N} \tag{8.2}$$

for $p,q = -N/2, \ldots, N/2 - 1$. The time series will be ordered from the first time point to the last. This definition of the one-dimensional discrete inverse Fourier transform will be utilized because MRI and fMRI measurements have the zero frequency centered and there is generally an even number of measurements. The discrete inverse Fourier transform $y(p\Delta t)$ at a given time point p in general consists of both a real and an imaginary part, but this imaginary part may be zero.

The one-dimensional forward discrete Fourier transformation procedure described in Equation 8.1 can be equivalently written as

$$(f_R + if_I) \;-\; (\bar{\Omega}_R + i\bar{\Omega}_I) \;\; (y_R + iy_I) \tag{8.3}$$

where $y_R = (y_R(1\Delta t), \ldots, y_R(N\Delta t))'$ and $y_I = (y_I(1\Delta t), \ldots, y_I(N\Delta t))'$, are the real and imaginary parts of the measured time series while $f_R = (f_R(1\Delta\nu), \ldots, f_R(N\Delta\nu))'$ and $f_I = (f_I(1\Delta\nu), \ldots, f_I(N\Delta\nu))'$ are the real and imaginary parts of the one-dimensional discrete Fourier transform of the time series (temporal frequencies). The Fourier matrix $\bar{\Omega} = (\bar{\Omega}_R + i\bar{\Omega}_I)$ in Equation 8.3 is given by

$$\bar{\Omega} = \begin{pmatrix} 1 & 1 & \cdots & 1 \\ 1 & W & & W^N \\ \vdots & \vdots & \ddots & \vdots \\ & & & W^{(N-1)N} \\ 1 & W^N & & W^{N*N} \end{pmatrix} \tag{8.4}$$

where $W = e^{-\frac{i2\pi}{N}}$ is an *a priori* known quantity from the sampling plan. The jkth element (row j and column k) of $\bar{\Omega}_R$ and $\bar{\Omega}_I$ are $\cos(2\pi(j-1)(k-1)/N)$ and $\sin(2\pi(j-1)(k-1)/N)$ where $j, k = 1, \ldots, N$.

Similarly, the one-dimensional discrete inverse Fourier transformation procedure described in Equation 8.2 can be equivalently written as

$$(y_R + iy_I) \;=\; (\Omega_R + i\Omega_I) \;\; (f_R + if_I) \tag{8.5}$$

where variables are as previously defined. The inverse Fourier matrix $\Omega = (\Omega_R + i\Omega_I)$ in Equation 8.5 is given by

$$\Omega = \begin{pmatrix} W^{(-\frac{N}{2})(-\frac{N}{2})} & W^{(-\frac{N}{2})(-\frac{N}{2}+1)} & \cdots & W^{(-\frac{N}{2})(\frac{N}{2}-1)} \\ W^{(-\frac{N}{2}+1)(-\frac{N}{2})} & W^{(-\frac{N}{2}+1)(-\frac{N}{2}+1)} & & W^{(-\frac{N}{2}+1)(\frac{N}{2}-1)} \\ & & \ddots & \vdots \\ & & & W^{(\frac{N}{2}-2)(\frac{N}{2}-1)} \\ W^{(\frac{N}{2}-1)(-\frac{N}{2})} & & & W^{(\frac{N}{2}-1)(\frac{N}{2}-1)} \end{pmatrix} \tag{8.6}$$

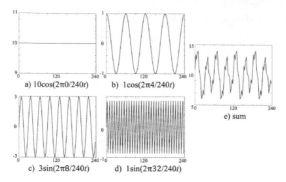

FIGURE 8.1
Discrete constituent parts and their sum for the one-dimensional function.

where $W = e^{\frac{i2\pi}{N}}$. The jkth element of Ω_R and Ω_I are $\cos(2\pi(j - N/2 - 1)(k - N/2 - 1)/N)$ and $\sin(2\pi(j - N/2 - 1)(k - N/2 - 1)/N)$.

The one-dimensional discrete Fourier transform can be illustrated with an example. A continuous one-dimensional signal $y(t)$ as in Equation 8.7 below

$$y(t) = 10\cos\left(2\pi\frac{0}{240}t\right) + \cos\left(2\pi\frac{4}{240}t\right) + 3\sin\left(2\pi\frac{8}{240}t\right) + \sin\left(2\pi\frac{32}{240}t\right) \qquad (8.7)$$

is sampled at time $t = (p - 1)\Delta t$ for $p = 1, \ldots, N$, $N = 96$, and $\Delta t = 2.5$ s (seconds) for a total of 240 s. The four discrete functions composing $y(p\Delta t)$ are presented in Figures 8.1a–d with their sum $y(p\Delta t)$ being in Figure 8.1e. The units for the horizontal axis in Figure 8.1 is seconds.

The one-dimensional discrete Fourier transform of $y(p\Delta t)$ for $p = 1, \ldots, N$ is given in Figure 8.2. After the one-dimensional discrete Fourier transform, the frequency spectrum was shifted so that the zero frequency is centered. What can be seen in Figure 8.2 is that the one-dimensional discrete Fourier transformation process selects out the constituent temporal frequencies. The units for the horizontal axes in Figure 8.2 is 1/s (cycles per second or Hz). In Figure 8.2 there are points (that have been connected) at the temporal frequencies that make up the time series. The real part of the temporal frequency spectrum in Figure 8.2a contains the cosine frequencies and the imaginary part in Figure 8.2b contains the sine frequencies. The magnitude in Figure 8.2c is a typical way of representing all the constituent frequencies with phase between the frequency parts in Figure 8.2d. There is a peak in the middle of Figure 8.2a (where the full vertical scale is limited to 96), corresponding to the zero cosine frequency (constant or baseline term) and at $4/240 \approx 0.0167$ Hz. We can also see points in Figure 8.2b are the $8/240 \approx 0.0333$ Hz and $32/240 \approx 0.1333$ Hz sine frequencies. The heights of the points are $NA_\nu/2$, where A_ν is the amplitude of the sinusoid at frequency ν except for the zero frequency, which has a height of NA_0. Also note that the real part is symmetric while the imaginary part is anti-symmetric.

A pictorial depiction of the one-dimensional discrete Fourier transform can be seen by displaying the time series as an image as in Figure 8.3. Again, after the discrete Fourier transform, the frequency spectrum was shifted so that the zero frequency is centered. In Figure 8.3c are the real (left) and imaginary (right) parts of the time series (intensity limited from 5 to 15), in Figure 8.3b are the real (left) and imaginary (right) parts of the Fourier matrix (intensity limited from −1 to 1), and in Figure 8.3a are the real (left) and imaginary (right) parts of the frequency spectrum (intensity limited from 0 to 96). The units for the vertical axis in Figure 8.3a is Hertz and for Figure 8.3c seconds. In this chapter, the greyscale for all images will use the convention that black will depict lower values and white,

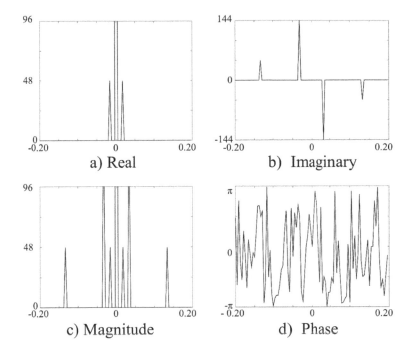

a) Real   b) Imaginary

c) Magnitude   d) Phase

FIGURE 8.2
Fourier transform of the time series.

higher values. Note that the intensity values of the horizontal stripes in Figure 8.3c (left) are brighter for higher values and depict the time series in Figure 8.1e.

Again, we can see that the one-dimensional discrete Fourier transform selects out the constituent temporal frequencies of the time series. The cosine frequencies are represented in Figure 8.3a (left) and the sines in Figure 8.3a (right).

In the same fashion, the inverse Fourier transform can be depicted as in Figure 8.4. The units for the axes in Figure 8.4a is seconds and Figure 8.4c Hertz. In Figure 8.4c are the real (left) and imaginary (right) parts to the frequency spectrum, in Figure 8.4b are the real

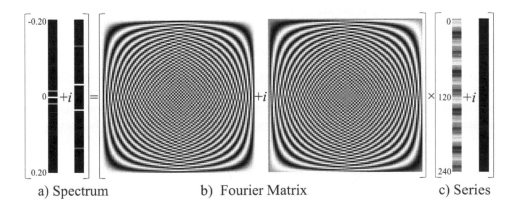

a) Spectrum   b) Fourier Matrix   c) Series

FIGURE 8.3
Matrix representation of one-dimensional discrete Fourier transform.

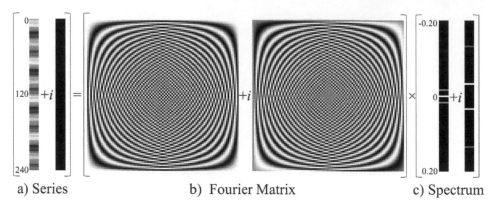

a) Series   b) Fourier Matrix   c) Spectrum

FIGURE 8.4
Matrix representation of one-dimensional discrete inverse Fourier transform.

(left) and imaginary (right) parts of the one-dimensional discrete inverse Fourier transform matrix (intensity limited from $-1/96$ to $1/96$), and in Figure 8.4a are the real (left) and imaginary (right) parts of the time series as originally specified.

We can see that when we have measured the constituent complex-valued frequencies, we can inverse Fourier transform reconstruct them into the measured time series. Although the one-dimensional discrete Fourier transform and discrete inverse Fourier transforms in Equation 8.3 and Equation 8.5 have a great conceptual interpretation, the implementations in Equation 8.1 and Equation 8.2 are computationally faster and generally how the transforms are performed. In Section 8.4, the discrete Fourier transform and discrete inverse Fourier transforms will be represented with an isomorphism that has an even better conceptual interpretation that is useful for statistical purposes. In MRI and fMRI, the measurements taken by the scanner are, to a good approximation, discrete complex-valued spatial frequencies and we reconstruct them into a discrete complex-valued image via the discrete inverse Fourier transform. The discrete Fourier transform and discrete inverse Fourier transform are reverse operations, meaning that you can one-dimensional discrete forward Fourier transform a time series to get temporal frequencies and then one-dimensional discrete inverse Fourier transform the temporal frequencies to get back the original time series. The discrete complex-valued frequencies can be obtained from the discrete complex-valued time series and vice versa provided there is no loss of data.

8.2.2   Two-Dimensional Fourier Transform

The mathematics for the two-dimensional discrete Fourier transform will be described similar to the way that it was done for the one-dimensional discrete Fourier transform, then an example involving an illustrative sample image will be presented. The two-dimensional discrete Fourier transform $F(q_x \Delta k_x, q_x \Delta k_x)$ of an image Y sampled at N_x horizontal and N_y vertical locations Δx and Δy distances apart is defined as in Equation 8.8 below

$$F(q_x \Delta k_x, q_y \Delta k_y) = \sum_{p_y=1}^{N_y} \sum_{p_x=1}^{N_x} Y(p_x \Delta x, p_y \Delta y) e^{-i2\pi \left(\frac{(p_x-1)(q_x-1)}{N_x} + \frac{(p_y-1)(q_y-1)}{N_y} \right)} \tag{8.8}$$

for $p_x, q_x = 1, \ldots, N_x$ and $p_y, q_y = 1, \ldots, N_y$. The differences in spatial frequencies between successive values Δk_x and Δk_y are $\Delta k_x = 1/(N_x \Delta x)$ and $\Delta k_y = 1/(N_y \Delta y)$. In order to satisfy the very important Shannon–Nyquist sampling criteria that allows the resolution of spatial frequencies, the image must be sampled at twice the highest constituent spatial frequency in each dimension. The highest spatial frequencies that can be resolved and hence the bounds on the two frequency axes are $k_{x,max} = 1/(2\Delta x)$ and $k_{y,max} = 1/(2\Delta y)$. If the image being discrete Fourier transformed is completely real-valued, then the discrete spatial frequencies have Hermitian symmetry where the bottom half is the conjugate transpose of the top half. Hermitian symmetry is generally not achieved in experimentally measured fMRI spatial frequencies. The two-dimensional discrete Fourier transform $F(q_x\Delta k_x, q_y\Delta k_y)$ at a given frequency pair $(q_x\Delta k_x, q_y\Delta k_y)$ in general consists of both a real and an imaginary part.

With the zero frequency of the frequency spectrum centered, the two-dimensional discrete inverse Fourier transform of $F(q_x\Delta k_x, q_y\Delta k_y)$ is defined as in Equation 8.9 below

$$Y(p_x\Delta x, p_y\Delta y) = \frac{1}{N_xN_y} \sum_{q_y=-N_y/2}^{N_y/2-1} \sum_{q_x=-N_x/2}^{N_x/2-1} F(q_x\Delta k_x, q_y\Delta k_y)e^{+i2\pi(\frac{p_xq_x}{N_x}+\frac{p_yq_y}{N_y})} \quad (8.9)$$

for $p_x, q_x = -N_x/2, \ldots, N_x/2 - 1$ and $p_y, q_y = -N_y/2, \ldots, N_y/2 - 1$. It should be noted that it is not necessarily the case that $N_x = N_y$ or that $\Delta x = \Delta y$. The image will be ordered from the first pixel or voxel to the last in both dimensions. This definition of the two-dimensional discrete inverse Fourier transform will be utilized because MRI and fMRI measurements have the zero frequency centered and generally an even array size in each dimension is measured. The two-dimensional discrete inverse Fourier transform of the image $Y(p_x\Delta x, p_y\Delta y)$, at a given voxel $(p_x\Delta x, p_y\Delta y)$ in general consists of both a real and an imaginary part, but this imaginary part may be zero.

The two-dimensional forward discrete Fourier transformation procedure described in Equation 8.8 can be equivalently written as

$$(F_R + iF_I) = (\bar{\Omega}_{yR} + i\bar{\Omega}_{yI}) \ (Y_R + iY_I) \ (\bar{\Omega}_{xR} + i\bar{\Omega}_{xI})^T \quad (8.10)$$

where T denotes the transpose, Y_R and Y_I are the real and imaginary parts of the measured image, and F_R and F_I are the real and imaginary parts of the Fourier transform of the image (spatial frequencies). If we have a real-valued image, then the imaginary part of the image Y_I is zero. The two-dimensional discrete Fourier matrices $\bar{\Omega}_x$ and $\bar{\Omega}_y$ in Equation 8.10 are the same as in the one-dimensional discrete Fourier transform in Equation 8.4 with appropriate dimensions.

The two-dimensional discrete inverse Fourier transformation procedure described in Equation 8.9 can be equivalently written as

$$(Y_R + iY_I) = (\Omega_{yR} + i\Omega_{yI}) \ (F_R + iF_I) \ (\Omega_{xR} + i\Omega_{xI})^T \quad (8.11)$$

where variables are all as previously defined. The discrete inverse Fourier matrices Ω_x and Ω_y in Equation 8.11 are the same as in the one-dimensional discrete inverse Fourier transform in Equation 8.6 with appropriate dimensions.

Similar to the one-dimensional discrete Fourier transform, the two-dimensional discrete Fourier transform can be illustrated with an example. A continuous two-dimensional image $Y(x, y)$ as in Equation 8.12 below

$$Y(x, y) = 10\cos\left(2\pi\frac{0}{240}x\right) + \frac{3}{2}\cos\left(2\pi\frac{8}{240}x\right)$$
$$+ \sin\left(2\pi\frac{24}{240}y\right) + \cos\left(2\pi\frac{16}{240}x + 2\pi\frac{16}{240}y\right) \quad (8.12)$$

is sampled at position $x = (p_x-1)\Delta x$ and $y = (p_y-1)\Delta y$ for $p_x = 1, \ldots, N_x, p_y = 1, \ldots, N_y,$

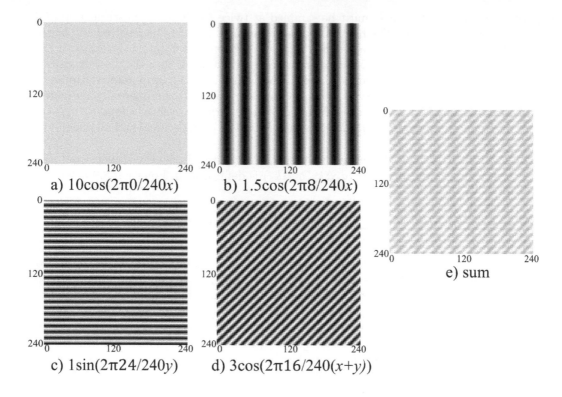

a) $10\cos(2\pi 0/240x)$

b) $1.5\cos(2\pi 8/240x)$

c) $1\sin(2\pi 24/240y)$

d) $3\cos(2\pi 16/240(x+y))$

e) sum

FIGURE 8.5
Discrete constituent parts and their sum for the two-dimensional image.

$N_x = 96$, $N_y = 96$, and $\Delta x = \Delta y = 2.5$ mm (millimeters) corresponding to a 240-mm field of view. The four discrete functions composing $Y(p_x\Delta x, p_y\Delta y)$ are presented in Figures 8.5a–d with their sum $Y(p_x\Delta x, p_y\Delta y)$ being in Figure 8.5e. The units for the axes in Figure 8.5 are millimeters.

What can be seen in Figure 8.6 is that the two-dimensional discrete Fourier transformation process again selects out the constituent spatial frequencies. After the two-dimensional discrete Fourier transform, the frequency spectrum was shifted so that the zero frequency is centered. In Figure 8.6c are the real (top) and imaginary (bottom) parts of the image (intensity limited from -15 to 15), in Figures 8.6b and d are the real (top) and imaginary (bottom) parts of the Fourier matrices (intensity limited from -1 to 1), and in Figure 8.6a are the real (top) and imaginary (bottom) parts of the frequency spectrum (intensity limited from 0 to 96^2). The units for the image in Figure 8.6c are millimeters and spatial frequencies in 8.6a are mm^{-1} (wave number). The transpose of $\bar{\Omega}_x$ is displayed in Figure 8.6d.

There are points at the spatial frequencies that make up the image. The real part of the spatial frequency spectrum contains the cosine frequencies and the imaginary part contains the sine frequencies. There is a peak in the middle of Figure 8.6a (top) corresponding to the zero cosine frequency (constant or baseline term), at the $8/240 \approx 0.0333$ mm^{-1} x spatial frequency, and at the $24/240 = 0.1000$ mm^{-1} both x and y spatial frequency. Also, we can see points in Figure 8.6a (bottom) at the $16/240 \approx 0.0667$ mm^{-1} spatial frequency. The heights of the points are $N_x N_y A_{k_x,k_y}/4$, where A_{k_x,k_y} is the amplitude of the planar sinusoid at frequency (k_x, k_y) except for the zero frequency, which has a height of $N_x N_y A_{0,0}$. It is important to note that the real part is symmetric while the imaginary

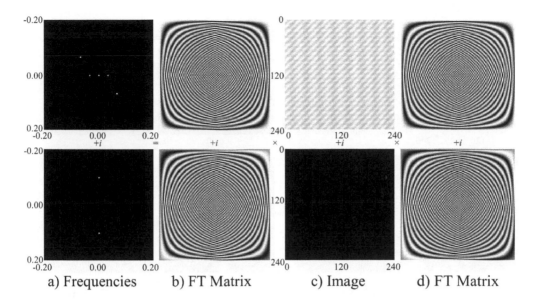

a) Frequencies b) FT Matrix c) Image d) FT Matrix

FIGURE 8.6
Matrix representation of two-dimensional discrete Fourier transform.

part is anti-symmetric, but for display purposes the intensities of the imaginary part were made completely positive.

In the same fashion, the inverse Fourier transform can be depicted as in Figure 8.7. In Figure 8.7c are the real (top) and imaginary (bottom) parts of the frequency spectrum, in Figure 8.7b and Figure 8.7d are the real (top) and imaginary (bottom) parts of the two-dimensional discrete inverse Fourier transform matrices (intensity limited from $-1/96$ to $1/96$ for Figure 8.7b and from $-1/96$ to $1/96$ for Figure 8.7d). Note the difference in appearance between Figure 8.6b and Figure 8.7b. In Figure 8.7a are the real (top) and imaginary (bottom) parts of the two-dimensional discrete inverse Fourier transformed image. The image being presented in Figure 8.7d is the transpose of Ω_x.

We can see that when we have measured the constituent complex-valued frequencies of an image, we can reconstruct them into the measured image. Although the discrete Fourier transform and discrete inverse Fourier transforms in Equation 8.10 and Equation 8.11 have a great conceptual interpretation, the implementations in Equation 8.8 and Equation 8.9 are computationally faster and most often how performed. In Section 8.4, the discrete Fourier transform and discrete inverse Fourier transforms will be represented with an isomorphism that has an even better conceptual interpretation and is useful for statistical analysis. In MRI and fMRI, the measurements taken by the scanner are, to a good approximation, complex-valued spatial frequencies and we reconstruct them into a complex-valued image via the discrete inverse Fourier transform.

The discrete Fourier transform and discrete inverse Fourier transform are reverse operations, meaning that you can two-dimensional discrete forward Fourier transform an image to get spatial frequencies then two-dimensional discrete inverse Fourier transform the spatial frequencies to get back the original image. The discrete complex-valued frequencies can be obtained from the discrete complex-valued time series and vice versa provided there is no loss of data. The discrete Fourier transform and discrete inverse Fourier transform are reverse operations provided there is no loss of data such as discarding phase images as is common in fMRI.

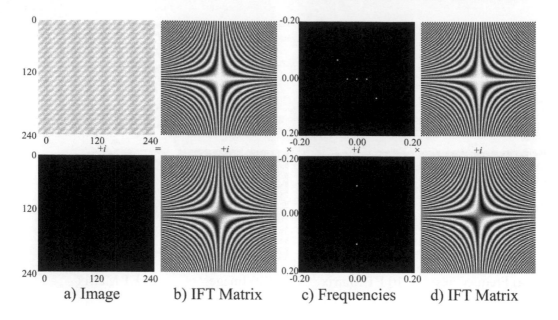

a) Image   b) IFT Matrix   c) Frequencies   d) IFT Matrix

FIGURE 8.7
Matrix representation of two-dimensional discrete inverse Fourier transform.

8.3   FMRI Acquisition and Reconstruction

It was important to review the one and two dimensional discrete Fourier transforms along with their inverses in Section 8.2 to prepare for measured fMRI data. In Section 8.2, it was seen that the two-dimensional discrete complex-valued spatial frequencies could be two dimensional discrete inverse Fourier transform reconstructed into a complex-valued image. In MRI and fMRI, the data measured by the scanner is, to a good approximation, the Fourier transform of the object being imaged (4). The spatial frequencies, called k-space, are measured by the scanner and inverse Fourier transform reconstructed into a complex-valued image. In this section, the MR signal equation is described along with a standard fMRI gradient echo–echo planar imaging (GRE-EPI) pulse sequence and the coverage of k-space (spatial frequencies).

8.3.1   The Signal Equation and k-Space Coverage

Without delving into the MRI physics, the continuous signal $s(k_x, k_y)$ to be measured by the scanner at time $t(k_x, k_y)$ for k-space location (k_x, k_y) is given by the following fairly general expression called the signal equation

$$s(k_x, k_y) = \int_{-\infty}^{\infty} \int_{-\infty}^{\infty} \rho(x,y) e^{-i2\pi(k_x x + k_y y)} dx dy$$

$$k_x(t(k_x, k_y)) = \frac{\gamma}{2\pi} \int_{0}^{t(k_x, k_y)} G_x(\tau) d\tau \qquad (8.13)$$

$$k_y(t(k_x, k_y)) = \frac{\gamma}{2\pi} \int_{0}^{t(k_x, k_y)} G_y(\tau) d\tau$$

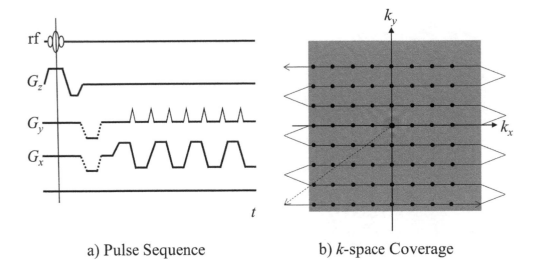

a) Pulse Sequence   b) k-space Coverage

FIGURE 8.8
Standard GRE-EPI pulse sequence and k-space coverage.

where

$$\rho(x,y) = \rho_0(x,y) \left(1 - e^{TR/T_1(x,y)}\right) e^{t(k_x,k_y)/T_2^*(x,y)} e^{i\gamma \Delta B(x,y)t(k_x,k_y)} . \qquad (8.14)$$

In Equation 8.14, $s(k_x, k_y)$ is the measured k-space signal at location (k_x, k_y), $\rho_0(x,y)$ is the proton spin density at position (x,y), $T_1(x,y)$ is the longitudinal relaxation at position (x,y), $T_2^*(x,y)$ is the transverse relaxation at position (x,y), $\Delta B(x,y)$ is the magnetic field inhomogeneity at position (x,y), TR is the time to repetition for measurement of the same slice, $\gamma = 2.67513 \times 10^8$ rad s$^{-1}T^{-1}$ is the gyromagnetic ratio of the H^1 nucleus, and $G_x(\tau)$ and $G_y(\tau)$ are magnetic field gradients. We can see that $\rho(x,y)$ is not strictly a function of position but varies depending on the time that a k-space point is to be measured, so the Fourier relationship is an approximate one.

This is the two-dimensional continuous Fourier encoding expression upon which the previous discrete two-dimensional discrete inverse Fourier transform image reconstruction in Section 8.2 was based. By changing the magnetic field gradients $G_x(t)$ and $G_y(t)$ through time t, the signal that is to be measured corresponds to the different locations in k-space.

The way that k-space is traversed and complex-valued data measured is with the use of magnetic field gradients that are in addition to but much smaller than the main magnetic field. The changing of the G_x and G_y magnetic field gradients can be graphically described using what is called a pulse sequence diagram. In a pulse sequence diagram, multiple gradient waveforms are presented where the gradient strength (either positive or negative) is on the vertical axis and time is on the horizontal axis.

A standard GRE-EPI pulse sequence diagram is presented in Figure 8.8a and corresponding k-space coverage trajectory in Figure 8.8b. In Figure 8.8a, a radio frequency pulse at the beginning of the first waveform trace is applied to tip the magnetization into the transverse plane (called excitation) and a gradient G_z selects the slice to be imaged in the second waveform trace. At this time, we are at the center of k-space, $(k_x, k_y) = (0, 0)$. After excitation and slice selection, gradient magnetic fields are applied as seen in the first negative

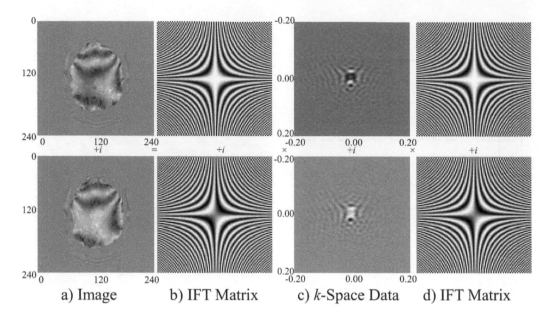

FIGURE 8.9
Matrix representation of fMRI image reconstruction.

trapezoidal-shaped changes in the G_y and G_x gradient waveforms (third and fourth traces). The initial negative trapezoidal G_y gradient takes us to the bottommost (most negative) k_y location while the initial negative trapezoidal G_x gradient takes us to the leftmost (most negative) k_x location. At this time, we are now at the bottom left corner of our k-space region. We apply a positive G_x gradient that takes us from left to right along the bottommost line of k-space. When we have reached the rightmost portion of k-space, we now apply a negative G_x gradient to take us from right to left in k-space and a positive triangular G_y gradient that takes up a k-space line. This process repeats until all of k-space is covered and complex-valued k-space array measurements are taken to fill in an array. The direction that we travel along lines of k-space is called the frequency encoding direction (k_x in this case) and the direction that we step up to travel along lines is called the phase encoding direction (k_y in this case). The trajectory that just described the coverage of k-space is illustrated in Figure 8.8b from bottom left to top left (with underlay being a lighter version of the real k-space array in Figure 8.9c).

The complex-valued k-space data (spatial frequencies) corresponding to an experimentally measured slice from a human fMRI scan (7) are displayed in Figure 8.9c. Note that the real part (cosines) and imaginary part (sines) of the k-space array is completely filled, unlike the simple illustrative example in Figure 8.7 of Section 8.2. The discrete inverse Fourier transform process is applied to the measured k-space data in Figure 8.9. The measured k-space data is in Figure 8.9c, the discrete inverse Fourier transform matrices are in Figure 8.9b and Figure 8.9d with the reconstructed (two-dimensional discrete inverse Fourier transformed) complex-valued image in Figure 8.9a. The two-dimensional discrete inverse Fourier transformation process as performed in Figure 8.7 of Section 8.2 reconstructs the experimentally measured k-space data into a complex-valued image. Note that the two discrete inverse Fourier transform matrices in Figure 8.7 are identical to those in Figure 8.9.

The reconstructed image in Figure 8.9a is in terms of real and imaginary parts. In most fMRI studies, the reconstructed images are transformed from Cartesian coordinates

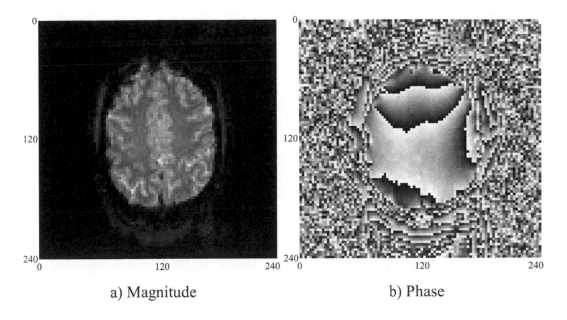

a) Magnitude b) Phase

FIGURE 8.10
Magnitude and phase of real and imaginary reconstructed image.

of real and imaginary to polar coordinates of magnitude and phase as seen in Figure 8.10. This transformation changes the statistical properties of the images. Further, in most fMRI studies, the phase half of the reconstructed image data as in Figure 8.10b is discarded and only the magnitude images as in Figure 8.10a are included in statistical analysis, in addition to processing performed on the images as will be briefly described in Section 8.4. From a statistical perspective, we should utilize all of our measured data as will be summarized in Section 8.5.

8.3.2 Nyquist Ghost k-Space Correction

When experimental fMRI k-space data is measured, there is a necessary processing step called Nyquist ghost correction that is performed on it in order to turn it into the k-space data seen in Figure 8.9c. A Nyquist ghost is a standard artifact that occurs in GRE-EPI due to a phase discrepancy between the odd and even frequency encode direction (right-to-left and left-to-right) lines of k-space.

The raw k-space measurements from an unprocessed experimental fMRI scan are presented in Figure 8.11c and are reconstructed into an image by the same discrete inverse Fourier transform process in Figure 8.11 and as in Figure 8.7 and Figure 8.9. Note the prominent ghost of the brain at the top and bottom of the real and imaginary images in Figure 8.11a. Images with the Nyquist ghost as presented are not suitable for either visual interpretation or statistical analysis. We can compare Figure 8.11a with the Nyquist ghost to the improved reconstructed image in Figure 8.10a, which is after a correction. Nyquist ghost correction will be described later in this Section.

The magnitude and phase of the reconstructed raw experimental k-space data in Figure 8.11 are presented in Figure 8.12. Again, note the prominent Nyquist ghost of the brain at the top and bottom of the magnitude image in Figure 8.12a as compared to the image

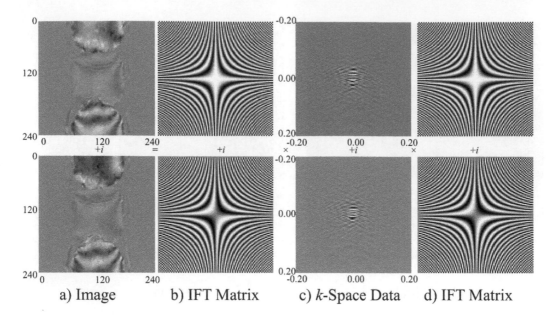

a) Image   b) IFT Matrix   c) k-Space Data   d) IFT Matrix

FIGURE 8.11
Reconstructed raw k-space data with Nyquist ghost.

after Nyquist ghost correction in Figure 8.10a. Images such as those in Figure 8.11a with the Nyquist ghost present are not suitable for either visual interpretation or statistical analysis.

Several methods have been developed to reduce such Nyquist ghosts (8, 9). Here, a method based upon the use of three navigator echoes, which allow the estimation and correction of the phase discrepancy (10) will be described. Before describing the three navigator echoes correction procedure, a little more explanation of the phase discrepancy and navigator echoes is in order. As previously noted, the Nyquist ghost occurs when there is a phase discrepancy between the odd and even lines of k-space. This phase discrepancy between the odd and even lines is illustrated in Figure 8.13a. Note that the array points are not aligned in the odd versus even lines in Figure 8.8a. The goal of the Nyquist ghost correction is to turn Figure 8.13a into Figure 8.8b.

In order to correct the phase discrepancy Δ, we first have to be able to estimate it. The Nyquist ghost phase discrepancy is echoes with the use of navigator. A navigator echo is a duplicate measurement of the entire $k_y = 0$ frequency encode line of k-space. In Figure 8.13b there are three navigator echoes presented, two from left to right and one from right to left. These navigator echoes could be measured at the beginning of the acquisition before the G_y phase encode gradient or in the middle of the acquisition when the center $k_y = 0$ line is being measured.

The process to experimentally estimate the phase discrepancy is to take each of the complex-valued navigator echo lines, labeled nav_1, nav_2, and nav_3, and one-dimensional discrete Fourier transform each of them to get NAV_1, NAV_2, and NAV_3. We also one-dimensional discrete Fourier transform each of the k-space lines for the entire array. Then we calculate

$$\hat{\omega}_0 = \text{angle}(NAV_3./NAV_1) \tag{8.15}$$

$$\hat{\phi} = NAV_2.*\exp(-i.*\hat{\omega}_0/2)./NAV_1 \tag{8.16}$$

$$\hat{\Delta} = \text{angle}(\text{median}(\text{real}(\hat{\phi})) + i\,\text{median}(\text{imag}(\hat{\phi}))) \tag{8.17}$$

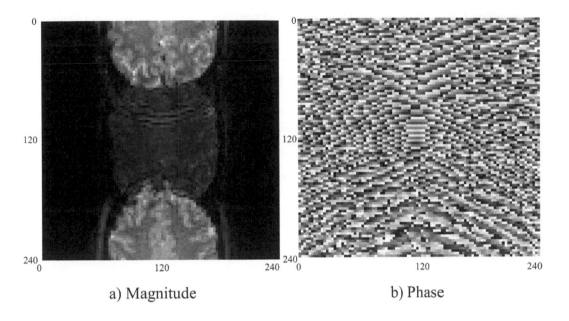

a) Magnitude   b) Phase

FIGURE 8.12
Magnitude and phase of reconstructed raw k-space data with Nyquist ghost.

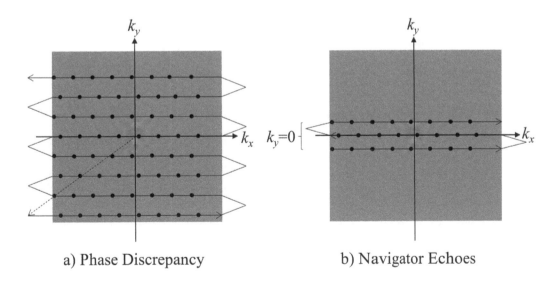

a) Phase Discrepancy   b) Navigator Echoes

FIGURE 8.13
Odd and even line phase discrepancy and navigator echoes.

where angle($\cdot$) indicates calculating the angle of its complex-valued argument, .$*$ indicates pointwise multiplication, ./ indicates pointwise division, median($\cdot$) returns the median of its vector argument, real($\cdot$) returns the real part of its argument, and imag($\cdot$) returns the imaginary part of its argument. Both $\hat{\Delta} = \text{angle}(\text{mean}((\hat{\phi})))$ and $\hat{\Delta} = \text{angle}(\text{median}((\hat{\phi})))$

have been utilized, but the median of the individual complex parts in Equation 8.17 works best experimentally.

Once the phase discrepancy is estimated to be $\hat{\Delta} = 0.4981$, the process to correct the k-space lines in Figure 8.13a is to one-dimensional discrete Fourier transform each row of k-space, and multiply the points in the odd lines by $\exp(-i\hat{\Delta})$ to shift them. Each of these corrected one-dimensional discrete Fourier transformed rows are now one-dimensional discrete inverse Fourier transformed with the result being the corrected rows of k-space as in Figure 8.9c. The corrected rows of k-space yield the real and imaginary images in Figure 8.9a and magnitude and phase in Figure 8.10 after image reconstruction.

As was demonstrated in this section, Nyquist ghost correction is a necessary processing operation in the display and analysis of fMRI experimental data. There are other processing operations performed on fMRI data that many find useful. These processing operations have been implemented to improve the statistical analysis, but the statistical ramifications of the processing are not well known. Any processing changes the statistical properties of the fMRI data. The AMMUST framework (11) is described in the next section and utilized to quantify the effect of select processing operations.

8.4   Image Processing

As was previously described, in MRI the data are Fourier encoded (4). The two-dimensional discrete inverse Fourier transform was utilized to reconstruct a complex-valued image in Section 8.3. In this section, the two-dimensional discrete inverse Fourier transform will be represented as a real-valued isomorphism involving the pre-multiplication of a vector of k-space data by a single inverse discrete Fourier transform matrix. Once this relationship is established, image processing can be represented as matrix multiplications and the resulting statistical properties can be quantified.

8.4.1   Reconstruction Isomorphism Representation

The discrete two-dimensional inverse Fourier transform in Equation 8.11 can be represented in terms of a single vector of spatial frequencies and a larger two-dimensional discrete inverse Fourier transform matrix (12). Let $f_R = \text{vec}(F'_R)$ and $f_I = \text{vec}(F'_I)$ be vectors of real and imaginary spatial frequency parts where $\text{vec}(\cdot)$ is defined to be the vectorization operator that stacks the columns of its matrix argument. The matrices F_R and F_I are the real and imaginary parts of the complex-valued spatial frequency matrix as described in Equation 8.11.

Then, we can perform a two-dimensional discrete inverse Fourier transformation as

$$
\begin{pmatrix} y_R \\ y_I \end{pmatrix} = \begin{pmatrix} \Omega_R & -\Omega_I \\ \Omega_I & \Omega_R \end{pmatrix} \begin{pmatrix} f_R \\ f_I \end{pmatrix}
$$

$$
y \qquad = \qquad \Omega \qquad f
$$

(8.18)

where $y_R = \text{vec}(Y'_R)$ contains the rows of the real part of the reconstructed image, $y_I = \text{vec}(Y'_I)$ contains the rows of the imaginary part of the reconstructed image, and Ω is given by

$$
\begin{aligned}
\Omega_R &= \left[(\Omega_{yR} \otimes \Omega_{xR}) - (\Omega_{yI} \otimes \Omega_{xI}) \right] \\
\Omega_I &= \left[(\Omega_{yR} \otimes \Omega_{xI}) + (\Omega_{yI} \otimes \Omega_{xR}) \right]
\end{aligned} ,
$$

(8.19)

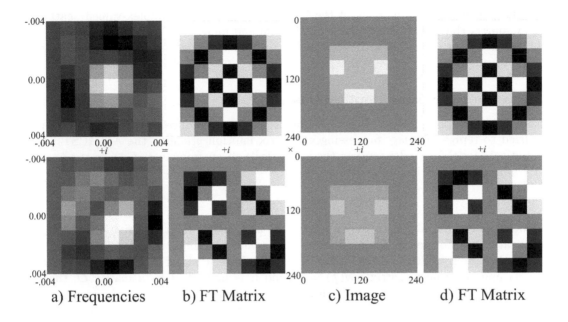

FIGURE 8.14
Image and spatial frequencies for illustrative isomorphism example.

where $\otimes$ is the Kronecker product that multiplies every element of its first matrix argument by its entire second matrix argument. In Equation 8.18, f and y are $2N_xN_y \times 1$ vectors and Ω is a $2N_xN_y \times 2N_xN_y$ matrix. The expression in Equation 8.19 was obtained by utilizing the result that the matrix multiplications $D = ABC$ can alternatively be expressed as $\text{vec}(D) = (C^T \otimes A)\text{vec}(B)$, so the two-dimensional discrete inverse Fourier transform in Equation 8.10 after attention to real and imaginary parts yields Equation 8.18 and Equation 8.19.

To illustrate the isomorphism representation of the two-dimensional discrete inverse Fourier transform, consider a small example involving the reconstruction of an 8×8 image as in Figure 8.14. In Figure 8.14c is the complex-valued 8×8 image, while Figure 8.14b and Figure 8.14d contain the two-dimensional discrete inverse Fourier transform matrices, in addition to Figure 8.14a containing the spatial frequencies.

In the isomorphism image reconstruction process, the spatial frequencies in Figure 8.14a are vectorized by stacking the rows of the real part of the frequencies on the rows of the imaginary part of the frequencies as displayed in Figure 8.15c and described in Equation 8.18. The two-dimensional discrete inverse Fourier transform matrices in Figure 8.14b and Figure 8.14d are utilized as in Equation 8.19 to form the single discrete inverse Fourier transformation matrix Ω as in Figure 8.15b. The vector of spatial frequencies f in Figure 8.15c is pre-multiplied by the larger two-dimensional discrete inverse Fourier transform matrix in Figure 8.15b to produce the vector of image voxel values y in Figure 8.15a. The vector of image voxel values in Figure 8.15a can be unstacked into the rows of the real part of the reconstructed image and the rows of the imaginary part of the reconstructed image as in Figure 8.14c.

The forward two-dimensional discrete Fourier transform can be performed in a similar fashion by multiplying the vector of stacked real and imaginary image rows by a single discrete forward Fourier matrix to obtain a vector of spatial frequencies that contains the rows of the real on top of the rows of the imaginaries to form a complex-valued spatial frequency matrix that can be unstacked into the real and imaginary frequency matrices.

This inverse Fourier matrix and vectorized array representation is useful in examining the effects of preprocessing as we will see next.

8.4.2   Image Processing Implications

The isomorphism representation of the two-dimensional discrete inverse Fourier transform was utilized to represent image processing as matrix multiplications in the AMMUST (A Mathematical Model for Understanding the STatistical) effects framework (11). In the AM-MUST framework, Equation 8.18 was extended as

$$y \;=\; O_I \;\; \Omega \;\; O_k \;\; f \tag{8.20}$$

to include k-space and image processing operations as a matrix multiplication O_k and O_I or all operations performed on the original measurements as a single matrix $O = O_I \Omega O_k$.

It is known from multivariate statistics that if we have a vector f with mean vector $\mathrm{E}(f) = f_0$ and covariance matrix $\mathrm{cov}(f) = \Gamma$, and it is multiplied by a matrix O, then the mean of the resulting vector $y = Of$ has a new mean $\mu = Of_0$ and covariance matrix $\Sigma = O\Gamma O^T$. This means that we can analytically quantify how image processing written as a matrix multiplication changes the mean, variance, and correlation structure of our data under ideal conditions. This change of the statistical properties of our data including potential induced correlation is purely from image processing and reconstruction and of no biological origin.

There are many processing operations performed on the fMRI data. A few common processing operations are listed in Table 8.1. These processing operations are performed on

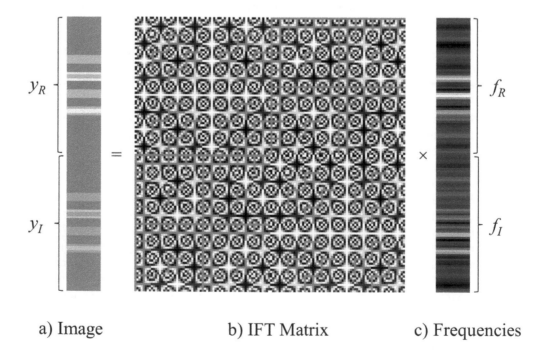

a) Image   b) IFT Matrix   c) Frequencies

FIGURE 8.15
Image reconstruction via isomorphism representation.

TABLE 8.1

Some k-space, reconstruction, and image processing operations.

Image Space	Reconstruction	k-space
Image Smoothing	*Inverse Fourier*	Nyquist Ghost
Static $B0$ Correction	IP SENSE	Non-Cartesian Gridding
Global Normalization	IP GRAPPA	Motion Correction
Image Registration	TP SENSE	Ramp Sampling
	TP GRAPPA	Homodyne Interpolation
		Zero Filling
		Apodization

the fMRI data before any statistical analysis, often without the knowledge of the person carrying out the analysis. The processes that are italicized in Table 8.1 are examined in more detail first, then the other processes will be elaborated upon.

Zero filling, also known as zero fill interpolation, is the process of taking the spatial frequency array corresponding to an image and placing it at the center of a much larger array with zeros around it. Philosophically, it is equivalent to signifying that these higher spatial frequencies have been measured to be zero. The zero fill operator matrix Z can be formed by taking an identity matrix and removing the columns that correspond to k-space locations that are zero filled. Then the Kronecker product between a two-dimensional identity matrix I_2 and this column removed identity matrix is taken to form Z.

Apodization of k-space data is another common processing operation that has the goal to mitigate ripples in images after reconstruction, called Gibbs ringing. Gibbs ringing is an artifact from the measurement of a finite central region of frequencies that extend beyond those measured. This is often called a spatial frequency truncation effect. One common apodization process is to use a Tukey filter. The Tukey apodization filter in Equation 8.21

$$T(k) \ = \ \begin{cases} 1 & |k| < k_c \\ \cos^2\left(\frac{\pi(|k|-k_c)}{2w}\right) & k_c \le |k| < k_c + w \\ 0 & k_c + w \le |k| \end{cases} \tag{8.21}$$

is defined in k-space where $k = (k_x^2 + k_y^2)^{-1/2}$ is the distance from the center of k-space. In this, k_c is the radius below which there is no apodization, and w is the distance over which there are no apodization transitions to complete filtering. The rows of this matrix T are placed into a diagonal matrix to form the apodization operator A. The function in Equation 8.21 produces a radially symmetric disk gradually decreasing to zero that is centered at the center of k-space.

The two-dimensional discrete inverse Fourier transformation image reconstruction process was described as an operator Ω for matrix multiplication earlier in this section. It is a necessary processing step in the generation of images for interpretation and statistical analysis. It therefore can't be skipped, unlike the other processing operations (except for Nyquist ghost correction).

Image smoothing or blurring is a procedure that many image processors and analysts use. Its purpose is to mitigate discrepancies between subjects' brains, increase signal within voxels relative to noise, and help with Gaussianity (normal distribution) assumptions when using random field theory for activation map thresholding. The Gaussian image smoothing kernel G defined in image space is described as

$$G(r) \ = \ \frac{1}{2\pi\sigma^2}\exp\left(-\frac{r^2}{2\sigma^2}\right) \tag{8.22}$$

where $r = (x^2 + y^2)^{-1/2}$ is the distance from the center of the voxel of interest and

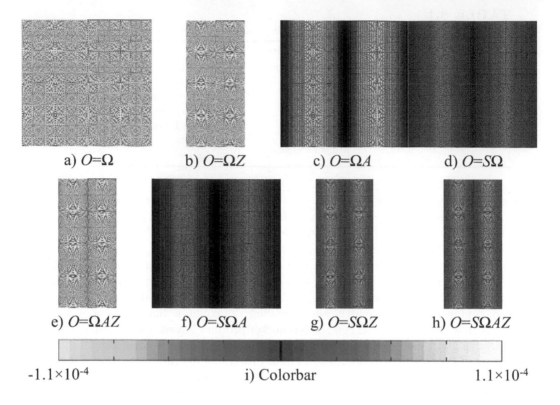

a) $O = \Omega$ b) $O = \Omega Z$ c) $O = \Omega A$ d) $O = S\Omega$

e) $O = \Omega AZ$ f) $O = S\Omega A$ g) $O = S\Omega Z$ h) $O = S\Omega AZ$

-1.1×10^{-4} i) Colorbar 1.1×10^{-4}

FIGURE 8.16
AMMUST processing operators, O.

$\sigma^2 = (\text{FWHM}/2)^2/(2\log 2)$ is the variance of a normal distribution, with FWHM denoting the full-width-at-half-max. The function in Equation 8.22 produces a radially symmetric Gaussian hill centered at the voxel of interest.

In order to gain some intuition into the processing and reconstruction operators, combinations of the four individual image processing operators Z, A, Ω, and S will be applied to an example for a typical size. Since we will want to reconstruct an image, the image reconstruction operator Ω will be included in all cases. The 96×96 k-space data in Figure 8.9c will be utilized as the mean image (f_0). When zero filling is applied, the central 64×64 portion of the k-space data in Figure 8.9c will be utilized and zero filled to 96×96. When apodization is applied, as in Equation 8.21, $k_c = 30$ and $w = 15$. All k-space data will be reconstructed to image space using the Ω operator as described in Equation 8.19 for 96×96 images. When smoothing is applied, a 2-voxel (5-mm) FWHM as described in Equation 8.22 will be used.

The processing and reconstruction operators are visually presented in Figure 8.16. The operator for only image reconstruction ($O = \Omega$) is given in Figure 8.16a; for zero filling from 64 to 96 and image reconstruction ($O = \Omega Z$) in Figure 8.16b; for apodization and image reconstruction ($O = \Omega A$) in Figure 8.16c; for image reconstruction and image smoothing ($O = S\Omega$) in Figure 8.16d; for zero filling, apodization, and image reconstruction ($O = \Omega AZ$) in Figure 8.16e; for apodization, image reconstruction, and image smoothing ($O = S\Omega A$) in Figure 8.16f; for zero filling, image reconstruction, and smoothing ($O = S\Omega Z$) in Figure 8.16g; and for zero filling, apodization, image reconstruction, and image smoothing ($O = S\Omega AZ$) in Figure 8.16h. All operator images are on the same intensity scale, $\pm 1.1 \times$

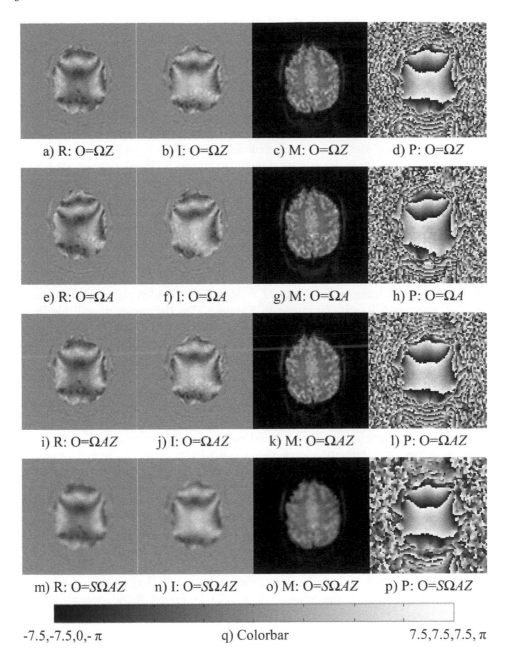

a) R: O=ΩZ   b) I: O=ΩZ   c) M: O=ΩZ   d) P: O=ΩZ

e) R: O=Ω*A*   f) I: O=Ω*A*   g) M: O=Ω*A*   h) P: O=Ω*A*

i) R: O=Ω*AZ*   j) I: O=Ω*AZ*   k) M: O=Ω*AZ*   l) P: O=Ω*AZ*

m) R: O=*S*Ω*AZ*   n) I: O=*S*Ω*AZ*   o) M: O=*S*Ω*AZ*   p) P: O=*S*Ω*AZ*

-7.5,-7.5,0,- π   q) Colorbar   7.5,7.5,7.5, π

FIGURE 8.17
Modified mean images from processing.

10^{-4} with the same color bar for the operators given in Figure 8.16i. Note the differences between the operators as some appear in the green hues while others in the purple hues.

The mean real and imaginary images, along with the magnitude and phase derived from them, are shown Figure 8.17 for the various processing pipelines. The mean images for reconstruction only ($O = \Omega$) are presented in Figure 8.9 and Figure 8.10 and thus not repeated here. The mean images for $O = S\Omega$, $O = S\Omega A$, and $O = S\Omega Z$ are also not shown as they are visually similar to the $O = S\Omega AZ$ mean images in Figures 8.17m–p, however, their magnitude of means are presented in Figure 8.18 as underlays.

The mean images after operator processing and reconstruction are visually presented in Figure 8.17. When applying zero filling from 64 to 96 and image reconstruction ($O = \Omega Z$) was performed, it yielded the mean real image in Figure 8.17a, the mean imaginary image in Figure 8.17b, the mean magnitude image in Figure 8.17c, and the mean phase image in Figure 8.17d. Note that the mean images after zero filling and reconstruction in Figures 8.17a–d appear slightly blurrier than those from solely reconstruction in Figure 8.9 and Figure 8.10. When applying apodization and image reconstruction ($O = \Omega A$) it yielded the mean real image in Figure 8.17e, the mean imaginary image in Figure 8.17f, the mean magnitude image in Figure 8.17g, and the mean phase image in Figure 8.17h. The mean images in Figures 8.17e–f appear similar but have slightly less detail than those in Figures 8.17a–d. When applying zero filling, apodization, and image reconstruction ($O = \Omega AZ$) it yielded the mean real image in Figure 8.17i, the mean imaginary image in Figure 8.17j, the mean magnitude image in Figure 8.17k, and the mean phase image in Figure 8.17l. The mean images in Figures 8.17i–l are virtually indistinguishable from those in Figures 8.17a–d. When all processing and reconstruction operations are applied, zero filling, apodization, image reconstruction, and image smoothing ($O = S\Omega AZ$), it yielded the mean real image in Figure 8.17m, the mean imaginary image in Figure 8.17n, the mean magnitude image in Figure 8.17o, and the mean phase image in Figure 8.17p. Note that the mean images in Figures 8.17m–p are noticeably blurrier that those from the other processing and reconstruction pipelines. All real and imaginary images are intensity limited in $[-7.5, 7.5]$, all magnitude images are intensity limited in $[0, 7.5]$, while all phase images are intensity limited in $[-\pi, \pi]$ as seen with the greyscale color bar in Figure 8.17q.

The correlations resulting from the various processing pipelines are displayed in Figure 8.18 assuming k-space measurements are uncorrelated. For the correlations, the entire $\Sigma = OO^T$ matrix was computed and then turned into a correlation matrix R. Since these are very large matrices, it is difficult to see the detail within them. To remedy the large matrices, the correlation between a given voxel and all others can be computed. The aforementioned processing and reconstruction operations yield the same correlation between any given voxel and all others. Therefore, the correlation between the center $(49, 49)$ voxel and all others will be displayed as a correlation image thresholded at $TH = 0.001$ and superimposed upon the corresponding magnitude image (intensity limited in $[0, 7.5]$) for visual reference. The resulting operator-induced correlation for only image reconstruction ($O = \Omega$) is given in Figure 8.18a; for zero filling from 64 to 96 and image reconstruction ($O = \Omega Z$) in Figure 8.18b; for apodization and image reconstruction ($O = \Omega A$) in Figure 8.18c; for image reconstruction and image smoothing ($O = S\Omega$) in Figure 8.18d; for zero filling, apodization, and image reconstruction ($O = \Omega AZ$) in Figure 8.18e; for apodization, image reconstruction, and image smoothing ($O = S\Omega A$) in Figure 8.18f; for zero filling, image reconstruction, and smoothing ($O = S\Omega Z$) in Figure 8.18g; and for zero filling, apodization, image reconstruction, and image smoothing ($O = S\Omega AZ$) in Figure 8.18h. All correlation images are on the same intensity scale, ± 1 with the same color bar for the correlations given in Figure 8.18i.

Note the differences in color and pattern between the various processing-induced correlations. It should be noted that $\Omega\Omega^\tau = \frac{1}{NxNy}I_{2NxNy}$ and that only reconstruction does not induce a correlation between voxels. When zero filling is performed, there is an extended sinc ripple correlation structure in Figures 8.18b, e, g and h. When image smoothing is applied, there is a central circular region of strongly induced correlation between the voxels. If one also thinks about what a combination of two individual operators in Figures 8.18b–d would look like, you can rationalize their combination in Figures 8.18e–g. It is also interesting to match the operators in Figure 8.16 to these correlations in Figure 8.18

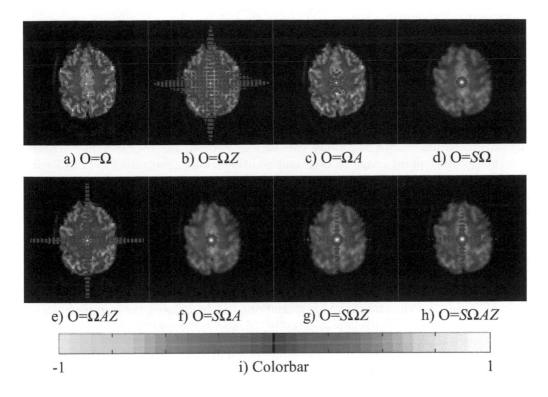

a) O=Ω b) O=ΩZ c) O=ΩA d) O=$S\Omega$

e) O=ΩAZ f) O=$S\Omega A$ g) O=$S\Omega Z$ h) O=$S\Omega AZ$

-1   i) Colorbar   1

FIGURE 8.18
Modified correlations from image processing.

to see that when the operators appear generally similar, the induced correlations are also generally similar. The Nyquist ghost correction in Section 8.3 also does not induce a correlation in and of itself.

Additional k-space operators could be or have been developed for the remaining k-space processing operations in Table 8.1. Non-Cartesian gridding is a process where k-space is originally not on the Cartesian grid, such as the spiral trajectory (13) or polar sampling (14). Cartesian gridding is a linear process and a Cartesian gridding operator C could be developed to bring this irregularly measured data to the rectangular grid. Motion correction is a process to adjust for minor subject movement over time. In a rigid body transformation, an individual image is translated and rotated to match another image. Motion correction is a linear process and an operator M could be developed to examine its statistical properties. Ramp sampling is a process where the frequency encoding lines of k-space (horizontal in examples in this chapter) are not flat but linearly angled toward the next line to be measured. Ramp sampling interpolation estimates the angled k-space onto horizontal lines. It was empirically found that ramp sampling interpolation induces a local sinc type correlation between the voxels (15) but it is linear and a matrix operator could be developed. Homodyne interpolation is a process where the Hermitian property of k-space is utilized. Half of the k-space array is measured plus a little bit more, called over-scan lines, and the remaining portion is interpolated (16). It was shown that Homodyne interpolation does not change the image mean or correlation, but does increase voxel variances (11) and produces a purely real-valued image.

The AMMUST framework was utilized to examine the effects of in-plane (IP) k-space subsampling and image reconstruction with the SENSE (SENSitivity Encoding) (17) and also the GRAPPA (GeneRalized Autocalibrating Partial Parallel Acquisition) (18) procedures. In the IP SENSE and GRAPPA reconstruction procedures, lines of k-space are skipped, which results in aliased reconstructed images. Additional full scans are measured, and with SENSE, the images are unaliased in image space with computed coil sensitivities while with GRAPPA, the missing lines of k-space are interpolated using a kernel. It was found that the use of SENSE and coil sensitivities induces a correlation between previously overlapping voxels (19) and that the use of GRAPPA with kernel interpolation also induces a correlation between previously overlapping voxels (20). These in-plane parallel image reconstruction procedures induce long-range correlations between the corresponding voxels in the previously overlapping image sections and additional image processing induces local correlation around these voxels. Care needs to be taken when interpreting these statistical results after either SENSE or GRAPPA in-plane acceleration reconstruction. Another avenue of image reconstruction that has been increasing in prominence is the selection of multiple slices for measurement resulting in k-space array images that are the sum of the arrays for the multiple slices. The measurement and separation of multiple overlapping slice images through-plane (TP) is called simultaneous multi-slice (SMS). The SENSE and GRAPPA unaliasing procedures are being applied TP and overlapping slice images separated. Since the same math and algorithms are being applied through-plane as in-plane, it is strongly believed that the same type of correlations will be induced through-plane into the separated slices and local correlations accentuated with image processing.

The effects of the remaining image space processing operations, have not been fully explored. Static $B0$ correction has been examined in simulations (21) but not for spatial statistical properties. Global normalization (22) and image registration (23) have yet to be examined and quantified. It is suspected that global normalization and image registration will induce a correlation between the voxels, but this has not been demonstrated. The quantification of image registration and global intensity normalization will provide an immense amount of knowledge to the field of fMRI as they are very common processing operations.

The AMMUST image processing framework was expanded (24) to stack k-space vectors from a time series of images, have a much larger matrix K of k-space processing to include different processing on each k-space vector, have a much larger matrix R that included reconstruction of each of the processed k-space vectors, and a much larger matrix I for processing on each of the reconstructed image vectors, then a matrix T that included first a permutation matrix so that the data is ordered by voxel and not by image along with a matrix that performs potentially different filtering of each of the time series as $O = TIRK$. It was found that time series processing such as frequency filtering induces a temporal correlation in voxel time series.

It was shown that different processing pipelines yield images with means and correlations that are modified in various ways including induced local correlation. The null hypothesis of no correlation between voxels is no longer valid when processing is involved. With the statistical properties of our image data changed in a known quantifiable manner, the processing operation matrices should be able to be incorporated into a statistical model for improved estimation and inference.

8.5   Additional Topics and Discussion

There are many topics that naturally arise from image reconstruction and processing. In this section, activation from complex-valued time series will be briefly described followed by a general discussion of the chapter.

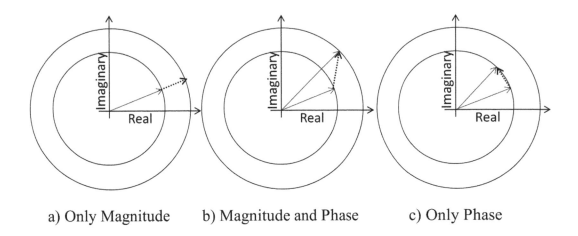

a) Only Magnitude   b) Magnitude and Phase   c) Only Phase

FIGURE 8.19
Complex-valued signal changes for activation.

8.5.1 Complex-Valued fMRI Activation

Since the image reconstruction methods yield complex-valued images, the time series within each voxel are also complex-valued. Models have been developed to detect three differential task-related changes in complex-valued fMRI time series.

Within complex-valued time series, there are three possible nonzero signal changes as shown in Figure 8.19. Figure 8.19a depicts a signal change only in the magnitude with a constant phase (CP), Figure 8.19b depicts a signal change in both the magnitude and phase (MP), and Figure 8.19c depicts a signal change only in the phase with a constant magnitude (CM). A CP activation model has been developed to detect statistically significant changes in the magnitude while specifying the phase is constant (25). An MP activation model has been developed where the full complex-valued (real-imaginary) data is utilized to detect statistically significant changes in both the magnitude and phase (26, 27).

For each voxel, the complex-valued measurement y_t at time t has been described as

$$
\begin{aligned}
y_t &= (\rho_t \cos\theta_t + \eta_{Rt}) + i(\rho_t \sin\theta_t + \eta_{It}) \\
\rho_t &= x_t'\beta = \beta_0 + \beta_1 x_{1t} + \dots + \beta_{q_1} x_{q_1 t} \\
\theta_t &= u_t'\gamma = \gamma_0 + \gamma_1 u_{1t} + \dots + \gamma_{q_2} x_{q_2 t}
\end{aligned}
\tag{8.23}
$$

where $t = 1, \dots, n$, $(\eta_{Rt}, \eta_{It})' \sim \mathcal{N}(0, \Sigma)$, x_t' is the tth row of an $n \times (q_1 + 1)$ design matrix X for the magnitude, u_t' is the tth row of an $n \times (q_2 + 1)$ design matrix U for the phase, and $\Sigma = \sigma^2 I_2$, while β and γ are magnitude and phase regression coefficient vectors respectively (26). Note that a separate design matrix U for the phase has been incorporated but X and U can be the same.

The usual fMRI activation model is a magnitude-only (MO) model where the phase images (and hence phase time series) are discarded (1). The MO model has been shown to be equivalent to a complex-valued activation model with unrestricted phase (28). A phase-only (PO) activation model, where the magnitude images (and hence the magnitude time series) are discarded, has been shown to detect statistically significant activation (12). When the phase images are discarded for the MO model, the MO model is not able to distinguish between CP signal changes in Figure 8.19a or MP signal changes in Figure 8.19b because it does not have rotational phase information. Similarly, when the magnitude images are discarded, the PO model is not able to distinguish between CM signal changes in Figure 8.19c or MP signal changes in Figure 8.19b because it does not have lengthening magnitude information.

Voxel time series after image reconstruction are complex-valued. Discarding the phase half of the data, as is generally done with the usual magnitude-only model, may not be optimal in terms of statistical modeling and detection of biological processes. Future statistical analyses should utilize the full complex-valued fMRI data, and there are many opportunities for the discovery of biological phenomenon such as oxygenation changes in the vasculature (29, 30) or direct detection of neuronal currents (31, 32).

8.5.2 Discussion

In this chapter, the foundations and fundamentals of image reconstruction and image processing have been examined. In Section 8.2, the one- and two-dimensional discrete forward and inverse transformations were discussed. Knowledge of the Fourier transform is essential to understanding the original k-space measurements taken by the MRI scanner. In Section 8.3, the signal equation and the traversal of k-space along a trajectory was explained with the use of a pulse-sequence diagram so that the process of original k-space measurement could be understood. Also in Section 8.3, the raw unprocessed k-space measurements were described along with the very important Nyquist ghost correction. In Section 8.4, the two-dimensional discrete inverse Fourier transform process for image reconstruction was written in terms of an isomorphism so that it could be utilized to examine and quantify the statistical effects of reconstruction and image processing. It was demonstrated that some common fMRI processing steps induce local correlation and it was described that other reconstruction techniques induce long-range correlations. In Section 8.5, additional related topics of complex-valued activation for the reconstructed complex-valued images.

In summary, there are many processes performed on fMRI data before typical analysis, generally with no knowledge retained about what the processing was. It is hoped that knowledge of processing operations can be utilized in the statistical analysis of fMRI data.

Acknowledgments

This work was supported by National Institutes of Health research grant R21NS087450.

This material was based upon work partially supported by the National Science Foundation under Grant DMS-1127914 to the Statistical and Applied Mathematical Sciences Institute. Any opinions, findings, and conclusions or recommendations expressed in this material are those of the author(s) and do not necessarily reflect the views of the National Science Foundation.

Bibliography

[1] Bandettini P.A., Wong E.C., Hinks R.S., Tikofsky R.S, and Hyde J.S. Time course EPI of human brain function during task activation. *Magn Reson Med*, 25(2):390–397, June 1992.

[2] Kwong K.K., Belliveau J.W., Chesler D.A., Goldberg I.E., Weisskoff R.M. Poncelet B.P. Kennedy D.N., Hoppel B.E., Cohen M.S., and Turner R. Oxygenation-sensitive contrast in magnetic resonance image of rodent brain at high magnetic fields. *Proc Natl Acad Sci USA*, 89(12):5675–5679, June 1992.

[3] Ogawa S., Tank D.W., Menon R.S., Ellermann J.M. ad Kim S.G., Merkle H., and Ugurbil K. Intrinsic signal changes accompanying sensory stimulation: Functional brain mapping with magnetic resonance imaging. *Proc Natl Acad Sci USA*, 89(13):5951–5955, July 1992.

[4] Kumar A., Welti D., and Ernst R.R. NMR Fourier zeugmatography. *J Magn Reson*, 18(1):69–83, April 1975.

[5] Shannon C.E. Communication in the presence of noise. In *Proc Institute of Radio Engineers*, volume 37, pages 10–21. John Wiley and Sons, January 1949.

[6] Nyquist H. Certain topics in telegraph transmission theory. *Trans AIEE*, 47:617–644, April 1928. Reprint as classic paper in *Proc. IEEE*, 90(2), Feb 2002.

[7] Karaman M.M., Bruce I.P., and Rowe D.B. Incorporating relaxivities to more accurately reconstruct MR images. *Magn Reson Imaging*, 33(4):374–384, 2014.

[8] Jesmanowicz A., Wong E.C., and Hyde J.S. Phase correction for EPI using internal reference lines. In *Proc Soc Magn Reson Med*, volume 12, page 1239, New York, New York, 1993. John Wiley and Sons.

[9] Jesmanowicz A., Wong E.C., and Hyde J.S. Self-correcting EPI reconstruction algorithm. In *Proc Soc Magn Reson Med*, volume 14, page 619, Nice, France, 1995. John Wiley and Sons.

[10] Nencka A.S., Hahn A.D., and Rowe D.B. The use of three navigator echoes in Cartesian EPI reconstruction reduces Nyquist ghosting. In *Proc Intl Soc Magn Reson Med*, volume 16, page 3032, Toronto, Canada, 2008. John Wiley and Sons.

[11] Nencka A.S., Hahn A.D., and Rowe D.B. A mathematical model for understanding the statistical effects of k-space (AMMUST-k) preprocessing on observed voxel measurements in fcMRI and fMRI. *J Neurosci Methods*, 181(2):268–282, 2009.

[12] Rowe D.B., Nencka A.S., and Hoffmann R.G. Signal and noise of Fourier reconstructed fMRI data. *J Neurosci Methods*, 159(2):361–369, 2007.

[13] Ahn C.B., Kim J.H., and Cho Z.H. High speed spiral scan echo planar NMR imaging-I. *IEEE Trans Med Imaging*, 5(1):2–7, 1986.

[14] Lauzon M.L. and Rutt B.K. Polar sampling in k-space: Reconstruction effects. *Magn Reson Med*, 40(5):769–782, November 1998.

[15] Deshpande G., LaConte S., Peltier S., and X. Hu. Integrated local correlation: A new measure of local coherence in fMRI data. *Hum Brain Mapp*, 30(1):13–23, 2009.

[16] Jesmanowicz A., Bandettini P.A., and Hyde J.S. Single-shot half k-space high-resolution gradient-recalled EPI for fMRI at 3 Tesla. *Magn Reson Med*, 40(5):754–762, November 1998.

[17] Pruessmann K.P., Weiger M., Scheidegger M.B., and Boesiger P. SENSE: Sensitivity encoding for fast MRI. *Magn Reson Med*, 42(5):952–962, November 1999.

[18] Griswold M.A., Jakob P.M. Heidemann R.M., Nittka M., Jellus V., Wang J., Kiefer B., and Haase A. Generalized autocalibrating partially parallel acquisitions (GRAPPA). *Magn Reson Med*, 47(6):1202–1210, June 2002.

[19] Bruce I.P., Karaman M.M., and Rowe D.B. A statistical examination of SENSE image reconstruction via an isomorphism representation. *Magn Reson Imaging*, 29(9):1267–1287, 2011.

[20] Bruce I.P. and Rowe D.B. Quantifying the statistical impact of GRAPPA in fcMRI data with a real-valued isomorphism. *IEEE Trans Med Imaging*, 33(2):495–503, 2014.

[21] Hahn A.D., Nencka A.S., and Rowe D.B. Improving robustness and reliability of phase-sensitive fMRI analysis using Temporal Off-Resonance Alignment of Single-Echo Timeseries (TOAST). *Neuroimage*, 44(3):742–752, 2009.

[22] Friston K.J., Ashburner J., Frith C.D., Poline J.B., Heather J.D., and Frackowiak R.S.J. Spatial registration and normalization of image. *Hum Brain Mapp*, 3(3):165–189, 1995.

[23] Jenkinson M., Bannister P., Brady M., and Smith S. Improved optimization for the robust and accurate linear registration and motion correction of brain images. *Neuroimage*, 17(2):825–841, 2002.

[24] Karaman M.M., Nencka A.S., Bruce I.P., and Rowe D.B. Quantification of the statistical effects of spatiotemporal processing of non-task fMRI data. *Brain Connect*, 4(9):649–661, 2014.

[25] Rowe D.B. and Logan B.R. A complex way to compute fMRI activation. *Neuroimage*, 32(3):1078–1092, 2004.

[26] Rowe D.B. Modeling both the magnitude and phase of complex-valued fMRI data. *Neuroimage*, 25(4):1310–1324, 2005.

[27] Rowe D.B. Magnitude and phase signal detection in complex-valued fMRI data. *Magn Reson Med*, 62(5):1356–1357, 2009.

[28] Rowe D.B. and Logan B.R. Complex fMRI analysis with unrestricted phase is equivalent to a magnitude-only model. *Neuroimage*, 24(2):603–606, 2005.

[29] Menon R.S. Postacquisition suppression of large-vessel BOLD signals in high-resolution fMRI. *Magn Reson Med*, 47(1):1–9, 2002.

[30] Nencka A.S. and Rowe D.B. Reducing the unwanted draining vein BOLD contribution in fMRI with statistical post-processing methods. *Neuroimage*, 37(1):177–188, 2007.

[31] Bodurka J., Jesmanowicz A., Hyde J.S., Xu H., Estowski L., and Li S.-J. Current-induced magnetic resonance phase imaging. *J Magn Reson*, 137(1):265–271, 1999.

[32] Chow L.S., Cook G.G., Whitby E., and Paley M.N. Investigating direct detection of axon firing in the adult human optic nerve using MRI. *Neuroimage*, 30(3):835–846, 2006.

9

Statistical Analysis on Brain Surfaces

Moo K. Chung and Seth D. Pollak

University of Wisconsin-Madison

Jamie L. Hanson

University of Pittsburgh

CONTENTS

9.1	Introduction		234
9.2	Surface Parameterization		235
	9.2.1	Local Parameterization by Quadratic Polynomial	236
	9.2.2	Surface Flattening	236
	9.2.3	Spherical Harmonic Representation	237
9.3	Surface Registration		239
	9.3.1	Affine Registration	239
	9.3.2	SPHARM Correspondence	239
	9.3.3	Diffeomorphic Registration	241
9.4	Cortical Surface Features		241
	9.4.1	Cortical Thickness	241
	9.4.2	Surface Area and Curvatures	242
	9.4.3	Gray Matter Volume	242
9.5	Surface Data Smoothing		243
	9.5.1	Diffusion Smoothing	243
	9.5.2	Iterated Kernel Smoothing	245
	9.5.3	Heat Kernel Smoothing	246
9.6	Statistical Inference on Surfaces		248
	9.6.1	General Linear Models	248
	9.6.2	Multivariate General Linear Models	249
	9.6.3	Small-n Large-p Problems	250
	9.6.4	Longitudinal Models	251
	9.6.5	Random Field Theory	252
	Bibliography		254

In this chapter, we review widely used statistical analysis frameworks for data defined along cortical and subcortical surfaces that have been developed in the last two decades. The cerebral cortex has the topology of a 2D highly convoluted sheet. For data obtained along curved non-Euclidean surfaces, traditional statistical analysis and smoothing techniques based on the Euclidean metric structure are inefficient. To increase the signal-to-noise ratio (SNR) and to boost the sensitivity of the analysis, it is necessary to smooth out noisy surface data. However, this requires smoothing data on curved cortical manifolds and assigning smoothing weights based on the geodesic distance along the surface. Thus, many cortical surface data analysis frameworks are differential geometric in nature (24). The smoothed surface data is then treated as smooth random fields and statistical inferences can be performed within Keith Worsley's random field theory (123, 122).

The methods described in this chapter are illustrated with the hippocampus surface dataset published in (26). Using this case study, we will determine if there is an effect of family income on the growth of hippocampus in children in detail. There are a total of 124 children and 82 of them have repeated magnetic resonance images (MRIs) two years later.

9.1 Introduction

The cerebral cortex has the topology of a 2D convoluted sheet. Most of the features that distinguish these cortical regions can only be measured relative to the local orientation of the cortical surface (33). As the brain develops over time, cortical surface area expands and its curvature changes (24). It is equally likely that such age-related changes with respect to the cortical surface are not uniform (30, 113). By measuring how geometric features such as the cortical thickness, curvature and local surface area change over time, statistically significant brain tissue growth or loss in the cortex can be detected locally at the vertex level.

The first obstacle in performing surface-based data analysis is a need for extracting cortical surfaces from MRI volumes. This requires correcting MRI field inhomogeneity artifacts. The most widely used technique is the nonparametric nonuniform intensity normalization method (N3) developed at the Montreal neurological institute (MNI), which eliminates the dependence of the field estimate on anatomy (108). The next step is the tissue classification into three types: gray matter, white matter and cerebrospinal fluid (CSF). This is critical for identifying the tissue boundaries where surface measurements are obtained. An artificial neural network classifier (71, 88) or Gaussian mixture models (53) can be used to segment the tissue types automatically. The Statistical Parametric Mapping (SPM) package[1] uses a Gaussian mixture with a prior tissue density map.

After the segmentation, the tissue boundaries are extracted as triangular meshes. In order to triangulate the boundaries, the marching cubes algorithm (76), level set method (102), the deformable surfaces method (35) or the anatomic segmentation using proximities (ASP) method (79) can be used. Brain substructures such as the brain stem and the cerebellum are usually automatically removed in the process. The resulting triangular mesh is expected to be topologically equivalent to a sphere. For example, the triangular mesh resulted from the ASP method consists of 40,962 vertices and 81,920 triangles with the average internodal distance of 3 mm. Figure 9.1 shows a representative cortical mesh obtained from ASP. Surface measurements such as cortical thickness can be automatically obtained at each mesh vertex. Subcortical brain surfaces such as amygdala and hippocampus are extracted similarly, but often done in a semi-automatic fashion with the marching cubes algorithm on manual edited subcortical volumes. In the hippocampus case study, the left and right hippocampi were manually segmented in the template using the protocol outlined in (99).

Comparing measurements defined across different cortical surfaces is not trivial due to the fact that no two cortical surfaces are identically shaped. In comparing measurements across different 3D whole brain images, 3D volume-based image registration such as Advanced Normalization Tools (ANTS) (6) is needed. However, 3D image registration techniques tend to misalign sulcal and gyral folding patterns of the cortex. Hence, 2D

---

[1]The SPM package is available at www.fil.ion.ucl.ac.uk/spm.

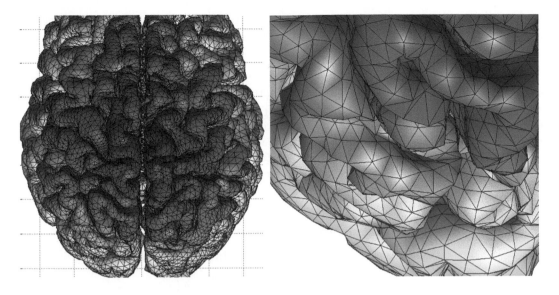

FIGURE 9.1
Left: The outer cortical brain surface mesh consisting of 81,920 triangles. Measurements are
defined at mesh vertices. Right: The part of the mesh is enlarged to show the convoluted
nature of the surface.

surface-based registration is needed in order to match measurements across different cortical
surfaces. Various surface registration methods have been proposed (28, 24, 115, 34, 82, 42).
Most methods solve a complicated optimization problem of minimizing the measure of
discrepancy between two surfaces. A much simpler spherical harmonic representation technique provide a simple way of approximately matching surfaces without time-consuming
numerical optimization (24).

Surface registration and the subsequent surface-based analysis usually require parameterizing surfaces. It is natural to assume the surface mesh to be a discrete approximation to
the underlying cortical surface, which can be treated as a smooth 2D Riemannian manifold.
Cortical surface parameterization has been done previously in (115, 67, 24). The surface
parameterization also provides surface shape features such as the Gaussian and mean curvatures, which measure anatomical variations associated with the deformation of the cortical
surface during, for instance, development and aging (33, 54, 67, 24).

9.2 Surface Parameterization

In order to perform statistical analysis on a surface, parameterization of the surface is often required (24). Brain surfaces are often mapped onto a plane or a sphere. Then surface
measurements defined on mesh vertices are also mapped onto the new domain and analyzed. However, almost all surface parameterizations suffer metric distortions, which in
turn influence the spatial covariance structure so it is not necessarily the best approach.

We model the cortical surface $\mathcal{M}$ as a smooth 2D Riemannian manifold parameterized by two parameters u^1 and u^2 such that any point $\mathbf{x} \in \mathcal{M}$ can be represented as

$$\mathbf{X}(u^1, u^2) = \{x_1(u^1, u^2), x_2(u^1, u^2), x_3(u^1, u^2) : (u^1, u^2) \in D \subset \mathbb{R}^2\}$$

for some parameter space $\mathbf{u} = (u^1, u^2) \in D \subset \mathbb{R}^2$ (16, 38, 30, 72). The aim of the parameterization is estimating the coordinate functions x_1, x_2, x_3 as smoothly as possible.

Both global and local parameterizations are available. A global parameterization, such as tensor B-splines and spherical harmonic representation, are computationally expensive compared to a local surface parameterization. A local surface parameterization in the neighborhood of point $\mathbf{x}$ can be obtained via the projection of the local surface patch onto the tangent plane $T_{\mathbf{x}}(\mathcal{M})$ (24, 67).

9.2.1 Local Parameterization by Quadratic Polynomial

A local parameterization is usually done by fitting a quadratic polynomial of the form

$$\mathbf{X}(u^1, u^2) = \beta_1 u^1 + \beta_2 u^2 + \beta_3 (u^1)^2 + \beta_4 u^1 u^2 + \beta_5 (u^2)^2 \tag{9.1}$$

in $(u^1, u^2) \in D \subset \mathbb{R}^2$. The data can be centered so there is no constant term in the quadratic form (9.1) (24). The coefficients β_i are usually estimated by the least squares method (67, 69, 24).

In estimating various differential geometric measures such as the Laplace–Beltrami operator or curvatures, it is not necessary to find global surface parameterization of $\mathcal{M}$. Local surface parameterization such as the quadratic polynomial fit is sufficient to obtain such geometric quantities (24). The drawback of the polynomial parameterization is that there is a tendency to weave the outermost mesh vertices to find vertices in the center. Therefore this is not advisable to directly fit (9.1) when one of the coordinate values rapidly changes.

9.2.2 Surface Flattening

Parameterizing cortical and subcortical surfaces with respect to simpler algebraic surfaces such as a unit sphere is needed to establish a standard coordinate system. However, polynomial regression type of local parameterization is ill-suited for this purpose. For the global surface parameterization, we can use *surface flattening* (3, 4), which is nonparametric in nature. The surface flattening parametrizes a surface by either solving a partial differential equation or optimizing its variational form.

Deformable surface algorithms naturally provide one-to-one maps from cortical surfaces to a sphere since the algorithm initially starts with a spherical mesh and deforms it to match the tissue boundaries (79). The deformable surface algorithms usually start with the second level of triangular subdivision of an icosahedron as the initial surface. After several iterations of deformation and triangular subdivision, the resulting cortical surface contains very dense triangle elements. There are many surface flattening techniques such as conformal mapping (4, 55, 62) quasi-isometric mapping (116), area preserving mapping (17), and the Laplace equation method (24).

For many surface flattening methods to work, the starting binary object has to be close to star-shape or convex. For shapes with a more complex structure, the methods may create numerical singularities in mapping to the sphere. Surface flattening can destroy the inherent geometrical structure of the cortical surface due to the metric distortion. Any structural or functional analysis associated with the cortex can be performed without surface flattening if the intrinsic geometric method is used (24).

9.2.3 Spherical Harmonic Representation

The spherical harmonic (SPHARM) representation[2] is a widely used subcortical surface parameterization technique (27, 24, 52, 55, 68, 104). SPHARM represents the coordinates of mesh vertices as a linear combination of spherical harmonics. SPHARM has been mainly used as a data reduction technique for compressing global shape features into a small number of coefficients. The main global geometric features are encoded in low degree coefficients while the noise components are in high degree spherical harmonics (55). The method has been used to model various subcortical structures such as ventricles (52), hippocampi (104) and cortical surfaces (27). The spherical harmonics have global support. So the resulting spherical harmonic coefficients contain the global shape features and it is not possible to directly obtain local shape information from the coefficients only. However, it is still possible to obtain local shape information by evaluating the representation at each fixed vertex, which gives the smoothed version of the coordinates of surfaces. In this fashion, SPHARM can be viewed as mesh smoothing (27, 29). In this section, we present a brief introduction of SPHARM within a Hilbert space framework.

Suppose there is a bijective mapping between the cortical surface $\mathcal{M}$ and a unit sphere S^2 obtained through a deformable surface algorithm (24). Consider the parameterization of S^2 by

$$\mathbf{X}(\theta, \varphi) = (\sin\theta\cos\varphi, \sin\theta\sin\varphi, \cos\theta),$$

with $(\theta, \varphi) \in [0, \pi) \otimes [0, 2\pi)$. The polar angle θ is the angle from the north pole and the azimuthal angle φ is the angle along the horizontal cross section. Using the bijective mapping, we can parameterize functional data f with respect to the spherical coordinates

$$f(\theta, \varphi) = g(\theta, \varphi) + \epsilon(\theta, \varphi), \tag{9.2}$$

where g is an unknown smooth coordinate function and ϵ is a zero mean random field, possibly Gaussian. The error function ϵ accounts for possible mapping errors. The unknown signal g is then estimated in the finite subspace of $\mathcal{L}^2(S^2)$, the space of square integrable functions in S^2, spanned by spherical harmonics in the least squares fashion (27).

Previous imaging and shape modeling research has used complex-valued spherical harmonics (19, 52, 55, 104). In practice, however, it is sufficient to use only real-valued spherical harmonics (32, 59), which is more convenient in setting up a real-valued stochastic model (9.2). The relationship between the real- and complex-valued spherical harmonics is given in (15, 59). The complex-valued spherical harmonics can be transformed into real-valued spherical harmonics using a unitary transform.

The spherical harmonic Y_{lm} of degree l and order m is defined as

$$Y_{lm} = \begin{cases} c_{lm} P_l^{|m|}(\cos\theta)\sin(|m|\varphi), & -l \leq m \leq -1, \\ \frac{c_{lm}}{\sqrt{2}} P_l^{|m|}(\cos\theta), & m = 0, \\ c_{lm} P_l^{|m|}(\cos\theta)\cos(|m|\varphi), & 1 \leq m \leq l, \end{cases}$$

where $c_{lm} = \sqrt{\frac{2l+1}{2\pi}\frac{(l-|m|)!}{(l+|m|)!}}$ and P_l^m is the *associated Legendre polynomial* of order m (32, 118), which is given by

$$P_l^m(x) = \frac{(1-x^2)^{m/2}}{2^l l!}\frac{d^{l+m}}{dx^{l+m}}(x^2-1)^l, x \in [-1,1].$$

---

[2]The SPHARM package is available at
www.stat.wisc.edu/~mchung/softwares/weighted-SPHARM/weighted-SPHARM.html.

The first few terms of the spherical harmonics are

$$Y_{00} = \frac{1}{\sqrt{4\pi}}, Y_{1,-1} = \sqrt{\frac{3}{4\pi}} \sin\theta \sin\varphi,$$

$$Y_{1,0} = \sqrt{\frac{3}{4\pi}} \cos\theta, Y_{1,1} = \sqrt{\frac{3}{4\pi}} \sin\theta \cos\varphi.$$

The spherical harmonics are orthonormal with respect to the inner product

$$\langle f_1, f_2 \rangle = \int_{S^2} f_1(\Omega) f_2(\Omega)\, d\mu(\Omega),$$

where $\Omega = (\theta, \varphi)$ and the Lebesgue measure $d\mu(\Omega) = \sin\theta d\theta d\varphi$. The norm is then defined as

$$\|f_1\| = \langle f_1, f_1 \rangle^{1/2}. \tag{9.3}$$

The unknown mean function g is estimated by minimizing the integral of the squared residual in $\mathcal{H}_k$, the space spanned by up to k-th degree spherical harmonics:

$$\widehat{g}(\Omega) = \arg\min_{h \in \mathcal{H}_k} \int_{S^2} \left| f(\Omega) - h(\Omega) \right|^2 d\mu(\Omega). \tag{9.4}$$

It can be shown that the minimization is given by

$$\widehat{g}(\Omega) = \sum_{l=0}^{k} \sum_{m=-l}^{l} \langle f, Y_{lm} \rangle Y_{lm}(\Omega), \tag{9.5}$$

the Fourier series expansion. The expansion (9.5) has been referred to as the *spherical harmonic representation* (27, 52, 55, 104, 103). This technique has been used in representing various brain subcortical structures such as hippocampi (104) and ventricles (52) as well as the whole brain cortical surfaces (27, 55). By taking each component of Cartesian coordinates of mesh vertices as the functional signal f, surface meshes can be parameterized as a function of θ and φ.

The spherical harmonic coefficients can be estimated in least squares fashion. However, for an extremely large number of vertices and expansions, the least squares method may be difficult to directly invert large matrices. Instead, the *iterative residual fitting (IRF)* algorithm (27) can be used to iteratively estimate the coefficients by partitioning the larger problem into smaller subproblems. The IRF algorithm is similar to the matching pursuit method (80). The IRF algorithm was developed to avoid the computational burden of inverting a large linear problem while the matching pursuit method was originally developed to compactly decompose a time-frequency signal into a linear combination of a pre-selected pool of basis functions. Although increasing the degree of the representation increases the goodness-of-fit, it also increases the number of estimated coefficients quadratically. So it is necessary to stop the iteration at the specific degree k, where the goodness-of-fit and the number of coefficients balance out. This idea was used in determining the optimal degree of SPHARM (27).

The limitation of SPHARM is that it produces the Gibbs phenomenon, i.e., ringing artifacts, for discontinuous and rapidly changing continuous measurements (27, 49). The Gibbs phenomenon can be effectively removed by weighting the spherical harmonic coefficients exponentially smaller, which makes the representation smooth out rapidly changing signals. The weighted version of SPHARM is related to isotropic diffusion smoothing (3, 20, 24, 28) as well as the diffusion wavelet transform (27, 29, 60).

9.3 Surface Registration

To construct a test statistic locally at each vertex across different surfaces, one must register the surfaces to a common template surface. Nonlinear cortical surface registration is often performed by minimizing objective functions that measure the global fit of two surfaces while maximizing the smoothness of the deformation in such a way that the cortical gyral or sulcal patterns are matched smoothly (28, 96, 115). There is also a much simpler way of aligning surfaces using SPHARM representation (27). Before any sort of nonlinear registration is performed, an affine registration is performed to align and orient the global brain shapes.

9.3.1 Affine Registration

Anatomical objects extracted from 3D medical images are aligned using affine transformations to remove the global size differences. Affine registration requires identifying corresponding landmarks either manually or automatically. The affine transform T of point $p = (p_1, \cdots, p_d)' \in \mathbb{R}^d$ to $q = (q_1, \cdots, q_d)'$ is given by

$$q = Rp + c,$$

where the matrix R corresponds to rotation, scaling and shear while c corresponds to translation. Although the affine transform is not linear, it can be made into a linear form by augmenting the transform. The affine transform can be rewritten as

$$\begin{pmatrix} q \\ 1 \end{pmatrix} = \begin{pmatrix} R & c \\ 0 \cdots 0 & 1 \end{pmatrix} \begin{pmatrix} p \\ 1 \end{pmatrix}. \tag{9.6}$$

Let

$$A = \begin{pmatrix} R & c \\ 0 \cdots 0 & 1 \end{pmatrix}.$$

Trivially, A is linear a linear operator. The matrix A is the most often used form for recording the affine registration.

Let $\mathbf{p}_i$ be the i-th landmark and its corresponding affine transformed points $\mathbf{q}_i$. Then we rewrite (9.6) as

$$\underbrace{\begin{pmatrix} \mathbf{q}_1 & \cdots & \mathbf{q}_n \end{pmatrix}}_{Q} = \begin{pmatrix} R & c \end{pmatrix} \underbrace{\begin{pmatrix} \mathbf{p}_1 & \cdots & \mathbf{p}_n \\ 1 & \cdots & 1 \end{pmatrix}}_{P}.$$

Then the least squares estimation is given as

$$\begin{pmatrix} \widehat{R} & \widehat{c} \end{pmatrix} = QP'(PP')^{-1}.$$

Then the points $\mathbf{p}_i$ are mapped to $\widehat{R}\mathbf{p}_i + \widehat{c}$, which may not coincide with $\mathbf{q}_i$ in general. In practice, landmarks are automatically identified from T1-weighted MRI.

9.3.2 SPHARM Correspondence

Using SPHARM, it is possible to approximately register surfaces with different mesh topology without any optimization. The crude alignment can be done by coinciding the first order ellipsoid meridian and equator in the SPHARM representation (52, 109). However,

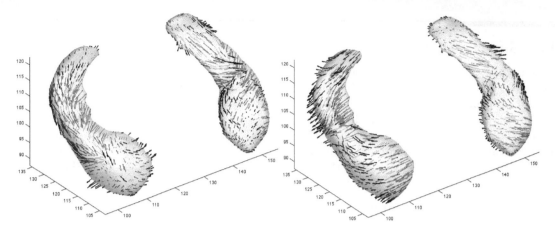

FIGURE 9.2
The displacement vector fields of registering from the hippocampus template to two individual surfaces. The displacement vector field is obtained from the diffeomorphic image registration (6). The variability in the displacement vector field can be analyzed using a multivariate general linear model (24).

this can be improved. Consider SPHARM representation of surface $H = (h_1, h_2, h_3)$ with spherical angles Ω given by

$$h_i(\Omega) = \sum_{l=0}^{k} \sum_{m=-l}^{l} h_{lm}^i Y_{lm}(\Omega),$$

where (v_1, v_2, v_3) are the coordinates of mesh vertices and SPHARM coefficient $h_{lm}^i = \langle v_i, Y_{lm} \rangle$. Consider another SPHARM representation $J = (j_1, j_2, j_3)$ obtained from mesh coordinate w_i:

$$j_i(\Omega) = \sum_{l=0}^{k} \sum_{m=-l}^{l} w_{lm}^i Y_{lm}(\Omega),$$

where $w_{lm}^i = \langle w_i, Y_{lm} \rangle$. Suppose the surface h_i is deformed to $h_i + d_i$ by the amount of displacement d_i. We wish to find d_i that minimizes the discrepancy between $h_i + d_i$ and j_i in the subspace $\mathcal{H}_k$ spanned using up to the k-th degree spherical harmonics. It can be shown that

$$\arg\min_{d_i \in \mathcal{H}_k} \left\| \widehat{h}_i + d_i - \widehat{j}_i \right\| = \sum_{l=0}^{k} \sum_{m=-l}^{l} (w_{lm}^i - v_{lm}^i) Y_{lm}(\Omega). \tag{9.7}$$

This implies that the optimal displacement of matching two surfaces is obtained by simply taking the difference between two SPHARM and matching coefficients of the same degree and order. Then a specific point $\widehat{h}_i(\Omega)$ in one surface corresponds to $\widehat{j}_i(\Omega)$ in the other surface. We refer to this point-to-point surface matching as the *SPHARM correspondence* (27). Unlike other surface registration methods (28, 96, 115), it is not necessary to consider an additional cost function that guarantees the smoothness of the displacement field since the displacement field $d = (d_1, d_2, d_3)$ is already a linear combination of smooth basis functions.

9.3.3 Diffeomorphic Registration

Diffeomorphic image registration is a recently popular technique for registering volume and surface data (6, 83, 117, 93, 127). From the affine transformed individual surfaces $\mathcal{M}_j$, an additional nonlinear surface registration to the template using the large deformation diffeomorphic metric mapping (LDDMM) framework can be performed (83, 117, 93, 127). In LDDMM, given the template surface $\mathcal{M}$, the diffeomorphic transformations, which are one-to-one, smooth forward, and inverse transformation, are constructed as follows. We estimate the diffeomorphism d between them as a smooth streamline given by the Lagrangian evolution:

$$\frac{\partial d}{\partial t}(x,t) = v \circ d(x,t)$$

with $d(x,0) = x$, $t \in [0,1]$ for time-dependent velocity field v. Note the surfaces $\mathcal{M}_j$ and $\mathcal{M}$ are the start and end points of the diffeomorphism, i.e., $\mathcal{M}_j \circ d(\cdot,0) = \mathcal{M}_j$ and $\mathcal{M}_j \circ d(\cdot,1) = \mathcal{M}$. By solving the evolution equation numerically, we obtain the diffeomorphism. By averaging the inverse deformation fields from the template $\mathcal{M}$ to individual subjects, we may obtain yet another final more refined template. The vector fields v are constrained to be sufficiently smooth to generate diffeomorphic transformations over finite time (40). Figure 9.2 shows the resulting displacement vector fields of warping from the template to two hippocampal surfaces. In the deformation-based morphometry, variability in the displacement is used to characterize surface growth and differences (24).

9.4 Cortical Surface Features

The human cerebral cortex has the topology of a 2D highly convoluted grey matter shell with an average thickness of 3 mm (24, 79). The outer boundary of the shell is called the *outer cortical surface* while the inner boundary is called the *inner cortical surface*. Various cortical surface features such as curvatures, local surface area and cortical thickness have been used in quantifying anatomical variations. Among them, cortical thickness has been more often analyzed than other features.

9.4.1 Cortical Thickness

Once we extract both the outer and inner cortical surface meshes, *cortical thickness* can be computed at each mesh vertex. The cortical thickness is defined as the distance between the corresponding vertices between the inner and outer surfaces (79). There are many different computational approaches in measuring cortical thickness. In one approach, the vertices on the inner triangular mesh are deformed to fit the outer surface by minimizing a cost function that involves bending, stretching and other topological constraints (24). There is also an alternate method for automatically measuring cortical thickness based on the Laplace equation (64).

The average cortical thickness for each individual is about 3 mm (58). Cortical thickness varies from 1 to 4 mm depending on the location of the cortex. In normal brain development, it is highly likely that the change of cortical thickness may not be uniform across the cortex. Since different clinical populations are expected to show different patterns of cortical thickness variations, cortical thickness has also been used as a quantitative index for characterizing clinical populations (28). Cortical thickness varies locally by region and is likely to be influenced by aging, development and disease (9). By analyzing how cortical thickness varies between clinical and non-clinical populations, we can locate regions on the brain related to a specific pathology.

9.4.2 Surface Area and Curvatures

As in the case of local volume changes in the deformation-based morphometry, the rate of cortical surface area expansion or reduction may not be uniform across the cortical surface (24). Suppose that cortical surface $\mathcal{M}$ is parameterized by the parameters $\mathbf{u} = (u^1, u^2)$ such that any point $\mathbf{x} \in \mathcal{M}$ can be written as $\mathbf{x} = \mathbf{X}(\mathbf{u})$. Let $\mathbf{X}_i = \partial \mathbf{X}/\partial u^i$ be the partial derivative vectors. The *Riemannian metric tensor* g_{ij} is then given by the inner product between two vectors $\mathbf{X}_i$ and $\mathbf{X}_j$, i.e.,

$$g_{ij}(t) = \langle \mathbf{X}_i, \mathbf{X}_j \rangle.$$

The tensor g_{ij} measures the amount of the deviation of the cortical surface from a flat Euclidean plane and can be used to measure lengths, angles and areas on the cortical surface.

Let $g = (g_{ij})$ be a 2×2 matrix of metric tensors. The total surface area of the cortex $\mathcal{M}$ is then given by

$$\int_D \sqrt{\det g} \, d\mathbf{u},$$

where $D = X^{-1}(\mathcal{M})$ is the parameter space (72). The term $\sqrt{\det g}$ is called the *surface area element* and it measures the transformed area of the unit square in the parameter space D via transformation $X : D \to \mathcal{M}$. The surface area element can be considered as the generalization of the Jacobian determinant, which is used in measuring local volume changes in the tensor-based morphometry (24).

Instead of using the metric tensor formulation, it is possible to quantify local surface area change in terms of the areas of the corresponding triangles in surface meshes. However, this formulation assigns the computed surface area to each face instead of each vertex. This causes a problem in both surface-based smoothing and statistical analysis, where values are required to be given on vertices. Interpolating scalar values on vertices from face values can be done by the weighted average of face values. It is not hard to develop surface-based smoothing and statistical analysis on face values, as a form of dual formulation, but the cortical thickness and the curvature metric are defined on vertices so we end up with two separate approaches: one for metrics defined on vertices and the other for metrics defined on faces. Therefore, the metric tensor approach provides a better unified quantitative framework for the subsequent statistical analysis.

The *principal curvatures* characterize the shape and location of the sulci and gyri, which are the valleys and crests of the cortical surfaces (10, 67, 69). By measuring the curvature changes, rapidly folding and cortical regions can be localized. The principal curvatures κ_1 and κ_2 can be represented as functions of β_i in quadratic surface (9.1) (16, 72).

9.4.3 Gray Matter Volume

Local volume can be computed using the determinant of the Jacobian of deformation and used in detecting regions of brain tissue growth and loss in brain development (24). Compared to the local surface area change, the local volume change measurement is more sensitive to small deformation of the brain. If a unit cube increases its sides by one, the surface area will increase by $2^2 - 1 = 3$ while the volume will increase by $2^3 - 1 = 7$. Therefore, the statistical analysis based on the local volume change is at least twice more sensitive compared to that of the local surface area change. So the local volume change should be able to pick out gray matter tissue growth patterns even when the local surface area change may not.

The gray matter can be considered as a thin shell bounded by two surfaces with varying cortical thickness. In most deformable surface algorithms like FreeSurfer, each triangle on the outer surface has a corresponding triangle on the inner surface. Let $\mathbf{p}_1, \mathbf{p}_2, \mathbf{p}_3$ be the three vertices of a triangle on the outer surface and $\mathbf{q}_1, \mathbf{q}_2, \mathbf{q}_3$ be the corresponding three vertices on the inner surface such that $\mathbf{p}_i$ is linked to $\mathbf{q}_i$. Then the volume of the triangular prism is given by the sum of the determinants

$$D(\mathbf{p}_1, \mathbf{p}_2, \mathbf{p}_3, \mathbf{q}_1) + D(\mathbf{p}_2, \mathbf{p}_3, \mathbf{q}_1, \mathbf{q}_2) + D(\mathbf{p}_3, \mathbf{q}_1, \mathbf{q}_2, \mathbf{q}_3)$$

where

$$D(\mathbf{a}, \mathbf{b}, \mathbf{c}, \mathbf{d}) = |\det(\mathbf{a} - \mathbf{d}, \mathbf{b} - \mathbf{d}, \mathbf{c} - \mathbf{d})|/6$$

is the volume of a tetrahedron whose vertices are $\{\mathbf{a}, \mathbf{b}, \mathbf{c}, \mathbf{d}\}$. Afterwards, the total gray matter volume can be estimated by summing the volumes of all triangular prisms (24).

9.5 Surface Data Smoothing

Cortical surface mesh extraction and cortical thickness computation are expected to introduce noise (28, 41, 79). To counteract this, surface-based data smoothing is necessary (3, 21, 20, 24). For 3D whole brain volume data, Gaussian kernel smoothing is desirable in many statistical analyses (39, 98). Gaussian kernel smoothing weights neighboring observations according to their 3D Euclidean distance. Specifically, Gaussian kernel smoothing of functional data or image $f(\mathbf{x}), \mathbf{x} = (x_1, \ldots, x_n) \in \mathbb{R}^n$ with *full width at half maximum* *(FWHM)* $= 4(\ln 2)^{1/2}\sqrt{t}$ is defined as the convolution of the Gaussian kernel with f:

$$F(\mathbf{x}, t) = \frac{1}{(4\pi t)^{n/2}} \int_{\mathbb{R}^n} e^{-(x-y)^2/4t} f(y) dy. \tag{9.8}$$

However, due to the convoluted nature of the cortex, whose geometry is non-Euclidean, we cannot directly use the formulation (9.8) on the cortical surface. For data that lie on a 2D surface, smoothing must be weighted according to the geodesic distance along the surface, which is not straightforward (3, 24, 31).

9.5.1 Diffusion Smoothing

By formulating Gaussian kernel smoothing as a solution of a diffusion equation on a Riemannian manifold, the Gaussian kernel smoothing approach can be generalized to an arbitrary curved surface. This generalization is called *diffusion smoothing* and was first introduced in the analysis of fMRI data on the cortical surface (3) and cortical thickness (31) in 2001.

It can be shown that Gaussian kernel smoothing (9.8) is the integral solution of the n-dimensional diffusion equation

$$\frac{\partial F}{\partial t} = \Delta F \tag{9.9}$$

with the initial condition $F(\mathbf{x}, 0) = f(\mathbf{x})$, where

$$\Delta = \frac{\partial^2}{\partial x_1^2} + \cdots + \frac{\partial^2}{\partial x_n^2}$$

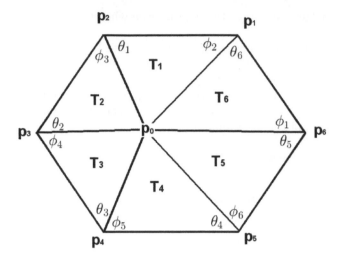

FIGURE 9.3
A typical triangulation in the neighborhood of $\mathbf{p} = \mathbf{p}_0$ in a surface mesh.

is the Laplacian in n-dimensional Euclidean space. Hence the Gaussian kernel smoothing is equivalent to the diffusion of the initial data $f(\mathbf{x})$ after time t.

Diffusion equations have been widely used in image processing as a form of noise reduction starting with Perona and Malik in 1990 (89). Numerous diffusion techniques have been developed for surface data smoothing (3, 24, 20, 21, 28, 65). When applying diffusion smoothing on curved surfaces, the smoothing somehow has to incorporate the geometrical features of the curved surface and the Laplacian Δ should change accordingly. The extension of the Euclidean Laplacian to an arbitrary Riemannian manifold is called the *Laplace–Beltrami operator* (5, 72). The approach taken in (3) is based on a local flattening of the cortical surface and estimating the planar Laplacian, which may not be as accurate as the cotan estimation based on the finite element method (FEM) given in (31). Further, the FEM approach completely avoids the use of surface flattening and parameterization; thus, it is more robust.

For given Riemannian metric tensor g_{ij}, the Laplace–Beltrami operator Δ is defined as

$$\Delta F = \sum_{i,j} \frac{1}{|g|^{1/2}} \frac{\partial}{\partial u^i} \left(|g|^{1/2} g^{ij} \frac{\partial F}{\partial u^j} \right), \tag{9.10}$$

where $(g^{ij}) = g^{-1}$ (5). Note that when g becomes an identity matrix, the Laplace–Beltrami operator reduces to the standard 2D Laplacian:

$$\Delta F = \frac{\partial^2 F}{\partial (u^1)^2} + \frac{\partial^2 F}{\partial (u^2)^2}.$$

Using FEM on the triangular cortical mesh, it is possible to estimate the Laplace–Beltrami operator as the linear weights of neighboring vertices using the cotan formulation, which is first given in (31).

Let $\mathbf{p}_1, \cdots, \mathbf{p}_m$ be m neighboring vertices around the central vertex $\mathbf{p} = \mathbf{p}_0$ (Figure 9.3). Then the estimated Laplace–Beltrami operator is given by

$$\widehat{\Delta F}(\mathbf{p}) = \sum_{i=1}^{m} w_i \big(F(\mathbf{p}_i) - F(\mathbf{p}) \big) \tag{9.11}$$

with the weights

$$w_i = \frac{\cot \theta_i + \cot \phi_i}{\sum_{i=1}^{m} \|T_i\|},$$

where θ_i and ϕ_i are the two angles opposite to the edge connecting $\mathbf{p}_i$ and $\mathbf{p}$, and $\|T_i\|$ is the area of the i-th triangle (Figure 9.3).

FEM estimation (9.11) is an improved formulation from the previous attempt in diffusion smoothing (3), where the Laplacian is simply estimated as the planar Laplacian after locally flattening the triangular mesh consisting of nodes $\mathbf{p}_0, \cdots, \mathbf{p}_m$ onto a flat plane. Afterwards, the finite difference (FD) scheme can be used to iteratively solve the diffusion equation at each vertex $\mathbf{p}$:

$$\frac{F(\mathbf{p}, t_{n+1}) - F(\mathbf{p}, t_n)}{t_{n+1} - t_n} = \widehat{\Delta} F(\mathbf{p}, t_n),$$

with the initial condition $F(\mathbf{p}, 0) = f(\mathbf{p})$ and fixed $\delta t = t_{n+1} - t_n$. After N iterations, the FD gives the diffusion of the initial data f after time $N\delta t$. If the diffusion were applied to Euclidean space, it would be approximately equivalent to Gaussian kernel smoothing with

$$\text{FWHM} = 4(\ln 2)^{1/2} \sqrt{N \delta t}.$$

For large meshes, computing the linear weights for the Laplace Beltrami operator takes a fair amount of time, but once the weights are computed, it is applied repeatedly throughout the iterations as a matrix multiplication. Unlike Gaussian kernel smoothing, diffusion smoothing is an iterative procedure.

9.5.2 Iterated Kernel Smoothing

Diffusion smoothing use FEM and FD, which are known to suffer numerical instability if sufficiently small step size is not chosen in the forward Euler scheme. To remedy the problem associated with diffusion smoothing, *iterative kernel smoothing*[3] was introduced (28). The method has been used in smoothing various cortical surface data: cortical curvatures (78, 47), cortical thickness (77, 14), hippocampus (105, 130), magnetoencephalography (MEG) (57) and functional-MRI (56, 63). This and its variations are probably the most widely used method for smoothing brain surface data at this moment. In iterated kernel smoothing, kernel weights are spatially adapted to follow the shape of the heat kernel in a discrete fashion.

The *n-th iterated kernel smoothing* of signal $f \in L^2(\mathcal{M})$ with kernel K_σ is defined as

$$K_\sigma^{(n)} * f(p) = \underbrace{K_\sigma * \cdots * K_\sigma}_{n \text{ times}} * f(p),$$

where σ is the bandwidth of the kernel. If K_σ is a heat kernel, we have the following iterative relation (28):

$$K_\sigma * f(p) = K_{\sigma/n}^{(n)} * f(p). \tag{9.12}$$

The relation (9.12) shows that kernel smoothing with large bandwidth σ can be decomposed into n repeated applications of kernel smoothing with smaller bandwidth σ/n. This idea can be used to approximate the heat kernel. When the bandwidth is small, the heat kernel behaves like the Dirac-delta function and, using the parametrix expansion (97, 119), we can approximate it locally using the Gaussian kernel.

---

[3]MATLAB package: www.stat.wisc.edu/~mchung/softwares/hk/hk.html.

Let $\mathbf{p}_1, \cdots, \mathbf{p}_m$ be m neighboring vertices of vertex $\mathbf{p} = \mathbf{p}_0$ in the mesh. The geodesic distance $d(\mathbf{p}, \mathbf{p}')$ between $\mathbf{p}$ and its adjacent vertex $\mathbf{p}_i$ is the length of edge between these two vertices in the mesh. Then the discretized and normalized heat kernel is given by

$$W_\sigma(\mathbf{p}, \mathbf{p}_i) = \frac{\exp\left(-\frac{d(\mathbf{p}, \mathbf{p}_i)^2}{4\sigma}\right)}{\sum_{j=0}^{m} \exp\left(-\frac{d(\mathbf{p} - \mathbf{p}_j)^2}{4\sigma}\right)}.$$

Note $\sum_{i=0}^{m} W_\sigma(\mathbf{p}, \mathbf{p}_i) = 1$. For small bandwidth, all the kernel weights are concentrated near the center, so we only need to worry about the first neighbors of a given vertex in a surface mesh. The discrete version of heat kernel smoothing on a triangle mesh is then given by

$$W_\sigma * f(p) = \sum_{i=0}^{m} W_\sigma(p, p_i) f(p_i).$$

The discrete kernel smoothing should converge to heat kernel smoothing as the mesh resolution increases. This is the form of the Nadaraya–Watson estimator (23) in statistics. Instead of performing a single kernel smoothing with large bandwidth $n\sigma$, we perform n iterated kernel smoothing with small bandwidth σ as follows $W_\sigma^{(n)} * f(p)$.

9.5.3 Heat Kernel Smoothing

The recently proposed *heat kernel smoothing* [4] framework constructs the heat kernel analytically using the eigenfunctions of the Laplace–Beltrami operator (101). This method avoids the need for the linear approximation used in iterative kernel smoothing that compounds the approximation error at each iteration. The method represents isotropic heat diffusion analytically as a series expansion so it avoids the numerical convergence issues associated with solving the diffusion equations numerically (3, 24, 65). This technique is different from other diffusion-based smoothing methods in that it bypasses the various numerical problems such as numerical instability, slow convergence, and accumulated linearization error.

Although recently there have been a few studies that introduce heat kernel in computer vision and machine learning (11), they mainly use heat kernel to compute shape descriptors (110, 18) or to define a multi-scale metric (37). These studies did not use heat kernel in smoothing functional data on manifolds. Further, most kernel methods in machine learning deal with the linear combination of kernels as a solution to penalized regressions, which significantly differs from the heat kernel smoothing framework, which does not have a penalized cost function. There are log-Euclidean and exponential map frameworks on manifolds, where the main interest is in computing the Fréchet mean along the tangent space (36, 43). Such approaches or related methods in (51), the Nadaya–Watson type of kernel regression, is reformulated to learn the shape or image means in a population. In the heat kernel smoothing framework, we are not dealing with manifold data but scalar data defined on a manifold, so there is no need for exploiting the manifold structure of the data itself.

Let Δ be the Laplace–Beltrami operator on $\mathcal{M}$. Solving the eigenvalue equation

$$\Delta \psi_j = -\lambda \psi_j, \tag{9.13}$$

we order eigenvalues

$$0 = \lambda_0 < \lambda_1 \leq \lambda_2 \leq \cdots,$$

and corresponding eigenfunctions $\psi_0, \psi_1, \psi_2, \cdots$ (97, 28, 75, 106). The eigenfunctions ψ_j

---

[4]MATLAB package: `brainimaging.waisman.wisc.edu/~chung/lb`.

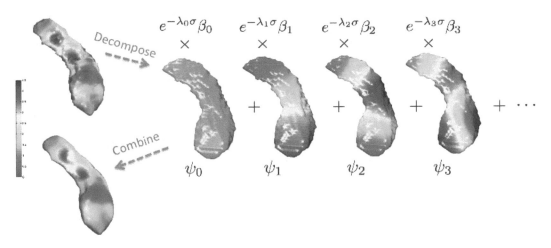

FIGURE 9.4
Schematic of hat kernel smoothing on a hippocampal surface. Given noisy functional on the surface, the Laplace–Beltrami eigenfunctions ψ_j are computed and their exponentially weighted Fourier coefficients $\exp -\lambda_j \sigma$ are multiplied as a form a regression. This process smoothes out the noisy functional signal with bandwidth σ.

form an orthonormal basis in $L^2(\mathcal{M})$, the space of square integrable functions in $\mathcal{M}$. Figure 9.4 shows the first four LB-eigenfunctions on a hippocampal surface.

There is extensive literature on the use of eigenvalues and eigenfunctions of the Laplace–Beltrami operator in medical imaging and computer vision (75, 92, 95, 94, 128, 129). The eigenvalues have been used in caudate shape discriminators (87). Qiu et al. used eigenfunctions in constructing splines on cortical surfaces (92). Reuter used the topological features of eigenfunctions (94). Shi et al. used the Reeb graph of the second eigenfunction in shape characterization and landmark detection in cortical and subcortical structures (107, 106). Lai et al. used the critical points of the second eigenfunction as anatomical landmarks for colon surfaces (73).

Using the eigenfunctions, *heat kernel* $K_\sigma(p, q)$ is defined as

$$K_\sigma(p, q) = \sum_{j=0}^{\infty} e^{-\lambda_j \sigma} \psi_j(p)\psi_j(q), \tag{9.14}$$

where σ is the bandwidth of the kernel. The heat kernel is the generalization of a Gaussian kernel. Then *heat kernel smoothing* of functional measurement Y is defined as

$$K_\sigma * Y(p) = \sum_{j=0}^{\infty} e^{-\lambda_j \sigma} \beta_j \psi_j(p), \tag{9.15}$$

where $\beta_j = \langle Y, \psi_j \rangle$ are Fourier coefficients (28). Kernel smoothing $K_\sigma * Y$ is taken as the estimate for the unknown mean signal θ. The degree for truncating the series expansion can be automatically determined using the forward model selection procedure (27). Figure 9.4 shows the heat kernel smoothing results with the bandwidth $\sigma = 0.5$ and $k = 500$ number of LB-eigenfunctions.

Unlike previous approaches to heat diffusion (3, 24, 65, 111), heat kernel smoothing avoids the direct numerical discretization of the diffusion equation. Instead, we discretize the basis functions of given manifold $\mathcal{M}$ by solving for the eigensystem (9.13) and obtain λ_j

and ψ_j. This provides more robust stable smoothing results compared to diffusion smoothing or iterated kernel smoothing approaches.

9.6   Statistical Inference on Surfaces

Surface measurements such as cortical thickness, curvatures, or fMRI responses can be modeled as random fields on the cortical surface:

$$Y(x) = \mu(x) + \epsilon(x), x \in \mathcal{M}, \tag{9.16}$$

where the deterministic part μ is the unknown mean of the observed functional measurement Y and ϵ is a mean zero random field. The functional measurements on the brain surface are often modeled using the general linear models (GLMs) or its multivariate version. Various statistical models are proposed for estimating and modeling the signal component $\mu(x)$ (67, 24, 27) but the majority of the methods are all based on GLM. GLMs have been implemented in the brain image analysis packages such as SPM and fMRISTAT.[5]

9.6.1   General Linear Models

We set up a GLM at each mesh vertex. Let y_i be the response variable, which is mainly coming from images and $\mathbf{x}_i = (x_{i1}, \cdots, x_{ip})$ to be the variables of interest and $\mathbf{z}_i = (z_{i1}, \cdots, z_{ik})$ to be nuisance variables corresponding to the i-th subject. Assume there are n subjects, i.e., $i = 1, \cdots, n$. We are interested in testing the significance of variables $\mathbf{x}_i$ while accounting for nuisance covariates $\mathbf{z}_i$. Then we set up GLM

$$y_i = \mathbf{z}_i \boldsymbol{\lambda} + \mathbf{x}_i \boldsymbol{\beta} + \epsilon_i$$

where $\boldsymbol{\lambda} = (\lambda_1, \cdots, \lambda_k)'$ and $\boldsymbol{\beta} = (\beta_1, \cdots, \beta_p)'$ are unknown parameter vectors to be estimated. We assume ϵ to be the usual zero mean Gaussian noise, although the distributional assumption is not required for the least squares estimation. We test hypotheses

$$H_0 : \boldsymbol{\beta} = 0 \text{ vs. } H_1 : \boldsymbol{\beta} \neq 0.$$

Subsequently the inference is done by constructing the F-statistic with p and $n - p - k$ degrees of freedom. GLMs have been used in quantifying cortical thickness, for instance, in child development (24, 28) and amygdala shape differences in autism (24).

In the hippocampus case study, the first T1-weighted MRI scans are taken at 11.6 ± 3.7 years for $n = 124$ children using a 3T GE SIGNA scanner. Variables **age** and **gender** are available. We also have variable **income**, which is a binary dummy variable indicating whether the subjects are from high- or low-income families. A total of 124 children and adolescents are from high- ($> 75000\$; n = 86$) and low-income ($< 35000\$, n = 38$) parents, respectively. In addition to this cross-sectional data, longitudinal data was available for 82 of these subjects ($n = 66, > 75000\$; n = 16, < 35000\$$). The second MRI scan was acquired for these 82 subjects about 2 years later at 14 ± 3.9 years. For now, we will simply ignore the correlation between the scans within a subject and will treat them as independent.

On the template surface, we have the displacement vector fields of mapping from the template to individual subjects (Figure 9.2). We take the length of the surface displacement, denoted as **deformation**, with respect to the template as the response variable. The

---

[5]The fMRISTAT package is available at `www.math.mcgill.ca/keith/fmristat`.

FIGURE 9.5
F-statistics maps on testing the significant hippocampus shape difference on the income level while controlling for age and gender. The arrows show the deformation differences between the groups (high income—low income). The fixed effects result (left) is obtained by treating the repeat scans as independent. The mixed effects result (right) is obtained by explicitly modeling the covariance structure of the repeat scans with a subject. Both results are not statistically significant under multiple comparisons even at 0.1 level.

displacement measures the shape difference with respect to the template. However, since the length measurement is noisy, surface-based smoothing is necessary. We have used heat kernel smoothing to smooth out noise with the bandwidth 1 and 500 LB-eigenfunctions. Then we set up the GLM:

$$\texttt{deformation} = \lambda_1 + \lambda_2 \texttt{age} + \beta_1 \texttt{income} + \epsilon$$

and test for the significance of β_1 at each mesh vertex. Figure 9.5-left shows the F-statistic result on testing β_1.

9.6.2 Multivariate General Linear Models

The multivariate general linear models (MGLMs) have been also used in modeling multivariate imaging features on brain surfaces. These models generalize a widely used univariate GLM by incorporating vector valued responses and explanatory variables (2, 46, 123, 125, 112, 24). Hotelling's T^2 statistic is a special case of MGLM that has been used primarily for inference on surface shapes, deformations and multi-scale features (22, 24, 48, 66, 70, 114). An example of this approach is Hotelling's T^2-statistic applied in determining the 3D brain morphology of HIV/AIDS patients (74). Hotelling's T^2-statistic is also applied to 2D deformation tensor at each mesh vertex on the hippocampal surface as a way to characterize Alzheimer's diseases (120).

Suppose there are a total of n subjects and p multivariate features of interest at each voxel. For MGLM to work, n should be significantly larger than p. Let $\mathbf{J}_{n \times p} = (J_{ij})$ be the measurement matrix, where J_{ij} is the measurement for subject i and for the j-th feature. The subscripts denote the dimension of the matrix. All the measurements over subjects for the j-th feature are denoted as $\mathbf{x}_j = (J_{1j}, \cdots, J_{nj})'$. The measurement vector for the i-th subject is denoted as $\mathbf{y}_i = (J_{i1}, \cdots, J_{ip})'$. $\mathbf{y}_i$ is expected to be distributed identically and independently over subjects. Note that

$$\mathbf{J} = (\mathbf{x}_1, \cdots, \mathbf{x}_p) = (\mathbf{y}_1, \cdots, \mathbf{y}_n)'.$$

We may assume the covariance matrix of $\mathbf{y}_i$ to be

$$\text{Cov}(\mathbf{y}_1) = \cdots = \text{Cov}(\mathbf{y}_n) = \mathbf{\Sigma}_{p \times p} = (\sigma_{kl}).$$

With these notations, we now set up the following MGLM at each mesh vertex:

$$\mathbf{J}_{n \times p} = \mathbf{X}_{n \times k} \mathbf{B}_{k \times p} + \mathbf{Z}_{n \times q} \mathbf{G}_{q \times p} + \mathbf{U}_{n \times p} \mathbf{\Sigma}_{p \times p}^{1/2}. \tag{9.17}$$

$\mathbf{X}$ is the matrix of contrasted explanatory variables while $\mathbf{B}$ is the matrix of unknown coefficients. Nuisance covariates are in matrix $\mathbf{Z}$ and the corresponding coefficients are in matrix $\mathbf{G}$. The components of Gaussian random matrix $\mathbf{U}$ are independently distributed with zero mean and unit variance. Symmetric matrix $\mathbf{\Sigma}^{1/2}$ is the square root of the covariance matrix accounting for the spatial dependency across different voxels. In MGLM (9.17), we are usually interested in testing hypotheses

$$H_0 : \mathbf{B} = 0. \ vs. \ H_1 : \mathbf{B} \neq 0.$$

The parameter matrices in the model are estimated via the least squares method. The multivariate test statistics such as Lawley–Hotelling trace or Roy's maximum root are used to test the significance of $\mathbf{B}$. When there is only one voxel, i.e., $p = 1$, these multivariate test statistics collapse to Hotelling's T^2 statistic (125, 24).

9.6.3   Small-n Large-p Problems

GLMs are usually fitted in each voxel separately. Instead of fitting the GLM at each voxel, one may be tempted to fit the model in the whole brain surface. For FreeSurfer meshes, we need to fit GLM over 300,000 vertices, which causes the *small-n large-p problem* (46, 100, 25).

Let $\mathbf{y}_j$ be the measurement vector at the j-th vertex. Assume there are n subjects and total p vertices in the surface. We have the same design matrix $\mathbf{Z}$ for all p vertices. Then we need to estimate the parameter vector $\boldsymbol{\lambda}_j$ in

$$\mathbf{y}_j = \mathbf{Z} \boldsymbol{\lambda}_j \tag{9.18}$$

for each j. Instead of solving (9.18) separately at each vertex, we combine all of them together so that we have this matrix equation

$$\underbrace{[\mathbf{y}_1, \cdots, \mathbf{y}_m]}_{\mathbf{Y}} = \mathbf{Z} \underbrace{[\boldsymbol{\lambda}_1, \cdots, \boldsymbol{\lambda}_m]}_{\boldsymbol{\Lambda}}. \tag{9.19}$$

The least squares estimation of the parameter matrix $\boldsymbol{\Lambda}$ is given by

$$\widehat{\boldsymbol{\Lambda}} = (\mathbf{Z}'\mathbf{Z})^{-1} \mathbf{Z}'\mathbf{Y}.$$

Note that $\mathbf{Z}$ is of size n by p and $\mathbf{Z}'\mathbf{Z}$ is only invertible when $n \ll p$. The least squares estimation does not provide robust parameter estimates for $n \ll p$, which is the usual case in surface modeling. For the small-n large-p problem, the GLM needs to be regularized using the L1-norm penalty (8, 7, 25, 45, 61, 81).

9.6.4 Longitudinal Models

So far we have only dealt with an imaging dataset where the parameters of the model are fixed and do not vary across subjects and scans. Such fixed-effects models are inadequate in modeling the within-subject dependency in longitudinally collected imaging data. However, mixed-effects models can explicitly model such dependency (44, 84, 85, 91). There are three advantages of the mixed-effects model over the usual fixed-effects model. It explicitly models individual growth patterns and accommodates an unequal number of follow-up image scans per subject and unequal time intervals between scans.

The longitudinal outcome Y_i from the i-th subject is modeled using the mixed-effects model (84) as

$$Y_i = X_i\beta + Z_i\gamma_i + e_i, \tag{9.20}$$

where β is the fixed effects shared by all subjects. γ_i is the subject-specific random effects and $e_i \sim N(0, \sigma^2)$ is independent and identically distributed noise. X_i and Z_i are the design matrices corresponding to the fixed and random effects respectively for the i-th subject. We assume $\gamma_i \sim N(0, \Gamma)$ and $\epsilon_i \sim N(0, \Sigma_i)$ with some covariance matrices Γ and Σ_i. Hierarchically we are modeling (9.20) as

$$Y_i|\gamma_i \sim N(X_i\beta + Z_i\gamma_i, \Sigma_i), \ \gamma_i \sim N(0, \Gamma).$$

Γ accounts for covariance among random effect terms. The within-subject variability between the scans is expected to be smaller than between-subject variability and explicitly modeled by Σ_i. The covariance of γ_i and ϵ are expected to have block diagonal structure such that there is no correlation among the scans of different subjects while there is high correlation between the scans of the same subject:

$$\mathbb{V}\begin{pmatrix} \gamma_i \\ \epsilon_i \end{pmatrix} = \begin{pmatrix} \Gamma & 0 \\ 0 & \Sigma_i \end{pmatrix}.$$

Subsequently, the overall covariance of Y_i is given by

$$\mathbb{V}Y_i = Z_i\Gamma Z_i' + \Sigma_i.$$

The random-effect contribution is $Z_i\Gamma Z_i'$ while the within-subject contribution is Σ_i.

The parameters and the covariance matrices can be estimated via the restricted maximum likelihood (REML) method (44, 91). The most widely used tools for fitting the mixed effects model are in the `nlme` library in the R statistical package (91). However, there is no need to use R to fit the mixed effects model. Keith Worsley has implemented the REML procedure in the `SurfStat` package [6] (24, 124).

Here we briefly explain how to set up a longitudinal mixed-effect model in practice. In the usual fixed-effect model, we have a linear model containing the fixed-effect term $\mathbf{age}_i$ for the i-th subject:

$$y_i = \beta_0 + \beta_1 \mathbf{age}_i + \epsilon_i, \tag{9.21}$$

where ϵ_i is assumed to follow independent Gaussian. In (9.21), every subject has identical growth trajectory $\beta_0 + \beta_1\mathbf{age}$, which is unrealistic. Biologically, each subject is expected to have its own unique growth trajectory. So we assume each subject to have its own intercept $\beta_0 + \gamma_{i0}$ and slope $\beta_1 + \gamma_{i1}$:

$$y_i = \beta_0 + \gamma_{i0} + (\beta_1 + \gamma_{i1})\mathbf{age}_i + \epsilon_i. \tag{9.22}$$

---

[6]The MATLAB package is available at `www.math.mcgill.ca/keith/surfstat`.

FIGURE 9.6
F-statistics maps testing the interaction between the income level and age while controlling for gender in a linear mixed-effects model. The arrows show the deformation differences between the groups (high income—low income). Significant regions are only found in the tail and midbody regions of the right hippocampus.

It is reasonable to assume the random vector $\gamma = (\gamma_{i0}, \gamma_{i1})$ to be multivariate normal. The model (9.22) can be decomposed into fixed- and random-effect terms:

$$y_i = (\beta_0 + \beta_1 \mathbf{age}_i) + (\gamma_{i0} + \gamma_{i1} \mathbf{age}_i) + \epsilon_i. \tag{9.23}$$

The fixed-effect term $\beta_0 + \beta_1 \mathbf{age}_i$ models the linear growth of the population while the random-effect term $\gamma_{i0} + \gamma_{i1} \mathbf{age}_i$ models the subject specific growth variations. Incorporating additional factors and interaction terms are done similarly.

In the hippocampus case study, the first MRI scans are taken at 11.6 ± 3.7 years while the second scans are taken at 14 ± 3.9 years. We are interested in determining the effects of income level on the shape of the hippocampus. In Section 9.6.1, we treated the second scans as if they came from independent subjects and modeled them using the fixed-effects model. Now we explicitly incorporate the dependency of repeated scans of the same subject. It is necessary to explicitly model the within-subject variability that is expected to be smaller than between-subject variability. This can be done by introducing a random-effect term. The resulting F-statistic maps are given in Figure 9.5-right. However, we did not detect any region that is affected by income. Thus, we tested the age and income interaction and found the regions of strong interaction (Figure 9.6).

9.6.5 Random Field Theory

Since we need to set up a GLM on every mesh vertex, it becomes a *multiple comparisons* problem. Correcting for multiple comparisons is crucial in determining overall statistical significance in correlated test statistics over the whole surface. For surface data, various methods are proposed: Bonferroni correction, random field theory (121, 123), false discovery rates (FDR) (12, 13, 50), and permutation tests (86). Among many techniques, the random field theory is probably the most natural in relation to surface data smoothing since it is able to explicitly control the amount of smoothing.

The generalization of a continuous stochastic process in $\mathbb{R}^n$ to a higher dimensional abstract space is called a *random field* (1, 24, 39, 66, 126). In the random field theory

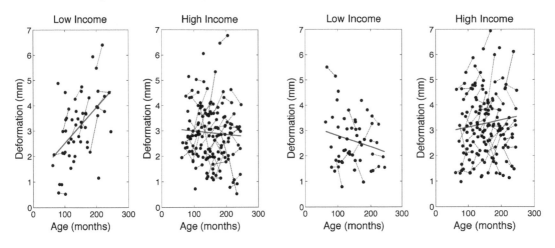

FIGURE 9.7

The plots showing income level-dependent growth differences in the posterior (left) and midbody (right) regions of the right hippocampus. The red lines are the linear regression lines. Scans within a subject are identified by dotted lines.

(121, 123), measurement Y at position $x \in \mathcal{M}$ is modeled as

$$Y(x) = \mu(x) + \epsilon(x),$$

where μ is the unknown signal to be estimated and ϵ is the measurement error. The measurement error at each fixed x can be modeled as a random variable. Then the collection of random variables $\{\epsilon(x) : x \in \mathcal{M}\}$ is called a *random field* or stochastic process. A measure-theoretic definition is given in (1).

Detecting the regions of statistically significant $\lambda(x)$ for some $x \in \mathcal{M}$ can be done via thresholding the maximum of a random field defined on the cortical surface (123, 122). For instance, T random field on the surface $\mathcal{M}$ is defined as

$$T(x) = \sqrt{n} \frac{M(x)}{S(x)},$$

where M and S are the sample mean and standard deviation over the n subjects. Under the null hypothesis

$$H_0 : \mu(x) = 0 \text{ for all } x \in \mathcal{M},$$

$T(x)$ is distributed as a student's T with $n - 1$ degrees of freedom at each voxel x. The p value of the local maxima of the T field will give a conservative threshold compared to FDR (123).

For sufficiently high threshold y, we can show that

$$P\left(\max_{x \in \mathcal{M}} T(\mathbf{x}) \geq y \right) \approx \sum_{i=0}^{3} \phi_i(\mathcal{M}) \rho_i(y), \tag{9.24}$$

where ρ_i is the i-dimensional EC-density and the Minkowski functional ϕ_i are

$$\phi_0(\mathcal{M}) = 2, \ \phi_1(\mathcal{M}) = 0, \ \phi_2(\mathcal{M} = \|\mathcal{M}\|, \ \phi_3(\mathcal{M}) = 0$$

and $\|\mathcal{M}\|$ is the total surface area of $\mathcal{M}$ (Worsley, 1996a). When diffusion or heat kernel smoothing with given FWHM is applied on surface $\mathcal{M}$, the 0-dimensional and 2-dimensional EC-density becomes

$$\rho_0(y) = \int_y^\infty \frac{\Gamma(\frac{n}{2})}{((n-1)\pi)^{1/2}\Gamma(\frac{n-1}{2})} \left(1 + \frac{y^2}{n-1}\right)^{-n/2} dy,$$

$$\rho_2(y) = \frac{1}{\text{FWHM}^2} \frac{4\ln 2}{(2\pi)^{3/2}} \frac{\Gamma(\frac{n}{2})}{(\frac{n-1}{2})^{1/2}\Gamma(\frac{n-1}{2})} y\left(1 + \frac{y^2}{n-1}\right)^{-(n-2)/2}.$$

The excursion probability, which is the probability of obtaining false positives for the one-sided alternate hypothesis, is approximated by the following formula:

$$P\left(\max_{x \in \mathcal{M}} T(x) \geq y\right) \approx 2\rho_0(y) + \|\mathcal{M}\|\rho_2(y).$$

For smoothing cortical thickness (24, 28), an FWHM of between 20 to 30 mm is recommended. This FWHM reflects the spatial frequency associated with the sulcal pattern. For measurements on smaller subcortical structures such as hippocampus and amygdala, significantly smaller FWHM is recommended. For amygdala and hippocampus, 0.5–1 mm would be sufficient.

In the hippocampus case study, we did not detect any statistically significant group difference at 0.01 level after correcting for multiple comparisons in both the left and right hippocampi. However, we obtained highly focalized regions of group difference in the growth rate, the interaction between income level and age, in the right hippocampus (corrected p value = 0.03). The posterior region is enlarging while the midbody and the anterior parts are shrinking in children from low-income families (Figure 9.6 and 9.7). On the other hand, the pattern is opposite for children from high-income families. Note that the right hippocampus is involved in the active maintenance of associations with spatial information (90). Future studies investigating the relationship between family socioeconomic status and spatial information processing measures are warranted.

Bibliography

[1] R.J. Adler and J.E. Taylor. *Random Fields and Geometry*. Springer Verlag, 2007.

[2] T.W. Anderson. *An Introduction to Multivariate Statistical Analysis*. Wiley, 2nd edition, 1984.

[3] A. Andrade, F. Kherif, J. Mangin, K.J. Worsley, A. Paradis, O. Simon, S. Dehaene, D. Le Bihan, and J-B. Poline. Detection of fMRI activation using cortical surface mapping. *Human Brain Mapping*, 12:79–93, 2001.

[4] S. Angenent, S. Hacker, A. Tannenbaum, and R. Kikinis. On the Laplace-Beltrami operator and brain surface flattening. *IEEE Transactions on Medical Imaging*, 18:700–711, 1999.

[5] G.B. Arfken. *Mathematical Methods for Physicists*. Academic Press, 5th edition, 2000.

[6] B.B. Avants, C.L. Epstein, M. Grossman, and J.C. Gee. Symmetric diffeomorphic image registration with cross-correlation: Evaluating automated labeling of elderly and neurodegenerative brain. *Medical Image Analysis*, 12:26–41, 2008.

[7] O. Banerjee, L. El Ghaoui, and A. d'Aspremont. Model selection through sparse maximum likelihood estimation for multivariate Gaussian or binary data. *The Journal of Machine Learning Research*, 9:485–516, 2008.

[8] O. Banerjee, L.E. Ghaoui, A. d'Aspremont, and G. Natsoulis. Convex optimization techniques for fitting sparse Gaussian graphical models. In *Proceedings of the 23rd International Conference on Machine Learning*, page 96, 2006.

[9] P. Barta, M.I. Miller, and A. Qiu. A stochastic model for studying the laminar structure of cortex from MRI. *IEEE Transactions on Medical Imaging*, 24:728–742, 2005.

[10] A. Bartesaghi and G. Sapiro. A system for the generation of curves on 3d brain images. *Human Brain Mapping*, 14(1):1–15, 2001.

[11] Mikhail Belkin, Partha Niyogi, and Vikas Sindhwani. Manifold regularization: A geometric framework for learning from labeled and unlabeled examples. *The Journal of Machine Learning Research*, 7:2399–2434, 2006.

[12] Y. Benjamini and Y. Hochberg. Controlling the false discovery rate: A practical and powerful approach to multiple testing. *J. R. Stat. Soc, Ser. B*, 57:289–300, 1995.

[13] Y. Benjamini and D. Yekutieli. The control of the false discovery rate in multiple testing under dependency. *Annals of Statistics*, pages 1165–1188, 2001.

[14] J. Bernal-Rusiel, M. Atienza, and J. Cantero. Detection of focal changes in human cortical thickness: Spherical wavelets versus Gaussian smoothing. *NeuroImage*, 41:1278–1292, 2008.

[15] M.A. Blanco, M. Florez, and M. Bermejo. Evaluation of the rotation matrices in the basis of real spherical harmonics. *Journal of Molecular Structure: THEOCHEM*, 419:19–27, 1997.

[16] W.M. Boothby. *An Introduction to Differential Manifolds and Riemannian Geometry*. Academic Press, London, 2nd edition, 1986.

[17] C. Brechbuhler, G. Gerig, and O. Kubler. Parametrization of closed surfaces for 3d shape description. *Computer Vision and Image Understanding*, 61:154–170, 1995.

[18] M. M. Bronstein and I. Kokkinos. Scale-invariant heat kernel signatures for non-rigid shape recognition. In *IEEE Conference on Computer Vision and Pattern Recognition (CVPR)*, pages 1704–1711, 2010.

[19] T. Bulow. Spherical diffusion for 3D surface smoothing. *IEEE Transactions on Pattern Analysis and Machine Intelligence*, 26:1650–1654, 2004.

[20] A. Cachia, J.-F. Mangin, D. Riviére, F. Kherif, N. Boddaert, A. Andrade, D. Papadopoulos-Orfanos, J.-B. Poline, I. Bloch, M. Zilbovicius, P. Sonigo, F. Brunelle, and J. Régis. A primal sketch of the cortex mean curvature: A morphogenesis based approach to study the variability of the folding patterns. *IEEE Transactions on Medical Imaging*, 22:754–765, 2003.

[21] A. Cachia, J.-F. Mangin, D. Riviére, D. Papadopoulos-Orfanos, F. Kherif, I. Bloch, and J. Régis. A generic framework for parcellation of the cortical surface into gyri using geodesic Voronoï diagrams. *Image Analysis*, 7:403–416, 2003.

[22] J. Cao and K.J. Worsley. The detection of local shape changes via the geometry of Hotelling's T2 fields. *Annals of Statistics*, 27:925–942, 1999.

[23] P. Chaudhuri and J.S. Marron. Scale space view of curve estimation. *The Annals of Statistics*, 28:408–428, 2000.

[24] M.K. Chung. *Computational Neuroanatomy: The Methods*. World Scientific, 2013.

[25] M.K. Chung. *Statistical and Computational Methods in Brain Image Analysis*. CRC Press, 2013.

[26] M.K. Chung, J.L. Hanson, R.J. Davidson, and S.D. Pollak. Effect of family income on hippocampus growth: Longitudinal study. *17th Annual Meeting of the Organization for Human Brain Mapping*, (2697), 2011.

[27] M.K. Chung, R. Hartley, K.M. Dalton, and R.J. Davidson. Encoding cortical surface by spherical harmonics. *Statistica Sinica*, 18:1269–1291, 2008.

[28] M.K. Chung, S. Robbins, and A.C. Evans. Unified statistical approach to cortical thickness analysis. *Information Processing in Medical Imaging (IPMI), Lecture Notes in Computer Science*, 3565:627–638, 2005.

[29] M.K. Chung, S.M. Schaefer, C. M. van Reekum, L.P. Schmitz, M. Sutterer, and R.J. Davidson. A unified kernel regression on manifolds detects aging-related changes in the amygdala and hippocampus. *MICCAI, Lecture Notes in Computer Science (LNCS)*, 8674:789–796, 2014.

[30] M.K. Chung, K.J. Worsley, S. Robbins, T. Paus, J. Taylor, J.N. Giedd, J.L. Rapoport, and A.C. Evans. Deformation-based surface morphometry applied to gray matter deformation. *NeuroImage*, 18:198–213, 2003.

[31] M.K. Chung, K.J. Worsley, J. Taylor, J. Ramsay, S. Robbins, and A.C. Evans. Diffusion smoothing on the cortical surface. *NeuroImage*, 13:S95, 2001.

[32] R. Courant and D. Hilbert. *Methods of Mathematical Physics*. Interscience, New York, English edition, 1953.

[33] A.M. Dale and B. Fischl. Cortical surface-based analysis I. Segmentation and surface reconstruction. *NeuroImage*, 9:179–194, 1999.

[34] C. Davatzikos. Spatial transformation and registration of brain images using elastically deformable models. *Comput. Vis. Image Understanding*, 66:207–222, 1997.

[35] C. Davatzikos and R.N. Bryan. Using a deformable surface model to obtain a shape representation of the cortex. *Proceedings of the IEEE International Conference on Computer Vision*, 9:2122–2127, 1995.

[36] B.C. Davis, P.T. Fletcher, E. Bullitt, and S. Joshi. Population shape regression from random design data. *International Journal of Computer Vision*, 90:255–266, 2010.

[37] F. de Goes, S. Goldenstein, and L. Velho. A hierarchical segmentation of articulated bodies. *Computer Graphics Forum*, 27:1349–1356, 2008.

[38] M.P. do Carmo. *Riemannian Geometry*. Prentice-Hall, Inc., 1992.

[39] E.R. Dougherty. *Random Processes for Image and Signal Processing*. IEEE Press, 1999.

[40] P. Dupuis, U. Grenander, and M. I. Miller. Variational problems on flows of diffeomorphisms for image matching. *Quarterly of Applied Math.*, 56:587–600, 1998.

[41] B. Fischl and A.M. Dale. Measuring the thickness of the human cerebral cortex from magnetic resonance images. *Proceedings of the National Academy of Sciences (PNAS)*, 97:11050–11055, 2000.

[42] B. Fischl, M.I. Sereno, R. Tootell, and A.M. Dale. High-resolution intersubject averaging and a coordinate system for the cortical surface. *Hum. Brain Mapping*, 8:272–284, 1999.

[43] P.T. Fletcher, C. Lu, S.M. Pizer, and S. Joshi. Principal geodesic analysis for the study of nonlinear statistics of shape. *IEEE Transactions on Medical Imaging*, 23:995–1005, 2004.

[44] J. Fox. *An R and S-Plus Companion to Applied Regression.* Sage Publications, Inc., 2002.

[45] J. Friedman, T. Hastie, and R. Tibshirani. Sparse inverse covariance estimation with the graphical lasso. *Biostatistics*, 9:432, 2008.

[46] K.J. Friston, A.P. Holmes, K.J. Worsley, J.-P. Poline, C.D. Frith, and R.S.J. Frackowiak. Statistical parametric maps in functional imaging: A general linear approach. *Human Brain Mapping*, 2:189–210, 1995.

[47] C. Gaser, E. Luders, P.M. Thompson, A.D. Lee, R.A. Dutton, J.A. Geaga, K.M. Hayashi, U. Bellugi, A.M. Galaburda, J.R. Korenberg, D.L. Mills, A.W. Toga, and A.L. Reiss. Increased local gyrification mapped in Williams syndrome. *NeuroImage*, 33:46–54, 2006.

[48] C. Gaser, H.-P. Volz, S. Kiebel, S. Riehemann, and H. Sauer. Detecting structural changes in whole brain based on nonlinear deformations: Application to schizophrenia research. *NeuroImage*, 10:107–113, 1999.

[49] A. Gelb. The resolution of the Gibbs phenomenon for spherical harmonics. *Mathematics of Computation*, 66:699–717, 1997.

[50] C.R. Genovese, N.A. Lazar, and T. Nichols. Thresholding of statistical maps in functional neuroimaging using the false discovery rate. *NeuroImage*, 15:870–878, 2002.

[51] Samuel Gerber, Tolga Tasdizen, and Ross Whitaker. Dimensionality reduction and principal surfaces via kernel map manifolds. In *IEEE 12th International Conference on Computer Vision (ICCV)*, pages 529–536, 2009.

[52] G. Gerig, M. Styner, D. Jones, D. Weinberger, and J. Lieberman. Shape analysis of brain ventricles using SPHARM. In *MMBIA*, pages 171–178, 2001.

[53] C.D. Good, I.S. Johnsrude, J. Ashburner, R.N.A. Henson, K.J. Friston, and R.S.J. Frackowiak. A voxel-based morphometric study of ageing in 465 normal adult human brains. *NeuroImage*, 14:21–36, 2001.

[54] L.D. Griffin. The intrinsic geometry of the cerebral cortex. *Journal of Theoretical Biology*, 166:261–273, 1994.

[55] X. Gu, Y.L. Wang, T.F. Chan, T.M. Thompson, and S.T. Yau. Genus zero surface conformal mapping and its application to brain surface mapping. *IEEE Transactions on Medical Imaging*, 23:1–10, 2004.

[56] D.J. Hagler Jr., A.P. Saygin, and M.I. Sereno. Smoothing and cluster thresholding for cortical surface-based group analysis of fMRI data. *NeuroImage*, 33:1093–1103, 2006.

[57] J. Han, J.S. Kim, C.K. Chung, and K.S. Park. Evaluation of smoothing in an iterative lp-norm minimization algorithm for surface-based source localization of meg. *Physics in Medicine and Biology*, 52:4791–4803, 2007.

[58] C.C. Henery and T.M. Mayhew. The cerebrum and cerebellum of the fixed human brain: Efficient and unbiased estimates of volumes and cortical surface areas. *Journal of Anatomy*, 167:167–180, 1989.

[59] H.H.H. Homeier and E.O. Steinborn. Some properties of the coupling coefficients of real spherical harmonics and their relation to gaunt coefficients. *Journal of Molecular Structure: THEOCHEM*, 368:31–37, 1996.

[60] A.P. Hosseinbor, W.H. Kim, N. Adluru, A. Acharya, H.K. Vorperian, and M.K. Chung. The 4d hyperspherical diffusion wavelet: A new method for the detection of localized anatomical variation. *MICCAI, Lecture Notes in Computer Science (LNCS)*, 8675:65–72, 2014.

[61] S. Huang, J. Li, L. Sun, J. Ye, A. Fleisher, T. Wu, K. Chen, and E. Reiman. Learning brain connectivity of Alzheimer's disease by sparse inverse covariance estimation. *NeuroImage*, 50:935–949, 2010.

[62] M. K. Hurdal and K. Stephenson. Cortical cartography using the discrete conformal approach of circle packings. *NeuroImage*, 23:S119–S128, 2004.

[63] H.J. Jo, J.-M. Lee, J.-H. Kim, Y.-W. Shin, I.-Y. Kim, J.S. Kwon, and S.I. Kim. Spatial accuracy of fMRI activation influenced by volume- and surface-based spatial smoothing techniques. *NeuroImage*, 34:550–564, 2007.

[64] S.E. Jones, B.R. Buchbinder, and I. Aharon. Three-dimensional mapping of cortical thickness using Laplace's equation. *Human Brain Mapping*, 11:12–32, 2000.

[65] A.A. Joshi, D. W. Shattuck, P. M. Thompson, and R. M. Leahy. A parameterization-based numerical method for isotropic and anisotropic diffusion smoothing on non-flat surfaces. *IEEE Transactions on Image Processing*, 18:1358–1365, 2009.

[66] S.C. Joshi. Large deformation diffeomorphisms and Gaussian random fields for statistical characterization of brain sub-manifolds. Ph.D. thesis. Washington University, St. Louis, 1998.

[67] S.C. Joshi, J. Wang, M.I. Miller, D.C. Van Essen, and U. Grenander. On the differential geometry of the cortical surface. *Vision Geometry IV, Vol. 2573, Proceedings of the SPIE 1995 International Symposium on Optical Science, Engineering and Instrumentation*, pages 304–311, 1995.

[68] A. Kelemen, G. Szekely, and G. Gerig. Elastic model-based segmentation of 3-d neuroradiological datasets. *IEEE Transactions on Medical Imaging*, 18:828–839, 1999.

[69] N. Khaneja, M.I. Miller, and U. Grenander. Dynamic programming generation of curves on brain surfaces. *IEEE Transactions on Pattern Analysis and Machine Intelligence*, 20:1260–1265, 1998.

[70] W.H. Kim, D. Pachauri, C. Hatt, M.K. Chung, S. Johnson, and V. Singh. Wavelet based multi-scale shape features on arbitrary surfaces for cortical thickness discrimination. In *Advances in Neural Information Processing Systems*, pages 1250–1258, 2012.

[71] K. Kollakian. Performance analysis of automatic techniques for tissue classification in magnetic resonance images of the human brain. Technical Report Master's thesis, Concordia University, Montreal, Quebec, Canada, 1996.

[72] E. Kreyszig. *Differential Geometry*. University of Toronto Press, 1959.

[73] Z. Lai, J. Hu, C. Liu, V. Taimouri, D. Pai, J. Zhu, J. Xu, and J. Hua. Intra-patient supine-prone colon registration in CT colonography using shape spectrum. In *Medical Image Computing and Computer-Assisted Intervention—MICCAI 2010*, volume 6361 of *Lecture Notes in Computer Science*, pages 332–339, 2010.

[74] N. Lepore, C.A. Brun, M.C. Chiang, Y.Y. Chou, R.A. Dutton, K.M. Hayashi, O.L. Lopez, H.J. Aizenstein, A.W. Toga, J.T. Becker, and P.M. Thompson. Multivariate statistics of the Jacobian matrices in tensor based morphometry and their application to HIV/AIDS. *Lecture Notes in Computer Science*, pages 191–198, 2006.

[75] B. Lévy. Laplace-Beltrami eigenfunctions towards an algorithm that "understands" geometry. In *IEEE International Conference on Shape Modeling and Applications*, page 13, 2006.

[76] W.E. Lorensen and H.E. Cline. Marching cubes: A high resolution 3D surface construction algorithm. In *Proceedings of the 14th Annual Conference on Computer Graphics and Interactive Techniques*, pages 163–169, 1987.

[77] E. Luders, K.L. Narr, P.M. Thompson, D.E. Rex, R.P. Woods, H. DeLuca, L. Jancke, and A.W. Toga. Gender effects on cortical thickness and the influence of scaling. *Human Brain Mapping*, 27:314–324, 2006.

[78] E. Luders, P.M. Thompson, K.L. Narr, A.W. Toga, L. Jancke, and C. Gaser. A curvature-based approach to estimate local gyrification on the cortical surface. *NeuroImage*, 29:1224–1230, 2006.

[79] J.D. MacDonald, N. Kabani, D. Avis, and A.C. Evans. Automated 3-D extraction of inner and outer surfaces of cerebral cortex from MRI. *NeuroImage*, 12:340–356, 2000.

[80] S. Mallat and Z. Zhang. Matching pursuits with time-frequency dictionaries. *IEEE Transactions on Signal Processing*, 41:3397–3415, 1993.

[81] R. Mazumder and T. Hastie. Exact covariance thresholding into connected components for large-scale graphical LASSO. *The Journal of Machine Learning Research*, 13:781–794, 2012.

[82] M.I. Miller, A. Banerjee, G.E. Christensen, S.C. Joshi, N. Khaneja, U. Grenander, and L. Matejic. Statistical methods in computational anatomy. *Statistical Methods in Medical Research*, 6:267–299, 1997.

[83] M.I. Miller and A. Qiu. The emerging discipline of computational functional anatomy. *Neuroimage*, 45(1S):S16–S39, 2009.

[84] J.K. Milliken and S.D. Edland. Mixed effect models of longitudinal Alzheimer's disease data: A cautionary note. *Statist. Med.*, 19:1617–1629, 2000.

[85] G. Molenberghs and G. Verbeke. *Models for Discrete Longitudinal Data.* Springer, 2005.

[86] T.E. Nichols and A.P. Holmes. Nonparametric permutation tests for functional neuroimaging: A primer with examples. *Human Brain Mapping*, 15:1–25, 2002.

[87] M. Niethammer, M. Reuter, F. Wolter, S. Bouix, N. Peinecke, M. Koo, and M.E. Shenton. Global medical shape analysis using the Laplace-Beltrami spectrum. *Lecture Notes in Computer Science*, 4791:850, 2007.

[88] M. Ozkan, B.M. Dawant, and R.J. Maciunas. Neural-network-based segmentation of multi-modal medical images: a comparative and prospective study. *IEEE Transactions on Medical Imaging*, 12:534–544, 1993.

[89] P. Perona and J. Malik. Scale-space and edge detection using anisotropic diffusion. *IEEE Trans. Pattern Analysis and Machine Intelligence*, 12:629–639, 1990.

[90] C. Piekema, R.P.C. Kessels, R.B. Mars, K.M. Petersson, and G. Fernández. The right hippocampus participates in short-term memory maintenance of object-location associations. *NeuroImage*, 33:374–382, 2006.

[91] J.C. Pinehiro and D.M. Bates. *Mixed Effects Models in S and S-Plus.* Springer, 3rd edition, 2002.

[92] A. Qiu, D. Bitouk, and M.I. Miller. Smooth functional and structural maps on the neocortex via orthonormal bases of the Laplace-Beltrami operator. *IEEE Transactions on Medical Imaging*, 25:1296–1396, 2006.

[93] A. Qiu and M.I. Miller. Multi-structure network shape analysis via normal surface momentum maps. *NeuroImage*, 42:1430–1438, 2008.

[94] M. Reuter. Hierarchical shape segmentation and registration via topological features of Laplace-Beltrami eigenfunctions. *International Journal of Computer Vision*, 89:287–308, 2010.

[95] M. Reuter, F.-E. Wolter, M. Shenton, and M. Niethammer. Laplace-Beltrami eigenvalues and topological features of eigenfunctions for statistical shape analysis. *Computer-Aided Design*, 41:739–755, 2009.

[96] S.M. Robbins. Anatomical standardization of the human brain in Euclidean 3-space and on the cortical 2-manifold. Technical Report PhD thesis, School of Computer Science, McGill University, Montreal, Quebec, Canada, 2003.

[97] S. Rosenberg. *The Laplacian on a Riemannian Manifold.* Cambridge University Press, 1997.

[98] A. Rosenfeld and A.C. Kak. *Digital Picture Processing.* Academic Press, 1982.

[99] B.D. Rusch, H.C. Abercrombie, T.R. Oakes, S.M. Schaefer, and R.J. Davidson. Hippocampal morphometry in depressed patients and control subjects: Relations to anxiety symptoms. *Biological Psychiatry*, 50:960–964, 2001.

[100] J. Schäfer and K. Strimmer. A shrinkage approach to large-scale covariance matrix estimation and implications for functional genomics. *Statistical Applications in Genetics and Molecular Biology*, 4:32, 2005.

[101] S. Seo, M.K. Chung, and H.K. Vorperian. Heat kernel smoothing using Laplace-Beltrami eigenfunctions. In *Medical Image Computing and Computer-Assisted Intervention—MICCAI 2010*, volume 6363 of *Lecture Notes in Computer Science*, pages 505–512, 2010.

[102] J.A. Sethian. *Level Set Methods and Fast Marching Methods: Evolving Interfaces in Computational Geometry, Fluid Mechanics, Computer Vision and Material Science.* Cambridge University Press, 2002.

[103] L. Shen and M.K. Chung. Large-scale modeling of parametric surfaces using spherical harmonics. In *Third International Symposium on 3D Data Processing, Visualization and Transmission (3DPVT)*, 2006.

[104] L. Shen, J. Ford, F. Makedon, and A. Saykin. Surface-based approach for classification of 3d neuroanatomical structures. *Intelligent Data Analysis*, 8:519–542, 2004.

[105] L. Shen, A. Saykin, M.K. Chung, H. Huang, J. Ford, F. Makedon, T.L. McHugh, and C.H. Rhodes. Morphometric analysis of genetic variation in hippocampal shape in mild cognitive impairment: Role of an il-6 promoter polymorphism. In *Life Science Society Computational Systems Bioinformatics Conference*, 2006.

[106] Y. Shi, I. Dinov, and A. W. Toga. Cortical shape analysis in the Laplace-Beltrami feature space. In *12th International Conference on Medical Image Computing and Computer Assisted Intervention (MICCAI 2009)*, volume 5762 of *Lecture Notes in Computer Science (LNCS)*, pages 208–215, 2009.

[107] Y. Shi, R. Lai, K. Kern, N. Sicotte, I. Dinov, and A. W. Toga. Harmonic surface mapping with Laplace-Beltrami eigenmaps. In *11th International Conference on Medical Image Computing and Computer Assisted Intervention (MICCAI 2008)*, volume 5242 of *Lecture Notes in Computer Science (LNCS)*, pages 147–154, 2008.

[108] J.G. Sled, A.P. Zijdenbos, and A.C. Evans. A nonparametric method for automatic correction of intensity nonuniformity in MRI data. *IEEE Transactions on Medical Imaging*, 17:87–97, 1988.

[109] M. Styner, I. Oguz, S. Xu, C. Brechbuhler, D. Pantazis, J. Levitt, M. Shenton, and G. Gerig. Framework for the statistical shape analysis of brain structures using SPHARM-PDM. In *Insight Journal, Special Edition on the Open Science Workshop at MICCAI*, 2006.

[110] J. Sun, M. Ovsjanikov, and L. J. Guibas. A concise and provably informative multi-scale signature based on heat diffusion. *Comput. Graph. Forum*, 28:1383–1392, 2009.

[111] T. Tasdizen, R. Whitaker, P. Burchard, and S. Osher. Geometric surface smoothing via anisotropic diffusion of normals. In *Geometric Modeling and Processing*, pages 687–693, 2006.

[112] J.E. Taylor and K.J. Worsley. Random fields of multivariate test statistics, with applications to shape analysis. *Annals of Statistics*, 36:1–27, 2008.

[113] P.M. Thompson, J.N. Giedd, R.P. Woods, D. MacDonald, A.C. Evans, and A.W Toga. Growth patterns in the developing human brain detected using continuum-mechanical tensor mapping. *Nature*, 404:190–193, 2000.

[114] P.M. Thompson, D. MacDonald, M.S. Mega, C.J. Holmes, A.C. Evans, and A.W Toga. Detection and mapping of abnormal brain structure with a probabilistic atlas of cortical surfaces. *Journal of Computer Assisted Tomography*, 21:567–581, 1997.

[115] P.M. Thompson and A.W. Toga. A surface-based technique for warping 3-dimensional images of the brain. *IEEE Transactions on Medical Imaging*, 15:1–16, 1996.

[116] B. Timsari and R. Leahy. An optimization method for creating semi-isometric flat maps of the cerebral cortex. In *The Proceedings of SPIE, Medical Imaging*, 2000.

[117] M. Vaillant, A. Qiu, J. Glaunès, and M.I. Miller. Diffeomorphic metric surface mapping in subregion of the superior temporal gyrus. *NeuroImage*, 34:1149–1159, 2007.

[118] G. Wahba. *Spline Models for Observational Data*. SIAM, New York, 1990.

[119] F.-Y. Wang. Sharp explicit lower bounds of heat kernels. *Annals of Probability*, 24:1995–2006, 1997.

[120] Y. Wang, Y. Song, P. Rajagopalan, T. An, K. Liu, Y.Y. Chou, B. Gutman, A.W. Toga, P.M. Thompson, and the Alzheimer's Disease Neuroimaging Initiative. Surface-based TBM boosts power to detect disease effects on the brain: An N= 804 ADNI study. *NeuroImage*, 56:1993–2010, 2011.

[121] K.J. Worsley. Local maxima and the expected Euler characteristic of excursion sets of χ^2, f and t fields. *Advances in Applied Probability*, 26:13–42, 1994.

[122] K.J. Worsley, M. Andermann, T. Koulis, D. MacDonald, and A.C. Evans. Detecting changes in nonisotropic images. *Human Brain Mapping*, 8:98–101, 1999.

[123] K.J. Worsley, S. Marrett, P. Neelin, A.C. Vandal, K.J. Friston, and A.C. Evans. A unified statistical approach for determining significant signals in images of cerebral activation. *Human Brain Mapping*, 4:58–73, 1996.

[124] K.J. Worsley, J.E. Taylor, F. Carbonell, M.K. Chung, E. Duerden, B. Bernhardt, O. Lyttelton, M. Boucher, and A.C. Evans. SurfStat: A MATLAB toolbox for the statistical analysis of univariate and multivariate surface and volumetric data using linear mixed effects models and random field theory. *NeuroImage*, 47:S102, 2009.

[125] K.J. Worsley, J.E. Taylor, F. Tomaiuolo, and J. Lerch. Unified univariate and multivariate random field theory. *NeuroImage*, 23:S189–195, 2004.

[126] A.M. Yaglom. *Correlation Theory of Stationary and Related Random Functions Vol. I: Basic Results*. Springer-Verlag, 1987.

[127] X. Yang, A. Goh, and A. Qiu. Locally linear diffeomorphic metric embedding (LLDME) for surface-based anatomical shape modeling. *NeuroImage*, 56:149–161, 2011.

[128] H. Zhang, O. van Kaick, and R. Dyer. Spectral methods for mesh processing and analysis. In *EUROGRAPHICS*, pages 1–22, 2007.

[129] H. Zhang, O. van Kaick, and R. Dyer. Spectral mesh processing. *Computer Graphics Forum*, 29:1865–1894, 2010.

[130] H. Zhu, J.G. Ibrahim, N. Tang, D.B. Rowe, X. Hao, R. Bansal, and B.S. Peterson. A statistical analysis of brain morphology using wild bootstrapping. *IEEE Transactions on Medical Imaging*, 26:954–966, 2007.

10

Neuroimage Preprocessing

Stephen C. Strother

Rotman Research Institute, Baycrest and Biophysics Department, University of Toronto

Nathan Churchill

St. Michael's Hospital, Toronto, Ontario, Canada

CONTENTS

10.1	Introduction	264
10.2	Principles for Studying and Optimizing Preprocessing Pipelines	267
	10.2.1 The Utility of Simulated Datasets	268
	10.2.2 Quantifying the Impact of Preprocessing Changes	268
	10.2.3 The Neuroscientific Importance of Preprocessing Choices	269
10.3	Metrics for Evaluating Neuroimaging Pipelines	270
	10.3.1 Pseudo-ROC Curves	271
	10.3.2 Cluster Overlap Metrics	271
	10.3.3 Intra-Class Correlation Coefficient	272
	10.3.4 Spatial Pattern Reproducibility Using Correlations	275
	10.3.5 Similarity Metric Ranking Approaches	275
	10.3.6 Prediction Metrics	276
	10.3.7 Combined Prediction versus Spatial Reproducibility Metrics	276
10.4	Preprocessing Pipeline Testing in the Literature	277
	10.4.1 Between-Subject, MRI Brain Registration	277
	10.4.2 Preprocessing for fMRI Resting-State Analysis	278
	10.4.3 Preprocessing for fMRI Task-Based Analysis	280
10.5	A Case Study: Optimizing fMRI Task Preprocessing	280
	10.5.1 Optimization with Prediction and Reproducibility Metrics	280
	10.5.2 An Individual Subject's (P, R) Curves	284
	10.5.3 fMRI Preprocessing Pipeline Optimization with (P, R) Curves	285
	10.5.3.1 Some fMRI Datasets	285
	10.5.3.2 Selecting Optimal Preprocessing Pipeline Steps	286
	10.5.4 Fixed and Individually Optimized Preprocessing Pipelines	289
	10.5.5 Independent Tests of Pipeline Optimization Results	292
10.6	Discussion of Open Problems and Pitfalls	293
	Bibliography	295

10.1   Introduction

In neuroimaging, preprocessing typically refers to the set of operations or steps following image data acquisition, which involve image and signal processing procedures designed to denoise, clean and normalize data for subsequent analysis, thereby improving the quality of results (e.g., for details with functional Magnetic Resonance Imaging (fMRI) see (147) and Chapter 6). As shown in Figure 10.1, neuroimaging studies can be thought of as an experimental pipeline with decisions made at multiple stages, from subject selection through data analysis, including the evaluation of pipeline efficacy using performance metrics. The preprocessing of neuroimaging data occurs at stage 4 within this pipeline; moreover, this stage can be considered a "sub-pipeline", in which multiple preprocessing decisions are made.

Both the overall experimental pipeline (stages 1–5), and the set of choices comprising the preprocessing sub-pipeline (i.e., stage 4) represent complex neuroimaging workflows. These may be managed within a growing number of neuroinformatics tools for manipulating neuroimaging workflows, which will only be peripherally addressed in this chapter, but are discussed in more detail in (158, 120, 60). The subsequent stages 6–7 of Figure 10.1 represent a quantitative evaluation framework for testing and optimizing earlier stages, including preprocessing pipeline choices. Stage 6 reflects the fact that neuroimaging experiments tend to occur as discrete scanning sessions linked to a single subject, and stages 1-5 may be optimized per scan session/subject. Stage 7 reflects the goal of many neuroimaging experiments to produce robust, generalizable group results from multiple scanning sessions, either for within-subject scans (e.g., test-retest changes in intervention studies), between-subject scans, or both. Preprocessing for multi-session analysis in stage 7 is often based on using a fixed set of stage 4 preprocessing choices for all scan sessions, which we will demonstrate may not be optimal for fMRI later in this chapter (e.g., (30, 35)).

Choices in any one or more of the multiple cascaded stages of Figure 10.1 may significantly interact with each other, and the steps chosen for a preprocessing pipeline at stage 4. Therefore, the performance of any particular preprocessing pipeline studied in isolation from the other experimental stages should only be generalised with great caution. Unfortunately, much of the neuroimaging preprocessing literature to date has focused on developments driven by an even more limited viewpoint than considering stage 4 in isolation, while implicitly hoping for generalisable results. Typically, new methodological preprocessing approaches have been introduced for only one or perhaps several preprocessing steps per paper. These reports are often accompanied by limited test data that demonstrates that the new approach is "better than" one or more of the existing approaches, and the chosen performance metrics are rarely consistent across studies. In addition, there is often no attempt made to address interactions with other preprocessing steps or the impact of different metrics, let alone the other stages of the overall experimental pipeline. For example, the impact of young versus old or disease cohorts (stage 1), different experimental designs such as block or single event stimuli for functional neuroimaging (stage 2), differences in data acquisition techniques such as MRI acquisition pulse sequences (stage 3), and different analysis model choices such as univariate versus multivariate approaches (stage 5).

Such research, although limited to one or a few preprocessing step choices, is important to understand the properties of particular algorithms and their software implementations. However, ignoring interactions with other preprocessing steps and/or the surrounding experimental stages has made it very difficult to know if reported methodological improvements in the literature are likely to generalise to overall improvements in preprocessing and/or experimental pipeline performance. We believe such generalisability must be explicitly tested

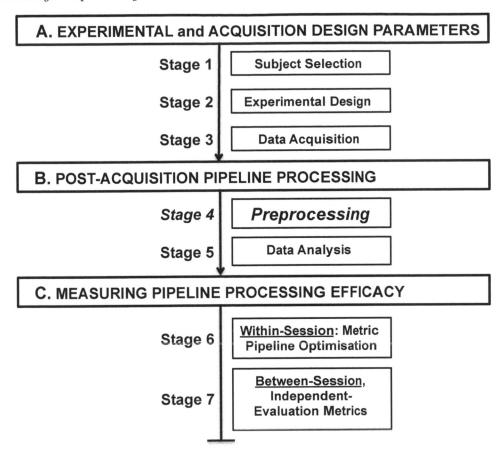

FIGURE 10.1
The multi-stage experimental pipeline for neuroimaging studies (see text for details).

both within stage 4 and for interaction effects with the other experimental stages in Figure 10.1. Such research focused on understanding the generalisable impact of what are often very complex neuroimaging pipeline hierarchies and how they may be optimised for particular neuroscientific questions is in its infancy. It seems likely that this research field would benefit significantly from insights provided by an influx of researchers from the broader statistical commmunity. In this view, we agree with Keith Worsley that neuroimaging may be statistics' "agricultural field trials of the 21st century" (155). In addition, preprocessing pipelines and their overall experimental interactions are a somewhat overlooked aspect of these 21st-century field trials in the literature to date.

The five neuroimaging modalities covered in this book (i.e., PET: Chapter 2, MRI: Chapter 3, DT1: Chapter 4, fMRI: Chapter 6, EEG: Chapter 7) involve preprocessing pipelines of varying levels of complexity. The most complex experimental and preprocessing pipelines are produced by modalities with relatively high sampling rates in both space and time, together with within-session, time-dependent experimental designs (step 2 of Figure 1). Of these five modalities, this includes explicit experimental designs for fMRI, EEG, and some dynamic PET studies. Furthermore, during the last decade even more complex, multi-modality experimental and preprocessing pipelines have been implemented for simultaneous EEG-MEG, fMRI-EEG and fMRI/MRI-PET experiments. There is a growing literature using multi-modal pipelines as illustrated for EEG-fMRI in Figure 10.2 (70, 116). Note that

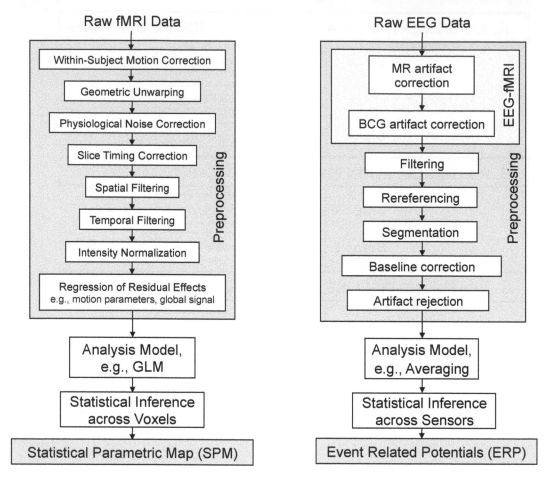

FIGURE 10.2
Examples of multi-modal preprocessing steps for fMRI (left) and simultaneous EEG-fMRI (right) experiments (see text for details).

the EEG pipeline here is specific for active tasks leading to event-related potentials (ERPs), and would be changed for recording spontaneous resting-state datasets. This Figure shows the complex, multiple image and signal processing steps that exist for these modalities within the preprocessing stage 4 of Figure 10.1. Furthermore, a simultaneous fMRI-EEG experiment ideally requires these two preprocessing pipelines to be jointly optimized, a critical piece of which is noted in the first two steps of the EEG pipeline. These steps reflect removal of EEG artifacts caused by magnetic gradient switching and small movements of the EEG electrodes due to ballistocardiographic (BCG) effects in the strong fixed magnetic field of the MRI scanner (70). To our knowledge, no group has attempted to jointly optimize such a multi-modal neuroimaging experiment.

Due to the complexity of modality-specific preprocessing pipelines, it will not be possible to address even the major individual details of preprocessing pipelines for the functional modalities reviewed in this book. Please refer to the chapters on the specific neuroimaging techniques for individual preprocessing approaches and the resulting modality-specific pipelines. Our goal in this chapter is to introduce and discuss issues that we believe are important for all modalities when evaluating preprocessing pipelines, and to illustrate these with case studies, particularly from fMRI, the modality where we have performed most of our own research on preprocessing pipeline optimization.

10.2 Principles for Studying and Optimizing Preprocessing Pipelines

Our approach to studying and optimizing preprocessing pipelines is guided by three general principles:

1. Simulated datasets, while potentially useful, provide only a rough guide for optimizing preprocessing pipelines, particularly for functional neuroimaging studies.

2. Seemingly small changes within a preprocessing pipeline may lead to large changes in the output.

3. New insights into human brain function may be obscured by poor or limited choices in the preprocessing pipeline particularly as a function of age and disease.

We discuss these principles in the context of the overall experimental stages in Figure 10.1, and a more detailed description of steps within stage 4, drawing on both the recent literature and our own research. We primarily focus on fMRI with occasional references to the other functional modalities because most of the limited literature exploring the generalizable impact of neuroimaging pipeline choices has focused on these neuroimaging modalities. This situation may be rapidly changing as neuroimaging researchers start to recognize the limitations of only using default settings in neuroimaging software packages, such as AFNI (2), FSL (75), SPM (144), FreeSurfer (52), EEG-Lab (48), etc. In addition, all functional neuroimaging techniques tend to rely on structural MRI scans to provide an anatomical template with a labeled map of the brain, making many neuroimaging pipeline results somewhat dependent on MRI image processing. For a description of the dependence of fMRI studies on MRI see (147). Such anatomical alignment and labeling tools are being continually refined as outlined in Chapters 4 and 5 and illustrated in the evaluation studies of Klein and colleagues (82, 84, 83) (see below).

Principle 1 stems from our belief that we generally do not have a sufficiently comprehensive understanding of the generative signal and noise models that produce neuroimaging datasets, particularly for functional data as a function of experimental tasks, age and/or disease.

Principle 2 may be tested using purely statistical criteria, and is therefore more directly accessible for statisticians to address without the need to manipulate stages 1–3 in Figure 10.1. However, much work remains to be done in studying and understanding the statistical properties of pipeline choices, particularly their bias–variance tradeoffs in highly ill-posed neuroimaging samples. For example, in fMRI and MRI, the number observations (e.g., N scanning sessions) across time (T time samples per session), $NT \ll$ number of variables (e.g., P spatial brain voxels), but for EEG/ERP, typically the number of spatial electrode locations $\ll NT$.

Principle 3 reminds us that satisfying purely mathematical/statistical criteria is necessary and important, but it does not imply an optimal or even necessarily useful neuroscientific or clinical outcome for the experiment.

10.2.1   The Utility of Simulated Datasets

Simulations in which truly activated signal voxels are generated with a known model allow different pipeline choices to be ranked from plots of true positive rates (TPR) versus false positive rates (FPR) using standard signal detection metrics based on receiver operating characteristic (ROC) curves (143, 153). However, while useful, simulated datasets often do not reflect all aspects of real datasets. Therefore they produce absolute TPR vs. FPR optimization results, which are biased, potentially leading users to select pipelines that produce results conforming to the assumptions of the simulation model. For a quantitative example of the limitations of simulated models, see (5) in which the information content of a simple simulation is shown to be much less than that of real data, although a comparison of different analysis techniques yielded a similar ranking (in terms of extracted information) for both real and simulated data. Hence, we believe performance measurement using real-world datasets should currently be given preference to focusing on ROC measurements from simulations, although the two may often be complementary and provide a more compelling research result. Disagreements between simulated and experimental performance may also help to better understand where simulations fail to model important properties of brain function. However, simulation studies may be improved in the future, as significant progress is being made in building more accurate generative models for fMRI (e.g., (40, 73)), PET (74), and EEG and MEG (39, 173). In addition, new data simulators have recently been developed that reflect the complex nonlinear dynamics of the underlying neuronal signals (42, 127), together with sophisticated simulations of the fMRI acquisition process (45, 46). For an extensive, but incomplete review of fMRI simulation studies see (167).

One related approach for studying the impact of preprocessing pipeline choices is to use hybrid datasets, where known simulated signals are superimposed on real datasets. Using either simulations alone or hybrid fMRI simulations, a series of studies have supported our second principle that "seemingly small changes within a preprocessing pipeline may lead to large effects on the output." They have shown that significant differences in signal detection performance should be expected for different preprocessing choices (e.g., (56, 140, 35)) and data analysis approaches (e.g., (92, 96, 97, 7, 92, 4, 31)). Moreover, (174) showed that ROC curve results for multiple machine learning analysis models applied to an extension of the simulations of (96) including simple BOLD signal parameters may reflect expected performance in real fMRI data. Finally, (9) introduced a more sophisticated evaluation approach based on comparing parametric bootstrap results from simulations with parameters estimated from real data to nonparametric distributions of the real data. However, Bellec's work was developed on data acquired with a 1.5T MRI, and to our knowledge has yet to be applied to data from a 3T MRI, which will have a significantly different signal and noise structure for the same experimental task (69, 160). All of these developments may help to significantly boost the role of simulated datasets in the study of preprocessing pipelines in the future.

10.2.2   Quantifying the Impact of Preprocessing Changes

Our second general principle that "seemingly small changes within a preprocessing pipeline may lead to large effects on the output" suggests that researchers should identify the best approaches and algorithms for each of the preprocessing steps in Figure 10.2 and then use fixed pipelines containing these steps. The knowledge generated by such a greedy search approach may then be tabulated as best-practice guidelines as was done for ERP measurement with EEG in (116). Many researchers appear to implicitly believe that major software packages reflect such good-enough, if not exactly best-practice, guidelines. While clearly a valuable approach to establish baseline standards in a field, this viewpoint assumes that such a fixed set of algorithms and software implementations exists for an established set of preprocessing steps. The neuroimaging literature shows that this may be far from the case, a situation that has recently been surveyed for fMRI in (24). For such a standardized pipeline to exist it would be necessary to show that it either applied broadly across interactions with many choices in stages 1–3 and 5–7 of Figure 10.1, or that separate standard pipelines exist for distinct subsets of these interacting choices. For example, should the same fixed pipeline be used for all subjects of a relatively homogeneous cohort performing a single functional task, or even across test-retest scan sessions within a single subject? Our own evidence for fMRI preprocessing pipelines indicates that such fixed pipelines may be far from optimal (30, 35). Furthermore, there are typically multiple algorithmic approaches and software implementations possible for each step of the preprocessing pipelines in Figure 10.2 as clearly seen on the Neuroimaging Informatics Tools and Resources Clearinghouse (NITRC) website (http://www.nitrc.org/). In neuroimaging there are a dizzying array of preprocessing and analysis algorithms and software options to choose from. We believe a minimal requirement for implementing fixed pipeline standards is a quantitative demonstration that there are sets of different algorithmic approaches and their software implementations that have equivalent performance, and some level of generality. Given the complexity of the space of choices outlined in stage 4 of Figure 10.1, and the interactions with other stages being explored in the literature, new approaches are required if we are to make substantial progress in optimizing preprocessing pipelines in the next decade.

This second principle may be evaluated using purely statistical measures. For example, there is growing quantitative evidence using multiple performance metrics that output statistical parametric maps (SPMs) are far from robust to a wide range of preprocessing choices (156, 147, 25, 35, 12). In contrast, studies examining a range of multi-stage interactions for cortical thickness measurements with MRI involving stages 3, 4 and 5 from Figure 10.1 have suggested that the outputs are somewhat robust to some changes in pipeline steps (63). Similar studies have been undertaken for fMRI task experiments across scanning sites (17), and such multi-site effects in clinical research studies are an increasingly important area within which to understand the impact of different preprocessing and other pipeline choices. Unfortunately, studying such effects has become more complicated in recent years. In 2006 (147) reviewed the large number of preprocessing possibilities for MRI and fMRI and some of the known interactions with the other experimental stages. Since then, the relevant literature and related software available on NITRC and other online sources has dramatically increased. Yet studies that attempt to analyze the impact of a range of either the overall experimental stages or a comprehensive set of preprocessing and analysis choices remain rare, largely due to the difficultly of performing them.

10.2.3 The Neuroscientific Importance of Preprocessing Choices

The third general principle that "new insights into human brain function may be obscured by poor choices in the preprocessing pipeline particularly as a function of age and disease" follows from the second. If small changes in the preprocessing pipeline lead to large changes in the output, then it is likely that some, and possibly many, such choices obscure or bias the picture of human brain function and/or disease that emerges from a given experiment. In particular, these changes may significantly interact with the population being studied, e.g., as a function of age and/or disease. However, this principle critically relies on the definition of pipeline performance metrics (i.e., in stages 6 and 7 of Figure 10.1) used to assess output changes as a function of pipeline choices. Ideally, the chosen metrics should act as quantitative proxies for the likelihood of new insights into human brain function, although this is typically a difficult connection to make. While recent results addressing the second principle show that the multivariate distribution of SPM patterns significantly varies across preprocessing pipeline choices in fMRI, this tells us nothing about related insights into human brain function. The SPMs are significantly different but which ones are better? This is a much harder problem to address, and is linked to the idea of what it means to produce useful scientific information in the context of a specific neuroscientific question or set of questions. One possibility here is to directly measure information content of output SPMs as proposed by (5), but it is not at all clear in what sense this is neuroscientifically better than metrics of pattern overlap, or intra-class correlation coefficients, which we review below.

If we are focusing on neuroscience questions linked to age- and disease-dependent brain function, we could choose performance metrics based on our ability to predict such age-dependent or disease effects. For those of us focused on statistical modeling, such choices should be made in collaboration with neuroscientists who have the required domain knowledge. For example, (55) showed that the extent to which the variability of fMRI predicts subjects' ages when comparing young versus older normal controls is significantly influenced by the chosen preprocessing pipeline. Similarly, (124) noted that if fMRI is to provide a better endophenotype than behavior for measuring the penetrance of genes into the underlying neural mechanisms, it must be relatively state-independent and have high test-retest reliability. This study found that while group fMRI results could be highly reliable, the signal-to-noise (SNR) levels of a subject's test-retest results could be quite unreliable, suggesting that in 2007 fMRI was not very suitable as a phenotypic measure for genetic studies. Reference (118) recently extended this work to show that the specific task domain may have a large impact on resulting levels of reliability in BOLD measurements. In particular, specific emotional, motivational and cognitive tasks were all found to have excellent between-subject, group-level reliability (i.e., intraclass correlation coefficients, ICC) "of 0.89–0.98 at the whole brain level and 0.66–0.97 within target ROIs." However, within-subject test-retest reliability was only fair to good for the motivational and cognitive tasks (ICC = 0.44–0.62) and very poor for the emotional task (ICC = −0.02–0.16).

Given one or more metrics that may be linked to Principles 2 and 3, it is then possible to optimize the choice of preprocessing pipeline steps in the hope that this will lead to new insights into human brain function. We believe it is important to recognize that while the underlying pipelines generated may be dependent on the metrics chosen, we do not necessarily require stable sets of pipeline steps, but rather seek optimization metrics that are consistently linked to important aspects of brain function. No single metric is likely to fulfill all or even most of the goals that may be derived from Principles 1 and 2. We believe that this represents another largely unexplored statistical research area in which metrics, or more likely ensembles of metrics, need to be developed and tested to establish viable quantitative proxies for improved neuroimaging insights into human brain function and clinical utility. Such metric ensembles need to be developed and studied as a function of age, and used to assess the suitability of different neuroimaging techniques to provide improved biomarkers for assessing clinical treatments and disease diagnosis.

10.3 Metrics for Evaluating Neuroimaging Pipelines

A wide variety of possible data-driven performance metrics have been proposed for the evaluation of neuroimaging pipelines. In the following section we summarise earlier discussions of such metrics from (5, 149) in light of more recent work. See (68) for a general discussion on measuring reliability in medical informatics studies. Many researchers have focused on metrics measuring spatial pattern reliability and agreement between independent repetitions of a neuroimaging experiment, and the link to signal detection and related SNRs. This is partly based on the need for reliable results noted above, and the recognition that smaller inferential p values when testing a null hypothesis do not necessarily imply a stronger likelihood of getting the same result in another replication of the same experiment (37, 113). Such p values are effectively training-set estimates that will be upwardly biased if used to optimize preprocessing pipelines; this leads to pipelines that are over-fitted to idiosyncrasies of the training data, reducing the generalizability of results to new data; some of the issues of such "circular analysis" are discussed in (88). Furthermore, replication has long been considered a fundamental criterion for a result to be considered as supporting strong scientific inference (117, 37, 76), making it a central focus of metric development for evaluating neuroimaging pipelines.

10.3.1 Pseudo-ROC Curves

One set of metrics derives from purely data-driven attempts to apply pseudo-ROC approaches without ground truth being available. Reference (58, 101) estimated pseudo-ROC curves and a reliability map from SPMs of repeated trials of an fMRI experiment based on a multinomial model of individual voxels. Reference (95) extended this empirical-ROC generation framework with a Bayesian technique for selecting the optimal operating point on the resulting ROC curve. An alternative procedure for generating empirical ROC curves that required a control state run to estimate false-positive rates together with a standard experimental task run was proposed by (111). Reference (28) estimated pseudo-ROC curves of multiple linear regression analysis applied to the datasets of stroke patients and healthy participants. The pseudo-curve approach was also implemented by obtaining a test statistic and thresholding it at different levels, and (99) reformulated the problem to eliminate this requirement of threshold choice. Finally, (148, 90) proposed a threshold-free, pseudo-ROC approach using combined prediction and reproducibility metrics that will be described in more detail below.

Spatial signal detection and associated values clearly provide a useful quantitative measure for Principle 2, but their link to Principle 3 is less clear since they are typically directly tied to a SNR definition associated with a particular analysis model, e.g., t-tests at individual voxels defined by a general linear model (16). If this model does not capture critical elements of brain structure (e.g., the off-diagonal voxel–voxel covariance terms needed to detect and track brain networks), then maximizing its signal detection power is unlikely to generally improve neuroscientific insights. It is well established in the literature that different models provide different ROC and associated SNR performance levels in given fMRI datasets (92). An important future research task is to understand the limitations of such model-specific approaches. Much of the fMRI neuroimaging literature continues to focus on optimizing GLM signal detection and reliability (79, 123, 12) without addressing potential limitations of this modeling approach when trying to understand brain function, and disease. However, one area in which ROC curves may directly reflect Principle 3 is when they are defined by true- and false-positive tradeoffs for biomarkers of disease detection and treatment response as we discuss below in Section 10.3.6.

10.3.2 Cluster Overlap Metrics

An alternative to signal detection and ROC curves is to measure cluster overlap of voxel-label maps, produced by thresholding an SPM and labeling voxels as active or inactive, making the results potentially nonlinear functions of thresholding levels (47). This complicates pipeline assessment, by requiring the selection of an appropriate statistical threshold for assigning voxels as active or inactive. For a discussion of statistical thresholding approaches for SPMs Chapter 6 and 13. Using highly averaged results as ground truth (93) performed a reliability assessment of binary maps using the kappa statistic (36). Other measures of overlap of spatial activation labels are the Jaccard coefficient (JC) (72) and the Dice coefficient (DC) (44). The Dice coefficient, which is asymptotically related to the kappa statistic (179), has been widely used for evaluation in functional neuroimaging (124, 129). The DC tends to produce a larger measure of overlap than the JC, with $DC = \frac{2JC}{(JC+1)}$, but JC is preferable as outlined below; for example, it directly measures the fractional overlap of SPMs (i.e., the ratio of intersection/union of thresholded voxels) (137). Let v_j and v_l represent the number of SPM image voxels identified as significantly activated in jth and lth replications of a fMRI experiment, respectively, with $v_{j,l}$ the number in both the jth and the lth replications. Then the Jaccard coefficient may be defined as

$$JC = \frac{v_{j,l}}{v_j + v_l - v_{j,l}} \tag{10.1}$$

Reference (100) notes that JC can be viewed as the conditional probability that a voxel is identified as activated in both the jth and the lth replications, given that it is identified as activated in at least one of the two replications. DC does not offer such a set-theoretic interpretation and also suffers from the undesirable property of aliasing (i.e., different input values may result in very similar DC values). In contrast, the JC is the best among a range of similarity indices in the context of comprehensively measuring social stability (131) while its complement from unity is a true distance metric (94). However, the JC is a pairwise reliability measure so that we get $\binom{M}{2}$ overlap measures for all pairs from M studies, which cannot be obviously combined into a single measure of activation reliability. Maitra addresses this and develops a description of the $\binom{M}{2}$ overlap measures, using a spectral decomposition of the similarity matrix of JC measures together with a testing strategy for identifying outliers. To our knowledge, this approach has not been used in any pipeline studies to date. Note that the pairwise overlap metrics DC and JC are most appropriate when replicated fMRI data are acquired on the same subject under the same experimental condition or for multiple subjects under the same experimental conditions. Spatial overlap metrics provide a clear quantitative measure for Principle 2, and their link to Principle 3 is perhaps stronger than for pseudo-ROC curves. The need for optimizing SPM overlap seems strongest in test-retest studies where a claim about spatial location in one scan session should be generalizable to a repeated scanning session, i.e., location reliability. Furthermore, spatial reliability is needed to support the use of regional spatial location and naming standards with which between-subject, group neuroimaging results are typically interpreted (180, 49). However, while a great deal of modern neuroimaging research has focused on between-subject reliability of structural neuroimaging (e.g., MRI registration and region naming (83)) it is less clear that this optimization should be as generally applied to between-subject reliability of fMRI SPMs, which may not reflect underlying individual subjects' activation patterns, e.g., (104, 105, 151). This is an area of much needed research since the extent to which nonoptimal preprocessing pipelines contribute to group heterogeneity has only just started to be explored.

10.3.3 Intra-Class Correlation Coefficient

In contrast, another widely used measure of neuroimaging reliability, the intra-class correlation coefficient (ICC) (11, 139, 20, 124, 165, 54), does not require thresholding, and is used for the more stringent requirement to match voxel signal levels rather than just their thresholded spatial locations. For an overview of ICC and related measures of reliability in fMRI together with a comprehensive summary of the related functional literature over the last 2 decades, see (11). Reference (11) provides four reasons to use reliability metrics: "1) highly unreliable results are generally considered to have little scientific value, 2) within-subject reliability is becoming important for clinical and diagnostic applications such as localization of language function for surgical treatment of intractable epilepsy, and measurement of before and after treatment effects, 3) highly reliable results are required to support attempts to use fMRI evidence, e.g., in the courtroom, and 4) to support data sharing of reliable results amongst researchers."

Reference (139) defines ICC as the correlation of a measurement on a "target" (a subject's scan in neuroimaging), and another "judge's" measurement on that target, i.e., a repeated within-subject scan. Repeated measurements on targets are performed by "judges" under a variety of sampling assumptions leading to one- and two-way variance decompositions. ICC is often equivalent to a variance ratio of the variance of interest between subjects' scans, σ_s^2, divided by σ_s^2 plus other variance components so that it lies between 0 and 1. Let $Y_{i,j}$ represent brain-voxel activity for the jth repeat measurement ($j = 1, \ldots, k$) on the ith subject ($i = 1, \ldots, n$), then consider the case of a two-way random-effects model with

$$Y_{i,j} = \mu + s_i + r_j + (sr)_{i,j} + e_{i,j} \tag{10.2}$$

where μ is a fixed effect and s, r, (sr) and e are random effects with zero mean and variances, σ_s^2, σ_r^2, σ_{sr}^2, and σ_e^2. Under these assumptions, the covariance of within-subject, test-retest scans is given by σ_s^2 since all the other covariance cross terms cancel or are assumed to sum to 0, giving a so-called $ICC(2,1)$ (139), which may be calculated as a normalized covariance given by

$$ICC(2,1) = \frac{\sigma_s^2}{\sigma_s^2 + \sigma_r^2 + \sigma_{sr}^2 + \sigma_e^2}. \tag{10.3}$$

Here $ICC(2,1)$ takes values near 1 when $\sigma_s^2 \gg$ any combination of σ_r^2, σ_{sr}^2, and σ_e^2, and near 0 when any combination of σ_r^2, σ_{sr}^2 and $\sigma_e^2 \gg \sigma_s^2$.

Within-subject repeat scan variance, interaction effects between subjects, and repeat scans and image noise should all be low for the between-subject measurement patterns to be considered as being highly reliable. $ICC(2,1)$ may be consistently estimated by

$$ICC(2,1) = \frac{BMS - EMS}{BMS + (k-1)EMS + k(RMS - EMS)/n} \tag{10.4}$$

where BMS is the between-subjects' mean square, EMS is the error mean square, and RMS is the repeated measures mean square. However, ICC exists in a variety of forms as defined by (139), and using different versions on the same sampling data may lead to quite different results (109).

$ICC(3,1)$ is the other key form for neuroimaging, which satisfies Eqn. (10.2), but treats repeated measurements as fixed rather than random effects. It is calculated as a normalized covariance given by

$$ICC(3,1) = \frac{\sigma_s^2 - \sigma_{sr}^2/(k-1)}{\sigma_s^2 + \sigma_{sr}^2 + \sigma_e^2}, \tag{10.5}$$

which may be consistently estimated by

$$ICC(3,1) = \frac{BMS - EMS}{BMS + (k-1)EMS}. \tag{10.6}$$

The interaction term between targets=subjects and (multiple judges)=(repeat scans) is included and discussed in (139), but has been consistently assumed to be zero, without clear justification, in related derivations in the functional neuroimaging literature (181, 165).

Reference (124) focused on test-retest fMRI results for a single subject calling $ICC(2,1)$

with $k{=}2$ and $n{=}1$, ICC_{within}, a measure of generalizable measurement agreement since it includes repeated measures as random effects. In contrast, when assessing the consistency of test-retest measurements across subjects, they argued that $ICC(3,1)$, called $ICC_{between}$ for $n >1$ was more appropriate for assessing endophenotypes, by treating systematic differences between repeat scans as fixed effects. This removes σ_r^2 from consideration in the denominator, generally leading to larger absolute ICC values. Similarly, (20, 12) have argued for the use of $ICC(3,1)$ in consistency studies of single-site fMRI datasets. Moreover, (130, 181) have suggested the use of $ICC(3,1)$ to remove possible systematic test-retest effects due to learning effects and/or fatigue in assessing the reliability of resting-state studies where independent component analysis is used to generate proxy patterns of possible brain networks. Reference (169) followed Zuo and used $ICC(3,1)$ to show that only a small number of extracted intrinsic connectivity networks (ICN) "demonstrated at least fair within-subject reliability at both the voxel- and network-level ($ICC \geq 0.40$)," their values being in rough agreement with the survey of (11). In addition, they found that "group-level reproducibility did not predict within-subject reliability, and that voxels with the highest connectivity strength within an ICN did not predict those with the highest reliability." Recently, the reliability of other regional measures such as regional homogeneity measured with ReHo (183, 175) have been examined together with the impact of a range of preprocessing and experimental conditions. For example, (12) used $ICC(3,1)$ to explore the impact of multiple stages of the experimental pipeline as outlined in Figure 10.1 on task-based fMRI studies. Regardless of which ICC measures are used, all studies that focus on multiple spatial locations (e.g., single voxels or volumes of interest) have the problem of summarizing the consensus spatial distribution of ICCs, which (11) has shown can be highly variable. This issue of which ICC is far from resolved, since (54) argued for a random-effects model to estimate between-site reliability of an fMRI experiment using $ICC(2,1)$, and (165) calculated the within- and between-subject $ICC(2,1)$ for some regions of interest and compared them. ICC is also being more widely used to assess multi-site reliability of structural MRI studies. Reference (22) shows that using state-of-the-art processing pipelines across 8 sites with 3T scanners produces between- and within-site ICCs of >0.90 for many structural regions. Finally, multiple authors have measured and reported both $ICC(2,1)$ and $ICC(3,1)$ in their studies to test the underlying model assumptions, e.g., (139, 118).

Recently a variant of ICC, the image intra-class correlation ($I2C2$) coefficient, has been proposed (138). It is designed to explicitly generalize ICC to deal with high-dimensional image vectors and addresses the case of $ICC(1,1)$ where σ_r^2, σ_{sr}^2, and σ_e^2 are all treated as a single error term in a one-way ANOVA, which takes the form

$$Y_{i,j}(v) = S_i(v) + E_{i,j}(v), \qquad (10.7)$$

where v is a voxel and Y, S and E are all Vx1 neuroimages with $S_i(v)$ the true images, which are independent across subjects, and $E_{i,j}(v)$ the error images, which are assumed independent of S and across subjects and replicates. Replacing the σ_s^2, and σ_e^2 variances with the traces of covariances, K_s and K_e, leads to a definition of a multivariate $ICC(1,1)$ as

$$I2C2 = \frac{trace(K_S)}{trace(K_S) + trace(K_E)}. \qquad (10.8)$$

Reference (138) describes efficient software tools for estimation, which are available online, and demonstrates that this definition is relatively robust to a misspecified model when the data follows a structure for which $ICC(2,1)$ or $ICC(3,1)$ might be more appropriate.

As for the other quantitative metrics outlined above ICC and $I2C2$ clearly provide useful measures for Principle 2, and possibly a more direct link than pseudoROC and cluster overlap measures to Principle 3. As argued in (124), for experiments that wish to treat the between-subject variation in neuroimaging measurements as true subject-dependent traits, it will always be better to have higher rather than lower values of these metrics, making this a potentially useful optimization target for preprocessing pipelines.

10.3.4 Spatial Pattern Reproducibility Using Correlations

Using unthresholded SPM patterns, (150, 156) proposed measurement of the Pearson product-moment correlation (R) between paired voxel values of SPMs from pairs of repeated, independent datasets, and others have suggested using Spearman or Concordance measures of spatial similarity (92, 169). Such an unthresholded measure may or may not agree with the above thresholded overlap metrics or ICC. For example, we have shown that with fMRI pipeline optimization for individual subjects at a fixed spatial smoothing scale, the average between-subject value of R significantly drops while that of the Jaccard overlap for activation patterns thresholded at a false discovery rate (FDR) of 0.05 rises significantly (30). This appears consistent with (123)'s observation that spatial location reliability is easier to attain than signal-level reliability, potentially due to global SPM effects.

Reference (11) argues against using Pearson correlations and in favor of ICC as a spatial similarity metric, while (124) has pointed out that at least for within-subject, test-retest comparisons of SPMs, R and $ICC(2,1)$ are very similar. This may be seen with reference to Figure 10.3a. For two independent, standard-normalized SPMs with mean 0 and $\sigma = 1$, the major signal axis of their voxel-by-voxel scatter plot has variance $(1+R)$ with variance of the uncorrelated minor noise axis given by $(1-R)$ (148). Following Eqn. (10.3) for $ICC(2,1)$ the shared variance of the two normalized SPM patterns is given by $\sigma_s^2 = (1+R)-(1-R) = 2R$, and the total variance by, $\sigma_s^2+\sigma_r^2+\sigma_e^2 = (1+R)+(1-R) = 2$. Therefore, $ICC(2,1)$ of two normalized, within-subject SPMs is equal to R, the Pearson correlation of the voxel-by-voxel scatter plot. Consequently, the difference between R and $ICC(2,1)$ in a comparison of two independent SPMs appears to depend on the impact of Gaussian normalization across space on the SPM's voxel signal levels. Reference (123) examined the scatter plot structure of unnormalized, test-retest SPMs for multiple tasks in terms of spatial pattern changes and global effects. For GLM t-values across multiple brain regions and tasks, they found primarily small-scale spatial pattern changes (e.g., due to partial voluming), and significant amplitude changes that appeared due to global effects. Overall, there does not seem to be a clear case for choosing ICC over R for within-subject, test-retest comparisons. If there are >2 within-subject scanning sessions, ICC may provide a more natural way to incorporate the multiple sessions, although this needs to be compared with the spectral decomposition approach of (100). Related multivariate approaches such as multi-dimensional scaling (MDS) (14, 15) as a visualization technique that provides a 2D representation of the relative distance between SPMs (or other outputs) may also be useful here.

10.3.5 Similarity Metric Ranking Approaches

There are other evaluation approaches that rank the pipelines and analysis techniques under evaluation, based on their overall agreement with other pipelines. Williams index (168) is an evaluation metric that uses a user-defined similarity measure (e.g. JC, DC, kappa, ICC, etc.) to compare label maps or continuous maps provided by the pipelines. The pipelines are ranked by their degrees of similarity to other methods. The method more similar to the other methods gets the best rank. The 2D representation of the relative distance between SPMs for MDS can use any of the common distance/agreement measures mentioned above although the use of the RV coefficient to compare matrix similarity (128, 80) in the 3-way extension of MDS, called DISTATIS, may have some advantages. (1) STAPLE is an evaluation technique developed by (164), initially for MRI measures of the relative quality of multi-label maps provided by the analysis techniques under evaluation and a reference standard through an Expectation Maximization (EM) framework. The methods are ranked by their similarity to the estimated reference. In order to calculate the reference, STAPLE, as for other techniques above, assumes, different analysis techniques make independent errors (false positive and false negative errors), which is not necessarily true. Generally, in the evaluation techniques that assess the consistency of outputs if a method outperforms other methods and detects a region of brain that other methods do not, this method is likely not to be rated as the best. Therefore, an evaluation measure that provides an individual metric for each method might be more useful. We refer readers to (15) for a comprehensive comparison of William's index, STAPLE and MDS on evaluation of several common methods of brain tissue segmentation, and a discussion of the pros and cons of metrics based on common agreement.

Most of the evaluation techniques listed above attempt to quantify common agreement between the SPMs or label maps calculated from independent datasets of the same task. The processing method that generates more reproducible SPMs also generates a larger area under its pseudo-ROC curve or a larger reproducibility/reliability metric using the test-retest approach indicating its superior performance. However, these metrics assume that the analysis techniques are making only random, independent errors on independent fMRI datasets, and therefore they will be unable to detect a consistent, pipeline-dependent bias. For example, a method that always declares some specific voxels as targets with fixed signal levels independent of the input data will have maximum reliability but very low accuracy or high bias. While clearly an extreme example, all models will have some, possibly spatially dependent bias when estimating statistical parameters in finite samples. Therefore, we have argued that for optimizing neuroimaging pipelines, such spatial similarity metrics should be combined with another metric that is not subject to the same bias (see below).

10.3.6 Prediction Metrics

There is another category of evaluation methods based on the prediction (P) or generalization error of a model for labeled scans grouped into experimental conditions/classes reflecting stable brain states or disease (145, 79, 90, 91, 108, 89, 146, 149) and unlabeled scans using probability density functions (PDF, (64)) or other coherent structure (3, 157). Such predictive techniques have come to be widely known as "mind reading" or multivariate pattern analysis (MVPA) in the fMRI literature (106, 115, 86). These evaluation methods use cross-validation subsampling to divide a dataset into independent training and test sets, fitting a model to the training set and testing its ability to predict either class labels or data structure in the test set. The prediction error mainly depends on the quality of the training and testing datasets, and how well the model being tested can predict the class labels or PDF. These measures may potentially reflect the quality of preprocessing approaches that remove artifacts and noise from the datasets (29, 146, 178). Reference (5) proposes a related subsampling approach by assessing the ability of spatial SPM patterns in training data to predict spatio-temporal structure in preprocessed, high-dimensional functional data, using a modified mutual information metric. Prediction has an advantage over the other metrics in that it may be the desired neuroscientific outcome measure when developing diagnostic biomarkers for disease, and treatment effects or cognitive states, providing an effective merger of a quantitative metric for Principle 2 with the desired goal for Principle 3.

10.3.7 Combined Prediction versus Spatial Reproducibility Metrics

The NPAIRS method (81, 148) combines two of the above metrics using both correlation-based spatial reproducibility and subsampled prediction or mutual information for the quantitative evaluation of neuroimage analysis techniques. NPAIRS relies on simultaneously measuring spatial pattern reproducibility of SPMs and their temporal prediction or mutual information of scan labels. Using PET scans, Kjems and Strother argued that temporal versus spatial metric curves can capture the bias–variance tradeoffs of an analysis model, and this was then extended to fMRI pipelines (90, 149, 30). This approach has been extended further to the comparison of non-linear BOLD hemodynamic models estimated in a Bayesian framework using Markov chain Monte Carlo techniques, with a Kullback–Leibler measure to estimate reproducibility (73). We have focused on this framework in our research because it incorporates the class of problems directly addressed by using a prediction metric (i.e., diagnostics), while simultaneously measuring the quality of the associated spatial pattern for purposes of visualization and interpretation (125). In addition, when within-subject pattern similarity is measured with the Pearson correlation coefficient (R), it is closely linked to the desirable goal of optimizing $ICC(2, 1)$. As a case study of pipeline optimization and metric choice in fMRI, in Section 10.5 below we focus on using the NPAIRS framework, with split-half subsampled prediction (P) and spatial reproducibility (R) metrics. We focus on using P and R to optimize within-subject, within-scanning-session pattern reliability and examine the impact of this within-session optimization on between-subject SPM pattern traits such as spatial location and behavioral predictions.

But first, we briefly review several other published MRI and fMRI studies that have performed large-scale examinations of interactions between experimental stages and/or steps within preprocessing pipelines. These studies provide examples of the ways in which the literature currently addresses the three general principles outlined above using a range of performance metrics. We emphasize that while the number of such studies is increasing, they remain relatively rare, and an understanding of the critical choices and interactions for preprocessing optimization pipelines in general, while perhaps best developed in fMRI and MRI, is in its infancy in all areas of neuroimaging.

10.4 Preprocessing Pipeline Testing in the Literature

10.4.1 Between-Subject, MRI Brain Registration

In several seminal papers, Klein et al. (82, 84) studied the use of MRI to establish a common spatial reference frame across subjects' brains to compare positions and sizes of brain regions. This was the first literature study of such size and scope in MRI neuroimaging that evaluated interactions between various experimental stages in Figure 10.1. This included the impact of 14 different non-linear deformation algorithms used to register subjects' brains to a common template space. The papers focused on group analysis of multiple brains with stages 5–7 of Figure 10.1 (i.e., analysis and optimization metrics) replaced by manual brain labeling. This provides a silver standard for evaluating metrics defined between subjects' labeled brains at stage 8. The 14 algorithms were evaluated using four different datasets and labeling protocols (i.e., interactions with stages 1 and 3) on which 8 different overlap metrics and 4 analysis methods were tested at stage 8. Their overlap metrics included Dice and Jaccard coefficients and related variants. While they controlled preprocessing per dataset including brain extraction by skull-stripping, and file formatting to allow all the algorithms to be consistently run on each dataset, they acknowledge the potential unknown impact of the limited set of preprocessing choices used. For example, it is unclear what the optimal preprocessing steps are for a particular algorithm with respect to skull stripping or not, interpolation approaches or not, bias-field corrected or uncorrected, intensity normalized or non-normalized, etc. Nevertheless they concluded "the relative performances of the registration methods under comparison appeared to be little affected by the choice of subject population, labeling protocol, and type of overlap measure." Furthermore, the best performing algorithms generalized across three different metric analysis approaches, i.e., permutation tests, one-way ANOVA tests, and indifference-zone ranking (82). In their subsequent study, they showed that particular processing pipeline steps produced superior overlap metrics for manually labeled brain regions across different brains: "1. de-skulling aids volume registration methods; 2. custom-made optimal average templates improve registration over direct pairwise registration." Due to measurement challenges, they were unable to show a clear advantage of either volume-based or surface-based nonlinear brain alignment (84).

As noted in Section 10.2.2 other recent work shows the high reliability possible for multiple structural, regional measurements from multi-site, 3T MRI studies; $ICC > 0.90$ for many regions (22). When combined with Klein's results, there appears to be converging evidence that structural MRI may be quite suitable for measuring stable neural traits in large multi-site studies, provided careful attention is paid to scanner stability and the types of processing pipelines employed. A far different picture emerges from the existing fMRI literature.

10.4.2 Preprocessing for fMRI Resting-State Analysis

Resting-state fMRI (rs-fMRI) is a functional neuroimaging approach in which brain function is measured in the absence of any explicit cognitive task. Rs-fMRI is rapidly growing in popularity due to the ease of implementation, however it carries some unique challenges in data preprocessing. Unlike task-based neuroimaging, there is no available information about the brain state at different time points. Therefore, when analyzing the data, the goal is not to characterize a specific task-related SPM, but rather the repertoire of brain states, which the brain explores while at rest. This is done by identifying brain regions whose BOLD signal tends to co-vary in time, either by performing linear decomposition into spatial and temporal components (e.g. PCA or ICA), or by measuring the functional connectivity between brain regions (e.g. seed-based connectivity studies, clustering and network analyses Chapter 14).

This has significant consequences when attempting to evaluate preprocessing pipelines, as there is no single SPM on which to assess the reliability of results, nor are there condition/class labels on which to measure prediction accuracy. In light of these challenges, studies of rs-fMRI preprocessing often advocate for a conservative approach that minimizes any potential artifacts. An example of this issue is in the correction of residual motion effects. Even after performing rigid-body motion correction, higher-motion subjects tend to have distinct differences in functional connectivity values, which may be attributable to residual motion artifacts, including decreased long-range connections and reduced network modularity (133, 161). This is typically controlled by regressing out head motion parameters as nuisance covariates, which has been demonstrated to have a small (57) but potentially positive impact on the detection of brain networks (166, 6). However, some studies have advocated more extensive correction approaches, including expanding the GLM-based regression model into derivatives and quadratic terms (134), with the goal of maximizing explained variance and removing any connectivity differences between high- and low-motion groups. Alternatively, (122) proposed that researchers discard volumes with abnormally high motion prior to analysis, although this may lead to discontinuities in the BOLD time series (26). Intriguingly, it has recently been shown that there may be a neurobiological basis causing certain subjects to have higher motion (85, 177), suggesting that it may be more difficult to disentangle motion artifacts from the BOLD signal than previously suspected.

Another prominent area of research is in the "global signal" effect present in resting-state fMRI; this is a low-frequency BOLD modulation that occurs throughout most/all of the brain, particularly in densely vascularized grey matter, leading to strong, uniformly positive correlations in connectivity analysis, but potentially impairing the ability to localize specific functional networks, and anti-correlations between brain regions. It is unknown whether it has a primarily physiological or neuronal basis (13, 136). Although global signal regression improves spatial specificity of connectivity analyses (166), it has been long established that this procedure creates potentially spurious anti-correlations between brain regions (51, 107, 77). However, alternative approaches have demonstrated a physiological component of the signal that can be specifically removed (27); moreover, (107) in PET and (23) in fMRI showed that the global response can be dissociated from other components of brain signal using PCA, allowing researchers to isolate it in this manner.

A handful of papers have also examined interactions between preprocessing steps in rs-fMRI, with a focus on the relative ordering of preprocessing algorithms. Reference (78) established that performing motion correction, followed by physiological correction and slice-timing correction, provides the optimal ordering that minimizes temporal variance in BOLD data, which they assumed to be a desirable outcome. Reference (7) used simulation studies with verifiable ground truth to demonstrate that discarding high-motion outliers (as proposed in (122)) should be performed prior to other temporal preprocessing steps. Finally, (62) demonstrated that bandpass filtering of BOLD data (used to remove high-frequency physiological effects), should not be performed prior to regression of nuisance covariates, as they will otherwise re-introduce high-frequency artifacts into fMRI data, although (166) found that the relative ordering of these steps had a marginal effect on functional connectivity measurements. Reference (166, 6) also examine the effects of adding multiple pipeline steps, including bandpass filtering, the removal of outlier volumes, global signal regression, and motion parameter regression. They both concluded that more extensive preprocessing tended to improve the specificity of detected functional networks.

Finally, a number of groups have sought to exploit the tendency for some ICA components to express features that are reliably associated with artifact, using classifiers that are manually trained to identify these features, e.g., (159, 41). Most recently, these approaches have been significantly extended for resting-state fMRI datasets using state-of-the-art high-speed fMRI acquisition sequences with a much larger range of artifactual features and types of classifiers (132, 61). Critically, all these approaches rely on the time-consuming (and potentially user-dependent) need to manually build a generalizable training set of artifactual features, which may be used with independent test data to classify ICA components as containing predominantly signal or artifact. In addition, in the task-setting ICA component denoising is not necessarily optimal (79, 35), and it remains to be seen if it is surpassed by other techniques for rs-fMRI, e.g. (3).

As shown in the preceding paragraphs, studies of preprocessing pipelines in rs-fMRI do not have any agreed-upon metric measures of data quality. A number of studies (6, 27, 166) have evaluated pipelines based on their ability to increase connections within expected brain networks, while minimizing connections outside of these regions. However, this presupposes knowledge of the true network structure in the brain, limiting its validity, and precluding researchers from using these pipelines to identify novel connectivity relationships. Other studies (135) have focused on selecting pipelines that minimize between-group differences. However, this approach must be used with care, to avoid removing the very between-group differences in BOLD response that we seek to detect. Similarly, (78) considers reduced temporal variance to be an important preprocessing goal, whereas other studies have considered variance, under certain constraints, to be a strong correlate of brain function (176, 55). Moreover, variance ratios in the form of ICC metrics have recently being used to test network extraction in resting-state studies, e.g., (181, 169). As an alternative, there has been some work in establishing quantitative metrics of prediction and reliability for functional resting-state networks, e.g., (10, 157), although they are currently focused on optimizing clustering models. While all of these methods offer promising solutions to pipeline development and selection, there remains much work to be done in validating these measures before routinely applying them for resting-state pipeline optimization. Nevertheless, there is much ongoing research, e.g., focused on the extent to which rs-fMRI may be able to match MRI as a biomarker for large multi-site studies of functional brain traits because of the relative simplicity of resting-state compared to task-based fMRI experiments.

10.4.3   Preprocessing for fMRI Task-Based Analysis

For a fixed preprocessing pipeline using $ICC(3,1)$, Reference (12) demonstrated significant effects on reliability of multiple experimental factors including experimental task design, type of cognitive task, contrast type and test-retest interval. In addition, they found significant interactions of task and design, and design and contrast. As a caveat, they note that $ICC(3,1)$ includes both within- and between-subject variability, and they are therefore not able to pinpoint the primary impact of their experimental manipulations. In our view, this reflects one of the limitations of relying purely on ICC measures to understand and optimize preprocessing pipeline choices.

Recently, (24) explicitly outlined the variability of fMRI results driven by a range of choices in the preprocessing pipeline and analysis steps. As others have demonstrated during the last 15 years (e.g., (30, 35, 90, 91, 67, 121)), Carp establishes that small changes within a preprocessing pipeline and analysis steps may lead to large effects on the output. He suggests that this leads to a much higher risk of inflated false positives than has been generally appreciated in the field. This occurs despite the extensive work on controlling multiple comparisons in SPMs of brain activations output at step 5 of Figure 10.1 (e.g., (112); Chapter 13). Carp's arguments draw strongly on the work of John Ioannidis with respect to the typically low power of many neuroscience experiments (19), and the selective reporting of the most significant test results (71). These issues have been repeatedly raised across the history of neuroimaging (162, 163), but remain a significant and unresolved issue for functional neuroimaging. Carp suggests that problems stemming from the influence of wide ranges of pipeline choices being used in relatively low-powered experiments may be mitigated by constraining the flexibility of analytic choices or by abstaining from selective analysis reporting.

We agree with the need to eliminate selective pipeline and analysis reporting and similar highly biased experimental approaches such as double dipping (88, 87). To start to address this problem, we strongly endorse the associated need for comprehensive and systematic reporting of neuroimaging pipeline steps called for in (119, 25), coupled with software and data sharing efforts, which allow other investigators to directly test results in the literature (120). This would allow the variability across meta-analysis of multiple studies with their individual preprocessing pipelines to be more clearly understood and potentially attributed to more than small sample sizes and failure to appropriately correct for multiple comparisons (163).

10.5   A Case Study: Optimizing fMRI Task Preprocessing

10.5.1   Optimization with Prediction and Reproducibility Metrics

The two data-driven metrics we use to optimize pipeline choices automatically are spatial reproducibility (R, Figure 10.3A) and temporal prediction accuracy (P, Figure 10.3B) es-

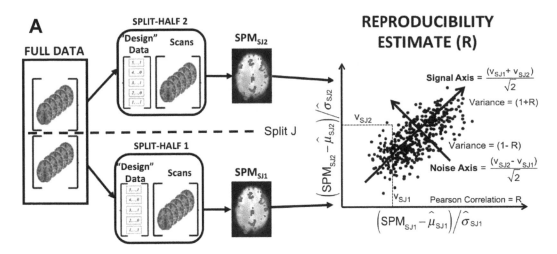

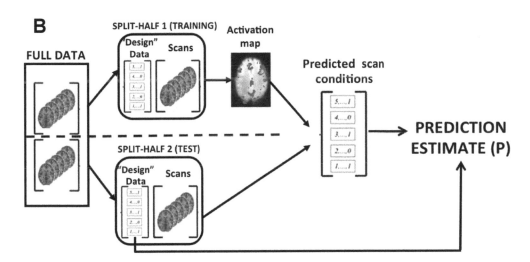

FIGURE 10.3

NPAIRS subsampling procedure producing training and test, split-half datasets from which are produced (A) spatial reproducibility metric, R, between split-half statistical parametric maps (SPM), and (B) a prediction metric, P, of the accuracy with which the training set SPM values are able to correctly predict the test set's scan condition/class labels.

timated in the NPAIRS split-half, subsampling framework (148, 151). Below we discuss the details of these metrics and the importance of jointly optimizing P and R using (P, R), pseudo-receiver operating characteristic plots, based on the supplementary material outlined in (35).

Reproducibility estimates how consistent extracted SPM voxel values are for a given data analysis pipeline, across (pseudo-)independent datasets. For this measure, we compute an SPM independently on each of two split-half datasets (SJ1, SJ2) from the dataset of interest as illustrated in Figure 10.3A. If split-half subsampling occurs across separate runs within a subject or between subjects, then it can be treated as truly independent. However, if splitting occurs within a single subject's fMRI run, the pseudo-independence qualifier is most appropriate even though all preprocessing and analysis occurs separately in the two split-half datasets, e.g., the temporal detrending order of Legendre polynomials is separately optimized in each split. All our results below reflect such within-run, pseudo-independent splits. Following splitting and independent application of preprocessing and analysis steps on each split-half dataset, we produce a scatter plot, where each point represents a single voxel's paired signal values from normalized SPMs $((v_{SJ1}, v_{SJ2})$, Figure 10.3A). The Pearson correlation of these two SPM vectors provides the global measure of SPM reproducibility (R), which is monotonically related to the global signal-to-noise ratio $(gSNR)$ of the image (146), a form of global effect size,

$$gSNR = \sqrt{\frac{2R}{1-R}}. \tag{10.9}$$

In addition, as noted in Section 10.3.4 R measures $ICC(2,1)$ for within subject test-retest splits. We may also use this splitting framework to obtain a conservative, reproducible, Z-scored SPM that maximizes the estimated global signal. If each split-half SPM has spatial mean zero $(\hat{\mu}_{SJ1} = 0)$ and values normalized by their global spatial standard deviation $(\hat{\sigma}_{SJ1})$, the major axis (1st component of a PCA on the scatterplot, accounting for most variance, i.e., $(1+R)$) is along the line of identity. This is the signal axis (see Figure 10.3A), and we project pairwise voxel values onto this line, in order to obtain the most-reproducible signal pattern. For SPMs in vector form X_1 and X_2, the transformed vector X_{sig} is

$$X_{sig} = [X_1|X_2][\frac{1}{\sqrt{2}}, \frac{1}{\sqrt{2}}]. \tag{10.10}$$

The minor axis (i.e., 2nd component of a PCA on the scatter-plot), which is orthogonal to the signal axis, forms the noise axis. The projection onto this axis measures the non-reproducible signal at each voxel giving noise vector X_{noi}:

$$X_{noi} = [X_1|X_2][-\frac{1}{\sqrt{2}}, \frac{1}{\sqrt{2}}]. \tag{10.11}$$

The standard deviation of X_{noi} provides a global estimate of SPM noise, which we use to normalize X_{sig}, to obtain a reproducible, Z-scored SPM $(rSPM(Z))$ from a single data split as

$$rSPM(Z) = \frac{X_{sig}}{std(X_{noi})}. \tag{10.12}$$

Note that when many subsampled data splits are available, an estimate of each voxel's variance may be calculated as the mean voxel variance across splits using Eqn. (10.11) (e.g., (33)). These methods provide our measure of model reproducibility, as well as a robust $rSPM(Z)$ that maximizes the reproducible spatial signal across split-halves of the dataset of interest. This measure is robust to dataset heterogeneity as the influence of localized outlying spatial variance values down-weights the results across more unequal split halves from heterogeneous datasets through division by $std(X_{noi})$ in Eqn. (10.12).

Prediction quantifies how well a model generalizes to independent results. It requires that we build a classification model on a training dataset, and then measure its ability to correctly classify scans in independent test data as shown for split-half datasets in Figure 10.3B. After building the training classifier with model parameters θ, we estimate the conditional probability $p(C_k|X; \theta)$ of correct assignment to class C_k with $k = 1 \ldots K$, for a test datapoint X. In this framework, we attempt to predict cognitive state (class C_k) based on brain measures (fMRI image vector X; a mind-reading model), defined by the Bayes posterior probability rule

$$p(C_k|X; \theta) = \frac{p(X|C_k; \theta)p(C_k)}{p(X)}. \tag{10.13}$$

This is the generalized form of our prediction measurement. We replace unknown, fixed prior probability $p(X)$ with a normalizing constant, to ensure that the sum across posterior class probabilities is unity

$$p(C_k|X;\theta) = \frac{p(X|C_k;\theta)p(C_k)}{\sum_{i=1}^{K} p(X|C_i;\theta)p(C_i)}. \qquad (10.14)$$

Prior class probability $p(C_k)$ is set as the proportion of scans in C_k. The conditional probability, $p(X|C_k;\theta)$ is dependent on the chosen analysis model. For our current results, we tested both multivariate and univariate classification models: 1) a regularized Canonical Variates Analysis (CVA, which is equivalent to a Linear Discriminant Analysis on an optimized Principal Component subspace (89, 148, 151)), and 2) a Gaussian Naive Bayes (GNB) model applied independently at every voxel (81, 106, 178).

A regularized CVA provides a linear transformation matrix L_{train} into the training discriminant subspace, where L_{train} is normalized so that training variance in this subspace is unity. In this case, the conditional likelihood of X_{test} for known class C_k reduces to the multivariate normal function:

$$p(X_{test}|C_k;\theta) = \frac{1}{\sqrt{2\pi}} exp\left(-\frac{1}{2}||L_{train}^T(X_{test} - \bar{X}_{train}^k)||^2\right) \qquad (10.15)$$

where $\bar{X}_{train}^k$ is the training mean of data from class C_k. For details of this classifier, refer to (90, 102, 89), and for related models used in an NPAIRS (P, R) framework, see (125, 151). Note that the use of split-half subsampling with multivariate prediction models in highly ill-posed datasets (i.e., NT $\ll$ V) is strongly supported by the recent statistical literature, showing that it helps to stabilize estimates of discriminant variable weights (e.g., SPMs) while bounding false positive rates (103).

GNB is a predictive GLM, which allows us to measure classification accuracy on independent data. It is one of the most widely used predictive models in the fMRI literature (115). GNB uses a training dataset to fit a Gaussian distribution on the BOLD signal in each experimental condition, performed independently for every brain voxel. Classification is then performed on a test dataset. For every test-data brain volume, we measure the probability that the training model will assign the volume to its true experimental condition, measured as the joint conditional likelihood over all brain voxels x_v $(v = 1...V)$, denoted:

$$p(X|C_k;\theta) = \prod_{v=1}^{V} (2\pi\sigma_{train}^2)^2 exp\left(-\frac{(x_v - \bar{x}_{train})^2}{2\sigma_{train}^2}\right) \qquad (10.16)$$

where $\bar{x}_{train}$ is the training mean, and σ_{train}^2 is the training variance of the vth voxel, for data obtained from class C_k.

Using Eqn. (10.14) with Eqns. (10.15) or (10.16) allows us to estimate the posterior probability $p(C_k|X_{test};\theta)$ for each class and scan, and a scan is assigned to the class with the largest $p(C_k|X_{test};\theta)$. The mean $p(C_k|X_{test};\theta)$ is computed over all scans in a given test split. Alternatively, for direct comparison with classifier models that do not have a probabilistic formulation, such as the widely used support vector machines (SVM) (91), we may simply replace likelihood $p(C_k|X_{test};\theta)$ with the percentage of scans allocated to this class. Then we compute the median of these mean values (across all test split-halves if there is more than one), as a measure robust to outlying splits, to produce a single prediction measure, P.

Although optimizing model prediction and reproducibility are goals for any neuroscientific experiment (76), it may be difficult if not impossible to simultaneously optimize both metrics in neuroimaging datasets with NT $\ll$ V. This appears due to prediction and reproducibility capturing important tradeoffs in model parameterization, and thus it may be undesirable to strictly optimize one metric alone, such as prediction. Models driven by R optimization tend to have more stable SPM patterns, but are often less sensitive to the stimulus-coupled brain response (i.e., poor generalizability) (34). As a simple illustration of this issue, an analysis model that ignores the data input and outputs a fixed SPM pattern will be perfectly reproducible ($R = 1$), but has no ability to predict brain states. By contrast, models driven by P may be highly predictive of the stimulus condition, but tend to extract unstable (less-reproducible) discriminant patterns (29, 172). As an example, a discriminant model that is driven by a small number of the most task-coupled brain voxels may be highly predictive of class structure ($P \approx 1$), but will have low reproducibility, as all other voxel values will tend to be more variable across splits. Standard analysis models and experimental data rarely provide such extreme results but lie somewhere in between. However, it has been previously shown that the choice of optimization criteria significantly alters results, potentially identifying different or only partial brain networks, with varying signal strengths, predictive task coupling and spatial extents (149, 151, 125, 152, 34). These papers examine the structure of (P, R) curves, and joint optimization based on Euclidean distance from (P=1, R=1), $D(P,R)$. They show that $D(P,R)$ provides an effective compromise between the two possible extremes of model parameterizations focusing on finding P_{MAX} or R_{MAX} as a function of model regularization. We briefly illustrate such (P, R) tradeoffs in the next section.

10.5.2 An Individual Subject's (P, R) Curves

We use a dataset from (65), which examined face and object representation in the human ventral temporal cortex. The data was obtained from the PyMVPA web site at http://www.pymvpa.org. In the experimental paradigm, the subjects viewed gray-scale images of eight object categories (bottle, cat, chair, face, house, scissors, scrambled, shoe) while performing a one-back repetition detection task. Each of the 8 stimuli was presented separately in 8 x 24 second blocks, separated by rest periods in each experimental run. For each subject, 12 experimental runs were conducted, for a total of 864 category scans per subject. For further details see (65). Preprocessing of the fMRI time series data comprised the following steps for each subject: (1) The functional images were skull-stripped, (2) corrections were made for rigid-body movement, (3) different versions of the dataset were created by spatially smoothing with 0, 3, 6, 9, 12, 15 mm FWHM isotropic Gaussian filters, (4) the time series were linearly de-trended and standardized within each run, (5) the scans were masked with subject-specific masks (mask vt.nii) provided with the dataset (307–675 ventral temporal voxels per subject, voxel size = 3.5 x 3.75 x 3.75 mm). For the analysis we used scans from two conditions (i.e., bottle vs. face), which gave 216 scans for analysis of subject 4 for a preprocessed data matrix of 216 scans x 675 voxels.

As described in (151), based on (126), the preprocessed dataset was analysed using several different classifiers with linear decision surfaces. In Figure 10.4, we show the (P, R) results for a SVM, regularized by varying the margin parameter C across $2^{[-30:0:5:30]}$ after scaling by the average eigenvalue of the input data covariance matrix. Figure 10.4 shows the resulting (P, R) curves as a function of three Gaussian spatial smoothing filters with their minimum distances to (P=1, R=1), D0, D6, and D12. All the curves show regions of tradeoff between R and P, where an increase in R (or P) is associated with decreased P (or R), at intermediate levels of regularization. We also observe extended portions of the curves where R (or P) remain constant while P (or R) decreases as a function of reduced (or increased) regularization; this tends to occur at the ends of the curves, i.e., extreme regularization values.

Focusing only on P_{MAX} suggests a 6-mm Gaussian filter is best with little difference between 0- and 12-mm filters, while R_{MAX} indicates little difference between 6- and 12-mm

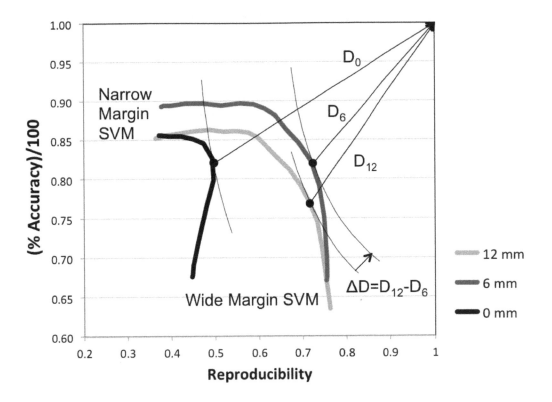

FIGURE 10.4
Prediction versus reproducibility curves as a function of regularization and spatial smoothing for a SVM classifier applied to two conditions for one subject from the dataset of (65) (see text for details).

filters. Clearly a 6-mm filter offers the best overall (P, R) across regularization settings; this curve is selected by minimizing the distance to $(P=1, R=1)$. For further discussion of choosing the point on a (P, R) curve, that minimizes the distance to $(1, 1)$ as an empirical operating point for regularization see (148, 151, 125, 90), and for comparison with P_{MAX} and R_{MAX} within subjects across task contrasts and linear classifiers, see (34). The optimization approaches and results outlined below are based on choosing models, and where applicable, their regularization parameters that minimize the Euclidean distance, $D(P, R)$, of their regularized operating point from $(1, 1)$.

10.5.3 fMRI Preprocessing Pipeline Optimization with (P, R) Curves

10.5.3.1 Some fMRI Datasets

We collected data from 27 young, healthy volunteers (15 female, ages 21–33 yrs, median 25 yrs) with retest session data from 20/27 (12 female, ages 22–33 yrs, median 25 yrs), and a test-retest interval of 2–23 months with a median of 6 months. Further details are available in (31).

For each test and retest scan session, participants performed two runs, consisting of four different tasks with task repeats, separated by approximately 10 minutes. Here we focus on preprocessing pipeline optimization for two of these four tasks: a forced-choice recognition task (REC) and the widely used clinical behavioural Trail-Making Test (TMT). For the REC task, following a 20-s instruction period, alternating task and baseline scanning blocks of 24 s were presented 4 times (run length = 212 s). During the task blocks, participants were presented every 3 s with a previously encoded Boston naming figure, side-by-side with two other previously unseen figures (semantic and perceptual foils) on a projection screen, and were asked to manipulate a cursor onto the location of the original figure, using the tablet and stylus. During control blocks, participants moved the cursor onto a fixation cross, presented at random intervals of 1–3 s. We analyzed the contrast between recall and control conditions, as a strong, robust block-design contrast. For the TMT task, following a 20-s instruction period, there were two stimulus types: TaskA, in which numbers 1–14 were pseudo-randomly displayed on a viewing screen, and TaskB, in which numbers 1–7 and letters A–G were displayed. Subjects used the tablet and stylus to draw a line connecting items in sequence (TaskA:1-2-3-4-...) or (TaskB: 1-A-2-B-...), connecting as many as possible in a 20-s block interval, while maintaining accuracy. A Control stimulus was presented after each task block, in which participants repeatedly traced a line from the center of the screen to a series of randomly presented dots. For a single run, each participant performed a 40-scan epoch of TaskA-Control-TaskB-Control twice. We analyzed the contrast between TaskB and TaskA conditions, as a relatively subtle block-design contrast of brain states (31).

Behavioral testing showed that learning and/or habituation had occurred between run1 and run2, particularly for the TMT task. Therefore we optimized preprocessing pipelines separately per run using a single split within each run. For REC, each split-half dataset within a run contained two alternating task and baseline scanning blocks for a total of 24 scans. For TMT, each split-half dataset within a run contained 10 TaskA and 10 TaskB scans for a total of 20 scans from a single 40-scan epoch of TaskA-Control-TaskB-Control. Pipeline optimization was repeated separately for each run in each test and retest scanning session, producing 4 sets of optimized pipelines and discriminant SPMs for each subject and task. For further task and experimental details, see (31).

10.5.3.2 Selecting Optimal Preprocessing Pipeline Steps

All the preprocessing steps outlined below are also described in (31) and were applied separately to each of the split-half datasets used to estimate the (P, R) performance metrics. This is required, as preprocessing steps applied across pseudo-independent task blocks of a single fMRI run (e.g. the regression of nuisance covariates) violate independence between split halves, leading to over-estimation of (P, R) metrics and the selection of over-fitted pipeline models that include all possible preprocessing steps.

The following preprocessing steps were fixed and not varied as part of the preprocessing pipeline optimization. Spatial smoothing was applied after Step 4, slice-timing correction, in the fixed order of preprocessing steps outlined below, using a 3D isotropic Gaussian kernel of 6 mm (AFNI 3d merge algorithm). The kernel size was fixed to avoid possible confounds due to its large impact on SPM reproducibility (see Figure 10.4). We then applied a subject-specific, non-neuronal tissue mask (i.e., vasculature, sinuses and ventricles) after spatial smoothing but before Step 5, temporal detrending. The PHYCAA+ algorithm (32) was used to estimate subject-specific masks, to account for inter-subject differences in vasculature. If these voxels are not excluded/down-weighted prior to analysis, they produce false-positive activations, and biased estimates of spatial reproducibility, particularly for multivariate analysis models.

We optimized our pipelines over all possible ON/OFF combinations for the fixed order of preprocessing steps outlined below. Turning each of these 8 binary step choices ON or OFF together with the 6 possible temporal detrending orders in step 5 provided 6 x 28 = 1536 possible preprocessing pipelines. Exploring different orderings is also possible but becomes much more computationally demanding.

Step 1. Motion correction (MC) [OFF/ON]: The AFNI 3dvolreg algorithm was used to transform each image to the volume with minimum estimated displacement, to correct for rigid-body head motion. Motion correction was tested in the pipeline, as its effects vary by dataset: it reduces motion artifact, particularly in children and older groups, and clinical datasets (18, 50), may produce biased results in cases of large BOLD response and relatively small head movements (53), and its effects vary with analysis model (178, 146).

Step 2. Censoring of outlier brain volumes (CENS) [OFF/ON]: We removed outlier time points caused by abrupt head motion, and replaced them by interpolating from adjacent volumes (code available at: www.nitrc.org/projects/spikecor_fmri). The censoring step is a robust alternative to typical scrubbing algorithms (122, 161), which is fully automated, and does not create discontinuities in the data (21, 31).

Step 3. Physiological correction; RETROICOR (RET) [OFF/ON]: We applied RETROICOR (59), using AFNI's 3dretroicor software. This parametric model uses external measures of respiration and heartbeat. This step is optimized, as its impact on signal detection has been shown to vary, as a function of subject and dataset (31, 35).

Step 4. Slice-timing correction (STC) [OFF/ON]: We corrected for timing offsets between axial slices due to EPI acquisition, by using AFNI's 3dTshift. This step was tested, as it remains unclear if it provides a significant benefit to signal detection in block designs (141).

Step 5. Temporal detrending (DET) [Order 0 to 5]: We regressed out low-frequency fluctuations from fMRI data, by fitting a Legendre polynomial of order N (0 to 5) in a denoising GLM, which also included steps 6 to 8. Detrending provides non-specific noise correction, including head motion, scanner drift, and physiological noise (98). Different orders were tested as the optimal order varies as a function of subject and task design (30, 31, 154).

Step 6. Motion parameter regression (MPR) [OFF/ON]: We performed PCA on the motion parameter estimates (output from 3dvolreg in step 2), and identified the 1-k PCs that accounted for >85% of motion variance. These components were regressed from the data in the denoising GLM model. This step was tested in the pipeline, as its effects vary by dataset: it controls residual motion artifact (53, 98, 50), but it may also reduce experimental power, particularly in block designs with large BOLD response and low head motion (30, 43, 114).

Step 7. Including the task design as a regressor (TASK) [OFF/ON]: We convolved the task paradigm with AFNI's standard SPMG1 HRF function (afni.nimh.nih.gov/pub/dist/doc/ program_help/3dDeconvolve.html). This regressor was included in the denoising GLM model with steps 5, 6, and 8 because when the nuisance regressors (5, 6, and 8) are correlated with the task paradigm, step 7 may protect against over-estimation of noise variance, and over-regression of task-related signal.

Step 8. Global signal regression using PCA (GSPC1) [OFF/ON]: We performed PCA on the fMRI data and regressed out the PC1 time-series, as part of the denoising GLM, because PC1 tends to be highly correlated with global signal effects (23), and residual motion artefacts (171). This approach minimizes the distortion of signal independent of global effects, unlike simple regression of mean BOLD signal (107, 77). The exact mechanism underlying global modulation remains unclear, but it may constitute physiological noise (13), neuronal response (136), or a mixture of both. The magnitude of global signal expression is subject-dependent (66, 170), indicating the importance of adaptively estimating it across subjects.

Step 9. Physiological correction; data-driven model (PHY+) [OFF/ON]: We used the multivariate data-driven PHYCAA+ model (32) (code available at: www.nitrc.org/projects/ phycaa_plus) to identify physiological noise components in the data, which were then regressed out of the fMRI data. It has been previously demonstrated that this step significantly improves the prediction and reproducibility of fMRI task analyses (32).

In order to select an optimal fixed pipeline set (i.e., a single, optimal set of the above

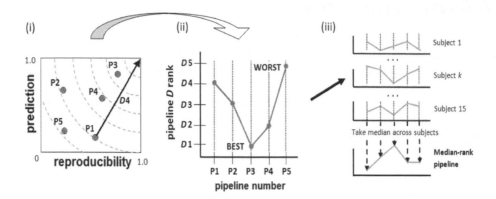

FIGURE 10.5

Steps in identifying the optimal set of fixed pipelines across all subjects (see text for details).
Reproduced with modifications from Figure 2 of (30).

preprocessing choices applied to all subjects), for a group of N subjects across K preprocessing pipelines, we apply a 2-step selection procedure: (a) for each scanning session and
task run in each subject we ranked all K tested pipelines based on $D(P,R)$ metrics, and
(b) from these sets of ranked pipelines, we identified the subset of pipelines that were not
statistically distinguishable from having the highest rankings across all subjects. Our approach identifies the fixed pipeline set that most consistently minimizes $D(P,R)$, across N
subjects. Because (P,R) distributions are heterogeneous across both subjects and pipelines,
we employ a rank-normalization procedure to test for a significant ordering in pipeline performance (30):

1. For each of the N subjects, separately rank the pipelines 1-K, by their $D(P,R)$
   value, with smaller D and higher rank indicating better pipeline performance.
   Illustrated for 5 pipelines from one subject in Figure 10.5 (i) and (ii).

2. For each pipeline k = 1 ... K, compute the median ranking across all N subjects'
   ranks, for a measure of relative performance, and identify the pipeline with highest
   median ranking, denoted PPL_{opt} (see Figure 10.5 (iii)).

3. Perform the Friedman multiple-treatment test (38), a non-parametric analogue
   to repeated-measures ANOVA. This procedure tests for consistent differences in
   pipelines' D-value ranks from step 2 (treatment response) across multiple subjects
   (samples).

4. If the Friedman test identifies a significant pipeline ordering, we performed a
   multiple-comparison based on the sum of ranks for each pipeline (38, 43). This
   estimates the critical-difference (CD) interval: the range at which a difference
   between pipeline rank-sums is greater than expected, based on the standard error
   of the rank-sum distribution. Pipelines with rank-sum differences less than the
   CD interval are not significantly distinguishable.

5. We identify the set of L pipelines that are within the CD interval from the PPL_{opt}
   pipeline, at $p = 0.05$, and thus not significantly worse than PPL_{opt}.

This method is used to identify a set of L optimal fixed pipelines with 95% confidence,
for a given set of subjects (30, 35, 31). Furthermore, in addition to the median-rank pipeline,

we use the D_{MIN}-optimal pipelines identified for each subject. Thus the above procedure provides both an optimal fixed pipeline (FIX) and individually optimized pipelines (IND).

For IND optimization, we require an additional step to account for task-coupled motion, which generates artifact that is task-correlated and reproducible, and thus not controlled by optimizing (P, R) metrics. We used an automated quantitative procedure to reject pipelines corrupted with motion artifact when choosing IND pipelines; for a detailed description see Supplemental document 4 in (35). Briefly we assume that step 1, motion correction (MC) provides at least minimal compensation for motion and pipelines with MC ON will not have worse spatial motion artifacts than with MC OFF. Therefore, we identify the set of individually optimized pipelines with MC forced ON, denoted INDMC, and compute the distribution of a measure of edge artifacts across this set of subject SPMs. Subjects with original IND pipelines with MC OFF, and significantly greater edge artifacts than the set of INDMC pipelines, are forced to replace the IND pipeline with their INDMC pipeline, which is the best-performing pipeline with MC ON.

10.5.4 Fixed and Individually Optimized Preprocessing Pipelines

For each combination of pipeline steps, the preprocessed data were analyzed in the NPAIRS split-half framework described above for a single split per run. We tested pipeline optimization for the two predictive analysis models described above: univariate (GNB) and multivariate (CVA). For the GNB model applied to each scanning session and run, we tested 6 x 28 = 1536 possible preprocessing pipelines. The CVA applied to each data split was regularised by transforming each split dataset into a reduced principal component subspace, of PCs 1-k with k = 1 ... 10. Each value of k may be thought of as a different PCA denoising and regularization step, leading to a 10-fold increase in the number of preprocessing pipelines for our multivariate model, compared with each GNB pipeline. As a result, for CVA we tested 15,360 combinations of preprocessing pipelines and CVA models for each scanning session and run. We analyzed each split and preprocessing-pipeline-model combination to produce (P, R) metrics for the resulting SPMs, and recorded the optimal individual (IND) and fixed (FIX) pipeline combinations based on DMIN, and the pipelines that produced P_{MAX} and R_{MAX} per session and run. In addition, we compared these results with a conservative fixed set of preprocessing steps (CON), which included 6-mm spatial smoothing, no non-neuronal tissue masking, and steps 1–6 from Section 10.5.3.2.

Figure 10.6 illustrates the preprocessing steps selected (i.e., ON (white) and OFF (black)) for the TMT task with the CVA model for just run2 in the test scanning session. The left panel of Figure 10.6 shows the steps selected (9 possible steps labeled along left vertical edge) for the 97 pipelines that are within the CD interval from the PPL_{opt} pipeline. Statistically indistinguishable pipelines are ordered horizontally from the pipeline with the least number of steps turned on, MIN, to the most, MAX. The best performing FIX pipeline consists of the following steps turned ON: MC, 3rd-order DET, and PHY+. Note that the MIN pipeline includes no steps turned ON, and the MAX pipeline includes only 7/9 steps ON with MPR and GSPC1 turned OFF. Furthermore, the CONS pipeline cannot be statistically distinguished from FIX in this dataset based on (P, R) values. We do not claim that the SPMs from these pipelines are statistically identical, only that their median $D(P, R)$s cannot be statistically distinguished in this experiment. The statistical differences between the SPMs from these pipelines may then be further studied, for example, using the DISTATIS multidimensional scaling technique (1). We have demonstrated this in (35).

The smaller right-hand panel of Figure 10.6 shows the individually optimized pipeline steps (IND) selected for run2 of each of the 27 subjects' test scanning sessions. Optimal pipelines based on the minimum $D(P, R)$ range from 3/9 steps turned ON for the first 3 subjects on the far left—all with MC and PHY+ ON, and CENS, RET and DET >0 vary-

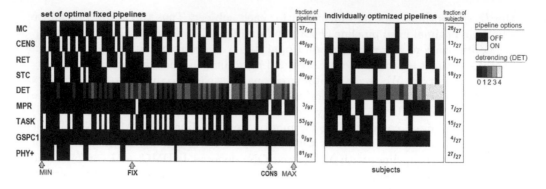

FIGURE 10.6

For 27 subjects performing the trails making task (TMT) analyzed with the CVA model we display the preprocessing steps selected ON (white) and OFF (black) for (left panel) 97 fixed group pipelines that cannot be statistically distinguished based on (prediction, reproducibility) metrics, and (right panel) 27 individually optimized pipelines (see text for details).

ing per subject-to 8/9 steps turned ON for the single subject on the far right, with only MPR OFF. Note that the detrending order also increases from left to right, varying from 0-1 to 4. This suggests that despite the additional steps being turned on across subjects from left to right, they remain unable to compensate for increasing levels of complex, low-frequency noise, resulting in higher-order detrending polynomials also being turned ON to try to compensate. This is an example of the large noise heterogeneity (both structured and unstructured) that occurs in BOLD fMRI datasets, which we show below cannot be adequately addressed using single, fixed sets of preprocessing pipeline steps across all subjects, or even runs within a subject.

In Figure 10.6 the smallest panels, labeled "fraction of pipelines" list the number of times a particular step is turned ON in the panel to the left for run2 of the test session for (1) the 97 pipelines, which are statistically indistinguishable from the $D(P,R)$ values of the FIX pipeline, and (2) each of the 27 subjects. Using these rankings we can compare the ranked importance of the various pipeline steps under the two objectives of FIX and IND pipeline optimization.

Figure 10.7 shows plots of prediction (P) versus global SNR ($gSNR$) for different preprocessing pipelines adapted from (31). Plots include both run1 and run2 from each subject's test session (2 runs x 27 subjects = 54 points) for the $CONS$ pipeline (red), the FIX pipeline (green), and three individually optimized pipelines per run, IND, for three optimization objectives (Prediction (IND–P, dark blue), $D(P,R)$ (IND–D, medium blue), Reproducibility (IND–R, light blue). Large icons show average (P, $gSNR$) coordinates for all runs and subjects per pipeline, for each experimental task and analysis model, with 1 Standard Deviation ellipses (enclosing $\approx$68% of data points). Dashed lines indicate chance (random guessing) for prediction, i.e., 0.5 for the two-class classifiers. For both tasks (left column: recognition (REC), right column, Trail-Making Test (TMT)), analysis models include (upper row: univariate Gaussian Nave Bayes (GNB), lower row: multivariate, PCA-regularized canonical variates analysis (CVA).

For all tasks and models, the $CONS$ pipeline (red) is outperformed on average by all other pipelines. The FIX pipeline (green) shows a clear improvement on average relative to $CONS$ for both P and $gSNR$, but it still underperforms relative to one or more of the three IND pipeline sets of results. The three IND pipelines on average follow the same

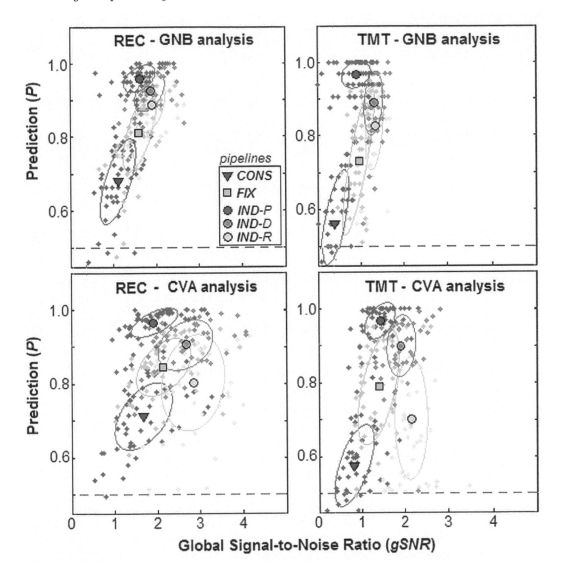

FIGURE 10.7

Plots of prediction (P) versus global SNR ($gSNR$) for 5 different preprocessing pipelines: (1) conservative ($CONS$, Red), (2) optimal fixed across all subjects (FIX, Green), (3) individually optimized (IND) for each subject's run and scanning session using optimization metrics, (3a. light blue) maximum reproducibility ($IND-R$), (3b. medium blue) minimum distance from (P=1, R=1) ($IND-D$), and (3c. dark blue) maximum prediction ($IND-P$). The left column shows the Recognition task results (REC), and the right column the Trail-Making Task results (TMT). The upper row illustrates results for a univariate, Gaussian Naive Bayes (GNB) predictive model, and the lower row for a multivariate, PCA-regularized canonical variates analysis (CVA) predictive model. See text for further details.

(P, R) tradeoff seen in Figure 10.4. As expected, on average, $IND-P$ produces higher Ps and lower Rs than $IND-D$ or $IND-R$, and $IND-R$ produces higher Rs and lower Ps than $IND-D$ and $IND-P$, and $IND-D$ lies intermediate between the other two. In particular, on average, $IND-D$ almost maintains the same $gSNR$ value as $IND-R$ while providing a clear boost in P. This behavior suggests that preprocessing steps may act as surrogate regularizers for an analysis model, by modifying the signal and noise structure of the data being modeled. This point is made most clearly by the GNB results, which have no form of explicit regularization, other than the modified sets of preprocessing steps selected to optimize $IND–P$, $IND–D$ and $IND–R$.

Finally, we note that these results are produced as part of Step 6 in Figure 10.1, and only demonstrate that the subsampling and optimization framework we have described performs as expected. On average across runs and subjects there are real, sometimes large improvements in our optimization metrics when selecting different preprocessing pipelines across different tasks and analysis models. This does not demonstrate that our (P, R) framework produces independently generalizable improvements in the extracted SPMs associated with these different pipelines since the P and $gSNR$ values reported in Figure 10.7 are training set outputs, produced by exploiting the flexibility of analytic choices rather than constraining them as proposed in (24). Such lack of constraint might be expected to produce upwardly biased results that may not generalize. However, using simulations we have shown that despite this possible bias, our framework results in real gains in effect sizes and resulting signal detection for a fixed sample (35). Furthermore, we have shown significant improvement in independent tests of activation pattern overlap of the resulting SPMs for these optimized pipeline results (35, 31). Below we examine the performance of such independent generalization tests for our (P, R) optimization framework.

10.5.5   Independent Tests of Pipeline Optimization Results

The sets of preprocessing steps chosen for each run and subject are optimized independently of each other. As a result, comparing SPMs across runs and/or subjects in Step 7 of Figure 10.1 is an independent test of the generalization performance of our framework. Therefore, we assess the generalizable reliability of outputs from our optimized pipelines across tasks and analysis models, by measuring the average Jaccard overlap of single-session and run task SPMs both between and within subjects. Underlying pattern heterogeneity and increased false negatives caused by low SNR may limit both the detection and generalization of the activation patterns that underlie the majority of fMRI studies in the literature today. If our pipeline optimization framework reinforces the consistency of active brain regions between subjects, and within subjects for test-retest scan sessions, we believe that new more generalizable insights into human brain function may be achieved.

Figure 10.8 (adapted from (31)) reflects the improvement of Jaccard activation overlap of the discriminant SPM activation patterns obtained with $IND–D$ optimized preprocessing pipelines plotted against the conservative fixed pipeline, CON, which included 6-mm spatial smoothing, non-neuronal tissue masking, and steps 1–6 from Section 10.5.3.2. The left-hand plot 8(a) summarizes the distribution of within-subject, Jaccard values for test-retest overlap between each subjects' SPMs, computed separately for test-retest tasks from each of run1 and run 2. The right-hand plot 8(b) summarizes the distribution of between-subject, Jaccard overlap values, computed as the average overlap of each subject's SPM with all other subjects' SPMs, computed separately for each of the four instances of each task from test and retest scans in run1 and 2.

For each task (REC and TMT) and both analysis models (GNB and CVA) the distribution of Jaccard overlap values is summarized with the grand mean and 1 Standard Deviation ellipses (enclosing $\approx 68\%$ of data points). For all combinations of task and analysis model, the $IND–D$ pipelines increase the overlap for the majority of runs and subjects. Increases tend to be larger for the weaker TMT contrast compared to the stronger REC contrast, the relative strengths of which are seen in their $gSNR$ values in Figure 10.7. Finally, activation pattern overlap increases are largest for the GNB model results for the weaker contrast of

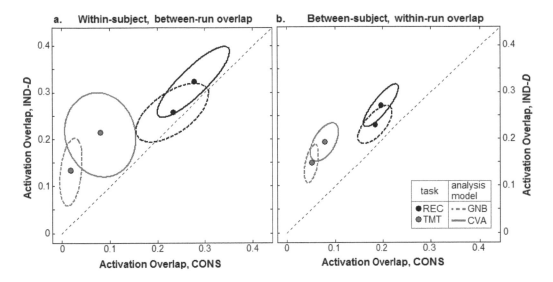

FIGURE 10.8

Plots of Jaccard activation overlap for individually optimized preprocessing pipelines using the minimum distance from $(P{=}1, R{=}1)$ $(IND\ D)$ versus the conservative pipeline defined in the text $(CONS)$ for (a) within-subject test-retest runs, and (b) between-subject overlap of run1-to-run1 and run2-to-run2. Individual session, run, and subject data points are not shown, but are summarized with 1 Standard Deviation ellipses (enclosing ≈68% of data points). Results are shown for the Recognition (REC) and Trail-Making Task (TMT) for predictive models using Gaussian Naive Bayes (GNB) and PCA-regularized canonical variates analysis (CVA). See text for further details.

TMT, i.e., the distribution ellipses for both within- and between-subject overlap are rotated towards the vertical from the 45 line of identity.

These results generalize our earlier publications to show that our framework improves independent measures of SPM activation overlap both between and within subjects across multiple tasks and data analysis models. Moreover, in (31) we show that these improved results extend to fast, single-event experimental designs and that the improved activation pattern overlap across subjects between runs also sometimes reflects greater, generalizable predictive power for simple measures of task behavior.

10.6   Discussion of Open Problems and Pitfalls

There are many open problems in the research area of neuroimaging preprocessing pipelines addressed in this chapter. This research field remains in its infancy, with much that remains unknown regarding the generalisable impact of preprocessing choices as part of what are often very complex neuroimaging pipeline hierarchies, and how they may be optimised for particular neuroscientific questions. While there seems to be a growing consensus that the field of neuroimaging should understand the impact of preprocessing pipeline choices, particularly on analysis power and result reliability, there is little agreement as to how this should be done. For example, what analysis models and other interaction effects should be considered, which optimization metrics should be used, and how should we go about assessing the impact on neuroscientific, treatment and disease-dependent problems? It is very easy to publish a paper in the neuroimaging literature that not only doesn't clearly describe the preprocessing and overall pipeline choices and methodology, but makes no attempt to understand the impact of these choices. As a result, the existing literature is only of limited utility in addressing the impact of preprocessing pipeline choices. This is partly hampered by a lack of agreed-upon guidelines for reporting and testing of pipelines, although literature (119) and standardisation initiatives within the Organisation for Human Brain Mapping (http://www.humanbrainmapping.org/files/2016/COBIDASreport.pdf) will improve this situation. We hope that the growing interest and urgent need to understand and improve the reliability of neuroimaging results, together with making the associated data and software readily available for independent testing, will change this situation during the next decade. As this change occurs, it seems likely that this research field will significantly benefit from insights provided by an influx of researchers from the broader statistical commmunity, particularly working in collaboration with relevant neuroscience and medical domain experts.

While we have outlined a number of possible optimization metrics for this effort, many other possibilities exist. This represents a largely unexplored statistical research area in neuroimaging in which metrics, or more likely ensembles of metrics, need to be developed and tested to establish viable quantitative proxies for optimizing preprocessing choices, which are linked to improved neuroimaging insights into human brain function. Such metric ensembles need to be developed and studied as a function of age and disease, and to evaluate the suitability of different neuroimaging techniques in providing improved biomarkers for assessing clinical treatments and disease diagnosis. We believe ROC measures will continue to be important for simulation studies, but for the foreseeable future will be of less importance than data-driven approaches. Based on studies of data-driven metrics reviewed in this chapter, we would propose a metric ensemble consisting of several forms of ICC including pattern reproducibility and $I2C2$, prediction measures generated in a cross-validation-style resampling framework, and Jaccard overlap. Our own experience and publications using (prediction, reproducibility) metric pairs indicate that while measures of reliability and reproducibility are important, they may not reflect the predictive generalizability of the resulting group data. Moreover, they may not be able to be simultaneously optimized, although (151) shows how reproducibility may be increased without sacrificing prediction using linear discriminants. These tradeoffs need to be better characterized and understood, along with tradeoffs in the choice of different reliability metrics and forms of $ICC/I2C2$ including the spectral summary of pairwise similarity metrics (100).

We believe the literature strongly supports the use of predictive resampling techniques for adaptive denoising approaches, designed to optimize preprocessing pipelines. Currently the three leading approaches for adaptive denoising all use classification techniques, but in rather different ways: one is directed at rs-fMRI (132), and two are directed at task-based studies; one for GLM denoising at the voxel level (79) and one for more general univariate or multivariate predictive tradeoffs with pattern reproducibility (31). It remains an open question, which if any of these approaches is more generally optimal for task-based fMRI, let alone rs-fMRI.

In general, the literature has only recently started to address how all of these metrics and preprocessing denoising approaches relate to one another. We believe they should all be simultaneously evaluated in both univariate and multivariate analysis frameworks across multiple data types (i.e., resting state, task, contrasts, etc.) to understand the impact of different SNR models, and as a function of the many experimental interactions outlined in Figure 10.1. However, as we have described, there are many pitfalls that must be considered and tested before producing generalizable results, which relate to such interactions with the rest of the neuroimaging experiment. For example, the task domain and test-retest interval may significantly impact such results among many other factors such as subjects' ages and clinical status. If these and other related issues have been considered at all, they have yet to be addressed in the literature in ways that suggest the results may be even partially generalizable. To accelerate such work, a broad range of publically accessible standardized datasets needs to become available, and be coupled with easy-to-use, flexible software testing environments (e.g., (60, 182)). Finally, both our colleagues and funding agencies need to recognize that such standardized, methodological work cannot proceed unless it is funded, and that such work is an important component of validating the utility of neuroimaging in neuroscience and for assessing clinical treatments and disease.

Bibliography

[1] H. Abdi, J. P. Dunlop, and L. J. Williams. How to compute reliability estimates and display confidence and tolerance intervals for pattern classifiers using the bootstrap and 3-way multidimensional scaling (DISTATIS). *Neuroimage*, 45(1):89–95, 2009.

[2] AFNI. AFNI and SUMA information central, 2014.

[3] B. Afshin-Pour, C. Grady, and S. Strother. Evaluation of spatio-temporal decomposition techniques for group analysis of fMRI resting state datasets. *Neuroimage*, 87:363–82, 2014.

[4] B. Afshin-Pour, S. M. Shams, and S. Strother. A hybrid LDA-gCCA model for fMRI data classification and visualization. *IEEE Trans Med Imaging*, 34(5):1031–41, 2015.

[5] B. Afshin-Pour, H. Soltanian-Zadeh, G. A. Hossein-Zadeh, C. L. Grady, and S. C. Strother. A mutual information-based metric for evaluation of fMRI data-processing approaches. *Hum Brain Mapp*, 32(5):699–715, 2011.

[6] A. Andronache, C. Rosazza, D. Sattin, M. Leonardi, L. D'Incerti, and L. Minati. Impact of functional MRI data preprocessing pipeline on default-mode network detectability in patients with disorders of consciousness. *Frontiers in Neuroinformatics*, 7:16, 2013.

[7] Tom Ash, John Suckling, Martin Walter, Cinly Ooi, Claus Tempelmann, Adrian Carpenter, and Guy Williams. Detection of physiological noise in resting state fMRI using machine learning. *Human Brain Mapping*, 34(4):985–998, 2013.

[8] C. F. Beckmann, M. DeLuca, J. T. Devlin, and S. M. Smith. Investigations into resting-state connectivity using independent component analysis. *Philosophical transactions of the Royal Society of London. Series B, Biological Sciences*, 360(1457):1001–13, 2005.

[9] P. Bellec, V. Perlbarg, and A. C. Evans. Bootstrap generation and evaluation of an fMRI simulation database. *Magn Reson Imaging*, 27(10):1382–96, 2009.

[10] P. Bellec, P. Rosa-Neto, O. C. Lyttelton, H. Benali, and A. C. Evans. Multi-level bootstrap analysis of stable clusters in resting-state fMRI. *Neuroimage*, 51(3):1126–39, 2010.

[11] C. M. Bennett and M. B. Miller. How reliable are the results from functional magnetic resonance imaging? *Ann N Y Acad Sci*, 1191:133–55, 2010.

[12] C. M. Bennett and M. B. Miller. fMRI reliability: Influences of task and experimental design. *Cognitive, Affective and Behavioral Neuroscience*, 13(4):690–702, 2013.

[13] R. M. Birn, J. B. Diamond, M. A. Smith, and P. A. Bandettini. Separating respiratory-variation-related fluctuations from neuronal-activity-related fluctuations in fMRI. *Neuroimage*, 31(4):1536–48, 2006.

[14] I. Borg and P.J.F. Groenen. *Modern Multidimensional Scaling: Theory and Applications*. Springer, London, 2005.

[15] S. Bouix, M. Martin-Fernandez, L. Ungar, M. Nakamura, M. S. Koo, R. W. McCarley, and M. E. Shenton. On evaluating brain tissue classifiers without a ground truth. *Neuroimage*, 36(4):1207–24, 2007.

[16] F. D. Bowman. Brain imaging analysis. *Annu Rev Stat Appl*, 1:61–85, 2014.

[17] G. G. Brown, D. H. Mathalon, H. Stern, J. Ford, B. Mueller, D. N. Greve, G. McCarthy, J. Voyvodic, G. Glover, M. Diaz, E. Yetter, I. B. Ozyurt, K. W. Jorgensen, C. G. Wible, J. A. Turner, W. K. Thompson, and S. G. Potkin. Multisite reliability of cognitive bold data. *Neuroimage*, 54(3):2163–75, 2011.

[18] E. T. Bullmore, M. J. Brammer, S. Rabe-Hesketh, V. A. Curtis, R. G. Morris, S. C. Williams, T. Sharma, and P. K. McGuire. Methods for diagnosis and treatment of stimulus-correlated motion in generic brain activation studies using fMRI. *Hum Brain Mapp*, 7(1):38–48, 1999.

[19] K. S. Button, J. P. Ioannidis, C. Mokrysz, B. A. Nosek, J. Flint, E. S. Robinson, and M. R. Munafo. Power failure: Why small sample size undermines the reliability of neuroscience. *Nature Reviews. Neuroscience*, 14(5):365–76, 2013.

[20] A. Caceres, D. L. Hall, F. O. Zelaya, S. C. Williams, and M. A. Mehta. Measuring fMRI reliability with the intra-class correlation coefficient. *Neuroimage*, 45(3):758–68, 2009.

[21] K. L. Campbell, O. Grigg, C. Saverino, N. Churchill, and C. L. Grady. Age differences in the intrinsic functional connectivity of default network subsystems. *Frontiers in Aging Neuroscience*, 5:73, 2013.

[22] T. D. Cannon, F. Sun, S. J. McEwen, X. Papademetris, G. He, T. G. van Erp, A. Jacobson, C. E. Bearden, E. Walker, X. Hu, L. Zhou, L. J. Seidman, H. W. Thermenos, B. Cornblatt, D. M. Olvet, D. Perkins, A. Belger, K. Cadenhead, M. Tsuang, H. Mirzakhanian, J. Addington, R. Frayne, S. W. Woods, T. H. McGlashan, R. T. Constable, M. Qiu, D. H. Mathalon, P. Thompson, and A. W. Toga. Reliability of neuroanatomical measurements in a multisite longitudinal study of youth at risk for psychosis. *Hum Brain Mapp*, 35(5):2424–34, 2014.

[23] F. Carbonell, P. Bellec, and A. Shmuel. Global and system-specific resting-state fMRI fluctuations are uncorrelated: Principal component analysis reveals anti-correlated networks. *Brain Connectivity*, 1(6):496–510, 2011.

[24] J. Carp. On the plurality of (methodological) worlds: estimating the analytic flexibility of fMRI experiments. *Frontiers in neuroscience*, 6:149, 2012.

[25] J. Carp. The secret lives of experiments: methods reporting in the fMRI literature. *Neuroimage*, 63(1):289–300, 2012.

[26] J. Carp. Optimizing the order of operations for movement scrubbing: Comment on power et al. *Neuroimage*, 76:436–8, 2013.

[27] C. Chang and G. H. Glover. Effects of model-based physiological noise correction on default mode network anti-correlations and correlations. *Neuroimage*, 47(4):1448–59, 2009.

[28] E. E. Chen and S. L. Small. Test-retest reliability in fMRI of language: Group and task effects. *Brain and Language*, 102(2):176–85, 2007.

[29] X. Chen, F. Pereira, W. Lee, S. Strother, and T. Mitchell. Exploring predictive and reproducible modeling with the single-subject FIAC dataset. *Hum Brain Mapp*, 27(5):452–61, 2006.

[30] N. W. Churchill, A. Oder, H. Abdi, F. Tam, W. Lee, C. Thomas, J. E. Ween, S. J. Graham, and S. C. Strother. Optimizing preprocessing and analysis pipelines for single-subject fMRI. I. Standard temporal motion and physiological noise correction methods. *Human Brain Mapping*, 33(3):609–27, 2012.

[31] N. W. Churchill, R. Spring, B. Afshin-Pour, F. Dong, and S. C. Strother. An automated, adaptive framework for optimizing preprocessing pipelines in task-based functional MRI. *PLoS One*, 10(7):e0131520, 2015.

[32] N. W. Churchill and S. C. Strother. PHYCAA+: An optimized, adaptive procedure for measuring and controlling physiological noise in bold fMRI. *Neuroimage*, 82C:306–325, 2013.

[33] Nathan Churchill, Robyn Spring, Herve Abdi, Natasa Kovacevic, Anthony R. McIntosh, and Stephen Strother. *The Stability of Behavioral PLS Results in Ill-Posed Neuroimaging Problems*, volume 56 of *Springer Proceedings in Mathematics and Statistics*, pages 171–183. Springer-Verlag, 2013.

[34] Nathan W. Churchill, Grigori Yourganov, and Stephen C. Strother. Comparing within-subject classification and regularization methods in fMRI for large and small sample sizes. *Human Brain Mapping*, 35(9):4499–517, 2014.

[35] N.W. Churchill, G. Yourganov, A. Oder, F. Tam, S.J. Graham, and S.C. Strother. Optimizing preprocessing and analysis pipelines for single-subject fMRI: 2. Interactions with ICA, PCA, task contrast and inter-subject heterogeneity. *PLoS One*, 7(2):e31147, 2012.

[36] J. Cohen. A coefficient of agreement for nominal scales. *Educ Psychol Meas*, 20: 3746, 1960.

[37] J. Cohen. The earth is flat (p ¡ .05). *American Psychologist*, 49(12):997–1003, 1994.

[38] W. J. Conover. *Practical Nonparametric Statistics*. John Wiley and Sons, New York, 1998.

[39] O. David and K. J. Friston. A neural mass model for MEG/EEG: Coupling and neuronal dynamics. *Neuroimage*, 20(3):1743–55, 2003.

[40] O. David, I. Guillemain, S. Saillet, S. Reyt, C. Deransart, C. Segebarth, and A. Depaulis. Identifying neural drivers with functional MRI: An electrophysiological validation. *PLoS Biology*, 6(12):2683–97, 2008.

[41] F. De Martino, F. Gentile, F. Esposito, M. Balsi, F. Di Salle, R. Goebel, and E. Formisano. Classification of fMRI independent components using IC-fingerprints and support vector machine classifiers. *Neuroimage*, 34(1):177–94, 2007.

[42] G. Deco, V. K. Jirsa, and A. R. McIntosh. Emerging concepts for the dynamical organization of resting-state activity in the brain. *Nature Reviews. Neuroscience*, 12(1):43–56, 2011.

[43] J. Demsar. Statistical comparisons of classifiers over multiple datasets. *Journal of Machine Learning Research*, 7:1–30, 2006.

[44] L.R. Dice. Measures of the amount of ecologic association between species. *Ecology*, 26:297302, 1945.

[45] I. Drobnjak, D. Gavaghan, E. Suli, J. Pitt-Francis, and M. Jenkinson. Development of a functional magnetic resonance imaging simulator for modeling realistic rigid-body motion artifacts. *Magnetic Resonance in Medicine: Official Journal of the Society of Magnetic Resonance in Medicine / Society of Magnetic Resonance in Medicine*, 56(2):364–80, 2006.

[46] I. Drobnjak, G. S. Pell, and M. Jenkinson. Simulating the effects of time-varying magnetic fields with a realistic simulated scanner. *Magnetic Resonance Imaging*, 28(7):1014–21, 2010.

[47] K. J. Duncan, C. Pattamadilok, I. Knierim, and J. T. Devlin. Consistency and variability in functional localisers. *Neuroimage*, 46(4):1018–26, 2009.

[48] EEGLAB. Eeglab, 2014.

[49] S. B. Eickhoff, D. Bzdok, A. R. Laird, F. Kurth, and P. T. Fox. Activation likelihood estimation meta-analysis revisited. *Neuroimage*, 59(3):2349–61, 2012.

[50] J. W. Evans, R. M. Todd, M. J. Taylor, and S. C. Strother. Group specific optimisation of fMRI processing steps for child and adult data. *Neuroimage*, 50(2):479–90, 2010.

[51] I. Ford. Confounded correlations: Statistical limitations in the analysis of interregional relationships of cerebral metabolic activity. *J Cereb Blood Flow Metab*, 6(3):385–8, 1986.

[52] FreeSurfer. FreeSurfer software suite, 2014.

[53] L. Freire and J. F. Mangin. Motion correction algorithms may create spurious brain activations in the absence of subject motion. *Neuroimage*, 14(3):709–722, 2001.

[54] L. Friedman, H. Stern, G. G. Brown, D. H. Mathalon, J. Turner, G. H. Glover, R. L. Gollub, J. Lauriello, K. O. Lim, T. Cannon, D. N. Greve, H. J. Bockholt, A. Belger, B. Mueller, M. J. Doty, J. He, W. Wells, P. Smyth, S. Pieper, S. Kim, M. Kubicki, M. Vangel, and S. G. Potkin. Test-retest and between-site reliability in a multicenter fMRI study. *Human Brain Mapping*, 29(8):958–72, 2008.

[55] D. D. Garrett, N. Kovacevic, A. R. McIntosh, and C. L. Grady. Blood oxygen level-dependent signal variability is more than just noise. *The Journal of Neuroscience: The Official Journal of the Society for Neuroscience*, 30(14):4914–21, 2010.

[56] M. Gavrilescu, M. E. Shaw, G. W. Stuart, P. Eckersley, I. D. Svalbe, and G. F. Egan. Simulation of the effects of global normalization procedures in functional MRI. *Neuroimage*, 17(2):532–42, 2002.

[57] M. Gavrilescu, G. W. Stuart, A. Waites, G. Jackson, I. D. Svalbe, and G. F. Egan. Changes in effective connectivity models in the presence of task-correlated motion: An fMRI study. *Human Brain Mapping*, 21(2):49–63, 2004.

[58] C. R. Genovese, D. C. Noll, and W. F. Eddy. Estimating test-retest reliability in functional MR imaging. i: Statistical methodology. *Magn Reson Med*, 38(3):497–507, 1997.

[59] G. H. Glover, T. Q. Li, and D. Ress. Image-based method for retrospective correction of physiological motion effects in fMRI: Retroicor. *Magn Reson Med*, 44(1):162–167, 2000.

[60] K. Gorgolewski, C. D. Burns, C. Madison, D. Clark, Y. O. Halchenko, M. L. Waskom, and S. S. Ghosh. Nipype: A flexible, lightweight and extensible neuroimaging data processing framework in Python. *Frontiers in Neuroinformatics*, 5:13, 2011.

[61] L. Griffanti, G. Salimi-Khorshidi, C. F. Beckmann, E. J. Auerbach, G. Douaud, C. E. Sexton, E. Zsoldos, K. P. Ebmeier, N. Filippini, C. E. Mackay, S. Moeller, J. Xu, E. Yacoub, G. Baselli, K. Ugurbil, K. L. Miller, and S. M. Smith. ICA-based artefact removal and accelerated fMRI acquisition for improved resting state network imaging. *Neuroimage*, 95:232–47, 2014.

[62] M. N. Hallquist, K. Hwang, and B. Luna. The nuisance of nuisance regression: spectral misspecification in a common approach to resting-state fMRI preprocessing reintroduces noise and obscures functional connectivity. *Neuroimage*, 82:208–25, 2013.

[63] X. Han, J. Jovicich, D. Salat, A. van der Kouwe, B. Quinn, S. Czanner, E. Busa, J. Pacheco, M. Albert, R. Killiany, P. Maguire, D. Rosas, N. Makris, A. Dale, B. Dickerson, and B. Fischl. Reliability of MRI-derived measurements of human cerebral cortical thickness: The effects of field strength, scanner upgrade and manufacturer. *Neuroimage*, 32(1):180–94, 2006.

[64] L. K. Hansen, J. Larsen, F. A. Nielsen, S. C. Strother, E. Rostrup, R. Savoy, N. Lange, J. Sidtis, C. Svarer, and O. B. Paulson. Generalizable patterns in neuroimaging: How many principal components? *Neuroimage*, 9(5):534–44, 1999.

[65] J. V. Haxby, M. I. Gobbini, M. L. Furey, A. Ishai, J. L. Schouten, and P. Pietrini. Distributed and overlapping representations of faces and objects in ventral temporal cortex. *Science*, 293(5539):2425–30, 2001.

[66] Hongjian He and Thomas T. Liu. A geometric view of global signal confounds in resting-state functional MRI. *Neuroimage*, 59(3):2339–2348, 2012.

[67] J. B. Hopfinger, C. Buchel, A. P. Holmes, and K. J. Friston. A study of analysis parameters that influence the sensitivity of event-related fMRI analyses. *Neuroimage*, 11(4):326–33, 2000.

[68] G. Hripcsak and D. F. Heitjan. Measuring agreement in medical informatics reliability studies. *J Biomed Inform*, 35(2):99–110, 2002.

[69] S.A. Huettel, A.W. Song, and G. McCarthy. *Functional Magnetic Resonance Imaging, 2nd Ed.* Sinauer Assoc., Inc., Sunderland, MA, 2009.

[70] R. J. Huster, S. Debener, T. Eichele, and C. S. Herrmann. Methods for simultaneous EEG-fMRI: an introductory review. *The Journal of Neuroscience: The Official Journal of the Society for Neuroscience*, 32(18):6053–60, 2012.

[71] J. P. Ioannidis, M. R. Munafo, P. Fusar-Poli, B. A. Nosek, and S. P. David. Publication and other reporting biases in cognitive sciences: Detection, prevalence, and prevention. *Trends in Cognitive Sciences*, 18(5):235–41, 2014.

[72] P. Jaccard. Etude comparative de la distribution florale dans unem portion des alpes et des jura. *Bull Soc Vaudoise Sci Nat*, 37:547579, 1901.

[73] D. J. Jacobsen, L. K. Hansen, and K. H. Madsen. Bayesian model comparison in nonlinear bold fMRI hemodynamics. *Neural Comput*, 20(3):738–55, 2008.

[74] S. Jan, G. Santin, D. Strul, S. Staelens, K. Assie, D. Autret, S. Avner, R. Barbier, M. Bardies, P. M. Bloomfield, D. Brasse, V. Breton, P. Bruyndonckx, I. Buvat, A. F. Chatziioannou, Y. Choi, Y. H. Chung, C. Comtat, D. Donnarieix, L. Ferrer, S. J. Glick, C. J. Groiselle, D. Guez, P. F. Honore, S. Kerhoas-Cavata, A. S. Kirov, V. Kohli, M. Koole, M. Krieguer, D. J. van der Laan, F. Lamare, G. Largeron, C. Lartizien, D. Lazaro, M. C. Maas, L. Maigne, F. Mayet, F. Melot, C. Merheb, E. Pennacchio, J. Perez, U. Pietrzyk, F. R. Rannou, M. Rey, D. R. Schaart, C. R. Schmidtlein, L. Simon, T. Y. Song, J. M. Vieira, D. Visvikis, R. Van de Walle, E. Wieers, and C. Morel. Gate: A simulation toolkit for PET and SPECT. *Phys Med Biol*, 49(19):4543–61, 2004.

[75] M. Jenkinson, C. F. Beckmann, T. E. Behrens, M. W. Woolrich, and S. M. Smith. Fsl. *Neuroimage*, 62(2):782–90, 2012.

[76] D. Jewett. What's wrong with single hypotheses? Why it is time for strong-inference-plus, 2005.

[77] T. Johnstone, K. S. Ores Walsh, L. L. Greischar, A. L. Alexander, A. S. Fox, R. J. Davidson, and T. R. Oakes. Motion correction and the use of motion covariates in multiple-subject fMRI analysis. *Hum Brain Mapp*, 27(10):779–88, 2006.

[78] T. B. Jones, P. A. Bandettini, and R. M. Birn. Integration of motion correction and physiological noise regression in fMRI. *Neuroimage*, 42(2):582–90, 2008.

[79] K. N. Kay, A. Rokem, J. Winawer, R. F. Dougherty, and B. A. Wandell. Glmdenoise: A fast, automated technique for denoising task-based fMRI data. *Frontiers in Neuroscience*, 7:247, 2013.

[80] F. Kherif, J. B. Poline, S. Meriaux, H. Benali, G. Flandin, and M. Brett. Group analysis in functional neuroimaging: Selecting subjects using similarity measures. *Neuroimage*, 20(4):2197–208, 2003.

[81] U. Kjems, L. K. Hansen, J. Anderson, S. Frutiger, S. Muley, J. Sidtis, D. Rottenberg, and S. C. Strother. The quantitative evaluation of functional neuroimaging experiments: mutual information learning curves. *Neuroimage*, 15(4):772–86, 2002.

[82] A. Klein, J. Andersson, B. A. Ardekani, J. Ashburner, B. Avants, M. C. Chiang, G. E. Christensen, D. L. Collins, J. Gee, P. Hellier, J. H. Song, M. Jenkinson, C. Lepage, D. Rueckert, P. Thompson, T. Vercauteren, R. P. Woods, J. J. Mann, and R. V. Parsey. Evaluation of 14 nonlinear deformation algorithms applied to human brain MRI registration. *Neuroimage*, 46(3):786–802, 2009.

[83] A. Klein and J. Tourville. 101 labeled brain images and a consistent human cortical labeling protocol. *Frontiers in Neuroscience*, 6:171, 2012.

[84] Arno Klein, Satrajit S. Ghosh, Brian Avants, B. T. T. Yeo, Bruce Fischl, Babak Ardekani, James C. Gee, J. J. Mann, and Ramin V. Parsey. Evaluation of volume-based and surface-based brain image registration methods. *Neuroimage*, 51(1):214–220, 2010.

[85] X. Z. Kong, Z. Zhen, X. Li, H. H. Lu, R. Wang, L. Liu, Y. He, Y. Zang, and J. Liu. Individual differences in impulsivity predict head motion during magnetic resonance imaging. *PloS One*, 9(8):e104989, 2014.

[86] N. Kriegeskorte, R. Goebel, and P. Bandettini. Information-based functional brain mapping. *Proc Natl Acad Sci U S A*, 103(10):3863–8, 2006.

[87] N. Kriegeskorte, M. A. Lindquist, T. E. Nichols, R. A. Poldrack, and E. Vul. Everything you never wanted to know about circular analysis, but were afraid to ask. *Journal of Cerebral Blood Flow and Metabolism: Official Journal of the International Society of Cerebral Blood Flow and Metabolism*, 30(9):1551–7, 2010.

[88] N. Kriegeskorte, W. K. Simmons, P. S. Bellgowan, and C. I. Baker. Circular analysis in systems neuroscience: The dangers of double dipping. *Nature Neuroscience*, 12(5):535–40, 2009.

[89] R. Kustra and S. Strother. Penalized discriminant analysis of [150]-water PET brain images with prediction error selection of smoothness and regularization hyperparameters. *IEEE Trans Med Imaging*, 20(5):376–87, 2001.

[90] S. LaConte, J. Anderson, S. Muley, J. Ashe, S. Frutiger, K. Rehm, L. K. Hansen, E. Yacoub, X. Hu, D. Rottenberg, and S. Strother. The evaluation of preprocessing choices in single-subject bold fMRI using NPAIRS performance metrics. *Neuroimage*, 18(1):10–27, 2003.

[91] S. LaConte, S. Strother, V. Cherkassky, J. Anderson, and X. Hu. Support vector machines for temporal classification of block design fMRI data. *Neuroimage*, 26(2):317–29, 2005.

[92] N. Lange, S. C. Strother, J. R. Anderson, F. A. Nielsen, A. P. Holmes, T. Kolenda, R. Savoy, and L. K. Hansen. Plurality and resemblance in fMRI data analysis. *Neuroimage*, 10(3 Pt 1):282–303, 1999.

[93] T. H. Le and X. P. Hu. Retrospective estimation and correction of physiological artifacts in fMRI by direct extraction of physiological activity from MR data. *Magn Reson Med*, 35(3):290–298, 1996.

[94] M. Levandowsky and D. Winter. Distance between sets. *Nature*, 234:3435, 1971.

[95] M. Liou, H. R. Su, J. D. Lee, P. E. Cheng, C. C. Huang, and C. H. Tsai. Functional MR images and scientific inference: Reproducibility maps. *Journal of Cognitive Neuroscience*, 15(7):935–45, 2003.

[96] A. S. Lukic, M. N. Wernick, and S. C. Strother. An evaluation of methods for detecting brain activations from functional neuroimages. *Artif Intell Med*, 25(1):69–88, 2002.

[97] A. S. Lukic, M. N. Wernick, D. G. Tzikas, X. Chen, A. Likas, N. P. Galatsanos, Y. Yang, F. Zhao, and S. C. Strother. Bayesian kernel methods for analysis of functional neuroimages. *IEEE Trans Med Imaging*, 26(12):1613–24, 2007.

[98] T. E. Lund, K. H. Madsen, K. Sidaros, W. L. Luo, and T. E. Nichols. Non-white noise in fMRI: Does modelling have an impact? *Neuroimage*, 29(1):54–66, 2006.

[99] R. Maitra. Assessing certainty of activation or inactivation in test-retest fMRI studies. *Neuroimage*, 47(1):88–97, 2009.

[100] R. Maitra. A re-defined and generalized percent-overlap-of-activation measure for studies of fMRI reproducibility and its use in identifying outlier activation maps. *Neuroimage*, 50(1):124–35, 2010.

[101] R. Maitra, S. R. Roys, and R. P. Gullapalli. Test-retest reliability estimation of functional mri data. *Magnetic Resonance in Medicine: Official Journal of the Society of Magnetic Resonance in Medicine / Society of Magnetic Resonance in Medicine*, 48(1):62–70, 2002.

[102] K. V. Mardia, J. T. Kent, and J. M. Bibby. *Multivariate Analysis*. Academic Press, San Diego, 1979.

[103] N. Meinshausen and P. Bhlmann. Stability selection. *Journal of the Royal Statistical Society: Series B (Statistical Methodology)*, 72(4):417473, 2010.

[104] M. B. Miller, C. L. Donovan, J. D. Van Horn, E. German, P. Sokol-Hessner, and G. L. Wolford. Unique and persistent individual patterns of brain activity across different memory retrieval tasks. *Neuroimage*, 48(3):625–35, 2009.

[105] M. B. Miller, J. D. Van Horn, G. L. Wolford, T. C. Handy, M. Valsangkar-Smyth, S. Inati, S. Grafton, and M. S. Gazzaniga. Extensive individual differences in brain activations associated with episodic retrieval are reliable over time. *Journal of Cognitive Neuroscience*, 14(8):1200–14, 2002.

[106] T.M. Mitchell, R. Hutchinson, R.S. Niculescu, F. Pereira, X. Wang, M. Just, and S. Newman. Learning to decode cognitive states from brain images. *Mach. Learn.*, 57:145175, 2004.

[107] J. R. Moeller and S. C. Strother. A regional covariance approach to the analysis of functional patterns in positron emission tomographic data. *J Cereb Blood Flow Metab*, 11(2):A121–35, 1991.

[108] N. Morch, L.K. Hansen, S.C. Strother, C. Svarer, D.A. Rottenberg, B. Lautrup, R. Savoy, and O.B. Paulson. *Nonlinear versus Linear Models in Functional Neuroimaging: Learning Curves and Generalization Crossover*, Lecture Notes in Computer Science, vol. 1230, pages 259–270. Springer-Verlag, New York, 1997.

[109] R. Muller and P. Buttner. A critical discussion of intraclass correlation coefficients. *Statistics in Medicine*, 13(23-24):2465–76, 1994.

[110] K. Murphy, R. M. Birn, D. A. Handwerker, T. B. Jones, and P. A. Bandettini. The impact of global signal regression on resting state correlations: are anti-correlated networks introduced? *Neuroimage*, 44(3):893–905, 2009.

[111] R. R. Nandy and D. Cordes. New approaches to receiver operating characteristic methods in functional magnetic resonance imaging with real data using repeated trials. *Magnetic Resonance in Medicine: Official Journal of the Society of Magnetic Resonance in Medicine / Society of Magnetic Resonance in Medicine*, 52(6):1424–31, 2004.

[112] T. Nichols and S. Hayasaka. Controlling the familywise error rate in functional neuroimaging: a comparative review. *Stat Methods Med Res*, 12(5):419–46, 2003.

[113] R. Nuzzo. Scientific method: Statistical errors. *Nature*, 506(7487):150–2, 2014.

[114] J.M. Ollinger, T.R. Oakes, Alexander A.L., F. Haeberli, K.M. Dalton, and R.J. Davidson. The secret life of motion covariates. In *NeuroImage*, volume 47, page S122, 2009.

[115] F. Pereira, T. Mitchell, and M. Botvinick. Machine learning classifiers and fMRI: a tutorial overview. *Neuroimage*, 45(1 Suppl):S199–209, 2009.

[116] T. W. Picton, S. Bentin, P. Berg, E. Donchin, S. A. Hillyard, Jr. Johnson, R., G. A. Miller, W. Ritter, D. S. Ruchkin, M. D. Rugg, and M. J. Taylor. Guidelines for using human event-related potentials to study cognition: Recording standards and publication criteria. *Psychophysiology*, 37(2):127–52, 2000.

[117] J. R. Platt. Strong inference: Certain systematic methods of scientific thinking may produce much more rapid progress than others. *Science*, 146(3642):347–353, 1964.

[118] M. M. Plichta, A. J. Schwarz, O. Grimm, K. Morgen, D. Mier, L. Haddad, A. B. Gerdes, C. Sauer, H. Tost, C. Esslinger, P. Colman, F. Wilson, P. Kirsch, and A. Meyer-Lindenberg. Test-retest reliability of evoked bold signals from a cognitive-emotive fMRI test battery. *Neuroimage*, 60(3):1746–58, 2012.

[119] R. A. Poldrack, P. C. Fletcher, R. N. Henson, K. J. Worsley, M. Brett, and T. E. Nichols. Guidelines for reporting an fMRI study. *Neuroimage*, 40(2):409–14, 2008.

[120] J. B. Poline, J. L. Breeze, S. Ghosh, K. Gorgolewski, Y. O. Halchenko, M. Hanke, C. Haselgrove, K. G. Helmer, D. B. Keator, D. S. Marcus, R. A. Poldrack, Y. Schwartz, J. Ashburner, and D. N. Kennedy. Data sharing in neuroimaging research. *Frontiers in Neuroinformatics*, 6:9, 2012.

[121] J. B. Poline, S. C. Strother, G. Dehaene-Lambertz, G. F. Egan, and J. L. Lancaster. Motivation and synthesis of the FIAC experiment: Reproducibility of fMRI results across expert analyses. *Hum Brain Mapp*, 27(5):351–9, 2006.

[122] J. D. Power, K. A. Barnes, A. Z. Snyder, B. L. Schlaggar, and S. E. Petersen. Spurious but systematic correlations in functional connectivity MRI networks arise from subject motion. *Neuroimage*, 59(3):2142–54, 2012.

[123] M. Raemaekers, S. du Plessis, N. F. Ramsey, J. M. Weusten, and M. Vink. Test-retest variability underlying fMRI measurements. *Neuroimage*, 60(1):717–27, 2012.

[124] M. Raemaekers, M. Vink, B. Zandbelt, R. J. van Wezel, R. S. Kahn, and N. F. Ramsey. Test-retest reliability of fMRI activation during prosaccades and antisaccades. *Neuroimage*, 36(3):532–42, 2007.

[125] P. M. Rasmussen, L. K. Hansen, K. H. Madsen, N.W. Churchill, and S.C. Strother. Pattern reproducibility, interpretability, and sparsity in classification models in neuroimaging. *Pattern Recognition*, 45(6):2085–2100, 2012.

[126] P.M. Rasmussen. Mathematical modeling and visualization of functional neuroimages. Thesis, Technical University of Denmark, 2012.

[127] P. Ritter, M. Schirner, A. R. McIntosh, and V. K. Jirsa. The virtual brain integrates computational modeling and multimodal neuroimaging. *Brain Connectivity*, 3(2):121–45, 2013.

[128] P. Robert and Y. Escoufier. A unifying tool for linear multivariate statistical methods: The RV-coefficient. *Appl. Statist.*, 25:257–265, 1976.

[129] S. A. Rombouts, F. Barkhof, F. G. Hoogenraad, M. Sprenger, and P. Scheltens. Within-subject reproducibility of visual activation patterns with functional magnetic resonance imaging using multislice echo planar imaging. *Magnetic Resonance Imaging*, 16(2):105–13, 1998.

[130] V. Rousson, T. Gasser, and B. Seifert. Assessing intrarater, interrater and test-retest reliability of continuous measurements. *Stat Med*, 21(22):3431–46, 2002.

[131] S. Ruddell, S. Twiss, and P. Pomeroy. Measuring opportunity for sociality: Quantifying social stability in a colonially breeding phocid. *Anim. Behav.*, 74:1357–1368, 2007.

[132] G. Salimi-Khorshidi, G. Douaud, C. F. Beckmann, M. F. Glasser, L. Griffanti, and S. M. Smith. Automatic denoising of functional MRI data: Combining independent component analysis and hierarchical fusion of classifiers. *Neuroimage*, 90:449–68, 2014.

[133] T. D. Satterthwaite, D. H. Wolf, J. Loughead, K. Ruparel, M. A. Elliott, H. Hakonarson, R. C. Gur, and R. E. Gur. Impact of in-scanner head motion on multiple measures of functional connectivity: Relevance for studies of neurodevelopment in youth. *Neuroimage*, 60(1):623–32, 2012.

[134] Theodore D. Satterthwaite, Mark A. Elliott, Raphael T. Gerraty, Kosha Ruparel, James Loughead, Monica E. Calkins, Simon B. Eickhoff, Hakon Hakonarson, Ruben C. Gur, Raquel E. Gur, and Daniel H. Wolf. An improved framework for confound regression and filtering for control of motion artifact in the preprocessing of resting-state functional connectivity data. *Neuroimage*, 64:240–256, 2013.

[135] Theodore D. Satterthwaite, Daniel H. Wolf, Kosha Ruparel, Guray Erus, Mark A. Elliott, Simon B. Eickhoff, Efstathios D. Gennatas, Chad Jackson, Karthik Prabhakaran, Alex Smith, Hakon Hakonarson, Ragini Verna, Christos Davatzikos, Raquel E. Gur, and Ruben C. Gur. Heterogeneous impact of motion on fundamental patterns of developmental changes in functional connectivity during youth. *Neuroimage*, 83:45–57, 2013.

[136] M. L. Scholvinck, A. Maier, F. Q. Ye, J. H. Duyn, and D. A. Leopold. Neural basis of global resting-state fMRI activity. *Proceedings of the National Academy of Sciences of the United States of America*, 107(22):10238–43, 2010.

[137] D. W. Shattuck, S. R. Sandor-Leahy, K. A. Schaper, D. A. Rottenberg, and R. M. Leahy. Magnetic resonance image tissue classification using a partial volume model. *Neuroimage*, 13(5):856–76, 2001.

[138] H. Shou, A. Eloyan, S. Lee, V. Zipunnikov, A. N. Crainiceanu, N. B. Nebel, B. Caffo, M. A. Lindquist, and C. M. Crainiceanu. Quantifying the reliability of image replication studies: The image intraclass correlation coefficient (i2c2). *Cogn Affect Behav Neurosci*, 13(4):714–24, 2013.

[139] P. E. Shrout and J. L. Fleiss. Intraclass correlations: Uses in assessing rater reliability. *Psychol Bull*, 86(2):420–8, 1979.

[140] P. Skudlarski, R. T. Constable, and J. C. Gore. Roc analysis of statistical methods used in functional MRI: Individual subjects. *Neuroimage*, 9(3):311–29, 1999.

[141] R. Sladky, K. J. Friston, J. Trostl, R. Cunnington, E. Moser, and C. Windischberger. Slice-timing effects and their correction in functional MRI. *Neuroimage*, 58(2):588–94, 2011.

[142] S. M. Smith, K. L. Miller, G. Salimi-Khorshidi, M. Webster, C. F. Beckmann, T. E. Nichols, J. D. Ramsey, and M. W. Woolrich. Network modelling methods for fMRI. *Neuroimage*, 54(2):875–91, 2011.

[143] J. A. Sorenson and X. Wang. ROC methods for evaluation of fMRI techniques. *Magnetic Resonance in Medicine: Official Journal of the Society of Magnetic Resonance in Medicine / Society of Magnetic Resonance in Medicine*, 36(5):737–44, 1996.

[144] SPM. Statistical parametric mapping, 2014.

[145] M. Stone. Cross-validatory choice and assessment of statistical predictions. *Journal of the Royal Statistical Society. Series B (Methodological)*, 36(2):111–147, 1974.

[146] S. Strother, A. Oder, R. Spring, and C. Grady. The NPAIRS computational statistics framework for data analysis in neuroimaging. *19th International Conference on Computational Statistics*, Keynote, Invited and Contributed Papers:111–120, 2010.

[147] S. C. Strother. Evaluating fMRI preprocessing pipelines. *IEEE Eng Med Biol Mag*, 25(2):27–41, 2006.

[148] S. C. Strother, J. Anderson, L. K. Hansen, U. Kjems, R. Kustra, J. Sidtis, S. Frutiger, S. Muley, S. LaConte, and D. Rottenberg. The quantitative evaluation of functional neuroimaging experiments: The NPAIRS data analysis framework. *Neuroimage*, 15(4):747–71, 2002.

[149] S.C. Strother, S. La Conte, L. Kai Hansen, J. Anderson, J. Zhang, S. Pulapura, and D. Rottenberg. Optimizing the fMRI data-processing pipeline using prediction and reproducibility performance metrics: I. A preliminary group analysis. *Neuroimage*, 23 Suppl 1:S196–207, 2004.

[150] S.C. Strother, N. Lange, J.R. Anderson, K.A. Schaper, K. Rehm, L.K. Hansen, and D.A. Rottenberg. Activation pattern reproducibility: Measuring the effects of group size and data analysis models. *Hum Brain Mapp*, 5:312–316, 1997.

[151] S.C. Strother, P.M. Rasmussen, N.W. Churchill, and L.K. Hansen. Stability and reproducibility in fMRI analysis. In I. Rish, G.A. Cecchi, A. Lozano, and A. Niculescu-Mizil, editors, *Practical Applications of Sparse Modeling*, Neural Information Processing Series, pages 99–121, Boston, 2014. MIT Press.

[152] S.C. Strother, S. Sharaf, and C. Grady. A hierarchy of cognitive brain networks revealed by multivariate performance metrics, 2015.

[153] J. A. Swets. Measuring the accuracy of diagnostic systems. *Science*, 240(4857):1285–93, 1988.

[154] J. Tanabe, D. Miller, J. Tregellas, R. Freedman, and F. G. Meyer. Comparison of detrending methods for optimal fMRI preprocessing. *Neuroimage*, 15(4):902–7, 2002.

[155] J. Taylor, A. Evans, and K. Friston. A tribute to: Keith Worsley, 1951–2009. *Neuroimage*, 46(4):891–4, 2009.

[156] C. Tegeler, S. C. Strother, J. R. Anderson, and S. G. Kim. Reproducibility of bold-based functional MRI obtained at 4 T. *Hum Brain Mapp*, 7(4):267–83, 1999.

[157] B. Thirion, G. Varoquaux, E. Dohmatob, and J. B. Poline. Which fMRI clustering gives good brain parcellations? *Frontiers in Neuroscience*, 8:167, 2014.

[158] A. Toga and J. Van Horn. Neuroimaging workflow design and data mining, 25 July, 2014 2012.

[159] J. Tohka, K. Foerde, A. R. Aron, S. M. Tom, A. W. Toga, and R. A. Poldrack. Automatic independent component labeling for artifact removal in fMRI. *Neuroimage*, 39(3):1227–45, 2008.

[160] C. Triantafyllou, R. D. Hoge, G. Krueger, C. J. Wiggins, A. Potthast, G. C. Wiggins, and L. L. Wald. Comparison of physiological noise at 1.5 T, 3 T and 7 T and optimization of fMRI acquisition parameters. *Neuroimage*, 26(1):243–50, 2005.

[161] K. R. Van Dijk, M. R. Sabuncu, and R. L. Buckner. The influence of head motion on intrinsic functional connectivity MRI. *Neuroimage*, 59(1):431–8, 2012.

[162] O. Vitouch and J. Gluck. "Small group petting": Sample sizes in brain mapping research. *Human Brain Mapping*, 5(1):74–77, 1997.

[163] T. D. Wager, M. A. Lindquist, T. E. Nichols, H. Kober, and J. X. Van Snellenberg. Evaluating the consistency and specificity of neuroimaging data using meta-analysis. *Neuroimage*, 45(1 Suppl):S210–21, 2009.

[164] S. K. Warfield, K. H. Zou, and W. M. Wells. Simultaneous truth and performance level estimation (staple): An algorithm for the validation of image segmentation. *IEEE Transactions on Medical Imaging*, 23(7):903–21, 2004.

[165] X. Wei, S. S. Yoo, C. C. Dickey, K. H. Zou, C. R. Guttmann, and L. P. Panych. Functional MRI of auditory verbal working memory: Long-term reproducibility analysis. *Neuroimage*, 21(3):1000–8, 2004.

[166] A. Weissenbacher, C. Kasess, F. Gerstl, R. Lanzenberger, E. Moser, and C. Windischberger. Correlations and anticorrelations in resting-state functional connectivity MRI: A quantitative comparison of preprocessing strategies. *Neuroimage*, 47(4):1408–16, 2009.

[167] M. Welvaert and Y. Rosseel. A review of fMRI simulation studies. *PLoS One*, 9(7):e101953, 2014.

[168] G.W. Williams. Comparing the joint agreement of several raters with another rater. *Biometrics*, 32:619627, 1976.

[169] K. M. Wisner, G. Atluri, K. O. Lim, and 3rd Macdonald, A. W. Neurometrics of intrinsic connectivity networks at rest using fMRI: Retest reliability and cross-validation using a meta-level method. *Neuroimage*, 76:236–51, 2013.

[170] C. W. Wong, V. Olafsson, O. Tal, and T. T. Liu. The amplitude of the resting-state fMRI global signal is related to EEG vigilance measures. *Neuroimage*, 83:983–90, 2013.

[171] R. P. Woods, M. Dapretto, N. L. Sicotte, A. W. Toga, and J. C. Mazziotta. Creation and use of a Talairach-compatible atlas for accurate, automated, nonlinear intersubject registration, and analysis of functional imaging data. *Human Brain Mapping*, 8(2-3):73–79, 1999.

[172] O. Yamashita, M. A. Sato, T. Yoshioka, F. Tong, and Y. Kamitani. Sparse estimation automatically selects voxels relevant for the decoding of fMRI activity patterns. *Neuroimage*, 42(4):1414–29, 2008.

[173] N. Yeung, R. Bogacz, C. B. Holroyd, and J. D. Cohen. Detection of synchronized oscillations in the electroencephalogram: An evaluation of methods. *Psychophysiology*, 41(6):822–32, 2004.

[174] G. Yourganov, T. Schmah, N. W. Churchill, M. G. Berman, C. L. Grady, and S. C. Strother. Pattern classification of fMRI data: Applications for analysis of spatially distributed cortical networks. *Neuroimage*, 96:117–32, 2014.

[175] Y. Zang, T. Jiang, Y. Lu, Y. He, and L. Tian. Regional homogeneity approach to fMRI data analysis. *Neuroimage*, 22(1):394–400, 2004.

[176] Y. F. Zang, Y. He, C. Z. Zhu, Q. J. Cao, M. Q. Sui, M. Liang, L. X. Tian, T. Z. Jiang, and Y. F. Wang. Altered baseline brain activity in children with ADHD revealed by resting-state functional MRI. *Brain and Development*, 29(2):83–91, 2007.

[177] L. L. Zeng, D. Wang, M. D. Fox, M. Sabuncu, D. Hu, M. Ge, R. L. Buckner, and H. Liu. Neurobiological basis of head motion in brain imaging. *Proceedings of the National Academy of Sciences of the United States of America*, 111(16):6058–62, 2014.

[178] J. Zhang, L. Liang, J. R. Anderson, L. Gatewood, D. A. Rottenberg, and S. C. Strother. A Java-based fMRI processing pipeline evaluation system for assessment of univariate general linear model and multivariate canonical variate analysis-based pipelines. *Neuroinformatics*, 6(2):123–34, 2008.

[179] A. P. Zijdenbos, R. Forghani, and A. C. Evans. Automatic "pipeline" analysis of 3-d MRI data for clinical trials: Application to multiple sclerosis. *IEEE Transactions on Medical Imaging*, 21(10):1280–91, 2002.

[180] K. Zilles and K. Amunts. Centenary of Brodmann's map–conception and fate. *Nat Rev Neurosci*, 11(2):139–45, 2010.

[181] X. N. Zuo, C. Kelly, J. S. Adelstein, D. F. Klein, F. X. Castellanos, and M. P. Milham. Reliable intrinsic connectivity networks: Test-retest evaluation using ICA and dual regression approach. *Neuroimage*, 49(3):2163–77, 2010.

[182] Xi-Nian Zuo, Jeffrey S. Anderson, Pierre Bellec, Rasmus M. Birn, Bharat B. Biswal, Janusch Blautzik, John C. S. Breitner, Randy L. Buckner, Vince D. Calhoun, F. Xavier Castellanos, Antao Chen, Bing Chen, Jiangtao Chen, Xu Chen, Stanley J. Colcombe, William Courtney, R. Cameron Craddock, Adriana Di Martino, Hao-Ming Dong, Xiaolan Fu, Qiyong Gong, Krzysztof J. Gorgolewski, Ying Han, Ye He, Yong He, Erica Ho, Avram Holmes, Xiao-Hui Hou, Jeremy Huckins, Tianzi Jiang, Yi Jiang, William Kelley, Clare Kelly, Margaret King, Stephen M. LaConte, Janet E. Lainhart, Xu Lei, Hui-Jie Li, Kaiming Li, Kuncheng Li, Qixiang Lin, Dongqiang Liu, Jia Liu, Xun Liu, Yijun Liu, Guangming Lu, Jie Lu, Beatriz Luna, Jing Luo, Daniel Lurie, Ying Mao, Daniel S. Margulies, Andrew R. Mayer, Thomas Meindl, Mary E. Meyerand, Weizhi Nan, Jared A. Nielsen, David OConnor, David Paulsen, Vivek Prabhakaran, Zhigang Qi, Jiang Qiu, Chunhong Shao, Zarrar Shehzad, Weijun Tang, Arno Villringer, Huiling Wang, Kai Wang, Dongtao Wei, Gao-Xia Wei, Xu-Chu Weng, Xuehai Wu, Ting Xu, Ning Yang, Zhi Yang, Yu-Feng Zang, Lei Zhang, Qinglin Zhang, Zhe Zhang, Zhiqiang Zhang, Ke Zhao, Zonglei Zhen, Yuan Zhou, Xing-Ting Zhu, and

Michael P. Milham. An open science resource for establishing reliability and reproducibility in functional connectomics. *Scientific Data*, 1, 2014.

[183] Xi-Nian Zuo, Ting Xu, Lili Jiang, Zhi Yang, Xiao-Yan Cao, Yong He, Yu-Feng Zang, F. Xavier Castellanos, and Michael P. Milham. Toward reliable characterization of functional homogeneity in the human brain: Preprocessing, scan duration, imaging resolution and computational space. *Neuroimage*, 65:374–386, 2013.

11

Linear and Nonlinear Models for fMRI Time Series Analysis

Tingting Zhang

Department of Statistics, University of Virginia

Haipeng Shen

Innovation and Information Management, School of Business, Faculty of Business and Economics, University of Hong Kong

Fan Li

Department of Statistical Science, Duke University

CONTENTS

11.1	Introduction	309
11.2	The GLM: Single-Level Analysis	310
11.3	Modeling the Hemodynamic Response Function in the Time Domain	312
	11.3.1 Parametric Models	313
	11.3.2 Nonparametric Models	317
	11.3.3 Comparison of Different HRF Estimation Methods	320
11.4	Hemodynamic Response Estimation in the Frequency Domain	321
11.5	Multi-Subject Analysis	322
	11.5.1 Semi-Parametric Approaches	323
11.6	Nonlinear Models	323
	11.6.1 The Balloon Model	324
	11.6.2 Volterra Series Model	325
	11.6.3 Bi-Exponential Nonlinear Model	326
	11.6.4 Volterra Series Models for Multi-Subject Data	326
11.7	Summary and Future Directions	327
	Bibliography	328

11.1 Introduction

Functional magnetic resonance imaging (fMRI) measures brain activity through detecting blood oxygen-level changes in blood vessels; it provides non-invasive measurements of human brain activity with high spatial resolution (see Chapter 6 for details). FMRI has become the most popular neuroimaging technology for studying brain functions in psychology research. In typical psychology experiments, several subjects from a target human population (either a population of normal human subjects or a population of patients with a certain disease) are recruited; each subject completes a protocol comprised of one or several experimental conditions, while his or her brain activity is measured by fMRI. Then a series of preprocessing steps—as described in Chapter 10—are performed. The ensuing fMRI data can be used in statistical analysis for different research goals: (i) measuring brain response in a designed condition, (ii) locating activated brain regions, (iii) comparing brain activity under different conditions, (iv) determining brain networks, and (v) predicting brain disease development. This chapter focuses on estimation methods for the first three goals. Chapters 14, 16, and 17 review existing statistical models and methods for prediction problems using human brain data.

The most common approach to measuring brain activity in designed conditions and constructing brain activation maps is the massive univariate analysis that analyzes the fMRI time series, one voxel at a time, usually within the framework of the general linear model (GLM). The GLM is inherently a linear regression model, treating the BOLD response as a linear combination of predictors plus error. Statistical inference of brain activity is performed using linear regression techniques such as the least squares estimation, Student t-tests, and analysis of variance. More details on single-level analysis of one session of fMRI data are discussed in Section 11.2. Many GLMs use a hemodynamic response function (HRF) to characterize the brain activity in an experimental condition. Modeling and estimation of the HRF are reviewed in Section 11.3. Group-level analysis involving multi-subject fMRI data is reviewed in Section 11.5. Nonlinear models are reviewed in Section 11.6.

11.2   The GLM: Single-Level Analysis

The GLM was first proposed by (25), with the following standard form

$$\mathbf{Y} = \mathbf{X}\boldsymbol{\beta} + \boldsymbol{\varepsilon}, \tag{11.1}$$

where $\mathbf{Y} = (Y_1, \ldots, Y_T)'$ is the fMRI time series of one given voxel of one subject and T is the number of fMRI scans, $\mathbf{X}$ is the design matrix, and $\boldsymbol{\varepsilon} = (\varepsilon_1, \ldots, \varepsilon_T)'$ is the vector of error terms. We assume $\boldsymbol{\varepsilon}$ follows a multivariate Gaussian distribution $N(0, \sigma^2 \cdot \boldsymbol{\Sigma})$, where σ^2 is the variance and $\boldsymbol{\Sigma}$ is a $T \times T$ correlation matrix. The first few scans after the experiment onset are usually discarded due to the magnetic saturation effect on the collected data. The design matrix $\mathbf{X}$ depends on the sequence of stimuli evoked during scanning and the model parameters $\boldsymbol{\beta}$ describe the task effect. For example, in a block design with alternating blocks of task and rest, the design matrix has two columns: the first column of ones corresponds to the intercept, and the second column with elements taking values one and zero corresponds to experimental conditions of the task and rest. The associated GLM is

$$\begin{pmatrix} Y_1 \\ Y_2 \\ \vdots \\ Y_T \end{pmatrix} = \begin{pmatrix} 1 & X_{11} \\ 1 & X_{21} \\ 1 & \vdots \\ 1 & X_{T1} \end{pmatrix} \times \begin{pmatrix} \beta_0 \\ \beta_1 \end{pmatrix} + \begin{pmatrix} \varepsilon_1 \\ \varepsilon_2 \\ \vdots \\ \varepsilon_T \end{pmatrix},$$

where the two parameters β_0 and β_1 characterize the brain activity at rest and in response to the task, respectively.

When the experiment involves more than two, say K, different conditions, the ensuing design matrix in the GLM would have $K + 1$ ($K > 2$) columns, where the first column is for the baseline intercept and each of the rest K columns characterizes the presence of one of the K conditions during the experiment. If the error terms $\boldsymbol{\varepsilon}$ are independent and identically distributed, that is, $\boldsymbol{\Sigma}$ is $\mathbf{I}_T$, a $T \times T$ identity matrix, $\boldsymbol{\beta}$ can be estimated by the ordinary least squares (OLS) estimate:

$$\hat{\beta}_o = (\mathbf{X}'\mathbf{X})^{-1}\mathbf{X}'\mathbf{Y}.$$

Time series data are often assumed to have autocorrelated errors, and consequently β is evaluated by the generalized least squares (GLS) estimate (73):

$$\hat{\beta}_g = (\mathbf{X}'\boldsymbol{\Sigma}^{-1}\mathbf{X})^{-1}\mathbf{X}'\boldsymbol{\Sigma}^{-1}\mathbf{Y}.$$

The correlation matrix $\boldsymbol{\Sigma}$ is usually unknown, but can be estimated as follows: First obtain residuals $e = \mathbf{Y} - \mathbf{X}\hat{\beta}_o$ from the OLS procedure, then fit autoregressive model AR(1) or AR(2) to e (67), estimate $\boldsymbol{\Sigma}$ based on the assumed AR model, estimate σ^2 by $e'\hat{\boldsymbol{\Sigma}}^{-1}e/(n - K - 1)$, and finally re-estimate β using the estimated $\hat{\sigma}^2$ and $\hat{\boldsymbol{\Sigma}}$.

Within the GLM framework, many scientific questions can be answered through hypothesis testing on β. We focus on the tests of linear contrasts of β with the null being $H_0 : c'\beta = d$, where the constant vector c' can take various forms depending on the problem. For example, if the goal is to study the brain response to a single condition (say the kth), then all the elements except for the one corresponding to β_k in c' equal zero. If the goal is to compare brain activities under different conditions, then the sum of c equals zero. The Student's t-test statistic is commonly used in these hypothesis tests:

$$t = \frac{c'\hat{\beta}_g - d}{\sqrt{S^2(c'\hat{\beta}_g)}},$$

where $S^2(c'\hat{\beta}_g)$ is the variance estimate of $c'\hat{\beta}_g$.

Since $\mathbf{V}(c'\hat{\beta}_g) = c'\mathbf{V}(\hat{\beta}_g)c = \sigma^2 \cdot c'(\mathbf{X}'\boldsymbol{\Sigma}^{-1}\mathbf{X})^{-1}c$ and $S^2(c'\hat{\beta}_g) = \hat{\sigma}^2 \cdot c'(\mathbf{X}'\hat{\boldsymbol{\Sigma}}^{-1}\mathbf{X})^{-1}c$, the test statistic follows the Student-t distribution with a degree of freedom ν approximated by the Satterthwaite method (72):

$$\nu = \frac{\text{tr}(\mathbf{R}\hat{\boldsymbol{\Sigma}})^2}{\text{tr}((\mathbf{R}\hat{\boldsymbol{\Sigma}})^2)},$$

where $\mathbf{R} = \mathbf{I}_T - \mathbf{X}(\mathbf{X}'\hat{\boldsymbol{\Sigma}}^{-1}\mathbf{X})^{-1}\mathbf{X}'\hat{\boldsymbol{\Sigma}}^{-1}$, and $\text{tr}(\cdot)$ stands for the trace of a matrix. Note that the tests discussed above are performed independently and repeatedly at each voxel. As such, the activation map construction, based on tests performed at spatially distributed voxels, is a multiple hypothesis testing problem, which is discussed in detail in Chapter 13.

Ill-posed problems due to non-invertible or close-to-singular matrix $\mathbf{X}'\mathbf{X}$ frequently arise in practice. Two common approaches can address this issue. One is to use the Moore–Penrose pseudoinverse (63, 66). Let $\mathbf{X}'\mathbf{X} = U\begin{pmatrix} S & 0 \\ 0 & 0 \end{pmatrix}U'$, where S is a diagonal matrix with no zero diagonal elements, and U is an orthogonal matrix. Then the pseudoinverse $(\mathbf{X}'\mathbf{X})^- = U'\begin{pmatrix} S^{-1} & 0 \\ 0 & 0 \end{pmatrix}U$. Another widely adopted method is through the Tikhonov regularization (77), of which a common example is the ridge regression: $\hat{\beta}_r = (\mathbf{X}'\mathbf{X} + \lambda \cdot \mathbf{I}_T)^{-1}\mathbf{X}'\mathbf{Y}$, where λ is a small positive constant. The ridge estimate is the minimizer of $||\mathbf{Y} - \mathbf{X}\beta||_2^2 + \lambda \cdot ||\beta||_2^2$, where the regularization term $\lambda \cdot ||\beta||_2^2$ controls the norm of β estimates. A generalized Tikhonov regularized estimate is $\hat{\beta}_t = (\mathbf{X}'\mathbf{X} + \lambda \cdot \boldsymbol{\Gamma}'\boldsymbol{\Gamma})^{-1}\mathbf{X}'\mathbf{Y}$, which is the minimizer of $||\mathbf{Y} - \mathbf{X}\beta||_2^2 + \lambda \cdot ||\boldsymbol{\Gamma}\beta||_2^2$. The matrix $\boldsymbol{\Gamma}$ is given according to the restriction to be imposed on β (more discussions in Section 11.3). The regularization parameter λ can be chosen by several methods, including cross validation, generalized cross validation (GCV) (82), and estimated mean squared error (91). The above estimators are summarized in Table 11.2.

TABLE 11.1
Summaries of notations in Section 11.2.

Notation	Description
$\mathbf{Y}$	a single fMRI time series data
$\hat{\beta}_o$	The OLS estimate
$\hat{\beta}_g$	The GLS estimate
$\hat{\beta}_r$	The ridge regression estimate
$\hat{\beta}_t$	The generalized Tikhonov regularized estimate

FMRI data often contain low-frequency signals due to physiological noise (7, 52, 74) or other fluctuation factors such as heartbeat, respiration, and head motion. Different approaches have been developed to remove such nuisance signal from fMRI data, including adding a low-order polynomial term of time t or splines (48, 60, 89) to the GLM, and using wavelet transforms (85) or discrete cosine basis functions—a high-pass filter—to filter noise (29, 88). With measurements of subjects' heartbeat and breathing, cardiac and respiratory signals can be removed by using their recorded time series as additional predictors. For example, (34) used Fourier series to characterize such signals. In both aforementioned approaches, additional, say m, predictors with associated design matrix $\mathbf{M}$ are included in the linear model (11.1) to account for the low-frequency signal:

$$\mathbf{Y} = \mathbf{M}\gamma + \mathbf{X}\beta + \varepsilon,$$

and the aforementioned regression methods are applied to estimate β.

Head motion, even small, can create time-dependent nonlinear changes in the magnetic field, and consequently can cause different effects on the fMRI data collected at different brain locations and collection time (32). Such motion, if not corrected in the brain activation map construction, increases the risk of false discoveries (23, 39), especially when the motion is correlated with the presence of experiment conditions. Different approaches have been developed for motion correction; these include realigning time series of images to a reference image using six parameters for translations and rotations, and incorporating estimated motion vectors into the linear model. Reference (43) compared three motion-correction processing methods: (i) correcting motion using the standard realignment, (ii) using motion estimates as nuisance covariates in the GLM after realigning the images, and (iii) including the motion estimates as nuisance covariates in the GLM without first realigning the images. Based on the comparison results, (43) proposed a general strategy that is widely applicable to different fMRI designs.

11.3 Modeling the Hemodynamic Response Function in the Time Domain

For block designs with sufficiently long—usually larger than 4s—blocks, the aforementioned t-tests can be directly applied to evaluate and compare average brain activity in different conditions. In event-related designs or block designs with short inter-stimulus intervals (ISI), typical GLMs involve convolutions between stimulus functions with a hemodynamic response function (HRF), as follows:

$$Y(t) = \int_0^m h(u) \cdot s(t-u)dt + \varepsilon(t), \tag{11.2}$$

where $s(t-u)$ is the stimulus function defined based on the paradigm of conditions used in the experimental session, $h(u)$ is the HRF used to characterize brain activity of the voxel in response to the stimulus, and $[0, m]$ defines the domain of the HRF. In a block design, the stimulus function $s(t)$ is a step function, equalling 1 during the stimulus, and 0 otherwise.

In an event-related design, $s(t)$ is a sum of delta functions: $s(t) = \sum_i \delta(t, t_i)$, where the t_is are the stimulus-evoked times, and $\delta(t, t_i) = 1$ if $t = t_i$ and 0 otherwise. Then the GLM (11.2) reduces to $Y(t) = \sum_i h(t - t_i) + \varepsilon(t)$. An example of this model is shown in Figure 11.1(b): here the stimulus—evoked at 0, 5, and 10 s—elicits HRFs of identical shape at the three time points, and the observed BOLD signal at time t, $Y(t)$, is the sum of the HRFs' values at the three time points plus error. The GLM (11.2) defines a linear time invariant (LTI) system; this system assumes the response to a single impulse (the HRF) is unchanged across time, and it also restricts the linear relationship between the system input (neuronal impulse) and output (BOLD response). The LTI assumption enables statistical inference of brain activity within the GLM framework.

When the experimental session involves more than two, say K, different conditions, the GLM is extended to

$$Y(t) = \mathbf{P}(t)\,\mathbf{d} + \sum_{k=1}^K \int_0^m h_k(u) \cdot s_k(t-u)dt + \varepsilon(t), \tag{11.3}$$

where $s_k(t)$ characterizes the experimental paradigm for the kth condition with $s_k(t) = 1$ if the kth condition is evoked at time t and 0 otherwise, and the term $\mathbf{P}(t)\mathbf{d}$ accounts for the low-frequency drift. A common choice of $\mathbf{P}(t)\mathbf{d}$ is a low-order polynomial of t with $\mathbf{P}(t) = (1, t, t^2)$ and $\mathbf{d} = (d_1, d_2, d_3)'$. The HRFs h_k are generally assumed to vary across conditions, voxels, and subjects. We point out that the variation in hemodynamic responses across different conditions of the same voxel is due to the different patterns of neuronal spike trains associated with each condition, though the hemodynamic response to each neuronal spiking is identical. Since it is impossible to explicitly characterize neuronal spiking activities under different experimental conditions within the GLM, we associate one unique HRF with each condition to account for the differences between experimental conditions.

The central task in the GLM framework is to estimate the HRFs. Once an estimate of h_k is obtained, its height (the maximum value), time to peak, and width (difference of t's at half maximum of h_k), as illustrated in Figure 11.1(a), can be used as low-dimensional summaries of the magnitude, reaction time, and duration, respectively, of the voxel's activity in the kth condition. Color-coded brain activation maps can be then be created (17, 18) based on these summaries of all voxels.

Estimation of HRFs can be carried under the time domain or the frequency domain, and below we describe several common methods under each domain.

11.3.1 Parametric Models

In the time domain, both parametric and nonparametric approaches to HRF estimation have been developed. Parametric models tend to use fewer parameters but can be restrictive, whereas nonparametric models are more flexible but often have estimates with large variability.

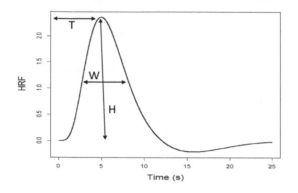

(a) The HRF

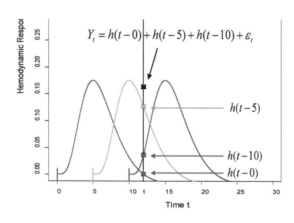

(b) The GLM

FIGURE 11.1
(a) Height (H), time-to-peak (T), and width (W) of the HRF. (b) The GLM for an event-related design, where the stimulus is evoked at 0, 5, and 10 s.

The simplest parametric model of HRFs is the Poisson model, first proposed by (27):

$$h(t) = \lambda^t e^{-\lambda}/t!,$$

where λ and t are positive integers. This model uses only one parameter to characterize magnitude and delay of HRF. Reference (3) proposed a more refined model:

$$h(t) = t^b e^{-t/c}, \quad t > 0,$$

where b and c are known realistic values that reflect the brain's hemodynamic responses. Given this HRF, (3) identified the optimum ISI for event-related designs with a fixed ISI. A more widely used HRF model is the difference of two Gamma density functions (89):

$$h(t) = A \left(b_1^{a_1} \frac{t^{a_1-1} \exp\{-b_1 \cdot t\}}{\Gamma(a_1)} - c \cdot b_2^{a_2} \frac{t^{a_2-1} \exp\{-b_2 \cdot t\}}{\Gamma(a_2)} \right), \tag{11.4}$$

where A determines HRF amplitude, b_1, a_1, b_2, and a_2 together determine HRF shape and scale, and c determines the poster-stimulus undershoot of the HRF. In the simplest case, only the magnitude A is treated as a free parameter, and the other parameters are fixed at $a_1 = 6, a_2 = 16, b_1 = b_2 = 1$, and $c = 1/6$, giving a "canonical" HRF (88). Another related model discussed in (33, 47) assumes:

$$h(t) = (t/d_1)^{a_1} \exp\{-(t - d_1)/b_1\} - c(t/d_2)^{a_2} \exp\{-(t - d_2)/b_2\}, \qquad (11.5)$$

where $a_1 = 6, a_2 = 12, b_1 = b_2 = 0.9, c = 0.35$, and $d_1 = a_1 b_1$ and $d_2 = a_2 b_2$ are the time to the peak and the time to the under-shoot, respectively.

Given an explicit form of the HRF with only magnitude unknown, the GLM (11.3) is converted to the linear regression model (11.1) with β consisting of drift coefficients and HRFs' magnitudes. This approach is also called the linear fit of canonical HRF. A more flexible approach—the nonlinear fit of HRF—treats all the parameters in (11.4) as unknown, and allows them to vary across stimuli, voxels, and subjects to accommodate variation of brain activity. The parameters can be estimated by minimizing the sum of squared error (SSE):

$$\text{SSE} = \sum_{t=1}^{T} \left(Y(t) - \mathbf{P}(t)\,\mathbf{d} - \sum_{k=1}^{K} \int_0^t h_k(u) \cdot s_k(t - u)dt \right)^2.$$

The SSE is a nonlinear function of the unknown parameters, and the optimization of SSE requires iterative algorithms such as the Newton–Raphson or the Levenberg–Marquardt algorithm.

An intermediate approach between the linear and nonlinear fit of Gamma functions uses the canonical HRF and its first-order derivative as basis functions, and estimates their coefficients by the least squares regression (12, 24, 41, 47). This approach is based on the following HRF model,

$$h(t) = A \cdot h_0(t - \delta),$$

where A and δ characterize the amplitude and delay of $h(t)$ with respect to the canonical HRF $h_0(t)$, respectively. Using the first-order Taylor expansion, $h(t) \approx A \cdot h_0(t) - \delta \cdot A \cdot h_0^{(1)}(t)$, model (11.2) is reduced to a linear regression with the design matrix consisting of the temporal terms associated with the drift and the convolutions of the stimulus paradigm with $h_0(t)$ and $h_0^{(1)}(t)$. Similarly, one can also incorporate a dispersion term into the GLM by using the Taylor approximation of the HRF model:

$$h(t) = A \cdot h_0((t - \delta)/W) \approx A \cdot h_0(t) - \delta \cdot A \cdot h_0^{(1)}(t) - A \cdot (W - 1) \cdot t \cdot h_0^{(1)}(t).$$

Using the canonical HRF's derivatives as functional bases provides more flexibility to detect the variation of brain activity.

Estimates of the latency, dispersion, and magnitude parameters involve the amplitude parameter of the canonical HRF and thus are not completely separate. To address this limitation, (47) proposed to choose functional bases by singular value decomposition (SVD), which can better approximate the HRF $h_0(t - \delta)$ for $-\Delta \leq \delta \leq \Delta$ given a reference h_0. The domain range for δ, Δ, typically equals 4.5. The idea is to first form a matrix $\mathbf{H}$ consisting of the values of $h_0(t - \delta)$ at equally spaced t (columns) and δ (rows). Then approximate $\mathbf{H}$ based on the SVD $\mathbf{H} = \mathbf{UDV}'$, where $\mathbf{U} = (\mathbf{u}_1, \mathbf{u}_2, \ldots)$, $\mathbf{V} = (\mathbf{v}_1, \mathbf{v}_2, \ldots)$ are orthonormal matrices, and $\mathbf{D} = \text{diag}(D_1 \geq D_2 \geq \ldots)$ is the diagonal matrix. Since $\mathbf{H}$ is close to $\mathbf{u}_1 D_1 \mathbf{v}_1' + \mathbf{u}_2 D_2 \mathbf{v}_2'$, approximate the HRF with $\mathbf{u}_1$ and $\mathbf{u}_2$ as two basis functions:

$$h_0(t - \delta) \approx w_1(\delta)\, U_1(t) + w_2(\delta)\, U_2(t),$$

where $U_1(t)$ and $U_2(t)$ are functional representations of $\mathbf{u}_2$ and $\mathbf{u}_2$, respectively, and functions of delay, $w_1(\delta)$ and $w_2(\delta)$, are approximated by $D_1\mathbf{v}_1$ and $D_2\mathbf{v}_2$, respectively. It has been shown that the variability of $h_0(t-\delta)$ can be explained by functional bases $U_1(t)$ and $U_2(t)$ more than by $h_0(t)$ and $h_0^{(1)}(t)$.

Plug $U_1(t)$ and $U_2(t)$, as the functional bases for the HRF, into the GLM, and obtain the least squares estimates of the coefficients $\hat{\beta}_1$ and $\hat{\beta}_2$ for $U_1(t)$ and $U_2(t)$, respectively. Then the delay parameter δ is estimated by solving $r(\hat{\delta}) = \hat{r}_C$, where

$$r(\delta) = w_2(\delta)/w_1(\delta), \quad \text{and} \quad \hat{r}_C = \hat{\beta}_2/\hat{\beta}_1 \cdot (1 + 1/T_1^2).$$

In the above equation, T_1—the t-statistic based on $\hat{\beta}_1$ for testing $\beta_1 = 0$—is used to correct for the effect of $\hat{\beta}_1$ being close to zero. The sampling distribution of the estimate of the delay parameter, $(\hat{\delta} - \mathbf{E}(\hat{\delta}))/\hat{\mathrm{sd}}(\hat{\delta})$, is approximated by a t-distribution with the regression degree of freedom. The variance of $\hat{\delta}$ is estimated using the Taylor expansion of its functional form $\hat{\delta} = r^{-1}(\hat{r}_c)$ given $\hat{\beta}_1$ and $\hat{\beta}_2$.

Reference (87) took a different basis-based parametric approach and characterized the HRF by four half-cosines with six free parameters in total. In order to obtain HRF estimates with a realistic shape, the authors imposed prior constraints on the model parameters within a Bayesian framework, and used the variational Bayes method to estimate the GLM. However, under this model, as well as Poisson and nonlinear Gamma fit models, the complicated relationship between the parameters and quantitative characteristics, such as height, width, and time-to-peak, of the HRFs renders interpretation of the model parameters difficult. To address this issue, (50) proposed an inverse logit model for the HRF:

$$h(t|\boldsymbol{\theta}) = \alpha_1 L\left((t - \delta_1)/W_1\right) + \alpha_2 L\left((t - \delta_2)/W_2\right) + \alpha_3 L\left((t - \delta_3)/W_3\right), \qquad (11.6)$$

where $L(x) = e^x/(1 + e^x)$ is the inverse logit function, and $\boldsymbol{\theta} = (\alpha_1, \delta_1, W_1, \alpha_2, \delta_2, W_2, \alpha_3, \delta_3, W_3)$. The three logit functions used in the model characterize different aspects of hemodynamic responses: The first describes the increase in brain activity after activation, the second describes the decrease after peak and undershoot, and the third describes the return to the baseline. For each inverse logit function, the parameter α indicates the direction and amplitude of changes, δ represents the shift, and W characterizes the extent of changes. In order to satisfy the requirement that the HRF equals zero at the boundary of the domain, the authors proposed to set $\alpha_3 = |\alpha_2| - |\alpha_1|$, and

$$\alpha_2 = \alpha_1 \frac{L(-\delta_3/W_3) - L(-\delta_1/W_1)}{L(-\delta_3/W_3) + L(-\delta_2/W_2)}.$$

Model (11.2) is rewritten as $Y(t) = f(t|\boldsymbol{\theta}) + \varepsilon(t)$, and the parameters are estimated by minimizing the SSE:

$$\mathrm{SSE}(\boldsymbol{\theta}) = (\mathbf{Y} - \mathbf{F}_{\boldsymbol{\theta}})'\boldsymbol{\Sigma}^{-1}(\mathbf{Y} - \mathbf{F}_{\boldsymbol{\theta}}),$$

where $\mathbf{Y} = (Y(1), \ldots, Y(T))'$, $\mathbf{F}_{\boldsymbol{\theta}} = (f(1|\boldsymbol{\theta}), \ldots, f(T|\boldsymbol{\theta}))'$, and $\boldsymbol{\Sigma}$ is the variance matrix of error terms. For an unknown $\boldsymbol{\Sigma}$, with the assumption of AR(1) for error terms, the SSE has an analytic form as

$$\mathrm{SSE}(\boldsymbol{\theta}, \phi) = z_1^2(1 - \phi^2) + \sum_{t=2}^{T}(z_t - \phi z_{t-1})^2,$$

where $z_t = Y(t) - f(t|\boldsymbol{\theta})$ and $-\phi$ is the superdiagonal entry values of $\boldsymbol{\Sigma}^{-1}$. An iterative Levenberg–Marquardt algorithm has been developed to minimize $\mathrm{SSE}(\boldsymbol{\theta}, \phi)$ with respect to $\boldsymbol{\theta}$.

11.3.2 Nonparametric Models

Nonparametric methods generally fall into two categories: either the value of the HRF at each unit time point is treated as a free parameter, or the HRF is represented by a linear combination of many functional bases, such as spline and wavelet bases, which are commonly used in nonparametric curve fitting and can flexibly approximate various functional shapes. The former approach can be viewed as a special case of the latter: in the finite impulse response (FIR) method (19, 44), the continuous HRF is discretized and represented by a set of delta functions in event-related designs or by a set of step functions in block designs. Reference (19, 16) pointed out that, with the discretization interval of the HRF being shorter than the fMRI sampling interval (TR), the resulting HRF estimates can have "a finer temporal resolution than that of the raw fMRI measurements" (19, p. 110).

Similar to parametric approaches, the GLM (11.3) with nonparametric specification for the HRFs is reduced to the linear regression model (11.1) with a much larger number of free parameters. The ensuing parameter estimation is straightforward, but the estimates usually have large variances. A variety of approaches have been proposed to address this issue. Capitalizing on the smooth property of the HRF (10), (36) proposed a regularized nonparametric HRF estimate, *smooth FIR* (SFIR), which has much smaller variance than standard least squares estimates and satisfies a boundary condition that the HRF values at the two ends equal zero. To illustrate, consider a simple situation with only one condition presented ($K = 1$) and without drift terms in the GLM. The idea is to assume a Gaussian prior for β: $\beta \sim \mathrm{N}(\mathbf{0}, \mathbf{R}^{-1})$, and estimate β by its posterior mode. The posterior distribution of β under independent Gaussian errors is

$$
\begin{aligned}
p(\beta|\mathbf{X}, \mathbf{Y}) &\propto \exp\{-(\mathbf{Y} - \mathbf{X}\beta)^2/2\sigma^2\} \cdot \exp\{-\beta' \mathbf{R}\beta/2\} \\
&\propto \exp\{-(\beta - \hat{\beta}_{\mathrm{MP}})'(\mathbf{X}'\mathbf{X} + \sigma^2 \mathbf{R})(\beta - \hat{\beta}_{\mathrm{MP}})/2\sigma^2\},
\end{aligned}
$$

where $\hat{\beta}_{\mathrm{MP}} = (\mathbf{X}'\mathbf{X} + \sigma^2 \mathbf{R})^{-1}\mathbf{X}'\mathbf{Y}$ is the posterior mode, and $\mathbf{R} = \Omega^{-1}$ with $\Omega_{ij} = v \exp\{-h(i - j)^2/2\}$ for $1 \leq i, j \leq d$ (d is the number of HRF parameters). The matrix $\mathbf{R}$ imposes a restriction on the HRF estimate such that the βs corresponding to temporally close points take similar values and the ensuing HRF estimate is smooth.

The regularization matrix can be further modified to satisfy the boundary condition that the βs corresponding to $h(0)$ and $h(m)$ equal zero. Specifically, let $\tilde{\Omega}$ be a $(d + 2) \times (d + 2)$ matrix with entries $\tilde{\Omega}_{ij} = v \exp(-h(i - j)^2/2)$ for $0 \leq i, j \leq d + 1$ and $\tilde{\mathbf{R}}$ be the central $d \times d$ submatrix of $\tilde{\Omega}^{-1}$. Then the SFIR estimator of β is

$$
\hat{\beta}_{\mathrm{MP}} = (\mathbf{X}'\mathbf{X} + \sigma^2 \tilde{\mathbf{R}})^{-1}\mathbf{X}'\mathbf{Y}.
$$

The hyper-parameter h controls the smoothness of the estimator. The larger h is the smaller the correlation between parameters. Reference (36) proposed to use $h = (TR/7)^2$. The extent of regularization in $\hat{\beta}_{\mathrm{MP}}$ is determined by the ratio of the variance σ^2 and the hyper-parameter v, σ^2/v. The parameters σ^2 and v are chosen by maximizing the density of $\mathbf{Y}$ given all the parameters:

$$
p(\mathbf{Y}|\mathbf{X}, \sigma^2, v, h) = (2\pi\sigma^2)^{-T/2} \left(\frac{\sigma^{2d}|\tilde{\mathbf{R}}|}{|\mathbf{X}'\mathbf{X} + \sigma^2\tilde{\mathbf{R}}|} \right)^{1/2} \cdot \exp\left\{ -\frac{\mathbf{Y}'(\mathbf{Y} - \mathbf{X}\hat{\beta}_{\mathrm{MP}})}{2\sigma^2} \right\}.
$$

Optimization of $p(\mathbf{Y}|\mathbf{X}, \sigma^2, v, h)$ is through iterative procedures. For example, given σ^2, one can obtain optimizers of $\hat{\beta}_{\mathrm{MP}}$ and v, based on which one updates the estimate of σ^2 (5, Sec. 10.4).

Reference (58, 59) proposed to use a discrete second-derivative matrix as the regularization matrix R in smooth FIR:

$$R = \frac{\tau^2}{\sigma^2} \cdot \frac{1}{\text{TR}^4} \begin{pmatrix} 5 & -4 & 1 & 0 & & \cdots & & 0 \\ -4 & 6 & -4 & 1 & 0 & & & \\ 1 & -4 & 6 & -4 & 1 & 0 & & \\ 0 & 1 & -4 & 6 & -4 & 1 & 0 & \vdots \\ & \ddots & \ddots & \ddots & \ddots & \ddots & \ddots & \ddots \\ \vdots & & 0 & 1 & -4 & 6 & -4 & 1 & 0 \\ & & & 0 & 1 & -4 & 6 & -4 & 1 \\ & & & & 0 & 1 & -4 & 6 & -4 \\ 0 & & \cdots & & & 0 & 1 & -4 & 5 \end{pmatrix},$$

where the hyper-parameter τ^2 adjusts the weight of the prior relative to noise; τ^2 is usually estimated by its posterior mode.

Reference (13, 14) developed a Tikhonov-regularization and generalized-cross-validation (Tik-GCV)-based method for HRF estimation. Let $\text{diag}(\mathbf{A}, \mathbf{B})$ denote the square matrix $[\mathbf{A}\ \mathbf{0}\ ;\ \mathbf{0}\ \mathbf{B}]$, and $\otimes$ denote the Kronecker product operation. For GLMs (11.3) with multiple conditions and drift predictors, define a regularization matrix $R = \text{diag}(\mathbf{0}, \mathbf{I}_K \otimes \Omega^{-1})$, where $\mathbf{0}$ is a zero vector corresponding to the drift parameters $\mathbf{d}$, and $\Omega = (L'L)^{-1}$ with

$$L = \begin{pmatrix} -2 & 1 & 0 & \cdots & & \cdots & 0 \\ 1 & -2 & 1 & 0 & & \ddots & \vdots \\ 0 & \ddots & \ddots & -\ddots & \ddots & & \vdots \\ \vdots & \ddots & \ddots & \ddots & \ddots & & 0 \\ \vdots & & \ddots & 0 & 1 & -2 & 1 \\ 0 & \cdots & & \cdots & 0 & 1 & -2 \end{pmatrix}.$$

Then $\boldsymbol{\beta}$ is estimated by $\hat{\boldsymbol{\beta}} = (\mathbf{X}'\mathbf{X} + \lambda R)^{-1}\mathbf{X}'\mathbf{Y}$, where the penalty parameter λ is chosen by the GCV.

A second class of methods represent HRFs by continuous functional base. Reference (90) proposed a two-level HRF estimation procedure using cubic smoothing splines for block designs. First, reformulate model (11.3) for $K = 1$ as

$$Y(t) = \sum_{j:0 \leq u_j \leq t} s(t - u_j)h(u_j)\delta_j + d(t) + \varepsilon(t), \quad t = 1, \ldots, T,$$

where $\delta_j = u_j - u_{j-1}$ and $d(t)$ stands for the low-frequency drift. The matrix representation of the above equation is:

$$\mathbf{Y} = \mathbf{S}\mathbf{h} + \mathbf{d} + \varepsilon, \tag{11.7}$$

where

$$\mathbf{S} = \begin{bmatrix} s(0) & 0 & \cdots & 0 \\ s(1) & s(0) & \cdots & 0 \\ \vdots & \vdots & \ddots & \vdots \\ s(m-1) & s(m-2) & \cdots & s(0) \\ s(m) & s(m-1) & \cdots & s(1) \\ \vdots & \vdots & \ddots & \vdots \\ s(T-1) & s(T-2) & \cdots & s(T-m) \end{bmatrix}, \quad \mathbf{h} = \begin{bmatrix} h(1) \\ \vdots \\ h(m) \end{bmatrix}, \quad \text{and } \mathbf{d} = \begin{bmatrix} d(1) \\ \vdots \\ d(T) \end{bmatrix}.$$

For $K > 1$, the matrix representation of model (11.3) becomes

$$\mathbf{Y} = \mathbf{S}_1\mathbf{h}_1 + \ldots + \mathbf{S}_K\mathbf{h}_K + \mathbf{d} + \varepsilon,$$

and the matrix formulation (11.7) continues to hold, with $\mathbf{S} = [\mathbf{S}_1, \ldots, \mathbf{S}_K]$ and $\mathbf{h} = [\mathbf{h}_1, \ldots, \mathbf{h}_K]'$. The basic idea of this curve-fitting method is to estimate $\mathbf{E}(y|x) = g(x)$ given a sample of random pairs $\{(x_i, y_i)\}_{i=1}^n$ by minimizing the penalized sum of squared errors (PSSE):

$$\frac{1}{n}\sum_{i=1}^n \{y_i - g(x_i)\}^2 + \lambda \int_a^b \{g^{(2)}(x)\}^2 dx, \quad \lambda > 0,$$

over the class of all twice-differentiable functions g. Define $\mathbf{g} = (g(x_1), g(x_2), \ldots, g(x_n))'$ and $\mathbf{y} = (y_1, \ldots, y_n)'$. According to Theorem 2.1 of (37), the above PSSE can be written in a matrix form as

$$(\mathbf{y} - \mathbf{g})'(\mathbf{y} - \mathbf{g})/n + \lambda \cdot \mathbf{g}'\mathbf{K}\mathbf{g},$$

where $\mathbf{K}$ is a known symmetric matrix whose elements are functions of $X_{i+1} - X_i$ for $1 \le i \le n-1$. Then the minimizer of the PSSE is $\hat{\mathbf{g}} = \mathbf{C}\mathbf{y}$, where $\mathbf{C}$ is a known smoothing-spline matrix related to $\mathbf{K}$. Based on this result, if $\mathbf{h}$ in (11.7) is known, a nonparametric cubic smoothing spline estimate of $\mathbf{d}$ is

$$\hat{\mathbf{d}}(\mathbf{h}) = \mathbf{C}_d(\mathbf{Y} - \mathbf{S}\mathbf{h}),$$

where $\mathbf{C}_d$ is an $n \times n$ smooth matrix of the cubic smoothing splines. Plugging $\hat{\mathbf{d}}(\mathbf{h})$ into (11.7), we get a one-level smoothing-spline estimate of $\mathbf{h}$:

$$\hat{\mathbf{h}}_{\text{rough}} = (\tilde{\mathbf{S}}'\hat{\boldsymbol{\Sigma}}^{-1}\tilde{\mathbf{S}})^{-1}\tilde{\mathbf{S}}'\hat{\boldsymbol{\Sigma}}^{-1}\tilde{\mathbf{Y}},$$

where $\tilde{\mathbf{Y}} = (\mathbf{I} - \mathbf{C}_d)\mathbf{Y}$, $\tilde{\mathbf{S}} = (\mathbf{I} - \mathbf{C}_d)\mathbf{S}$, and $\hat{\boldsymbol{\Sigma}}$ is the estimated correlation matrix of $\boldsymbol{\varepsilon}$. Applying the cubic spline smoothing to the rough one-level estimate gives a two-level HRF estimate:

$$\hat{\mathbf{h}}_{\text{smooth}} = \mathbf{C}_h\hat{\mathbf{h}}_{\text{rough}},$$

where $\mathbf{C}_h$ denotes the cubic spline smoothing matrix of $\mathbf{h}$. With the HRF estimates, the one-level and two-level estimates of the drift components are

$$\hat{\mathbf{d}}_{\text{rough}} = \mathbf{C}_d(\mathbf{Y} - \mathbf{S}\hat{\mathbf{h}}_{\text{rough}}) \quad \text{and} \quad \hat{\mathbf{d}}_{\text{smooth}} = \mathbf{C}_d(\mathbf{Y} - \mathbf{S}\hat{\mathbf{h}}_{\text{smooth}}).$$

Reference (78) applied functional data analysis techniques to HRF estimation using B-spline expansions and Tikhonov regularization. To illustrate, take the simplest single-condition model (11.2) as an example. Let $h(t) = \sum_{j=1}^J c_j\phi_j(t) = \boldsymbol{\phi}_t'\mathbf{c}$, where $\{\phi_j(t)\}_{j=1}^J$ is a set of known basis functions, typically, B-spline bases (21), $\mathbf{c} = (c_1, \ldots, c_J)'$, and $\boldsymbol{\phi}_t' = (\phi_1(t), \ldots, \phi_J(t))$. Model (11.2) is converted to a linear regression model

$$Y(t) = \sum_{j=1}^J c_j \int_0^m \phi_j(u) \cdot s(t-u)du + \varepsilon(t).$$

The matrix representation is

$$\mathbf{Y} = \boldsymbol{\Phi}\mathbf{c} + \boldsymbol{\varepsilon},$$

where $\boldsymbol{\Phi}$ is the design matrix with elements consisting of $\int_0^m \phi_j(u) \cdot s(t-u)du$ for $j = 1, \ldots, J$ and $t = 1, \ldots, T$. The basis coefficients are estimated by minimizing the PSSE with a penalty on the second-order derivative of the HRF:

$$\text{PSSE}(\mathbf{c}) = \sum_{t=1}^T \left\{ Y(t) - \sum_{j=1}^J c_j \int_0^m \phi_j(u) \cdot s(t-u)du \right\}^2$$

$$+ \lambda \cdot \int_0^m \left\{ \sum_{j=1}^J c_j\phi_j^{(2)}(u) \right\}^2 du.$$

The penalty term is to avoid overfitting and to smooth the least squares estimate of the HRF. The penalty is quadratic of $\mathbf{c}$, and thus the matrix representation of PMSE($\mathbf{c}$) is

$$\text{PSSE}(\mathbf{c}) = (\mathbf{Y} - \mathbf{\Phi c})'(\mathbf{Y} - \mathbf{\Phi c}) + \lambda \cdot \mathbf{c}'\mathbf{P}\mathbf{c},$$

where $\mathbf{P}(j_1, j_2) = \int_0^m \int_0^m \phi_{j_1}^{(2)}(u_1)\phi_{j_2}^{(2)}(u_2)du_1 du_2$. The minimizer $\hat{\mathbf{c}}_\lambda$ of PSSE($\mathbf{c}$) is $(\mathbf{\Phi}'\mathbf{\Phi} + \lambda \cdot \mathbf{P})^{-1}\mathbf{\Phi}'\mathbf{Y}$. The penalty parameter λ is selected through GCV by minimizing

$$\text{GCV}(\lambda) = \frac{1}{T}\frac{\hat{\varepsilon}_\lambda'\hat{\varepsilon}_\lambda}{\left(1 - \text{tr}\mathbf{S}_\lambda/T\right)^2},$$

where $\hat{\varepsilon}_\lambda = \mathbf{Y} - \mathbf{\Phi}\hat{\mathbf{c}}_\lambda$ are the regression residuals, and $\mathbf{S}_\lambda = \mathbf{\Phi}(\mathbf{\Phi}'\mathbf{\Phi} + \lambda \cdot \mathbf{P})^{-1}\mathbf{\Phi}'$.

In the experimental designs with multiple conditions and drift terms, the matrix representation of model (11.3) becomes

$$\mathbf{Y} = \mathbf{Dd} + \mathbf{\Phi}_1\mathbf{c}_1 + \ldots + \mathbf{\Phi}_K\mathbf{c}_K + \varepsilon.$$

Let $\mathbf{\Phi} = [\mathbf{D}, \mathbf{\Phi}_1, \ldots, \mathbf{\Phi}_K]$, $\mathbf{c} = [\mathbf{d}, \mathbf{c}_1, \ldots, \mathbf{c}_K]'$, and the penalty matrix $\mathbf{\Psi} = \text{diag}[\mathbf{0}, \mathbf{I}_K \otimes \mathbf{P}]$. The estimate of $\mathbf{c}$ becomes $(\mathbf{\Phi}'\mathbf{\Phi} + \lambda \cdot \mathbf{\Psi})^{-1}\mathbf{\Phi}'\mathbf{Y}$.

11.3.3   Comparison of Different HRF Estimation Methods

Reference (50) conducted extensive comparisons of five HRF modeling approaches for estimating response amplitude/height (H), time-to-peak (T), and full-width at half-max (W) of the HRF using simulated and real data. The five HRF estimation methods are inverse logit model (IL), nonlinear fit on two Gamma functions, the canonical HRF with temporal derivatives, Smooth FIR (SFIR), and linear fit of Gamma functions, i.e., the standard canonical HRF model. We first elaborate the procedure for estimating H, T, and W based on the HRF estimate $\hat{h}(t)$ from different approaches. For models without closed forms of H, T, and W, the estimates of T and H are

$$\hat{T} = \min\{t|\hat{h}^{(1)}(t) = 0, \ \hat{h}^{(2)}(t) < 0\} \ \text{ and } \ \hat{H} = \hat{h}(\hat{T}).$$

Let $t_u = \min\{t|t > \hat{T}, \ \hat{h}(t) < \hat{H}/2\}$ and $t_l = \max\{t|t < \hat{T}, \ \hat{h}(t) < \hat{H}/2\}$. For continuous $\hat{h}$, $\hat{W} = t_u - t_l$. For discretized $\hat{h}$, W is estimated through a linear interpolation: $\hat{W} = (t_{u-1} + \triangle_u) - (t_{l+1} + \triangle_l)$, where

$$\triangle_l = \frac{h(t_{l+1}) - 0.5\text{H}}{h(t_{l+1}) - h(t_l)} \ \text{ and } \ \triangle_u = \frac{h(t_{u-1}) - 0.5\text{H}}{h(t_{u-1}) - h(t_u)}.$$

For the canonical HRF with its derivative, to correct for the bias due to the confounding effects between the magnitude and delay terms, (12) used $\text{sign}(\hat{\beta}_m)\sqrt{\hat{\beta}_m^2 + \hat{\beta}_d^2}$ as an amplitude estimate, where $\hat{\beta}_m$ and $\hat{\beta}_d$ are the coefficient estimates for the canonical HRF and its temporal derivative, respectively. For the IL (11.6), $\hat{\text{H}} = \alpha_1$, $\hat{\text{T}} = \delta_1 + W_1 c$, and $\hat{\text{W}} = \delta_2 - \delta_1 - W_2 \log(2\alpha_2/\alpha_1 - 1)$, where c is a constant, chosen by the analyst, and can be set to, for example, $\log((0.99^{-1} - 1)^{-1})$.

Through simulations, (50) demonstrated that each of the five methods introduces some bias in estimation, and the largest biases occurred in all the models for HRFs with an extended width. However, the IL did not have bias in estimating H, and had only a small bias in estimating T. The SFIR had the highest power in detecting the true effects, though it has less interpretable model parameters. In general, IL and SFIR performed better than the other HRF estimation methods. The authors suggested that SFIR is more suitable for modeling shorter-duration events, while IL is better for fMRI designs with longer and more variable epochs. In terms of computational expenses, SFIR beats IL, especially for GLMs with multiple stimuli and weak signal-to-noise ratio.

Extending (50), (49) introduced procedures based on bootstrap and the sign permutation test for inference of summary statistics and investigated independence of parameter estimates by seven models: the canonical HRF used in SPMs, the canonical HRF plus its temporal derivative, the canonical HRF plus its temporal and dispersion derivatives, the model based on the finite impulse response basis set, SFIR, two Gamma functions with six parameters, and IL. They determined the amount of dependence among estimates of H, T, and W, and found that the IL has the least parameter dependence based on simulation studies and real data analysis.

In the GLMs for fMRI with different conditions, if each HRF is represented by multiple functional bases, it is not self-evident how to choose proper statistics for comparing brain activations in different conditions. Different approaches have been proposed. For example, SPM tests all the basis coefficients via an F-test. Reference (24) suggested comparing estimates of the main basis coefficient. Reference (12) recommended comparing the norm of the coefficients for the canonical HRF and its derivatives. Reference (49) compared low-dimensional characteristics, e.g. height and width, of HRF estimates. Alternatively, one can perform joint estimation and detection within a Bayesian framework (15, 53, 54, 80).

11.4   Hemodynamic Response Estimation in the Frequency Domain

In contrast to the vast literature of the time-domain estimation methods, only a few methods have been proposed to estimate the HRF in the frequency domain (2, 45, 57, 83, 84). The basic idea of these frequency domain methods is to first transform the original fMRI voxel time series into the frequency domain using discrete Fourier transform, and then develop a statistical model for the obtained Fourier coefficients. In comparison to the time domain approaches, the frequency domain methods are less sensitive to the temporal correlation assumption of the error process, since the Fourier coefficients are approximately uncorrelated across frequencies (57).

One of the first frequency domain approaches is given by (45), which models the HRF with a two-parameter gamma function, and allows the parameters to change across voxels. Reference (2) stated that the approach of (45) was reported to have the issue of parameter identifiability. Reference (57) addressed the issue with a fixed HRF for experimental designs with periodic stimuli.

Recently, (2) and (84) developed nonparametric HRF estimation methods in the frequency domain, which are more flexible than the earlier parametric frequency-domain methods. In particular, (2) considered voxel-specific estimation for event-related designs with a single stimulus, and demonstrated how the method can be used to test the linearity assumption (11.2). Numerical studies are used to illustrate the nice estimation performance of the method. However, the method analyzed one voxel time series at a time, and ignored the fact that fMRI data are naturally spatially dependent, for example, many fMRI studies reveal spatially contiguous activation regions with rather sharp edges (84). This motivated (84) to extend the approach of (2) to develop a multiscale adaptive smoothing model (MASM), that incorporates spatial and temporal dependence using time series across all voxels in a three-dimensional (3D) brain volume in a multi-scale and adaptive fashion. (Note that (83) reports some preliminary results of (84).) Below we provide some details of the MASM and the corresponding estimation algorithm as described in (84).

Let $\mathcal{D} \subset R^3$ denote the 3D brain volume of a single subject. For voxel $\mathbf{d} \in \mathcal{D}$, $Y(t, \mathbf{d})$ is the recorded BOLD signal, which according to the LTI theory, is the convolution of the stimuli functions $\mathbf{S}(t) = (s_1(t), \ldots, s_K(t))^T$ and the corresponding HRFs $\mathbf{H}(t, \mathbf{d}) = (h_1(t, \mathbf{d}), \ldots, h_K(t, \mathbf{d}))^T$, plus an error process $\epsilon(t, \mathbf{d})$. The time-domain convolution model can be written as

$$Y(t, \mathbf{d}) = \mathbf{H}(\cdot, \mathbf{d}) \otimes \mathbf{S}(t) + \epsilon(t, \mathbf{d}) = \int <\mathbf{H}(\mathbf{d}, t - u), \mathbf{S}(u)> du + \epsilon(t, \mathbf{d}). \qquad (11.8)$$

MASM first transforms model (11.8) into the frequency domain, by applying discrete Fourier transform to both sides of the model. Denote the discrete Fourier coefficients of $Y(t, \mathbf{d})$, $\mathbf{H}(t, \mathbf{d})$, $\mathbf{S}(t)$, and $\epsilon(t, \mathbf{d})$, as $\phi_{\mathbf{Y}}(f_m, \mathbf{d})$, $\phi_{\mathbf{H}}(f_m, \mathbf{d})$, $\phi_{\mathbf{S}}(f_m)$, and $\phi_\epsilon(f_m, \mathbf{d})$, respectively, at the fundamental frequencies $f_m = m/T$ for $m = 0, \cdots, T - 1$. Then, MASM assumes that

$$\phi_Y(f, \mathbf{d}) = <\phi_{\mathbf{H}}(f, \mathbf{d}), \phi_{\mathbf{S}}(f)> + \phi_\epsilon(f, \mathbf{d}),$$

where $\phi_{\mathbf{H}}(f, \mathbf{d}) = (\phi_{h_1}(f, \mathbf{d}), \ldots, \phi_{h_K}(f, \mathbf{d}))^T$, $\phi_{\mathbf{S}}(f) = (\phi_{s_1}(f), \ldots, \phi_{s_K}(f))^T$. Considering a fixed voxel $\mathbf{d}$ and $K = 1$, the above model reduces to the model considered by (2). To incorporate spatial and temporal dependence, MASM then borrows strength across nearby voxels and time points by assuming that for each stimulus $s_k(t)$ and each voxel d, $\phi_{h_k}(f, \mathbf{d})$ is close to $\phi_{h_k}(f', \mathbf{d}')$ for some neighboring voxels $\mathbf{d}'$ and frequencies f'.

To estimate $\phi_{\mathbf{H}}(f, \mathbf{d})$, the MASM combines all information at fundamental frequencies $f_m \in \eta_f(r) = (f - r, f + r) \cap \{m/T : m = 0, 1, \ldots, T - 1\}$ with $r > 0$ and voxels $\mathbf{d}' \in B(\mathbf{d}, s)$, where $B(\mathbf{d}, s)$ is a spherical neighborhood of voxel $\mathbf{d}$ with radius $s \geq 0$ to construct an approximation equation as follows:

$$\phi_Y(f_m, \mathbf{d}') \approx <\phi_{\mathbf{H}}(f, \mathbf{d}), \phi_{\mathbf{S}}(f_m)> + \phi_\epsilon(f_m, \mathbf{d}').$$

Then, the MASM estimates $\phi_{H_k}(f, \mathbf{d})$ for $k = 1, \cdots, K$ using weighted least squares of K local weighted functions $L_{[-k]}(\phi_{H_k}(f, \mathbf{d}); r, s)$ defined as

$$\sum_{f_m \in \eta_f(r)} \sum_{\mathbf{d}' \in B(\mathbf{d}, s)} |\phi_{Y[-k]}(f_m, \mathbf{d}') - \phi_{h_k}(f, \mathbf{d})\phi_{s_k}(f_m)|^2 \tilde{w}_k(\mathbf{d}, \mathbf{d}', f, f_m; r, s)$$

for $k = 1, \cdots, K$, where $\phi_{Y[-k]}(f_m, \mathbf{d}') = \phi_Y(f_m, \mathbf{d}') - \sum_{l \neq k} \phi_{h_k}(f_m, \mathbf{d}')\phi_{s_k}(f_m)$. The weight $\tilde{w}_k(\mathbf{d}, \mathbf{d}', f, f_m; r, s)$ characterizes the physical distance between $(f, \mathbf{d})$ and $(f_m, \mathbf{d}')$ and the similarity between $\phi_{h_k}(f, \mathbf{d})$ and $\phi_{h_k}(f_m, \mathbf{d}')$. The spatial radii s and the frequency radii r are carefully chosen in a multiscale adaptive fashion, following the Propagation-Separation approach of (76). With $\hat{\phi}_{h_k}(f, \mathbf{d})$ for $k = 1, \cdots, K$, the HRF estimates are obtained as

$$\tilde{h}_k(t, \mathbf{d}) = \frac{1}{T} \sum_{m=0}^{T-1} \hat{\phi}_{h_k}(f_m, \mathbf{d}) \exp(i2\pi t f_m) \text{ for any } \mathbf{d} \in \mathcal{D} \text{ and } t. \qquad (11.9)$$

The numerical studies of (84) suggest that the MASM performs significantly better than the SFIR of (36), the inverse logit model of (50), and the canonical HRF.

11.5   Multi-Subject Analysis

For multi-subject fMRI data analysis, the simplest approach is to estimate each subject's HRF independently using the same method (e.g., one of those in the previous section), and then summarize the results to account for the variability across subjects (1, 40). However, this approach may be inefficient when the data from each individual have a low signal-to-noise ratio (SNR). Another approach is to assume a two-level, mixed-effect or random-effect model (4, 26, 31, 64) for multi-subject fMRI data, which will be discussed in detail in Chapter 12. This section focuses on several recently proposed semi-parametric methods, which assume common characteristics of the HRFs shared across a population to improve the estimation efficiency of each subject's HRF estimate.

11.5.1 Semi-Parametric Approaches

Reference (92) proposed a general model for the HRFs in multi-subject settings, assuming that the HRFs for a fixed voxel under stimulus k share a common functional form but can differ in magnitude and latency across subjects,

$$h_{i,k}(t) = A_{i,k} \cdot f_k(t + D_{i,k}), \qquad (11.10)$$

where $A_{i,k}$ and $D_{i,k}$ represent magnitude and latency of the chosen voxel's reaction to the kth stimulus of subject i, respectively. No parametric assumption, except differentiability, is imposed on the population-average HRF $f_k(t)$. Model (11.10) enables "borrowing" information across subjects while allowing for subject-specific characteristics. Also, the nonparametric nature of f_k provides maximum flexibility in modeling heterogeneous brain activities across regions and stimuli. Model (11.10) was also considered in (24) and (41), where $f_k(t)$ is fixed as the canonical HRF.

To estimate the semi-parametric model (11.10), Reference (92) suggested representing $f_k(t)$ by *cubic B-spline basis*: $f_k(t) = \sum_{l=1}^{L} a_{l,k} b_l(t)$, where the basis functions $b_l(t)$ are chosen based on a partition $\Pi_q = (t_0 = 0, t_1, \cdots, t_q = m)$ of the interval $[0, m]$. Applying this B-spline representation to its first-order Taylor expansion, model (11.10) reduces to a bilinear form:

$$Y_i(t) = \mathbf{P}_i(t) \cdot \mathbf{d}_i + \sum_{k=1}^{K} \sum_{l=2}^{L-1} \omega_{l,k}^i \, \rho_{l,k}^i(t) + \sum_{k=1}^{K} \sum_{l=2}^{L-1} \nu_{l,k}^i \, \varrho_{l,k}^i(t) + \varepsilon_i(t), \qquad (11.11)$$

where $\rho_{l,k}^i(t) = \int_0^m b_l(u) s_{i,k}(t - u) du$ and $\varrho_{l,k}^i(t) = \int_0^m b_l'(u) s_{i,k}(t - u) du$ are known functions; $\omega_{l,k}^i = A_{i,k} \cdot a_{l,k}$ and $\nu_{l,k}^i = A_{i,k} \cdot D_{i,k} \cdot a_{l,k}$. Though $A_{i,k}$, $D_{i,k}$ and $a_{l,k}$ are not directly identifiable, they are estimable by additional moment constraints. To estimate the parameters of the reduced model (11.11), (92) developed a fast algorithm that avoids knot selection and iterative estimation by imposing a penalty on the roughness of the HRFs. The crucial penalty parameter can be selected by minimizing the average MSE, or based on GCV (70, 71) or the restricted maximum likelihood (REML) method (86).

11.6 Nonlinear Models

The existence of nonlinearities in evoked responses, particularly in event-related fMRI designs, has been widely recognized in the literature, e.g. (11, 28, 30, 62, 75, 79, 81). The degree of nonlinearity, induced by nonlinearity both in the vascular response and at the neuronal level, usually varies across brain regions and stimuli, and shorter intervals between stimuli lead to stronger nonlinearity (8, 20, 51, 79). Though the importance of adjusting for nonlinearity in estimating hemodynamic responses has been demonstrated (for a compelling example, see (81)), quantification of nonlinearity is challenging. Below we review three models for a nonlinear system.

11.6.1   The Balloon Model

The nonlinear behavior of the hemodynamic response to a stimulus has been shown to depend on several physiological parameters, including cerebral blood volume (CBV), cerebral blood flow (CBF) and deoxyhemoglobin concentration. The Balloon model is a biomechanical model first proposed by (11). Denote the baseline (at rest) deoxyhemoglobin content, blood volume fraction, and oxygen extraction fraction by Q_0, V_0, and E_0, respectively. Define normalized total deoxyhemoglobin content $q = Q/Q_0$ and normalized blood volume fraction $v = V/V_0$.

The BOLD signal S consists of intravascular S_i and extravascular S_e signals:

$$S = (1 - V)S_e + VS_i. \tag{11.12}$$

Taking a small change on both sides of the above equation and dividing the ensuing equation by (11.12) gives:

$$\frac{\triangle S}{S} = \frac{V_0}{1 - V_0 + \gamma V_0} \left\{ \frac{1 - V_0}{V_0} \frac{\triangle S_e}{S_e} + \frac{\triangle S_i}{S_e} + (1 - v)(1 - \gamma) \right\}, \tag{11.13}$$

where $\gamma = S_i/S_e$ is the intrinsic signal ratio.

With numerical derivations of changes in intravascular and extravascular signals $\triangle S_e/S_e$ and $\triangle S_i/S_e$ based on the results in (65) and (6), model (11.13) becomes

$$\frac{\triangle S}{S} = V_0 \left\{ k_1(1 - q) + k_2(1 - \frac{q}{v}) + k_3(1 - v) \right\}, \tag{11.14}$$

where $\frac{\triangle S}{S}$ is the BOLD signal in $Y(t)$, and parameters k_1, k_2, and k_3 are obtained from the results in the literature (6, 65).

The next step is to model the transient change of the venous volume and deoxyhemoglobin content. The Balloon model assumes that with neglected capillary contribution, the venous vessel's biomechanical function in receiving output from a capillary bed is similar to a swelling balloon. The changes in normalized deoxyhemoglobin content and venous volume depend on the CBF into and out of the receiving venous compartment, denoted by F_{in} and F_{out}, respectively, and the ensuing models are

$$\frac{dq(t)}{dt} = \frac{1}{\tau_0} \left\{ f_{\text{in}}(t) \frac{E(f_{\text{in}}, E_0)}{E_0} - f_{\text{out}}(V) \frac{q(t)}{v(t)} \right\}, \tag{11.15}$$

$$\frac{dv(t)}{dt} = \frac{1}{\tau_0} (f_{\text{in}}(t) - f_{\text{out}}(v)),$$

where $\tau_0 = V_0/F_0$ is the mean transit time through the venous compartment at rest, $f_{\text{in}}(t) = F_{\text{in}}(t)/F_0$ and $f_{\text{out}}(t) = F_{\text{out}}(V)/F_0$ are normalized CBF flows, and $E(f_{\text{in}}, E_0)$ is the net extraction of O_2, depending on the O_2 extraction at rest and CBF inflow.

Using the result in (9), the dependence of E on E_0 and f_{in} is

$$E(f_{\text{in}}, E_0) = 1 - (1 - E_0)^{1/f_{\text{in}}}. \tag{11.16}$$

The model for f_{out} is a combination of a power law (38) and a linear component (56, 55), corresponding to steady states and transition period between two steady states, respectively:

$$f_{\text{out}} = 1 + \lambda_1(v - 1) + \lambda_2(v - 1)^b, \tag{11.17}$$

where $\lambda_1 = 0.2$, $\lambda_2 = 4.0$ and $b = 0.47$. In summary, Equations (11.14), (11.15), (11.16), and (11.17) jointly model the connection between the BOLD signal with changes in blood deoxyhemoglobin and volume.

Reference (30) extended the above model by adding equations to characterize the connection of the inflow to the neuronal activity $s(t)$:

$$\frac{df_{\text{in}}(t)}{dt} = q(t)$$

$$\frac{dq(t)}{dt} = \phi\, s(t) - q(t)/\tau_s - (f_{\text{in}}(t) - 1)/\tau_f,$$

where ϕ, τ_s and τ_f are unknown parameters, representing the efficacy with which neuronal activity causes an increase in signal, the rate of signal decay, and the rate of autoregulatory feedback from the blood flow, respectively. For each voxel, the six parameters—E_0, τ_0, b, ϕ, τ_s and τ_f—can be estimated by minimizing the least squared difference between the first- and second-order kernels of the proposed differential equations and the empirical kernels.

11.6.2 Volterra Series Model

For the Balloon model including equations for neuronal activity $s(t)$, the output of the dynamic system $Y(t)$ can be mathematically represented as a nonlinear convolution of neuronal activity $s(t)$, as an input, with Volterra series:

$$Y(t) = Q_0(t) + \sum_d^\infty \int_0^t \cdots \int_0^t Q_d(t, u_1, \ldots, u_d) \times s(u_1) \ldots s(u_d) du_1 \ldots du_d,$$

(11.18)

where Q_d is the dth-order Volterra kernels. The formulation of Q_d, $d = 1, \ldots, \infty$, depend on the Balloon model formulation and model parameters.

Reference (28) proposed to use a truncated second-order Volterra series to characterize nonlinearity in evoked hemodynamic responses as follows:

$$Y(t) = \mathbf{P}(t)\,\mathbf{d} + \int_0^m h(u) \cdot s(t-u) du$$

$$+ \int_0^m Q(u_1, u_2) \cdot s(t - u_1) \cdot s(t - u_2) du_1 du_2 + \varepsilon_i(t). \quad (11.19)$$

Based on model (11.19), an extension of the GLM and an approximation of the Balloon model, one can bypass estimating the Balloon model and use estimation techniques for the GLM to evaluate both the linear and nonlinear brain responses in different conditions, as explained below.

Representing h and Q by a set of functional bases, say cubic B-splines $b_l(t)$ for $l = 1, \ldots, L$:

$$h(t) = \sum_{l=1}^L c_l^1 b_l(t) \quad \text{and} \quad Q(t_1, t_2) = \sum_{l_1=1}^L \sum_{l_2=1}^L c_{l_1 l_2}^2 b_{l_1}(t_1) b_{l_2}(t_2).$$

Then model (11.19) reduces to a linear regression

$$Y(t) = \mathbf{P}(t)\,\mathbf{d} + \sum_{l=1}^L c_l^1 x_l(t) + \sum_{l_1=1}^L \sum_{l_2=1}^L c_{l_1 l_2}^2 x_{l_1}(t) x_{l_2}(t) + \varepsilon_i(t),$$

where $x_l(t) = \int_0^m b_l(u) \cdot s(t-u) du$. One can apply standard linear regression techniques described in Section 11.2 to estimate the coefficients and the Volterra kernels.

Handbook of Neuroimaging Data Analysis

11.6.3   Bi-Exponential Nonlinear Model

Reference (81) proposed to use a bi-exponential function of the position of stimulus in the time sequence to approximate the nonlinear behavior of the HRF in height, onset, and peak delay. Specifically, for brain response in the visual and motor cortex to stimuli separated 1 s apart, they assumed that

$$
\begin{aligned}
m(x) &= 1.7141e^{-2.1038x} + 0.4932e^{-0.0770x} \\
d(x) &= -13.4097e^{-1.0746x} + 4.8733e^{0.1979x} \\
p(x) &= 37.5445e^{-2.6760x} - 3.204e^{-0.2120x} + 5.6344,
\end{aligned}
$$

where x is the position of the stimulus (e.g. 1 stands for the first time a stimulus occurs in the time sequence), $m(x)$ is the stimulus magnitude, as a proportion of the canonical HRF height with no stimulation history, $d(x)$ is the onset delay of the canonical SPM99 HRF in s, and $p(x)$ is the time to peak. Then the refined model for hemodynamic response, modified by using the nonlinear models for $m(x)$, $p(x)$ and $d(x)$, is

$$
h(t) = \frac{m(x)}{\max\left\{ \frac{(t-d(x))^{p(x)-1}e^{-\lambda t}}{\int_0^\infty t^{p(x)-1}e^{-\lambda t}dt} \right\}} \left\{ \frac{(t-d(x))^{p(x)-1}e^{-\lambda t}}{\int_0^\infty t^{p(x)-1}e^{-\lambda t}dt} - \frac{1}{6}\frac{(t-d(x))^{15}e^{-\lambda t}}{\int_0^\infty t^{15}e^{-\lambda t}dt} \right\}, \quad (11.20)
$$

where λ is a constant scaling parameter equal to the TR divided by sampling resolution. Simulation studies showed that the modified linear model and the Balloon model gave similar results, both beating the conventional linear model.

11.6.4   Volterra Series Models for Multi-Subject Data

Reference (81) pointed out several limitations of model (11.20). First, it may not be flexible enough to characterize nonlinear behavior of brain activity across all regions, especially for those exhibiting noncanonical saturation effects. Second, the bi-exponential formulas may only be suitable for designs with ISI around 1 s. In comparison, the Volterra-series-based nonlinear models are flexible to accommodate nonlinear behavior across different brain regions, stimulus types, and inter-stimulus interval lengths. But these models are also more difficult to estimate given limited available data. Reference (93) proposed a new semi-parametric Volterra series model for multi-subject data to alleviate these issues. They rewrote model (11.19) to allow the drift term $\mathbf{P}$, the HRF h, and the second-order Volterra kernel Q to be subject-dependent:

$$
Y_i(t) = \mathbf{P}_i(t) \cdot \mathbf{d}_i + \sum_{k=1}^{K} \int_0^m h_{i,k}(u) \cdot s_{i,k}(t-u)du
$$

$$
+ \sum_{k_1,k_2=1}^{K} \int_0^m Q_{i,k_1k_2}(u_1,u_2) \cdot s_{i,k_1}(t-u_1) \cdot s_{i,k_2}(t-u_2)du_1 du_2 + \varepsilon_i(t).
$$

$$(11.21)$$

To simultaneously model population-wide and subject-specific characteristics of brain activity, and to "borrow information" across subjects, (93) assumed a semi-parametric form for both h and Q:

$$
\begin{aligned}
Q_{i,k_1k_2}(t_1,t_2) &= M_{i,k_1k_2} \cdot Q_{k_1k_2}(t_1,t_2), \\
h_{i,k}(t) &= A_{i,k} \cdot f_k(t+D_{i,k}).
\end{aligned} \quad (11.22)
$$

The above model assumes that the interaction pattern between hemodynamic responses of a given pair of stimuli is identical, but differs in intensity across subjects. No parametric assumption except for differentiability is imposed on f_k and $Q_{k_1 k_2}$.

Model (11.22) is similar to the semi-parametric linear model (11.10) discussed in Section 11.5.1, and (93) proposed an analogous spline-basis-based regularized strategy to estimate model (11.22). Specifically, one can represent the bivariate function $Q_{k_1 k_2}(t_1, t_2)$ by cubic spline bases:

$$Q_{k_1 k_2}(t_1, t_2) = \sum_{l_1 l_2 = 1}^{L} Z_{k_1 k_2 l_1 l_2} \cdot b_{l_1}(t_1) \cdot b_{l_2}(t_2).$$

Nonlinearity is known to disappear if events are spaced at least 5 seconds apart (61), and therefore $Q_{k_1 k_2}(t_1, t_2) = 0$ for $|t_1 - t_2| \geq 5$. Let $\mathcal{L}^2 = \{(l_1, l_2) : 1 \leq l_1, l_2 \leq L; |l_1 - l_2| \geq 4 + 5/m \cdot (L - 2)\}$ and $\mathcal{K}^2 = \{(k_1, k_2): \text{there exists at least one } (u_1, u_2) \in (0, m)^2 \text{ such that } s_{i,k_1}(t - u_1) = s_{i,k_2}(t - u_2) = 1 \text{ for at least one subject } i \text{ and } |u_1 - u_2| < 5\}$. Then the nonlinear functional model (11.21) is transformed to a bilinear model:

$$
\begin{aligned}
Y_i(t) = \ & \mathbf{P}_i(t) \cdot \mathbf{d}_i + \sum_{k=1}^{K} \sum_{l=2}^{L-1} \omega_{i,kl} \cdot \rho_{i,kl}(t) + \sum_{k=1}^{K} \sum_{l=2}^{L-1} \phi_{i,kl} \cdot \varrho_{i,kl}(t) \\
& + \sum_{(k_1, k_2) \in \mathcal{K}^2} \sum_{(l_1, l_2) \in \mathcal{L}^2} \nu_{i,k_1 k_2 l_1 l_2} \cdot \psi_{k_1 k_2 l_1 l_2}(t) + \varepsilon_i(t),
\end{aligned}
\tag{11.23}
$$

where $\omega_{i,kl} = A_{i,k} \cdot a_{kl}$, $\phi_{i,kl} = A_{i,k} \cdot D_{i,k} \cdot a_{kl}$, $\nu_{i,k_1 k_2 l_1 l_2} = M_{i,k_1 k_2} \cdot Z_{k_1 k_2 l_1 l_2}$, $\rho_{i,kl}(t) = \int_0^m b_l(u) \cdot s_{i,k}(t - u)du$, $\varrho_{i,kl}(t) = \int_0^m b_l(u) \cdot s_{i,k}(t - u)du$, and $\psi_{k_1 k_2 l_1 l_2}(t) = \int_0^m \int_0^m b_{l_1}(u_1) \cdot b_{l_2}(u_2) \cdot s_{i,k_1}(t - u_1) \cdot s_{i,k_2}(t - u_2)du_1 du_2$ are known functions. Parameters of model (11.23) can be estimated through a penalized approach similar to that in Section 11.5.1.

11.7   Summary and Future Directions

The statistical models and methods described in this chapter are for voxel-wise analysis of fMRI data. Since spatially close voxels tend to have similar fMRI data, it is advantageous to incorporate the spatial information of fMRI data into HRF estimation and construction of the brain activation map. Reference (53, 54) developed a brain-parcellation-based Bayesian model for simultaneous brain activation detection and HRF estimation. This model is based on the assumption that the HRFs of voxels within the same parcel share a common functional shape but with different magnitudes. Reference (80) extended this Bayesian model by using a spatially adaptive prior for the hemodynamic response magnitudes of voxels within the same parcel. Reference (15) further developed a joint parcellation-detection-estimation procedure for simultaneously identifying brain parcels of voxels with homogeneous hemodynamic properties, detecting activated brain regions, and estimating HRFs in within-subject fMRI analysis. Recently, (22) proposed a Bayesian model for joint activation detection and HRF estimation for multi-subject fMRI studies.

FMRI data in the region of interest or estimated HRFs of activated voxels can be used in scalar-on-image (SI) regressions for predicting subjects' outcomes such as mental states. Several statistical models (35, 42, 69, 46) have been proposed for SI regressions using fMRI data. SI regressions with alternative imaging modalities, such as diffusion tensor imaging (68), have also been discussed.

Bibliography

[1] GK Aguirre, E Zarahn, and M D'Esposito. The variability of human, BOLD hemodynamic responses. *NeuroImage*, 8:360–369, 1998.

[2] P Bai, Y Truong, and X Huang. Nonparametric estimation of hemodynamic response function: A frequency domain approach. *Lecture Notes-Monograph Series*, pages 190–215, 2009.

[3] PA Bandettini and RW Cox. Event-related fMRI contrast when using constant inter-stimulus interval: Theory and experiment. *Magnetic Resonance in Medicine*, 43:540–548, 2000.

[4] CF Beckmann, M Jenkinson, and SM Smith. General multi-level linear modelling for group analysis in fMRI. *NeuroImage*, 20:1052–1063, 2003.

[5] CM Bishop. *Neural Networks for Pattern Recognition*. Oxford University Press, New York, NY, 1995.

[6] JL Boxerman, PA Bandettini, KK Kwong, JR Baker, T Davis, BR Rosen, and RM Weisskoff. The intravascular contribution to fMRI signal change: Monte Carlo modeling and diffusion-weighted studies in vivo. *Magnetic Resonance in Medicine*, 34:4–10, 1995.

[7] J Brosch, T Talavage, J Ulmer, and J Nyenhuis. Simulation of human respiration in fMRI with a mechanical model. *IEEE Transactions on Biomedical Engineering*, 49:700–707, 2002.

[8] RL Buckner. Event-related fMRI and the hemodynamic response. *Human Brain Mapping*, 6:373–377, 1998.

[9] RB Buxton and LR Frank. A model for the coupling between cerebral blood flow and oxygen metabolism during neural stimulation. *Journal of Cerebral Blood Flow and Metabolism*, 17:64–72, 1997.

[10] RB Buxton, K Uludag, DJ Dubowitz, and TT Liu. Modeling the hemodynamic response to brain activation. *Neuroimage*, 23:220–233, 2004.

[11] RB Buxton, EC Wong, and LR Frank. Dynamics of blood flow and oxygenation changes during brain activation: The balloon model. *Magnetic Resonance in Medicine*, 39:855–864, 1998.

[12] VD Calhoun, MC Stevens, GD Pearlson, and KA Kiehl. fMRI analysis with the general linear model: Removal of latency-induced amplitude bias by incorporation of hemodynamic derivative terms. *NeuroImage*, 22:252–257, 2004.

[13] R Casanova, S Ryali, J Serences, L Yang, R Kraft, PJ Laurienti, and JA Maldjian. The impact of temporal regularization on estimates of the BOLD hemodynamic response function: A comparative analysis. *NeuroImage*, 40(4):1606–1618, 2008.

[14] R Casanova, L Yang, WD Hairston, PJ Laurienti, and JA Maldjian. Evaluating the impact of spatio-temporal smoothness constraints on the BOLD hemodynamic response function estimation: An analysis based on Tikhonov regularization. *Physiological Measurement*, 30(5):37–51, 2009.

[15] L Chaari, F Forbes, T Vincent, and P Ciuciu. Adaptive hemodynamic-informed parcellation of fMRI data in a variational joint detection estimation framework. In *15th Proceedings MICCAI, LNCS*, 7512:180–188, 2012.

[16] P Ciuciu, JB Poline, G Marrelec, J Idier, C Pallier, and H Benali. Unsupervised robust nonparametric estimation of the hemodynamic response function for any fMRI experiment. *IEEE Transactions on Medical Imaging*, 22:1235–1251, 2003.

[17] RW Cox. AFNI software for analysis and visualization of functional magnetic resonance neuroimages. *Computers and Biomedical Research*, 29:162–173, 1996.

[18] RW Cox and JS Hyde. Software tools for analysis and visualization of fMRI data. *NMR in Biomedicine*, 10:171–178, 1997.

[19] A Dale. Optimal experimental design for event-related fMRI. *Human Brain Mapping*, 8:109–114, 1999.

[20] AM Dale and RL Buckner. Selective averaging of rapidly presented individual trials. *Human Brain Mapping*, 5:323–340, 1997.

[21] C de Boor. *A Practical Guide to Splines*. Springer, Berlin, 1978.

[22] D Degras and MA Lindquist. A hierarchical model for simultaneous detection and estimation in multi-subject fMRI studies. *NeuroImage*, 98:61–72, 2014.

[23] AS Field, YF Yen, JH Burdette, and AD Elster. False cerebral activation on bold functional MR images: Study of low-amplitude motion weakly correlated to stimulus. *AJNR Am J Neuroradiol*, 21:1388–1396, 2000.

[24] KJ Friston, P Fletcher, O Josephs, A Holmes, MD Rugg, and R Turner. Event-related fMRI: Characterizing differential responses. *NeuroImage*, 7:30–40, 1998.

[25] KJ Friston, A Holmes, KJ Worsley, J-P Poline, CD Frith, and RSJ Frackowiak. Statistical parametric mapping: A general linear approach. *Human Brain Mapping*, 2:189–210, 1995.

[26] KJ Friston, AP Holmes, CJ Price, C Büchel, and JK Worsley. Multisubject fMRI studies and conjunction analyses. *NeuroImage*, 10:385–396, 1999.

[27] KJ Friston, P Jezzard, and R Turner. Analysis of functional MRI time-series. *Human Brain Mapping*, 1:153–171, 1994.

[28] KJ Friston, O Josephs, G Rees, and R Turner. Nonlinear event-related responses in fMRI. *Magn Reson Med*, 39:41–52, 1998.

[29] KJ Friston, O Josephs, E Zarahn, AP Holmes, S Rouquette, and JB Poline. To smooth or not to smooth? *NeuroImage*, 12:196–208, 2000.

[30] KJ Friston, A Mechelli, E Turner, and CJ Price. Nonlinear responses in fMRI: The Balloon model, Volterra kernels, and other hemodynamics. *NeuroImage*, 12:466–477, 2000.

[31] KJ Friston, W Penny, C Phillips, S Kiebel, G Hinton, and J Ashburner. Classical and Bayesian inference in neuroimaging: Theory. *NeuroImage*, 16:465–483, 2002.

[32] KJ Friston, S Williams, R Howard, RS Frackowiak, and R Turner. Movement-related effects in fMRI time-series. *Magn Reson Med*, 35:346–355, 1996.

[33] GH Glover. Deconvolution of impulse response in event-related BOLD fMRI. *NeuroImage*, 9:416–429, 1999.

[34] GH Glover, TQ Li, and D Ress. Image-based method for retrospective correction of physiological motion effects in fMRI: RETROICOR. *Magnetic Resonance in Medicine*, 44:162–167, 2000.

[35] J Goldsmith, L Huang, and CM Crainiceanu. Smooth scalar-on-image regression via spatial Bayesian variable selection. *Journal of Computational and Graphical Statistics*, 23:46–64, 2014.

[36] C Goutte, FA Nielsen, and LK Hansen. Modeling the hemodynamic response in fMRI using smooth FIR filters. *IEEE Transactions on Medical Imaging*, 19:1188–1201, 2000.

[37] PJ Green and BW Silverman. *Nonparametric Regression and Generalized Linear Models: A Roughness Penalty Approach*. Chapman & Hall, London, 1994.

[38] RL Grubb, ME Raichle, JO Eichling, and MM Ter-Pogossian. The effects of changes in PCO_2 on cerebral blood volume, blood flow, and vascular mean transit time. *Stroke*, 5:630–639, 1974.

[39] JV Hajnal, R Myers, A Oatridge, JE Schwieso, IR Young, and GM Bydder. Artifacts due to stimulus-correlated motion in functional imaging of the brain. *Magn Reson Med*, 3:283–291, 1994.

[40] DA Handwerker, JM Ollinger, and D'Esposito M. Variation of BOLD hemodynamic responses across subjects and brain regions and their effects on statistical analyses. *NeuroImage*, 21:1639–1651, 2004.

[41] R Henson, C.J Price, MD Rugg, R Turner, and KJ Friston. Detecting latency differences in event-related BOLD responses: Application to words versus nonwords and initial versus repeated face presentations. *NeuroImage*, 15:83–97, 2002.

[42] L Huang, J Goldsmith, PT Reiss, DS Reich, and CM Crainiceanu. Bayesian scalar-on-image regression with application to association between intracranial DTI and cognitive outcomes. *NeuroImage*, 83:210–223, 2013.

[43] T Johnstone, KS Ores Walsh, LL Greischar, AL Alexander, AS Fox, RJ Davidson, and TR Oakes. Motion correction and the use of motion covariates in multiple-subject fMRI analysis. *Human Brain Mapping*, 27:779 –788, 2006.

[44] N Lange, SC Strother, JR Anderson, FA Nielsen, AP Holmes, T Kolenda, R Savoy, and L. K. Hansen. Plurality and resemblance in fMRI data analysis. *NeuroImage*, 10:282–303, 1999.

[45] N Lange and SL Zeger. Non-linear Fourier time series analysis for human brain mapping by functional magnetic resonance imaging (with discussion). *Journal of the Royal Statistical Society: Series C (Applied Statistics)*, 46(1):1–29, 1997.

[46] F Li, T Zhang, Q Wang, MZ Gonzalez, EL Maresh, and JA Coan. Spatial Bayesian variable selection and grouping in high-dimensional covariate spaces with application to fMRI. *Annals of Applied Statistics*, 9:687–713, 2015.

[47] CH Liao, KJ Worsley, JB Poline, JAD Aston, GH Duncan, and AC Evans. Estimating the delay of the fMRI response. *NeuroImage*, 16:593–606, 2002.

[48] MA Lindquist. The statistical analysis of fMRI data. *Statistical Science*, 23:439–464, 2008.

[49] MA Lindquist, JM Loh, LY Atlas, and TD Wager. Modeling the hemodynamic response function in fMRI: Efficiency, bias and mis-modeling. *NeuroImage*, 45:187–198, 2009.

[50] MA Lindquist and TD Wager. Validity and power in hemodynamic response modelling: A comparison study and a new approach. *Human Brain Mapping*, 28:764–784, 2007.

[51] H Liu and J Gao. An investigation of the impulse functions for the nonlinear bold response in functional MRI. *Magn. Reson. Imaging*, 18:931–938, 2000.

[52] H Luo and S Puthusserypady. Analysis of fMRI datawith drift: Modified general linear model and Bayesian estimator. *IEEE Transactions on Biomedical Engineering*, 55:1504–1511, 2008.

[53] S Makni, P Ciuciu, J Idier, and JB Poline. Joint detection-estimation of brain activity in functional MRI: A multichannel deconvolution solution. *IEEE Transactions on Signal Process*, 53(9):3488–3502, 2005.

[54] S Makni, J Idier, T Vincent, B Thirion, G Dehaene-Lambertz, and P Ciuciu. A fully Bayesian approach to the parcel-based detection-estimation of brain activity in fMRI. *NeuroImage*, 41(3):941–969, 2008.

[55] JB Mandeville, JJA Marota, C Ayata, G Zaharchuk, MA Moskowitz, BR Rosen, and et al. Evidence of a cerebrovascular postarteriole windkessel with delayed compliance. *Journal of Cerebral Blood Flow and Metabolism*, 19(6):679–689, 1999.

[56] JB Mandeville, JJA Marota, BE Kosovsky, JR Keltner, R Weissleder, BR Rosen, and RM Weisskoff. Dynamic functional imaging of relative cerebral blood volume during rat forepaw stimulation. *Magnetic Resonance in Medicine*, 39:615–624, 1996.

[57] JL Marchini and BD Ripley. A new statistical approach to detecting significant activation in functional MRI. *NeuroImage*, 12(4):366–380, 2000.

[58] G Marrelec, H Benali, P Ciuciu, M Pelegrini-Issac, and JB Poline. Robust estimation of the hemodynamic response function in event-related BOLD fMRI using basic physiological information. *Human Brain Mapping*, 19:1–17, 2003.

[59] G Marrelec, H Benali, P Ciuciu, and JB Poline. Bayesian estimation of the hemodynamic of the hemodynamic response function in functional MRI. *AIP Conference Proceedings*, 617:229–247, 2001.

[60] VS Mattay, JA Frank, AKS Santha, JJ Pekar, JH Duyn, AC McLaughlin, and DR Weinberger. Whole brain functional mapping with isotropic MR imaging. *Radiology*, 201:399–404, 1996.

[61] FM Miezin, L Maccotta, JM Ollinger, SE Petersen, and RL Buckner. Characterizing the hemodynamic response: Effects of presentation rate, sampling procedure, and the possibility of ordering brain activity based on relative timing. *NeuroImage*, 11(6):735–759, 2000.

[62] KL Miller, WM Luh, TT Liu, A Martinez, T Obata, EC Wong, LR Frank, and RB Buxton. Nonlinear temporal dynamics of the cerebral blood flow response. *Human Brain Mapping*, 13:1–12, 2001.

[63] EH Moore. On the reciprocal of the general algebraic matrix. *Bulletin of the American Mathematical Society*, 26(9):394–395, 1920.

[64] JA Mumford and TE Nichols. Modeling and inference of multisubject fMRI data. *IEEE Engineering in Medicine and Biology Magazine*, 25(2):42–51, 2006.

[65] S Ogawa, RS Menon, DW Tank, S-G Kim, H Merkle, JM Ellerman, and K Ugurbil. Functional brain mapping by blood oxygenation level-dependent contrast magnetic resonance imaging: A comparison of signal characteristics with a biophysical model. *Biophysical Journal*, 64:803–812, 1993.

[66] R Penrose. A generalized inverse for matrices. *Proceedings of the Cambridge Philosophical Society*, 51:406–413, 1955.

[67] P Purdon and R Weisskoff. Effects of temporal autocorrelation due to physiological noise and stimulus paradigm on voxel-level false-positive rates in fMRI. *Human Brain Mapping*, 6:239–249, 1998.

[68] PT Reiss, L Huo, Y Zhao, C. Kelly, and RT Ogden. Wavelet-domain regression and predictive inference in psychiatric neuroimaging. *Annals of Applied Statistics*, 9:1076–1101, 2015.

[69] PT Reiss, M Mennes, E Petkova, L Huang, MJ Hoptman, BB Biswal, SJ Colcombe, XN Zuo, and MP Milham. Extracting information from functional connectivity maps via function-on-scalar regression. *NeuroImage*, 56:140–148, 2011.

[70] PT Reiss and RT Ogden. Functional principal component regression and functional partial least squares. *Journal of the American Statistical Association*, 102:984–996, 2007.

[71] PT Reiss and RT Ogden. Smoothing parameter selection for a class of semiparametric linear models. *Journal of the Royal Statistical Society, Series B*, 71:505–523, 2009.

[72] F Satterthwaite. An approximate distribution of estimates of variance components. *Biometrics Bulletin*, 2:110–114, 1946.

[73] SR Searle. *Linear Models*. John Wiley & Sons, 1971.

[74] A Smith, B Lewis, U Ruttinmann, and et al. Investigation of low frequency drift in fMRI signal. *NeuroImage*, 9:526–533, 1999.

[75] DA Soltysik, KK Peck, KD White, B Crosson, and RW Briggs. Comparison of hemodynamic response non-linearity across primary cortical areas. *NeuroImage*, 22:1117–1127, 2004.

[76] Karsten Tabelow, Jörg Polzehl, Henning U Voss, and Vladimir Spokoiny. Analyzing fMRI experiments with structural adaptive smoothing procedures. *NeuroImage*, 33(1):55–62, 2006.

[77] AN Tikhonov and VY Arsenin. *Solution of Ill-Posed Problems*. Winston, New York, 1977.

[78] VA Vakorin, R Borowsky, and GE Sarty. Characterizing the functional MRI response using Tikhonov regularization. *Statistics in Medicine*, 26(21):3830–3844, 2007.

[79] AL Vazquez and DC Noll. Nonlinear aspects of the BOLD response in functional MRI. *NeuroImage*, 7:108–118, 1998.

[80] T Vincent, L Risser, and P Ciuciu. Spatially adaptive mixture modeling for analysis of fMRI time series. *IEEE Transactions on Medical Imaging*, 29(4):1059–1074, 2010.

[81] TD Wager, A Vazquez, L Hernandez, and DC Nollb. Accounting for nonlinear BOLD effects in fMRI: Parameter estimates and a model for prediction in rapid event-related studies. *NeuroImage*, 25:206–218, 2005.

[82] G Wahba. *Spline Models for Observational Data*. SIAM, Philadelphia, 1990.

[83] J Wang, H Zhu, J Fan, K Giovanello, and W Lin. Adaptively and spatially estimating the hemodynamic response functions in fMRI. In *Medical Image Computing and Computer-Assisted Intervention–MICCAI 2011*, pages 269–276. Springer, 2011.

[84] J Wang, H Zhu, J Fan, K Giovanello, and W Lin. Multiscale adaptive smoothing models for the hemodynamic response function in fMRI. *The Annals of Applied Statistics*, 7(2):904–935, 2013.

[85] JB Weaver, Y Xu, DM Healy, and LD Cromwell. Filtering noise from images with wavelet transforms. *Magnetic Resonance in Medicine*, 21:288–295, 1991.

[86] SN Wood. Fast stable restricted maximum likelihood and marginal likelihood estimation of semiparametric generalized linear models. *Journal of the Royal Statistical Society, Series B*, 73:3–36, 2011.

[87] MW Woolrich, TE Behrens, and SM Smith. Constrained linear basis sets for HRF modelling using variational Bayes. *Neuroimage*, 21:1748–1761, 2004.

[88] KJ Worsley and KJ Friston. Analysis of fMRI time-series revisited again. *NeuroImage*, 2:173–181, 1995.

[89] KJ Worsley, CH Liao, J Aston, V Petre, GH Duncan, F Morales, and A Evans. A general statistical analysis for fMRI data. *NeuroImage*, 15:1–15, 2002.

[90] CM Zhang, Y Jiang, and T Yu. A comparative study of one-level and two-level semiparametric estimation of hemodynamic response function for fMRI data. *Statistics in Medicine*, 26:3845–3861, 2007.

[91] T Zhang, F Li, L Beckes, C Brown, and JA Coan. Nonparametric inference of hemodynamic response for multi-subject fMRI data. *NeuroImage*, 63:136–145, 2012.

[92] T Zhang, F Li, L Beckes, and JA Coan. A semi-parametric model of the hemodynamic response for multi-subject fMRI data. *NeuroImage*, 75:1754–1765, 2013.

[93] T Zhang, F Li, M Gonzalez, E Maresh, and JA Coan. A semi-parametric nonlinear model for event-related fMRI. *NeuroImage*, 97:178–187, 2014.

12

Functional Neuroimaging Group Studies

Bertrand Thirion

Université Paris Saclay, France

CONTENTS

12.1	Introduction	335
12.2	Variability of Brain Shape and Function	337
12.3	Mixed-Effects and Fixed-Effects Analyses	339
12.4	Group Analysis for Functional Neuroimaging	339
	12.4.1 Problem Setting and Notations	340
	12.4.2 Estimation	341
	12.4.3 Statistical Inference	342
	12.4.4 The Random Effects t-Test	343
12.5	Taking into Account the Spatial Context in Statistical Inference	344
12.6	Type I Error Control with Permutation Testing	346
12.7	Illustration of Various Inference Strategies on an Example	347
12.8	Conclusion	350
	Bibliography	350
	A Table of the Most Standard Statistical Tests in Neuroimaging	354

Multi-subject statistical analysis is an essential step of neuroimaging studies, as it makes it possible to draw conclusions that hold with a prescribed confidence level for the population under study. The use of the linear assumption to model activation signals in brain images and their modulation by various factors has opened the possibility to rely on relatively simple estimation and statistical testing procedures. Specifically, the analysis of functional neuroimaging signals is typically carried out on a per-voxel basis, in the so-called *mass univariate framework*. However, the lack of power in neuroimaging studies has incited neuroscientists to develop new procedures to improve this framework: various solutions have been set up to take into account the spatial context in statistical inference or to deal with violations of distributional assumptions of the data. In this chapter, we review the general framework for group inference, the ensuing mixed-effects model design and its simplifications, together with the various solutions that have been considered to improve the standard mass-univariate testing framework.

12.1 Introduction

Most neuroimaging statistical problems deal with the comparison of sets of images that embed some features of brain structure or function across a group of subjects, with the purpose to detect some common effects across individuals or some differences across

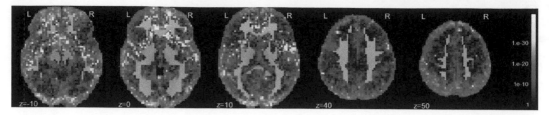

FIGURE 12.1

Illustration of the deviation from normality of functional neuroimaging datasets: The map shows the p-value of a test (5) rejecting the normality of the distribution of the z-transformed activation statistics across subjects using the dataset presented in Section 12.7 ($n = 573$). Note that the Gaussian hypothesis is significantly rejected in *all* cortical regions.

sub-populations. The most standard framework consists of comparing images on a voxel-by-voxel basis (or a vertex-by-vertex basis if the data is sampled on a mesh), after resampling in a common spatial referential, such as the referential defined by the Montreal NeuroImaging Institute (MNI) template (18), or a coordinate system on the cortical surface (7). This resampling is assumed to correct for pose and shape differences across individuals, so that possible remaining differences are related to the anatomical or functional feature of interest and not to a mere residual of between-subject registration. Note that this is an assumption, and that the limitations of the brain image coregistration procedures used for this purpose are key to understanding the difficulty of statistical inference in the context of neuroimaging group studies.

The simplest statistical inference procedure for image-based data, a.k.a. mass-univariate inference, relies on the computation of a statistic in each voxel. This statistic is then compared to a reference distribution that represents its likely values when the hypothetical effect is absent. If the actual statistic value is extreme with respect to this distribution—*its p-value is low*—one can conclude that it would not likely be observed under the null hypothesis, hence it is more likely explained by some alternative hypothesis: *the null hypothesis is rejected.* This inference scheme is known as *classical statistics.* In practice, the question of how low the p-value should be to support the conclusion is dealt with arbitrarily, and the corresponding choice ($p < .05$ is commonly accepted in the literature) cannot be justified within the framework of classical statistics (23).

The choice of the decision statistic is very important. As many studies involve comparing the mean value of a given image-derived feature across populations, parametric tests such as Student's t-test or Fisher's F-test are natural choices. However, they are limited in two regards:

- In order to yield accurate p-values, they involve a *normality hypothesis* which is most often violated (see e.g. Fig. 12.1). Depending on how whether standard parametric statistics are robust enough to these deviations or other non-parametric statistics should be used instead is an important question. In practice, however, the price to pay in terms of computation cost or loss of sensitivity when using non-parametric inference is most often too high, while accurate p-values for t or F tests can be obtained under weak assumptions by *permutation testing* (2, 21).

- The individual values that are compared in the test are not directly observed; instead, they are estimated with some level of uncertainty from the acquired data. Taking into account the uncertainty in the statistical evaluation leads formally to a *mixed-effects model* (19). Such a model is more accurate, but also arguably more expensive than a simple random-effects model. We review the mixed-effects formulation and its simplifications in detail in Section 12.4.

Then comes an issue that is standard in statistics, yet particularly problematic in neuroimaging, namely that of *multiple comparisons*: since many tests are performed simultaneously, the risk of observing low p-values by chance, and hence making false detections, is high. Depending on the statistical guarantees required for the analysis, a suitable correction can be implemented, such as family-wise error control (of which the *Bonferroni correction*, that corrects the significance level by the number of tests performed, is the simplest example) or false discovery rate control. Again, the most reliable procedure is given by permutation tests (see Section 12.6).

One of the major issues with image-based statistical works is that they are often carried out on small samples of subjects; for instance, most cognitive neuroimaging findings are based on cohorts of no more than 20 subjects, due to the cost of data acquisition and processing. This situation leads to a degradation of the ability to separate signal from noise; in practice, enforcing a strict control on type I errors jeopardizes the analyst's ability to detect the actual signal. As a result, neuroimaging studies typically suffer from a *lack of power* and a *low reproducibility* of the findings (3).

For the sake of sensitivity, the mass-univariate setting can be enhanced by taking into account the image context in statistical inference. Smoothing is a relatively standard image analysis procedure, but it strongly biases the shape of the signal of interest and thus can only be used sparingly. However, one can also consider the continuous structure of the signals of interest embedded in the images by focusing on the size of the connected components of supra-threshold areas for a given detection threshold. Assuming that the distribution of such sizes under the global null hypothesis (that there is no effect present in any region of the image) is known, observing larger region sizes typically indicates the presence of an effect in these regions. This type of cluster-level inference has become a standard in neuroimaging (26); it has also been extended to general procedures that avoid the prior selection of a cluster-forming threshold (31). This and other related procedures are discussed in Section 12.5. We give a brief account of permutation testing approaches in Section 12.6 and conclude in Section 12.7 by illustrating examples.

12.2   Variability of Brain Shape and Function

Variability of Brain Organization and Brain Imaging

Neuroimaging group analyses test the effect of some external variables of interest on the image signal, i.e., they compare the amount of signal explained by the variables of interest to the residual signal after fitting a linear model. They do thus measure the fraction of between-subject signal variance that is correlated with the target variable. However, this variability has a complex nature, as it encodes functional and structural features unique to each individual, together with limitations or various signal corruptions in the imaging process; in any case, it cannot be simply conceptualized as an additive random noise in the observations. Building suitable imaging features that achieve some robustness against the observed between-subject variability is thus an important challenge. It requires some effort to understand and capture the information of interest in the presence of distortions and structured noise.

Variability of Brain Shape

The variability of brain size and shape is mostly observed through anatomical imaging and various computational geometric procedures that measure the thickness, the regularity of

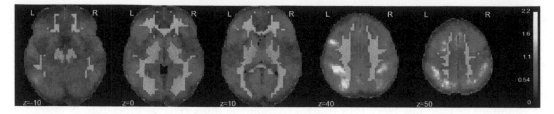

FIGURE 12.2
Illustration of the relative magnitude of within- and between-subject variability, through
the average between/within variance ratio of the dataset presented in Section 12.7. While
the ratio is close to 1 in many regions, it is larger in regions that display a non-zero mean
effect across the population (compare with Fig. 12.5).

the cortical surface, or some of its singularities (sulcal pits, sulci fundi, etc.). The variability
of such features readily poses a challenge for the comparison of brains from different subjects:
what makes each brain location unique from an anatomical perspective? Or put differently,
how can we warp each individual brain such that the localized individual features can be
considered as corresponding to each other? The best approaches so far consist of first compensating for differences in image pose and brain size through linear transformations, then
using high-dimensional diffeomorphic registration to align individual gray matter outlines
approximately (1), and then to perform statistical analysis, yet in the absence of further
guarantee on the identity of the tested structures. This framework results in an uncertainty
of about 10 mm on the actual voxel correspondences, which can be taken as a blur on the
results of any group analysis (17, 33). Current attempts to improve upon this situation rely
on surface-based mapping (8)—yet without any formal guarantee of accurately aligning
cyto-architectural areas nor functional areas—or using functional localizer experiments to
define individual regions of interest (22).

A point relevant for all neuroimaging studies is that these differences are not modeled,
because current imaging contrasts are not sufficient to disambiguate the nature and organization of all brain areas; it is also clear that human cortical folding cannot be mapped
diffeomorphically across individuals (28). The corresponding variability is thus inaccurately
handled as unstructured noise.

Variability of Brain Function

Besides the variability of brain shape and anatomical organization, different subjects may
display different brain activation patterns, which can be interpreted as differences in functional organization, cognitive strategies, attention, or more simply signal-to-noise ratio of
the imaging data related, e.g., to presence of motion or various acquisition artifacts. In the
absence of external data, these sources of variability cannot be identified easily, nor can they
be removed. They are often considered as an additional additive noise. Note that functional
MRI is not a quantitative modality, in the sense that the corresponding measurements of the
BOLD (blood oxygen-level-dependent) signal are not expressed in absolute physical units.
Practitioners have found that expressing the signal fluctuations as a percentage of the baseline level was a practical measure, yet it is unclear how invariant the resulting quantification
is to parameters of no interest in statistical analysis, such as MR sequence parameters (see
an illustration in Fig. 12.5(b)).

A rough measure of the between-subject variability can be given by comparing the
amount of between-subject variance to average amount of within-subject variance (related
to observation noise). An example of such a ratio is displayed in Fig. 12.2.

12.3   Mixed-Effects and Fixed-Effects Analyses

Fig. 12.2 illustrates a situation that arises frequently in functional neuroimaging, namely the presence of two or more heterogeneous sources of variance in the data. One is related to the observation process and can be viewed as noise, while the other is related to the difference between individuals, as discussed in Section 12.2. We refer to the corresponding variance estimates as *first-level* and *second-level* variance respectively. These sources of variability are jointly embedded in the observations, yet the terms are separated in the so-called mixed-effects modeling framework (19), which we will discuss in detail in Section 12.4.

It should be noted that one can neglect the second-level variance, which leads to a *fixed-effects model*: Such a model typically assumes that a given effect has been observed in all subjects as if it were a repeated measurement on the same individual, and thus aims at deriving the mean effect and variance of this common effect without considering cross-subject fluctuations. Such an inference cannot be used as a population-level inference, given that it ignores the variability that stems from the subject effect (10). The difference between fixed- and mixed-effects inference is illustrated in Fig. 12.3.

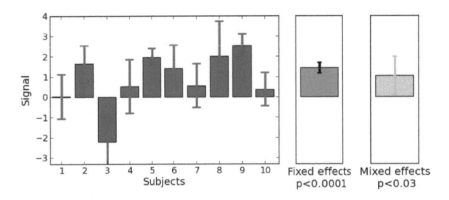

FIGURE 12.3

Illustration of the difference between fixed- and mixed-effects inference: Given $n = 10$ observations associated with a given level of uncertainty (left), one can perform a fixed-effects inference that ignores cross-subject variability of the observations and thus leads to population effects with tight uncertainty (middle), or consider this variance and then obtain wider uncertainty estimates (right). Only the mixed effects model yields a valid inference on the population from which the observations were sampled.

12.4   Group Analysis for Functional Neuroimaging

In this section, we first review the two-level linear model for functional neuroimaging, then discuss the estimation of the mixed-effects model and ensuing statistical tests.

12.4.1 Problem Setting and Notations

As a preliminary note, it is assumed here that some functional data are observed at a given set of brain locations across multiple subjects, be these locations cortical surface nodes or voxels (see, e.g. (36)).

To clarify the notations, we use bold capital letters for matrices (e.g. $\mathbf{A}$), bold letters for vectors (e.g. $\mathbf{a}$), and small letters for scalars (e.g. a). We denote $\mathcal{N}(\boldsymbol{\mu}, \boldsymbol{\Sigma})$ the multivariate Gaussian distribution with mean $\boldsymbol{\mu}$ and covariance matrix $\boldsymbol{\Sigma}$. Given n scalars $(\sigma_1^2, .., \sigma_n^2)$, we denote by $\mathrm{diag}(\sigma_1^2, .., \sigma_n^2)$ the $(n \times n)$ diagonal matrix with these scalars on the diagonal. We use the following conventions: $n_{subjects}$ denotes the number of subjects, p the number of voxels, and $m(s)$ the number of scans of a given dataset in a given subject s.

Within-Subject Model

We consider a study performed over $n_{subjects}$ subjects. For any subject $s \in [n_{subjects}]$, let $\mathbf{Y}_s$ denote a set of observations (fMRI scans) obtained in this subject. To simplify the statistical formalism, each fMRI scan is *flattened* to a vector representation, where each coordinate of the vector represents the fMRI activity in a given voxel within a suitable brain mask. Let us denote $m(s)$ the length of the time series in subject s. $\mathbf{Y}_s$ is thus a matrix of shape $(m(s) \times p)$ and the value $\mathbf{Y}_s$ (i, j) for $1 \leq i \leq m(s)$ and $1 \leq j \leq p$ represents the fMRI signal at voxel j acquired at time i. Let $(\mathbf{X}_s)_{s \in [n_{subjects}]}$ represent the design matrices that model experimental and nuisance effects that are likely to be reflected in brain signals. The first-level general linear model (GLM) takes the form

$$\forall s \in [n_{subjects}], \mathbf{Y}_s = \mathbf{X}_s \mathbf{B}_s + \boldsymbol{E}_s, \tag{12.1}$$

where $(\mathbf{B}_s)$ for each $s \in [n_{subjects}]$ is a matrix of shape $(n_{reg} \times p)$ that yields the coefficients associated with the columns of the design matrix and $(\boldsymbol{E}_s)$ is the unmodeled signal, considered as observation noise.

In practice, only a certain combination of the parameters is of interest and will be the subject of further inference. For instance, the effects associated with motion parameters may not be considered in population-level analyses, while the activation signal associated with a combination of experimental conditions quantifies a cognitive response of interest that one wishes to compare with individual characteristics. To keep the setting clear yet comprehensive, we thus rewrite Equation 12.1 with split design and parameter matrices to introduce a distinction between the coefficients of interest and those modeling other effects. Note that singling out a parameter of interest amounts to defining a *contrast*, i.e., a linear combination of the effects $\mathbf{B}_s$, that represents it in the data. Here we assume that the design matrix is written in a form such that the contrast of interest corresponds to the first column $\mathbf{x}_s^i$ of $\mathbf{X}_s$, i.e., $\mathbf{X}_s = [\mathbf{x}_s^i, \mathbf{X}_s^r]$; correspondingly, the effects of interest and residual effects are written as $\mathbf{b}_s$ and $\mathbf{B}_s^r$, i.e., $\mathbf{X}_s \mathbf{B}_s = \mathbf{x}_s^i \mathbf{b}_s + \mathbf{X}_s^r \mathbf{B}_s^r$, thus yielding

$$\forall s \in [n_{subjects}], \mathbf{Y}_s = \mathbf{x}_s^i \mathbf{b}_s + \mathbf{X}_s^r \mathbf{B}_s^r + \boldsymbol{E}_s \tag{12.2}$$

Group-Subject Model

At the population level, it is expected that the contrasts of interest, observed across subjects, could potentially be explained by subject-dependent variables, such as age, behavioral tests, or genetic variables. We note $\mathbf{B} = [\mathbf{b}_1, .., \mathbf{b}_{n_{subjects}}]^T$ the $(n_{subjects} \times n_{voxels})$ matrix that represents the contrasts of interest measured across individuals. The explanatory variables of interest are grouped in a second-level design matrix $\mathbf{Z}$, that has a shape $(n_{subjects} \times n_{factors})$

$$\mathbf{B} = \mathbf{Z}\boldsymbol{\beta} + \mathcal{E}, \tag{12.3}$$

where β is a matrix of shape $(n_{factors} \times n_{voxels})$ that represents the population-level effects. Without loss of generality, Eq. (12.1)–(12.3) can thus be rewritten

$$\mathbf{Y}_s = \mathbf{x}_s^i \mathbf{b}_s + \mathbf{X}_s^r \mathbf{B}_s^r + \boldsymbol{E}_s, \ \forall s \in [n_{subjects}] \qquad (12.4)$$

$$\mathbf{B} = [\mathbf{b}_1, .., \mathbf{b}_{n_{subjects}}]^T = \mathbf{Z}\beta + \boldsymbol{\mathcal{E}}. \qquad (12.5)$$

Finally, the question of interest is where in the brain a certain combination of the factors of interest yields a positive effect on average in the population, i.e., whether $\mathbf{c}^T \beta > 0$, where $\mathbf{c}$ is a suitable vector of contrast on the population-level effects: if $\mathbf{Z}$ contains three variates related to the sex, the age of the subjects, and an intercept, setting $\mathbf{c} = (0, 1, 0)$ will display the effect of age on the observed BOLD signal. Different types of contrasts correspond to different statistical questions; see Appendix A for an overview. The problem consists thus in estimating with what confidence one can reject the null hypothesis $\mathbf{c}^T \beta = 0$. Since the model defined in Eqs. (12.4)–(12.5) can be handled at each brain location independently, we proceed with voxel-level analysis (estimation and statistical analysis). We keep focusing on voxel-level probabilistic assessment till the end of this section; taking into account the joint signal distribution over voxels is deferred to Section 12.5. Such an approach is typical of a mass-univariate modeling framework.

Interpretation as a Mixed Effects Model

It is straightforward to observe that Eqs. (12.1)–(12.3) can be concatenated into one equation:

$$\forall s \in [n_{subjects}], \mathbf{Y}_s = \mathbf{x}_s^i \mathbf{Z}|_s \beta + \mathbf{x}_s^i \boldsymbol{\mathcal{E}}|_s + \mathbf{X}_s^r \mathbf{B}_s^r + \boldsymbol{E}_s, \qquad (12.6)$$

where $\mathbf{Z}|_s$ and $\boldsymbol{\mathcal{E}}|_s$ denote the restrictions of matrix $\mathbf{Z}$ and $\boldsymbol{\mathcal{E}}$ to their row s. This means that the observed data $\mathbf{Y}_s$ are actually composed of four different effects:

- The effect of interest to be tested: $\mathbf{x}_s^i \mathbf{Z}|_s \beta$

- A random subject effect $\mathbf{x}_s^i \boldsymbol{\mathcal{E}}|_s$

- Some within-subject effects of no interest, or nuisance effects, handled as fixed effects $\mathbf{X}_s^r \mathbf{B}_s^r$

- Some observation noise $\boldsymbol{E}_s$

The model thus comprises both fixed and random effects, hence the reference to a *mixed effects* model.

12.4.2   Estimation

It is then generally assumed that the two random components are Gaussian distributed, with unknown variance: Let $(\boldsymbol{\Lambda}_s)$ be the $(m \times m)$ variance-covariance matrix of the noise $\boldsymbol{E}_s$ in a subject $s \in [1..n_{subjects}]$ and $\boldsymbol{\Delta}$ be the variance-covariance matrix $(n_{subjects} \times n_{subjects})$ of the random effects. While $\boldsymbol{\Lambda}_s$ are often taken as the covariance matrix of a (temporal) auto-regressive process scaled by an unknown variance, $\boldsymbol{\Delta}$ can be assumed to be diagonal, as $(\boldsymbol{\mathcal{E}}|_1, .., \boldsymbol{\mathcal{E}}|_{n_{subjects}})$ have been sampled independently. In the absence of additional information on the population structure, $\boldsymbol{\Delta} = \gamma^2 \mathbf{I}_{n_{subjects}}$, where $\mathbf{I}_n$ is the $n \times n$ identity matrix. The estimation of the parameters of the mixed-effects model, is carried out, generally following the maximum likelihood principle. The parameters to estimate are

$$\Theta = \left((\boldsymbol{\Lambda}_s, \mathbf{B}_s^r), \beta, \gamma^2 \right), \qquad (12.7)$$

given $(\mathbf{Y}_s, \mathbf{X}_s)_{s=1..n}$ and $\mathbf{Z}$:

$$\hat{\Theta} = \text{argmax}_\Theta \, \mathcal{N}(\mathbf{B}; \mathbf{Z}\beta, \gamma^2 \mathbf{I}_{n_{subjects}}) \prod_{s=1}^{n_{subjects}} \mathcal{N}(\mathbf{Y}_s; \mathbf{x}_s^i \beta_s + \mathbf{X}_s^r \mathbf{B}_s^r, \Lambda_s).$$

For the sake of simplicity and computation efficiency, a two-step method is frequently used. First, the estimation of the effects of interest and their covariance, independently in each subject, and then the estimation of the population parameters. Standard solutions include: the EM algorithm (41, 29, 36), Bayesian methods (39) or the Gauss–Newton method or some variant thereof (35). Here we follow the EM approach. First level model fit (for notational simplicity, we assume that Λ_s is given, while it is voxel-specific and data-dependent in practice). The first-level estimation procedure yields estimates $(\hat{\mathbf{b}}_s, \hat{\sigma}_s^2)$ of the individual effects and associated variance. These estimates have a large number of degrees of freedom $m(s) - n_{reg}$. The group model (Eq. (12.4)–(12.5)) then boils down to

$$\hat{\mathbf{b}}_s = \mathbf{b}_s + \mathbf{e}_s, \; \forall s \in [n_{subjects}]$$
$$\mathbf{B} = [\mathbf{b}_1, .., \mathbf{b}_{n_{subjects}}]^T = \mathbf{Z}\beta + \mathcal{E},$$

where for each subject, $s \in [n_{subjects}], \sigma_s^2 = var(\mathbf{e}_s)$ is estimated with a large number of degrees of freedom, hence we assume that it is exact, while $\gamma^2 = var(\mathcal{E})$ is unknown, and has to be estimated. (β, γ^2) can then be estimated so that they maximize the likelihood

$$\mathcal{L}(\hat{\mathbf{B}}; \beta, \gamma^2) = \mathcal{N}\left(\hat{\mathbf{B}}; \mathbf{Z}\beta, \gamma^2 \mathbf{I}_{n_{subjects}} + \text{diag}(\sigma_1^2, .., \sigma_{n_{subjects}}^2)\right) \qquad (12.8)$$

using an EM algorithm: If we denote $\mathcal{N}(\tilde{\mathbf{b}}_s, \tilde{\mathbf{s}}_s^2)$ the variational distribution of $\mathbf{b}_s$, and $\tilde{\mathbf{B}} = [\tilde{\mathbf{b}}_1, .., \tilde{\mathbf{b}}_{n_{subjects}}]$, the EM algorithm consists of iterating the two steps:

$$\text{E-step: } \forall s \in [1..n_{subjects}], \; \tilde{\mathbf{s}}_s^2 = \left(\frac{1}{\gamma^2} + \frac{1}{\sigma_s^2}\right)^{-1}, \; \tilde{\mathbf{b}}_s = \tilde{\mathbf{s}}_s^2 \left(\frac{\hat{\mathbf{b}}_s}{\gamma^2} + \frac{\beta}{\sigma_s^2}\right) \qquad (12.9)$$

$$\text{M-step: } \beta = (\mathbf{Z}^T \mathbf{Z})^{-1} \mathbf{Z}^T \tilde{\mathbf{B}}, \; \gamma^2 = \frac{1}{n_{subjects}} \left(\sum_{s=1}^{n_{subjects}} \tilde{\mathbf{s}}_s^2 + \|\tilde{\mathbf{B}} - \mathbf{Z}\beta\|^2\right). \qquad (12.10)$$

It converges toward a local maximum of the likelihood function. Although there is no guarantee that the reached maximum is global, it can be observed in many cases that the likelihood function only has one maximum. See, e.g., in Fig. 12.4 the likelihood as a function of the parameters (β, v) obtained for the toy data used in Fig. 12.3.

It should be noted that the above algorithm may not yield the optimal estimators for the variance parameters; for instance, more accurate estimates of the variance may be obtained by using restricted maximum likelihood approaches (14).

12.4.3   Statistical Inference

The main question remains to test whether the effect of interest is different from zero, i.e., $\mathbf{c}^T \beta > 0$ or $\mathbf{c}^T \beta \neq 0$. It is not obvious what test should be used. A Wald statistic can be computed as

$$t = \frac{\mathbf{c}^T \beta}{\sqrt{\mathbf{c}^T (\mathbf{Z}^T W^{-1} \mathbf{Z})^{-1} \mathbf{c}}}, \qquad (12.11)$$

where $W(\gamma) = \gamma^2 \mathbf{I}_{n_{subjects}} + \text{diag}(\sigma_1^2, .., \sigma_{n_{subjects}}^2)$. However, it does not conform exactly to a student distribution under the null hypothesis (39, 4). This means that non-parametric

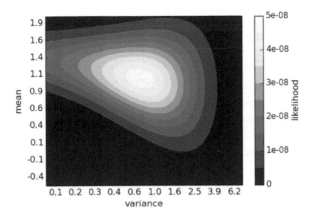

FIGURE 12.4
Likelihood of the observations displayed in Fig. 12.3 as a function of the parameters (β, γ^2) (which are 2 scalars in that case). It can be seen that the likelihood function has a unique maximum.

methods have to be used to obtain an estimate of its null distribution. Given this requirement, the likelihood ratio statistic—which is known as the most powerful test for the model—is the best possible test. This is defined as follows

$$\Lambda = 2 \left(\sup_{\beta,\gamma} \log \mathcal{L}(\hat{\mathbf{B}}; \beta, W(\gamma)) - \sup_{\beta:c^T\beta=0,\gamma} \log \mathcal{L}(\hat{\mathbf{B}}; \beta, W(\gamma)) \right), \qquad (12.12)$$

where $\mathcal{L}$ is defined as in Equation (12.8). Λ is readily computed by running the EM algorithm twice, once in a constrained mode ($c^T\beta = 0$), once in an unconstrained mode.

12.4.4 The Random Effects t-Test

A crude approximation of the above model consists of neglecting the first-level variance; this amounts to assuming that the individual variances are identical ($\sigma_1^2 = \cdots = \sigma_{n_{subjects}}^2$) and clearly simplifies the ensuing statistical inference. The assumption yields a simple *random-effects model*:

$$\mathbf{B} = \mathbf{Z}\beta + \mathcal{E}', \qquad (12.13)$$

where $\mathcal{E}' \sim \mathcal{N}(0, g^2 \mathbf{I}_{n_{subjects}})$, for which the decision statistic is simply

$$t_{RFX} = \frac{\mathbf{c}^T\beta}{\sqrt{g^2 \mathbf{c}^T (\mathbf{Z}^T\mathbf{Z})^{-1}\mathbf{c}}}, \qquad (12.14)$$

where β and g^2 are easily estimated with a least-squares fit. Note that it is the most widely used model for population-level inference in neuroimaging. Similarly, an F statistic can be defined when one is interested in unsigned or multi-dimensional contrasts. In that case, and under the Gaussian hypothesis, the log-likelihood ratio defined in Equation (12.12) is a monotonous function of the Fisher statistic, or, if one is interested in signed effects, of the t statistic. We provide an empirical comparison of the t statistic with the full mixed-effects model in Section 12.7. We also provide a table of the most frequently used statistical models (one-sample t-test, two-sample t-test, paired-t-test, two-way ANOVA) in Table A.1. See (25) for a more complete discussion on this topic.

12.5 Taking into Account the Spatial Context in Statistical Inference

Multiple Comparisons in the Mass Univariate Framework

So far, we have considered each voxel independently, which is a requirement at the estimation stage, given the computation cost of the procedure, but comes at the price of low sensitivity when performing statistical inference. In the mass-univariate setting, one statistical test is performed per voxel, and the significance is then simply corrected for the complete set of tests: Corrected p-values can be obtained by family-wise error-rate (FWER) control, i.e., the probability of performing one false detection, or the generally more lenient false discovery rate (FDR) control, that controls for the proportion of false discoveries among the detections. We discuss the practical computation of FWER-corrected thresholds in Section 12.6. For false discovery rate control, a simple reference procedure is presented in the neuroimaging context in (13) and is most often used due to its simplicity, but non-parametric alternatives exist too (12).

This framework is usually adopted for its convenience and relatively simple interpretability. However, it is limited by its inability to take into account the image structure of the data: the observed cross-subject variability in brain organization limits the relevance of a voxel-by-voxel description of the data, and yields a reduced sensitivity to detect the true effects.

Spatial Regularization

The first way to take into account the image structure in statistical modeling is to regularize spatially the data: such a regularization is most often implemented with data smoothing in the volume or on the surface, with a 6- to 10-mm full width at half maximum (fWHM) kernel. Choosing the adapted kernel directly leads to a bias–variance tradeoff: larger kernels reduce the variance yet yield a large bias on the activation peaks position. Smoothing can instead be replaced by Markov random field modeling (6) or anisotropic smoothing (32), but such procedures are not widely used by practitioners due to the lack of efficient implementations.

Inference on the Region Size

A second possibility to take into account the image structure in the inference procedure is to consider the size of the connected components of the set of supra-threshold voxels (27), assuming that the chosen threshold (e.g. t > 3 for a t-test) correctly separates the peaks of the image from the background noise. The size statistic has to be compared with the reference distribution of the size of such structures obtained when no effect is present, i.e., under the null hypothesis. Such a distribution is in general not known a priori, but it can be approximated under the hypothesis that the data follows a known parametric distribution (26), or more simply by permutation (15); see Section 12.6. The latter approach is typically recommended. The underlying intuition is that, especially when looking at population-level statistics maps, narrow regions are unlikely to represent truly active brain structures. There are, however, two major drawbacks with such an approach: First, it depends on an arbitrary cluster-forming threshold, so that the result does not outline intrinsic characteristics of the data. Second, varying this threshold potentially yields abrupt changes in the detected areas, thus leading to a fundamental instability of the results.

It is, however, relatively popular in the neuroimaging community due to its enhanced sensitivity, which is attributable to the weakness of the test. It rejects the global null hypothesis for the whole cluster, but does not make it possible to localize the activation within the supra-threshold cluster. Hence, it should only be used with high cluster-forming thresholds (38).

Cluster-Mass Testing and Threshold-Free Cluster Enhancement (TFCE)

An improvement upon these procedures consists of combining the size and height of peak regions in a common statistic (27, 16). However, more general combinations of extent and height can be used alternatively, resulting in the so-called threshold-free cluster enhancement (TFCE) procedure (31). Let $v \to \phi(v)$ be the statistical map obtained from the univariate inference step; the TFCE statistic is computed as follows:

$$\text{TFCE(v)} = \int_0^{\phi(v)} z^H a_v(z)^E dz, \qquad (12.15)$$

where $a(z)$ is the size of the connected component of the map thresholded at level z that contains voxel v, and H and E are two non-negative numbers. In general $H = 2$ and $E = .5$, as in the reference TFCE implementation in (31). Note that $H = 1$ and $E = 0$ yields a cluster mass statistic, while $H = 0$ and $E = 1$ yields a generalized cluster size statistic integrated over thresholds. The setting $H = 2, E = .5$ was designed to represent optimally paraboloid-shaped activations.

Randomized Parcellation-Based Inference (RPBI)

A recent alternative to TFCE has been proposed in (20), and consists of using multiple parcellations of the brain volume. A *parcellation* is a partition of the brain volume into connected regions *(parcels)*, for instance 1000 or 2000 parcels; taking parcel-based averages of the input signal can be understood as a dimension reduction procedure well suited for smooth images. The parcel-based signal estimation can be carried in a group of subjects in parallel, and thus used in a population model, where standard statistics (e.g. Eq. (12.11)) are computed, resulting in a parcel-based statistical map. While the signal compression brought by parcel-based analysis is certainly beneficial to sensitivity, the resulting map is heavily biased by the initial parcellation specification: to avoid such biases, one has to marginalize out the parcellation choice. The RPBI statistic is thus defined as the voxel-based sum of binary variables testing whether the decision statistic in the parcel including the voxel was above a given threshold or not, computed over a large enough number of initial parcellations. Let $\mathcal{P}$ be a set of parcellations, and V be the set of voxels under consideration. Given a voxel v and a parcellation P, the parcel-based thresholding function θ_t is defined as:

$$\theta_t(v, P) = \begin{cases} 1 \text{ if } F(\Phi_P(v)) > t \\ 0 \text{ otherwise} \end{cases} \qquad (12.16)$$

where $\Phi_P : V \to P$ is the mapping function that associates each voxel with a parcel from the parcellation P. For a predefined test, F returns the F-statistic associated with the average signal of a given parcel (a t or other statistic is also possible). Finally, the aggregating statistic at a voxel v is given by the counting function C_t:

$$C_t(v, \mathcal{P}) = \sum_{P \in \mathcal{P}} \theta_t(v, P). \qquad (12.17)$$

$C_t(v, \mathcal{P})$ represents the number of times the voxel v was part of a parcel associated with a statistical value larger than t across the folds of the analysis conducted on the set of parcellations $\mathcal{P}$. The parameter t should be set to ensure a Bonferroni-corrected control at $p < 0.1$ in each of the parcel-level analyses. In practice, the results are weakly sensitive to mild variations of t.

12.6   Type I Error Control with Permutation Testing

Permutation testing is the reference approach for statistical inference, as it provides valid and accurate p-values under a typically restricted set of hypotheses. In practice, many statistical procedures described previously (cluster size, TFCE, RPBI) do not have a well-defined distribution under the null hypothesis. Even the family-wise error control in the presence of correlations does not have a perfectly known distribution and is correctly approximated only in some peculiar settings (40). While some reference distributions could be estimated or even simulated under the assumption of Gaussian noise with pre-defined covariance—a framework popularized as Gaussian Random Field Theory (27)—the hypotheses involved in such a procedure are relatively strong, hence easily proved wrong. In practice, the smoothness of the signal can vary across regions of an image, challenging the stationarity of the random field (30). In general, permutation-based inference is thus the reference solution to obtain accurate control of type I errors.

As the statistics in (12.11) and (12.12) are pivotal, the advised procedure is to use the Westfall–Young kind of correction for multiple comparisons (37). Family-wise error rate (FWER) corrected p-values are obtained by sampling the null distribution of the maximal statistic through a permutation scheme. Let us consider a statistical map $v \to \phi(v)$ in which the values follow the same distribution under the null hypothesis (this is called the *pivotality* hypothesis). We assume that the data are exchangeable under the null hypothesis: for instance, we compare the mean effect across two populations of subjects, and do not have any other covariate in the data model. Then, for any permutation π of the data (i.e., permutation of the rows of the second-level design matrix $\mathbf{Z}$), one can compute the corresponding map $v \to \phi^{\pi}(v)$, by applying the usual estimation procedure. Assuming that J such permutation-based estimations are carried out, the critical threshold for the family-wise-error-corrected p-value α is given by

$$\phi^c(\alpha) = \mathcal{Q}_{1-\alpha}\left((\max_{v\in[1..p]}\phi^{\pi_j}(v))_{j\in[1..J]}\right), \qquad (12.18)$$

where $\mathcal{Q}_{1-\alpha}$ stands for the $(1-\alpha)$ quantile of the values (in this case, across permutations $(\pi_1, \cdots, \pi_J)$). The significance at level α of the non-permuted statistics $\phi(v)$ is thus assessed by thresholding of these values against ϕ^c_α. Slightly more complex strategies can be used to control the false discovery rate instead (12).

A particular case is when there is no regressor in the model, i.e., the test only consists of the intercept, meaning that the inference is about the mean effect being larger than zero. In that case, permuting the data does not change the statistic. What is done instead is to further assume that the distribution under the null hypothesis is symmetric, and then swap the sign of the observations: for n observations, this generates up to 2^n resamplings.

Limitations of Permutation Testing

There are, however, some cases where the use of permutation tests requires some care, because simplistic implementations lead to inaccurate results. One such case is when models include some covariates on top of the effects of interest: due to the possible correlations between the contrasts of interest and other covariates, the permutation test needs to be handled with specific procedures (9, 38).

Another obvious limitation of permutation testing is its cost: to guarantee a stable estimate of ϕ^c_α in Eq. (12.18), it is necessary to use $J \gg \frac{1}{\alpha^2}$ permutations.

12.7   Illustration of Various Inference Strategies on an Example

The difference between the output of a mixed-effects and a fixed-effects model is usually large in neuroimaging settings, where inter-subject variability is large with respect to intra-subject variability (see e.g. Fig. 12.3). As neuroimaging studies are concerned with population-level validity of the findings, fixed-effects inference is ruled out.

By contrast, the difference between the mixed-effects statistic and the random effects statistic (i.e., the test in Eq. (12.14) as opposed to the test in Eq. (12.12)) is more subtle and deserves some consideration. Neglecting first-level variance is actually equivalent to assuming that it is equal in all subjects, i.e., that all individual observations are equally reliable (see Section 12.4.4). Hence any difference between truly mixed-effects or random-effects models clearly outlines the impact of the difference in reliability across samples in the population.

Two examples are described next.

- The first example in Fig. 12.5(a) shows the results of permutation-controlled $\alpha = .05, J = 10^4$ mixed-effects and random-effects tests to detect the mean effect of a computation task (from which a simple reading effect has been subtracted (24)), based on a sample of $n = 30$ subjects.

- The second example in Fig. 12.5(b) is a two-sample test that targets the differential effect of two MRI scanners on the activation obtained after completing the same task as the first example. More precisely, it aims at detecting the difference between a Siemens Trio 3T scanner (data acquired in 2007–2008 (36)) and a 3T Bruker scanner (data acquired in 2003–2006 (24)). The acquisition parameters of the Bruker dataset are TR = 2400 ms, TE = 60 ms, the matrix size is 64×64, FOV = 24 $cm \times$ 24 cm. Each volume consisted of n_a 3-mm- or 4-mm-thick axial slices without gap, where n_a varied from 26 to 40 according to the session. A session comprised 130 scans. The acquisition parameters of the Trio dataset are TR = 2400 ms, TE = 30 ms, the matrix size is 64×64, FOV = 19.2 $cm \times$ 19.2 cm and the number of slices is 40, while the session duration is unchanged. Note that the two sets of images have been preprocessed in the same way. The cohort studied comprises $n_1 = 72$ and $n_2 = 78$ subjects from the Bruker and Siemens scanners respectively.

In both cases, random- and mixed-effects maps are presented, after thresholding at the $p < .05$ level, corrected for multiple comparisons by a permutation procedure. While the two kinds of maps are clearly similar, one can observe more sensitivity in the case of mixed-effects inference in the two settings:

- In the one-sample test, more active voxels are found in the ventral striatum, the anterior cingular cortex, the insula, and the right intraparietal sulcus, which are regions of the dorsal attentional network. Moreover, the detected regions display more significant p-values.

- In the two-sample test, the effects observed in the mixed-effects analysis are wider and more significant in all regions, although this is more visible in the ventral striatum and the thalamus.

Note that the superiority of the mixed model in terms of sensitivity has already been reported in the literature (39, 29, 36, 4).

Next, to assess the impact of spatially aware statistics, we perform a qualitative comparison of the maps obtained through the different spatial models, coupled with the mixed-effects statistic: peak significance (no spatial context), cluster size significance, TFCE and RPBI. The results are shown in Fig. 12.6, and are based on the one-sample test example.

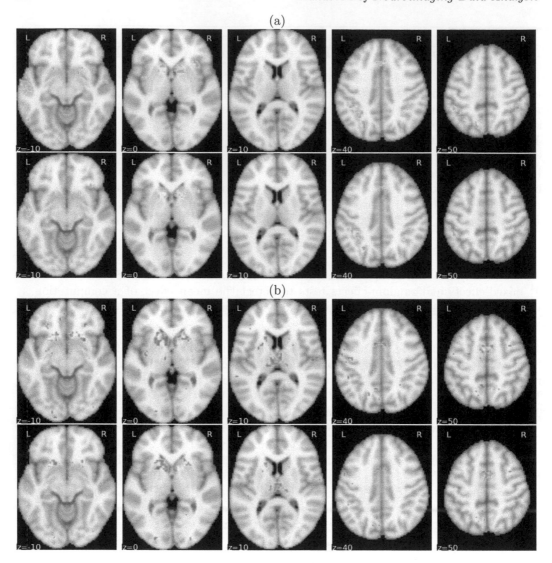

FIGURE 12.5
Difference between a mixed-effects model and a random-effects model for statistical analysis. (a) One-sample test that aims at detecting the mean effect on a computation task (from which a simple reading effect has been subtracted (24)), based on a sample of $n_{subjects} = 30$ subjects; top: mixed-effects model; bottom: random-effects model. Both maps are thresholded at the $p < 0.05$ level, corrected for multiple comparisons, using permutation testing. (b) Illustration of the difference on a two-sample test, where the differential effect of two MRI scanners on the activation obtained in the same task as (a). The mixed- (top) and random- (bottom) effects maps, both thresholded at $p < 0.05$, FWER-corrected by permutation, show the same effects, but again the mixed-effects inference is more sensitive.

They clearly show that for a given significance level ($p < 0.05$, corrected for multiple comparison through an max-control permutation-based procedure), spatially informed approaches yield more active voxels than voxel-level inference, a gain that can be interpreted as a higher sensitivity. For instance, some temporal or frontal activation foci are not found

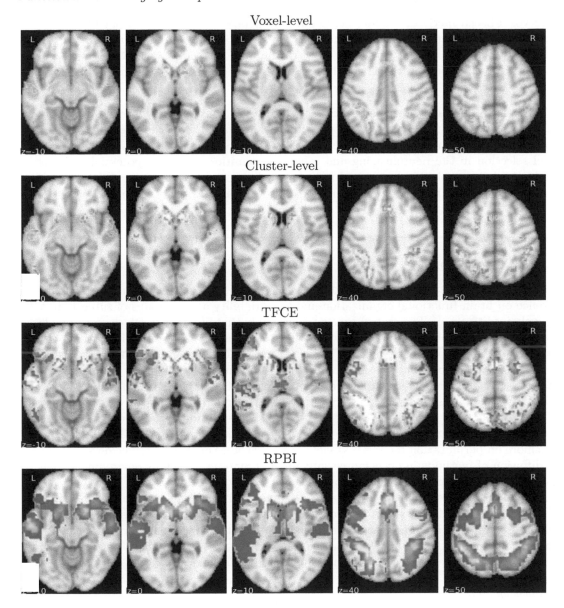

FIGURE 12.6
Impact of including the spatial context in the neuroimaging statistical inference procedure: the four maps above represent the activation related to the one-sample test presented in Fig. 12.5(a), thresholded at a significance level of $p < .05$, corrected for multiple comparison through an F-max permutation scheme. Voxel-level, cluster-level, TFCE and RPBI present increasing amounts of activations.

in the voxel-based approach, but are detected with the other approaches. It should be noted, however, that this increase in sensitivity is mitigated by two important drawbacks:

- Only the voxel-level inference provides a guarantee of the active status of each detected voxel, as the other approaches only provide a rejection of the null hypothesis for some (unknown) voxel in the detected areas.

- Non-voxel-based methods require additional parameters. For instance, in our procedure, we relied on a cluster-forming threshold of $p < 10^{-4}$, the E and H parameters of the TFCE (we kept the default $E = .5$, $H = 2$), the internal threshold of the RPBI approach, as well as the number of atlases and their resolution. See Section 12.5 for a description of these parameters.

Importantly, it has been reported that the increased sensitivity afforded by spatially aware method comes with higher reproducibility, i.e., a higher chance of reproducing the results on another sample of images (see e.g. (20)), which is acknowledged as a criterion for model selection in the neuroimaging and other communities (34). This potentially means that in the low-sample/low-power regime common to most neuroimaging studies, the increase in sensitivity afforded by these methods offsets their drawbacks (see also (35)).

12.8 Conclusion

The flexibility afforded by the linear model has made it possible for neuroimagers to make inference on brain functional organization and its variability across groups of subjects in a relatively intuitive way. The central concept is arguably that of *contrast*, which corresponds to stating a precise question to identify the effect of variables on brain activation signals. Current gains in computational efficiency have led to the development of permutation-based approaches, which are typically more accurate and make it possible to draw inference from quantities that do not have any known or simple distribution under the null hypothesis.

In practice, neuroscientists often rely on the solution offered by the software that they use most often, such as SPM, FSL or AFNI, which make different choices: SPM relies mostly on parametric statistics, while FSL gives access to the popular TFCE statistic and relies more on permutation tools; AFNI yields a wider choice of statistics. The development and diffusion of methods in open-source software and more user-friendly environments will condition the adoption of the more advanced tools by the community.

Methodological research still has to address some hard challenges, such as the design of more efficient methods to take into account the spatial structure in the images while performing probabilistic inference. Also, the same level of statistical rigor needs to be achieved in more complex statistical analyses that consider the signals from multiple regions, such as multivariate pattern analysis and functional connectivity analysis.

Acknowledgment

We are very thankful to Philippe Pinel who provided the data used in the comparison. This work was also possible thanks to the software developments carried out by Benoit da Mota, Virgile Fritsch and Gaël Varoquaux. Many thanks also to Ana Luisa Grilo Pinho, who suggested the addition of the statistical tests table. We acknowledge funding from the Human Brain Project, the Digiteo project iConnectom, and the Microsoft Research (MediLearn project).

Bibliography

[1] John Ashburner and Karl J. Friston. Diffeomorphic registration using geodesic shooting and Gauss-Newton optimisation. *Neuroimage*, 55(3):954–967, Apr 2011.

[2] E. Bullmore, M. Brammer, S. C. Williams, S. Rabe-Hesketh, N. Janot, A. David, J. Mellers, R. Howard, and P. Sham. Statistical methods of estimation and inference for functional MR image analysis. *Magn Reson Med*, 35(2):261–277, Feb 1996.

[3] Katherine S. Button, John P. A. Ioannidis, Claire Mokrysz, Brian A. Nosek, Jonathan Flint, Emma S. J. Robinson, and Marcus R. Munafò. Power failure: Why small sample size undermines the reliability of neuroscience. *Nat Rev Neurosci*, 14(5):365–376, May 2013.

[4] Gang Chen, Ziad S. Saad, Audrey R. Nath, Michael S. Beauchamp, and Robert W. Cox. fMRI group analysis combining effect estimates and their variances. *Neuroimage*, 60(1):747–765, Mar 2012.

[5] R. D'Agostino and E. S. Pearson. Testing for departures from normality. *Biometrika*, 60:613–622, 1973.

[6] X. Descombes, F. Kruggel, and D. Y. von Cramon. Spatio-temporal fMRI analysis using Markov random fields. *IEEE Trans Med Imaging*, 17(6):1028–1039, Dec 1998.

[7] B. Fischl, M. I. Sereno, R. B. Tootell, and A. M. Dale. High-resolution intersubject averaging and a coordinate system for the cortical surface. *Hum Brain Mapp*, 8(4):272–284, 1999.

[8] Bruce Fischl, Niranjini Rajendran, Evelina Busa, Jean Augustinack, Oliver Hinds, B. T. Thomas Yeo, Hartmut Mohlberg, Katrin Amunts, and Karl Zilles. Cortical folding patterns and predicting cytoarchitecture. *Cereb Cortex*, 18(8):1973–1980, Aug 2008.

[9] D. Freedman and D. Lane. A nonstochastic interpretation of reported significance levels. *Journal of Business & Economic Statistics*, 1(4):292–98, 1983.

[10] K. J. Friston, A. P. Holmes, and K. J. Worsley. How many subjects constitute a study? *Neuroimage*, 10(1):1–5, Jul 1999.

[11] K. J. Friston, K. J. Worsley, R. S. Frackowiak, J. C. Mazziotta, and A. C. Evans. Assessing the significance of focal activations using their spatial extent. *Hum Brain Mapp*, 1(3):210–220, 1994.

[12] Y.C. Ge, S. Dudoit, and T.P. Speed. Resampling-based multiple testing for microarray data analysis. *Test*, 12:1–77, 2003.

[13] Christopher R. Genovese, Nicole A. Lazar, and Thomas Nichols. Thresholding of statistical maps in functional neuroimaging using the false discovery rate. *Neuroimage*, 15(4):870–878, Apr 2002.

[14] David A. Harville. Maximum likelihood approaches to variance component estimation and to related problems. *Journal of the American Statistical Association*, 72(358):320–338, 1977.

[15] Satoru Hayasaka and Thomas E. Nichols. Validating cluster size inference: Random field and permutation methods. *Neuroimage*, 20(4):2343–2356, Dec 2003.

[16] Satoru Hayasaka and Thomas E. Nichols. Combining voxel intensity and cluster extent with permutation test framework. *Neuroimage*, 23(1):54–63, Sep 2004.

[17] P. Hellier, C. Barillot, I. Corouge, B. Gibaud, G. Le Goualher, D. L. Collins, A. Evans, G. Malandain, N. Ayache, G. E. Christensen, and H. J. Johnson. Retrospective evaluation of intersubject brain registration. *IEEE Trans Med Imaging*, 22(9):1120–1130, Sep 2003.

[18] J. C. Mazziotta, A. W. Toga, A. Evans, P. Fox, and J. Lancaster. A probabilistic atlas of the human brain: Theory and rationale for its development. The International Consortium for Brain Mapping (ICBM). *Neuroimage*, 2(2):89–101, Jun 1995.

[19] Robert A. McLean, William L. Sanders, and Walter W. Stroup. A unified approach to mixed linear models. *The American Statistician*, 45(1):54–64, 1991.

[20] Benoit Da Mota, Virgile Fritsch, Gaël Varoquaux, Tobias Banaschewski, Gareth J. Barker, Arun L. W. Bokde, Uli Bromberg, Patricia Conrod, Jürgen Gallinat, Hugh Garavan, Jean-Luc Martinot, Frauke Nees, Tomas Paus, Zdenka Pausova, Marcella Rietschel, Michael N. Smolka, Andreas Ströhle, Vincent Frouin, Jean-Baptiste Poline, Bertrand Thirion, and the I. M. A. G. E. N. consortium. Randomized parcellation based inference. *Neuroimage*, 89:203–215, Apr 2014.

[21] Thomas E. Nichols and Andrew P. Holmes. Nonparametric permutation tests for functional neuroimaging: A primer with examples. *Hum Brain Mapp*, 15(1):1–25, Jan 2002.

[22] Alfonso Nieto-Castañón and Evelina Fedorenko. Subject-specific functional localizers increase sensitivity and functional resolution of multi-subject analyses. *Neuroimage*, 63(3):1646–1669, Nov 2012.

[23] Regina Nuzzo. Scientific method: Statistical errors. *Nature*, 506(7487):150–152, Feb 2014.

[24] Philippe Pinel, Bertrand Thirion, Sébastien Meriaux, Antoinette Jobert, Julien Serres, Denis Le Bihan, Jean-Baptiste Poline, and Stanislas Dehaene. Fast reproducible identification and large-scale databasing of individual functional cognitive networks. *BMC Neurosci*, 8:91, 2007.

[25] Russell A. Poldrack, Jeanette A. Mumford, and Thomas E. Nichols. *Handbook of Functional MRI Data Analysis*. Cambridge University Press, Cambridge, New York, 2011.

[26] J. B. Poline and B. M. Mazoyer. Analysis of individual positron emission tomography activation maps by detection of high signal-to-noise-ratio pixel clusters. *J Cereb Blood Flow Metab*, 13(3):425–437, May 1993.

[27] J. B. Poline, K. J. Worsley, A. C. Evans, and K. J. Friston. Combining spatial extent and peak intensity to test for activations in functional imaging. *Neuroimage*, 5(2):83–96, Feb 1997.

[28] Denis Rivière, Jean-François Mangin, Dimitri Papadopoulos-Orfanos, Jean-Marc Martinez, Vincent Frouin, and Jean Régis. Automatic recognition of cortical sulci of the human brain using a congregation of neural networks. *Med Image Anal*, 6(2):77–92, Jun 2002.

[29] Alexis Roche, Sébastien Mériaux, Merlin Keller, and Bertrand Thirion. Mixed-effect statistics for group analysis in fMRI: a nonparametric maximum likelihood approach. *Neuroimage*, 38(3):501–510, Nov 2007.

[30] Gholamreza Salimi-Khorshidi, Stephen M. Smith, and Thomas E. Nichols. Adjusting the neuroimaging statistical inferences for nonstationarity. *Med Image Comput Comput Assist Interv*, 12(Pt 1):992–999, 2009.

[31] Stephen M. Smith and Thomas E. Nichols. Threshold-free cluster enhancement: Addressing problems of smoothing, threshold dependence and localisation in cluster inference. *Neuroimage*, 44(1):83–98, Jan 2009.

[32] Andres Fco. Solé, Shing-Chung Ngan, Guillermo Sapiro, Xiaoping Hu 0001, and Antonio López. Anisotropic 2d and 3d averaging of fMRI signals. *IEEE Trans. Med. Imaging*, 20(2):86–93, 2001.

[33] Peter Stiers, Ronald Peeters, Lieven Lagae, Paul Van Hecke, and Stefan Sunaert. Mapping multiple visual areas in the human brain with a short fMRI sequence. *Neuroimage*, 29(1):74–89, Jan 2006.

[34] Stephen C. Strother, Jon Anderson, Lars Kai Hansen, Ulrik Kjems, Rafal Kustra, John Sidtis, Sally Frutiger, Suraj Muley, Stephen LaConte, and David Rottenberg. The quantitative evaluation of functional neuroimaging experiments: The NPAIRS data analysis framework. *Neuroimage*, 15(4):747–771, Apr 2002.

[35] Bertrand Thirion, Philippe Pinel, Sébastien Mériaux, Alexis Roche, Stanislas Dehaene, and Jean-Baptiste Poline. Analysis of a large fMRI cohort: Statistical and methodological issues for group analyses. *Neuroimage*, 35(1):105–120, Mar 2007.

[36] Alan Tucholka, Virgile Fritsch, Jean-Baptiste Poline, and Bertrand Thirion. An empirical comparison of surface-based and volume-based group studies in neuroimaging. *Neuroimage*, 63(3):1443–1453, Nov 2012.

[37] P. H. Westfall and S.S. Young. *Resampling-Based Multiple Testinging: Examples and Methods for p-Value Adjustment*. Wiley-Interscience, 1993.

[38] Anderson M. Winkler, Gerard R. Ridgway, Matthew A. Webster, Stephen M. Smith, and Thomas E. Nichols. Permutation inference for the general linear model. *Neuroimage*, 92:381–397, May 2014.

[39] Mark W. Woolrich, Timothy E. J. Behrens, Christian F. Beckmann, Mark Jenkinson, and Stephen M. Smith. Multilevel linear modelling for fMRI group analysis using Bayesian inference. *Neuroimage*, 21(4):1732–1747, Apr 2004.

[40] K. J. Worsley. An improved theoretical p value for SPMS based on discrete local maxima. *Neuroimage*, 28(4):1056–1062, Dec 2005.

[41] K. J. Worsley, C. H. Liao, J. Aston, V. Petre, G. H. Duncan, F. Morales, and A. C. Evans. A general statistical analysis for fMRI data. *Neuroimage*, 15(1):1–15, Jan 2002.

A Table of the Most Standard Statistical Tests in Neuroimaging

TABLE A.1

Example of statistical models used in neuroimaging. The first column describes the model, the second column describes how the data are ordered in the outcome vector, the third column shows the design matrix, and the last column illustrates the hypothesis tests and corresponding contrasts. Note, in the ANOVA example F-tests are used for all contrasts, whereas t-tests are used for the other examples. We use the notations from Eq. (12.13). This table is adapted from (25).

Test Description	Order of Data	$Z\beta$	Hypothesis test
One-sample t-test, 5 observations	B_1 B_2 B_3 B_4 B_5	$\begin{pmatrix} 1 \\ 1 \\ 1 \\ 1 \\ 1 \end{pmatrix} \begin{pmatrix} \beta_1 \end{pmatrix}$	H_0: Overall mean $= 0$ H_0: $\beta_1 = 0$ H_0: $c^T\beta = 0$ $c = [1]$
Two-sample t-test, 5 observations $Y_1 \cdots Y_5$ in group 1 and 5 observations $Y_6 \cdots Y_{10}$ in group 2.	B_1 B_2 B_3 B_4 B_5 B_6 B_7 B_8 B_9 B_{10}	$\begin{pmatrix} 1 & 0 \\ 1 & 0 \\ 1 & 0 \\ 1 & 0 \\ 1 & 0 \\ 0 & 1 \\ 0 & 1 \\ 0 & 1 \\ 0 & 1 \\ 0 & 1 \end{pmatrix} \begin{pmatrix} \beta_1 \\ \beta_2 \end{pmatrix}$	H_0: equal mean in both groups H_0: $\beta_1 - \beta_2 = 0$ H_0: $c^T\beta = 0$ $c = [1 \ -1]$
Paired t-test, 5 paired measures of two responses B and B', corresponding e.g. to two experimental conditions observed in 5 subjects.	B_1 B'_1 B_2 B'_2 B_3 B'_3 B_4 B'_4 B_5 B'_5	$\begin{pmatrix} 1 & 1 & 0 & 0 & 0 & 0 \\ -1 & 1 & 0 & 0 & 0 & 0 \\ 1 & 0 & 1 & 0 & 0 & 0 \\ -1 & 0 & 1 & 0 & 0 & 0 \\ 1 & 0 & 0 & 1 & 0 & 0 \\ -1 & 0 & 0 & 1 & 0 & 0 \\ 1 & 0 & 0 & 0 & 1 & 0 \\ -1 & 0 & 0 & 0 & 1 & 0 \\ 1 & 0 & 0 & 0 & 0 & 1 \\ -1 & 0 & 0 & 0 & 0 & 1 \end{pmatrix} \begin{pmatrix} \beta_{\text{diff}} \\ \beta_1 \\ \beta_2 \\ \beta_3 \\ \beta_4 \\ \beta_5 \end{pmatrix}$	H_0: B' is equal to B H_0: $\beta_{\text{diff}} = 0$ H_0: $c^T\beta = 0$ $c = [1\,0\,0\,0\,0\,0]$
Two way ANOVA. Factor B has two levels and factor B' has 3 levels. There are 2 observations for each B/B' combination.	$B_1B'_1(1)$ $B_1B'_1(2)$ $B_1B'_2(1)$ $B_1B'_2(2)$ $B_1B'_3(1)$ $B_1B'_3(2)$ $B_2B'_1(1)$ $B_2B'_1(2)$ $B_2B'_2(1)$ $B_2B'_2(2)$ $B_2B'_3(1)$ $B_2B'_3(2)$	$\begin{pmatrix} 1 & 1 & 1 & 0 & 1 & 0 \\ 1 & 1 & 1 & 0 & 1 & 0 \\ 1 & 1 & 0 & 1 & 0 & 1 \\ 1 & 1 & 0 & 1 & 0 & 1 \\ 1 & 1 & -1 & -1 & -1 & -1 \\ 1 & 1 & -1 & -1 & -1 & -1 \\ 1 & -1 & 1 & 0 & -1 & 0 \\ 1 & -1 & 1 & 0 & -1 & 0 \\ 1 & -1 & 0 & 1 & 0 & -1 \\ 1 & -1 & 0 & 1 & 0 & -1 \\ 1 & -1 & -1 & -1 & 1 & 1 \\ 1 & -1 & -1 & -1 & 1 & 1 \end{pmatrix} \begin{pmatrix} \beta_{\text{mean}} \\ \beta_{Y1} \\ \beta_{Y'1} \\ \beta_{Y2} \\ \beta_{Y1\,Y'1} \\ \beta_{Y1\,Y'2} \end{pmatrix}$	F-tests for all contrasts $c^T\beta = 0$ H_0: Overall mean is 0 $c = [1\,0\,0\,0\,0\,0]$ H_0: Main B effect is 0 $c = [0\,1\,0\,0\,0\,0]$ H_0: Main B' effect is 0 $c = \begin{bmatrix} 0\,0\,1\,0\,0\,0 \\ 0\,0\,0\,1\,0\,0 \end{bmatrix}$ H_0: B/B' interaction is 0 $c = \begin{bmatrix} 0\,0\,0\,0\,1\,0 \\ 0\,0\,0\,0\,0\,1 \end{bmatrix}$

13

Corrections for Multiplicity in Functional Neuroimaging Data

Nicole A. Lazar

University of Georgia

CONTENTS

13.1   Introduction 355
13.2   Control of Familywise Error Rate 356
13.3   Control of False Discovery Rate 360
13.4   Accounting for Spatial Dependence 362
13.5   Summary 364
        Bibliography 365

Due to the large size of neuroimaging data, the question of corrections for multiple testing cannot be ignored if valid inference is to be obtained. Historically, familywise error rate has been a relevant quantity to control; for massive datasets, however, this criterion is generally too conservative. More recently, methods for the control of the false discovery rate have gained in popularity in the functional neuroimaging literature, as they are better able to accommodate a large number of tests carried out simultaneously. Bayesian procedures and approaches that explicitly account for the spatial correlation in brain image data are also becoming more prevalent.

13.1   Introduction

The problem of adjusting for multiple testing in functional magnetic resonance imaging was one of the first points of entry for statisticians in this area of application (10). When imaging psychologists tried to apply well-known corrections for multiple comparisons (note: not multiple testing; the two are not the same—a distinction that I will clarify below) such as the Bonferroni correction, they found that these were ineffective, eliminating virtually all signals, even those in areas that "should have been" activated by a particular task or stimulus.

   A simple example will highlight the difficulty. A typical image might consist of 30 slices covering the brain, and each slice might be of resolution 64×64. At each of the $30 \times 64 \times 64 = 122,880$ voxels individually, a simple model such as a linear model or t-test is fit to check for differences in activation levels between a control and task condition. But then how to decide which voxels show a statistically significant difference? At each voxel the null hypothesis of no difference is tested. If the null hypothesis is true at every location, we would expect on average to find $\alpha \times 30 \times 64 \times 64$ "active" voxels—which, if α is taken as the traditional 0.05, comes out to 6,144 apparently active, but actually null, voxels. Clearly, this is not reasonable, and so some adjustment must be made.

One commonly used adjustment in the psychology literature is the Bonferroni correction. This procedure is easy both to implement and to understand, hence its popularity. If one wishes to claim an overall Type I error probability of α, and one is performing v hypothesis tests, then instead of testing each at level α, test at level α/v. In the previous example, this would correspond to setting a much more stringent significance level—instead of 0.05, one would use $0.05/122,880 = 0.000000407$. In terms of equivalent Z scores, this means that instead of a threshold of 1.96 for a two-sided hypothesis test, the level for significance would be 5.07, which is obviously much harder to attain. Furthermore, one can show that with this adjustment, the probability of any falsely rejected null hypothesis out of the v is controlled at the original level α.

Historically, statisticians had addressed the question of multiple comparisons in the context of *post hoc* testing for analysis of variance. In this setting, a researcher carries out an analysis of variance, rejecting the overall null hypothesis that all group means are the same. The multiple comparison approach then investigates further, looking at all pairwise differences, or possibly contrasts involving more than two groups, to find where the differences are (e.g. (20) but there are many others). There is a need to adjust for this after-the-fact peek at the data, so that the overall familywise error rate (FWER) is preserved at the stated level. The Bonferroni correction is an example of such a procedure. A vast literature exists; see (17) and the copious references therein, for a summary of the field through the late 1970s. Note that, although related, this is not quite the same as the statistical question of current interest. Hence solutions that were satisfactory in the context of examining contrasts for the smaller problems of the 1950s–1980s are not necessarily appropriate in the modern setting of testing multiple hypotheses simultaneously.

13.2   Control of Familywise Error Rate

To understand some of the issues involved in error rate control, I consider some simple examples. In the following example, I generate fake p-values, uniformly distributed between 0 and 1, for a single 64×64 slice of data. The results are shown in Figure 13.1. The first panel is a histogram of all of the p-values, and indeed they look to be uniformly distributed. The second panel is a heat map of the p-values in the 64×64 grid; there is no apparent pattern as indeed there should not be, since these data were generated without any consideration of spatial structure. The third panel shows the points in the map that are less than 0.05 in value, i.e. voxels that would be declared significant if no correction for multiplicity were applied. In this particular simulation, there are 206 such voxels, which happens to be just a little bit above the nominal 0.05 (it's 0.0503, to be a bit more precise). Again, there is no apparent pattern—you can see mostly "singletons" or individual points without others nearby, but of course there are a few connected "clusters" as well. Finally, in the last panel, we can see the rejected null hypotheses when the Bonferroni correction is applied— all apparent significant results have been eliminated. In this example, that is in fact the correct outcome, since all of these p-values were generated from the null hypothesis.

I repeated this experiment 5000 times and for each I recorded how many apparent "discoveries" (all of them false, of course) there are when no correction was applied, and how many when the Bonferroni correction is used. On average, with no correction, there are 204.517 false rejections, which is 0.04993 of the 4096–very close to the nominal level.

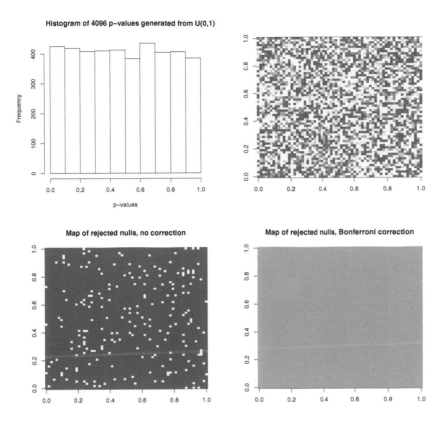

FIGURE 13.1
Simulation results from Uniform(0,1). The top left panel shows a histogram of the p-values, while the top right represents these p-values in a heat map. On the bottom are the points that cross the 0.05 threshold without (left) and with (right) Bonferroni correction.

There are, however, simulations with as few as 158 (just under 4%) or as many as 254 (just over 6%) spurious rejections.

By contrast, with the Bonferroni correction, most of the time—4768/5000 = 0.9536—none of the p-values survive the thresholding. Since we controlled at 0.05, this is the expected result—5% of the time, there is at least one false rejection. In fact, out of the other 232 cases, in 226 there is just a single false rejection, and 6 times there are two. With the null data, then, we get results that are consistent with the claims of each approach, which is not surprising but nonetheless attaching some solid numbers helps to reinforce the ideas.

What happens if I add in some "true activation"? Now I will put a patch of activation in the grid—an area of 100 points where the p-values are uniformly distributed between 0 and 0.01, say, to represent voxels drawn from an alternative hypothesis. The sparsity of the true signal is representative of functional magnetic resonance imaging (fMRI) data, where the proportion of task-responsive voxels is generally quite low. Results are presented in Figure 13.2, in the same format as the first example. Note that in the histogram of p-values, there is now a small but obvious spike in the lowest bar, since I added in 100 p-values from a different, more concentrated, distribution. In a real data analysis, a p-value distribution like this would be indicative of some voxels coming from the alternative—but of course we

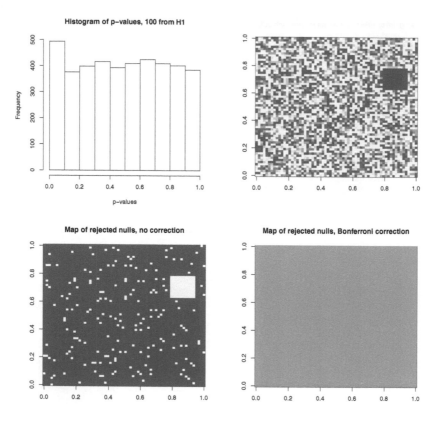

FIGURE 13.2

Simulation results from 100 voxels taken from the alternative and the rest from the null hypothesis. The top left panel shows a histogram of the p-values, while the top right represents these p-values in a heat map. On the bottom are the points that cross the 0.05 threshold without (left) and with (right) Bonferroni correction.

would not know which ones, and the task is to pick those out from the noise. The heat map shows the patch of activation quite clearly. The thresholded map with no correction picks out the entire patch, and many other points as well; in this particular simulation, 284 voxels pass the 0.05 threshold, although just 100 actually come from the truly active set; there are many more false rejections than true ones, and this is the price one pays for no multiplicity adjustment in imaging data. On the other hand, when the Bonferroni correction is applied, again all activation is wiped out, including the truly active set; this is the price one pays for control of the familywise error rate with the conservative Bonferroni adjustment, when the number of tests is very big (and 4096 is not all that large).

As before, I repeated this procedure 5000 times. On average, there are 300 discoveries; since there are always 100 true discoveries to be found, we have on average an "extra" 200 discoveries. And there are always more false findings than true ones—the minimum number of rejections is 254, and the maximum is 351. The Bonferroni correction does not fare well in this scenario. In 4174 of the 5000 cases, no null hypothesis is rejected; in 753 cases there is one rejection; in 65 cases there are two rejections; in six cases there are three rejections;

and in two cases there are four rejections. At best we find four of the 100 truly active points, a small percentage of what is there to find, so most true discoveries are missed.

Although work on multiple comparisons, or simultaneous inference, continued throughout the 1980s (see, for example, (15); (16); (23); (24); and many others), a major philosophical and statistical shift began in the 1990s, with the earliest examples of what we would now call "Big Data" problems. The testing question in the neuroimaging, or more generally large data, setting, is similar but not quite the same, hence some researchers refer instead to "multiple testing" here. There is no overall null hypothesis to be rejected, leading to detailed pairwise or groupwise comparisons. Rather, a separate hypothesis test is conducted at each voxel. There is still a need to control the FWER at some stated level, but it does not derive now from performing multiple *post hoc* comparisons, but instead from performing multiple tests simultaneously. An infamous study (5) highlights the problem in the imaging context. Bennett and colleagues put a dead salmon into an MR scanner, and showed it images of humans in different social situations. After standard analysis, but without control for multiple testing, areas of apparent activation were found inside the brain of the dead fish. Clearly these are all false positives, but the work cleverly shows the importance of taking into account the fact that many tests of (dependent) hypotheses are being tested at once. Early efforts to improve control of the FWER for fMRI data include (12), (28), and (29). For an excellent overview and comparison of FWER control methods in the neuroimaging context, see (18).

Forman et al. (1995) (12) note that activation of interest is likely to occur in clusters of voxels, and that supposedly activated voxels that are isolated from other areas of activation are more likely to be spurious. That is, there is a spatial correlation in the imaging data that should be accounted for in any adjustment for multiplicity. They propose a Monte Carlo simulation scheme based on the extent of spatial dependence, and combine this with a threshold adjustment for each voxel. Under their procedure, a cluster as a whole is considered to be significantly activated if it consists of at least k contiguous (touching) voxels, each of which is significant at a specified level α. k would typically be set by the investigator, and is a function of the spatial structure. Clusters that are smaller than k in extent will not be detected. The advantage of this approach is that it explicitly includes the spatially dependent nature of the activation; on the other hand, the researcher does need to decide in advance on the size of the clusters of interest. In an exploratory study in particular, this might be hard to do, or scientists might be reluctant to miss promising leads by setting the cluster size too small.

Worsley ((28); (29)) proposes a random field approach to control the familywise error rate. His method is based on high-level mathematical analysis of what a field with a particular distribution (Gaussian, t, etc.) "should" look like in a null state. Departures from the null state are captured by excursion sets, regions where the field is more extreme than expected. The Euler characteristic describes the behavior of the excursion set. In three dimensions, to which fMRI are usually reduced via a voxel-wise statistical analysis prior to thresholding, the Euler characteristic counts the number of "connected components" less the number of "holes" (27). As the threshold increases, the holes gradually disappear, as do most of the components, which are peaks in the random field. The Euler characteristic at high thresholds counts the number of peaks in the brain map. Eventually, if the threshold is high enough, just a single peak will remain, and this is the maximum value of the test statistic in the image as a whole. Worsley shows more formally that, at high enough thresholds, this is essentially equivalent to control of FWER. The random field approach as typically used in neuroimaging relies on Gaussian fields, and is one of the reasons that researchers usually spatially smooth their data as part of the standard analysis.

A resampling approach to FWER control is described by Nichols and Holmes (19). As in the use of random fields, the goal is to learn about the extreme tail behavior of the distribution of test statistics, and more specifically of the maximum test statistic. Whereas random field theory depends on distributional assumptions, resampling is a non-parametric, simulation-based analysis, in which data are resampled under the null hypothesis of no activation at any voxel. If the resamples are drawn with replacement, one obtains a bootstrap distribution; if drawn without, which results in a permuting of data labels, this yields a permutation distribution. The resampling methods proceed under the assumption that there is no difference in activation levels between task and control conditions, hence the labels on the stimuli of "task" and "control" are themselves irrelevant. When one samples the labels with or without replacement, corresponding to bootstrap and permutation respectively, a new value of the test statistic can be calculated. Continuing in this way over multiple resamples, the empirical distribution of the test statistics under the null hypothesis of no activation is built up and tail behavior, particularly of extremes, can be examined.

Nichols and Hayasaka (18) provide an excellent and detailed overview of Bonferroni and related methods, random field, and resampling analyses, as well as a comparison of the strengths, weaknesses, and performance of each in the context of functional neuroimaging data. A major advantage of the Bonferroni-type procedures, which include a variety of so-called step-up and step-down methods, is that they are valid under no additional assumptions, or under mild assumptions of positive dependence among tests. When the joint distribution of the tests, or equivalently their dependence structure, is unknown, as will typically be the case for neuroimaging data, it is desirable to impose a minimum of assumptions. On the other hand, these approaches are often conservative, and don't take advantage of the spatial structure in the data at all.

Random field theory, by contrast, makes very strong assumptions on the data, particularly regarding the smoothness of the image, stationarity of the data, and normal distribution. Any of these may be questionable, although they are rarely questioned in practice. When the data are not sufficiently smooth, or if one is not willing to accept the assumptions needed for random field theory, permutation test or bootstrap can be used to obtain an empirical distribution for the maximal statistic. These methods are computationally intense, particularly for correlated data, which need to be handled carefully. Furthermore, they need to be developed on a case-by-case basis, taking into consideration the specifics of the null and alternative hypotheses, and the test statistic in question, as well as the attributes of the data that are relevant to preserve under resampling. Through examination of real and simulated data, (18) conclude that random field theory is generally conservative, unless the data are very smooth. The permutation test approach is most sensitive, but it is computationally intensive.

13.3   Control of False Discovery Rate

A radically different approach to the question is to control not the FWER but another quantity called the false discovery rate (FDR; (1)). Whereas FWER is the probability of making even one false rejection, and hence is extremely conservative especially when the number of tests is vast, FDR is the expected proportion of falsely rejected hypotheses among all of the rejected. This is, in theory, a much more generous standard, which furthermore adapts to the features of the data, such as the overall strength of the signal. Benjamini and Hochberg (1) introduce the basic control procedure; Genovese, Lazar, and Nichols (13) introduce FDR control to the fMRI community. Since the initial works, many variations and improvements for FDR control have been proposed for imaging and genetics data in particular.

The basic procedure is as follows. One starts by ordering all of the p-values obtained from the hypothesis tests, from smallest to largest. Let $p_{(i)}$ denote the ith smallest p-value. To control the FDR at level q, find the largest i for which

$$p_{(i)} \leq i\frac{q}{v},$$

where v is the number of tests. Note that the inequality may not hold for any i, which provides insight into the increased sensitivity of the FDR control procedure: there will be rejected hypotheses if and only if the inequality holds for some i. By contrast, $i = 1$ yields the Bonferroni threshold adjustment when q is the overall significance level desired in a FWER setting, that is, the smallest p-value is compared to the Bonferroni standard. The Bonferroni procedure, therefore, leads to some rejected hypotheses if and only if the inequality holds specifically for $i = 1$. Finally, $i = v$ gives q, the unadjusted level. Of course, this correspondence holds if we take q to be equivalent to α, but there is no particular reason to do so. Ideally, q would be chosen by the researcher to represent the proportion of false rejections he or she would be willing to tolerate, and as such this value should be highly context-dependent. In practice, many people seem to default to $q = 0.05$ but in some of my own work, I've found much lower q values to be helpful.

I repeated the experiment described in Section 13.2 using the basic false discovery rate control procedure, with $q = 0.05$. For the null data, and 5000 repetitions, there are 4763 cases in which no nulls are rejected, compared to 4768 in the simulation with the Bonferroni correction. In most instances, the procedure correctly fails to make any discoveries. On the other hand, in keeping with the more generous nature of the FDR control technique, there are slightly more discoveries—as many as three in three of the simulations. Overall the proportion of false discoveries is 0.0518, which is close to the nominal $q = 0.05$.

To better understand the behavior of the FDR procedure and its sensitivity to both the choice of q and the strength of the alternative, I conducted a series of simulations following the same general scheme as outlined above: 100 p-values from the "alternative" taken to be uniform on $(0, \delta)$, for $\delta = 0.001$ (an easier case) and $\delta = 0.01$ (a harder case), with the other 3996 uniform on $(0, 1)$; and values of q that represent varying levels of tolerance for false discoveries, $q = 0.01, 0.05, 0.15$. In all cases the simulation is repeated 5000 times.

Starting with the easy case, when $q = 0.01$, a low level of tolerance for false rejections, the overall number of discoveries is also small in general: at most, there are 9 "rejections" and a majority of the time there are none. Correspondingly, then, the number of false discoveries is also small, at most 2, and on average 0.0426 false discoveries. The false discovery proportion, calculated as the proportion of falsely rejected nulls out of all rejected, assuming at least one null is rejected is 0.0355 in this simulation, which is slightly higher than the claimed $q = 0.01$.

With $q = 0.05$, a *de facto* default level of tolerance for many researchers, the performance is better. On average there are 100 discoveries, although the smallest number is 32 and the largest is 115, so there is strong variability in the results. The number of false rejections ranges from 0 to 15, with an average of 4.387. And the proportion itself ranges from 0 to 0.13, with an average of 0.0428. Although the average is close to the required level, it is interesting that even in this easy case, the actual (observed, simulated) false discovery proportion can be as high as 0.13.

If one is willing to tolerate a higher proportion of false discoveries, and take $q = 0.15$, there are always spurious rejections. Over the 5000 simulations, the number of rejections range from 104 to 140, with an average of 117.4. The 100 true alternatives are correctly identified. Since there were only 100 true draws from the alternative, some of the discoveries must be false. As expected, the number of false discoveries is from 4 to 40, with an average of 17.4 and this results in a false discovery proportion anywhere from 0.038 to 0.29. The average for this scenario is 0.147, which is indeed very close to the claimed $q = 0.15$.

Not surprisingly, the more difficult scenario, in which the alternative draws are less extreme, has poorer performance overall. When $q = 0.01$ there are very few discoveries, at most three out of the 100, and so if even one of them is wrong, the proportion of false discoveries will be high. Indeed, the average proportion of false discoveries, when any nulls are rejected at all, is 0.2. For $q = 0.05$, the situation isn't much better. At most there are five rejections and the average false discovery proportion is 0.31. In both of these cases, the median is 0 and the maximum is 1, showing that the distribution of values is very skewed.

Under the most tolerant level of $q = 0.15$, the number of discoveries ranges from 0 to 26, so still most of the 100 are missed; the average number of rejections is just 1.5. The false discovery proportion (FDP) is 0.27 on average, much higher than claimed, although the median is close at 0.17. And again the maximum FDP value is 1.

From this small simulation, it is apparent that the independence version of the FDR control procedure should be used with caution. If the signal is weak, which might easily be true for single-subject fMRI data for example, even this more sensitive thresholding approach may not be sufficient. A generous q may help, but one runs the risk of actually achieving a false discovery rate that is very different from the claimed. When the signal is strong, as might be the case for data combined across multiple subjects, FDR performs as claimed. However, if the value of q is too high, false discoveries will be guaranteed.

Control of the false discovery rate, along with the simple FDR-controlling procedure of Benjamini and Hochberg, proved revolutionary for many large data endeavors, including neuroimaging, because it allowed for a shift in the perception of an appropriate error rate to control. Many researchers have developed variations on the basic method, for example to accommodate general dependence among the tests, or to improve power, or to establish alternatives especially to connect with Bayesian ideas.

Benjamini and Yekutieli (4) show that in fact the original procedure continues to control FDR even for dependent tests, particularly under a condition called *positive regression dependence*.

A simple modification to the original procedure allows for general dependence among the p-values, the usual situation for neuroimaging. The comparison becomes

$$p_{(i)} \leq \frac{i}{vc(v)}q.$$

Here $c(v) = \sum_{i=1}^{v} \frac{1}{i}$ under arbitrary dependence.

Proposals to improve the power of the FDR procedure have focused mostly on attempts to obtain an initial estimate of the number of true null hypotheses in the collection, and hence reduce the number of tests in the comparison. Benjamini and Hochberg (2) give an adaptive FDR control method, that first estimates the number of true nulls and then applies the usual procedure taking this number into account. Alternatives are given in (11), (26), and (3). Most of these continue to control the false discovery rate under independence of the tests; some do so under dependence as well.

13.4   Accounting for Spatial Dependence

An important characteristic of fMRI data—and hence of the hypotheses that are simultaneously tested at each voxel—is their spatial dependence. It is crucial to account for this structure, as both the Forman et al. and Worsley random field approaches explicitly do. The simpler Bonferroni and basic FDR procedures do not, however, consider spatial structure per se, although both can be valid even in the face of lack of independence. An immediate effect of the spatial dependence is that the number of tests to correct or control for is not obvious. If the voxels were completely independent, then control would be related directly to the number of voxels, v. If they were completely dependent, then there would be just one real piece of data. The truth is somewhere between these extremes. It is not, however, necessarily easy to estimate the "true" number of hypotheses that are being tested; see (18) for an example. In practice, then, the focus has been more on modeling.

In recent years especially, there have been increased efforts to control for multiplicity in the presence of spatial dependence specifically, as opposed to general dependence such as in the work on false discovery rate cited in Section 13.3. The problem is more subtle than it might appear at first, since the voxels themselves have no intrinsic physiological or neurological meaning. Rather they are simply an artifact of how the data are obtained in the scanner. For this reason some, such as (9) and (8), have argued that in fact any correction for multiplicity that is based on the voxels is, in some sense, "missing the essential point." Instead, one should consider and perform inference on features of the map, such as peaks, harking back to the random field theory of Worsley. These authors also point out, correctly, that the sharp null hypothesis of "no activation" at the voxel level is almost surely false everywhere, since the brain is always and everywhere activated. (This reasoning may or may not carry over to the question of group comparisons—are group-averaged brains also always and everywhere differentially activated, for example?) Interestingly, Chumbley and Friston (9) attribute this in part to the spatial smoothing that is (usually) applied to the subject-level data, rather than to any characteristic of the data themselves. In my opinion, this actually reverses the problem, since spatial smoothing is an attempt to modify the data to fit a model (random field theory), when we should be looking for models that are suitable for the data. The spatial dependence likely exists, however, with or without smoothing.

Nonetheless, there is perhaps promise in the approaches that attempt to use spatial information directly. Even if these are still based on the voxel unit, they recognize that tagging a single voxel as "active" or "not active" is not of that much scientific interest or relevance. Recent examples of this perspective include (14) and (7).

Heller and colleagues (14) propose a cluster FDR procedure that operates on connected groups of voxels rather than on individuals. They define clusters empirically, iteratively joining together voxels that are maximally correlated over their time course. In this way, a set of nonoverlapping clusters is found. Each cluster is summarized by the average time course of its voxels, which is taken to be the "signal" for that cluster. Based on this average time course, a statistical test for cluster-level activation is carried out. The Benjamini–Hochberg procedure for FDR control can then be applied to each of the clusters. The authors note that the number of clusters might still be large, although it will be much smaller than the number of voxels, hence the multiple testing problem is mitigated to a large extent. Also, the inference is on the cluster, not its individual components. As a consequence, if a cluster is concluded to be task-related, this says nothing about the individual voxels in the cluster, beyond the fact that some of them must also be task-related (although how many and which ones are unknown). On the other hand, if a cluster is deemed not task-related, then none of its component voxels should be either.

A completely different direction is given in (7), which extends the work of Scott and Berger (21),(22) on Bayesian multiplicity control to incorporate simple spatial information. A difficulty with the use of Bayesian methods in the analysis of fMRI data is the size of the data and number of parameters needed for a complete spatio-temporal model. Brown et al. (7) avoid this problem by fitting a conditional autoregressive (CAR) model to the statistical maps obtained after a "first level" analysis, that is, collapsing over time. The CAR model is built around small, local neighborhoods, with a prior placed on the probability that a voxel is significant, i.e., that the null hypothesis of no difference between task and rest condition at that voxel should be rejected. Voxels share information, so that rejections in the form of high posterior probabilities of significance are naturally clustered. The procedure is self-calibrating, as shown by both Scott and Berger and by Brown and colleagues, which means that one can consider the posterior probabilities of activation with no need for additional correction for multiplicities.

13.5   Summary

For multiple testing problems in functional neuroimaging contexts, as this review shows, a researcher should consider the following dimensions.

1. **Control of FWER or of FDR.** There is ample evidence, even before the advent of massive datasets, that control of the familywise error rate is a conservative approach to the multiplicity problem. Very large numbers of tests exacerbate the issue. While dependence reduces the number of simultaneous hypotheses, it is harder to model; the tradeoff between these two aspects of the problem deserves further attention. Control of the false discovery rate has gained in popularity in neuroimaging, as well as in many other scientific contexts. With the extensions of the original FDR-control procedure (1) to account for various dependence structures, clusters, and Bayesian interpretations, the usefulness of this approach has expanded. Still, one should consider carefully which rate to control, and at what level. The latter question, in particular, has tended to be ignored, with many researchers relying on the "default" 0.05 level even for FDR control. Since FDR admits an intuitive interpretation, there is no reason to adhere to traditional (albeit themselves arbitrary) levels of control.

2. **Frequentist or Bayes.** Naturally enough, most work on the multiplicity problem has been from a frequentist perspective, as point-null hypothesis testing is not of interest to the traditional Bayesian. Further, many Bayesian procedures are self-calibrating, thereby obviating the need for active error control even in the face of a large number of hypotheses. Some recent Bayesian work has turned to the multiple testing problem ((21), (22)) and there is a strong literature on Bayesian interpretations of false discovery rate procedures (see, for example (25)).

3. **Independent voxels or dependent clusters.** It is evident that the true number of independent pieces of information in an fMRI study of a single brain—and hence the true size of the multiple testing problem—is somewhere between one (complete dependence) and the number of voxels in the dataset (complete independence). Historically, analysis including control for multiplicity, has been on the voxel level, that is, assuming spatial independence. This was widely recognized as an approximation and simplification, but how to deal with the spatial correlation has still not been resolved, although there are various proposals in the literature. One can account for spatial dependence at the modeling stage, as in, for example, (30), which takes a geostatistical approach, or (6), which carries out a spatiotemporal analysis on regions of interest. Alternatively, as in (7), one can model the data independently at each voxel and incorporate the dependence later, at the point of accounting for multiple testing issues. Yet another approach considers clusters of voxels, again either at the modeling stage or at the thresholding stage, as in (14). Overall, it is probably better to acknowledge the spatial

dependence rather than to ignore it altogether, although the latter approach often works "well enough" in practice and is generally easier to implement.

The challenge of thresholding, or more generally, assessing statistical and scientific significance in functional neuroimaging data, is not easy to overcome. Multiple testing corrections are typically necessary to avoid over-optimistic and unrealistic conclusions from what are often not very strong data. Most neuroimaging researchers are by now aware of the need for some sort of control, although it is still possible to find published articles that do not correct for multiplicity. There is room for further developments, especially in the direction of some more recent approaches that take advantage of the specific spatial structure in the data. Such methods do not rely on general dependence, and hence should be more powerful. They also acknowledge directly that the voxel is not a meaningful unit of analysis, but rather that collections of voxels that cluster in spatially relevant ways may yield more substantive and reproducible scientific results.

Bibliography

[1] Y. Benjamini and Y. Hochberg. Controlling the false discovery rate: A practical and powerful approach to multiple testing. *Journal of the Royal Statistical Society, Series B*, 57:289–300, 1995.

[2] Y. Benjamini and Y. Hochberg. On the adaptive control of the false discovery rate in multiple testing with independent statistics. *Journal of Educational and Behavioral Statistics*, 25:60–83, 2000.

[3] Y. Benjamini, A.M. Krieger, and D. Yekutieli. Adaptive linear step-up procedures that control the false discovery rate. *Biometrika*, 93:491–507, 2006.

[4] Y. Benjamini and D. Yekutieli. The control of the false discovery rate in multiple testing under dependency. *Annals of Statistics*, 29:1165–1188, 2001.

[5] C. Bennett, A. Baird, M. Miller, and G. Wolford. Neural correlates of interspecies perspective taking in the post-mortem Atlantic salmon: An argument for proper multiple comparisons correction. *Journal of Serendipitous and Unexpected Results*, 1:1–5, 2010.

[6] F.D. Bowman. Spatiotemporal models for region of interest analyses of functional neuroimaging data. *Journal of the American Statistical Association*, 102:442–453, 2007.

[7] D.A. Brown, N.A. Lazar, G.S. Datta, W. Jang, and J.E. McDowell. Incorporating spatial dependence into Bayesian multiple testing of statistical parametric maps in functional neuroimaging. *NeuroImage*, 84:97–112, 2014.

[8] J. Chumbley, K. Worsley, G. Flandin, and K. Friston. Topological FDR for neuroimaging. *NeuroImage*, 49:3057–3064, 2010.

[9] J.R. Chumbley and K.J. Friston. False discovery rate revisited: FDR and topological inference using Gaussian random fields. *NeuroImage*, 44:62–70, 2009.

[10] W.F. Eddy, M. Fitzgerald, C. Genovese, N. Lazar, A. Mockus, and J. Welling. The challenge of functional magnetic resonance imaging. *Journal of Computational and Graphical Statistics*, 8:545–558, 1999.

[11] B. Efron, R. Tibshirani, J.D. Storey, and V. Tusher. Empirical Bayes analysis of a microarray experiment. *Journal of the American Statistical Association*, 96:1151–1160, 2001.

[12] S.D. Forman, J.D. Cohen, M. Fitzgerald, W.F. Eddy, M.A. Mintun, and D.C. Noll. Improved assessment of significant activation in functional magnetic resonance imaging (fMRI): Use of a cluster-size threshold. *Magnetic Resonance in Medicine*, 33:636–647, 1995.

[13] C.R. Genovese, N.A. Lazar, and T.E. Nichols. Thresholding of statistical maps in functional neuroimaging using the false discovery rate. *NeuroImage*, 15:870–878, 2002.

[14] R. Heller, D. Stanley, D. Yekutieli, N. Rubin, and Y. Benjamini. Cluster-based analysis for fMRI data. *NeuroImage*, 33:599–608, 2006.

[15] Y. Hochberg. A sharper Bonferroni procedure for multiple tests of significance. *Biometrika*, 75:800–802, 1988.

[16] Y. Hochberg and A.C. Tamhane. *Multiple Comparison Procedures*. John Wiley & Sons, Hoboken, NJ, USA, 1987.

[17] R.G. Miller. *Simultaneous Statistical Inference*. Springer-Verlag, New York, NY, USA, second edition, 1981.

[18] T.E. Nichols and S. Hayasaka. Controlling the familywise error rate in functional neuroimaging: A comparative review. *Statistical Methods in Medical Research*, 12:419–446, 2003.

[19] T.E. Nichols and A.P. Holmes. Nonparametric permutation tests for functional neuroimaging: A primer with examples. *Human Brain Mapping*, 15:1–25, 2001.

[20] H. Scheffé. A method for judging all contrasts in the analysis of variance. *Biometrika*, 40:87–104, 1953.

[21] J.G. Scott and J.O. Berger. An exploration of aspects of Bayesian multiple testing. *Journal of Statistical Planning and Inference*, 136:2144–2162, 2006.

[22] J.G. Scott and J.O. Berger. Bayes and empirical-Bayes multiplicity adjustment in the variable selection problem. *Annals of Statistics*, 38:2587–2619, 2010.

[23] J.P. Shaffer. Multiple hypothesis testing. *Annual Review of Psychology*, 46:561–584, 1995.

[24] R.J. Simes. An improved Bonferroni procedure for multiple tests of significance. *Biometrika*, 73:751–754, 1986.

[25] J.D. Storey. The positive false discovery rate: A Bayesian interpretation and the q-value. *Annals of Statistics*, 31:2013–2035, 2003.

[26] J.D. Storey, J.E. Taylor, and D. Siegmund. Strong control, conservative point estimation and simultaneous conservative consistency of false discovery rates: A unified approach. *Journal of the Royal Statistical Society, Series B*, 66:187–205, 2004.

[27] K.J. Worsley. Detecting activation in fMRI data. *Statistical Methods in Medical Research*, 12:401–418, 2003.

[28] K.J. Worsley, A.C. Evans, S. Marrett, and P. Neelin. A three-dimensional statistical analysis for CBF activation studies in human brain. *Journal of Cerebral Blood Flow and Metabolism*, 12:900–918, 1992.

[29] K.J. Worsley, S. Marrett, P. Neelin, A.C. Vandal, K.J. Friston, and A.C. Evans. A unified statistical approach for determining significant signals in images of cerebral activation. *Human Brain Mapping*, 4:58–73, 1995.

[30] J. Ye, N.A. Lazar, and Y. Li. Sparse principal component analysis and geostatistical analysis in clustering fMRI time series. *Journal of Neuroscience Methods*, 199:336–345, 2011.

14

Functional Connectivity Analyses for fMRI Data

Ivor Cribben

University of Alberta School of Business

Mark Fiecas

University of Warwick

CONTENTS

14.1	Introduction	369
14.2	Methods and Measures for FC	371
	14.2.1 Setup	371
	14.2.2 Cross-Correlation and Partial Cross-Correlation	371
	14.2.3 Stability Selection	372
	14.2.4 Cross-Coherence and Partial Cross-Coherence	373
	14.2.5 Mutual Information	374
	14.2.6 Principal and Independent Components Analyses	374
	14.2.7 Time-Varying Connectivity	375
14.3	Simulation Study	376
	14.3.1 Results	378
14.4	Functional Connectivity Analysis of Resting-State fMRI Data	381
	14.4.1 Data Description and Preprocessing	381
	14.4.2 Overview of the Estimation Procedure	382
	14.4.3 Results	382
14.5	Future Directions and Open Problems	384
	Bibliography	389

14.1 Introduction

Traditional neuroimaging studies focus on "activation studies" to determine distributed patterns of brain activity that are associated with specific tasks or psychological states. These studies are usually referred to as functional segregation studies (41). In other words, researchers are interested in segregating or partitioning the brain into regions based on their activation while performing certain tasks. However, recently, in order to thoroughly understand brain function, researchers have begun to study the interaction of brain regions, as a great deal of neural processing is performed by an integrated network of several brain regions. This is sometimes referred to as functional integration. The first discussion on segregation and integration of brain function occurred at the International Medical Congress meeting held in London on August 4, 1881. Friedrich Goltz, a German physiologist, argued for the idea that brain function was dependent on related pathways and connections and not on the idea of localization (84). This discussion continues to this day. That is, can physiological changes from specific tasks or psychological states be explained more effectively by functional localization, the interregional functional connections (42), or their combination?

In the analysis of functional magnetic resonance imaging (fMRI), positron emission tomography (PET), electroencephalography (EEG) and magnetoencephalography (MEG) time series data, functional connectivity (FC) is the name given to studying this interaction, correlation or dependence between signals observed from spatially remote brain regions (44, 38). The idea of FC originally arose in the analysis of spike train data by (46). FC is a measure of dependence or "relatedness" but does not comment on how the dependence is mediated. It is sometimes referred to as undirected association. Estimating the FC between pre-defined brain regions or voxels allows for the characterization of interregional neural interactions during particular experimental tasks or merely from spontaneous brain activity while subjects are being scanned at rest. Using fMRI and PET, researchers have been able to create maps of FC with distinct spatial distributions of temporally correlated brain regions called functional networks. One of the aims of the Human Connectome Project (94), an ambitious project sponsored by the National Institutes of Health (NIH), is the investigation of the functional connections of the human brain. While FC is used to measure undirected association, effective connectivity is defined as the influence that one brain region or system has over another one. It can be thought of in a circuit diagram format and is sometimes referred to as directed association.

As well as functional and effective connectivity, researchers are also interested in structural connectivity, the study of how brain regions are anatomically linked. It is thought that by using diffusion tensor imaging (DTI) to measure the anatomical fiber structure between brain regions, the anatomical infrastructure can be learned and used to support the measurement of effective connectivity patterns (4, 16, 94). In addition, structural connectivity has the power to be very useful in the estimation of FC by using the information in the measured structural connectivity as a prior in a concrete Bayesian framework. This may enable more precise model estimation and efficient model comparison.

It has been shown that neurological disorders disrupt the FC or structural properties of the brain (50, 75). However, it is still unknown whether they are the cause or consequence of the disease. The estimation of FC and linking this structure to disorders is a good starting point for the treatment of the disease. Reference (19) investigated the link between FC and schizophrenia with the objective of finding biomarkers for the disorder. References (15, 51, 93, 98) have found different static FC network structures in subjects with Alzheimer's disease compared to healthy subjects. We expect this area of research to expand in future years due to the many unsolved open statistical problems.

The main focus of this chapter is statistical methods that estimate static FC where the time series data within each brain region is stationary or the experimental condition does not change over time. To address recent evidence that FC is evolving over time, we also describe some of the new techniques for estimating time-varying FC. There are numerous methods for quantifying FC, and each method has its own strengths and weaknesses (106, 38). We also apply the various FC metrics to both simulated data and to a resting-state fMRI dataset, whilst additionally making comparisons between them. Finally, we discuss the statistical issues in estimating FC and the many open statistical problems.

TABLE 14.1
A table of abbreviations for the FC metrics.

FC Metric	Abbreviation
Cross-correlation	CCor
Partial correlation - unregularized	PCCor
Partial correlation - glasso	PCCorG
Partial correlation - glasso with refit	PCCorGR
Partial correlation - glasso with refit and subsampling	PCCorGRS
Cross-coherence	CCoh
Partial coherence - unregularized	PCCoh
Partial coherence - shrinkage	PCCohS

14.2 Methods and Measures for FC

14.2.1 Setup

Before we detail the methods used to estimate or measure FC, it is important to first describe the data for which we estimate the FC. After the preprocessing steps (see Chapter 10) are completed, we can estimate the FC between selected voxel time courses or between time courses from pre-defined regions of interest. Brain regions, commonly referred to as regions of interest or ROIs, are usually defined using an anatomical atlas (e.g., the Automated Anatomical Labeling (AAL) atlas or the Brodmann atlas) or using a data-driven technique (21). Obviously, the FC patterns change as the definition of the boundaries of the regions change, and in fact, (92) showed that the inferred dependency structure is impacted by how the ROIs are constructed. Other methods that can be used to define a brain region include Group Independence Component Analysis (ICA; (18)), which decomposes multi-subject neuroimaging data into functional brain regions. After defining the ROI using any of the procedures above, we average the time courses across the voxels within that region and carry that data forward to an FC analysis. By defining brain regions we not only reduce the computational burden but also introduce statistical benefits due to the reduction in variability. Finally, a seed-based FC analysis is a simple approach that compares the FC between a "seed" region or voxel in the brain with all of the other regions/voxels throughout the brain. A seed-based FC analysis allows one to investigate the spatial variation of FC with respect to the seed.

Throughout this chapter, we will let $\mathbf{X}(t)$, $t = 1, \ldots, T$ be a P-variate time series representing the data. Each dimension corresponds to a time series from a voxel or an ROI. An FC analysis then is simply an investigation of the dependency structure of $\mathbf{X}(t)$. A list of the FC metrics or measures of dependency we will use in this chapter is in Table 14.1.

14.2.2 Cross-Correlation and Partial Cross-Correlation

Cross-correlation (CCor) is the simplest and most popular metric for computing the FC between two distinct dimensions of $\mathbf{X}(t)$, say, $X_j(t)$ and $X_k(t)$, $j \neq k$. CCor quantifies the (lagged) temporal dependencies between two signals. Estimating CCor and its asymptotic distribution for statistical inference have been well investigated in the literature. One often performs statistical inference after applying the Fisher-Z transform, defined by $Z(\widehat{\rho}) = -0.5 \log((1 + \widehat{\rho})/(1 - \widehat{\rho}))$, on the sample cross-correlation $\widehat{\rho}$ because the asymptotic variance of $Z(\widehat{\rho})$ is not a function of the population-level cross-correlation (14).

CCor quantifies only the marginal linear dependence between $X_j(t)$ and $X_k(t)$; it is possible that the linear association between $X_j(t)$ and $X_k(t)$ is driven by a third signal $X_i(t)$. In order to remove the effects of $X_i(t)$ when investigating the dependency between $X_j(t)$ and $X_k(t)$, one should compute the *partial* cross-correlation (PCCor) between the two signals. The PCCor between $X_j(t)$ and $X_k(t)$, after removing the effects of a third signal $X_i(t)$, proceeds by first regressing each of $X_j(t)$ and $X_k(t)$ on $X_i(t)$, and then cross-correlating the residuals (with possibly some lag). This approach, however, can be very inefficient if one is interested in all pairwise PCCors between tens or hundreds of signals. One could, instead, extract all pairwise values of PCCor from the precision matrix (the inverse of the covariance matrix) of the data (103, 70). Because PCCor quantifies the "direct" relationship between two signals (conditional on the other signals observed), PCCor is a valuable metric for estimating brain networks (70, 24). However, when the dimensionality of $\mathbf{X}(t)$ is large and the time courses are relatively short, estimates of the precision matrix can be numerically unstable, or, if the sample covariance matrix is not full rank, impossible to obtain. Specifically, let S be the sample covariance matrix of $\mathbf{X}(t)$. Whenever the dimensionality P is much larger than the length of the time series T, then S is not of full rank. This problem can be alleviated via, e.g., regularization procedures. One example that has been used for FC analyses is the graphical lasso (glasso) (92, 24, 38, 25). Given the sample covariance matrix S, we can obtain an estimate of the precision matrix, $\widehat{\Omega}$, that maximizes the penalized log-likelihood

$$\log(\det(\widehat{\Omega})) - \operatorname{tr}(S\widehat{\Omega}) - \lambda||\widehat{\Omega}||_1, \qquad (14.1)$$

where the matrix norm $||\cdot||_1$ is the sum of the absolute values of the elements of the matrix, including the diagonal elements. The sparsity of the graph is controlled by a parameter λ and is often estimated using cross-validation. Smith et al. (92) showed, using simulated fMRI time courses that the PCCor obtained from a ℓ_1-regularized precision matrix yielded high sensitivity to estimating the connectivity of a network structure, and (38) showed, using a test-retest resting-state fMRI dataset, that an ℓ_1-regularized estimate of the precision matrix yielded estimates of PCCor that had higher test-retest reliability than the non-regularized estimates. There have been numerous theoretical and methodological developments for estimating large-scale covariance matrices (see, e.g., (85) for a thorough review), and because it is often the case that the dimensionality P is much larger than the length of the time series T, some type of regularization procedure for estimating the precision matrix should be used whenever possible. We carefully point out that the graphical lasso assumes that the precision matrix is sparse. Sparsity is a modeling assumption that is difficult to check, and this assumption may not hold for fMRI data (49).

14.2.3   Stability Selection

The precision matrix obtained from an ℓ_1-regularized precision matrix for estimating the connectivity of a network structure can also be represented by a graphical model. Graphical models display the dependency structure of $\mathbf{X}(t)$ using a graph G. A graph, $G = (V, E)$, consists of a set of vertices V and corresponding edges E that connect pairs of vertices. They may be defined as either undirected or directed with respect to how the edges connect one vertex to another. Directed graphs infer directionality between variables (or vertices) while undirected graphs do not. This chapter deals with undirected graphs exclusively. In this case, each vertex represents a time series and edges encode dependencies between the variables.

As we have mentioned above, we can estimate a precision matrix or an undirected graph using the graphical lasso (40). Here, an edge and missing edge between two vertices in the graph indicates a partial correlation and conditional independence between brain signals respectively. We first estimate the undirected graph for a path of λ values and choose the value of λ (sparsity of the graph) using 10–fold cross-validation (see Section 14.3). After estimating the graph using cross-validation and identifying non-zero edges, the model is refit without the sparse inducing (or l_1) constraint while keeping the zero elements in the matrix fixed to reduce bias and improve the model selection performance (24, 25). As the graphical lasso is known to estimate a number of false positive edges in the estimated undirected graphs, a subsampling stability selection approach (74) can be performed to rectify the problem. Here, the goal is to control the family-wise type I multiple testing error by looking at the selection probabilities of every edge under subsampling. In this setup, the data are subsampled many times and we choose all edges that occur in a large fraction of the resulting selection sets. This is very similar to the procedure carried out in (24). The graphical glasso estimate, the graphical lasso estimate with a refit, and the graphical lasso estimate with a refit and the subsampling procedure are referred to as PCCorG, PCCorGR, and PCCorGRS, respectively.

14.2.4 Cross-Coherence and Partial Cross-Coherence

Cross-coherence (CCoh) quantifies the linear relationship between two signals that are being driven by frequencies (or frequency bands) of interest, i.e., CCoh is the analog of CCor in the frequency domain. Reference (79) showed that the CCoh between two signals is asymptotically equivalent to the cross-correlation between the two signals after being passed through a band-pass filter. Partial cross-coherence (PCCoh) quantifies the linear relationship between two signals conditional on the other signals observed, and it is the analog of PCCor in the frequency domain. These metrics are frequently used in imaging modalities that have a fine temporal resolution, e.g., EEG (95, 73, 36) and MEG (95, 2, 54), though it has been successfully used in fMRI studies (96, 89, 20).

Estimating CCoh or PCCoh requires an estimate of the spectral density matrix $\mathbf{f}(\omega)$, which can be interpreted as the covariance matrix of the data in the frequency domain (12). If we assume that $\mathbf{X}(t)$ is weakly stationary (i.e., the first two moments of the data are time-invariant), then we can estimate the spectral density matrix by computing the discrete Fourier transform of the data to obtain the vector of Fourier coefficients $\mathbf{d}_X(\omega) = T^{-1/2} \sum_{t=1}^{T} \mathbf{X}(t) \exp(-i2\pi\omega t)$, which are used to construct the periodogram matrix $\mathbf{I}(\omega) = \mathbf{d}_X(\omega)\mathbf{d}_X(\omega)^*$, where (*) denotes the complex conjugate transpose. Though $\mathbf{I}(\omega)$ is an asymptotically unbiased estimator for $\mathbf{f}(\omega)$, its variance does not decrease in large samples, and thus, it is not a consistent estimator (12). However, if we pick a smoothing kernel $K_T(\cdot)$ whose smoothing span M_T is such that that $M_T \to \infty$ and $M_T/T \to 0$ when $T \to \infty$ and use it to smooth each element of $\mathbf{I}(\omega)$ over frequencies to obtain the smoothed periodogram matrix $\widehat{\mathbf{f}}(\omega)$, i.e., $\widehat{f}_{jk}(\omega) = \int_{-0.5}^{0.5} K_T(\omega - \alpha)I_{jk}(\alpha)d\alpha$, then $\widehat{\mathbf{f}}(\omega)$ is a consistent estimator for $\mathbf{f}(\omega)$ (12). There are other well-established nonparametric estimators that are consistent estimators for $\mathbf{f}(\omega)$, such as Welch periodograms and the multitaper estimator, and we refer to the cited works for details (83, 28, 73, 106). The cross-coherence between the j-th and k-th dimension of $\mathbf{X}(t)$ is denoted by $\rho_{jk}^2(\omega) = |f_{jk}(\omega)|^2/(f_{jj}(\omega)f_{kk}(\omega))$, and given an estimate $\widehat{\mathbf{f}}(\omega)$ of the spectral density matrix, we can estimate the cross-coherence by plugging in the appropriate entries of the matrix. Since CCoh is the frequency domain analog of the square of CCor, we report instead $|\rho_{jk}(\omega)|$ so that it is comparable to CCor. Just as with CCor, the Fisher-Z transform of $|\rho_{jk}(\omega)|$ is used to stabilize the asymptotic variance for statistical inference (14, 79).

PCCoh is a function of the inverse of the spectral density matrix, and so its estimates may be impossible to obtain whenever $P \gg T$ (89, 9, 37). Thus, we similarly advocate the use

of regularized estimators of the spectral density matrix in order to obtain stable estimates
of PCCoh (73, 37). Moreover, (38) used resting-state fMRI data to show that PCCoh
was impossible to estimate without regularization because the length of the time courses
were too short. We now describe the shrinkage procedure developed by (9) to regulate
smoothed periodogram matrices in order to estimate PCCoh whenever $P >> T$; partial
coherence estimates obtained from the smoothed periodogram matrix after shrinkage will
be denoted as PCCohS. Briefly, the philosophy of the shrinkage procedure is to push the
eigenvalues of $\widehat{\mathbf{f}}(\omega)$ towards their mean, say $\mu(\omega)$. If the extreme eigenvalues are closer
to each other, then this will result in a smaller condition number, which is defined to be
the ratio of the maximum eigenvalue to the minimum eigenvalue. The shrinkage estimator
accomplishes this goal by taking the convex combination between the smoothed periodogram
and the scaled identity matrix, whose scale is precisely $\mu(\omega)$. The shrinkage estimator for
the spectral density matrix is $\widehat{\mathbf{f}}_S(\omega) = W(\omega)\mu(\omega)\mathrm{Id} + (1 - W(\omega)\widehat{\mathbf{f}}(\omega)$, where Id is the $P \times P$
identity matrix and $W(\omega)$, called the shrinkage weight, is a scalar that determines the
amount of regularization at frequency ω (note that $W(\omega) = 0$ denotes no regularization and
$W(\omega) = 1$ denotes full regularization); $W(\omega)$ functions similar to the penalty parameter
λ in Equation (14.1). The optimal $W(\omega)$ is derived using a quadratic loss function. We
refer the reader to (9) and (37) for details on estimating the optimal $W(\omega)$. The resulting
$\widehat{\mathbf{f}}_S(\omega)$ is numerically more stable than $\widehat{\mathbf{f}}(\omega)$, making it possible to estimate partial coherence
whenever the dimensionality is large.

14.2.5 Mutual Information

Nonlinear dependencies between functional brain time series can be captured using mutual
information (MI; (90)). This metric quantifies how much information is being shared by
the two time series of signals; if the two signals are independent, then the MI between
the two signals is zero, and if one is a function of the other, then the MI between the
two signals is infinity. MI is estimated via the joint probability density function between
the two signals and the marginal probability density functions of each of the two signals.
To estimate MI non-parametrically, one can proceed to estimate these probability density
functions using kernel-smoothed histograms (61, 13, 82). However, one should be cautious
using such estimators whenever the time courses are short (13). Alternatively, one can obtain
a parametric estimator by using the data to estimate the parameter of the model and then
analytically or numerically estimate MI. For example, if one assumes that the signals are
Gaussian, then MI reduces to a monotonic function of the CCoh between the two signals,
which greatly simplifies computations (see, e.g., (89) and (106)).

While capturing nonlinear dependencies is, indeed, attractive, (55) argued that resting-
state fMRI data obtained from healthy subjects are almost Gaussian, and so pursuing
non-linear methods may not be practical. In contrast, the utility of MI for FC analyses in
different modalities and in different subject groups is shown in other works (60, 78, 30).

14.2.6 Principal and Independent Components Analyses

One of the earliest works on the analysis of FC for neuroimaging data was performed via
principal components analysis (PCA) on a PET dataset collected from subjects performing
verbal tasks (44). PCA aims to identify the "principal directions" of the covariance structure of the data through the eigenvectors of the covariance or the correlation matrix. In
practice, the first few principal components are extracted and each principal component is
then declared to be a functionally connected network (7, 104). That is, the first few principal
components that explain the majority of the variation of the data are interpreted as the func-

tionally connected "systems" of the brain. However, the dimension of the spatial covariance matrix is in the hundreds of thousands, which can make the storage amount prohibitive. Reference (104) derived a computational simplification for computing the singular value decomposition (SVD) of the large-scale covariance matrix. If the spatial extent of the principal components is small or if other sources of noise are present, e.g., physiological noise and other artifacts, PCA and SVD may fail to identify the correct sources of the signal (6, 7, 104). Moreover, (104) pointed out two important problems when using PCA for FC analyses. First, statistical inference for declaring functionally connected regions is difficult because of the lack of theoretical results for obtaining the p-values of the local maxima of the principal components. Second, picking the number of components requires qualitative assessment.

Similar in spirit to PCA and SVD, independent components analysis (ICA) is used to detect regions with related spatial or temporal characteristics (72, 7). Whereas PCA extracts the relevant spatial or temporal features of the covariance matrix of the data such that the resulting components are uncorrelated, ICA attempts to identify the independent "source signals" that generated the data (71, 7). The probabilistic ICA, developed by (8) and part of the FSL software for analyzing fMRI data, extends the standard ICA by accounting for the variation in the signals. Since then, there have been many methodological extensions, particularly for obtaining the independent components given a group of subjects (52, 17, 22).

In the absence of *a priori* information on regions of interest, PCA, SVD, ICA, and their variants are good for exploring the correlation structure of the high-dimensional data instead of investigating all pairwise correlations. These methods have been used in various imaging modalities, including fMRI (7, 104) and EEG (80, 22). Analyses using these methods have reduced dimensionality because they are performed on the principal components that explain most of the variability in the data. The functionally connected regions discovered by these methods can then be used to derive scientific hypotheses for future investigations.

14.2.7 Time-Varying Connectivity

The methods for estimating FC in Sections 14.2.2–14.2.6 assume that the data is stationary over time, that is, the dependence or the FC between brain regions or voxels remains constant through the experimental time course. However, while this assumption is convenient for estimation and computation purposes as it keeps the connectivity analysis from becoming vastly more complex, it presents a simplified version of a highly integrated and dynamic phenomenon. Evidence of the non-stationary behavior of brain activity has been observed in high temporal resolution EEG data (3, 69, 81), task-based fMRI experiments (24, 25, 31, 39, 35, 87), and even prominently in resting-state data (32, 34).

This has led to a surge in activity in developing new statistical methods for estimating the time-varying connectivity. The popular Psychophysiological Interactions (PPI) method introduced by (43) is one of the first methods to consider time-varying FC, which investigates whether the correlation between the time series from two brain regions differs depending on psychological context. In other words, the method is interested in whether a significant interaction exists between the psychological state and the functional coupling between two brain regions. More recently (47) introduced the statistical parametric network analysis, which allows researchers to study the dynamics of functional networks under different levels of cognitive demand. Bassett et al. (5) explored the changes in the organization (modularity) of graphs for human learning using predefined data windows spanning multiple temporal scales (e.g. days, hours and minutes) during motor learning. Here the width of the window is chosen by the researcher prior to analysis, where it would ideally be determined from the data itself.

Often, in many experimental settings it is difficult to specify the nature, timing and duration of the psychological processes or states being studied *a priori*. Hence, new techniques have been introduced to estimate the time-varying nature of FC in terms of connectivity strength (edge thickness in an undirected graph) and directionality (sign of edge in an undirected graph). Reference (20, 64, 59) investigated the non-stationary behavior of resting-state connectivity using a sliding window approach and a time-frequency coherence analysis based on the wavelet transform, an independent component analysis (ICA) and a correlation (CCor) analysis, respectively. Reference (1, 53, 62, 88) also considered using a Group ICA (18) to decompose multi-subject resting-state data into functional brain regions, a sliding window and k-means clustering of the windowed correlation matrices to study whole brain time-varying networks. Reference (67) studied whole brain dynamic FC using a sliding window and a principal component analysis (PCA) technique for resting-state data. Reference (66) introduced a data-driven multivariate method called higher-order singular value decomposition (HOSVD), a tensor decomposition, to model whole-brain networks from group-level dynamic FC data using a sliding window.

Of course, all the static methods mentioned in Sections 14.2.2–14.2.6 have a natural time-varying analogue in conjunction with a sliding window. The sliding window technique begins at the first time point, a window (or block) of a fixed number of time points, say k, is selected, and all data points within the window are used to estimate the FC. The window is then shifted a certain number of time points and the FC is estimated on the new dataset. By shifting the window to the end of the experimental time course, researchers can estimate the time-varying FC. While the sliding window method can be used to observe time-varying FC in neuroimaging studies as described in many works above, (58) has outlined limitations of the method. For example, the choice of window size is very important as different window sizes can lead to different FC patterns. Another pitfall is that the technique gives equal weight to all observations of k time points in the past and 0 weight to all others (68). Hence, in order to estimate the time-varying FC without the use of the sliding window, (24, 25, 23) introduced a novel data-driven method called Dynamic Connectivity Regression (DCR) for detecting FC change points between brain regions where the number of change points and their location are not known in advance. After finding the change points, DCR estimates a graph or series of relationships between the brain regions for data in the temporal partition that falls between pairs of adjacent change points. It is assumed that the graph does not change within each partition. This methodology has been extended to allow for the case when the number of brain regions considered is greater than the number of time points ($P > T$). The new method estimates changes in the community network structure between brain regions in order to depict the time-varying FC (26). Another work that bypasses the sliding window methodology is (105). They proposed a novel dynamic Bayesian variable partition model (DBVPM) that estimates and models multivariate dynamic functional interactions using a unified Bayesian framework. DBVPM first detects the temporal boundaries of piecewise quasi-stable functional interaction patterns, which are then modeled by representative signature patterns and whose temporal transitions are characterized by finite-state transition machines.

14.3   Simulation Study

In order to assess the performance of the FC metrics described above, a simulation study was performed. We focused on simulated data akin to the single-subject analysis. The simulations illustrate the application of the FC metrics to correlated multivariate normal (MVN) data with $P = 10$ time series (or nodes) and $P = 90$ time series. For the simulation with 10 time series we simulated 1000 datasets with $T = 200$ data points, while for the simulation

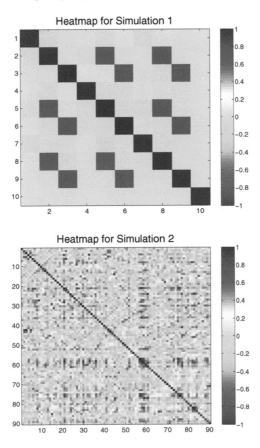

FIGURE 14.1
The true correlation structure for the multivariate normal dataset with 10 and 90 brain signals, respectively.

with 90 time series we simulated 100 datasets with 200 data points. The MVN data represents the whitened data or the residuals from fitting an autoregressive integrated moving average model to the univariate time series to remove the autocorrelation structure inherent in neuroimaging time series. For the simulation with 10 time series, we simulated an MVN dataset with a sparse correlation structure, that is, the dataset had only 6 edges of strength 0.7. For the simulation with 90 time series, we simulated the data using the correlation matrix estimated from the first subject of the resting-state dataset in Section 14.4. The objective of each simulation is to understand and provide insight into the various FC metrics. A heat map in Figure 14.1 depicts the true correlation structure for each simulation.

We applied all the FC metrics contained in Table 14.1. For PCCorG, PCCorGR, and PCCorGRS, we estimated the undirected graph for a path of λ values (0.01 to 1) and choose the value of λ (sparsity of the graph) using 10-fold cross-validation. For the stability selection procedure (Section 14.2.3) for inference on the edges, we used a threshold of 0.95 based on 100 subsamples. Furthermore, for the FC metrics in the frequency domain (CCoh, PCCoh, and PCCohS) we estimated the spectral density matrix by smoothing the periodogram matrix using a 31-point Hann window. From the smoothed periodogram matrix we extracted CCoh and then averaged it in the frequency band [0.01 0.10] Hertz as in the real data analysis in Section 14.4. To estimate PCCohS, we regularized the smoothed periodogram matrix via

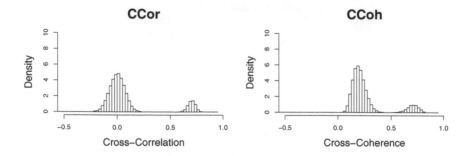

FIGURE 14.2
The distribution of CCor and CCoh for 1000 simulations of a multivariate normal dataset with 10 brain signals and a sparse FC structure.

shrinkage to yield invertible matrices and averaged the resulting estimate over the frequency band [0.01 0.10] Hertz. For the 90 time series setting, the smoothed periodogram matrix was impossible to invert, and so we could not extract PCCoh. However, after regularization via shrinkage, we were able to extract PCCohS.

14.3.1 Results

The estimates of all the FC metrics for the MVN simulations with 10 time series are shown in Figures 14.2 and 14.3. These figures display a histogram of all the FC metrics computed between all 10 ROIs, resulting in 45 values per FC metric per simulation. In Figure 14.2, we see the two FC metrics quantifying marginal dependencies, CCor and CCoh. The distribution of the CCor has a bimodal structure: a null distribution around 0 and a cluster of values around the true correlations of 0.7. The distribution of the CCoh is very similar to the distribution of CCor, but because all values of CCoh are strictly positive by definition, the null distribution is centred around 0.2.

Figure 14.3 shows the distribution of the regularized and unregularized estimates of partial correlation and of regularized and unregularized estimates of partial coherence. For PCCor, there is again evidence of a bimodal distribution: a null distribution around 0 and a cluster of values around 0.4. The lower values (compared to the CCor values in Figure 14.2) are due to the fact that conditional correlations are being estimated.

The graphical lasso and its variants impose sparsity on the precision matrix, thus, yielding more null partial correlations. Notice that PCCorGRS has the most null values, followed by PCCorGR and then PCCorG. Also, we see slightly larger values for the true non-null correlations for PCCorGRS and PCCorGR compared to PCCorG as they both include a refitting procedure without the l_1 regularization, which reduces bias. Now, focusing on the frequency domain metrics, PCCoh and PCCohS, we see that PCCoh has only positive values and there is no evidence of a clear separation of the two distributions as is the case for the PCCor. The distribution of PCCohS appears to have smaller values than PCCoh at the larger values of PCCoh, but at the same time the null values are closer to 0.

The estimates of all the FC metrics for the MVN simulations with 90 time series are shown in Figures 14.4 and 14.5. These figures display a histogram of all the FC metrics computed between all 90 ROIs, resulting in 4005 values per FC metric per simulation. In Figure 14.4, we see the two FC metrics quantifying marginal dependencies, CCor and CCoh. Here, the bimodal structure seen in the simulation with 10 time series and a sparse

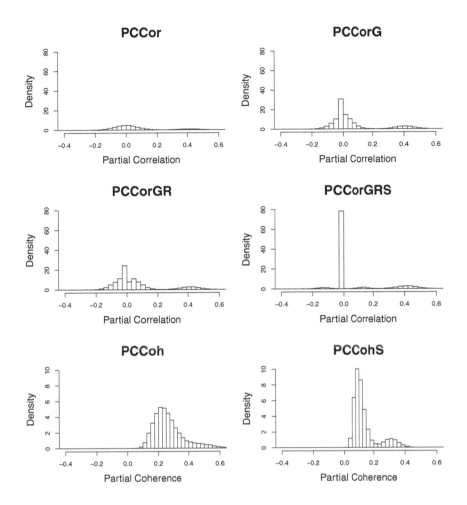

FIGURE 14.3
The distribution of PCCor, PCCorG, PCCorGR, PCCorGRS, PCCoh and PCCohS for 1000 simulations of a multivariate normal dataset with 10 brain signals and a sparse FC structure.

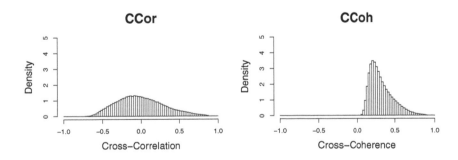

FIGURE 14.4
The distribution of CCor and CCoh for 100 simulations of a multivariate normal dataset with 90 brain signals and an FC network structure obtained from the resting-state fMRI data in Section 14.4.

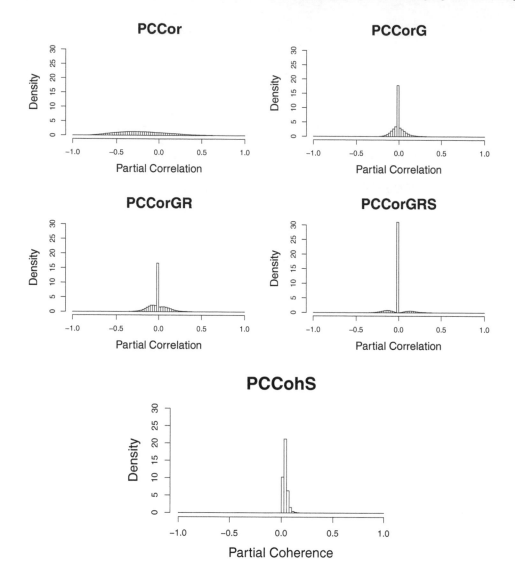

FIGURE 14.5
The distribution of PCCor, PCCorG, PCCorGR, PCCorGRS and PCCohS for 100 simulations of a multivariate normal dataset with 90 brain signals and an FC network structure obtained from the resting-state fMRI data in Section 14.4.

correlation structure is not evident. The distribution of the CCoh is again very similar to the distribution of CCor but the values are strictly positive by definition.

Figure 14.5 shows the distribution of the regularized and unregularized estimates of partial correlation and of regularized and unregularized estimates of partial coherence for the MVN simulation with 90 time series. Again, as for CCor and CCoh, there is no evidence of a bimodal distribution for PCCor for this simulation. In addition, PCCorGRS has many null values due to the fact that the subsampling procedure is used. Comparing PCCorGR and PCCorGRS to PCCorG, we see that there is a more distinct separation between the null and non-null values as PCCorG did not include a refitting procedure, and so the values are smaller. We could not obtain PCCoh because the smoothed periodogram matrix was not invertible. Though we were able to obtain PCCohS, it is difficult to determine from the empirical distribution those estimates that are different from the null distribution.

14.4   Functional Connectivity Analysis of Resting-State fMRI Data

14.4.1   Data Description and Preprocessing

We used a resting-state fMRI dataset of $N = 25$ participants (mean age 29.44 ± 8.64, 10 males) that is publicly available at NITRC (http://www.nitrc.org/projects/nyu_trt). A Siemens Allegra 3.0-Tesla scanner was used to obtain three resting-state scans for each participant, though for this analysis, we only analyze the third scan. Each scan consisted of $T = 197$ contiguous EPI functional volumes with a time repetition (TR) = 2000 ms; time echo (TE) = 25 ms; flip angle (FA) = $90°$; 39 number of slices, matrix = 64×64; field of view (FOV) = 192 mm; and voxel size $3 \times 3 \times 3$ mm^3. During each scan, each participant was asked to relax and remain still with eyes open during the scan. For spatial normalization and localization, a high-resolution T1-weighted magnetization prepared gradient echo sequence was obtained (MPRAGE, TR = 2500 ms; TE — 4.35 ms; inversion time — 900 ms; FA = $8°$, number of slices = 176; FOV = 256 mm).

The data were preprocessed using both FSL (http://www.fmrib.ox.ac.uk) and AFNI (http://afni.nimh.nih.gov/afni). The images were (1) motion corrected using FSL's `mcflirt` (rigid body transform; cost function normalized correlation; reference volume the middle volume) and then (2) normalized into the Montreal Neurological Institute space using FSL's `flirt` (affine transform; cost function mutual information). FSL's `fast` was then used to (3) obtain a probabilistic segmentation of the brain to acquire white matter and cerebrospinal fluid (CSF) probabilistic maps, thresholded at 0.99. AFNI's `3dDetrend` was then used to (4) remove the nuisance signals, namely the six motion parameters, white matter and CSF signals, and the global signal. Finally, using FSL's `fslmaths`, the volumes were (5) spatially smoothed using a Gaussian kernel with FWHM = 6 mm.

We performed both ROI-level connectivity analysis and a voxel-wise connectivity analysis. The ROIs were constructed using the automatic anatomical atlas (AAL) (99). This atlas parcellates each hemisphere into 45 regions for a total of 90 regions in the analysis. Each regional mean time course was obtained for each subject by averaging over all of the voxels within that region. Each regional time course was then detrended and standardized to unit variance, and then we applied a 4th order Butterworth filter with passband at [0.01 0.10] Hertz.

To investigate how FC varies over space, we also conducted a voxel-wise connectivity analysis over the whole brain. The voxel-wise analysis is a bivariate analysis, performed on the seed with each of the 228,483 voxels that cover the brain. We used a seed in the posterior cingulate cortex (PCC), defined as a cube having side length 6 mm centered at MNI coordinates $(−6, 58, 28)$. The PCC has been shown numerous times to be a vital component of the "default mode network" (DMN) (51, 39). This is the seed coordinate used in previous works (91, 38). The seed time course was defined as the average of the voxel time courses within the cube, that was then detrended, standardized to unit variance, and then filtered with a 4th order Butterworth filter with passband at [0.01 0.10] Hertz. For the voxel-wise connectivity analysis, we restricted our FC analysis only to CCor and CCoh. It would be unreasonable to use, e.g., PCCor or PCCoh, in a full-brain voxel-wise connectivity analysis because all conditional dependencies between the seed and any other voxel in the brain will be practically zero as a result of removing the effects of the time course of a voxel that is a spatial neighbor of the seed.

14.4.2 Overview of the Estimation Procedure

For the time domain metrics, we first computed the 90×90 sample covariance matrix per subject, then averaged the matrices over the subjects, as this is the MLE for the covariance matrix for the population. From the averaged sample covariance matrix, we extracted CCor. To estimate PCCor, we took the inverse of the sample covariance matrices averaged over all subjects and then extracted PCCor. We used different variants of the graphical lasso to obtain a penalized estimate of PCCor using the averaged sample covariance matrix. For the graphical lasso, we estimated the undirected graph for a path of λ values and chose the value of λ (sparsity of the graph) using 10-fold cross-validation. For the stability selection procedure (Section 0.2.3) for inference on the edges, we used a threshold of 0.95 based on 100 runs.

For the frequency domain metrics, to compute CCoh and PCCoh, for each subject we estimated the spectral density matrix by smoothing the periodogram matrix using a 31-point Hann window. From the smoothed periodogram matrix we extracted CCoh and then averaged it in the frequency band [0.01 0.10] Hertz because these are the relevant frequencies that contribute to resting-state FC (89). However, we could not obtain PCCoh because the smoothed periodogram matrix was not invertible. We used two methods to circumnavigate this. First, following how we estimated the PCCor and its variants, we averaged the smoothed periodogram matrices over all subjects. This increased the effective sample size enough to yield invertible matrices to obtain unregularized estimates of PCCoh. Second, we followed (38) by regularizing each subject's smoothed periodogram matrix via shrinkage to yield invertible matrices, extracted the partial coherence per subject, and then averaged these partial coherences over the subjects. Finally, we averaged the resulting estimate over the frequency band [0.01 0.10] Hertz to obtain the regularized estimates PCCohS. We used this approach because the shrinkage method was developed only for single-subject analyses.

14.4.3 Results

The estimates of each FC metric for the ROI-level analysis are shown in Figures 14.6 and 14.8. These figures display a histogram of all the FC metrics computed between all 90 ROIs, resulting in 4005 values per FC metric. In Figure 14.6, we see the two FC metrics quantifying marginal dependencies, CCor and CCoh. The distribution of the CCor include many negative values because of the preprocessing step that includes the removal of the global signal. (A discussion of the controversy surrounding this preprocessing step can be found in (77) and (76)). By the definition of CCoh, all values are strictly positive. To illustrate that both metrics quantify the same linear association, we plot the absolute value of CCor against CCoh, as shown in Figure 14.7. The two are highly correlated (r = 0.926), and the two values are almost identical at the larger values. Moreover, note that most of the values of CCor near 0.0 have values of CCoh near 0.27, suggesting that these values of CCoh are likely null.

Figure 14.8 shows the distribution of regularized and unregularized estimates of both partial correlation and partial coherence. Note that the regularized estimates are shifted more towards the null. Specifically, the graphical lasso and its variants impose sparsity on the precision matrix, thus yielding more null partial correlations. Moreover, PCCorGRS has the most null values, followed by PCCorGR and then PCCorG. Comparing PCCorGR and PCCorGRS, we see that there is a more distinct separation between the null and non-null

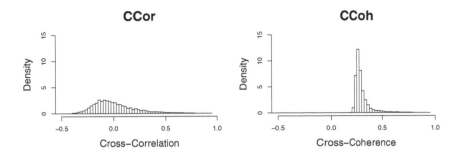

FIGURE 14.6
The distribution of the 4005 FC estimates using CCor and CCoh on the resting-state fMRI data.

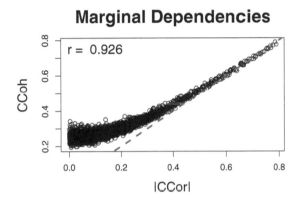

FIGURE 14.7
CCoh plotted against the absolute value of CCor for the resting-state fMRI data. The identity line is the dashed blue line.

values for PCCorGRS than PCCorGR. We can see more clearly the effect of regularization for obtaining estimates of the conditional dependencies in Figure 14.9. For partial coherence, we see the effect of *linear* shrinkage on PCCoh. Specifically, PCCohS seems to be a linear transformation of PCCoh, with PCCohS smaller than PCCoh at the larger values of PCCoh. Unlike the lasso penalty, the shrinkage procedure does not enforce sparsity, and so the estimates of PCCohS are biased towards the null whereas the regularized estimates of partial correlation have exactly null values.

Now we compare the marginal dependencies with the conditional dependencies. In Figure 14.10, we see empirically the relationship between CCor with each of the estimates of partial correlations. We see that the values of CCor shrink towards zero, and more specifically, many of the negative values of CCor either have values near zero or even exactly zero (per PCCorG, PCCorGR, and PCCorGRS) for partial correlation. This suggests that two ROIs that are negatively correlated marginally may in fact be uncorrelated conditionally, i.e., the dependencies between these two ROIs are being driven by the other ROIs in the analysis. Comparing CCoh with each of PCCoh and PCCohS, we see a similar shrinking effect towards the null for the partial coherence estimates.

The results of the voxel-wise seed analyses are shown in Figures 14.11 and 14.12. For the voxel-wise seed analysis using both CCor and CCoh, the seed was strongly positively

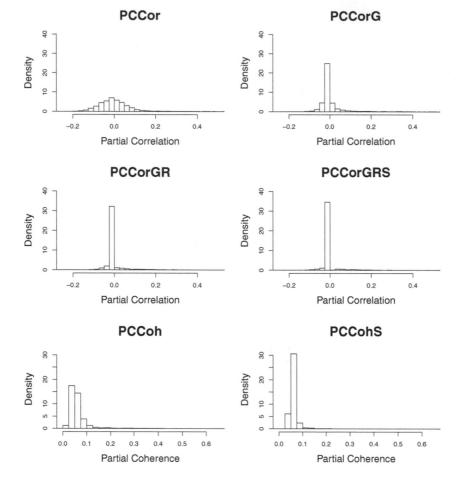

FIGURE 14.8
The distribution of the 4005 FC estimates of the conditional dependency methods (PCCor, PCCorG, PCCorGR, PCCorGRS, PCCoh, and PCCohS) on the resting-state fMRI data.

correlated with the medial prefrontal cortex, which is in agreement with previous studies (51, 39). Moreover, if we look at the strength of FC, then both CCor and CCoh have converging results, as shown in Figure 14.13.

14.5   Future Directions and Open Problems

Statistical methods for FC analyses are still in their infancy, and statisticians will continue to face daunting challenges for developing statistical theories and methodologies. With the explosion of the number of neuroimaging datasets being collected, statisticians are in great demand to play a bigger role in analyzing these datasets. Top journals for neuroimaging analyses (e.g., *NeuroImage, Human Brain Mapping*) regularly publish articles written by statisticians contributing new statistical methodology to the field. As the datasets become richer and ever more massive, the number of open *statistical* problems for FC analyses will

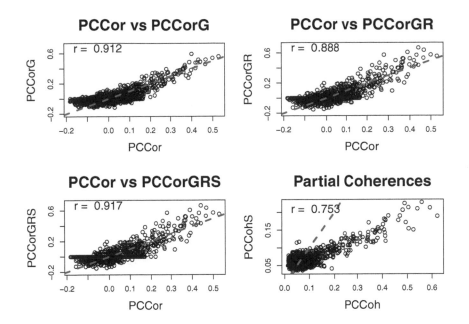

FIGURE 14.0
The unregularized estimates of the conditional dependencies plotted against each of their regularized counterparts in the resting-state fMRI data. The correlation coefficient r between each FC metric pair is computed between those values whose regularized estimate is non-null.

also increase. We now give an overview of some of the current open problems for estimating FC.

Currently, there exists no unified modeling framework for estimating FC or brain networks similar to the General Linear Model (GLM; (45)) for fMRI activation data. In fact, (48) noted that there is no principled methodological framework for estimating networks or FC. However, due to the fact that fMRI and the statistical methods used to analyze the data are still in their infancy, it is unlikely that an optimal estimating procedure for quantifying FC exists. Hence, different statistical estimation techniques will be required for different experimental contexts. If the FC between brain regions is represented using a graph structure, it is possible to summarize the graph or network using a descriptive statistic such as small-worldness, degree distribution, and modularity, to name just a few (see Chapter 17 for more details). Also, while an FC analysis may be an end in itself, in order to understand brain networks, their complex organization, and how they link to behavior and possible disorders, a multivariate framework is essential. In particular, there is growing interest in statistical methods that link the effects of multiple variables of interest including behavior and disease status on the overall network structure using exponential random graph models (ERGMs). For more details, see Chapter 17. These methods are still in their infancy, have not been widely utilized in the field, and also require further research work. However, the benefit of applying this type of analysis will lead to a greater understanding of the link between behavior and functional brain connectivity.

Methods for estimating dynamic FC are now being developed at an accelerated pace and the results from these analyses are very promising. However, many challenges remain. For example, the question of how to combine information across subjects is an open

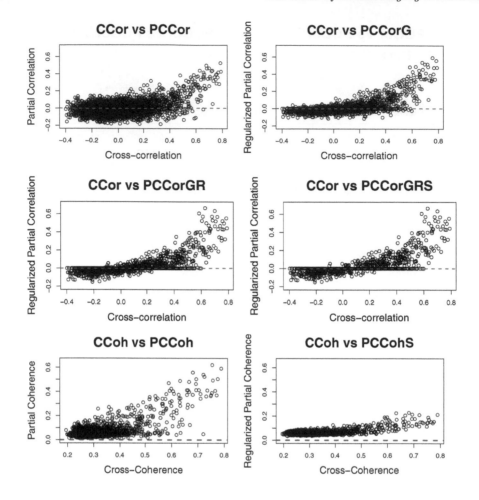

FIGURE 14.10
The marginal dependencies plotted against their corresponding conditional dependencies for the resting-state fMRI data. The zero line is the dashed blue line.

problem. In many cases it is assumed that within each partition the FC or graph structure is the same across subjects (102, 24, 25, 26). In addition, some studies focus exclusively on data at the subject level (27) because group average analyses risk removing unique individual patterns of activity and analyzing individual-subject data allows us to understand the sources of inter-subject variability in brain activity. It is assumed that by averaging across the group, the noise effects are removed. However, by analyzing individual subjects and the variations between them, the results may provide insight to the understanding of the mind–brain relationship (101). In other words, while analyses of multi-subject data are highly valuable for understanding general cognitive processes, the study of variation between individuals may be able to provide us with a deeper knowledge of the workings of the brain. The question of whether to combine information across subjects (to analyze group average results) or whether to look at individual patterns of brain activity is critical for individual treatment of brain disorders and very topical in the field of neuroscience (see (101) for a discussion). From a statistical perspective, our current investigations into multi-subject analyses not reported here has shown that accounting for inter-subject variability will lead to higher statistical power. Evaluating how time-varying functional connectivity

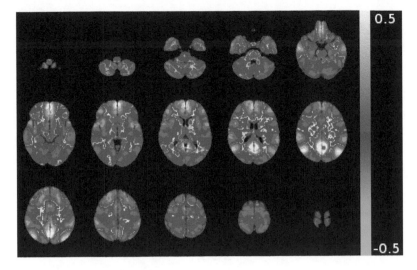

FIGURE 14.11
Voxel-wise FC analysis of the resting-state fMRI data, with the seed (represented by a purple dot) in the PCC, using CCor.

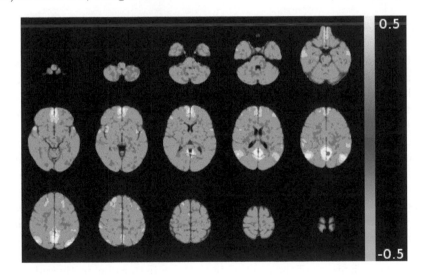

FIGURE 14.12
Voxel-wise FC analysis of the resting-state fMRI data, with the seed (represented by a purple dot) in the PCC, using CCoh.

structures relate to brain dysfunction or disorders such as depression and Alzheimer's disease will require methodological extensions of this framework. Robustness and goodness of fit measures will be very important in this setup.

It has been shown that neurological disorders disrupt the FC or structural properties of the brain (50, 75). Reference (15, 51, 93, 98) found different static FC structures in subjects with Alzheimer's disease compared to healthy subjects. However, in order to gain a true understanding of brain disorders, we may have to study the difference between the time-varying connectivity for control subjects and the subjects with the disorder. This comparison

Voxel–Level Marginal Dependencies

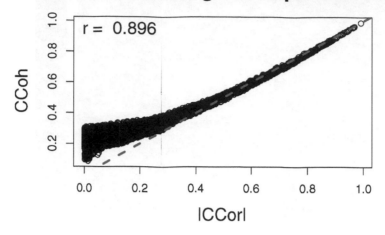

FIGURE 14.13

CCoh plotted against the absolute value of CCor from the resting-state fMRI data. The identity line is the dashed blue line.

could then be used to identify indicators of disease. This work has already begun in earnest. For example, using a sliding-window analysis, Reference (88) evaluated changes in the time-varying FC of patients with schizophrenia. They discovered that there was less connectivity between the selected brain regions and that there were significant differences in the time-varying connectivity between patients with schizophrenia compared to healthy subjects. Reference (56) linked dynamic brain states with depression by considering changes in static FC. They intimated that the differences could be due to a bias towards more frequent down states and predicted a decrease in mood-related brain dynamics due to the diminished capacity for self-induced mood changes. For Alzheimer's disease, Reference (62) studied the differences in time-varying FC between patients and controls by estimating a graph summary statistic, the modularity metric (86), and using a sliding-window approach. This paper is the first to report changes in Alzheimer's patients beyond FC in a resting-state experiment. Considering temporal features of FC may therefore provide a more accurate description of Alzheimer's disease, potentially leading not only to a better understanding of the large-scale characterizations of the disease, but also to better diagnostic and prognostic indicators. We expect this area to grow rapidly in the next few years due to the fact that there are many unsolved open statistical problems.

There are very few statistical models that account for both the spatial and temporal dependence in functional neuroimaging data for FC analyses. This is primarily because of the challenge of estimating a large-scale spatial covariance matrix whenever the number of voxels in the images are in the hundreds of thousands. One simple way to reduce the dimensionality of the problem is to restrict the data to a more tractable number of voxels (e.g., analyze only the voxels within regions of interests), or summarize the voxel-level data to ROI-level data so that the spatial dimension of the problem reduces to the number of ROIs. Despite the simplified data structure, the spatial correlations of the voxels within the ROIs can affect statistical inference (63). Working with the large-scale spatial covariance matrix of the data is a major hurdle to overcome for properly modeling the spatial correlation between the voxels in fMRI data (10, 33, 63). We refer to Chapter 19 for a more thorough discussion on spatial statistical models.

Structural connectivity is commonly measured using diffusion tensor imaging (DTI, see Chapter 3). Briefly, DTI measures the diffusivity of water molecules in the brain, and we infer the orientation of white matter fiber tracts in the brain by quantifying the diffusion patterns of the water molecules (65). Presumably, structural connectivity and FC are somehow related, and evidence for this assertion has been shown in recent studies using resting-state fMRI and DTI (100, 29, 57, 97, 107). Thus, there is a growing interest in jointly analyzing fMRI and DTI data in order to gain a better understanding of the structure–function relationships of the brain. Previous works have taken various approaches on how to combine fMRI and DTI data (see, e.g., (107) for a thorough overview). For instance, some works used fMRI to assist DTI analysis by first using a task-based fMRI experiment to identify the spatial locations with pronounced hemodynamic activity, and these locations are then used as a seed point for the DTI tractography. Conversely, some works used DTI to assist fMRI analysis by using the structural information for a more informed FC analysis. Recently, Reference (11) developed a hierarchical model that fused fMRI and DTI data in order to compute FC from the fMRI data while weighting the results using the structural connectivity obtained from the DTI data, and they showed that accounting for the structural information improves the sensitivity of the results of FC analyses. More research needs to be conducted in order to obtain a better understanding of the structure–function relationships of the brain.

New techniques can be adopted and tested from other fields such as electrophysiology (LFP analysis) or computer science (pattern recognition) , which offer a wide variety of tools for dynamic data analysis. There is also an inherent value in having researchers examine dynamics to identify patterns that may exist in the data and are not detected by most algorithms. However, there has to be a balance between data transparency and overwhelming the reader with information, which may hinder interpretation (1). Hence, we require fruitful methods that find ways of reducing dimensionality while maintaining key features.

Addressing computational issues due to the high-dimensional nature of the data is an important problem requiring input from statisticians. Many researchers are now developing statistical software, packages, and Graphic User Interfaces (GUI) that allow non-statisticians to utilize the current and up-to-date statistical methods. Again, we believe this pattern should continue and all researchers should make their code freely available. However, caution must prevail. Many researchers in neuroimaging may find that the new methods developed by statisticians are difficult to understand and hence may see the techniques as "black boxes" and use the default parameter settings in the package. We believe that it is our duty as statisticians to inform and educate researchers of the benefits but more importantly the drawbacks, of our newly developed statistical methods. In summary, we believe the neuroimaging field would benefit greatly from an influx of statistical knowledge and expertise. This collaboration will also engender new statistical areas and methodologies.

Bibliography

[1] E. A. Allen, E. Damaraju, S. M. Plis, E. B. Erhardt, T. Eichele, and V. D. Calhoun. Tracking whole-brain connectivity dynamics in the resting state. *Cerebral Cortex*, 2012. In press.

[2] J. Alonso, J. Poza, M. Mananas, S. Romero, A. Fernández, and R. Hornero. MEG connectivity analysis in patients with Alzheimer's disease using cross mutual information and spectral coherence. *Ann. Biomed Eng*, 39:524–536, 2011.

[3] Amos Arieli, Alexander Sterkin, Amiram Grinvald, and Ad Aertsen. Dynamics of ongoing activity: Explanation of the large variability in evoked cortical responses. *Science*, 273(5283):1868–1871, 1996.

[4] D. S. Bassett and E.T. Bullmore. Human brain networks in health and disease. *Current Opinion in Neurobiology*, 2:340–347, 2009.

[5] D. S. Bassett, N. F. Wymbs, M. A. Porter, P. J. Mucha, J. M. Carlson, and S. T. Grafton. Dynamic reconfiguration of human brain networks during learning. *Proceedings of the National Academy of Sciences USA*, 108:7641–7646, 2011.

[6] R. Baumgartner, L. Ryner, W. Richter, and R. Summers. Comparison of two exploratory data analysis methods for fMRI: Fuzzy clustering vs. principal component analysis. *Magnetic Resonance Imaging*, 18:89–94, 2000.

[7] C. Beckmann, M. DeLuca, J. Devlin, and S. Smith. Investigations into resting-state connectivity using independent component analysis. *Philosophical Transactions: Biological Sciences*, 360:1001–1013, 2005.

[8] C. Beckmann and S. Smith. Probabilistic independent component analysis for functional magnetic resonance imaging. *IEEE Trans. Med. Imaging*, 23:137–152, 2004.

[9] H. Böhm and R. von Sachs. Shrinkage estimation in the frequency domain of multivariate time series. *Journal of Multivariate Analysis*, 100:913–935, 2009.

[10] F. Bowman, B. Caffo, S. Bassett, and C. Kilts. A Bayesian hierarchical framework for spatial modeling of fMRI data. *NeuroImage*, 39:146–156, 2008.

[11] F. Bowman, L. Zhang, G. Derado, and S. Chen. Determining functional connectivity using fMRI data with diffusion-based anatomical weighting. *NeuroImage*, 62:1769–1779, 2012.

[12] D. Brillinger. *Time Series: Data Analysis and Theory*. Society for Industrial and Applied Mathematics, 2001.

[13] D. Brillinger. Second-order moments and mutual information in the analysis of time series. *Recent Advances in Statistical Methods*, pages 64–76, 2002.

[14] P. Brockwell and R. Davis. *Time Series: Theory and Methods*. Springer, 1998.

[15] Randy L. Buckner, Jorge Sepulcre, Tanveer Talukdar, Fenna M. Krienen, Hesheng Liu, Trey Hedden, Jessica R. Andrews-Hanna, Reisa A. Sperling, and Keith A. Johnson. Cortical hubs revealed by intrinsic functional connectivity: Mapping, assessment of stability, and relation to Alzheimer's disease. *The Journal of Neuroscience*, 29(6):1860–1873, 2009.

[16] E.T. Bullmore and O. Sporns. Complex brain networks: Graph theoretical analysis of structural and functional systems. *Nature Review Neurosciences*, 10:186–198, 2009.

[17] V. Calhoun, J. Liu, and T. Adali. A review of group ICA for fMRI data and ICA for joint inference of imaging, genetic, and ERP data. *NeuroImage*, 45:S163S172, 2009.

[18] V.D. Calhoun, T. Adali, G.D. Pearlson, and J.J. Pekar. A method for making group inferences from functional MRI data using independent component analysis. *Human Brain Mapping*, 14(3):140–151, 2001.

[19] Vince D. Calhoun, Tom Eichele, and Godfrey Pearlson. Functional brain networks in schizophrenia: A review. *Frontiers in Human Neuroscience*, 3:17, 2009.

[20] C. Chang and G.H. Glover. Time-frequency dynamics of resting-state brain connectivity measured with fMRI. *NeuroImage*, 50:81–89, 2010.

[21] R. Cameron Craddock, G. Andrew James, Paul E. Holtzheimer, Xiaoping P. Hu, and Helen S. Mayberg. A whole brain fMRI atlas generated via spatially constrained spectral clustering. *Human Brain Mapping*, 33(8):1914–1928, 2012.

[22] C. Crainiceanu, B. Caffo, S. Luo, V. Zipunnikov, and N. Punjabi. Population value decomposition: A framework for the analysis of image populations. *Journal of the American Statistical Association*, 106:775–790, 2011.

[23] I. Cribben. Detecting dependence change points in multivariate time series with applications in neuroscience and finance. PhD thesis, Columbia University, 2012.

[24] I. Cribben, R. Haraldsdottir, L. Y. Atlas, T. D. Wager, and M. A. Lindquist. Dynamic connectivity regression: Determining state-related changes in brain connectivity. *NeuroImage*, 61:907–920, 2012.

[25] I. Cribben, T. D. Wager, and M. A. Lindquist. Detecting functional connectivity change points for single subject fMRI data. *Frontiers in Computational Neuroscience*, 7:143, 2013.

[26] I. Cribben and Y. Yu. Estimating whole brain dynamics using spectral clustering. arXiv preprint arXiv:1509.03730, 2015.

[27] J. Cummine, I. Cribben, C. Luuand E. Kim, R. Bahktiari, G. Georgiou C.A., and C.A. Boliek. Understanding the role of speech production in reading: Evidence for a print-to-speech neural network using graphical analysis. *Neuropsychology*, 2015.

[28] Ming Dai and Wengsheng Guo. Multivariate spectral analysis using Cholesky decomposition. *Biometrika*, 91:629–643, 2004.

[29] J. Damoiseaux and M. Greicius. Greater than the sum of its parts: A review of studies combining structural connectivity and resting-state functional connectivity. *Brain Structure and Function*, 213:525–533, 2009.

[30] Olivier David, Diego Cosmelli, and Karl J. Friston. Evaluation of different measures of functional connectivity using a neural mass model. *NeuroImage*, 21:659–673, 2004.

[31] Stefan Debener, Markus Ullsperger, Markus Siegel, and Andreas K Engel. Single-trial EEG–fMRI reveals the dynamics of cognitive function. *Trends in Cognitive Sciences*, 10(12):558–563, 2006.

[32] Pascal Delamillieure, Gaëlle Doucet, Bernard Mazoyer, Marie-Renée Turbelin, Nicolas Delcroix, Emmanuel Mellet, Laure Zago, Fabrice Crivello, Laurent Petit, Nathalie Tzourio-Mazoyer, et al. The resting state questionnaire: an introspective questionnaire for evaluation of inner experience during the conscious resting state. *Brain Research Bulletin*, 81(6):565–573, 2010.

[33] G. Derado, F. Bowman, and C. Kilts. Modeling the spatial and temporal dependence in fMRI data. *Biometrics*, 66:949–957, 2010.

[34] Gaëlle Doucet, Mikaël Naveau, Laurent Petit, Laure Zago, Fabrice Crivello, Gaël Jobard, Nicolas Delcroix, Emmanuel Mellet, Nathalie Tzourio-Mazoyer, Bernard Mazoyer, et al. Patterns of hemodynamic low-frequency oscillations in the brain are modulated by the nature of free thought during rest. *NeuroImage*, 59(4):3194–3200, 2012.

[35] Tom Eichele, Stefan Debener, Vince D. Calhoun, Karsten Specht, Andreas K. Engel, Kenneth Hugdahl, D. Yves von Cramon, and Markus Ullsperger. Prediction of human errors by maladaptive changes in event-related brain networks. *Proceedings of the National Academy of Sciences*, 105(16):6173–6178, 2008.

[36] M. Fiecas and H. Ombao. The generalized shrinkage estimator for the analysis of functional connectivity of brain signals. *The Annals of Applied Statistics*, 5:1102–1125, 2011.

[37] M. Fiecas, H. Ombao, C. Linkletter, W. Thompson, and J. Sanes. Functional connectivity: Shrinkage estimation and randomization test. *NeuroImage*, 5:1102–1125, 2010.

[38] M. Fiecas, H. Ombao, D. van Lunen, R. Baumgartner, A. Coimbra, and D. Feng. Quantifying temporal correlations: A test-retest evaluation of functional connectivity in resting-state fMRI. *NeuroImage*, 65:231–241, 2013.

[39] Michael D. Fox, Abraham Z. Snyder, Justin L. Vincent, Maurizio Corbetta, David C. Van Essen, and Marcus E. Raichle. The human brain is intrinsically organized into dynamic, anticorrelated functional networks. *Proceedings of the National Academy of Sciences of the United States of America*, 102(27):9673–9678, 2005.

[40] J. H. Friedman, Ta Hastie, and Ra Tibshirani. Sparse inverse covariance estimation with the graphical lasso. *Biostatistics*, 9:432–441, 2007.

[41] K. J. Friston. Functional and effective connectivity: A review. *Brain Connectivity*, 1:13–36, 2011.

[42] K. J. Friston and C. Buechel. Functional connectivity. In R.S.J. et al. Frackowiak, editor, *Human Brain Function (2nd Edition)*. Academic Press, 2003.

[43] K. J. Friston, C. Buechel, G. R. Fink, J. Morris, E. Rolls, and R. J. Dolan. Psychophysiological and modularity interactions in neuroimaging. *NeuroImage*, 6:218–229, 1997.

[44] K. J. Friston, C. D. Firth, R. F. Liddle, and R. S. J. Frackowiak. Functional connectivity: The principal component analysis of large (PET) datasets. *Journal of Celebral Blood Flow and Metabolism*, 13:5–14, 1993.

[45] Karl J. Friston, Andrew P. Holmes, Keith J. Worsley, J-P. Poline, Chris D. Frith, and Richard S.J. Frackowiak. Statistical parametric maps in functional imaging: A general linear approach. *Human Brain Mapping*, 2(4):189–210, 1994.

[46] G.L. Gerstein and D.H. Perkel. Simultaneously recorded trains of action potentials: Analysis and functional interpretation. *Science*, 164:828–830, 1969.

[47] C.E. Ginestet and A.E. Simmons. Statistical parametric network analysis of functional connectivity dynamics during a working memory task. *NeuroImage*, 55:688–704, 2011.

[48] Cedric E. Ginestet, Thomas E. Nichols, Ed T. Bullmore, and Andrew Simmons. Brain network analysis: Separating cost from topology using cost-integration. *PloS One*, 6(7):e21570, 2011.

[49] J. Gonzalez, Z. Saad, D. Handwerker, S. Inati, N. Brenowitz, and P. Bandettini. Whole-brain, time-locked activation with simple tasks revealed using massive averaging and model-free analysis. *Proceedings of the National Academy of Sciences*, 109:5487–5492, 2012.

[50] Michael D. Greicius, Vesa Kiviniemi, Osmo Tervonen, Vilho Vainionpää, Seppo Alahuhta, Allan L. Reiss, and Vinod Menon. Persistent default-mode network connectivity during light sedation. *Human Brain Mapping*, 29(7):839–847, 2008.

[51] Michael D. Greicius, Gaurav Srivastava, Allan L. Reiss, and Vinod Menon. Default-mode network activity distinguishes Alzheimer's disease from healthy aging: Evidence from functional MRI. *Proceedings of the National Academy of Sciences of the United States of America*, 101(13):4637–4642, 2004.

[52] Y. Guo and G. Pagnoni. A unified framework for group independent component analysis for multi-subject fMRI data. *NeuroImage*, 42:1078–1093, 2008.

[53] Daniel A. Handwerker, Vinai Roopchansingh, Javier Gonzalez-Castillo, and Peter A. Bandettini. Periodic changes in fMRI connectivity. *NeuroImage*, 63(3):1712–1719, 2012.

[54] A. Hillebrand, G. Barnes, J. Bosboom, H. Berendse, and C. Stam. Frequency-dependent functional connectivity wiwith resting-state networks: An atlas-based MEG beamformer solution. *NeuroImage*, 59:3909–3921, 2012.

[55] J. Hlinka, M. Palus, M. Vejmelka, D. Mantini, and M. Corbetta. Functional connectivity in resting-state fMRI: Is linear correlation sufficient? *NeuroImage*, 54:2218–2225, 2011.

[56] Paul E. Holtzheimer and Helen S. Mayberg. Deep brain stimulation for psychiatric disorders. *Annual Review of Neuroscience*, 34:289–307, 2011.

[57] C. Honey, O. Sporns, L. Cammoun, X. Gigandet, J. Thiran, R. Meull, and P. Hagmann. Predicting human resting-state functional connectivity from structural connectivity. *PNAS*, 106:2035–2040, 2009.

[58] R. Matthew Hutchison, Thilo Womelsdorf, Elena A. Allen, Peter A. Bandettini, Vince D. Calhoun, Maurizio Corbetta, Stefania Della Penna, Jeff Duyn, Gary Glover, Javier Gonzalez-Castillo, et al. Dynamic functional connectivity: Promises, issues, and interpretations. *NeuroImage*, 2013.

[59] R. Matthew Hutchison, Thilo Womelsdorf, Joseph S. Gati, Stefan Everling, and Ravi S. Menon. Resting-state networks show dynamic functional connectivity in awake humans and anesthetized macaques. *Human Brain Mapping*, 34(9):2154–2177, 2013.

[60] J. Jeong, J. Gore, and B. Peterson. Mutual information analysis of the EEG in patients with Alzheimer's disease. *Clinical Neurophysiology*, 112:827–835, 2001.

[61] Harry Joe. Estimation of entropy and other functional of a multivariate density. *Ann. Inst. Statist. Math*, 41:683–697, 1989.

[62] David T. Jones, Prashanthi Vemuri, Matthew C. Murphy, Jeffrey L. Gunter, Matthew L. Senjem, Mary M. Machulda, Scott A. Przybelski, Brian E. Gregg, Kejal Kantarci, and David S. Knopman. Non-stationarity in the "resting brains modular" architecture, 2012.

[63] H. Kang, H. Ombao, C. Linkletter, N. Long, and D. Badre. Spatio-spectral mixed-effects model for functional magnetic resonance imaging data. *Journal of the American Statistical Association*, 107:568–577, 2012.

[64] Vesa Kiviniemi, Tapani Vire, Jukka Remes, Ahmed Abou Elseoud, Tuomo Starck, Osmo Tervonen, and Juha Nikkinen. A sliding time-window ICA reveals spatial variability of the default mode network in time. *Brain Connectivity*, 1(4):339–347, 2011.

[65] D. Le Bihan. Looking into the functional architecture of the brain with diffusion MRI. *Nature Review Neuroscience*, 4:469–480, 2003.

[66] N. Leonardi and D. Van de Ville. Identifying network correlates of brain states using tensor decompositions of whole-brain dynamic functional connectivity. In *Pattern Recognition in Neuroimaging (PRNI), 2013 International Workshop on*, pages 74–77, June 2013.

[67] Nora Leonardi, Jonas Richiardi, Markus Gschwind, Samanta Simioni, Jean-Marie Annoni, Myriam Schluep, Patrik Vuilleumier, and Dimitri Van De Ville. Principal components of functional connectivity: A new approach to study dynamic brain connectivity during rest. *NeuroImage*, 83:937–950, 2013.

[68] M. A. Lindquist, X. Yuting, M.B. Nebel, and B.S. Caffo. Evaluating dynamic bivariate correlations in resting-state fMRI: A comparison study and a new approach. *NeuroImage*, 2014.

[69] Scott Makeig, Stefan Debener, Julie Onton, and Arnaud Delorme. Mining event-related brain dynamics. *Trends in Cognitive Sciences*, 8(5):204–210, 2004.

[70] G. Marrelec, A. Krainik, H. Duffau, M. Pélégrini-Issac, S. Lehéricy, J. Doyon, and H. Benali. Partial correlation for functional brain interactivity investigation in functional MRI. *Neuroimage*, 32:228–237, 2006.

[71] M. McKeown, L. Hansen, and T. Sejnowski. Independent component analysis of functional MRI: What is signal and what is noise? *Current Opinion in Neurobiology*, 13:620–629, 2003.

[72] M. McKeown, T. Jung, S. Makeig, G. Brown, S. Kindermann, T. Lee, and T. Sejnowski. Spatially independent activity patterns in functional MRI data during the stroop color-naming task. *PNAS*, 95:803–810, 1998.

[73] T. Medkour, A. Walden, and A. Burgess. Graphical modelling for brain connectivity via partial coherence. *Journal of Neuroscience Methods*, 180:374–383, 2009.

[74] N. Meinshausen and P. Bühlmann. Stability selection. *Journal of the Royal Statistical Society B*, 72:417–473, 2010.

[75] Vinod Menon. Large-scale brain networks and psychopathology: A unifying triple network model. *Trends in Cognitive Sciences*, 15(10):483–506, 2011.

[76] Michael Fox, Dongyang Zhang, Abraham Snyder, and Marcus Raichle. The global signal and observed anticorrelated resting state brain networks. *Journal of Neurophysiology*, 101:3027–3283, 2009.

[77] Kevin Murphy, Rasmus M. Birn, Daniel A. Handwerker, Tyler B. Jones, and Peter A. Bandettini. The impact of global signal regression on resting state correlations: Are anti-correlated networks introduced? *NeuroImage*, 44:893–905, 2009.

[78] S. Na, S. Jin, S. Kim, and B. Ham. EEG in schizophrenic patients: Mutual information analysis. *Clinical Neurophysiology*, 113:1954–1960, 2002.

[79] H. Ombao and S. van Bellegem. Coherence analysis: A linear filtering point of view. *IEEE Transactions on Signal Processing*, 56:2259–2266, 2008.

[80] Hernando Ombao, Rainder von Sachs, and Wensheng Guo. SLEX analysis of multivariate nonstationary time series. *Journal of the American Statistical Association*, 100:519–531, 2005.

[81] Julie Onton, Marissa Westerfield, Jeanne Townsend, and Scott Makeig. Imaging human EEG dynamics using independent component analysis. *Neuroscience & Biobehavioral Reviews*, 30(6):808–822, 2006.

[82] L. Paninski. Estimation of entropy and mutual information. *Neural Computation*, 15.6:1191–1253, 2003.

[83] D. B. Percival and A. T. Walden. *Spectral Analysis for Physical Applications: Multitaper and Conventional Univariate Techniques*. Cambridge University Press, 1993.

[84] C.G. Phillips, S. Zeki, and H.B. Barlow. Localization of function in the cerebral cortex. past, present and future. *Brain*, 107:327–361, 1984.

[85] M. Pourahmadi. Covariance estimation: The GLM and regularization perspectives. *Statistical Science*, 26:369–387, 2011.

[86] Mikail Rubinov and Olaf Sporns. Weight-conserving characterization of complex functional brain networks. *Neuroimage*, 56(4):2068–2079, 2011.

[87] Sepideh Sadaghiani, Guido Hesselmann, and Andreas Kleinschmidt. Distributed and antagonistic contributions of ongoing activity fluctuations to auditory stimulus detection. *The Journal of Neuroscience*, 29(42):13410–13417, 2009.

[88] Ünal Sakoğlu, Godfrey D. Pearlson, Kent A. Kiehl, Y. Michelle Wang, Andrew M. Michael, and Vince D. Calhoun. A method for evaluating dynamic functional network connectivity and task-modulation: Application to schizophrenia. *Magnetic Resonance Materials in Physics, Biology and Medicine*, 23(5-6):351–366, 2010.

[89] R. Salvador, J. Suckling, C. Schwarzbauer, and E. Bullmore. Undirected graphs of frequency-dependent functional connectivity in whole-brain networks. *Philosophical Transactions of the Royal Society B*, 360:937–946, 2005.

[90] C.E. Shannon and W. Weaver. A mathematical theory of communications. Technical report, *Bell Systems Technical Journal*, 1948.

[91] Zarrar Shehzad, A. M. Clare Kelly, Philip T. Reiss, Dylan G. Gee, Kristin Gotimer, Lucina Q. Uddin, Sang Han Lee, Daniel S. Marguilies, Amy Krain Roy, Bharat B. Biswal, Eva Petkova, F. Xavier Castellanos, and Michael P. Milham. The resting brain: Unconstrained yet reliable. *Cerebral Cortex*, 19:2209–2229, 2009.

[92] Stephen M. Smith, Karia L. Miller, Gholamreza Salimi-Khorshidi, Matthew Webster, Christian F. Beckmann, Thomas E. Nichols, Joseph D. Ramsey, and Mark W. Woolrich. Network modelling methods for fMRI. *NeuroImage*, 54:875–891, 2011.

[93] Christian Sorg, Valentin Riedl, Mark Mhlau, Vince D. Calhoun, Tom Eichele, Leonhard Ler, Alexander Drzezga, Hans Frstl, Alexander Kurz, Claus Zimmer, and Afra M. Wohlschlger. Selective changes of resting-state networks in individuals at risk for Alzheimer's disease. *Proceedings of the National Academy of Sciences*, 104(47):18760–18765, 2007.

[94] O. Sporns, G. Tononi, and R. Kötter. The human connectome: A structured description of the human brain. *Plos Computational Biology*, 1:245–251, 2005.

[95] R. Srinivasan, W. Winter, J. Ding, and P. Nunez. EEG and MEG coherence: Measures of functional connectivity at distinct spatial scales of neocortical dynamics. *Journal of Neuroscience Methods*, 166:41–52, 2007.

[96] F. Sun, L. Miller, and M. D'Esposito. Measuring interregional functional connectivity using coherence and partial coherence analyses of fMRI data. *NeuroImage*, 21:647–658, 2004.

[97] K. Supekar, L. Uddin, K. Prater, H. Amin, M. Greicius, and V. Menon. Development of functional and structural connectivity with the default mode network in young children. *Neuroimage*, 52:290–301, 2010.

[98] Kaustubh Supekar, Vinod Menon, Daniel Rubin, Mark Musen, and Michael D. Greicius. Network analysis of intrinsic functional brain connectivity in Alzheimer's disease. *PLoS Computational Biology*, 4(6):e1000100, 2008.

[99] N. Tzourio-Mazoyer, B. Landeau, D. Papathanassiou, F. Crivello, O. Etard, N. Delcroix, B. Mazoyer, and M. Joliot. Automated anatomical labeling of activations in SPM using a macroscopic anatomical parcellation of the MNI MRI single-subject brain. *NeuroImage*, 15:273–289, 2002.

[100] M. van den Heuvel, R. Mandl, J. Luigjes, and H. Hulshoff Pol. Microstructural organization of the cingulum tract and the level of default mode functional connectivity. *Journal of Neuroscience*, 28:10844–10851, 2008.

[101] J. D. Van Horn, S. T. Grafton, and M. B. Miller. Individual variability in brain activity: a nuisance or an opportunity? *Brain Imaging and Behavior*, 2:327–334, 2008.

[102] G. Varoquaux, A. Gramfort, J. B. Poline, B. Thirion, R. Zemel, and J. Shawe-Taylor. Brain covariance selection: Better individual functional connectivity models using population prior. *Advances in Neural Information Processing Systems*, pages 2334–2342, 2010.

[103] J. Whittaker. *Graphical Models in Applied Multivariate Statistics*. Wiley: Chichester England and New York, 1990.

[104] Keith Worsley, Jen-I. Chen, Jason Lerch, and Alan C. Evans. Comparing functional connectivity via thresholding correlations and singular value decomposition. *Philosophical Transactions of the Royal Society B*, 360:913–920, 2005.

[105] Jing Zhang, Xiang Li, Cong Li, Zhichao Lian, Xiu Huang, Guocheng Zhong, Dajiang Zhu, Kaiming Li, Changfeng Jin, Xintao Hu, Junwei Han, Lei Guo, Xiaoping Hu, Lingjiang Li, and Tianming Liu. Inferring functional interaction and transition patterns via dynamic Bayesian variable partition models. *Human Brain Mapping*, pages n/a–n/a, 2013.

[106] D. Zhou, W. Thompson, and G. Siegle. MATLAB toolbox for functional connectivity. *NeuroImage*, 47:1590–1607, 2009.

[107] D. Zhu, T. Zhang, X. Jiang, X. Hu, H. Chen, N. Yang, J. Lv, J. Han, L. Guo, and T. Liu. Fusing DTI and fMRI data: A survey of methods and applications. *NeuroImage*, page in press, 2013.

15

Multivariate Decompositions in Brain Imaging

Ani Eloyan

Department of Biostatistics, Brown University

Vadim Zipunnikov, Juemin Yang, and Brian Caffo

Department of Biostatistics, Bloomberg School of Public Health, Johns Hopkins University

CONTENTS

15.1	Introduction	399
15.2	Principal Component Analysis and Singular Value Decomposition	400
	15.2.1 Singular Value Decomposition	401
	15.2.2 Principal Components Analysis	401
	15.2.3 PCA in Brain Imaging	402
15.3	Structured PCA Models	403
	15.3.1 Calculation of High-Dimensional PCA	404
15.4	Independent Component Analysis	405
	15.4.1 ICA in Brain Imaging	406
	15.4.2 Homotopic Group ICA	407
	15.4.3 Computation of High-Dimensional ICA	407
15.5	Discussion of Other Methods	409
15.6	Acknowledgements	411
	Bibliography	411

15.1   Introduction

In brain imaging studies, information rich, high-dimensional data are collected, typically along with important research or clinical demographic and covariate information. A variety of brain imaging modalities exist to probe brain structure, function, and chemical composition. One, in particular, is functional magnetic resonance imaging (fMRI), where four-dimensional images of the brain are collected with three dimensions corresponding to space and the fourth to time. In structural magnetic resonance imaging (sMRI) and other static imaging techniques, such as static positron emission tomography (PET), the images are three dimensional for each session. In each case, another dimension is created with multiple scanning sessions per subject, such as in a longitudinal or crossover study.

Given the size and complexity of brain imaging datasets, dimension reduction and decomposition techniques are routinely used when analyzing these data. Decomposition methods are often used for different purposes in analyzing brain imaging data, with each purpose requiring different processing and data organization. Consider two conceptual examples. In the first, four-dimensional fMRI arrays are transformed into a two dimensional matrix by vectorizing the 3D brain image for each time point and concatenating the vectors over (within session scanning) time ((29) provides details on programming common fMRI analyses in the R software (77)). The resulting two dimensional matrix is typically high-dimensional, with hundreds of time points and tens of thousands of non-background voxels. In the second, sMRI three dimensional images for each subject are often vectorized to construct a subject-by-voxel matrix. In both cases the result is a rectangular matrix, with vectorized space along the columns (or rows). One can then take a principal component decomposition of these data to identify and visualize the directions of variability in the data.

Principal component analysis (PCA) (53) is easily the most well known and used decomposition and dimension reduction method. PCA is based on the second-order moments of observed variables where principal components are the eigenvectors associated with the covariance or correlation matrix. In our two conceptual examples, the variability under consideration is spatial variability over intra-subject scanning time in fMRI and over inter-subject sampling variability in sMRI. The dimension of the large covariance matrix is reduced by PCA, while not losing important variability information in the data. The so-called principal components provide the best rank $K(< p)$ linear approximation to the (p-dimensional) data where "best" is defined in terms of preserving variability. Thus, one would interpret the first principal component in the conceptual fMRI example as the spatial map that is a linear combination of observed maps for this subject that explains the largest amount of temporal variability in the fMRI intensities. The similar first spatial component in the conceptual sMRI example explains the largest amount of sampling (inter-subject) variability in the imaging intensities.

Building on these two examples, PCA and the closely related singular value decomposition (SVD), can be used to explore variation in structural and functional images. In addition to PCA, other related decompositions and factor analytic techniques have similar goals and applicability. Given this starting point, these tools can be used in a staggeringly diverse collection of ways that include the investigation of processing, biological, spatial, temporal, sampling, and structured (based on the design) variability. The methods can be approached from formal generative probabilistic models or data-level mathematical thinking. One particular probabilistic model, independent components analysis (ICA), has had a major impact on the field of fMRI. This model is often expressed as a linear reorganization of the SVD and unobserved independent factors. This chapter aims to give broad directions in this difficult landscape of powerful tools.

The approach taken in this chapter is to broadly introduce decomposition methods as used primarily in structural and functional brain imaging using a diverse collection of examples for context. Admittedly, the chapter is biased toward the authors' contributions complimented with a didactic overview of the area. The chapter is organized as follows. Section 15.2 provides an introduction to principal component analysis and the singular value decomposition. In Section 15.3, recent methods developed for PCA analysis of structured data are discussed. The independent component analysis model is described in Section 15.4. Finally, Section 15.5 concludes the chapter by providing a non-exhaustive list of other decomposition methods.

15.2 Principal Component Analysis and Singular Value Decomposition

Suppose that we have a sample of images $\mathbf{Y}_i, i = 1, \ldots, n$. In a hypothetical example, each image $\mathbf{Y}_i$ may be a three dimensional array, e.g. a structural MRI, of dimension

$p = p_1 \times p_2 \times p_3$ collected for subject $i = 1, \ldots, n$. However, depending on the research question and the assumptions on the noise structure between voxels in the brain, one may be willing to partially disregard the spatial structure of the voxels and regard the three dimensional image $\mathbf{Y}_i$ as a long vector. In other words, suppose every image is unfolded into a $p \times 1$ dimensional vector containing voxels in a pre-defined order preserved across all images. In this section, we will refer to the $p \times n$ sample data matrix as defined by $\mathbf{Y} = (\mathbf{Y}_1, \ldots, \mathbf{Y}_n)$, where each column i contains the unfolded image for subject $i = 1, \ldots, n$. This representation of the observed images is particularly beneficial for the application of two dimensional matrix decomposition methods discussed in this chapter. Tensor decomposition techniques developed for analyzing the observed images directly are briefly mentioned in Section 15.5.

15.2.1 Singular Value Decomposition

The singular value decomposition (SVD) of a rank r matrix $\mathbf{Y}$ can be written as

$$\mathbf{Y} = \mathbf{V}\mathbf{\Sigma}\mathbf{U}' = \sum_{l=1}^{r} \sigma_l \mathbf{v}_l \mathbf{u}'_l, \tag{15.1}$$

where the $p \times r$ matrix $\mathbf{V} = [\mathbf{v}_1, \ldots, \mathbf{v}_r]$ contains orthonormal left singular vectors $\mathbf{v}_i \in \mathbb{R}^p$, the $n \times r$ matrix $\mathbf{U} = [\mathbf{u}_1, \ldots, \mathbf{u}_r]$ contains orthonormal right singular vectors $\mathbf{u}_i \in \mathbb{R}^n$, and the $r \times r$ matrix $\mathbf{\Sigma}$ is diagonal with the ordered singular values $\sigma_1 \geq \sigma_2 \geq \ldots \geq \sigma_r > 0$ on the diagonal.

The SVD delivers the best rank K approximation of the matrix $\mathbf{Y}$ in the Frobenius norm, $\|\mathbf{Y}\|_F = \sqrt{\sum_{i,j} Y_{ij}^2}$. The truncated SVD representation $\mathbf{Y}^{(K)} = \sum_{k=1}^{K} \sigma_k \mathbf{v}_k \mathbf{u}'_k$ is the solution to the problem

$$\min_{rank(\mathbf{Y}^{(K)})=K} \left\| \mathbf{Y} - \mathbf{Y}^{(K)} \right\|_F.$$

In addition, SVD is closely related to the generalized Moore–Penrose inverse of $\mathbf{Y}$ which is defined as a unique $n \times p$ matrix $\mathbf{Y}^{(-)}$ such that (i) $\mathbf{Y}\mathbf{Y}^{(-)}\mathbf{Y} = \mathbf{Y}$, (ii) $\mathbf{Y}^{(-)}\mathbf{Y}\mathbf{Y}^{(-)} = \mathbf{Y}^{(-)}$, (iii) $(\mathbf{Y}\mathbf{Y}^{(-)})' = \mathbf{Y}\mathbf{Y}^{(-)}$, (iv) $(\mathbf{Y}^{(-)}\mathbf{Y})' = \mathbf{Y}^{(-)}\mathbf{Y}$. The Moore–Penrose inverse of matrix $\mathbf{Y}$ with SVD as in (15.1) is given by $\mathbf{Y}^{(-)} = \mathbf{U}\mathbf{\Sigma}^{-1}\mathbf{V}'$.

Computationally, the SVD of an $n \times p$ matrix $\mathbf{Y}$ can be obtained with $np \cdot O(min(n,p))$ effort (38). Section 15.3.1 provides details on efficient SVD calculation in imaging applications dealing with a high-dimensional p.

15.2.2 Principal Components Analysis

Principal components analysis (PCA) is a very popular dimension reduction technique that is closely related to the SVD. The goal of PCA is to identify a set of orthonormal linear combinations of the data such that the projection onto each of the directions explains the most variability in the data in the corresponding space. The principal components are typically ordered based on the amount of variance explained in a decreasing order. Several algorithms have been developed for PCA with various computational properties. As a first step, a PCA algorithm de-means the vectors $\mathbf{Y}_i$ by subtracting the sample average image, $\boldsymbol{\mu} = \sum_{i=1}^{n} \mathbf{Y}_i/n$. For notational simplicity, we use the same notation, $\mathbf{Y}_i$, for the demeaned vectors. At the second step, the sample $p \times p$ covariance matrix $\widehat{\mathbf{K}}$ is calculated as $\widehat{\mathbf{K}} = \frac{1}{n}\sum_{i=1}^{n} \mathbf{Y}_i \mathbf{Y}'_i$. Note that if the SVD of the demeaned data matrix is defined as $\mathbf{Y} = \mathbf{V}\mathbf{\Sigma}\mathbf{U}'$, then the covariance operator $\widehat{\mathbf{K}}$ can be decomposed as $(1/n)\mathbf{V}\mathbf{\Sigma}^2\mathbf{V}'$, where the orthonormal columns of the $p \times r$ matrix $\mathbf{V}$ are referred to as *eigenimages* and the non-negative diagonal elements $\sigma_1^2/n \geq \sigma_2^2/n \geq .. \geq \sigma_r^2/n > 0$ are called *eigenvalues*. At the third step, the n-dimensional vectors of *eigenscores* are calculated as $\boldsymbol{\xi}_i = \mathbf{V}'\mathbf{Y}_i = (\mathbf{\Sigma}\mathbf{U}')_i$. In practice, only a few, say, K leading eigenimages are used to model original images as $\mathbf{Y}_i = \boldsymbol{\mu} + \sum_{k=1}^{K} \mathbf{v}_k \xi_{ik}$. The number of eigenimages, K, is typically chosen to make the proportion of variance explained, which can be computed as $(\sigma_1^2 + \ldots + \sigma_K^2)/(\sigma_1^2 + \ldots + \sigma_r^2)$, larger than a certain predefined threshold. Formal rank-selection approaches are based either on linear mixed-effects models as in Di et al. (23) or Bayesian models as in Everson and Roberts (30) and Minka (73). In addition, Reference (44) proposed a cross-validation approach for estimating the number of principal components.

Note that the independence of images $\mathbf{Y}_i$ translates geometrically into orthogonality of the rows of $\mathbf{U}$. For high-dimensional p, the brute-force eigenanalysis requires $O(p^3)$ operations to decompose the $p \times p$ covariance operator $\widehat{\mathbf{K}}$ and, as a result, is infeasible. For some cases, p is large enough that the task of calculating and storing $\widehat{\mathbf{K}}$ is impossible. Reference (42) provides a detailed discussion on the use of randomized algorithms to estimate low-dimensional approximations of matrices for massive datasets. Reference (79) proposes incremental updates of eigenbases in the context of visual tracking. Section 15.3.1 provides details on efficient calculation of PCA in p-linear time with sequential access to the data, which is especially crucial in imaging applications where ultra high-dimensional data are routinely analyzed.

Consistency of estimated PCs in the high-dimension low sample size (HDLSS) context depends on the spacing of the eigenvalues of the population covariance matrix (54) and sparcity of the leading PCs (81). Consistency of the n-length, right singular vectors of the high-dimensional sample data matrices has been investigated by Leek (66). Analytical asymptotic confidence intervals for PCs typically require the assumption of normally distributed data (34, 91) or existence and computation of fourth-order moments (59, 60, 76).

15.2.3 PCA in Brain Imaging

Many functional neuroimaging applications greatly benefit by using PCA while pre-processing the data to remove motion artifacts (69, 75) and for dimension reduction (46, 44, 61, 4, 2). In fMRI applications, PCA has been widely used for clustering visual networks (25), identifying functional networks in normally aging subjects (72), and studying functional brain connectivity in various populations (96, 67, 87, 19, 33, 86, 5, 22).

Among other imaging modalities, PCA has proved to be useful in gadolinium delayed-enhanced MRI for detecting and clustering lesions in multiple sclerosis patients (83, 82), for estimating cortical thickness (101), in the analysis of brain morphology (104, 15), in functional near-infrared spectroscopy (88), as well as in analyzing PET (97, 70) and single photon emission computed tomography (108) data. The use of PCA in the context of Canonical Variates Analysis is proposed by (98) to identify associations of fMRI and PET signals with outcomes of interest. Reference (56) discusses comparisons of the dimension reduction techniques using cross-validation.

PCA played an important role in the winning algorithm of the ADHD200 prediction competition (27) with the goal of developing methods that predict whether children have attention deficit hyperactivity disorder (ADHD) or are neurotypical by using fMRI scans along with other phenotypic data. The algorithm consisted of four parts, two of which used PCA for dimension reduction of the fMRI data before applying the chosen prediction algorithm such as the gradient boosting method. Particularly, the prediction sub-algorithm with the highest accuracy in the internal test set used PCA to identify voxels in the space exhibiting the highest variability over time and used those voxels as seed voxels in a typical connectivity analysis inherently identifying the residual motion in the fMRI scans.

The functional version of PCA (96) additionally assumes that principal components are smooth functions over the temporal or spatial domain. In imaging, functional PCA has been used for modeling white-matter tracts (35, 37, 36, 47, 103) and for smoothing dynamic PET signals (50).

15.3 Structured PCA Models

Many modern imaging studies acquire brain images over multiple visits according to nested, crossed or longitudinal designs. These sampling structures may be parsimoniously modeled and analyzed via structured PCA models discussed below.

Zipunnikov et al. (104) considered a two-way imaging design for a sample of repeated images $\mathbf{Y}_{ij}$, where $\mathbf{Y}_{ij}$ is a brain image of the ith subject, $i = 1, \ldots, I$ at visit $j, j = 1, \ldots, J$ with the total number of images $n = IJ$. The main idea is to model each image via a linear combination of latent processes $\mathbf{Y}_{ij} = \boldsymbol{\mu} + \mathbf{R}_i^{(1)} + \mathbf{R}_{ij}^{(2)}$, where $\boldsymbol{\mu}$ is the overall mean image, $\mathbf{R}_i^{(1)}$ is a subject-specific image deviation from the population mean, and $\mathbf{R}_{ij}^{(2)}$ is the visit-specific image deviation from the subject-specific mean. The latent processes $\mathbf{R}_i^{(1)}$ and $\mathbf{R}_{ij}^{(2)}$ may be modeled via estimating and decomposing their covariance matrices

$$\widehat{\mathbf{K}}_{R^{(1)}} = \frac{1}{I(J^2 - J)} \sum_{i=1}^{I} \sum_{j_1 \neq j_2} \mathbf{Y}_{ij_1} \mathbf{Y}'_{ij_2} \quad \text{and}$$

$$\widehat{\mathbf{K}}_{R^{(2)}} = \frac{1}{2I(J^2 - J)} \sum_{i=1}^{I} \sum_{j_1 \neq j_2} (\mathbf{Y}_{ij_1} - \mathbf{Y}_{ij_2})(\mathbf{Y}_{ij_1} - \mathbf{Y}_{ij_2})'.$$

Then each image is parsimoniously represented as $\mathbf{Y}_{ij} = \boldsymbol{\mu} + \sum_{k_1=1}^{K_1} \mathbf{v}_{k_1}^{(1)} \xi_{ik_1} + \sum_{k_2=1}^{K_2} \mathbf{v}_{k_2}^{(2)} \zeta_{ijk_2}$, where $\mathbf{v}_{k_1}^{(1)}$, ξ_{ik_1}, $\mathbf{v}_{k_2}^{(2)}$, and ζ_{ijk_2} are eigenimages and eigenscores for processes $\mathbf{R}_i^{(1)}$ and $\mathbf{R}_{ij}^{(2)}$, respectively. Shou et al. (84) defined the image intra-class correlation (I2C2) coefficient, a multivariate generalization of the classical intra-class correlation coefficient (ICC), as $\rho = trace(\widehat{\mathbf{K}}_{R^{(1)}})/[trace(\widehat{\mathbf{K}}_{R^{(1)}}) + trace(\widehat{\mathbf{K}}_{R^{(2)}})]$ and explored replication accuracy of three imaging modalities: MRI, resting-state fMRI, and diffusion tensor imaging (DTI).

Shou et al. (85) extended the approach and considered a comprehensive list of models that emerge from more general experimental designs and capture a wide variety of correlation structures. A general m-way crossed model is expressed via a linear combination of latent processes as $\mathbf{Y}_{i_1 i_2 \cdots i_m} = \boldsymbol{\mu} + \mathbf{R}_{\mathcal{T}_1} + \mathbf{R}_{\mathcal{T}_2} + \cdots + \mathbf{R}_{\mathcal{T}_d}$. The model uses d sub-index sets, $\{\mathcal{T}_1, \mathcal{T}_2, \ldots, \mathcal{T}_d\}$, that define the model structure. For example, the two-way model above is obtained with $d = 2$, $\{\mathcal{T}_1, \mathcal{T}_2\} = \{i, ij\}$. The latent processes $\mathbf{R}_{\mathcal{T}_i}$ are assumed to be mean zero and mutually uncorrelated, which guarantees their identifiability. Consequently, the total variability of imaging outcomes is decomposed into process-specific variations. Shou et al. (85) showed that the covariance matrices of the latent processes $\mathbf{R}_{\mathcal{T}_i}$ can always be estimated as $\widehat{\mathbf{K}}_{R_{\mathcal{T}_i}} = \mathbf{Y}' \mathbf{H}_{R_{\mathcal{T}_i}} \mathbf{Y}$, where $\mathbf{H}_{R_{\mathcal{T}_i}}$'s are design-specific matrices.

Another relatively new direction focuses on explicit modeling of longitudinal structures $(\mathbf{Y}_{ij}, T_{ij}, Z_{ij}), i = 1, \ldots, I, j = 1, \ldots, J_i$, where images $\mathbf{Y}_{ij}$ of subject i are obtained at times T_{ij} together with covariates Z_{ij}. Figure 15.1 gives an example of a longitudinal design in the study of demyelination in corpus callosum in multiple sclerosis (MS) patients and its association with the Expanded Disability Status Scale (EDSS) (106). To explicitly model longitudinal designs, Zipunnikov et al. (106) proposed the following model

$$\mathbf{Y}_{ij} = \boldsymbol{\mu} + \mathbf{R}_{i,0} + T_{ij} \mathbf{R}_{i,1} + Z_{ij} \mathbf{R}_{i,2} + \mathbf{R}_{ij}.$$

The proposed method separates brain regions changing linearly with time, modeled by $\mathbf{R}_{i,1}$, from those staying unchanged, modeled by $\mathbf{R}_{i,0}$, and those linearly correlated with Z_{ij}, modeled by $\mathbf{R}_{i,2}$. After estimating and decomposing relevant covariance matrices, each image is parsimoniously modeled via eigenvectors and eigenvalues as

$$\mathbf{Y}_{ij} = \boldsymbol{\mu} + \sum_{k_1=1}^{K_1} (\mathbf{v}_{k_1}^{R,0} + T_{ij}\mathbf{v}_k^{R,1} + Z_{ij}\mathbf{v}_k^{R,2})\xi_{ik_1} + \sum_{k_2=1}^{K_2} \mathbf{v}_{k_2}^{R}\zeta_{ijk_2}.$$

Lee et al. (64) adapted the longitudinal model above to separate registration errors from baseline and physiological longitudinal changes in morphometric RAVENS maps of MS patients. Based on the longitudinal model, Lee et al. (65) introduced a distance between longitudinal trajectories of two subjects i_1 and i_2 as $\mathbf{D}^2(i_1, i_2) = \sum_{k=1}^{K_1} (\xi_{i_1k} - \xi_{i_2k})^2$ and used the distance to cluster subjects based on trajectories in their longitudinally collected images.

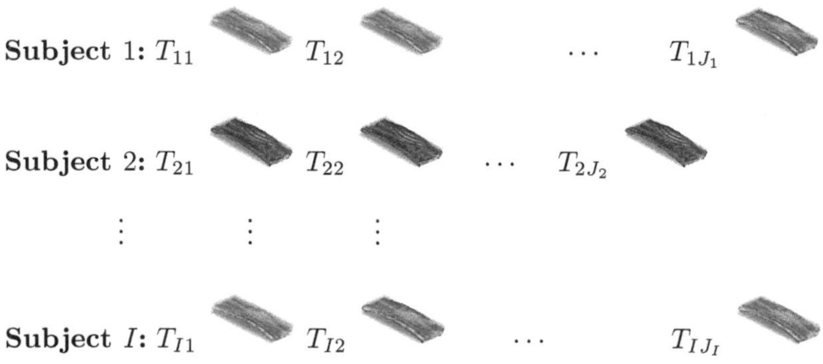

Subject 1: T_{11} T_{12} $\cdots$ T_{1J_1}

Subject 2: T_{21} T_{22} $\cdots$ T_{2J_2}

Subject I: T_{I1} T_{I2} $\cdots$ T_{IJ_I}

FIGURE 15.1
An example of the longitudinal design for the study presented by Zipunnikov et al. (106). Fractional Anisotropy (FA) in the Corpus Callosum of subjects diagnosed with MS. Darker color means lower FA.

15.3.1   Calculation of High-Dimensional PCA

In imaging applications, the brute-force PCA calculation is infeasible as it would require constructing, storing, and decomposing the $p \times p$ covariance operators. Therefore, References (105, 106) developed an algorithm that allows efficient calculation of the eigenimages and eigenvalues without either calculating or storing the covariance operators. All necessary calculations are done using sequential access to data and require only a number of operations linear in the number of parameters p. One of the main assumptions of the approach is that the sample size, n, is moderate so calculations of order $O(n^3)$ are feasible.

The algorithm starts with calculating $n \times n$ symmetric matrix $\mathbf{Y}'\mathbf{Y}$ and its spectral decomposition $\mathbf{Y}'\mathbf{Y} = \mathbf{U}\boldsymbol{\Sigma}\mathbf{U}'$. Note that for high-dimensional p the matrix $\mathbf{Y}$ cannot be loaded into the memory. The solution is to partition it into L slices as $\mathbf{Y}' = [(\mathbf{Y}^1)'|(\mathbf{Y}^2)'|\dots|(\mathbf{Y}^L)']$, where the size of the lth slice, $\mathbf{Y}^l$, is $(p/L) \times n$ and can be adapted to the available computer memory and optimized to reduce implementation time. The matrix $\mathbf{Y}'\mathbf{Y}$ is then calculated as $\sum_{l=1}^{L} (\mathbf{Y}^l)'\mathbf{Y}^l$ by streaming the individual blocks. After that, matrix $\mathbf{V}$ can be obtained as $\mathbf{V} = \mathbf{Y}\mathbf{U}\boldsymbol{\Sigma}^{-1}$. The actual calculations can be performed on the slices of the partitioned matrix $\mathbf{Y}$ as $\mathbf{V}^l = \mathbf{Y}^l\mathbf{U}\boldsymbol{\Sigma}^{-1}, l = 1, \dots, L$. The concatenated slices $[(\mathbf{V}^1)'|(\mathbf{V}^2)'|\dots|(\mathbf{V}^L)']$ form the matrix of the left singular vectors $\mathbf{V}'$.

An important extension of the algorithm above to the structured PCA models in Section 15.3 stems from the observation that after obtaining the SVD of $\mathbf{Y}$, each image $\mathbf{Y}_{i_1 i_2 \cdots i_m}$ can be represented as $\mathbf{Y}_{i_1 i_2 \cdots i_m} = \mathbf{V} \mathbf{\Sigma} \mathbf{U}_{i_1 i_2 \cdots i_m}$. Hence, the transformation $\mathbf{V}' \mathbf{Y}_{i_1 i_2 \cdots i_m} = \mathbf{\Sigma} \mathbf{U}_{i_1 i_2 \cdots i_m}$ is a (linear) *structure-, rank-, and variability*-preserving projection onto a low-dimensional space spanned by columns of matrix $\mathbf{\Sigma} \mathbf{U}'$.

15.4   Independent Component Analysis

Independent component analysis (ICA) (49) is a matrix decomposition method that attempts to uncover a latent linear representation of the data with independent latent factors (sources). The use of ICA as opposed to PCA is often required when the underlying sources are believed to be independent non-Gaussian random variables. While PCA was widely used in analyzing brain imaging data, it was most often implemented for data cleaning and preprocessing mainly due to the difficulties with the interpretation of the estimated components. Secondly, the estimated brain networks have a specific structure of sparsity. PCA requires the components to be orthogonal and there is no inherent structure of sparsity in the principal components. In ICA, sparsity of the independent components is modeled by specific choices of the non-Gaussian densities of underlying components or specific approximations for these densities.

The concept was first introduced in the context of signal processing by (55) who coined the name ICA. Since then several methods have been developed for estimating the variables in the ICA model. Suppose that in an fMRI setting, a three dimensional array (of dimension $V = v_1 \times v_2 \times v_3$) of intensities is collected for one subject at time points $t = 1, \ldots, T$. As in the previous example, the image is unfolded at each time point into a V-dimensional vector. We define by $\mathbf{Y}$ the observed matrix such that the tth row of $\mathbf{Y}$ corresponds to the unfolded three dimensional image collected at time point t. The intensity of the image at voxel v at time t is defined as $\mathbf{Y}(t, v)$. Note that we are changing the notation in this section to be consistent with the literature on ICA. The general ICA model can be written as

$$\mathbf{Y}(t, v) = \sum_{q=1}^{Q} \mathbf{A}(t, q) \mathbf{S}(q, v) + \epsilon(t, v) \tag{15.2}$$

where $\mathbf{A}$ is the $T \times Q$ mixing matrix and $\mathbf{S}$ is the $Q \times V$ matrix of independent components. The rows of $\mathbf{S}$ contain brain networks corresponding to various functions such as motor, vision, default brain network, etc. Finally, $\epsilon(t, v) \sim N(0, \sigma^2)$ are independent additive Gaussian noise components.

The theoretical work for estimating the parameters in model (15.2) primarily focuses on two directions based on the assumptions on the noise structure. Most early work on the methodology development was based on the assumption $\epsilon(t, v) = 0$ in the so-called noise-free model. Some of the commonly used methods are based on maximizing a measure of non-Gaussianity of the independent components such as the fourth-moment of the density. For instance, fastICA introduced by (48) estimates the unmixing matrix $\mathbf{W} = \mathbf{A}^{-1}$ by optimizing an approximation of the fourth-order moments of independent components. JADE, proposed by (13), is based on cumulant tensors.

A group of algorithms for ICA is based on maximum likelihood estimation of the unmixing matrix. Assuming that the components $\mathbf{S}(q, \cdot)$ are independent identically distributed vectors with density $f_q(\cdot)$, one may write the likelihood function for $\mathbf{W}$ as:

$$L(\mathbf{W}, f) = \prod_{t=1}^{T} \prod_{v=1}^{V} f_q(\sum_{q=1}^{Q} \mathbf{W}(t,q)\mathbf{Y}(q,v)). \qquad (15.3)$$

Product Density ICA, proposed by (90), uses a tilted Gaussian density for the independent components, a kernel density-based estimator was proposed by (10) and a semiparametric likelihood ICA introduced by (26) uses a semi-parametric model for the densities of independent components allowing for flexibility in modeling the underlying densities. Reference (71) approaches the problem of minimizing the dependence among the sources directly by employing distance covariance.

Whether ICA or PCA is the appropriate model for analyzing a dataset mainly depends on the assumptions one is willing to make about the structure of the underlying components. PCA only uses the second-order moments to estimate the components (assuming normally distributed components) while in ICA, higher-order moments are considered (assuming non-Gaussian, and statistically independent components). PCA (also referred to as whitening in this context) is typically used as a preprocessing step before applying ICA. If the underlying sources are non-Gaussian the resulting PCA estimates recover the original components often only up to a rotation matrix. As an example, one may consider the estimation of brain networks in fMRI experiments. Typically, relatively few voxels are activated within a brain network resulting in a high frequency of intensities around 0 (see Figure 15.2). Hence, the distribution of the intensities within the component is highly non-Gaussian. ICA utilizes higher-order moments of the distribution to estimate the underlying independent components along with the corresponding mixing matrix and has been very successful in the area of network estimation in fMRI among other applications.

15.4.1   ICA in Brain Imaging

The use of ICA in neuroimaging has been introduced by (12) who proposed a generalization of the model for group settings. The main idea is that the independent components are common across the group, whereas the mixing matrices are estimated for each subject. The group ICA model can be written as

$$\mathbf{Y}_i(t,v) = \sum_{q=1}^{Q} \mathbf{A}_i(t,q)\mathbf{S}(q,v), \qquad (15.4)$$

where i is the index for subjects. Based on the proposed algorithm, the group estimation consists of a first-level dimension reduction of each subject specific matrix using PCA, followed by temporal concatenation of the resulting matrices. A second-level dimension reduction is then applied to the group data to obtain the first Q principal components where Q is the pre-specified number of independent components to be estimated. Finally, ICA is applied to the resulting lower-dimensional matrix to estimate the independent brain networks and their corresponding temporal mixing patterns.

Since then, ICA has been used to uncover the underlying structure of networks in the brain from functional MRI data in experiments where subjects are performing certain tasks as well as when subjects are at rest (9), in electroencephalography (EEG) to study the function of the brain given more granular observations in time with lower spatial resolution ((20), (21)). Reference (40) proposed a unified framework for group ICA based on modeling the densities of the independent components using Gaussian mixture densities while estimating the mixture parameters and the matrices $\mathbf{W}$ for each subject via the EM algorithm. Three different structures of the unmixing matrices are considered for the estimation. Reference (39) derived the exact equations for estimation of the parameters in the EM algorithm. In addition, (41) proposed a new approach to group ICA where the population-level independent components are assumed to be the same for all observations in the population while the subject-level variation between the independent components is modeled in a second-level hierarchical model. Reference (93) introduced the Canonical ICA method for group ICA similar to the mixed-effects models where the subspace common to the group is estimated via canonical correlation analysis followed by an ICA-based pattern extraction.

The second set of models, primarily referred to as probabilistic ICA models, incorporate the additive Gaussian noise in (15.2). Reference (6) proposed a maximum likelihood estimation procedure for estimating the parameters in the ICA model along with the noise covariance. Reference (7) proposed a tensorial approach where the standard ICA model is generalized to higher dimensions. A model with autocorrelated noise was discussed by (63). An extensive overview and comparison of statistical methods for ICA is provided by (78). One overarching issue with estimation in the fMRI setting is the high dimensionality of the data. Next, we describe two approaches proposed in the literature to overcome the dimensionality problem.

15.4.2   Homotopic Group ICA

Homotopic group ICA (H-gICA) was introduced by (99) with the goal of improving group estimates of the networks as well as reducing the noise in the resulting sources via brain functional homotopy. The result follows from exploiting the high degree of synchrony in spontaneous activity that exists between geometrically corresponding interhemispheric regions (109). The approach utilizes the fact that most of the brain networks are symmetric in left and right hemispheres. The model is similar to the general group ICA formulation except that the common sources are estimated for a single hemisphere instead of the whole brain. The homotopic group ICA model can be written as

$$\mathbf{Y}_{i,j}(t,v) = \sum_{q=1}^{Q} \mathbf{A}_{i,j}(t,q)\mathbf{S}(q,v), \qquad (15.5)$$

where i is the index for subjects, j is the index for hemisphere, and $\mathbf{A}_{i,j}$ is the mixing matrix of the ith subject corresponding to hemisphere j. $\mathbf{Y}_{i,1}(\cdot, v)$ and $\mathbf{Y}_{i,2}(\cdot, v)$ are images of the corresponding voxels in the left and right hemispheres. In the H-gICA model, the independent components are assumed to be common across subjects and hemispheres, while how they mix to produce the signal can differ among both subjects and hemispheres. Compared with group ICA, H-gICA actually doubles the number of parameters while reducing the number of voxels in the estimated sources by half.

Under the setting of H-gICA, functional homotopy can be defined for each underlying network. The subject- and network-specific functional homotopy for the qth ICA-based network of subject i can be defined as $\mathbf{H}_i(q) = Cor(\mathbf{A}_{i,1}^{(q)}, \mathbf{A}_{i,2}^{(q)})$, where $\mathbf{A}_{i,j}^{(q)}$ is a vector of the time course modulating spatial map q. Relations between functional homotopy and ADHD disease status is studied by (99). Figure 15.2 shows four networks identified by H-gICA using fMRI scans for 20 children with ADHD overlaid on the MNI template brain.

15.4.3   Computation of High-Dimensional ICA

HDICA (28) is a likelihood-based semi-parametric approach for independent component analysis. It can be used to estimate the brain networks based on a large set of fMRI scans. The method is one of the few in the literature that proposes a unique estimation procedure

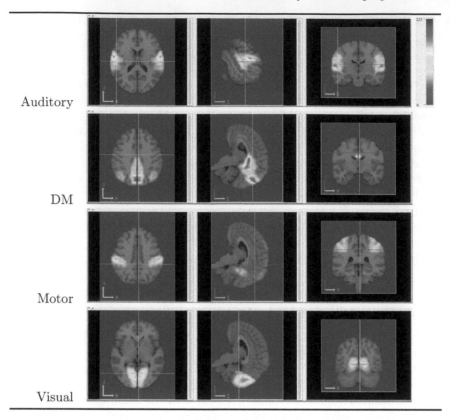

FIGURE 15.2
Four networks (Auditory, Default Mode (DM), Motor, and Visual) computed for 20 subjects using the H-gICA algorithm. The color bar on the right shows the values corresponding to each color, namely the highest intensities are colored in red followed by the yellow and blue for voxels with lower intensities.

to ensure full identifiability of the parameters in the ICA model. As discussed by (17), the parameters in the model (i.e., the mixing matrix and the independent components) are identifiable up to a permutation, scale, and sign. In HDICA, a set of constraints on the densities of the independent components is implored resulting in full identifiability of the parameters.

The densities of the underlying independent components $\mathbf{S}_q(\cdot)$ are modeled using a flexible mixture of Gaussian densities as given by

$$f_q(s) = \sum_{j=1}^{N_q} \frac{\theta_{qj}}{\sigma_q} \phi\left(\frac{s - \mu_{qj}}{\sigma_q}\right),$$

where N_q is the number of mixture densities, μ_{qj} and σ_q are the means and standard deviations of the mixture densities. Here the means and standard deviations of mixture elements are assumed to be known entities while the weights θ_{qj} are estimated using the EM algorithm. Further, a binning procedure is proposed to estimate the densities in the ultra-high-dimensional domain. Finally, each subject-specific unmixing matrix is updated using a Newton–Raphson optimization update for each iteration of the algorithm.

15.5   Discussion of Other Methods

Sparse PCA ((51), (107)) is often used to decompose datasets with a specific structure, particularly where the matrix of eigenscores is sparse, in other words, each resulting element in the dimension reduced matrix is only a linear combination of some of the principal components rather than all of them. The estimation of the parameters is based on a regression-type formulation of the SVD problem. Namely, as discussed in Section 15.2.1, the decomposition in (15.1) can be obtained by finding the solution to the following problem.

$$minimize \|\mathbf{Y} - \mathbf{UDV}^T\|_F \tag{15.6}$$

subject to $\mathbf{U}^T\mathbf{U} = \mathbf{I}_k$, $\mathbf{V}^T\mathbf{V} = \mathbf{I}_k$, and $\mathbf{D}$ is a diagonal matrix with non-negative values.

In the Sparse PCA framework, additional constraints are imposed on the vectors of principal scores such as the lasso type penalty or elastic net (107). In the generalized least-squares matrix decomposition framework proposed by (1) the Frobenius norm in (15.6) is extended to the more general $\mathbf{Q}$, $\mathbf{R}$ norm. The utility of the generalized matrix decomposition as compared with regular PCA is investigated by analyzing a task-based fMRI dataset. Reference (43) proposed a scale-invariate sparse PCA model for high-dimensional data. Reference (94) and (95) discuss the use of Sparse PCA in the context of fMRI analysis incorporating between subject variability, while (8) proposes a spatially constrained approximation for brain parcellation.

Principal Component Regression (PCR) is commonly used in prediction problems where the number of predictor variables is very high. In this framework, instead of using the full matrix of predictor variables in regression, the first few principal components of that matrix are modeled and, typically, the Ordinary Least Squares (OLS) estimates of the regression coefficients of those PCs are estimated. While some issues persist with the interpretation of the estimated parameters in the PCR model, a more important problem (as pointed out by (52)) is that some of the PCs with relatively smaller eigenvalues, which would be excluded from the analysis, may be as important as the PCs that would be included in the regression model due to having larger eigenvalues. In other words, large variability in a linear combination of the explanatory variables does not necessarily translate into better prediction. PCR is routinely used in neuroimaging (whether referred to as PCR or not) mainly where imaging data is used to predict a disease status where the number of predictors (e.g. connectivity maps) is high dimensional. For example, (89) describes the use of PCR for predicting functional responses across individuals.

Partial Least Squares (PLS) regression is a method similar to PCR in that a low-dimensional linear transformation of the predictor variables is used in regression as opposed to the full matrix. PLS differs from PCR in that the PCs are computed not only to identify directions of maximum variability in the space of the predictor variables but also to consider the directions that have high correlations with the outcome (31). A gentle tutorial for PLS and a comprehensive review of PLS-related literature and its applications in neuroimaging is given by (61).

Non-Negative Matrix Decomposition (NNMD) is another matrix decomposition technique introduced by (62) in the context of face recognition. The model of NNMD is similar to the model given by (15.2) with the restriction that the elements in $\mathbf{A}$ are non-negative. Using the same notation as in Section 15.4, the NNMD model assumes

$$\mathbf{Y}(t, v) \sim Poisson(\sum_{q=1}^{Q} \mathbf{A}(t, q)\mathbf{S}(q, v))$$

where $\mathbf{A}(t, q) \geq 0$ for $t = 1, \ldots, T$ and $q = 1, \ldots, Q$. By restricting the model to non-negative values of $\mathbf{A}$, the resulting basis components in NNMD represent parts of the original image which reproduce the original image in the additive model, rather than representations of the full image as in PCA and ICA.

Reference (3) deployed NNMD to analyze multimodal MRI, fMRI, and phenotypic data together to identify differences in typically developing children as compared with children with ADHD.

Projection Pursuit is a statistical method for identifying projections of multivariate data that are deemed most "interesting" (32). The argument against using the PCs for projection when, say, clustering multivariate data is the possible loss of information about the structure of clusters when projecting onto the PCs. Hence, a linear mapping algorithm is proposed to identify optimal projections. Reference (24) argued that the least "interesting" directions for projection are the directions where the projected data is most Gaussian. We point out the main steps of the heuristic arguments as follows. A projection is "interesting" if it is structured or non-random which can be quantified by its entropy defined as $\int f \log f$, i.e., the projection with smaller entropy is more structured. On the other hand, among the densities with a fixed mean and variance, the Gaussian density attains the highest entropy. As a result, the projections that maximize a measure of non-Gaussianity are most "interesting." This idea is also in line with many algorithms in ICA, where the heuristic justification of maximizing a measure of non-Gaussianity is based on maximizing the independence among the variables.

Factor Analysis (45) is a generative latent-variable model related to PCA that has been widely used in social sciences. Given an observed data matrix $\mathbf{Y}$, which has been de-meaned, the model can be expressed as follows:

$$\mathbf{Y} = \mathbf{AF} + \epsilon$$

where the factors $\mathbf{F}$ are uncorrelated and independent of the random Gaussian noise ϵ. The matrix $\mathbf{A}$ is the matrix of loadings. Define the covariance matrix of the observed data as $\mathbf{C}_y$ and the noise covariance as $\mathbf{C}_\epsilon$, then

$$\mathbf{C}_y = \mathbf{AA}^T + \mathbf{C}_\epsilon.$$

The parameters are identifiable up to an orthogonal transformation. In practice, the uniqueness is often imposed according to a set of assumptions based on the specific problem at hand. Several methods exist to estimate the factors and their loadings in the literature on factor analysis.

Reference (18) proposed the Population Value Decomposition (PVD) as a framework for the analysis of populations of images. As defined in Section 15.2, if $\mathbf{Y}_i$, is defined as the observed matrix for subject $i = 1, \ldots, n$, then the PVD decomposition can be written as

$$\mathbf{Y}_i = \mathbf{P}\mathbf{V}_i\mathbf{D}.$$

Reference (11) used PVD to study altered functional connectivity in patients with Alzheimer's disease. Closely related methods have been proposed by (100) and (102).

Tensor (multi-way) statistical analysis is an emerging area that offers powerful data-compression tools. In neuroscience, tensor analysis is adapted by methods requiring third- and fourth-order moment interactions such as JADE (13) or when brain images are stored and analyzed in their natural $3D$ (MRI) or $4D$ (fMRI) form (80). The two most popular ways to decompose a three-way tensor $\mathbf{Y} \in \mathbb{R}^{p_1 \times p_2 \times p_3}$ are as follows: (i) canonical decomposition/parallel factors (CANDECOMP/PARAFAC)(14) that represents

$$Y(i_1, i_2, i_3) = \sum_{k=1}^{K} \lambda_k v_{i_1 k} u_{i_2 k} w_{i_3 k},$$

where $\mathbf{v} \in \mathbb{R}^{p_1 \times K}, \mathbf{u} \in \mathbb{R}^{p_2 \times K}, \mathbf{w} \in \mathbb{R}^{p_3 \times K}, \lambda \in \mathbb{R}^K$, and (ii) Tucker decomposition (92) that represents

$$Y(i_1, i_2, i_3) = \sum_{k_1=1}^{K_1} \sum_{k_2=1}^{K_2} \sum_{k_3=1}^{K_3} \lambda_{k_1 k_2 k_3} v_{i_1 k_1} u_{i_2 k_2} w_{i_3 k_3},$$

where $\mathbf{v} \in \mathbb{R}^{p_1 \times K_1}, \mathbf{u} \in \mathbb{R}^{p_2 \times K_2}, \mathbf{w} \in \mathbb{R}^{p_3 \times K_3}, \lambda \in \mathbb{R}^{K_1 \times K_2 \times K_3}$. For more comprehensive reviews, see (58, 68, 74, 57, 16).

15.6 Acknowledgements

The project described was supported by grant number R01 EB012547 from the National Institute of Biomedical Imaging and Bioengineering.

Bibliography

[1] G.I. Allen, L. Grosenick, and J. Taylor. A generalized least-square matrix decomposition. *Journal of the American Statistical Association*, 109(505):145–159, 2014.

[2] A.H. Andersen and W.S. Rayens. Structure-seeking multilinear methods for the analysis of fMRI data. *NeuroImage*, 22(2):728–739, 2004.

[3] A. Anderson, P.K. Douglas, W.T. Kerr, V.S. Haynes, A.L. Yuille, J. Xie, Y.N. Wu, J.A. Brown, and M.S. Cohen. Non-negative matrix factorization of multimodal MRI, fMRI and phenotypic data reveals differential changes in default mode subnetworks in ADHD. *NeuroImage*, 2013.

[4] M. Barnathan, V. Megalooikonomou, C. Faloutsos, S. Faro, and F.B. Mohamed. TWave: High-order analysis of functional MRI. *NeuroImage*, 58(2):537–548, 2011.

[5] R. Baumgartner, C. Teichtmeister, M. Diemling, W. Backfrieder, M. Sámal, M. Barth, H. Bergmann, and E. Moser. Application of rotated principal components for a paradigm-free analysis of conventional gradient-echo fMRI data. *NeuroImage*, 3(3, Supplement):S50–, 996.

[6] C.F. Beckmann and S. M. Smith. Probabilistic independent component analysis for functional magnetic resonance imaging. *IEEE Trans. Med. Imaging*, 23:137–152, 2004.

[7] C.F. Beckmann and S. M. Smith. Tensorial extensions of independent component analysis for multisubject fMRI analysis. *Neuroimage*, 25:295–311, 2005.

[8] Alexis Benichoux and Thomas Blumensath. A spatially constrained low-rank matrix factorization for the functional parcellation of the brain. In *Signal Processing Conference (EUSIPCO), 2014 Proceedings of the 22nd European*, pages 26–30. IEEE, 2014.

[9] B.B. Biswal, M. Mennes, X.-N. Zuo, S. Gohel, C. Kelly, et al. Toward discovery science of human brain function. *Proc Natl Acad Sci U S A*, 107(10):4734–9, 2010.

[10] R. Boscolo, H. Pan, and V.P. Roychowdhury. Independent component analysis based on nonparametric density estimation. *Neural Networks, IEEE Transactions on*, 15 (1):55 –65, 2004.

[11] B.S. Caffo, C.M. Crainiceanu, G. Verduzco, S. Joel, S.H. Mostofsky, S.S. Bassett, and J.J. Pekar. Two-stage decompositions for the analysis of functional connectivity for fMRI with application to Alzheimer's disease risk. *NeuroImage*, 51(3):1140–1149, 2010.

[12] V.D. Calhoun, T. Adali, G.D. Pearlson, and J.J. Pekar. A method for making group inferences from functional MRI data using independent component analysis. *Human Brain Mapping*, 14:140–151, 2001.

[13] J.-F. Cardoso. Eigen-structure of the fourth-order cumulant tensor with application to the blind source separation problem. *Proc. IEEE Int. Conf. on Acoustics, Speech and Signal Processing (ICASSP'90)*, Albuquerque, New Mexico, pages 2655–2658, 1990.

[14] J.D. Carroll and J.-J. Chang. Analysis of individual differences in multidimensional scaling via an n-way generalization of "Eckart-Young" decomposition. *Psychometrika*, 35(3):283–319, 1970.

[15] K. Chen, E.M. Reiman, G.E. Alexander, D. Bandy, R. Renaut, W.R. Crum, N.C. Fox, and M.N. Rossor. An automated algorithm for the computation of brain volume change from sequential MRIs using an iterative principal component analysis and its evaluation for the assessment of whole-brain atrophy rates in patients with probable Alzheimer's disease. *NeuroImage*, 22(1):134–143, 2004.

[16] A. Cichocki, D. Mandic, A.H. Phan, C. Caiafa, G. Zhou, Q. Zhao, and L. De Lathauwer. Tensor decompositions for signal processing applications from two-way to multiway component analysis. arXiv preprint arXiv:1403.4462, 2014.

[17] P. Comon. Independent component analysis: A new concept? *Signal Processing*, 36 (3):287–314, 1994.

[18] C.M. Crainiceanu, B.S. Caffo, S. Luo, V.M. Zipunnikov, and N.M. Punjabi. Population value decomposition: A framework for the analysis of image populations. *Journal of the American Statistical Association*, 106(495), 2011.

[19] A.T. Curtis and R.S. Menon. Highcor: A novel data-driven regressor identification method for BOLD fMRI. *NeuroImage*, 98(0):184 – 194, 2014.

[20] A. Delorme and S. Makeig. EEGLAB: An open source toolbox for analysis of single-trial EEG dynamics including independent component analysis. *Journal of Neuroscience Methods*, 134(1):9–21, 2004.

[21] A. Delorme, T.J. Sejnowski, and S. Makeig. Enhanced detection of artifacts in EEG data using higher-order statistics and independent component analysis. *Neuroimage*, 34(4):1443–1449, 2007.

[22] O. Demirci, V.P. Clark, and V.D. Calhoun. A projection pursuit algorithm to classify individuals using fMRI data: Application to schizophrenia. *NeuroImage*, 39(4):1774 – 1782, 2008.

[23] C.-Z. Di, C.M. Crainiceanu, B.S. Caffo, and N.M. Punjabi. Multilevel functional principal component analysis. *The Annals of Applied Statistics*, 3(1):458, 2009.

[24] P. Diaconis and D. Freedman. Asymptotics of graphical projection pursuit. *The Annals of Statistics*, pages 793–815, 1984.

[25] C. Ecker, E. Reynaud, S.C. Williams, and M.J. Brammer. Detecting functional nodes in large-scale cortical networks with functional magnetic resonance imaging: A principal component analysis of the human visual system. *Human Brain Mapping*, 28(9): 817–834, 2007.

[26] A. Eloyan and S.K. Ghosh. A semiparametric approach to source separation using independent component analysis. *Computational Statistics & Data Analysis*, 58:383–396, 2013.

[27] A. Eloyan, J. Muschelli, M.B. Nebel, H. Liu, F. Han, T. Zhao, A.D. Barber, S. Joel, J.J. Pekar, S.H. Mostofsky, and B.S. Caffo. Automated diagnoses of attention deficit hyperactive disorder using magnetic resonance imaging. *Frontiers in Systems Neuroscience*, 6, 2012.

[28] A. Eloyan, C.M. Crainiceanu, and B.S. Caffo. Likelihood-based population independent component analysis. *Biostatistics*, 14(3):514–527, 2013.

[29] A. Eloyan, S. Li, J. Muschelli, J.J. Pekar, S.H. Mostofsky, and B.S. Caffo. Analytic programming with fMRI data: A quick-start guide for statisticians using r. *PloS one*, 9(2):e89470, 2014.

[30] R. Everson and S. Roberts. Inferring the eigenvalues of covariance matrices from limited, noisy data. *Signal Processing, IEEE Transactions on*, 48(7):2083–2091, 2000.

[31] L.E. Frank and J.H. Friedman. A statistical view of some chemometrics regression tools. *Technometrics*, 35(2):109–135, 1993.

[32] J.H. Friedman and J.W. Tukey. A projection pursuit algorithm for exploratory data analysis. 1973.

[33] M. Gabbay, C. Brennan, E. Kaplan, and L. Sirovich. A principal components-based method for the detection of neuronal activity maps: Application to optical imaging. *NeuroImage*, 11(4):313–325, 2000.

[34] M.A. Girshick. On the sampling theory of roots of determinantal equations. *The Annals of Mathematical Statistics*, 10(3):203–224, 1939.

[35] J. Goldsmith, J. Bobb, C.M. Crainiceanu, B.S. Caffo, and D.S. Reich. Penalized functional regression. *Journal of Computational and Graphical Statistics*, 20(4):830–851, 2010.

[36] J. Goldsmith, C.M. Crainiceanu, B.S. Caffo, and D.S. Reich. Penalized functional regression analysis of white-matter tract profiles in multiple sclerosis. *NeuroImage*, 57(2):431–439, 2011.

[37] J. Goldsmith, C.M. Crainiceanu, B.S. Caffo, and D.S. Reich. Longitudinal penalized functional regression for cognitive outcomes on neuronal tract measurements. *Journal of the Royal Statistical Society: Series C (Applied Statistics)*, 61(3):453–469, 2012.

[38] G.H. Golub and C.F. Van Loan. *Matrix Computations*. Johns Hopkins University, Press, Baltimore, MD, USA, pages 374–426, 1996.

[39] Y. Guo. A general probabilistic model for group independent component analysis and its estimation methods. *Biometrics*, 67(4):1532–1542, 2011.

[40] Y. Guo and G. Pagnoni. A unified framework for group independent component analysis for multi-subject fMRI data. *Neuroimage*, 42:1078–1093, 2008.

[41] Y. Guo and L. Tang. A hierarchical model for probabilistic independent component analysis of multi-subject fMRI studies. *Biometrics*, pages 1–12, 2013.

[42] N. Halko, P.G. Martinsson, and J.A. Tropp. Finding structure with randomness: Stochastic algorithms for constructing approximate matrix decompositions, 2009. URL http://arxiv. org/abs/0909.4061.

[43] F. Han and H. Liu. Scale-invariant sparse PCA on high-dimensional meta-elliptical data. *Journal of the American Statistical Association*, 109(505):275–287, 2014.

[44] L.K. Hansen, J. Larsen, F.Å. Nielsen, S.C. Strother, E. Rostrup, R. Savoy, N. Lange, J. Sidtis, C. Svarer, and O.B. Paulson. Generalizable patterns in neuroimaging: How many principal components? *NeuroImage*, 9(5):534 – 544, 1999.

[45] H.H. Harman. *Modern Factor Analysis*. University of Chicago Press, 2nd edition, 1967.

[46] S. Haufe, S. Dähne, and V.V. Nikulin. Dimensionality reduction for the analysis of brain oscillations. *NeuroImage*, 2014.

[47] L. Huang, J. Goldsmith, P.T. Reiss, D.S. Reich, and C.M. Crainiceanu. Bayesian scalar-on-image regression with application to association between intracranial DTI and cognitive outcomes. *NeuroImage*, 83(0):210–223, 2013.

[48] A. Hyvarinen and E. Oja. A fast fixed-point algorithm for independent component analysis. *Neural Computation*, 9(7):1483–1492, 1997.

[49] A. Hyvärinen, J. Karhunen, and E. Oja. *Independent Component Analysis*. John Wiley & Sons, 2001.

[50] C.-R. Jiang, J.A.D. Aston, and J.-L. Wang. Smoothing dynamic positron emission tomography time courses using functional principal components. *NeuroImage*, 47(1): 184–193, 2009.

[51] I.M. Johnstone and A.Y. Lu. Sparse principal components analysis. *Unpublished manuscript*, 2004.

[52] I.T. Jolliffe. A note on the use of principal components in regression. *Applied Statistics*, pages 300–303, 1982.

[53] I.T. Jolliffe. *Principal Component Analysis*. Wiley Online Library, 2005.

[54] S. Jung, J.S. Marron, et al. PCA consistency in high dimension, low sample size context. *The Annals of Statistics*, 37(6B):4104–4130, 2009.

[55] C. Jutten and J. Herault. Blind separation of sources, Part I: An adaptive algorithm based on neuromimethic architecture. *Signal Processing*, 24:1–10, 1991.

[56] Ferath Kherif, Jean-Baptiste Poline, Guillaume Flandin, Habib Benali, Olivier Simon, Stanislas Dehaene, and Keith J. Worsley. Multivariate model specification for fMRI data. *Neuroimage*, 16(4):1068–1083, 2002.

[57] B.N. Khoromskij. Tensors-structured numerical methods in scientific computing: Survey on recent advances. *Chemometrics and Intelligent Laboratory Systems*, 110(1): 1–19, 2012.

[58] T.G. Kolda and B.W. Bader. Tensor decompositions and applications. *SIAM Review*, 51(3):455–500, 2009.

[59] T. Kollo and H. Neudecker. Asymptotics of eigenvalues and unit-length eigenvectors of sample variance and correlation matrices. *Journal of Multivariate Analysis*, 47(2): 283–300, 1993.

[60] T. Kollo and H. Neudecker. Asymptotics of Pearson–Hotelling principal-component vectors of sample variance and correlation matrices. *Behaviormetrika*, 24:51–70, 1997.

[61] A. Krishnan, L.J. Williams, A.R. McIntosh, and H. Abdi. Partial least squares (PLS) methods for neuroimaging: A tutorial and review. *NeuroImage*, 56(2):455–475, 2011.

[62] D.D. Lee and H.S. Seung. Learning the parts of objects by non-negative matrix factorization. *Nature*, 401(6755):788–791, 1999.

[63] S. Lee, H. Shen, Y. Truong, M. Lewis, and X. Huang. Independent component analysis involving autocorrelated sources with an application to functional magnetic resonance imaging. *Journal of the American Statistical Association*, 106(495):1009–1024, 2011.

[64] S. Lee, V.M. Zipunnikov, B.S. Caffo, D.S. Reich, and D.L. Pham. Statistical image analysis of longitudinal ravens images: Methodology and case study. In *submitted*, 2013.

[65] S. Lee, V.M. Zipunnikov, N. Shiee, C.M. Crainiceanu, B.S. Caffo, and D.L. Pham. Clustering of high dimensional longitudinal imaging data. In *Pattern Recognition in Neuroimaging (PRNI), 2013 International Workshop on*, pages 33–36, 2013.

[66] J.T. Leek. Asymptotic conditional singular value decomposition for high-dimensional genomic data. *Biometrics*, 67(2):344–352, 2011.

[67] N. Leonardi, J. Richiardi, M. Gschwind, S. Simioni, J.-M. Annoni, M. Schluep, P. Vuilleumier, and D. Van De Ville. Principal components of functional connectivity: A new approach to study dynamic brain connectivity during rest. *NeuroImage*, 83 (0):937–950, 2013.

[68] H. Lu, K.N. Plataniotis, and A.N. Venetsanopoulos. A survey of multilinear subspace learning for tensor data. *Pattern Recognition*, 2011.

[69] H. Mandelkow, D. Brandeis, and P. Boesiger. Good practices in EEG-MRI: The utility of retrospective synchronization and PCA for the removal of MRI gradient artefacts. *NeuroImage*, 49(3):2287–2303, 2010.

[70] P.J. Markiewicz, J.C. Matthews, J. Declerck, and K. Herholz. Robustness of correlations between PCA of FDG-PET scans and biological variables in healthy and demented subjects. *NeuroImage*, 56(2):782–787, 2011.

[71] D.S. Matteson and R.S. Tsay. Independent component analysis via distance covariance. arXiv preprint arXiv:1306.4911, 2013.

[72] P. Metzak, E. Feredoes, Y. Takane, L. Wang, S. Weinstein, T. Cairo, Elton T.C. Ngan, and T.S. Woodward. Constrained principal component analysis reveals functionally connected load-dependent networks involved in multiple stages of working memory. *Human Brain Mapping*, 32(6):856–871, 2011.

[73] T.P. Minka. Automatic choice of dimensionality for PCA. In *NIPS*, volume 13, pages 598–604, 2000.

[74] M. Mørup. Applications of tensor (multiway array) factorizations and decompositions in data mining. *Wiley Interdisciplinary Reviews: Data Mining and Knowledge Discovery*, 1(1):24–40, 2011.

[75] J. Muschelli, M.B. Nebel, B.S. Caffo, A.D. Barber, J.J. Pekar, and S.H. Mostofsky. Reduction of motion-related artifacts in resting state fMRI using aCompCor. *NeuroImage*, 96(0):22–35, 2014.

[76] H. Ogasawara. Concise formulas for the standard errors of component loading estimates. *Psychometrika*, 67(2):289–297, 2002.

[77] R Core Team. *R: A Language and Environment for Statistical Computing*. R Foundation for Statistical Computing, Vienna, Austria, 2014.

[78] B.B. Risk, D.S. Matteson, D. Ruppert, A. Eloyan, and B.S. Caffo. An evaluation of independent component analyses with an application to resting-state fMRI. *Biometrics*, 70(1):224–236, 2014.

[79] David A. Ross, Jongwoo Lim, Ruei-Sung Lin, and Ming-Hsuan Yang. Incremental learning for robust visual tracking. *International Journal of Computer Vision*, 77(1-3):125–141, 2008.

[80] T. Schultz, A. Fuster, A. Ghosh, R. Deriche, L. Florack, L. Lek-Heng, et al. Higher-order tensors in diffusion imaging. *Visualization and Processing of Tensors and Higher Order Descriptors for Multi-Valued Data*. Dagstuhl Seminar 2011, 2013.

[81] D. Shen, H. Shen, and J.S. Marron. Consistency of sparse pca in high dimension, low sample size contexts. *Journal of Multivariate Analysis*, 115:317–333, 2013.

[82] R.T. Shinohara, C.M. Crainiceanu, B.S. Caffo, M.I. Gaitán, and D.S. Reich. Population-wide principal component-based quantification of blood–brain-barrier dynamics in multiple sclerosis. *NeuroImage*, 57(4):1430–1446, 2011.

[83] R.T. Shinohara, C.M. Crainiceanu, B.S. Caffo, M.I. Gaitán, and D.S. Reich. Population-wide principal component-based quantification of blood–brain-barrier dynamics in multiple sclerosis. *NeuroImage*, 57(4):1430–1446, 2011.

[84] H. Shou, A. Eloyan, S. Lee, V.M. Zipunnikov, A.N. Crainiceanu, M.B. Nebel, B.S. Caffo, M.A. Lindquist, and C.M. Crainiceanu. Quantifying the reliability of image replication studies: The image intraclass correlation coefficient (I2C2). *Cognitive, Affective, & Behavioral Neuroscience*, 13(4):714–724, 2013.

[85] H. Shou, V.M. Zipunnikov, C.M. Crainiceanu, and S. Greven. Structured functional principal component analysis. arXiv preprint arXiv:1304.6783, 2013.

[86] H. Shou, A. Eloyan, M.B. Nebel, A. Mejia, J.J. Pekar, S.H. Mostofsky, B.S. Caffo, M.A. Lindquist, and C.M. Crainiceanu. Shrinkage prediction of seed-voxel brain connectivity using resting state fMRI. *NeuroImage*, 2014.

[87] M. Sugiura, J. Watanabe, Y. Maeda, Y. Matsue, H. Fukuda, and R. Kawashima. Different roles of the frontal and parietal regions in memory-guided saccade: A PCA approach on time course of BOLD signal changes. *Human Brain Mapping*, 23(3):129–139, 2004.

[88] S. Tak and J.C. Ye. Statistical analysis of fNIRS data: A comprehensive review. *NeuroImage*, 85, Part 1(0):72–91, 2014.

[89] Bertrand Thirion, Gaël Varoquaux, Olivier Grisel, Cyril Poupon, and Philippe Pinel. Principal component regression predicts functional responses across individuals. In *Medical Image Computing and Computer-Assisted Intervention–MICCAI 2014*, pages 741–748. Springer, 2014.

[90] R. Tibshirani and T.J. Hastie. Independent components analysis through product density estimation. In *Advances in Neural Information Processing Systems*, pages 649–656, 2002.

[91] M.E. Tipping and C.M. Bishop. Probabilistic principal component analysis. *Journal of the Royal Statistical Society: Series B (Statistical Methodology)*, 61(3):611–622, 1999.

[92] L.R. Tucker. Some mathematical notes on three-mode factor analysis. *Psychometrika*, 31(3):279–311, 1966.

[93] Gaël Varoquaux, Sepideh Sadaghiani, Philippe Pinel, Andreas Kleinschmidt, Jean-Baptiste Poline, and Bertrand Thirion. A group model for stable multi-subject ICA on fMRI datasets. *Neuroimage*, 51(1):288–299, 2010.

[94] Gaël Varoquaux, Alexandre Gramfort, Fabian Pedregosa, Vincent Michel, and Bertrand Thirion. Multi-subject dictionary learning to segment an atlas of brain spontaneous activity. In *Information Processing in Medical Imaging*, pages 562–573. Springer, 2011.

[95] Gaël Varoquaux, Yannick Schwartz, Philippe Pinel, and Bertrand Thirion. Cohort-level brain mapping: Learning cognitive atoms to single out specialized regions. In *Information Processing in Medical Imaging*, pages 438–449. Springer, 2013.

[96] R. Viviani, G. Grön, and M. Spitzer. Functional principal component analysis of fMRI data. *Human Brain Mapping*, 24(2):109–129, 2005.

[97] B.J. Weder, K. Schindler, T.J. Loher, R. Wiest, M. Wissmeyer, P. Ritter, K. Lovblad, F. Donati, and J. Missimer. Brain areas involved in medial temporal lobe seizures: A principal component analysis of ictal SPECT data. *Human Brain Mapping*, 27(6): 520–534, 2006.

[98] Keith J. Worsley, Jean-Baptiste Poline, Karl J. Friston, and A.C. Evans. Characterizing the response of PET and fMRI data using multivariate linear models. *NeuroImage*, 6(4):305–319, 1997.

[99] J. Yang, A. Eloyan, A. Barber, M.B. Nebel, S.M. Mostofsky, J.J. Pekar, C.M. Crainiceanu, and B.S. Caffo. Homotopic group ICA for multi-subject brain imaging data. 2013.

[100] J. Ye. Generalized low rank approximations of matrices. *Machine Learning*, 61(1–3): 167–191, 2005.

[101] U. Yoon, J.-M. Lee, K. Im, Y.-W. Shin, B.H. Cho, I.Y. Kim, J.S. Kwon, and S.I. Kim. Pattern classification using principal components of cortical thickness and its discriminative pattern in schizophrenia. *NeuroImage*, 34(4):1405–1415, 2007.

[102] S. Yu, J. Bi, and J. Ye. Matrix-variate and higher-order probabilistic projections. *Data Mining and Knowledge Discovery*, 22(3):372–392, 2011.

[103] Y. Yuan, J.H. Gilmore, X. Geng, S. Martin, K. Chen, J.-L. Wang, and H. Zhu. FMEM: Functional mixed effects modeling for the analysis of longitudinal white matter tract data. *NeuroImage*, 84(0):753–764, 2014.

[104] V.M. Zipunnikov, B.S. Caffo, D.M. Yousem, C. Davatzikos, B.S. Schwartz, and C.M. Crainiceanu. Functional principal component model for high-dimensional brain imaging. *NeuroImage*, 58(3):772–784, 2011.

[105] V.M. Zipunnikov, B.S. Caffo, D.M. Yousem, C. Davatzikos, B.S. Schwartz, and C.M. Crainiceanu. Multilevel functional principal component analysis for high-dimensional data. *Journal of Computational and Graphical Statistics*, 20(4), 2011.

[106] V.M. Zipunnikov, S. Greven, B.S. Caffo, D.S. Reich, and C.M. Crainiceanu. Longitudinal high-dimensional data principal component analysis with application to diffusion tensor imaging of multiple sclerosis. 2011.

[107] H. Zou, T.J. Hastie, and R. Tibshirani. Sparse principal component analysis. *Journal of Computational and Graphical Statistics*, 15(2):265–286, 2006.

[108] G. Zuendorf, N. Kerrouche, K. Herholz, and J.-C. Baron. Efficient principal component analysis for multivariate 3D voxel-based mapping of brain functional imaging datasets as applied to FDG-PET and normal aging. *Human Brain Mapping*, 18(1): 13–21, 2003.

[109] X.-N. Zuo, C. Kelly, A. Di Martino, M. Mennes, D.S. Margulies, S. Bangaru, R. Grzadzinski, A.C. Evans, Y.-F. Zang, F.X. Castellanos, et al. Growing together and growing apart: Regional and sex differences in the lifespan developmental trajectories of functional homotopy. *The Journal of Neuroscience*, 30(45):15034–15043, 2010.

16

Effective Connectivity and Causal Inference in Neuroimaging

Martin A. Lindquist

Department of Biostatistics; Johns Hopkins University

Michael E. Sobel

Department of Statistics; Columbia University

CONTENTS

16.1	Introduction	419
16.2	Effective Connectivity	420
16.3	Models of Effective Connectivity	422
	16.3.1 Structural Equation Models	422
	16.3.2 Dynamic Causal Models	427
	16.3.3 Granger Causality	429
16.4	Effective Connectivity and Causation	432
16.5	Conclusions	435
	Bibliography	436

16.1 Introduction

Neuroscientists have long sought to identify brain regions responsible for specific mental functions. More recently, they have expressed great interest in identifying the interactions between regions in the generation of mental activity. In the neuroimaging literature, these two interests are often referred to as *functional specialization* and *functional integration*, respectively (17, 39).

Functional specialization refers to the existence of specific brain areas, often organized into distinct neuronal populations or cortical areas, that are specialized for different functions. In studies of functional specialization, the aim is to discover regionally specific effects and identify regions specialized for a particular task. In contrast, functional integration is the coordinated activation of large numbers of neurons within the distributed system of the cerebral cortex. Brain areas are assumed to be highly interconnected and process information in a distributed manner. In studies of functional integration, the aim is to analyze the interactions between different regions in a neuronal system and determine how experimental manipulations affect these interactions.

To date, the majority of research has focused on localization, often with the goal of constructing maps indicating regions of the brain that are activated by specific tasks (see, for example, Chapters 11, 12, and 13). However, although the brain adheres to certain principles of functional specialization, these alone cannot satisfactorily account for its operation. Instead, an understanding of the "connectivity" between regions in the production of mental activity is required.

"Connectivity" is an umbrella term that has been used to refer to a number of related aspects of brain organization. In the neuroimaging literature it is common to distinguish between anatomical, functional, and effective connectivity (58). Anatomical connectivity describes the physical connections between different brain regions and is typically studied using techniques such as diffusion tensor imaging (DTI; see Chapter 4). Functional connectivity is defined as the undirected association between two or more time series of measurements from different regions (see Chapter 14). Finally, effective connectivity is defined as the directed influence of one brain region on others (17).

Functional connectivity analyses seek to establish the statistical associations among regions, for example, the association between time series from two different regions (bivariate connectivity). Simple functional connectivity analyses usually compare associations between ROIs using data collected from subjects in a resting state or experimental state, or between a seed region of interest and voxels throughout the brain. Alternatively, multivariate decomposition techniques such as principal components analysis (PCA) or independent components analysis (ICA) are often used in functional connectivity studies; see Chapter 15.

Effective connectivity analyses, on the other hand, are model-dependent. Typically, a small set of regions and a hypothesized set of directed connections are specified *a priori*, and tests of fit are used to compare a small number of alternative models and assess the statistical significance of individual connections. Typically, the connections in these models are endowed with a causal interpretation. Because there are many possible models, the choice of regions and connections must be both anatomically and scientifically motivated; these choices, which are not always clear-cut given the current state of knowledge in the field, can dramatically affect the substantive conclusions drawn.

In this chapter we focus on the concept of effective connectivity. We begin by describing particular issues involved in analyses of effective connectivity. We then consider several modeling methods widely used in effective connectivity studies. Although these methods are often used to make inferences about causation, the connection between these inferences and any causal parameters that these methods might be estimating is not clear. We therefore introduce an approach to causal inference from the statistical literature that clarifies these matters and that illustrates how causal inferences in neuroscience rest on important assumptions (which may be incorrect) of which investigators are usually unaware. We illustrate these points by means of a simple example.

16.2   Effective Connectivity

It is often said that functional connectivity analyses summarize patterns of correlations among brain systems, while effective connectivity analyses model the mechanisms that generate these correlations. However, the distinction between functional and effective connectivity is not always clear (26), and differences in the interpretation of these concepts can lead to disagreement as to whether an analysis should be categorized as functional or effective connectivity. If the discriminating features between the two approaches are a directional model in which causal influences are specified and the ability exists to draw conclusions about direct or indirect connections, then many analyses using linear regression, for example, psychophysiological interaction (PPI) analysis (18), which simply tests for the significance of an interaction effect, would be included in the domain of effectivity connectivity. In contrast, (16) argues that the distinction lies in the fact that functional connectivity is an "observable phenomenan" that can be quantified using measures of statistical dependencies (e.g., correlations), while effective connectivity corresponds to the parameter of a model that seeks to explain these observed dependencies.

The most commonly used methods in effective connectivity studies include Structural Equation Models (SEMs) (38), Dynamic Causal Modeling (DCM) (14), and Granger causality (48). Both structural equation modeling and Granger causality have long histories in the behavioral and social sciences, while DCM is a model specifically tailored to the analysis of neuroimaging data. We describe these methods in the next section. First, we discuss several factors that are common to all analyses of effective connectivity and require consideration.

In most analyses of effective connectivity, a small set of regions with a proposed set of directed connections are specified *a priori*, and hypothesis tests are used to assess the statistical significance of individual connections. Granger causality is a notable exception, in which instead the pair-wise relationship between two time courses is studied, or more generally, the relationship between a pre-specified seed region and the rest of the brain. That said, most methods for studying effective connectivity depend on two models, a neuroanatomical model describing the areas of interest, and a model describing the directed connections between these areas. Often, the brain regions are represented as nodes in a directed graph, with an edge from region 1 to 2 representing a hypothesized influence of region 1 on 2, and the absence of an edge from node 1 to 2 representing the hypothesis of no influence from 1 to 2; see Fig. 16.1 for an example. Typically the nodes and edges are assumed *a priori* known, and the goal of the analysis is to determine the strength of the edges.

The data used in effective connectivity analyses often consist of the average signal across time within predetermined regions of interest (ROI); these can be specified using either structural or functional features. Structural ROIs can be specified using automated anatomical labeling of subject-specific data. Here, ROIs for each subject are specified based on their anatomy. Alternatively, one can use single-subject anatomical atlases, such as the AAL atlas (60) or Talairach atlas (59), or multi-subject probabilistic atlases, such as the LONI Probabilistic Brain Atlas (52). However, since normalization only partially accounts for inter-subject variability, considerable residual variability in the shape and location of regions defined based on anatomical markers will remain. Therefore, defining ROIs for each subject based on their specific anatomy is preferable, if possible. Alternatively, functional ROIs can be defined by performing a separate localizer task. This allows one to identify functionally distinct nodes in the brain that can be used for subsequent connectivity analysis. However, it is important that the data used to localize the regions be performed on a different dataset than that being analyzed in the later connectivity analysis. Finally, the results of previous studies on a similar topic can be useful in defining regions of interest. Meta-analysis of functional imaging studies is growing in interest (see Chapter 61) and may ultimately prove useful in this endeavor. The hope is that the results can be used to generate ROIs that are less sensitive to noise than those based on single-subject activations.

Typically, effective connectivity is assessed using data from a single subject. However, as when imaging is used to map the brain regions activated in response to an experimental stimulus, researchers are typically more interested in the ability to perform population-level inference. One solution to this problem is to simply aggregate the time courses across subjects and perform the analysis in the same manner as the single-subject case. However, this type of averaging across subjects can give rise to results that are difficult to interpret and not particularly meaningful. Hierarchical models could be used as an alternative to pooling together the single-subject results as above, though they have not found wide usage in the field to date. Another commonly used option is to fit separate models for each subject and thereafter enter the subject-specific parameter estimates of interest into a second-level analysis (e.g. a t-test) (31). It should be noted that this approach does not take into consideration subject-specific differences in variation when performing the test, and thus can give misleading results.

The brain activity about which inferences are drawn is another important issue. While researchers are typically interested in making inferences about the underlying neuronal activation, most imaging procedures provide indirect measures of these activations. Often researchers analyze these indirect (observed) measurements; whether or not this is reasonable may depend upon the methodology and imaging modality used in a particular investigation. For example, vector autoregression models associated with Granger causality are used to determine the temporal precedence between two different time courses. When the outcome is the blood-oxygenation-level-dependent (BOLD) response measured using functional magnetic resonance imaging (fMRI), these time courses measure the neuronal activity after being filtered through the hemodynamic response function (HRF). However, as the HRF is known to vary across different brain regions (see Chapter 6), it is impossible to determine whether the temporal precedence observed in one time course compared to another is due to neuronal activation or inter-regional HRF differences. Recently, researchers have attempted to circumvent this issue by first deconvolving the BOLD response and thereafter examining Granger causality using the resulting time courses.

While researchers have also called attention to other possible problems that might render causal inferences from effective connectivity models invalid, for example, spatio-temporal limitations in imaging modalities, they implicitly assume these inferences would be valid in the absence of these problems. However, as we shall see subsequently, whether or not such inferences are warranted depends on additional considerations such as how the effects are defined and whether or not these are identified.

16.3   Models of Effective Connectivity

In this section we discuss several commonly used modeling procedures for estimating the relationships between different regions, including structural equation models (SEM) (38), dynamic causal models (DCM) (14), and Granger causality mapping (48).

16.3.1   Structural Equation Models

Linear structural equation models (29, 4), also called mean and covariance structure models, model the relationship between outcomes η, some of which may be unobserved, to other outcomes and independent variables ξ, some of which may also be unobserved, through "structural" equations. The unobserved outcomes and independent variables are related to observed outcomes and independent variables, respectively, through measurement equations. Here we consider the special case of a covariance structure model, where the means are not of interest. Focusing on this case is both adequate for our purposes and in keeping with previous literature on the use of structural equation models to study effective connectivity.

The vector of latent outcomes η is related to its components and latent inputs ξ through the structural equation

$$\eta = \mathbf{B}\eta + \mathbf{\Gamma}\xi + \zeta, \tag{16.1}$$

where $\mathbf{\Gamma}$ is a matrix relating the latent inputs to outputs, the diagonal elements of $\mathbf{B}$ are $\mathbf{0}$, and ζ is a vector of errors independent of ξ with mean $\mathbf{0}$. The latent vectors η and ξ are connected to mean centered observed vectors $\mathbf{Y}$ and $\mathbf{X}$, respectively, through the measurement equations

$$\mathbf{Y} = \mathbf{\Lambda}_y \boldsymbol{\eta} + \boldsymbol{\varepsilon}, \tag{16.2}$$

$$\mathbf{X} = \mathbf{\Lambda}_x \boldsymbol{\xi} + \boldsymbol{\delta}, \tag{16.3}$$

where $\mathbf{\Lambda}_y$ and $\mathbf{\Lambda}_x$ are matrices of factor loadings, the $\mathbf{0}$ mean vector $\boldsymbol{\varepsilon}$ is independent of $\boldsymbol{\eta}$, $\boldsymbol{\xi}$ and $\boldsymbol{\delta}$, and the $\mathbf{0}$ mean vector $\boldsymbol{\delta}$ is independent of $\boldsymbol{\eta}$, $\boldsymbol{\xi}$ and $\boldsymbol{\varepsilon}$. The elements of the matrices $\mathbf{B}$ and $\mathbf{\Gamma}$ are typically interpreted as effects.

With these assumptions, the covariance matrix $\mathbf{\Sigma}$ of $(\mathbf{Y}', \mathbf{X}')'$ can be expressed as a function of the parameters $\mathbf{B}$, $\mathbf{\Gamma}$, $\mathbf{\Lambda}_y$, $\mathbf{\Lambda}_x$ and the covariance matrices $\mathbf{\Sigma}_{\zeta\zeta}$ of $\boldsymbol{\zeta}$, $\mathbf{\Sigma}_{\varepsilon\varepsilon}$ of $\boldsymbol{\varepsilon}$, and $\mathbf{\Sigma}_{\delta\delta}$ of $\boldsymbol{\delta}$. Assuming independent and identically distributed observations $(\mathbf{Y_i}', \mathbf{X}_i')'$ from the distribution of $(\mathbf{Y}', \mathbf{X}')'$, $i = 1, ...n$, the parameters of an identified model may be estimated by minimizing or maximizing an objective function $F(\mathbf{\Sigma}, \mathbf{S})$ with respect to the model parameters, where $\mathbf{S}$ is the sample covariance matrix. To identify the model, it is necessary to place constraints on the elements of the covariance and parameter matrices. These usually take the form of exclusion restrictions, setting parameters to take the value 0, or equality constraints, setting different parameters to take equal values.

To test whether or not a given overidentified model fits the data, a likelihood ratio test can be used. The resulting test statistic follows a χ^2 distribution in large samples, with degrees of freedom equal to the number of non-redundant elements in $\mathbf{\Sigma}$ minus the number of non-redundant parameters estimated. Such tests can also be used to compare two or more models that are parametrically nested; thus one can test the null hypothesis that one or more of the $\beta_{kk'}$ are 0 by comparing the model in which these parameters are set to 0 with the model in which they are not. However, non-nested models cannot be compared using such tests. In order to compare such models, alternative criteria, e.g., the Bayesian Information Criterion (BIC) or the Aikake information criterion (AIC), can be used.

The use of linear structural equation models (LSEMs) in neuroscience to describe how neural activity in one brain region affects activity in other regions dates back to the early 1990s and its initial use in positron emission tomography (PET) imaging (40, 38, 41). Subsequently, many papers using structural equation modeling in conjunction with fMRI data have been published (e.g., see (5) for an early example). The technique has found less usage in magneto-encephalography (MEG) and electroencephalography (EEG), though examples can be found (1).

In a typical application using fMRI data, K brain regions are chosen *a priori* and some function of the BOLD responses Y_{ivkt} for a given subject i at voxel v in region k in experimental period $t = 1, ..., T$ is chosen to represent the response Y_{ikt} in region k. For example, one might choose the average value of the response for the voxels in that region, $Y_{ikt} \equiv N_k^{-1} \sum_{v \in k} Y_{ivkt}$, where N_k is the number of voxels in region k.

Let $\mathbf{Y}_{it} = (Y_{i1t}, ..., Y_{iKt})'$. In single-subject analyses under a given condition (for example, resting state), the time series of values $\mathbf{Y}_{i1}, ..., \mathbf{Y}_{iT}$ are treated (inappropriately) as independent and identically distributed copies from the distribution of a random variable $\mathbf{Y}_i$ and used to estimate the covariance matrix $\mathbf{\Sigma}_{y_i y_i}$ of $\mathbf{Y}_i$. It is assumed that $\mathbf{\Lambda}_{y_i}$ is the $K \times K$ identity matrix, $\boldsymbol{\varepsilon}_i = \mathbf{0}$. There are no inputs $\boldsymbol{\xi}_i$, so $\mathbf{\Gamma}_i = \mathbf{0}$. Hence (16.1) can be re-expressed for subject i as

$$\mathbf{Y}_i = \mathbf{B}_i \mathbf{Y}_i + \boldsymbol{\zeta}_i. \tag{16.4}$$

Here $\mathbf{B}_I$ expresses the effective connectivity between regions. Rewriting $\mathbf{Y}_i$ as $(I - \mathbf{B}_i)^{-1} \boldsymbol{\zeta}_i$ yields

$$\mathbf{\Sigma}_{y_i y_i} = (I - \mathbf{B}_i)^{-1} \mathbf{\Sigma}_{\zeta_i \zeta_i} (I - \mathbf{B}_i')^{-1}. \tag{16.5}$$

Equation (16.5) is a system of $\frac{K(K+1)}{2}$ equations in the $\frac{K(K+1)}{2} + K(K-1)$ parameters in $\mathbf{\Sigma}_{\zeta_i \zeta_i}$ and $\mathbf{B}_i$. To identify the model, restrictions must be imposed on the parameters.

$$\mathbf{Y} = \mathbf{BY} + \zeta$$

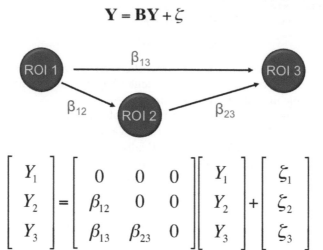

$$\begin{bmatrix} Y_1 \\ Y_2 \\ Y_3 \end{bmatrix} = \begin{bmatrix} 0 & 0 & 0 \\ \beta_{12} & 0 & 0 \\ \beta_{13} & \beta_{23} & 0 \end{bmatrix} \begin{bmatrix} Y_1 \\ Y_2 \\ Y_3 \end{bmatrix} + \begin{bmatrix} \zeta_1 \\ \zeta_2 \\ \zeta_3 \end{bmatrix}$$

FIGURE 16.1
An example of a simple recursive three-variable SEM. The edges between the three nodes are captured in the matrix **B**.

To express the idea that activity in region k' does not cause activity in region k for this subject, $\beta_{ikk'}$ in $\mathbf{B}_i$ is set to 0. Such restrictions are clearly warranted when two regions k and k' are not anatomically connected; in this case $\beta_{ikk'} = \beta_{ik'k} = 0$. When two regions are anatomically connected, substantive considerations in the context under investigation may (or many not) suggest that activity in region k does not affect activity in region k' and/or activity in region k' does not affect activity in region k. If the regions can be ordered so that $\mathbf{B}_i$ is lower triangular, and also $\mathbf{\Sigma}_{\zeta_i\zeta_i}$ is diagonal, the model is said to be recursive; see Fig. 16.1 for an example. Recursive models are identified and express the idea of unidirectional causal flow: if activity in region k' affects activity in region k, activity in region k does not affect activity in region k' (38, 5). It should be noted that models in which $\beta_{ikk'} \neq 0$ and $\beta_{ik'k} \neq 0$ have also been considered in the neuroimaging literature (45).

As the components of $\mathbf{Y}_{it}$ are measured simultaneously and activity in region k, even if it causes activity in region k', does not result in an immediate change in the value $Y_{ik't}$, Equation (16.1) does not admit a straightforward causal interpretation even in a recursive model (unless the measurements are actually temporally ordered). Worse, some authors have interpreted models where $\beta_{ikk'} \neq 0$ and $\beta_{ik'k} \neq 0$ as evidencing bi-directional causality, which is fundamentally at odds with the inherently asymmetric notion of causation.

Thus, in place of (16.1), (55) considers the discrete time dynamical linear system

$$\mathbf{Y}_{it} = \mathbf{B}_i \mathbf{Y}_{i,t-1} + \zeta_i, \tag{16.6}$$

which expresses the idea that it is activity in specific regions at time $t-1$ that generates activity at time t. Repeated iteration of (16.6) then gives:

$$\mathbf{Y}_{i,t+L} = \mathbf{B}_i^{L+1} \mathbf{Y}_{i,t-1} + \sum_{\ell=0}^{L} \mathbf{B_i}^j \zeta_i, \tag{16.7}$$

from which it is evident that if $\mathbf{B}_i^L$ converges to $\mathbf{0}$ as L increases without bound, then $\mathbf{Y}_{i,t+L}$ has limit $\tilde{\mathbf{Y}}_i = \lim_{L\to\infty} \mathbf{Y}_{i,t+L} = (\mathbf{I} - \mathbf{B})^{-1} \zeta_i$. Then, the system (16.7) has the unique equilibrium value $\tilde{\mathbf{Y}}_i$, and if the system is in equilibrium, Equation (16.7) holds with $\tilde{\mathbf{Y}}_i$ substituted into both sides of the equation. Thus, if the system under observation is in equilibrium, (16.1) applies. For recursive models, $\mathbf{B}_i^K = \mathbf{0}$ and the system reaches equilibrium after $K - 1$ periods. While the interpretation of (16.1) as the limit of (16.7) clarifies the meaning of $\mathbf{B}_i$ and resolves the ambiguities associated with time noted above, it is also important to note that the error $\boldsymbol{\zeta}_i$ in (16.7) is constant. While this would be reasonable were $\boldsymbol{\zeta}_i$ an input to the model that the researcher could hold constant over time, it is not reasonable to believe a stochastic error that is not under the researcher's control will take the same value over time.

LSEMs have also seen much application at the group level, most often after further aggregation of the data by averaging over voxels and stacking the time courses obtained from each subject (51, 28), so that a single model can be fit. Not surprisingly, aggregation in this manner can lead to findings that are at odds with findings from single-subject analyses (46). As an alternative, in many group-level studies, researchers perform t-tests across the individual subject-specific parameter estimates of each edge (32) under the null hypothesis that the parameters are equal across subjects; however, problems arise if the variances differ across subjects, and thus we do not recommend this approach. More preferable strategies include the construction of a multi-level SEM (53), or the embedding of the single subject models into a larger SEM, as in (42) and described below.

The strategy of embedding SEMs into a larger SEM can also be used to study contrasts and similarities in connectivity between baseline and experimental states that are temporally separated. Typically, this is accomplished by allowing the covariance matrix to depend on states, e.g., using the data obtained from the subject in the resting condition 1 to estimate a sample covariance matrix $\mathbf{S}_{i1}$ of $\boldsymbol{\Sigma}_{i1}$ and the data from the experimental state 2 to estimate a sample covariance matrix $\mathbf{S}_{i2}$ of $\boldsymbol{\Sigma}_{i2}$. In a full model, the model parameters vary by state. Using χ^2 tests, as previously described, the full model can then be compared with restricted models in which selected parameters of interest are set to 0 and/or constrained to take the same value in different states.

It seems evident that the SEMs above cannot handle dynamic relationships flexibly, severely limiting their value for studying effective connectivity. Further, treating the time series data $\mathbf{Y}_{it}$ as independent draws from a common distribution will lead to standard error estimates that are inconsistent. Nor do the models as used incorporate the association of the BOLD response with experimental inputs. While this is not necessarily problematic with resting-state data or comparisons of baseline and experimental states when these are "well separated" and "transitional observations" occurring after a state change are discarded, this approach would require discarding the majority of the data in studies using event-related designs. Nor can the models above be used to study changes in connectivity that result immediately and shortly after a change in state.

The foregoing concerns have motivated methodological researchers to develop truly dynamic SEMs (or related state-space models—see (7)) to study brain dynamics. Reference (32) propose a "unified SEM" model that combines the features of the conventional SEM above with a vector autoregression:

$$\mathbf{Y}_{it} = \mathbf{B}_i \mathbf{Y}_{i,t} + \sum_{\ell=1}^{L} \boldsymbol{\Phi}_{i\ell} \mathbf{Y}_{i,t-\ell} + \boldsymbol{\zeta}_{it}, \tag{16.8}$$

where $\mathbf{Y}_{i,t}$ is the BOLD response of subject i in period t, $\mathbf{B}_i$ has diagonal elements 0, $\boldsymbol{\Phi}_{i,t-\ell}$ is a parameter matrix for the lagged BOLD responses $\mathbf{Y}_{i,t-\ell}$, and $\boldsymbol{\zeta}_{it}$ is a $\mathbf{0}$ mean white noise vector with covariance matrix $\Sigma_{\zeta_i \zeta_i}$. Let $\boldsymbol{\Lambda}$ denote the covariance matrix of $(\mathbf{Y}_{i,t-1}, ..., \mathbf{Y}_{i,t-L})$ and $\boldsymbol{\theta}$ the vector of (free) parameters to be estimated. Treating $\boldsymbol{\Lambda}$ as unrestricted, the model is estimated using the implied covariance structure $\Sigma(\boldsymbol{\theta})$ of $(\mathbf{Y}_{it}, \mathbf{Y}_{i,t-1}, ..., \mathbf{Y}_{i,t-L})$ and optimizing a discrepancy function between the sample covariance matrix and $\Sigma(\boldsymbol{\theta})$ with respect to the parameters $\boldsymbol{\theta}$. To handle multiple subjects, in a second stage, (32) regresses estimated model parameters on covariates such as age and gender.

As before, the approach in (32) does not allow consideration of the association of the BOLD response with experimental inputs. Reference (19) recently proposed an "extended unified SEM" (euSEM) that generalizes the unified SEM and takes into account this association. The model bears some resemblance to the dynamic causal model (discussed in the next section) due to Friston and collaborators (14), but whereas the DCM models underlying neural activity, reference (19) models the BOLD response. For ease of exposition, we consider the following case of the model:

$$\mathbf{Y}_{it} = \mathbf{B}\mathbf{Y}_{it} + \sum_{\ell}^{L} \boldsymbol{\Phi}_{i\ell}\mathbf{Y}_{i,t-\ell} + \boldsymbol{\Gamma}_0 f_t(\bar{z}_t) + \boldsymbol{\Gamma}_1 f_{t-1}(\bar{z}_{t-1}) + \boldsymbol{\Delta}\mathbf{Y}_{i,t-1}f_{t-1}(\bar{z}_{t-1}) + \boldsymbol{\zeta}_{it}, \quad (16.9)$$

where $\boldsymbol{\Gamma}_0$ is a $K \times 1$ parameter vector to be estimated and $f_t(\bar{z}_t)$ is the convolution of the hemodynamic response function with the experimental sub-regimen $\bar{z}_t \equiv (z_1, ..., z_t)$, each component of which has value 1 when the stimulus is "on," 0 when "off"; $\boldsymbol{\Gamma}_1$ and $f_{t-1}(\bar{z}_{t-1})$ are defined similarly. The matrix $\boldsymbol{\Delta}$ is the parameter matrix due to the interaction of the lag one BOLD response vector with the convolved hemodynamic response function. The matrices $\mathbf{B}$ and $\boldsymbol{\Phi}_{\ell}$, $\ell = 1, ..., L$ describe the structure of connectivity among the K regions in the resting-state data. The terms $\boldsymbol{\Gamma}_0$ and $\boldsymbol{\Gamma}_1$ are interpreted as the effects of the experimental inputs (effective connectivity) and $\boldsymbol{\Delta}$ the modulating effects of the experimental inputs on effective connectivity. Note that interpreting $\boldsymbol{\Gamma}_1$ as the effect of the lagged stimulus $f_{t-1}(\bar{z}_{t-1})$ ignores the interaction between this and the lagged response $\mathbf{Y}_{i,t-1}f_{t-1}$ in (16.9).

Reference (19) emphasizes that while analyses of effective connectivity using structural equation models are apparently confirmatory, given the current preliminary stage of knowledge, exploratory approaches to discover which connections are present/absent are needed. The euSEM can be used in conjunction with either approach; in exploratory approaches, the Lagrange multiplier test is used to sequentially add model terms, similar in spirit to stepwise regression. Reference (43) discusses exploratory approaches based on Bayesian networks with both Gaussian and non-Gaussian data; one of the limitations of these approaches is the assumption that the $\mathbf{Y}_{it}$ are independent over time, although this assumption can be modified to allow for a moderate amount of autocorrelation, as might exist when measurements are taken infrequently. It seems that exploratory approaches to studying effective connectivity using regularization (for example L1 or L2 regularization) have not been considered to date.

A different line of generalization, which also incorporates experimental inputs, is the linear state-space model proposed by (25):

$$\mathbf{Y}_{it} = \boldsymbol{\alpha}_{iy} + \boldsymbol{\Lambda}_{iyt}\boldsymbol{\eta}_{it} + \boldsymbol{\varepsilon}_{iyt}, \quad (16.10)$$

$$\boldsymbol{\eta}_{it} = \boldsymbol{\alpha}_{i\eta} + \mathbf{B}_{i,t-1}\boldsymbol{\eta}_{i,t-1} + \boldsymbol{\zeta}_{i\eta t}, \quad (16.11)$$

where (16.10) and (16.11) are the observation and state equations, respectively. In (16.10), each component Y_{ikt} of the detrended (for scanner drift) BOLD response vector $\mathbf{Y}_{it}$ is the sum of the error ε_{ikt} and the "true response" $\alpha_{ik} + (\boldsymbol{\Lambda}_{iyt})_{kk}\eta_{ikt}$, where the kk element of the diagonal matrix $\boldsymbol{\Lambda}_{iyt}$, $(\Lambda_{iyt})_{kk} = f_{kt}(\bar{\mathbf{z}}_t)$, is the convolution of the (possibly region-specific) hemodynamic response function with the experimental sub-regimen $\bar{\mathbf{z}}_t \equiv (z_1, ..., z_t)$ previously defined and η_{ikt} is the amplitude in region k at time t, α_{ik} is a baseline "true response" and the vector of errors $\boldsymbol{\varepsilon}_{iyt}$ is assumed to follow a normal distribution with mean $\mathbf{0}$ and diagonal covariance matrix $\boldsymbol{\Sigma}_{\varepsilon_{it}\varepsilon_{it}}$.

An important contrast to previous work, as well as the subsequent papers by (32) and (19), all of which model the BOLD response only, is that (25) treats the BOLD response, in accordance with the interpretation that cerebral blood flow is a manifestation of the more fundamental neural activity of interest, as an indicator of activity. The relationship between these is modeled in the observation (or measurement) Equation (16.10). The state (or structural) Equation (16.11) describes the dependence of the unobserved amplitudes η_{ikt} on the period $t-1$ amplitude vector $\boldsymbol{\eta}_{i,t-1}$ through the time-varying connectivity matrix $\mathbf{B}_{i,t-1}$ and the error vector $\boldsymbol{\zeta}_{i\eta t}$, the latter assumed to follow a normal distribution with mean $\mathbf{0}$ and diagonal covariance matrix $\boldsymbol{\Sigma}_{\zeta_{it}\zeta_{it}}$.

The time-varying connectivity matrix, which governs the evolution of the state vector, has components $\beta_{i,k'k,t-1} = \sum_{k=1}^{K} \gamma_{k'k} f_{k,t-1}(\bar{z}_{t-1})$, allowing changes in latent amplitudes to depend on experimental stimuli through the resulting hemodynamic response. Thus, the model can be used with different experimental stimuli and with event-related designs. However, in this model, the fundamental connectivity parameters $\gamma_{k'k}$ are time invariant. This implies that connectivity varies over time only through the hemodynamic response to the experimental stimulus and does not qualitatively change in different states. If it is believed that the γ parameters should depend on the particular experimental stimuli, the model above cannot be used to study more than one experimental stimulus. However, it should be possible to extend the model to handle multiple stimuli. Note also that the implication that $\beta_{i,k'k,t-1} = 0$ if $f_{k,t-1}(\bar{z}_{t-1}) = 0$ for all k, as would be the case for resting state data or data collected during a resting period after the hemodynamic response to the stimulus had washed out. Thus, in this model, resting state constitutes a baseline from which the effects of experimental stimuli on latent amplitudes are measured.

16.3.2 Dynamic Causal Models

The measurements used in most effective connectivity analyses (e.g. SEM) are typically based upon observed time series data that are several layers removed from the underlying neuronal activity of interest, thereby limiting the scope for interpretation at this level. Dynamic Casual Modeling (DCM) (14) moves the analysis to the neuronal level by modeling the observed data as outputs of latent neuronal activity.

DCM uses a standard state-space design, and treats the brain as a deterministic non-linear dynamic system that receives inputs and produces outputs. It is based on a neuronal model of interacting cortical regions, supplemented with a forward model describing how neuronal activity is transformed into the observed response. Within the DCM framework, effective connectivity is parameterized in terms of the coupling among unobserved neuronal activity in different regions. This coupling is studied by applying experimental inputs and measuring the response in the regions under observation. Experimental inputs cause changes in effective connectivity at the neuronal level, which in turn cause changes in the observed data.

There are multiple variants of DCM, each tailored to the specific mechanisms that generated the observed data. However, they all follow a certain general form. The basic idea is based on the fact that both hemodynamic (fMRI) and electromagnetic (EEG/MEG) signals arise from a network of brain regions or neuronal populations. This network is described in DCM by a state-space model consisting of a set of two equations. The first equation describes how experimental manipulations $\mathbf{U}_t$ influence the dynamics of the latent neuronal states of the system $\mathbf{Z}_t$ at time t. This can be expressed using the following general equation $\dot{\mathbf{Z}}_t = f(\mathbf{Z}_t, \mathbf{U}_t, \theta)$, where $\dot{\mathbf{Z}}_t$ represents the rate of change of $\mathbf{Z}_t$, f describes its dynamics, and θ the unknown model parameters. The second equation maps the latent states to the observed data $\mathbf{Y}_t$ and can be expressed through the general equation $\mathbf{Y}_t = g(\mathbf{Z}_t, \phi)$. Here, g describes the mapping and ϕ the unknown model parameters.

DCM was originally introduced for application to fMRI data (14), and in this modality the method has found its widest usage. Since that time, extensions of the methodology have been made to include EEG and MEG data (10), where each variant takes the particular biophysical requirements of the specific modality into consideration. Here we describe in detail DCM for fMRI. Readers should, however, be aware that certain changes to the model are required before applying it to data from other modalities.

In the fMRI setting, a distinction is made between the neuronal level and the hemodynamic level. A bilinear model is used to represent the neuronal level and an extended Balloon model is used for the hemodynamic level. In a DCM there are J experimental inputs and K outputs, one output per each region included in the model. Each region has five state variables, four corresponding to the hemodynamic model and a fifth corresponding to neuronal activity. The goal of DCM is to estimate parameters at the neuronal level so that the modeled BOLD signals are as close as possible to the experimentally observed BOLD signals.

To illustrate, let us define the neuronal states as $\mathbf{Z}_{it} = (Z_{i1t}, \ldots Z_{iKt})^T$, where Z_{ikt} corresponds to the kth region under consideration at time t for subject i. As above, the effective connectivity model is given as $\dot{\mathbf{Z}}_{it} = f(\mathbf{Z}_{it}, \mathbf{U}_{it}, \theta)$, where f is a non-linear function describing the influences that $\mathbf{Z}_{it}$ and $\mathbf{U}_{it}$ exert upon changes in the neuronal states at time t. Using a bilinear Taylor approximation of f, the neuronal model is expressed in DCM as

$$\dot{\mathbf{Z}}_{it} \approx \left(\mathbf{A}_i + \sum_{j=1}^{J} U_{it}(j)\mathbf{B}_i^j\right)\mathbf{Z}_{it} + \mathbf{C}_i\mathbf{U}_{it}, \qquad (16.12)$$

where $U_{it}(j)$ is the jth of J observed inputs at time t. The matrices $\mathbf{A}_i$, $\mathbf{B}_i^j$ and $\mathbf{C}_i$ are subject specific and time invariant. $\mathbf{A}_i$ represents the first-order connectivity among regions in the absence of input. It specifies how regions are connected and whether these connections are uni- or bidirectional. $\mathbf{C}_i$ represents the extrinsic influence of inputs on neuronal activity. It specifies how inputs are connected to regions. Finally, the matrices $\mathbf{B}_i^j$ represent the change in coupling induced by the jth input. These specify how inputs change connections; see Fig. 16.2 for an example. Note that the neuronal model specified in Eq. (16.12) is deterministic.

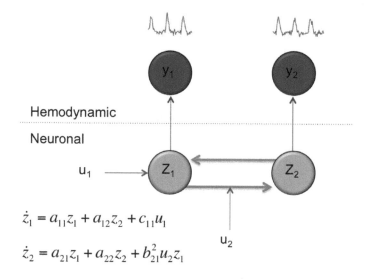

FIGURE 16.2
An example of a simple two-variable DCM.

A stochastic version of DCM exists as well (8), but has not found as much use in the field to date.

Neuronal activity causes changes in blood volume and deoxyhemoglobin that in turn cause changes in the observed BOLD response. The hemodynamics are described using an extended Balloon model (6, 15), which involves a set of hemodynamic state variables, state equations, and hemodynamic parameters ϕ. These state variables include the vasodilatory signal s_t, the inflow f_t, the blood volume v_t and the deoxygenation content q_t. The Balloon model consists of a sequence of differential equations that describes the coupling among the variables (s_t, f_t, v_t, q_t) based upon the hemodynamic parameters ϕ. Changes in the neuronal activity give rise to a vasodilatory signal, which in turn leads to increased blood flow and changes in blood volume and deoxygenation content.

Finally, the predicted BOLD response is a non-linear function of v_t and q_t. Combining the neuronal and hemodynamic states for each subject i as $\mathbf{X}_{it} = (\mathbf{Z}_{it}, s_{it}, f_{it}, v_{it}, q_{it})$ gives us the following state-space model:

$$\dot{\mathbf{X}}_{it} = f(\mathbf{X}_{it}, \mathbf{U}_{it}, \gamma) \tag{16.13}$$
$$\mathbf{Y}_{it} = g(\mathbf{X}_{it}, \gamma). \tag{16.14}$$

Here, all neurodynamic and hemodynamic parameters are contained in the vector $\gamma = (\theta\ \phi)^T$.

The estimation process is then carried out using Bayesian methods, where empirical priors are placed on ϕ and shrinkage priors are placed on the coupling parameters contained in θ. An optimization scheme is used to estimate parameters that maximize the posterior probability. The posterior density is then used to make inferences about the significance of the connections between various brain regions. As the posterior is intractable to compute, it is assumed that the posterior distribution of the parameters is Gaussian, whose mean and covariance are given by the moments of the posterior distribution. This is used to test the probability that a given parameter exceeds a chosen threshold.

Bayesian model selection can be used to determine whether the data favors one model over another. Let $\bar{\mathbf{Y}}_i = (\mathbf{Y}_{i1}, \mathbf{Y}_{i2}, \dots \mathbf{Y}_{iT})$. The model evidence is defined as

$$p(\bar{\mathbf{Y}}_i|M_1) = \int p(\bar{\mathbf{Y}}_i|\theta, M_1)p(\theta|M_1)d\theta. \tag{16.15}$$

This is the distribution of the observed data under a particular model M_1, which is used to compute the Bayes factor, allowing us to compare models M_1 and M_2:

$$B_{12} = \frac{p(\bar{\mathbf{Y}}_i|M_1)}{p(\bar{\mathbf{Y}}_i|M_2)}. \tag{16.16}$$

If B_{12} is large, model M_1 is more likely than model M_2. This type of comparison is only valid if the data are identical for all models. This means that only models that contain the same regions can be compared. Model selection cannot be used to address whether or not to include a particular area in the model.

DCM is a biophysical model with latent dynamics and a forward model that links it to the observed data. As mentioned above, DCM can be applied to a number of different imaging modalities, though the exact form of the equations needs to be refined accordingly. In general, DCM is quite computationally demanding and can only handle a limited number of regions (roughly 8).

16.3.3 Granger Causality

Granger causality mapping uses multivariate time series responses to determine temporal precedence among activity in different regions or voxels of the brain. It is based upon the use of the concept of Granger causality (23, 24), which has a long history in the field of economics.

Granger defines Ω_t as the stochastic history of the world through time t. Let $\mathbf{Y}_t$ denote a vector of random variables measured at time t, and let $\{\mathbf{Y_s} : \mathbf{s} \leq \mathbf{t}\}$ denote the entire history of these variables; $\mathbf{X}_t$ and $\{\mathbf{X_s} : \mathbf{s} \leq \mathbf{t}\}$ are defined similarly. In this setting, $\{\mathbf{Y_s} : \mathbf{s} \leq \mathbf{t}\}$ is said to Granger cause $\mathbf{X}_{t+1}$ if the history of $\mathbf{Y}_t$ contains information about the current value of $\mathbf{X}_t$, which is $\Pr(\mathbf{X}_{t+1} \in A \mid \Omega_t) \neq \Pr(\mathbf{X}_{t+1} \in A \mid \Omega_t - \{\mathbf{Y_s} : \mathbf{s} \leq \mathbf{t}\})$ for some (measurable) set A.

Reference (23) also discusses the concepts of feedback and "instantaneous" causation. Feedback occurs when $\{\mathbf{Y_s} : \mathbf{s} \leq \mathbf{t}\}$ Granger causes $\mathbf{X}_{t+1}$ and $\{\mathbf{Y_s} : \mathbf{s} \leq \mathbf{t}\}$ Granger causes $\mathbf{Y}_{t+1}$, i.e., the history of each variable contains information about the current value of the other variable. Similarly, $\mathbf{Y}_t$ is said to Granger cause $\mathbf{X}_t$ instantaneously if $\Pr(\mathbf{X}_t \in A \mid \Omega_{t-1} \cup \mathbf{Y}_t) \neq Pr(\mathbf{X}_t \in A \mid \Omega_{t-1})$ for some (measurable) set A. From this definition, it follows that if $\mathbf{Y}_t$ Granger causes $\mathbf{X}_t$ instantaneously, $\mathbf{X}_t$ Granger causes $\mathbf{Y}_t$ instantaneously. Note that as causation is universally conceptualized as an asymmetrical relationship, interpreting instantaneous Granger causation as a measure of whether or not two series cause one another instantaneously does not make sense.

Regardless of the merits of these definitions for assessing whether or not associations are causal (which we consider later), operationalization requires the replacement of Ω_t with a smaller information set and the assessment of the probability relationships above (or their counterparts in expectation) through the use of a model. In a single-subject analysis in neuroscience, that information set typically consists of measurements of a subject's brain activity made at discrete times in different brain locations. Note that measurements from other systems, e.g., the cardiovascular system, are not part of this information set.

A typical approach is to take a "seed" for each subject i, $X_{it}, t = 1, ..., T$, which is a time series of measurements for a particular brain location, and to take $\mathbf{Y}_{it}, t = 1, ..., T$ as the multivariate time series of all other measurements. Although the seed is typically univariate, the following exposition holds also for the case where X_{it} is multivariate. Together $\{X_{is}, \mathbf{Y}_{is} : s \leq t\}$ constitute the information set through time t.

One then assesses the elements of $\mathbf{Y}_{it}$ that are sources (Granger causes X_{it}) and targets (Granger caused by X_{it}). Assuming second-order stationarity, the two vector autoregressive models:

$$\mathbf{Y}_{it} = \sum_{\ell=1}^{L} \Phi_{i\ell y} \mathbf{Y}_{i,t-\ell} + \delta_{ity}, \qquad (16.17)$$

$$\mathbf{Y}_{it} = \sum_{\ell=1}^{L} (\Phi_{i\ell yy} \mathbf{Y}_{i,t-\ell} + \Phi_{i\ell yx} X_{i,t-\ell}) + \delta_{ityy}, \qquad (16.18)$$

where δ_{ity} is a mean $\mathbf{0}$ white noise error with variance–covariance matrix Σ_{i1yy} and δ_{ityy} is a mean $\mathbf{0}$ white noise error with variance–covariance matrix Σ_{i2yy}, can be compared to assess whether X_i Granger causes $\mathbf{Y}_i$. If $\Phi_{i\ell yx} \neq 0$ for one or more values of ℓ, or equivalently, if the errors δ_{ityy} are significantly less variable than the errors δ_{ity}, this is evidence that X_i Granger causes $\mathbf{Y}_i$. Similarly, one can assess whether or not $\mathbf{Y}_i$ Granger causes X_i. In theory, L may be infinite, but in practice, L is finite.

Reference (20, 21) proposed to quantify the "influences" above by comparing generalized variances in the autoregressive models (16.17) and

$$X_{it} = \sum_{\ell=1}^{L} \Phi_{i\ell x} X_{i,t-\ell} + \delta_{itx}, \qquad (16.19)$$

where δ_{itx} is a white noise error with variance σ_{i1xx}, with the generalized variance in the autorregressive model:

$$\mathbf{Z}_{it} = \sum_{\ell=1}^{L} \Phi_{i\ell} \mathbf{Z}_{i,t-\ell} + \boldsymbol{\delta}_{itz},$$ (16.20)

where $\boldsymbol{\delta}_{its}$ is a white noise error with mean $\mathbf{0}$ and covariance matrix

$$\boldsymbol{\Sigma}_{izz} = \left(\begin{array}{cc} \sigma_{izxx} & \boldsymbol{\Sigma}_{izxy} \\ \boldsymbol{\Sigma}_{izyx} & \boldsymbol{\Sigma}_{izyy} \end{array} \right).$$

He decomposed the ratio $F_{i,x,y} \equiv \ln(\sigma_{i1xx} \cdot |\boldsymbol{\Sigma}_{i1yy}|/|\boldsymbol{\Sigma}_{iz})$ into three components: (1) the "influence" of X_i on $\mathbf{Y}_i$, $F_{i,x \to y} \equiv \ln(|\boldsymbol{\Sigma}_{i1yy}|/|\boldsymbol{\Sigma}_{izyy}|)$, (2) the "influence" of $\mathbf{Y}_i$ on X_i, $F_{i,y \to x} \equiv \ln(\sigma_{i1xx}/\sigma_{izxx})$, (3) the instantaneous mutual influence $F_{ixy} \equiv \ln(\sigma_{izxx} \cdot |\boldsymbol{\Sigma}_{izyy}|/|\boldsymbol{\Sigma}_{izz}|)$. He also extended this decomposition to the frequency domain and to the case of conditional Granger causality (see below). As above, $L = \infty$ in theory, but finite in implementations.

In the past few years, neuroscientists have both advocated for and criticized the use of Granger causality. Whereas SEMs and DCMs tend to be confirmatory in nature, requiring the specifications of which regions are linked (not linked) directly, Granger causality is exploratory (48) and therefore potentially useful especially in the earlier stages of scientific inquiry.

In early applications (for example, (2, 30)), Granger causality was applied to local field potential data and data collected using EEG. Granger causality has also been applied to data collected using MEG (22). While these types of data have high temporal resolution, making Granger causality a potentially useful procedure, reference (3) notes that preprocessing steps used with EEG and MEG data may induce spurious associations.

The subsequent use of Granger causality with fMRI data raises additional issues stemming from the lower temporal resolution of fMRI, variation in the hemodynamic response function between brain regions, and the use of the BOLD response as an indirect measurement of neuronal activity. On the one hand, (48) argues that the problems stemming from the lower temporal resolution can be overcome if shorter TRs are used and they offer several supporting simulations. In contrast, (9) argues that Granger causality mapping should not be applied to fMRI data because the hemodynamic response function varies across brain regions, thereby potentially creating spurious results. Reference (9) further examines Granger causality using deconvolved fMRI time series data, providing results that (49) argues actually demonstrate the usefulness of Granger causality. Reference (11) suggests that Granger causality can be used with fMRI data at typical TRs even when hemodynamic response variability across regions is present. This argument is further extended by (50) to the group level, though their findings have been questioned by (54).

More generally, neuroscientists have acknowledged and are working to ameliorate a number of current limitations with applications of Granger causality to brain imaging data. As noted above, in testing for Granger causality, second-order stationary linear AR models are used. Because the stationarity assumption implies that the marginal distributions in period t have the same mean vector and covariance matrix, the models should not be applied to time series consisting of data from two or more different states (e.g., baseline and experimental). Like the majority of SEM models previously considered, the models are most useful for resting-state data. To apply the models to non-stationary series, either the data must be partitioned into components that are stationary and the models can be applied to these components, as advocated in (13), or a non-stationary model that accounts for transitions between (possibly) stationary periods must be constructed. Note that the former strategy is similar to that often used in structural equation modeling when the time series is composed of data from multiple states. Second, the linearity assumption, while convenient, is also suspect. A number of attempts to incorporate nonlinear relations have been made. As an example, (12) used nonparametric methods based on Fourier and wavelet transforms to assess Granger causality in the frequency domain; the resulting measures can then be transferred to the time domain. Reference (37) exploits the "kernel trick" to estimate the linear AR model in a (possibly much higher dimensional) feature space defined by functions of the original data. For a general treatment of this strategy, see the monograph (47).

While the study of effective connectivity is concerned broadly with the structure of causal relations among brain regions, Granger causality only indicates whether activity in a given brain location or set of locations Granger causes (does not Granger cause) activity in another location(s) and/or is (is not) Granger caused by activity in other locations. It does not, however, attempt to make a statement about how this occurs.

For example, suppose a researcher wants to know if activity in location $Y_{i1} \in \mathbf{Y}_i$ affects activity in location X_i directly and/or indirectly by causing activity in other locations, e.g., $Y_{i2}, ..., Y_{iK}$, which in turn directly and/or indirectly cause activity in location X_i. Suppose he considers the model:

$$X_{it} = \sum_{\ell=1}^{L} \mathbf{\Phi}_{i\ell xx} X_{i,t-\ell} + \sum_{\ell=1}^{L} \mathbf{\Phi}_{i\ell xy} Y_{i,t-\ell} + \delta_{itxx}, \qquad (16.21)$$

where δ_{itxx} is a white noise error with variance σ_{ixx}. The usual test for Granger causality compares the variances σ_{i1xx} in (16.19) with σ_{ixx}. While this indicates whether or not the inclusion of $\mathbf{Y}_i$ significantly reduces the variance, it tells us little about the role of Y_{i1}. Equivalently, this is a test of the null hypothesis: $\mathbf{\Phi}_{i\ell xy} = \mathbf{0}$ for $\ell = 1, ..., L$ in (16.21), rejection of which does not indicate which components of $\mathbf{Y}_i$ might be "causing" X_i nor in which manner.

The use of conditional Granger causality in neuroscience is based on the idea that if $\mathbf{Y}_i$ Granger causes X_i, and Y_{i1} is a "direct" Granger cause, then net of the other components of $\mathbf{Y}_i$, the coefficients $\phi_{i\ell xy_1}$ corresponding to the first element in $\mathbf{Y}_{i,t-\ell}$, $\ell = 1, ..., L$, should not all be 0. However, it is not clear to us in what sense Y_{i1}) directly causes X_i (see the next section). The influence measures proposed by (20) can also be extended to cover conditional Granger causality (21).

Group-level analyses using Granger causality pose additional challenges. One approach is to combine the results from single-subject analyses qualitatively, e.g., estimate the average influence of X on $\mathbf{Y}$, $\hat{F}_{x \to y} = n^{-1} \sum_{i=1}^{n} \hat{F}_{i,x \to y}$, or the proportion of subjects in which X_i Granger causes $\mathbf{Y}_i$. While this approach does not suffer from aggregation bias, it is limited and tells us little about the variability across subjects. A better approach would be to construct a hierarchical model for the single-subjects time series; although straightforward in principle, to the best of our knowledge, such an approach has yet to be taken.

16.4   Effective Connectivity and Causation

While statistical methods are useful for studying relationships among variables, additional assumptions, of which researchers are often unaware, are needed to justify interpreting these relationships as indicative of causation, and if these assumptions are not met, the resulting causal inferences will generally be invalid. Here we use "potential outcomes" notation (44), which is the standard notation used by statisticians working in the area of causal inference, to explicate additional assumptions under which the parameters of a method used to study effective connectivity warrant the interpretation as effects. These assumptions are in addition to those made when these methods are used for descriptive or predictive purposes. Though widely used in the statistical literature on causal inference, this notation has only recently been introduced into the neuroscience literature (34, 35, 33, 36, 57).

We have described three approaches to the study of effective connectivity. While there are significant differences among these, in all three cases, a given subset of brain regions are chosen and a model of the directed connections between these regions is estimated, with the coefficients for the connections typically interpreted as the effect of activity in one region upon activity in another. While researchers have warned that limitations on the temporal and/or spatial resolution of various imaging modalities may sometimes render such causal inferences invalid, they implicitly assume these inferences would be valid in the absence of these limitations. However, as the effects are not even defined, we do not see how it is possible to ascertain whether the model parameters associated with these approaches actually warrant a causal interpretation in some sense and/or the conditions under which these might warrant such an interpretation.

To discuss these issues, note first that as causal inference is the act of making inferences about causal relationships, and as there are different notions of causation, whether or not an inference is valid depends on the notion of the causal relation under consideration (56). Although there has been some discussion of causation in the neuroscience literature, there is little discussion of the relationship between the methods used to make causal inferences and the concept of causation itself. Dovetailing with the modern statistical literature on causation (27) we shall treat causation as a singular or unit relation that supports a counterfactual conditional relation. As an example, consider the statement that "the treatment caused John to die." As opposed to a general statement of the form "treatment causes death," the causal statement here is singular, referring only to John. Further, for Mary, this statement might not be true. Second, we require the statement to mean not only that John took the treatment and died, but also that had John not taken the treatment he would not have died. The latter clause is counterfactual because John took the treatment.

To represent this notion formally in the simple case above, let $Z_i = 0$ if unit i does not receive treatment, 1 if unit i receives treatment. For each unit i, there are two potential outcomes. $Y_i(0)$ is the outcome value when i does not receive treatment, say $Y_i(0) = 1$ if i dies, 0 if i lives. Similarly, $Y_i(1)$ is the outcome value when i receives treatment, with $Y_i(1) = 1$ if i dies, 0 if i lives. The unit effect may then be defined as $Y_i(1) - Y_i(0)$. This cannot be observed as only one of the potential outcomes $Y_i = Z_i Y_i(1) + (1 - Z_i)Y_i(0)$ is observed, depending on the value of Z_i. Consequently, interest often focuses on estimation of the average treatment effect $E(Y(1) - Y(0))$ (or the average treatment effect $E(Y(1) - Y(0) \mid \mathbf{X} = \mathbf{x})$), where $\mathbf{X}$ is a vector of covariates), with the expectation taken over the distribution of the unit effects in a population of units $\mathcal{P}$ (or subpopulation of units $\mathcal{P}_\mathbf{x}$).

To estimate the average treatment effect, one might use the difference in sample means: $n_1^{-1}\sum_{i:Z_i=1} Y_i - n_0^{-1}\sum_{i:Z_i=0} Y_i$ where n_1 is the number of subjects assigned to treatment and n_0 the number assigned to the control condition. In general, however, $n_1^{-1}\sum_{i:Z_i=1} Y_i$ estimates $E(Y \mid Z = 1) = E(Y(1) \mid Z = 1) \neq E(Y(1))$. Similarly, $n_0^{-1}\sum_{i:Z_i=0} Y_i$ estimates $E(Y(0) \mid Z = 0)$. However, in a randomized experiment, $Y(0), Y(1), \mathbf{X} \| Z$, where $\|$ denotes statistical independence. Thus, as a result of the randomization, for $z = 0, 1$, $E(Y(z))$ equals the conditional expectation $E(Y \mid Z = z)$ and as this is identified, $E(Y(z))$ is identified; thus the average treatment effect is identified. Similarly, for $z = 0, 1$, $E(Y(z) \mid \mathbf{X} = \mathbf{x}) = E(Y \mid Z = z, \mathbf{X} = \mathbf{x})$ and as this is identified, $E(Y(z) \mid \mathbf{X} = \mathbf{x})$ is identified; thus average treatment effects in subpopulations $\mathbf{X} = \mathbf{x}$ are also identified.

The foregoing example illustrates the key ideas behind the approach to causal inference that derives from the statistical literature. First, define the cause(s), then use the notation that captures the idea that the causal relation supports a counterfactual conditional to define the potential outcomes. Second, use the potential outcomes to define the causal estimand (causal parameter) of interest. Third, ask whether the estimand is identified from the type of data that has been collected. Fourth, if the causal parameter is identified, construct an estimate of it from these data.

By way of contrast, in the approach to causal inference taken in the effective connectivity literature we have described, parameters describing the association between outcomes in different brain regions are estimated and called effects, i.e., steps 1–3 above are missing. For the simple example above, the explicit approach we have advocated is not strictly necessary. In a randomized experiment, intuition suggests that the difference between sample means estimates a treatment effect, whereas in observational studies intuition suggests that control group and treatment group cases may differ systematically in ways related to the outcome(s). But in more complicated cases, the failure to use potential outcomes and define explicitly the causal estimands of interest typically leads researchers to make causal claims that are neither clear nor supported by the data. That appears to be the case with effective connectivity.

We illustrate this briefly using a simple example with brain regions $k = 1, ..., K$, a single subject, and second-order temporal dependence. We assume the data are collected while the subject is in a resting or constant experimental state. Suppose for now that neural activity is measured directly, resulting in the observed vectors $\mathbf{Y}_{it} = (Y_{i1t}, Y_{i2t}, ..., Y_{iKt})'$, $t = 1, ..., T$. Further assume that the temporal resolution is fine enough so that within a given period, activity in any given region k cannot generate activity in a different region k'.

The researcher estimates the model (16.17) for various values of L and determines that $L = 2$ is "adequate." He interprets this to mean that values of the response variable $\mathbf{Y}$ three or more periods prior to period t have no effect on the response at time t and he interprets the coefficients $\phi_{ikk'\ell}$ of $\mathbf{\Phi}_{i\ell}$ for $\ell = 1, 2$ as effects, more specifically, as a "direct" effect for subject i of a one-unit change in region k' on region k, ℓ periods later.

The interpretation above is vague and the foregoing approach leaves many questions unanswered. What is (are) the estimand(s) of interest? What conditions must be satisfied in order to interpret the model coefficients as effects? And what are these effects precisely? And are these the effects that an investigator is/should be interested in estimating?

To address these issues, we first define the cause(s) and causal estimand(s). We begin by defining potential outcomes $\mathbf{Y}_{it}(\mathbf{y}_{i,t-1}, \mathbf{y}_{i,t-2}, ...\mathbf{y}_{i1}) = \mathbf{Y}_{it}(\bar{\mathbf{y}}_{i,t-1})$, where the notation $\bar{\mathbf{y}}_{i,t-j} \equiv (\mathbf{y}_{i,t-j}, ..., \mathbf{y}_{i1})$ for $j = 1, ..., t-1$ and $t = 2, ...T$. Note that the potential outcomes in period t are allowed to depend only on previous $\mathbf{y}_{i,t-j}$, $j > 0$, but are not otherwise restricted. We now compare subject i in period t with himself/herself under different "treatment regimens" $\bar{\mathbf{y}}_{t-1}$ and $\bar{\mathbf{y}}_{t-1}^*$, defining the unit effects as $\mathbf{Y}_{it}(\bar{\mathbf{y}}_{t-1}) - \mathbf{Y}_{it}(\bar{\mathbf{y}}_{t-1}^*)$. As above, the unit effects cannot be observed, as only one of the potential outcomes can be observed. For $t = 2, ..., T$, we define the causal estimands

$$E(\mathbf{Y}_{it}(\bar{\mathbf{y}}_{,t-1}) - \mathbf{Y}_{it}(\bar{\mathbf{y}}_{t-1}^*)), \tag{16.22}$$

the average effect of $\bar{\mathbf{y}}_{t-1}$ vs. $\bar{\mathbf{y}}_{t-1}^*$ for unit i. Note that here, as opposed to the case above where probabilities and expectations are taken over the distribution of subjects, probabilities and expectations are computed over replications within subject i.

Recall the researcher's interpretation of the model (16.17): the nonzero coefficients are "direct effects" and lagged responses with $j \geq 3$ do not affect the outcomes. To see that such an interpretation may not be warranted, we now examine the second of these interpretations. If this interpretation were warranted, then

$$E(\mathbf{Y}_{it}(\mathbf{y}_{i,t-1}, \mathbf{y}_{i.t-2}, \bar{\mathbf{y}}_{i,t-3})) = E(\mathbf{Y}_{it}(\mathbf{y}_{i,t-1}, \mathbf{y}_{i.t-2}, \bar{\mathbf{y}}_{i,t-3}^*))$$

$$\equiv E(\mathbf{Y}_{it}(\mathbf{y}_{i,t-1}, \mathbf{y}_{i.t-2})) \tag{16.23}$$

for all $(\bar{\mathbf{y}}_{i,t-3}, \bar{\mathbf{y}}_{i,t-3}^*)$. Now, in general,

$$E(\mathbf{Y}_{it} \mid \mathbf{Y}_{i,t-1} = \mathbf{y}_{t-1}, \mathbf{Y}_{i,t-2} = \mathbf{y}_{t-2}, \bar{\mathbf{Y}}_{i,t-3} = \bar{\mathbf{y}}_{t-3}) =$$

$$E(\mathbf{Y}_{it}(\mathbf{y}_{t-1}, \mathbf{y}_{t-2}, \bar{\mathbf{y}}_{t-3}) \mid \mathbf{Y}_{i,t-1} = \mathbf{y}_{t-1}, \mathbf{Y}_{i,t-2} = \mathbf{y}_{t-2}, \bar{\mathbf{Y}}_{i,t-3} = \bar{\mathbf{y}}_{t-3}). \tag{16.24}$$

Thus, (16.17) implies

$$
\begin{aligned}
E(\mathbf{Y}_{it}(\mathbf{y}_{t-1}, \mathbf{y}_{t-2}, \bar{\mathbf{y}}_{t-3}) \mid \mathbf{Y}_{i,t-1} = \mathbf{y}_{t-1}, \mathbf{Y}_{i,t-2} = \mathbf{y}_{t-2}, \bar{\mathbf{Y}}_{i,t-3} = \bar{\mathbf{y}}_{t-3}) = \\
E(\mathbf{Y}_{it}(\mathbf{y}_{t-1}, \mathbf{y}_{t-2}, \bar{\mathbf{y}}_{t-3}^*) \mid \mathbf{Y}_{i,t-1} = \mathbf{y}_{t-1}, \mathbf{Y}_{i,t-2} = \mathbf{y}_{t-2}, \bar{\mathbf{Y}}_{i,t-3} = \bar{\mathbf{y}}_{t-3}^*).
\end{aligned} \quad (16.25)
$$

But clearly (16.25) does not imply (16.23).

It is also easy to see that if $\mathbf{Y}_{it}(\mathbf{y}_{i,t-1}, \bar{\mathbf{y}}_{i,t-2}) = \mathbf{Y}_{it}(\mathbf{y}_{i,t-1}, \bar{\mathbf{y}}_{i,t-2}^*)$ for all $(\bar{\mathbf{y}}_{i,t-2}, \bar{\mathbf{y}}_{i,t-2}^*)$, that is, the period t response is not affected by neural activity in periods prior to $t - 1$, this does not imply that (16.17) holds with $L = 1$. That is, the researcher's model is compatible with the case where the period t response is affected only by neural activity in the immediately preceding period. It follows immediately in this case that the interpretation of the nonzero coefficients in $\boldsymbol{\Phi}_{i2}$ as effects is unwarranted.

The foregoing analysis is intended only to briefly make the point that the commonly employed approaches to the study of effective connectivity do not necessarily yield results that merit a causal interpretation. It is also possible to state conditions under which such models do admit a causal interpretation. That said, it is not clear to us that the effects we have considered necessarily correspond to those that should be of greatest interest, and we hope our approach will help researchers to consider more carefully what effects they wish to define and estimate. As a simple example, it might be important and/or interesting to ask what would happen in subsequent periods $j > 0$ if neural activity in region k took value y_{tk} vs. y_{Tk}^*. The potential outcomes are then $\mathbf{Y}_{i,t+j}(y_{tk})$ and $\mathbf{Y}_{i,t+j}(y_{tk}^*)$ and the estimand of interest is $\mathbf{E}\mathbf{Y}_{i,t+j}(y_{tk}) - \mathbf{Y}_{i,t+j}(y_{tk}^*)$. A researcher might also be interested in how activity in region k affects activity in a specific region k' two periods later through specific pathways of interest; effects of this nature can also be defined and conditions for their identification and estimation can be developed. In a future manuscript we shall further develop these points.

Second, and just as important as the mathematical issues addressed above, we have assumed thus far that neural activity is measured directly. When this is not the case, researchers sometimes treat the observed measurements as indicators of latent neuronal activity and model the relationship among the latent variables, as in the state-space model of Ho et al. (25) or dynamic causal modeling. But researchers also model the observed measurements and impart a causal interpretation to the relationship among these even when these do not directly measure the underlying activity of interest, as in many of the applications using structural equation modeling or Granger causality mapping where fMRI data, for example, are analyzed. We do not believe the resulting parameters are interpretable as effects: whereas manipulating neuronal activity in region k may cause the activity in region k' to increase, manipulating the BOLD response in region k will not necessarily lead to an increase in the BOLD response in region k'.

16.5 Conclusions

In recent years, neuroscientists have become increasingly interested in brain connectivity, that is, the manner in which brain regions interact and how these interactions depend on experimental conditions. Various types of connectivity have been studied, including anatomical connectivity, functional connectivity, and effective connectivity, and various procedures have been used to study these types of connectivity.

In this chapter we have focused on the concept of effective connectivity, that is, the directed influence of one brain region on others, and the methods most commonly used to study these influences. First, we considered several issues that need to be addressed when performing analyses of effective connectivity. We then reviewed the methods most frequently used to study effective connectivity, focusing on structural equation modeling, dynamic causal modeling, and Granger causality modeling. While researchers using these methods typically interpret their findings as indicative of causation, they do not generally state what they mean by this term, nor do they define the effects they purport to be estimating or state conditions under which their interpretations are valid. Therefore we introduce an approach to causation and causal inference from the statistical literature, in which effects and conditions for their identification are explicitly defined and given, respectively. Using a simple example of the type used in the literature on Granger causality, we show how the usual analyses can be misleading. We also show how our approach can be used to define many other potentially interesting effects that researchers do not seem to consider in the literature on effective connectivity. By making explicit the assumptions involved when various experimental designs, imaging modalities and modeling procedures are used to make causal inferences, statisticians have the potential to make important contributions to the further development of neuroscience.

Bibliography

[1] Laura Astolfi, Febo Cincotti, Claudio Babiloni, Filippo Carducci, Alessandra Basilisco, Paolo Maria Rossini, Serenella Salinari, Donatella Mattia, Sergio Cerutti, D Ben Dayan, et al. Estimation of the cortical connectivity by high-resolution EEG and structural equation modeling: simulations and application to finger tapping data. *Biomedical Engineering, IEEE Transactions on*, 52(5):757–768, 2005.

[2] Corrado Bernasconi and Peter KoÈnig. On the directionality of cortical interactions studied by structural analysis of electrophysiological recordings. *Biological Cybernetics*, 81(3):199–210, 1999.

[3] Steven L Bressler and Anil K Seth. Wiener–Granger causality: A well-established methodology. *Neuroimage*, 58(2):323–329, 2011.

[4] Michael W Browne and Gerhard Arminger. Specification and estimation of mean-and covariance-structure models. In *Handbook of Statistical Modeling for the Social and Behavioral Sciences*, pages 185–249. Springer, 1995.

[5] Christian Büchel and KJ Friston. Modulation of connectivity in visual pathways by attention: Cortical interactions evaluated with structural equation modelling and fMRI. *Cerebral Cortex*, 7(8):768–778, 1997.

[6] R B Buxton, E C Wong, and L R Frank. Dynamics of blood flow and oxygenation changes during brain activation: The balloon model. *Magnetic Resonance in Medicine*, 39:855–864, 1998.

[7] Sy-Miin Chow, Moon-ho R Ho, Ellen L Hamaker, and Conor V Dolan. Equivalence and differences between structural equation modeling and state-space modeling techniques. *Structural Equation Modeling*, 17(2):303–332, 2010.

[8] J Daunizeau, KJ Friston, and SJ Kiebel. Variational Bayesian identification and prediction of stochastic nonlinear dynamic causal models. *Physica D: Nonlinear Phenomena*, 238(21):2089–2118, 2009.

[9] Olivier David, Isabelle Guillemain, Sandrine Saillet, Sebastien Reyt, Colin Deransart, Christoph Segebarth, and Antoine Depaulis. Identifying neural drivers with functional MRI: an electrophysiological validation. *PLoS Biology*, 6(12):e315, 2008.

[10] Olivier David, Stefan J Kiebel, Lee M Harrison, Jérémie Mattout, James M Kilner, and Karl J Friston. Dynamic causal modeling of evoked responses in EEG and MEG. *NeuroImage*, 30(4):1255–1272, 2006.

[11] Gopikrishna Deshpande, Priya Santhanam, and Xiaoping Hu. Instantaneous and causal connectivity in resting state brain networks derived from functional MRI data. *Neuroimage*, 54(2):1043–1052, 2011.

[12] Mukeshwar Dhamala, Govindan Rangarajan, and Mingzhou Ding. Estimating Granger causality from Fourier and wavelet transforms of time series data. *Physical Review Letters*, 100(1):018701, 2008.

[13] Mingzhou Ding, Steven L Bressler, Weiming Yang, and Hualou Liang. Short-window spectral analysis of cortical event-related potentials by adaptive multivariate autoregressive modeling: Data preprocessing, model validation, and variability assessment. *Biological Cybernetics*, 83(1):35–45, 2000.

[14] K J Friston, L Harrison, and W Penny. Dynamic causal modelling. *NeuroImage*, 19:1273–1302, 2003.

[15] K J Friston, A Mechelli, R Turner, and C J Price. Nonlinear responses in fMRI: The balloon model, Volterra kernels, and other hemodynamics. *NeuroImage*, 12:466–477, 2000.

[16] Karl J Friston. Functional and effective connectivity: A review. *Brain Connectivity*, 1(1):13–36, 2011.

[17] KJ Friston. Functional and effective connectivity in neuroimaging: A synthesis. *Human Brain Mapping*, 2:56–78, 1994.

[18] KJ Friston, C Buechel, GR Fink, J Morris, E Rolls, and RJ Dolan. Psychophysiological and modulatory interactions in neuroimaging. *Neuroimage*, 6(3):218–229, 1997.

[19] Kathleen M Gates, Peter Molenaar, Frank G Hillary, and Semyon Slobounov. Extended unified SEM approach for modeling event-related fMRI data. *NeuroImage*, 54(2):1151–1158, 2011.

[20] John Geweke. Measurement of linear dependence and feedback between multiple time series. *Journal of the American Statistical Association*, 77(378):304–313, 1982.

[21] John F Geweke. Measures of conditional linear dependence and feedback between time series. *Journal of the American Statistical Association*, 79(388):907–915, 1984.

[22] David W Gow Jr, Jennifer A Segawa, Seppo P Ahlfors, and Fa-Hsuan Lin. Lexical influences on speech perception: A Granger causality analysis of MEG and EEG source estimates. *Neuroimage*, 43(3):614–623, 2008.

[23] Clive WJ Granger. Investigating causal relations by econometric models and cross-spectral methods. *Econometrica: Journal of the Econometric Society*, pages 424–438, 1969.

[24] Clive WJ Granger. Testing for causality: A personal viewpoint. *Journal of Economic Dynamics and Control*, 2:329–352, 1980.

[25] Moon-Ho Ringo Ho, Hernando Ombao, and Robert Shumway. A state-space approach to modelling brain dynamics. *Statistica Sinica*, 15:407–425, 2005.

[26] B Horwitz. The elusive concept of brain connectivity. *NeuroImage*, 19:466–470, 2003.

[27] Guido W Imbens and Donald B Rubin. *Causal Inference in Statistics, Social, and Biomedical Sciences: An Introduction.* Cambridge University Press, 2014.

[28] G Andrew James, Mary E Kelley, R Cameron Craddock, Paul E Holtzheimer, Boadie W Dunlop, Charles B Nemeroff, Helen S Mayberg, and Xiaoping P Hu. Exploratory structural equation modeling of resting-state fMRI: Applicability of group models to individual subjects. *Neuroimage*, 45(3):778–787, 2009.

[29] Karl G Jöreskog. Factor analysis by least squares and maximum likelihood methods. 1977.

[30] Maciej Kamiński, Mingzhou Ding, Wilson A Truccolo, and Steven L Bressler. Evaluating causal relations in neural systems: Granger causality, directed transfer function and statistical assessment of significance. *Biological Cybernetics*, 85(2):145–157, 2001.

[31] Christian Herbert Kasess, Klaas Enno Stephan, Andreas Weissenbacher, Lukas Pezawas, Ewald Moser, and Christian Windischberger. Multi-subject analyses with dynamic causal modeling. *Neuroimage*, 49(4):3065–3074, 2010.

[32] Jieun Kim, Wei Zhu, Linda Chang, Peter M Bentler, and Thomas Ernst. Unified structural equation modeling approach for the analysis of multisubject, multivariate functional MRI data. *Human Brain Mapping*, 28(2):85–93, 2007.

[33] MA Lindquist. Functional causal mediation analysis with an application to brain connectivity. *Journal of the American Statistical Association*, 107:1297–1309, 2012.

[34] MA Lindquist and M.E. Sobel. Graphical models, potential outcomes and causal inference: Comment on Ramsey, Spirtes and Glymour. *NeuroImage*, 57:334–336, 2011.

[35] MA Lindquist and ME Sobel. Cloak and DAG: A response to the comments on our comment. *NeuroImage*, 76:446–449., 2013.

[36] X Luo, DS Small, CR Li, and PR Rosenbaum. Inference with interference between units in an fMRI experiment of motor inhibition. *Journal of the American Statistical Association*, 107:530–541, 2012.

[37] Daniele Marinazzo, Wei Liao, Huafu Chen, and Sebastiano Stramaglia. Nonlinear connectivity by Granger causality. *Neuroimage*, 58(2):330–338, 2011.

[38] A McIntosh and F Gonzalez-Lima. Structural equation modeling and its application to network analysis in functional brain imaging. *Human Brain Mapping*, 2:2–22, 1994.

[39] Anthony Randal McIntosh. Towards a network theory of cognition. *Neural Networks*, 13(8):861–870, 2000.

[40] AR McIntosh and F Gonzalez-Lima. Structural modeling of functional neural pathways mapped with 2-deoxyglucose: Effects of acoustic startle habituation on the auditory system. *Brain Research*, 547(2):295–302, 1991.

[41] AR McIntosh, CL Grady, Leslie G Ungerleider, JV Haxby, SI Rapoport, and B Horwitz. Network analysis of cortical visual pathways mapped with PET. *The Journal of Neuroscience*, 14(2):655–666, 1994.

[42] Andrea Mechelli, Will D Penny, Cathy J Price, Darren R Gitelman, and Karl J Friston. Effective connectivity and intersubject variability: Using a multisubject network to test differences and commonalities. *Neuroimage*, 17(3):1459–1469, 2002.

[43] Jeanette A Mumford and Joseph D Ramsey. Bayesian networks for fMRI: A primer. *NeuroImage*, 86:573–582, 2014.

[44] J Neyman. On the application of probability theory to agricultural experiments. Essay on principles. Section 9. *Statistical Science*, 5:465–472, 1990.

[45] Will D Penny, Klaas E Stephan, Andrea Mechelli, and Karl J Friston. Modelling functional integration: A comparison of structural equation and dynamic causal models. *Neuroimage*, 23:S264–S274, 2004.

[46] Joseph D Ramsey, Stephen José Hanson, Catherine Hanson, Yaroslav O Halchenko, Russell A Poldrack, and Clark Glymour. Six problems for causal inference from fMRI. *Neuroimage*, 49(2):1545–1558, 2010.

[47] Carl Edward Rasmussen and CKI Williams. *Gaussian Processes for Machine Learning*, The MIT Press, Cambridge, MA, pages 715–719, 2006.

[48] A Roebroeck, E Formisano, and R Goebel. Mapping directed influence over the brain using granger causality and fMRI. *NeuroImage*, 25:230–242, 2005.

[49] Alard Roebroeck, Elia Formisano, and Rainer Goebel. The identification of interacting networks in the brain using fMRI: Model selection, causality and deconvolution. *Neuroimage*, 58(2):296–302, 2011.

[50] Marleen B Schippers, Remco Renken, and Christian Keysers. The effect of intra-and inter-subject variability of hemodynamic responses on group-level Granger causality analyses. *Neuroimage*, 57(1):22–36, 2011.

[51] Ralf Schlösser, Thomas Gesierich, Bettina Kaufmann, Goran Vucurevic, Stefan Hunsche, Joachim Gawehn, and Peter Stoeter. Altered effective connectivity during working memory performance in schizophrenia: A study with fMRI and structural equation modeling. *Neuroimage*, 19(3):751–763, 2003.

[52] David W Shattuck, Mubeena Mirza, Vitria Adisetiyo, Cornelius Hojatkashani, Georges Salamon, Katherine L Narr, Russell A Poldrack, Robert M Bilder, and Arthur W Toga. Construction of a 3d probabilistic atlas of human cortical structures. *Neuroimage*, 39(3):1064–1080, 2008.

[53] Anders Skrondal and Sophia Rabe-Hesketh. *Generalized Latent Variable Modeling: Multilevel, Longitudinal, and Structural Equation Models*. CRC Press, 2004.

[54] Stephen M Smith, Peter A Bandettini, Karla L Miller, TEJ Behrens, Karl J Friston, O David, T Liu, Mark William Woolrich, and Thomas E Nichols. The danger of systematic bias in group-level fMRI-lag-based causality estimation. *Neuroimage*, 59(2):1228–1229, 2012.

[55] Michael E Sobel. Effect analysis and causation in linear structural equation models. *Psychometrika*, 55(3):495–515, 1990.

[56] Michael E Sobel. Causal inference in the social and behavioral sciences. In *Handbook of Statistical Modeling for the Social and Behavioral Sciences*, pages 1–38. Springer, 1995.

[57] Michael E Sobel and Martin A Lindquist. Causal inference for fMRI time series data with systematic errors of measurement in a balanced on/off study of social evaluative threat. *Journal of the American Statistical Association*, (just-accepted):00–00, 2014.

[58] Olaf Sporns. *Networks of the Brain*. MIT Press, 2011.

[59] Jean Talairach and Pierre Tournoux. Co-planar stereotaxic atlas of the human brain. 3-dimensional proportional system: An approach to cerebral imaging, 1988.

[60] Nathalie Tzourio-Mazoyer, Brigitte Landeau, Dimitri Papathanassiou, Fabrice Crivello, Olivier Etard, Nicolas Delcroix, Bernard Mazoyer, and Marc Joliot. Automated anatomical labeling of activations in SPM using a macroscopic anatomical parcellation of the MNI MRI single-subject brain. *Neuroimage*, 15(1):273–289, 2002.

[61] Wager TD, Lindquist M, and Kaplan L (2007). Meta-analysis of functional neuroimaging data: current and future directions. *Social cognitive and affective neuroscience*, 2(2):150–158.

17

Network Analysis

Cedric E. Ginestet

Department of Biostatistics, King's College London

Mark Kramer

Department of Mathematics and Statistics, Boston University

Eric D. Kolaczyk

Department of Mathematics and Statistics, Boston University

CONTENTS

17.1	Introduction	441
17.2	Network Construction	443
	17.2.1 Notation	443
	17.2.2 Vertex Set	444
	17.2.3 Edge Set	446
	17.2.4 Thresholding Networks	446
17.3	Descriptive Measures of Network Topology	447
	17.3.1 Characteristic Path Length	447
	17.3.2 Clustering Coefficient	448
	17.3.3 Degree Distribution	449
17.4	Network Models	450
	17.4.1 Erdős–Rényi Random Graphs	451
	17.4.2 Small-World Networks	452
	17.4.3 Preferential Attachment	453
	17.4.4 Exponential Random Graph Models	453
	17.4.5 Stochastic Block Models	454
17.5	Estimation and Comparison of Networks	455
	17.5.1 Statistical Parametric Networks	456
	17.5.2 Density-Integrated Topology	457
	17.5.3 Comparison of Weighted Networks	458
17.6	Conclusion	460
	Bibliography	460

17.1 Introduction

The brain is first and foremost a network of interacting biological elements. This simple observation motivates the use of network analytical methods in neuroimaging. Specifically, brain activity is assumed to originate with the interactions of a very large number of individual neurons. Moreover, other patterns of interaction have also generated some interest in the literature, including the network formed by neuroglial cells. These microscale networks are naturally of critical importance to aid our understanding of motor and cognitive brain functions. Neuroimaging and neuroanatomy have shown that such microscale networks are integrated into larger-scale networks of cortical and subcortical regions. It is these macroscale networks that will be discussed in the chapter at hand.

Methodologically, the analysis of such networks falls under the umbrella of an emerging field of statistical research, loosely referred to as complex network analysis. This field has rapidly gathered scientific momentum in the last two decades. In particular, this impetus has generated renewed interest in a classical branch of mathematics called graph theory. Researchers from a large variety of academic disciplines have joined their efforts to build a new body of literature, which is sometimes described as "network science." This interest in networks is itself subsumed under a larger multidisciplinary research framework commonly referred to as "complexity theory" (51). A vast number of scientists have found this network perspective theoretically useful, as attested by the considerable number of citations garnered by some of the seminal papers in this area (68, 3).

The adoption of this network perspective has also been motivated by the automated collection of large datasets in biomedicine and engineering. Consequently, this novel approach has led to the development of an adapted methodological machinery. Thus, a large number of statisticians and mathematicians have proposed new tools for the type of methodological questions associated with network data. In particular, some abstract problems in graph theory have gained increased practical relevance. These include graph coloring and clique enumeration, for instance (72, 50, 49). In turn, the study of real-world networks has also yielded new theoretical questions, thereby stimulating the research of pure mathematicians, such as graph theorists (14, 13).

The study of social networks has a long history in the social sciences, which dates back to the early and mid-twentieth century (67). By contrast, the emergence of modern network science can be traced back to the seminal description of small-world networks (68), and scale-free networks (3), about fifteen years ago. Some of these modern concepts have then been adopted in neuroimaging both at a theoretical (59, 60) and at an experimental level (23). This has led to a large array of research projects aiming at classifying macroscale brain networks according to their topological properties (2, 1, 38). The application of network analysis to neuroscience is particularly stimulated by the recent launching of a large NIH-funded consortium named the *Human Connectome Project*. A goal of this project is to compile and combine different neuroimaging datasets, in order to better understand the networks underlying brain activity in humans.

Most of the discussion in this chapter directly pertains to functional MRI (fMRI) data. This emphasis reflects the wide popularity of network analysis for this type of data (15). However, most of our discussion will also apply to other neuroimaging modalities, such as structural MRI (sMRI) and diffusion tensor imaging (DTI). Moreover, macro- and mesoscale networks have also been constructed on the basis of electrophysiological measures of brain activity, which include data collected using electroencephalogram (EEG), magnetoencephalogram (MEG), and electrocorticogram (ECoG). While the choice and meaning of vertices and edges for these different datasets may vary, the overall statistical approach for comparing topology across several families of networks can be applied in many different contexts. We will especially emphasize the methodological differences between these neuroimaging modalities, when considering network construction.

This chapter does not attempt to offer an exhaustive presentation of the rapidly evolving set of statistical methods that are now available for the analysis of network data. The interested reader is therefore invited to consult a textbook-level treatment of this emerging statistical sub-discipline, for a more specialized overview of the field (41). The chapter is organized as follows. In Section 17.2, we first introduce the general toolbox for building networks from neuroimaging datasets. In particular, we emphasize the critical importance of the choice of vertex sets and the choice of edge sets, in such constructions. In Section 17.3, we present the main aspects of graph topology that have gained popularity in the neuroimaging literature. In Section 17.4, we discuss some of the general models that have been used for representing real-world networks. Finally, in Section 17.5, the topic of statistical inference is covered. In this last topic, we discuss how to evaluate the properties of a single network by constructing a relevant null or reference distribution, and describe how to compare the topology of several groups of networks.

17.2   Network Construction

In this section, we introduce the graph-theoretical notation that will be used throughout the rest of the chapter. We then consider the three main decisions that have to be made when constructing a network from neuroimaging data. This process firstly involves the selection of a vertex set, which represents a set of regions of interest (ROIs). Secondly, one needs to choose the similarity metric, which will provide information about the strength of association between different nodes in the network. This provides us with an edge set. In general, such networks are weighted, and it is then necessary to apply a threshold in order to obtain binary graphs. We address these three issues in turn.

17.2.1   Notation

Network science is built upon graph theory, a branch of pure mathematics. While a network is composed of *nodes* that are connected with *links* a graph is composed of *vertices*, which are connected with *edges*. Apart from these differences in terminology, all networks are usually treated as graphs. Thus, in this chapter, the terms *graph* and *network* will be used interchangeably. The use of these two terms will only be varied for stylistic convenience.

Formally, a graph, G, is composed of a set of vertices, denoted $V(G)$. In neuroimaging, each vertex or node in $V(G)$ may represent an ROI in the brain, which will correspond to portions of gray matter. Every pair of vertices in $V(G)$ may be linked by an edge, which is represented by an element in $E(G)$, the edge set. For example, if the vertices v and u in $V(G)$ are linked, then the pair (v, u) belongs to $E(G)$. For notational convenience, we often use the shorthand vu for the pair (v, u). Altogether, therefore, a graph is defined as a pair of two sets, such that $G = (V, E)$.

Two vertices linked by an edge are said to be *adjacent*. Similarly, two edges are called adjacent if they have one vertex in common. By contrast, the relationship between edges and vertices is referred to as incidence. That is, the edge vu is said to be *incident* to both v and u. The *order* of a graph is its number of vertices, and will be denoted by $N_v = |V(G)|$. This should be distinguished from the *size* of a graph, which is defined as its number of edges, which will be denoted by $N_e = |E(G)|$. An *undirected* graph, G, is one in which the direction of the edges is ignored. That is, if one can reach v from u, then one can reach u from v. In such cases, we include either $vu \in E(G)$, or $uv \in E(G)$, but not both. Therefore, the maximal number of edges in an undirected graph is $N_v(N_v - 1)/2$. When G possesses the maximal number of edges, we say that this graph is *saturated*, or complete. The *density* of a graph is its size, normalized by the size of the saturated version of a graph of the same order. That is, the density of G is defined as follows:

$$\text{dens}(G) = \frac{N_e}{N_v(N_v - 1)/2}.$$

By contrast, a graph is said to be *directed*, when the presence of the edge, vu, only implies that u can be reached from v. A directed graph is sometimes referred to as a digraph. The directed edges of a digraph are commonly called *arcs*, and generally represented by arrows. The starting vertex, v, of an arc, vu, is its tail, whereas the vertex, u, to which the arrow points is called its head.

In addition, networks may be either *weighted* or *unweighted*. Weighted undirected graphs will be denoted by the triple $G = (V, E, W)$, where $W(G)$ is a set of weights, whose elements are indexed by the entries in $E(G)$, such that $w(e) = w(v, u)$, for some edge $e = vu$ and vertices $v, u \in V(G)$. The weight set can be represented as a symmetric matrix, W, whose diagonal elements are null. In general, it is common to assume that $w(e) \geq 0$, for every edge. The *density* of a weighted graph is its average weight. We will use this concept extensively, when discussing the thresholding of weighted networks in neuroimaging. Often, we will not draw an explicit distinction between a weighted and an unweighted network through our notation. However, which one we are referring to should be clear from the context.

17.2.2 Vertex Set

The first key decision that needs to be made when constructing a network from a neuroimaging dataset is the selection of the nodes of the network. Three main approaches have been adopted in the literature, which consist of (i) constructing voxel-wise networks, (ii) applying a parcellation atlas to the dataset, and (iii) using pre-existing sets of ROIs that have been reported as significant in previous activation studies. We here provide the reader with an overview of these three approaches, and defer to (62) for a comprehensive treatment of this topic.

The first strategy that one may consider when constructing macroscale brain networks is to treat each voxel as an individual node. This has the direct benefit of objectivity, since this approach does not require any extraneous information in order to aggregate the voxels into larger regions. In particular, this approach has the benefit of providing a uniform spatial distribution of the vertices over the images under scrutiny. This advantage should be especially contrasted with the use of a parcellation scheme, in which the regions may have different sizes. The voxel-wise strategy for constructing a network vertex set has been adopted by several groups of researchers. However, there is currently no consensus on the choice of the number of voxels. While some authors have used 3,400 nodes (45), others have incorporated all the voxels present in standard MRI images, yielding up to 140,000 vertices per individual network (23). Naturally, the computational cost of handling these large graphs constitutes the main limitation of this voxel-wise approach to network construction. In particular, certain topological metrics described in section 17.3 may not be computationally tractable for such large networks.

The second approach to vertex selection addresses these limitations by using an anatomical atlas or parcellation scheme, in order to identify a set of relevant ROIs. This atlas-based approach generally consists of approximately one hundred different brain regions, which are then treated as the network nodes. The Automated Anatomical Labeling (AAL) template, for instance, is comprised of 116 non-overlapping cortical, subcortical and cerebellar regions (63). This particular anatomical atlas has proved to be very popular in the literature with more than three-quarters of the published studies applying network analysis to neuroimaging, being based on the AAL atlas (62). One of the advantages of this approach to

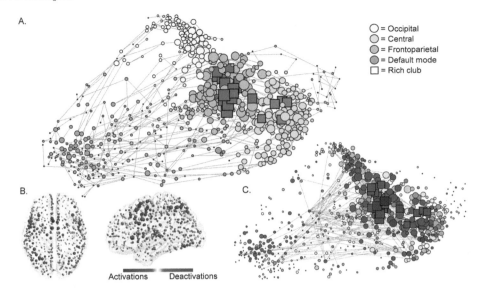

Activations   Deactivations

FIGURE 17.1

Meta-analytic functional coactivation network, based on data combining in excess of 1,600 neuroimaging studies published between 1985 and 2010. (A) A minimum spanning tree is used to locate nodes in relation to their topological proximity to each other. Different modules are coded by color, with the size of all nodes proportional to their weighted degree (strength). (B) Nodes in anatomical space, colored according to their number of activations and deactivations. (C) Nodes arranged in the same layout as A, and colored as in B. Note that the rich club concentrates most of the activations, whereas the periphery and particularly the default-mode network concentrates the deactivations. See (19), and Section 17.3.3 in this chapter, for a discussion of hubs.

network construction is that the resulting networks are directly comparable from one study to another.

Finally, a third strategy for vertex selection that has gained in popularity in the neuroimaging community is the use of existing activation studies for identifying sets of ROIs. In this approach, the entire cortical surface needs not be entirely covered by a set of anatomical regions. Rather, the rationale for adopting such an approach is that previous research has already shown that a small subset of regions are particularly salient to the phenomena of interest. In particular, this approach has paved the way for the construction of specific summary networks that combine the results of several previous studies in order to produce what may be called a *meta-analytical network* (19). An illustrative implementation of this approach has been provided in Figure 17.1.

Naturally, the choice of vertex set will be highly dependent on the type of neuroimaging modality that had been utilized for collecting the original dataset. The aforementioned three strategies are especially relevant for fMRI data, although they could equally be applied to DTI and sMRI. By contrast, for EEG, MEG, and ECoG data, it is common practice to utilize the sensors as network nodes. That is, the vertex set is associated with the standard space of sensors. Moreover, with electrophysiological data, one can also estimate the sources of the signal under scrutiny, and use these reconstituted sources as network vertices. Regardless of the modality, there is a marked tendency in the literature toward using several choices of vertex sets in order to evaluate the sensitivity of the results to a particular choice of parcellation scheme or set of ROIs (71).

17.2.3   Edge Set

Once a vertex set has been selected, one must decide how to quantify the strength of association between these different vertices. Which information should one attach to the presence or absence of an edge? There is a wide range of measures of association that have been proposed in the literature. In general, such a decision depends on the type of data at hand, and the spatial resolution of the selected vertex set.

For fMRI data, one can readily use the voxel-specific time series in order to compute a specific measure of association. If one has applied a parcellation scheme to the original data, then it is common practice to compare the mean regional time series of each pair of ROIs (35). Since the temporal resolution of fMRI is relatively poor, one often uses Pearson's correlation coefficient for evaluating linear associations between the mean regional time series of two distinct brain regions. By contrast, electrophysiological experiments based on EEG, MEG or ECoG data provide a much higher temporal resolution. For these datasets, one may hence choose to apply more sophisticated association metrics, including mutual information, coherence, Granger's causality, and synchronization likelihood (54, 47). When computing such measures of association, it is customary to analyze these associations for different frequency bands. In the case of fMRI data, several research groups have indeed reported that slow (< 0.1Hz) fluctuations are especially salient for the study of large-scale brain networks (2, 15, 35).

For sMRI, a somewhat different perspective is usually adopted. In such cases, one generally computes a correlation coefficient or another measure of association with respect to the entire sample of subjects. Several authors, for instance, have compared the average cortical thickness of different areas, by computing a correlation coefficient over a large sample of subjects (38, 17). Finally, for DTI, when possible, the number of tracts linking one region to another is computed, and utilized as a measure of association between two given cortical regions. This particular measure of inter-nodal association is sometimes referred to as *connection density* (36).

17.2.4   Thresholding Networks

In general, the above two-step procedure—the selection of a vertex set, followed by the use of a measure of association between the corresponding brain regions—will generate a weighted graph. However, most of the standard tools of graph theory have been developed for binary networks, in which each edge is either present or not. It is therefore often convenient to threshold the weighted graphs, in order to apply the standard graph-theoretical machinery. However, in doing so, we discard information about the original signal. Moreover, in most situations, the choice of a threshold is not unique, and such a decision may be difficult to justify.

Understandably, this topic has therefore generated a substantial amount of methodological discussion. A strategy that has gained currency in the neuroimaging literature is the use of a mass-univariate approach, in which a statistical test is performed for every possible edge in the network (38, 2, 35). One may, for instance, wish to test whether each individual correlation coefficient is significant. Naturally, since such tests are not independent, it is then necessary to correct for multiple comparison by using standard techniques such as the Bonferroni correction or the false discovery rate (FDR) (7, 53). In doing so, we are supplanting the choice of an arbitrary cutoff point by the use of a particular significance level, as illustrated in Figure 17.2. The question of the choice of threshold is also associated with other methodological issues, related to the comparison of networks across different experimental conditions. This topic will be covered in greater detail in Section 17.5.

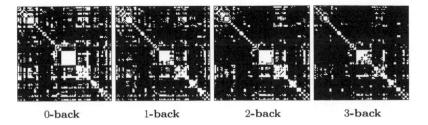

0-back   1-back   2-back   3-back

FIGURE 17.2

Thresholded correlation matrices for four levels of a working memory experimental manipulation, called the N-back task, over a vertex set based on the AAL template with 90 cortical and subcortical ROIs. Adjacency matrices become sparser with increasing working memory load. These adjacency matrices were obtained by constructing group mean networks, using a mass univariate approach based on z-tests with FDR correction (base rate $\alpha_0 = .05$). Zero entries are denoted in black and edges in white. (See (35) for details.)

17.3 Descriptive Measures of Network Topology

In this section, we provide an overview of the topological graph properties that have become popular in modern network science. Naturally, this represents a vast subject that encompasses a large array of graph theoretical concepts. Here, we solely cover three of the most common topological metrics that have been especially relevant for the study of neuroimaging data, namely, (i) the characteristic path length, (ii) the clustering coefficient, and (iii) the degree distribution. For a more extensive treatment of this topic from the perspective of network data analysis, the interested reader is referred to (41); for a comprehensive survey of the various topological metrics used in neuroimaging, one may consult (57).

17.3.1 Characteristic Path Length

One of the fundamental properties of a network is the *reachability* of a given vertex from another. A vertex is reachable from a starting vertex, when there exists a *path* between these two vertices. A path is here defined as a sequence of adjacent edges. A graph for which there exists a path between every pair of vertices is said to be *connected*. A disconnected graph can be decomposed into *connected components*. The connectedness of a graph is an inexpensive quantity to compute, as it can be checked in $O(N_v + N_e)$ time.

A closely related notion is the one of a *shortest path* between two vertices. In an undirected network, the length of a path is simply defined as the number of edges in the path. For a given pair of vertices, such a shortest path needs not be unique. However, this is rarely a source of concerns, as we are often mainly interested in the length of that path. When two vertices belong to two distinct connected components, the length of the shortest path linking them is set to infinity. Formally, for every pair $v, u \in V(G)$, let the length of the shortest path between these two nodes be denoted by $\rho(v, u)$. It is easy to see that this function is a proper metric, in the sense that it is non-negative, symmetric, and satisfies the triangle inequality, as well as the coincidence axiom (i.e. $\rho(v, u) = 0$, if and only if, $v = u$). Therefore, one may note that the space formed by the Cartesian product, $V(G) \times V(G)$, is a genuine metric space, when equipped with this shortest path metric.

For a weighted graph, $G = (V, E, W)$, the shortest path length can be similarly defined with respect to the weight set of G, such that the weighted shortest path between every two vertices, v and u, is the path that possesses the smallest total weight. The computation of shortest paths in a weighted graph is a well-studied problem in discrete mathematics, and can be solved efficiently using Dijkstra's algorithm. Given a single source vertex, one can compute the shortest paths to all the remaining vertices in the graph in $O(N_e + N_v \log N_v)$ time (21). A natural way to summarize the global speed of information transfer is to consider the average shortest path length, sometimes referred to as *characteristic path length*. This particular topological metric has been extensively used in the neuroimaging literature (64, 61).

The characteristic path length, however, suffers from one disadvantage. This quantity is infinite for disconnected graphs. Some authors have addressed this limitation by suggesting an equivalent formula for measuring the speed of information transfer. In this case, we compute the mean of the reciprocals of the pairwise shortest path lengths (44, 43). This alternative measure is termed the *global efficiency*, and is formally defined as follows,

$$\text{GloEff}(G) = \frac{1}{N_V(N_V - 1)} \sum_{v \in V} \sum_{\substack{u \in V \\ u \neq v}} \frac{1}{\rho(v, u)}, \tag{17.1}$$

where G can be either connected or disconnected. Furthermore, observe that this quantity also has the advantage of being normalized between 0 and 1. Global efficiency has proved to be a popular topological metric in the neuroimaging literature (57, 35, 34). Authors tend to favor global efficiency over the characteristic path length, since the former can be computed for disconnected graphs. This is a desirable property when varying the density of a family of graphs, as is commonly done when comparing network topology over several groups of subjects. We will return to this issue in Section 17.5.

17.3.2 Clustering Coefficient

A natural counterpoint to the characteristic path length is the clustering coefficient. Whereas the average shortest path quantifies global information transfer, the clustering coefficient is a local measure of the flow of information. For each individual vertex, this quantity is defined in terms of the number of triangles (i.e. cliques of three vertices of *complete* graphs, K_3) in the neighborhood of that vertex. The neighborhood of a vertex, $v \in V(G)$, is here denoted by $\partial_G(v)$ and is defined as the vertex subset of $V(G)$ formed by v and its first-order neighbors.

The number of triangles in the neighborhood of v is compared with the number of *connected triples* in $\partial_G(v)$. This quantity is defined as the vertex triples in $\partial_G(v)$ containing at least two edges. The clustering coefficient of a single vertex, $v \in V(G)$, is then defined as follows,

$$\text{cl}(v) = \frac{\tau_\Delta(v)}{\tau_3(v)},$$

where $\tau_\Delta(v)$ denotes the number of triangles in the neighborhood of v and $\tau_3(v)$ represents the number of connected triples in $\partial_G(v)$. Observe that this latter quantity is straightforwardly obtained as $\tau_3(v) = \binom{d_v}{2}$, where d_v is the degree of vertex v (see Section 17.3.3). Although the clustering coefficient is a measure of local connectivity, it is often important to summarize this local information into a global measure pertaining to the entire graph. In such cases, we compute the mean clustering coefficient over all the vertices in the graph. This gives

$$\text{cl}(G) = \frac{1}{|V'|} \sum_{v \in V'} \text{cl}(v), \tag{17.2}$$

where $V' \subseteq V(G)$ is the vertex subset of all vertices with at least two neighbors. The restriction to that particular vertex subset ensures that we are not dividing by zero, when computing the clustering coefficient of vertices with a single neighbor. Since its introduction by Watts and Strogatz (68), the clustering coefficient has attracted a considerable amount of interest. This trend has been reflected in the neuroimaging literature, in which both the clustering coefficient and the characteristic path length are routinely reported and analyzed (15, 57).

While the formulation of the clustering coefficient described above remains a popular choice in the literature, it should be noted that there exist at least two alternative methods. Firstly, certain graph theorists have suggested the use of the normalized version of the clustering coefficient, denoted by $\mathrm{cl}_T(G)$, which has the advantage of being comprised between 0 and 1 (12). In fact, this normalized topological measure predates the notion of clustering coefficient. This quantity is commonly referred to as *transitivity*, and is defined as follows,

$$\mathrm{cl}_T(G) = \frac{\sum_{v \in V'} \tau_3(v) \, \mathrm{cl}(v)}{\sum_{v \in V'} \tau_3(v)} = \frac{3\tau_\Delta(G)}{\tau_3(G)}.$$

where $\tau_\Delta(G) = (1/3) \sum_{v \in V'} \tau_\Delta(v)$, and $\tau_3(G) = \sum_{v \in V'} \tau_3(v)$ represent the total number of triangles in G, and the total number of connected triples in G, respectively. Transitivity can therefore be interpreted as the fraction of transitive connected triples.

Local connectivity can also be estimated using a quantity called *local efficiency* (44, 43). This measure is defined as the average of the global efficiencies of every vertex-specific neighborhood, such that

$$\mathrm{LocEff}(G) := \frac{1}{N_v} \sum_{v \in V} \mathrm{GloEff}(G_v),$$

with $\mathrm{GloEff}(\cdot)$ defined as in Equation (17.1), and where G_v denotes the induced subgraph of G formed by $\partial_G(v) - v$. That is, G_v is the subgraph of G, composed of all of the neighbors of v. Contrary to $\partial_G(v)$, observe that v does not belong to $V(G_v)$. As for its global counterpart, local efficiency has the advantage of being well-defined for every vertex in G. Note that this property is not shared by the clustering coefficient, since in Equation 17.2, we restricted our attention to vertices of degree two or larger. As for global efficiency, this is one reason that explains the popularity of this last measure in the neuroimaging literature (46, 1, 39).

17.3.3 Degree Distribution

The *degree*, $d(v)$, of a vertex $v \in V(G)$ is the number of vertices adjacent to v, or equivalently, the number of edges incident to v. Thus, the degree of a vertex provides some information about the level of connectedness of that node. For a vertex v in a directed graph, we usually distinguish between the *out-degree* (i.e. the number of arcs whose tail is v) and the *in-degree* (i.e. the number of arcs whose head is v). In such directed graphs, the distributions of the out- and in-degrees are often analyzed separately. In a weighted graph, the weighted degree of a node, v, denoted $d_W(v)$, is defined as the sum of the weights of the edges connecting v to other nodes, such that

$$d_W(v) = \sum_{u \in \partial_G(v)} w(v, u).$$

Observe that a weighted degree can, in general, take negative values. However, the weights of the networks commonly encountered in neuroimaging are often restricted to be non-negative (34), and in such settings the degrees are also non-negative.

Of particular interest is the *degree distribution* of a graph G. This is formally defined by treating $d(v)$ as a discrete random variable taking values in the non-negative integers, and considering the density of that random variable. That is, for every non-negative integer, δ, let

$$p_G(\delta) = \frac{1}{N_v} \sum_{v \in V} 1\{d(v) = \delta\},$$

where naturally, $\sum_{\delta \in \mathbb{N}} p_G(\delta) = 1$, and with $1\{f(v)\}$ denoting the indicator function, which takes a value of 1 if $f(v)$ is true, and 0 otherwise. In effect, $p_G(\delta)$ therefore represents the histogram of the set of vertex degrees. The in- and out-degree distributions can similarly be computed for directed graphs. One of the consistent findings of the last fifteen years of research in network science, has been that large real-world networks commonly exhibit heavy-tailed degree distributions, resembling *power laws* (3, 18). This has led to the characterization of a family of network models called *scale-free* networks. These ideas will be described in greater detail in Section 17.4. A power law degree distribution can be described for some positive number, $\beta \geq 0$, as follows,

$$p_G(\delta) \propto \frac{1}{\delta^\beta}.$$

Several research groups in neuroimaging have fitted the degree distribution of various macroscale brain networks, in order to estimate the value of that power law parameter (2, 4). This estimation is commonly done by estimating the slope coefficient of a regression model based on the log-log transformed degree distribution, using $\log p_G(\delta)$ as the response variable. This gives the following model

$$\log p_G(\delta) = C - \beta \log \delta + \varepsilon,$$

where C is the intercept and ε denotes the error term. However, some caution should be exerted when fitting a power law using a log-log regression, as these seemingly straightforward statistical analyses are increasingly treated with suspicion in the network science literature, and more sophisticated alternatives have been proposed (18).

Node degrees are also used for classifying vertices, thereby providing a summary of the network architecture. Specifically, a node with a high degree is referred to as a *hub*. Some network analytical strategy focuses on identifying a subset of nodes that is mainly composed of hubs. Such a vertex subset may be referred to as a "rich club" of nodes. This particular strategy has been used to good effect in neuroimaging, where the identification of a rich club provides a parsimonious representation of large-scale cortical networks (19), as illustrated in Figure 17.1, for the case of a meta-analytical study.

17.4   Network Models

The different network models that have been articulated in the network science literature can be loosely classified into three distinct categories:

1. *Mathematical models*, such as the Erdős–Rényi (ER) random graph model and the small-world model.

2. *Statistical models*, such as the exponential random graph models (ERGMs) and the stochastic block models (SBMs).

3. *Biophysical models*, which rely on systems of differential equations to model the patterns of interactions between individual neurons or brain regions.

In this section, we will solely cover the first two categories, by considering some of the main examples of mathematical and statistical models. The biophysical approach to network modeling is generally tackled using tools from the physics literature. The interested reader may consult the work of Friston et al. on this particular topic, for an overview (28).

In its most general sense, a network model will be defined as a probability density function (pdf) defined over a space of graphs, denoted $\mathcal{G}$, such that the set of discrete values taken by this pdf is given by

$$\left\{ \mathbb{P}_{\boldsymbol{\theta}}[G] : G \in \mathcal{G}, \ \boldsymbol{\theta} \in \boldsymbol{\Theta} \right\},$$

where $\boldsymbol{\theta}$ is a vector of parameters controlling the distribution of the random variable, G. Sometimes, and especially when considering a uniform distribution, this vector of parameters will be omitted. The complexity of this model depends on (i) the definition of the space of graphs, $\mathcal{G}$, and (ii) the choice of distribution, $\mathbb{P}_{\boldsymbol{\theta}}$ on $\mathcal{G}$. Observe, for instance, that one may define a uniform distribution on the space of all graphs with exactly two disconnected components. Although such a probabilistic model can be easily described, it would nonetheless be difficult to sample realizations from such a distribution. Other approaches to construct $\mathbb{P}_{\boldsymbol{\theta}}$ rely on a generative process, which usually constructs each element, $G \in \mathcal{G}$, through a sequence of recursive steps.

In this section, we will examine five classes of network models that have proved to be especially popular in the neuroimaging literature. These include the classical random graph model, due to Erdős and Rényi (25, 26), in Section 17.4.1, and the small-world and preferential attachment models in Sections 17.4.2 and 17.4.3, respectively. These two mathematical models have been particularly instrumental in the renewed popularity of network science in the past decade. Finally, we will consider two statistical approaches to network modeling: the exponential random graph model in Section 17.4.4, and the stochastic block model in Section 17.4.5.

17.4.1 Erdős–Rényi Random Graphs

In a series of papers, Erdős and Rényi posited the bases for the study of random graphs with a pre-specified number of vertices and edges (24, 25, 26). In this classical model, the space of graphs of interest is defined as the set of all graphs over N_v vertices and N_e edges. This space is denoted by $\mathcal{G}(N_v, N_e)$, and each graph, G, in that space has the following probability of occurring,

$$\mathbb{P}[G | N_v, N_e] = \binom{N_s}{2}^{-1},$$

where we have defined N_s as $N_v(N_v-1)/2$. That is, N_s denotes the maximal number of edges that a graph of order N_v may contain. A different, yet asymptotically equivalent, formulation of the classical ER model is one based on a random allocation of individual edges. In this alternative model, every edge has an independent probability of being present. Thus, we specify the number of vertices, N_v, accompanied by a probability $p \in [0, 1]$, such that the space of interest is denoted by $\mathcal{G}(N_v, p)$ (32). In this case, the probability of obtaining a graph G with N_e edges is given by

$$\mathbb{P}[G | N_v, p] = p^{N_e} (1 - p)^{N_s - N_e}.$$

Although these two models are asymptotically equivalent, this second model has perhaps gained greater currency in practice, mainly owing to its natural sampling mechanism. Indeed, from the aforementioned probability of any graph G with N_e edges, one can straightforwardly sample from the space $\mathcal{G}(N_v, p)$ by drawing N_e realizations from a Bernoulli random variable with probability p. Such a sampling scheme has therefore a computational complexity of order $O(|N_v|^2)$. However, there also exist more efficient sampling methods that are less costly computationally. Batagelj and Brandes, for instance, have proposed an algorithm that solely necessitates $O(|N_v| + |N_e|)$ steps, for sampling from such random graphs (6).

This framework can be extended to more general scenarios, where the graphs are not simply constrained to have a fixed order and size. Several other models of random graphs have been proposed in the literature, in which various aspects of the networks are fixed. One may, for instance, be interested in fixing the degree distribution of a family of graphs, and then consider all random graphs that possess the same number of vertices, the same number of edges, and the same degree distribution.

Once a particular reference model has been selected, it is then possible to compute a particular network topological metric for each of the realizations from that distribution. For instance, given a network of interest, G, we may control for its number of vertices and edges, by generating realizations from $\mathcal{G}(N_v, N_e)$. These random graphs will be denoted by G_j^*, for $j = 1, \ldots, m$, where m is chosen to be large. If we wish to evaluate how uncommon the clustering coefficient of G is, we can compare the value of $\mathrm{cl}(G)$ with the values of $\mathrm{cl}(G_j^*)$, for every realization from the reference model. This provides us with an indication of how unlikely the value $\mathrm{cl}(G)$ is under the assumption that G was drawn from $\mathcal{G}(N_v, N_e)$. More details on this topic can be found in Chapter 6 of (41).

17.4.2   Small-World Networks

Modern network science was spearheaded by the seminal paper of Watts and Strogatz (68), which highlighted the presence of topological similarities over large families of real-world networks. These included biological networks such as protein–protein interactions, utility networks such as the power grid, and transportation networks. All of these datasets have been characterized as non-random, since they tend to exhibit a greater level of local clustering than typical ER graph models. Despite these differences in local clustering, such networks nevertheless approximate the characteristic path lengths of random graph models. Therefore, real-world networks appear to strike a balance between the global efficiency of random graphs and the local clustering of *regular graphs*. (A graph is said to be k-regular, when all of its nodes have degree k.)

The network model proposed by Watts and Strogatz is a growth model, where the original graph is a regular graph. At each step of the process, randomly selected edges are randomly rewired. Naturally, if this process is continued for a sufficiently long time, the resulting graph corresponds to an ER model. Crucially, the transition window between a regular lattice and a random graph was found to contain an interesting family of networks that Watts and Strogatz called *small-world*. This particular choice of phrase was motivated by the proverbial small-world phenomenon, which was originally studied in the social sciences (48).

Small-world networks are thus characterized by a surprising combination of local clustering and small average shortest path length. This model of brain connectivity has been extensively applied to neuroimaging data. Macroscale cortical networks based on various parcellation schemes have been found to display a small-world network architecture in normal subjects (2). Moreover, these topological properties have also been found to be disrupted in patients suffering from Alzheimer's disease (61), and in psychiatric patients with schizophrenia (46). However, one should note that there is no unanimous consensus on the ubiquity of small-worldness in neuroscience. Some other researchers have challenged the claim that large-scale brain networks possess a small-world configuration (9, 30).

17.4.3 Preferential Attachment

This third mathematical model formalizes the notion in economics, which postulates that the "rich get richer." This concept has been used and studied in the social network literature for almost half a century (58). More recently, however, this general graph model has been described by Barabási and Albert, and applied to a large range of real-world datasets, including the World Wide Web (3).

Formally, the preferential attachment model is described by characterizing the patterns of connectivity of new vertices, as they are progressively being added to an existing network (11). Given a fixed sequence of vertices $v_1, v_2, \ldots$, we wish to construct a random graph process, G_t with $t = 0, 1, 2, \ldots$, where the starting point is the empty graph $G_0 = (V_0, E_0)$ with $V_0 = E_0 = \varnothing$. In its simplest form, the preferential attachment algorithm is based on the addition of a single vertex and a single edge at each time point. Thus, at each step of the process, we add a vertex v_t to V_{t-1}, and this vertex is linked to one of the existing vertices in V_{t-1}, which we may denote by $u \in V_{t-1}$. The probability of v_t being linked to u is then given by

$$p(v_t \sim u) = \frac{d_{G_{t-1}}(u)}{\sum_{v \in V_{t-1}} d_{G_{t-1}}(v)}.$$

That is, every new vertex, v_t, is connected to an existing one, u, with a probability that is proportional to the number of vertices already connected to u. This model can then be generalized to the addition of several edges at each time point. The interested reader may consult (11) for such extensions.

This preferential attachment growth process favors vertices with high degrees. We would therefore expect that graphs produced using this construction rule exhibit a skewed degree distribution, characterized by few nodes with a high degree, and many nodes with a small degree. Indeed, the degree distributions of preferential attachment networks tend to follow a power law (3). As a result, these models are sometimes referred to as *scale-free* networks, owing to the common use of this phrase in the physics literature. This preferential attachment model has been applied to several neuroimaging datasets, such as resting-state fMRI (65).

17.4.4 Exponential Random Graph Models

Small-world and scale-free networks are useful descriptive models that allow the characterization of real-world networks. However, these two classical models do not lend themselves to the estimation of parameters controlling the network process under scrutiny. A full probabilistic formulation that has gained traction in the modern literature on network science is the exponential random graph model (ERGM). In this model, we specify the probability of a particular graph, as a function of a set of constraints on the binary random variables controlling the presence or absence of an edge.

Formally, a random graph $G = (V, E)$ is associated with a set of binary random variables, Y_{ij}'s, where each Y_{ij} controls whether or not the edge between the ith and the jth nodes is present. Thus, the corresponding random adjacency matrix, $\mathbf{Y} = (\mathbf{Y_{ij}})$, completely determines the topology of G. An ERGM is then specified through the joint distribution of this family of binary variates for some vector of parameters, $\boldsymbol{\theta}$. This is commonly formulated as follows,

$$\mathbb{P}_{\boldsymbol{\theta}}[\mathbf{Y} = \mathbf{y}] = \frac{1}{\kappa(\boldsymbol{\theta})} \exp\left\{ \sum_{\mathbf{h} \in \mathcal{H}} \theta_{\mathbf{h}} \mathbf{g_h}(\mathbf{y}) \right\},$$

where each $h \subseteq (V \times V)$, is a *configuration* that stipulates a set of edges amongst the set of all possible edges in G. Each of these configurations may be interpreted as a network motif or constraint, whose probability of occurrence is controlled by the parameter θ_h. Thus, for every configuration, h, the function $g_h(\mathbf{y})$ is defined as

$$g_h(\mathbf{y}) = \prod_{(\mathbf{i},\mathbf{j}) \in \mathbf{h}} \mathbf{y_{ij}},$$

which takes a value of one, if and only if all the edges described in h are present. Clearly, if the parameter, θ_h, is non-zero, then the corresponding Y_{ij}'s are dependent upon each other. Finally, $\kappa(\boldsymbol{\theta})$ is a normalizing constant, which is given by

$$\kappa(\boldsymbol{\theta}) = \sum_{\mathbf{y}} \exp\left\{ \sum_{h \in \mathcal{H}} \theta_h g_h(\mathbf{y}) \right\},$$

where the summation is here conducted with respect to all the possible values of $\mathbf{y}$. That is, $\mathbf{y} \in \{0,1\}^{\mathbf{N_s}}$, where as before $N_s = N_v(N_v - 1)/2$.

The topology of an ERGM therefore critically depends on the choice of the set of induced subgraphs, $\mathcal{H}$, and on the values of their associated weights, $\boldsymbol{\theta}$. Different choices of $\mathcal{H}$ and $\boldsymbol{\theta}$ will yield different models. A *Bernoulli random graph model*, for instance, can be constructed by specifying each $h \in \mathcal{H}$, such that it only contains a single edge. This produces exactly $N_v(N_v - 1)/2$ configurations, whose corresponding configuration functions take the form,

$$g_h(\mathbf{y}) = \mathbf{g_{ij}}(\mathbf{y}) = \mathbf{y_{ij}},$$

for every $(i, j) \in E(G)$. For the Bernoulli random graph model, the original defining equation for the general ERGM simplifies to the following,

$$\mathbb{P}_{\boldsymbol{\theta}}[\mathbf{Y} = \mathbf{y}] = \frac{1}{\kappa(\boldsymbol{\theta})} \exp\left\{ \sum_{(\mathbf{i},\mathbf{j}) \in \mathbf{V^2}} \theta_{\mathbf{ij}} \mathbf{y_{ij}} \right\}.$$

Such a model, however, contains too many parameters, and it is hence customary to impose an additional homogeneity condition on the θ_{ij}'s. For example, we may recover the ER random graph by setting all the θ_{ij}'s to be equal.

This simple illustration of the versatility of the ERGM formulation also demonstrates its main weakness. This approach tends to lead to convoluted graph models, characterized by high-dimensional parameter spaces. Such models therefore necessitate complex optimization procedures, as these large parameter spaces need to be explored in an efficient manner. The methods that have been proposed in the literature to tackle this estimation problem include Markov chain Monte Carlo maximum likelihood estimation (31), and a stochastic version of the Newton–Raphson algorithm (56). A viable alternative to ERGMs, which has proved to be particularly useful in the network science literature, is the model described in the next section.

17.4.5   Stochastic Block Models

A natural way of approaching complex networks is to divide them into subgraphs. Stochastic block models (SBMs) provide a formal framework for performing such partitioning (16). Each subgraph in an SBM is expected to exhibit a comparable connectivity pattern. As such, SBMs can be conceived as a model-based approach to community detection. In classical community detection, one is mainly interested in identifying highly connected subgraphs. By contrast, in an SBM, one can detect homogeneous sets of nodes, which will contain vertices with similar in-group and out-group topological properties. Thus, the purpose of an SBM is to partition the vertex set into a *known* number of non-overlapping subsets.

For k subgraphs, the vertex set of a graph G will therefore be partitioned into a sequence $\{C_1, \ldots, C_k\}$, satisfying

$$V(G) = \bigcup_{l=1}^{k} C_l, \quad \text{and} \quad C_l \cap C_{l'} = \varnothing,$$

for every $l \neq l'$.

Probabilistically, an SBM can be described as a two-level hierarchical statistical model, comprised of a latent and an observed component. At the first level, the latent group affiliation is given by a set of k-dimensional vectors, $\mathbf{Z}_i$'s. For every vertex, one of these vectors is drawn from a multinomial distribution,

$$\mathbf{Z}_i \overset{\text{iid}}{\sim} \text{Multin}(1; \pi_1, \ldots, \pi_k),$$

for every $i = 1, \ldots, N_v$, and where the π_k's are the group proportions satisfying $0 \leq \pi_k \leq 1$, and $\sum_k \pi_k = 1$, determining the size of each vertex subsets. At the second level, the observed adjacency matrix is given by a set of Bernoulli random variables, Y_{ij}'s, that are conditionally independent, given the group affiliations for each edge, $\mathbf{z}_i$ and $\mathbf{z}_j$, obtained from the first level of the model. That is, we have

$$Y_{ij} | \mathbf{Z}_i = \mathbf{z}_i, \mathbf{Z}_j = \mathbf{z}_j \overset{\text{ind}}{\sim} \text{Bern}(\mathbf{z}_i^T \mathbf{B} \mathbf{z}_j),$$

where $\mathbf{B}$ is a $k \times k$ symmetric matrix of coefficients, controlling the within-group and between-group connectivity patterns. For example, a natural choice of $\mathbf{B}$ may be one composed of only two parameters, α and β, such that

$$\mathbf{B} = \begin{vmatrix} \alpha & \beta & \cdots & \beta \\ \beta & \alpha & \cdots & \beta \\ \vdots & \ddots & \ddots & \vdots \\ \beta & \beta & \cdots & \alpha \end{vmatrix}.$$

This example is a community partitioning model in which the diagonal entries, i.e. the within-group probabilities of two vertices being connected, is set to be much larger than the off-diagonal entries, which represent the probabilities of between-community edges. Also, observe that such SBMs can straightforwardly be extended to directed networks by selecting $\mathbf{B}$ to be non-symmetric.

SBMs have recently gained in popularity in the network science literature. These models appear to strike a reasonable balance between the naive homogeneity of ER models, and the sophistication of ERGMs, while remaining computationally tractable. Moreover, SBMs have been shown to be amenable to theoretical analysis (10, 20). This family of models has not yet been widely adopted in the analysis of neuroimaging data. However, this should constitute a fruitful modeling approach in the future.

17.5 Estimation and Comparison of Networks

In this final section, we consider the statistical network analysis of neuroimaging data, per se. Firstly, we present a mass-univariate approach, which is sometimes referred to as statistical parametric networks (SPNs). Secondly, we discuss the thorny issue of comparing the topology of networks that differ in density and evaluate the advantages of using density-integrated estimators of topological metrics. Thirdly, we consider the use of weighted topological metrics as a possible strategy for comparing several populations of macroscale cortical networks.

17.5.1   Statistical Parametric Networks

In modern network science, researchers commonly analyze a single network at a time (68). By contrast, neuroimagers are generally interested in comparing the properties of one or several families of *subject-specific* networks. As described in Section 17.2, these graphs are routinely standardized with respect to their vertex sets. Moreover, such networks tend to be weighted, where each weight represents the strength of association between each pair of vertices. Thus, subject-specific connectivity patterns can solely differ in terms of their edge weights. A family of n weighted networks may hence be denoted as follows, $\{G_1, \ldots, G_n\}$, where each $G_i = (V, E_i, W_i)$, for every $i = 1, \ldots, n$, since all the G_i's share the same vertex set.

It may be hard to visualize such a family of networks in order to comprehend the topological trends characterizing the sample. Consequently, it is natural to ask whether a *summary network* can be constructed in order to help with the visualization and comprehension of the properties of the dataset. Probably the most popular method for accomplishing such a task is the use of a mass-univariate approach, whereby each edge is subjected to an independent statistical test. This methodological approach is reminiscent of the well-known statistical parametric mapping (SPM) framework, which has been widely adopted in neuroimaging as the standard way of summarizing sets of MRI and electrophysiological images (27). Thus, by analogy, one may refer to this method as *statistical parameter mapping* (SPN).

This approach to summarizing families of networks can be naturally applied to single-group or multiple-group experimental designs. Here, the different groups of subject-specific networks could either represent the various levels of an experiments, or the different stages of a repeated measures design. As such, the implementation of an SPN solely requires the specification of an independent linear model for every edge in the networks of interest. As for SPM, the SPN framework suffers from one important methodological limitation. Since these tests are applied independently, one must correct for multiple comparison in order to control the probability of committing type I errors. As for other mass-univariate approaches, this correction can be performed using either the classical Bonferroni correction or the false discovery rate (FDR) (7, 53).

Given a particular experimental design, one can choose to summarize the data by constructing several condition-specific networks, as was conducted in Figure 17.3. This strategy

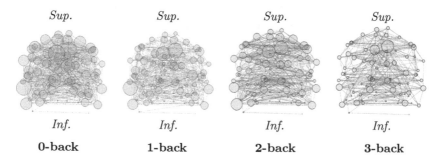

FIGURE 17.3
Graphical representations of mean SPNs, over four levels of a cognitive task. The mean SPNs for an N-back task in the coronal planes are here presented, after FDR correction (base rate $\alpha_0 = .05$). The locations of the nodes represent the stereotaxic centroids of the corresponding cortical regions. The orientation axis is indicated in italics: inferior–superior for the coronal section. The size of each node is proportional to its degree. (See (35) for a full description.)

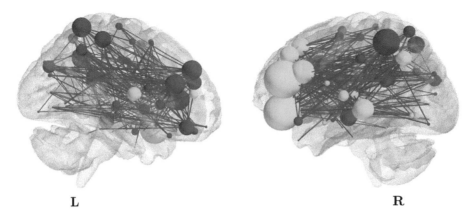

L   R

FIGURE 17.4
Visualization of a differential SPN, summarizing the effect of a cognitive factor. Left and right sagittal section of a negative differential SPN, representing the significantly "lost" edges, due to the N-back experimental factor. The presence of an edge is here determined by the thresholding of p-values at .01, uncorrected. (See (35) for further details.)

has the advantage of providing an overview of the network differences across experimental conditions. Such summary networks are sometimes referred to as *mean networks* (35). By contrast, one can also conduct an analysis of variance (ANOVA) for each edge, in order to produce a single network representing the number of edges that have been significantly lost or gained as a result of undergoing the experimental manipulation. One may interpret this particular type of approach as the identification of an "edge contrast," by analogy with the terminology used in the SPM literature. The graphs capturing these connectivity differences may then be termed *differential networks*. One such network is reported in Figure 17.4, for an experiment on working memory.

17.5.2   Density-Integrated Topology

We have already seen in Section 17.2.4 that weighted networks in neuroimaging are generally thresholded. Such binarized networks should be contrasted with the original datasets composed of real numbers. Understandably, the issue of thresholding networks has thus generated a substantial amount of academic discussions (66, 34). Unfortunately, this debate has not yet reached a satisfying conclusion, and there is currently no consensus on how to compare the topological properties of several families of networks that significantly differ in density.

This methodological problem stems from the fact that the topological properties reported in Section 17.3 are highly dependent on the density of the graphs under scrutiny. Here, the term density should be understood as either the number of edges for an unweighted graph, or as the mean weight for a weighted one. Therefore, any statistical comparison of the topological properties of several families of graphs cannot be interpreted without a consideration of the differences in the number of edges of these graphs. In some sense, we would like to be able to "control" for differences in the number of edges or mean weight, in the same way that we control for extraneous variables in a linear model. Note, however, that the relationship between the number of edges and the topological metrics studied in Section 17.3 is unlikely to be linear, and therefore a naive linear control of such differences in density is unlikely to be sufficient for our purpose (66).

One approach to this problem that has proved to be popular in the literature is the use of a *density-integrated* topological metric. This is applicable to an arbitrary topological metric. Without loss of generality, let us assume that this metric is denoted by a function, $T(\cdot)$, which takes a binary graph and returns a real number. Importantly, this topological metric should be well-defined for every graph density. That is, recall that the characteristic path length, for instance, is not finite for disconnected graphs. In such cases, it is preferable to take an equivalent metric, such as global efficiency, as introduced in Section 17.3, since the latter metric is finite for every graph density. Thus, given a weighted graph $G = (V, E, W)$, we can define the density-integrated version of T, as follows,

$$T_K(G) = \sum_{k=1}^{N_s} T(\gamma(G, k))p(k),$$

where as before, $N_s = N_v(N_v - 1)/2$, and $\gamma(G, k)$ is a thresholding function, which takes a weighted graph, G, and a density level, k, and returns a binary graph, whose edge set is composed of the k most weighted edges in the original graph, G. Naturally, the density cutoff points range from a single edge to the number of edges in the saturated network. Observe that each density level can be given a weighting, $p(k)$, that allows us to emphasize certain portions of the range of densities, and where we have $\sum_k p(k) = 1$. It has indeed been reported that certain connectivity patterns are more likely to occur at a low density level (2). This would therefore justify a partial integration of the density levels, which may be controlled by setting certain $p(k)$'s to zero.

At the heart of this thresholding conundrum is the question of topological invariance. Indeed, we wish to identify the topological properties of a graph that are invariant under certain transformations of the density of that network. However, which family of density transformations shall we choose? It can be shown that such topological invariance can be obtained by considering the family of all *monotonic transformations* of the graph weight set. That is, every rank-preserving function of the weight set will leave the resulting density-integrated topological property invariant (34). This theoretical result could be regarded as a definition of what we mean by separating density from topology. In this setting, a difference in density-integrated topology is therefore one that is invariant to any monotonic transformation of the weight set.

17.5.3   Comparison of Weighted Networks

The problem highlighted in the previous section could also be naturally tackled by directly considering the weighted versions of the topological metrics of interest. Many such weighted metrics have been proposed in the literature (57). Indeed, why shall we go through the process of thresholding a weighted network to compute a measure of topology, if such a quantity can be directly calculated using the original weighted network?

Unfortunately, despite its natural appeal, this approach has been shown to be limited in practice. Indeed, for certain weighted topological metrics, it can be shown that comparing such metrics is equivalent to comparing the mean weights of two families of networks. Such a result has been rigorously proved for the weighted version of global efficiency, for instance (34). This theorem has been demonstrated under mild assumptions on the range of the edge weights, which are likely to hold in most practical settings. Although such a result solely holds for the weighted global efficiency, it is likely that similar results could be derived for other weighted topological metrics.

The strong dependence of weighted topology on weighted density can also be illustrated in the context of community detection. In Figure 17.5, we have reported some simulated results that exemplify the problem associated with comparing the community structure of

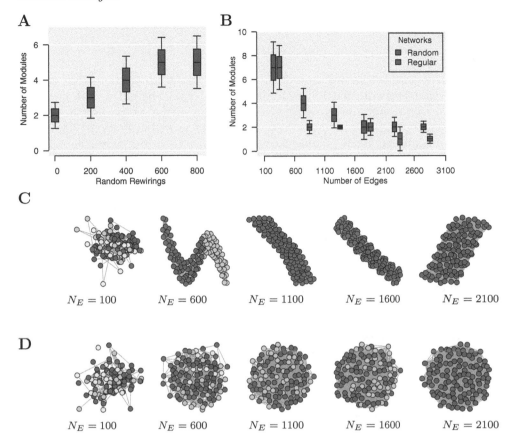

FIGURE 17.5

Topological randomness and number of edges predict number of modules. (A) Relationship between the number of random rewirings of a regular lattice and the number of modules in such a network. Here, the number of edges is kept constant throughout all rewirings. (B) Relationship between the number of edges in a network and its number of modules for both regular (i.e., lattice) and random graphs. This shows that the number of modules tends to decrease as more edges are added to both types of networks. (C-D) Modular structures of regular (C) and random (D) networks for different number of edges, N_E. These networks are represented using the algorithm of (40), with different colors representing different modules. In all simulations, the number of vertices is $N_V = 112$, as in (5) and (33).

weighted graphs that differ in density. This particular set of synthetic data replicates some previous results, originally reported in (5). As one can observe from Figure 17.5, the number of communities identified under each scenario is highly dependent on the number of edges in each graph. As the density increases, the number of modules decreases for both random and regular graphs. Such simulations show that the interpretation of the topological differences of several families of weighted graphs that differ in their average weights, should be reported with caution.

17.6   Conclusion

In this chapter, we have provided an overview of the different statistical methods that can be used for the analysis of network data in neuroimaging. It should be noted that much of the work to date is descriptive, in the sense that it is common for researchers in this field to compute several topological metrics, such as the ones presented in Section 17.3, for a family of networks of interest. Such topological properties are then routinely compared across different groups of subject-specific networks. However, we have seen in Section 17.5 that such comparisons are fraught with difficulties, since disentangling differences in density from differences in topology has proved to be particularly hard. One possible solution to this problem is the use of density-integrated topological metrics, which appear promising. By contrast, the use of full statistical network models, such as the ones described in Section 17.4, has received little attention in the neuroimaging literature. This may be partly explained by the inherent difficulties associated with the estimation of such models in practice. While this limitation is certainly true for ERGMs, the use of SBMs may give researchers a sufficient level of flexibility to extract meaningful information from network data.

An emerging challenge for network data analysis is the consideration of time-evolving networks, in what may be called *dynamic network analysis* (8, 52). Several models of time-evolving networks have been proposed in the literature, including dynamic ERGMs (42, 37), and dynamic SBMs (69, 70). The dynamics of dyads in evolving networks have also been studied successfully in social networks (55, 29). These methodological extensions to the time domain are especially relevant to neuroimaging data, since many such datasets can be characterized as evolving over time. This is naturally the case for networks constructed using fMRI data. For such datasets, however, the temporal resolution of these networks may not be sufficiently high, to exploit the full power of dynamic network analysis. By contrast, the use of electrophysiological recordings as in EEG, MEG and ECoG may reveal time-indexed changes in macroscale cortical connectivity. Dynamic network analysis has indeed already been successfully applied to a range of different datasets in neuroimaging, including fMRI (5) and EEG data (22).

Bibliography

[1] Sophie Achard and E. Bullmore. Efficiency and cost of economical brain functional networks. *PLOS Computational Biology*, 3:174–182, 2007.

[2] Sophie Achard, Raymond Salvador, Brandon Whitcher, John Suckling, and Ed Bullmore. A resilient, low-frequency, small-world human brain functional network with highly connected association cortical hubs. *J. Neurosci.*, 26(1):63–72, January 2006.

[3] A-L Barabasi and R. Albert. Emergence of scaling in random networks. *Science*, 286:509–512, 1999.

[4] Danielle S. Bassett, Andreas Meyer-Lindenberg, Sophie Achard, Thomas Duke, and Edward Bullmore. Adaptive reconfiguration of fractal small-world human brain functional networks. *Proceedings of the National Academy of Sciences of the United States of America*, 103(51):19518–19523, 2006.

[5] Danielle S. Bassett, Nicholas F. Wymbs, Mason A. Porter, Peter J. Mucha, Jean M.

Carlson, and Scott T. Grafton. Dynamic reconfiguration of human brain networks during learning. *Proceedings of the National Academy of Sciences*, 108(18):7641–7646, May 2011.

[6] V. Batagelj and U. Brandes. Efficient generation of large random networks. *Physical Review E*, 71(3):036113–036123, 2005.

[7] Yoav Benjamini and Yosef Hochberg. Controlling the false discovery rate: A practical and powerful approach to multiple testing. *Journal of the Royal Statistical Society. Series B (Methodological)*, 57(1):289–300, 1995.

[8] Tanya Y. Berger-Wolf and Jared Saia. A framework for analysis of dynamic social networks. In *Proceedings of the 12th ACM SIGKDD International Conference on Knowledge Discovery and Data Mining*, pages 523–528. ACM, 2006.

[9] Stephan Bialonski, Marie-Therese Horstmann, and Klaus Lehnertz. From brain to earth and climate systems: Small-world interaction networks or not? *Chaos: An Interdisciplinary Journal of Nonlinear Science*, 20(1):013134, 2010.

[10] Peter J. Bickel and Aiyou Chen. A nonparametric view of network models and Newman-Girvan and other modularities. *Proceedings of the National Academy of Sciences*, 106(50):21068–21073, 2009.

[11] B. Bollobas, O. Riordan, J. Spencer, and G. Tusnady. The degree sequence of a scale-free random graph process. *Random Struct. Alg.*, 18(3):279–290, 2001.

[12] B. Bollobas and O.M. Riordan. Mathematical results on scale-free random graphs. In S. Bornholdt and H.G. Schuster, editors, *Handbook of Graphs and Networks: From the Genome to the Internet*, pages 1–32. Wiley, London, 2003.

[13] Bela Bollobas and Oliver Riordan. The diameter of a scale-free random graph. *Combinatorica*, 24(1):5–34, 2004.

[14] Bela Bollobas and Oliver Riordan. Shortest paths and load scaling in scale-free trees. *Phys. Rev. E*, 69(3):36114–36124, 2004.

[15] E. Bullmore and Olaf Sporns. Complex brain networks: Graph theoretical analysis of structural and functional systems. *Nature Reviews Neuroscience*, 10(1):1–13, 2009.

[16] Alain Celisse, Jean-Jacques Daudin, and Laurent Pierre. Consistency of maximum-likelihood and variational estimators in the stochastic block model. *Electronic Journal of Statistics*, 6:1847–1899, 2012.

[17] Zhang J. Chen, Yong He, Pedro Rosa-Neto, Jurgen Germann, and Alan C. Evans. Revealing modular architecture of human brain structural networks by using cortical thickness from MRI. *Cerebral Cortex*, 18(10):2374–2381, 2008.

[18] Aaron Clauset, Cosma Rohilla Shalizi, and Mark E.J. Newman. Power-law distributions in empirical data. *SIAM Review*, 51(4):661–703, 2009.

[19] Nicolas A. Crossley, Andrea Mechelli, Petra E. Vertes, Toby T. Winton-Brown, Ameera X. Patel, Cedric E. Ginestet, Philip McGuire, and Edward T. Bullmore. Cognitive relevance of the community structure of the human brain functional coactivation network. *Proceedings of the National Academy of Sciences*, 110(28):11583–11588, 2013.

[20] Aurelien Decelle, Florent Krzakala, Cristopher Moore, and Lenka Zdeborova. Inference and phase transitions in the detection of modules in sparse networks. *Phys. Rev. Lett.*, 107(6):065701, August 2011.

[21] E.W. Dijkstra. A note on two problems in connexion with graphs. *Numerische Mathematik*, 1:269–271, 1959.

[22] Stavros I. Dimitriadis, Nikolaos A. Laskaris, Vasso Tsirka, Michael Vourkas, Sifis Micheloyannis, and Spiros Fotopoulos. Tracking brain dynamics via time-dependent network analysis. *Journal of Neuroscience Methods*, 193(1):145–155, 2010.

[23] Victor M. Eguiluz, Dante R. Chialvo, Guillermo A. Cecchi, Marwan Baliki, and A. Vania Apkarian. Scale-free brain functional networks. *Phys. Rev. Lett.*, 94(1):18102–18106, January 2005.

[24] P. Erdős and A. Rényi. On random graphs I. *Publ. Math. Debrecen*, 6:290–297, 1959.

[25] P. Erdős and A. Rényi. The evolution of random graphs. *Publ. Math. Inst. Hungar. Acad. Sci.*, 5:17, 1960.

[26] P. Erdős and A. Rényi. On the strength of connectedness of a random graph. *Acta Mathematica Hungarica*, 12(1–2):261–267, 1961.

[27] Karl J. Friston. Functional and effective connectivity in neuroimaging: A synthesis. *Human Brain Mapping*, 2(1–2):56–78, 1994.

[28] K.J. Friston and J. Ashburner. Generative and recognition models for neuroanatomy. *NeuroImage*, 23(1):21–24, September 2004.

[29] Lazaros K. Gallos, Diego Rybski, Fredrik Liljeros, Shlomo Havlin, and Hernan A. Makse. How people interact in evolving online affiliation networks. *Physical Review X*, 2(3):031014, 2012.

[30] Felipe Gerhard, Gordon Pipa, Bruss Lima, Sergio Neuenschwander, and Wulfram Gerstner. Extraction of network topology from multi-electrode recordings: Is there a small-world effect? *Frontiers in Computational Neuroscience*, 5:4–4, 2010.

[31] Charles J. Geyer and Elizabeth A. Thompson. Constrained Monte Carlo maximum likelihood for dependent data (with discussion). *Journal of the Royal Statistical Society. Series B. Methodological*, 54(3):657–699, 1992.

[32] Edgar N. Gilbert. Random graphs. *The Annals of Mathematical Statistics*, pages 1141–1144, 1959.

[33] C.E. Ginestet, A.P. Fournel, and A. Simmons. Statistical network analysis for functional MRI: Summary networks and group comparisons. *Frontiers in Computational Neuroscience*, 8(51):1–10, 2014.

[34] Cedric E. Ginestet, Thomas E. Nichols, Edward T. Bullmore, and A. Simmons. Brain network analysis: Separating cost from topology using cost-integration. *PLoS ONE*, 6(7):e21570, 2011.

[35] Cedric E. Ginestet and A. Simmons. Statistical parametric network analysis of functional connectivity dynamics during a working memory task. *NeuroImage*, 5(2):688–704, 2011.

[36] Patric Hagmann, Leila Cammoun, Xavier Gigandet, Reto Meuli, Christopher J. Honey, Van J. Wedeen, and Olaf Sporns. Mapping the structural core of human cerebral cortex. *PLoS Biol*, 6(7):159–169, July 2008.

[37] Steve Hanneke, Wenjie Fu, and Eric P. Xing. Discrete temporal models of social networks. *Electronic Journal of Statistics*, 4:585–605, 2010.

[38] Yong He, Zhang J. Chen, and Alan C. Evans. Small-world anatomical networks in the human brain revealed by cortical thickness from MRI. *Cereb. Cortex*, 17(10):2407–2419, October 2007.

[39] Yong He, Jinhui Wang, Liang Wang, Zhang J. Chen, Chaogan Yan, Hong Yang, Hehan Tang, Chaozhe Zhu, Qiyong Gong, Yufeng Zang, and Alan C. Evans. Uncovering intrinsic modular organization of spontaneous brain activity in humans. *PLoS ONE*, 4(4):1–18, April 2009.

[40] T. Kamada and S. Kawai. An algorithm for drawing general undirected graphs. *Information Processing Letters*, 31(1):7–15, 1989.

[41] E.D. Kolaczyk. *Statistical Analysis of Network Data: Methods and Models*. Springer-Verlag, London, 2009.

[42] Pavel N. Krivitsky and Mark S. Handcock. A separable model for dynamic networks. *Journal of the Royal Statistical Society: Series B (Statistical Methodology)*, 76(1):29–46, 2014.

[43] V. Latora and M. Marchiori. Economic small-world behavior in weighted networks. *The European Physical Journal B—Condensed Matter and Complex Systems*, 32(2):249–263, March 2003.

[44] Vito Latora and Massimo Marchiori. Efficient behavior of small-world networks. *Phys. Rev. Lett.*, 87(19):198701–198705, October 2001.

[45] Yang-Yu Liu, Jean-Jacques Slotine, and Albert-Laszlo Barabasi. Controllability of complex networks. *Nature*, 473(7346):167–173, 2011.

[46] Yong Liu, Meng Liang, Yuan Zhou, Yong He, Yihui Hao, Ming Song, Chunshui Yu, Haihong Liu, Zhening Liu, and Tianzi Jiang. Disrupted small-world networks in schizophrenia. *Brain*, 131(4):945–961, 2008.

[47] Sifis Micheloyannis, Michael Vourkas, Vassiliki Tsirka, Eleni Karakonstantaki, Kassia Kanatsouli, and Cornelis J. Stam. The influence of ageing on complex brain networks: A graph theoretical analysis. *Human Brain Mapping*, 30(1):200–208, 2009.

[48] S. Milgram. The small world problem. *Psychology Today*, 2:60–67, 1967.

[49] Ron Milo, Shalev Itzkovitz, Nadav Kashtan, Reuven Levitt, and Uri Alon. Response to comment on "Network motifs: Simple building blocks of complex networks" and "Superfamilies of evolved and designed networks." *Science*, 305(5687):1107d–, August 2004.

[50] Ron Milo, Shalev Itzkovitz, Nadav Kashtan, Reuven Levitt, Shai Shen-Orr, Inbal Ayzenshtat, Michal Sheffer, and Uri Alon. Superfamilies of evolved and designed networks. *Science*, 303(5663):1538–1542, March 2004.

[51] M. Mitchell. *Complexity: A Guided Tour*. Oxford University Press, 2009.

[52] Peter J. Mucha, Thomas Richardson, Kevin Macon, Mason A. Porter, and Jukka-Pekka Onnela. Community structure in time-dependent, multiscale, and multiplex networks. *Science*, 328(5980):876–878, 2010.

[53] Thomas Nichols and Satoru Hayasaka. Controlling the familywise error rate in functional neuroimaging: A comparative review. *Statistical Methods in Medical Research*, pages 419–446, October 2003.

[54] Ellie Pachou, Michael Vourkas, Panagiotis Simos, Dirk Smit, Cornelis Stam, Vasso Tsirka, and Sifis Micheloyannis. Working memory in schizophrenia: An EEG study using power spectrum and coherence analysis to estimate cortical activation and network behavior. *Brain Topography*, 21(2):128–137, December 2008.

[55] Mark T. Rivera, Sara B. Soderstrom, and Brian Uzzi. Dynamics of dyads in social networks: Assortative, relational, and proximity mechanisms. *Annual Review of Sociology*, 36:91–115, 2010.

[56] Herbert Robbins and Sutton Monro. A stochastic approximation method. *The Annals of Mathematical Statistics*, pages 400–407, 1951.

[57] Mikail Rubinov and Olaf Sporns. Complex network measures of brain connectivity: Uses and interpretations. *Neuroimage*, 52:1059–1069, 2010.

[58] H. Simon. On a class of skew distribution functions. *Biometrika*, 42:425–440, 1955.

[59] O. Sporns, G. Tononi, and G.M. Edelman. Theoretical neuroanatomy: Relating anatomical and functional connectivity in graphs and cortical connection matrices. *Cereb. Cortex*, 10(2):127–141, February 2000.

[60] Olaf Sporns, Dante R. Chialvo, Marcus Kaiser, and Claus C. Hilgetag. Organization, development and function of complex brain networks. *Trends in Cognitive Sciences*, 8(9):418–425, September 2004.

[61] C.J. Stam, B.F. Jones, G. Nolte, M. Breakspear, and Ph. Scheltens. Small-world networks and functional connectivity in Alzheimer's disease. *Cereb. Cortex*, 17(1):92–99, January 2007.

[62] Matthew L. Stanley, Malaak N. Moussa, Brielle M. Paolini, Robert G. Lyday, Jonathan H. Burdette, and Paul J. Laurienti. Defining nodes in complex brain networks. *Frontiers in Computational Neuroscience*, 7:169, 2013.

[63] N. Tzourio-Mazoyer, B. Landeau, D. Papathanassiou, F. Crivello, O. Etard, N. Delcroix, B. Mazoyer, and M. Joliot. Automated anatomical labeling of activations in SPM using a macroscopic anatomical parcellation of the MNI MRI single-subject brain. *NeuroImage*, 15(1):273–289, January 2002.

[64] Martijn P. van den Heuvel, Cornelis J. Stam, Rene S. Kahn, and Hilleke E. Hulshoff Pol. Efficiency of functional brain networks and intellectual performance. *J. Neurosci.*, 29(23):7619–7624, June 2009.

[65] M.P. van den Heuvel, C.J. Stam, M. Boersma, and H.E. Hulshoff Pol. Small-world and scale-free organization of voxel-based resting-state functional connectivity in the human brain. *NeuroImage*, 43(3):528–539, November 2008.

[66] Bernadette C. M. van Wijk, Cornelis J. Stam, and Andreas Daffertshofer. Comparing brain networks of different size and connectivity density using graph theory. *PLoS ONE*, 5(10):13701–13716, October 2010.

[67] Stanley Wasserman. *Social Network Analysis: Methods and Applications*, volume 8. Cambridge University Press, 1994.

[68] Duncan J. Watts and Steven H. Strogatz. Collective dynamics of "small-world" networks. *Nature*, 393(6684):440–442, June 1998.

[69] Eric P. Xing, Wenjie Fu, and Le Song. A state-space mixed membership blockmodel for dynamic network tomography. *The Annals of Applied Statistics*, 4(2):535–566, 2010.

[70] Kevin S. Xu and Alfred O. Hero III. Dynamic stochastic blockmodels: Statistical models for time-evolving networks. In *Social Computing, Behavioral-Cultural Modeling and Prediction*, pages 201–210. Springer, 2013.

[71] Andrew Zalesky, Alex Fornito, and Edward T. Bullmore. Network-based statistic: Identifying differences in brain networks. *NeuroImage*, 53(4):1197–1207, December 2010.

[72] Lihua Zhang, Wenli Xu, and Cheng Chang. Genetic algorithm for affine point pattern matching. *Pattern Recognition Letters*, 24(13):9–19, January 2003.

18

Modeling Change in the Brain: Methods for Cross-Sectional and Longitudinal Data

Philip T. Reiss

Department of Child and Adolescent Psychiatry, New York University

Ciprian M. Crainiceanu

Department of Biostatistics, Johns Hopkins University

Wesley K. Thompson

Department of Psychiatry, University of California, San Diego

Lan Huo

Department of Child and Adolescent Psychiatry, New York University

CONTENTS

18.1	Introduction	468
18.2	Notation and Road Map	468
18.3	Cross-Sectional and Longitudinal Designs	469
	18.3.1 Cross-Sectional Designs	470
	18.3.2 Single-Cohort Longitudinal Designs	470
	18.3.3 Multi-Cohort Longitudinal Designs	471
18.4	Region-Wise Linear Models for the Mean	472
	18.4.1 What Is v, and What Is t?	472
	18.4.2 Cross-Sectional Data	473
	18.4.3 Longitudinal Data	473
	18.4.3.1 Mixed-Effects Models	473
	18.4.3.2 Marginal Models	474
	18.4.4 Relative Efficiency for Estimating the Mean Function	474
	18.4.5 Complications Due to Misalignment	475
	18.4.6 Borrowing Information "Spatially"	477
18.5	Nonlinear Models for the Mean	477
	18.5.1 Polynomial Models	477
	18.5.2 Nonparametric and Semiparametric Models	477
	18.5.3 Analyses with Repeated Cross-Sectional Subsamples	480
18.6	Beyond Modeling the Mean	482
	18.6.1 Individual-Specific Curves	482
	18.6.2 Modeling Components of Change: Longitudinal Functional Principal Component Analysis	483
	18.6.3 Modeling the Entire Age-Specific Distribution	485
	18.6.4 Modeling Local Rates of Change	485
18.7	Discussion	486

Acknowledgments 487
Bibliography 487

18.1   Introduction

Change in the human brain takes many forms, which are studied in various biomedical fields including neurology, psychiatry, developmental psychology and radiology. Imaging studies, in particular using magnetic resonance imaging (MRI), have helped to characterize the typical course of brain development in early life (e.g., 30), and of brain aging in later life (e.g., 21). With regard to abnormalities of cognition, behavior, and brain health, the focus has depended on the age group being studied: research on disorders of the young has sought to pinpoint anomalies in neurodevelopmental trajectories (e.g., 86), whereas studies of adult disorders have emphasized modeling disease-related brain changes (e.g., 44).

An example of relating atypical brain development to a disorder is the study of Shaw et al. (79), who compared development of cortical thickness in 223 children with attention-deficit/hyperactivity disorder (ADHD) versus 223 typically developing controls. Cortical thickness in many parts of the cerebrum increases until adolescence, and then decreases. Applying quadratic growth models, Shaw et al. (79) found that peak thickness was attained later for the ADHD group in most of the cerebrum. This finding suggested that ADHD may arise from a delay in cortical maturation. Attempts to connect brain change with later-life disease progression are exemplified by numerous studies of the Alzheimer's Disease Neuroimaging Initiative (ADNI) cohort. For example, Hua et al. (40) applied tensor-based morphometry to measure brain atrophy in 684 members of the ADNI cohort who were scanned at baseline and one year later. They found faster atrophy in younger subjects, especially those presenting with amnestic mild cognitive impairment.

Although studies targeting different age ranges pose different sets of problems, one can nevertheless speak of a coherent perspective, which has recently emerged, of seeking to understand change in the brain across the lifespan (e.g., 83, 12). Insel (42) drew on this unifying perspective when he argued that, just as brain-based biomarkers can now signal the predementia phase of Alzheimer's disease, an emphasis on studying neurodevelopmental trajectories may help to identify young people who are at risk for behavioral disorders.

This chapter aims to present a set of approaches with broad applicability to the study of change in the brain, and to highlight some areas where new statistical methods may help to drive scientific advances. A recurring theme will be the question of what can be inferred from cross-sectional versus from longitudinal data. It is often argued that longitudinal data are required to understand trajectories of change in the brain (e.g., 82, 31). This is a fair claim, but one that needs to be unpacked. Statisticians have critical roles to play in elucidating the precise meaning of this claim, in critically examining both the advantages and the pitfalls of longitudinal data, and in developing analytic tools for extracting scientifically meaningful knowledge from both cross-sectional and longitudinal imaging studies.

18.2   Notation and Road Map

The generic data structure for this chapter entails a variable t representing age or time, covariates x, and a response y. We shall write t_i for the ith individual's age/time in a cross-sectional setting, and t_{ij} when referring to the jth observation for the ith individual in a longitudinal setting. Single or double subscripts are defined analogously for covariates, although $\boldsymbol{x}_i$ can be used in longitudinal contexts to denote a non-time-varying predictor. For the most part, we are concerned not with a single response per individual (or per scan), but rather with a set of responses defined at each of a set of brain regions $v \in \mathcal{V}$. When one thinks of these regions as being modeled individually, it is appropriate to use a third subscript to denote the response for region v, say y_{ijv}. Alternatively, when the set of responses across the brain are thought of as constituting a *functional response* in the sense of Ramsay and Silverman (65), a more appropriate notation treats v as the argument of a function, say $y_{ij}(v)$. In most of what follows, v appears in subscripts, but it occurs as a function argument in §18.4.5 and §18.6.2.

Most of this chapter will be concerned with variants of the general model

$$y_{ijv} = \mu_v(t_{ij}, \boldsymbol{x}_{ij}) + \varepsilon_{ijv} \text{ for } i = 1, \ldots, I, j = 1, \ldots, J_i, \tag{18.1}$$

where $J_i = 1$ for each i (and hence the j's can be dropped) in the cross-sectional case. The mean-zero errors ε_{ijv} are ordinarily uncorrelated across individuals, i.e., $\mathrm{Cor}(\varepsilon_{i_1 j_1 v_1}, \varepsilon_{i_2 j_2 v_2}) = 0$ whenever $i_1 \neq i_2$. We usually have some within-scan "spatial" correlation, i.e., $\mathrm{Cor}(\varepsilon_{ijv_1}, \varepsilon_{ijv_2}) \neq 0$ for some $v_1 \neq v_2$, although the "mass-univariate" tradition in neuroimaging is to fit a separate model for each v (but §18.4.6 below outlines some functional-response approaches and other attempts to progress beyond this tradition). For longitudinal data ($J_i > 1$ for at least some i), we generally have nonzero temporal correlation within subjects, i.e., $\mathrm{Cor}(\varepsilon_{ij_1 v}, \varepsilon_{ij_2 v}) \neq 0$ for $j_1 \neq j_2$. In the simplest case, the effect of the covariates $\boldsymbol{x}$ on y is linear and independent of t, so that (18.1) becomes

$$y_{ijv} = \boldsymbol{x}_{ij}^T \boldsymbol{\alpha}_v + f_v(t_{ij}) + \varepsilon_{ijv}.$$

We shall also consider linear effects of $\boldsymbol{x}$ that do depend on t: interaction models in the parametric case, and varying-coefficient models in the semiparametric case.

In what follows, we shall often use the term "trajectory" as a shorthand for the population average trajectory, i.e. the function $\mu_v(\cdot, \boldsymbol{x})$ in (18.1)—in other words, the mean of the quantity of interest y as a function of time or age, conditional on the covariates $\boldsymbol{x}$. This terminology is unsatisfying, inasmuch as *trajectory* connotes a pattern of change undergone by an individual (which we shall sometimes refer to as an "individual trajectory"). As discussed in §18.3, a mean function estimate, especially when derived from cross-sectional data, may be completely unrepresentative of the individual trajectories that are sampled. Nevertheless, it is common in the literature to refer to the mean function as a "trajectory"; moreover, understanding how to model the mean is a prerequisite for proceeding to more sophisticated techniques that may provide more insight into individual change.

We examine modeling of the mean function in §18.4 and §18.5, which consider linear and nonlinear models, respectively, while §18.6 presents several approaches that move beyond merely estimating the mean. Of course, choosing an appropriate model depends on the study design, so before we begin to discuss modeling, §18.3 compares several designs for studying change in the brain.

18.3   Cross-Sectional and Longitudinal Designs

In a critical review focusing on MRI research in psychiatry, Horga et al. (39) argue that closer attention to study design would improve the replicability and generalizability of neuroimaging findings, and would make it easier to identify developmental trajectories and infer causal mechanisms. They highlight ascertainment biases and other pitfalls that can afflict common designs, such as case-control designs with convenience samples, and they advocate "yoking" imaging studies to designs that are less vulnerable to such pitfalls, including longitudinal and epidemiological studies. In this section, drawing on the work of Thompson et al. (84), we discuss cross-sectional and longitudinal design types.

18.3.1 Cross-Sectional Designs

Lifespan inferences are often drawn from cross-sectional designs, in which each subject is studied once at a particular age, with age varying across subjects (e.g., 87, 5). In an unstructured cross-sectional study, each subject is observed only once and the between-subject ages have a random distribution over the age span of interest. Cross-sectional studies completely confound within- and between-subject sources of variation and hence presuppose negligible differences among age cohorts when making longitudinal inferences (77, 78). Developmental or longitudinal inferences from cross-sectional designs can also be misleading because they cannot capture individual variation in the shape of trajectories across subjects.

Kraemer et al. (47) issue a number of caveats about inferring developmental processes from cross-sectional studies. They write that, in order for the shape of individual trajectories to be inferred correctly from a curve of mean values derived from cross-sectional data, three conditions must be met:

1. Sampling from the population during the time span of interest must be random.

2. The error variance must be constant over time.

3. All individual trajectories must be parallel (vertical translates of each other).

One form of violation of condition 1 is the "healthy survivor" effect, in which individuals with a better outcome are more likely to be available for a sample. For example, if those with lower cognitive abilities are unable to participate in an aging study, then older subjects in particular may be unrepresentative of their age cohort, and thus estimated cognitive performance trajectories may be biased upward at later ages (84).

As an illustration of the impact of violations of condition 3, consider a simple model in which all individuals' measures y follow a common trajectory f, but with mutually independent random horizontal and vertical shifts, τ and Δ, each with mean zero. Assuming mean-zero measurement error, an individual observed at time t has conditional mean

$$E(y|t, \tau, \Delta) = f(t + \tau) + \Delta \qquad (18.2)$$

and thus the mean function is

$$t \mapsto E(y|t) = E_{\tau,\Delta}[E(y|t, \tau, \Delta)] = E_\tau[f(t + \tau)],$$

a "horizontally blurred" version of f. If f is marked by a period of monotonic change preceded and followed by plateaux, the above function will rise or fall less steeply than f and thus underestimate the rate of change, as in Figure 18.1. Thompson et al. (84) demonstrate this phenomenon for entorhinal cortical thinning in the Alzheimer's Disease Neuroimaging Initiative (ADNI) sample.

18.3.2 Single-Cohort Longitudinal Designs

Given the limitations of cross-sectional designs, single-cohort longitudinal designs (SLDs) have rightfully been considered the gold standard for understanding within-subject change over time at any stage of the lifespan. SLDs follow subjects in one cohort over the entire age span of interest, measuring outcomes periodically and repeatedly over time. Analyses of SLD data therefore base inferences about development or aging on data measuring within-subject

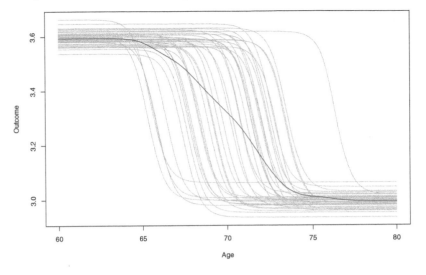

FIGURE 18.1
Forty simulated curves following a decline trajectory of form (18.2), with normally distributed horizontal shifts τ and vertical shifts Δ. The mean of the curves, shown in red, declines less steeply than any of the individual curves.

change. However, SLDs also have important methodological and practical limitations. SLDs completely confound aging and period effects (78). Significant practical problems hamper SLDs covering a broad span of ages. Relevant clinical tools and measures are virtually certain to change substantially over the course of, say, a 20- or 40-year single-cohort study, creating problems of within-study comparability. This is an especially important limitation in longitudinal imaging studies, where technology changes rapidly. Missing data and subject attrition are likely to be substantial and unlikely to be missing completely at random in the sense of Little and Rubin (54).

18.3.3   Multi-Cohort Longitudinal Designs

Multi-cohort longitudinal designs (MLDs) collect data from multiple age cohorts, each followed longitudinally for a given length of time—typically a small fraction of the total age range under study. A wider age range can thus be sampled in a shorter time span than in an SLD. Unlike cross-sectional studies, MLDs capture both between- and within-subject variation in outcomes by age, making it possible to disentangle the effects of age cohort or sampling variation from within-subject change. Most longitudinal imaging studies are in fact MLDs, though the potential impact of age cohort sampling differences is often glossed over.

Unstructured MLDs prespecify the age range of interest but generally recruit unbalanced or non-overlapping age cohorts. Unstructured MLDs are particularly common in late-life imaging studies where age at entry covers a very broad span, often 20–40 years (e.g., 58, 13, 73, 38). Unstructured designs cannot fully address the methodological challenges inherent in characterizing longitudinal trajectories, potentially lacking power to accurately estimate nonlinear trajectories, and are suboptimal for testing for sampling effects or the presence of significant between-subject variation in trajectory shapes. The result is often misleading or inaccurate determinations of the onset and course of disorders.

The deficiencies of unstructured MLDs can be largely avoided by a *structured* MLD, in which one carefully prespecifies the number of subjects in each age cohort and the degree of overlap of within-subject trajectories across neighboring cohorts. In a structured MLD, individuals enter the study at pre-selected ages (age cohorts) spanning the age range of interest, with each subject followed longitudinally over a shorter time span relative to the entire age range. Structured MLDs have been successfully applied to developmental research (e.g., 76, 56, 45), but have thus far been relatively underutilized for investigating the development and progression of brain disorders. An important special case of structured MLDs is the accelerated longitudinal design proposed by Bell (6, 7). In this design, each age cohort in effect undergoes a prospective longitudinal study of relatively short duration. A crucial design requirement is that the short cohort age spans overlap. If the age span were 55–76 years, one might follow 10 age cohorts yearly at ages 55–58, 57–60, etc. Each subject would thus be followed for three years, covering the 55–76 year age span with a two-year overlap between consecutive age cohorts. Thus, by choosing appropriate overlapping cohorts, trajectories across long age spans can be estimated within a much shorter time period. By including baseline age as a term in the model, the trajectories of subjects in overlapping age cohorts can be used to estimate and test for differences among age cohorts (84). Moreover, nonlinear developmental trajectories can potentially be estimated, even if the individual subject-level trajectories are too short to model as nonlinear.

18.4   Region-Wise Linear Models for the Mean

In this section and §18.5 we restrict our attention to the population mean trajectory $t \mapsto \mu_v(t|\boldsymbol{x})$ in (18.1). The aims of this section are to review models for $\mu_v(\cdot|\boldsymbol{x})$ that are linear with respect to time, and to explore some of the issues that they raise when modeling change in the brain.

18.4.1   What Is v, and What Is t?

Before attempting to fit model (18.1), one must clearly define the unit of analysis $v \in \mathcal{V}$ within the brain and the time variable t. In many studies the set $\mathcal{V}$ of brain locations is a singleton: one is interested in a single parameter such as hippocampal volume, or whole-brain average of an image-derived quantity. But in brain mapping research, $\mathcal{V}$ is a larger set such as

- a collection of prespecified regions of interest (ROIs);

- a set of voxels lying along a line or arc, as in recent papers that have considered fractional anisotropy $y(v)$ for v located along a white matter tract or from anterior to posterior along the corpus callosum (e.g. 33, 95, 94);

- tens of thousands of vertices throughout the cerebral cortex, as in cortical thickness studies;

- tens or hundreds of thousands of voxels forming a 3D grid into which the whole brain is divided; or

- in connectivity studies (see, e.g., §18.6.3), a set of pairs of ROIs, sometimes called "connections of interest."

The appropriate definition of the time variable depends on both the design and the scientific questions of interest. For studying development, t might represent age; for progression of neurological disease, t might refer to time since onset or since the baseline visit.

But rules of thumb are less important here than sensitivity to the given application: for example, in their analysis of multiple sclerosis-related brain atrophy, Jones et al. (44) show that it is advantageous to use age, rather than time since the baseline visit, as the predictor of interest.

18.4.2   Cross-Sectional Data

In the cross-sectional setting we have $J_i \equiv 1$ in (18.1). Ignoring for now the dependence on v, the model reduces to

$$y_i = \mu(t_i, \boldsymbol{x}_i) + \varepsilon_i \tag{18.3}$$

for $i = 1, \ldots, I$. For example, suppose the vector $\boldsymbol{x}_i$ is simply an indicator variable x_i for sex or for presence of a genotype or diagnosis, and that μ is assumed to be linear with respect to time, with both the intercept and the slope depending on x_i. Model (18.3) is then the interaction model

$$y_i = \beta_0 + \beta_1 x_i + \beta_2 t_i + \beta_3 x_i t_i + \varepsilon_i = \begin{cases} \beta_0 + \beta_2 t_i + \varepsilon_i, & \text{if } x_i = 0; \\ (\beta_0 + \beta_1) + (\beta_2 + \beta_3) t_i + \varepsilon_i, & \text{if } x_i = 1. \end{cases} \tag{18.4}$$

18.4.3   Longitudinal Data

Guillaume et al. (34) note that the two most popular modeling approaches for longitudinal neuroimaging data are both variants of ordinary least squares (OLS): naïve OLS (N-OLS) and summary statistics OLS (SS-OLS). N-OLS incorporates subject-specific intercepts in the model. SS-OLS extracts subject-specific parameters of interest, such as slopes with respect to age, and fits a group OLS model to these summary statistics. As also noted by Guillaume et al. (34), the recent increased popularity of longitudinal neuroimaging studies has prompted a number of authors to apply more sophisticated longitudinal analysis methods to brain imaging data. Such methods can be divided into two classes of models widely employed by biostatisticians since the 1980s for analyzing longitudinal data: mixed-effects models and marginal models.

18.4.3.1   Mixed-Effects Models

Linear mixed-effects models (48) incorporate random effects to account for the dependence among repeated measures. In the case of model (18.1), we can capture the correlation among residuals $\varepsilon_{ij}, \varepsilon_{ik}$ $(j \neq k)$ for the ith individual by assuming that for each i, j, $\varepsilon_{ij} = u_i + e_{ij}$, where the random effects u_i are independent random variables with the $N(0, \sigma_u^2)$ distribution and error terms e_{ij} are independent of the u_i's. For simplicity we omit covariates $\boldsymbol{x}$ and assume the simple linear model $E(y_{ij}|t_{ij}) = \beta_0 + \beta_1 t_{ij}$; the main effects of covariates and their interactions with time can readily be added as in (18.4). The above decomposition of ε_{ij} gives rise to the so-called random intercept model

$$y_{ij} = (\beta_0 + u_i) + \beta_1 t_{ij} + e_{ij}. \tag{18.5}$$

The most straightforward error specification is that the e_{ij}'s are mutually independent with an identical $N(0, \sigma_e^2)$ distribution, implying the "compound symmetry" assumption of constant correlation for any pair of observations from the same individual: specifically, $\text{Cor}(\varepsilon_{ij}, \varepsilon_{ik}) = \frac{\sigma_u^2}{\sigma_u^2 + \sigma_e^2}$ whenever $j \neq k$. Guillaume et al. (34) offer evidence that the compound symmetry assumption is unrealistic for longitudinal neuroimaging studies. Two more-flexible alternatives are to assume autocorrelated errors, or else to retain the assumption of uncorrelated errors but replace (18.5) with the random intercept/random slope (RIRS) model

$$y_{ij} = (\beta_0 + u_{0i}) + (\beta_1 + u_{1i})t_{ij} + e_{ij}, \tag{18.6}$$

in which the ordered pair (u_{0i}, u_{1i}) is assumed to arise from a bivariate normal distribution with mean zero. Voxel-by-voxel linear mixed-effects models have recently been studied by Chen et al. (14) and Bernal-Rusiel et al. (9) for longitudinal neuroimaging data, but the general approach was previously employed by a number of authors (e.g. 79).

18.4.3.2 Marginal Models

Marginal models account for dependence among repeated measures without explicitly modeling subject-specific effects. Available methods for longitudinal neuroimaging data include those of Skup et al. (81), who proposed a generalized method of moments approach (GMM; 35) suitable for covariates that may be time-varying, and Li et al. (50), who pursued a generalized estimating equations (GEE; 52) approach. Both of these papers improve upon separate voxelwise analyses by means of multiscale adaptive smoothing of the images. Guillaume et al. (34) propose a more computationally efficient marginal approach in which OLS parameter estimation is combined with a sandwich estimator for the standard errors. They consider small-sample adjustments and conduct extensive simulations comparing variants of their method with N-OLS, SS-OLS, and linear mixed-effects models.

18.4.4 Relative Efficiency for Estimating the Mean Function

Consider longitudinally observed responses that are assumed to follow the random-effects model

$$y_{ij} = \mu(t_{ij}) + u_i + e_{ij}, \tag{18.7}$$

a generalization of the random-intercept linear model (18.5). Many authors have emphasized the need for longitudinal data when inferring neurodevelopmental trajectories (e.g., 39), but as noted above, inferences drawn from longitudinal data often center not on individual trajectories of change but on the mean function $\mu(t)$. If $\mu(t)$ is the target of inference, then what can we say about the advantage of longitudinal as opposed to cross-sectional data?

To aid intuition, in the left panel of Figure 18.2 we present a "spaghetti plot" of the cortical thickness measures for one vertex, from the longitudinal sample described in §18.5.3 below. Rather than imposing parametric assumptions on $\mu(\cdot)$ in (18.7), we fitted a nonparametric mixed model to these data (see §18.5.2). The right panel displays the values $\hat{\mu}(t_{ij}) + \hat{e}_{ij}$ from the fitted model, i.e., the data with the predicted random effects $\hat{u}_i$ "subtracted off." These values are much less noisy than those in the left panel, suggesting that the longitudinal design allows for more precise estimation of $\mu(\cdot)$.

To frame the issue more formally, suppose that, instead of observing I individuals at $J_1, \ldots, J_I$ time points respectively, we had observed $N \equiv \sum_{i=1}^{I} J_i$ separate individuals cross-sectionally. In this cross sectional scenario, instead of (18.7), the model might be specified as

$$y_i^* = \mu(t_i^*) + \varepsilon_i^*, \tag{18.8}$$

$i = 1, \ldots, N$; note that since the cross-sectional design does not allow us to separate between- from within-subject error (as in the subject effect u_i vs. the noise term e_{ij} as in (18.7)), both are lumped together in the error term ε_i^*. Suppose further that the same collection of N time points occur in both scenarios, i.e., $(t_{ij} : i = 1, \ldots, I, j = 1, \ldots, J_i)$ and $(t_i^* : i = 1, \ldots, N)$ are identical except for possible reordering. If we let $v_{long}(t)$ and $v_{cs}(t)$ denote the variance in estimating $\mu(t)$ by models (18.7) and (18.8), respectively, then the gain from the longitudinal design can be measured as $\text{avg}[v_{long}(t)]/\text{avg}[v_{cs}(t)]$, where "avg"

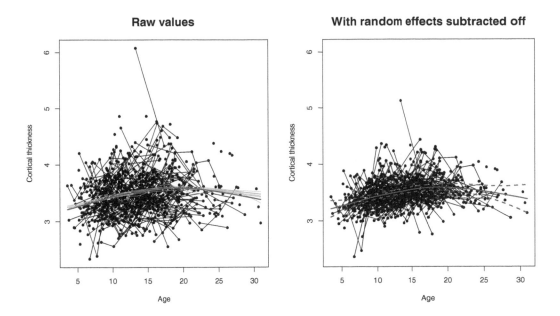

FIGURE 18.2
At left, the raw cortical thickness values at one vertex for a longitudinal sample (the curves are discussed below in §18.5.3). Upon fitting model (18.7), the estimated values with random subject effects removed, i.e. $\hat{\mu}(t_{ij})+\hat{e}_{ij}$, are as shown at right; the red curves represent $\hat{\mu}(t)\pm 2$ standard errors.

may denote either the average over all N time points or the integral over the time range of interest. Borrowing a term from the experimental design literature (e.g., 43), we can refer to this ratio as the I-efficiency of the longitudinal design relative to the cross-sectional design.

We have undertaken a preliminary theoretical study of this relative I-efficiency, defined using the average over the N time points, under the assumption that $\mu(\cdot)$ is modeled as a linear combination of K basis functions $b_1(t), \ldots, b_K(t)$—for example, the polynomials $b_1(t) \equiv 1, b_2(t) = t, \ldots, b_K(t) = t^{K-1}$, or (unpenalized) B-splines. The I-efficiency can be shown to depend on the within-subject correlation $\rho = \sigma_u^2/(\sigma_u^2 + \sigma_e^2)$, where σ_u^2 and σ_e^2, the random effects variance and noise variance in (18.7), respectively, are assumed to be estimable without error. As $\rho \to 1$, irrespective of the choice of basis functions or time points, the I-efficiency approaches $\bar{J}/K$, where $\bar{J} = \sum_{i=1}^{I} J_i/I$. Thus if the average number of observations per subject is lower than the number of basis functions, the I-efficiency will be less than 1 for high ρ; but it is above 1 for small ρ, at least for all scenarios we have observed. Intuitively, by using a longitudinal design to disentangle between- from within-subject variation, one can estimate $\mu(\cdot)$ more efficiently in this particular sense, but only if the between-subject proportion of the variance is sufficiently large and the repeated observations are sufficiently sparse. Note, however, that the I-efficiency has not been examined for estimation by penalized basis functions (see §18.5.2).

18.4.5 Complications Due to Misalignment

All longitudinal studies contain multiple major sources of error that are modeled statistically using subject-specific effects or treated as noise. Longitudinal *imaging* studies have a specific source of error that can probably be mitigated, but cannot be eliminated: *registration*, the process of aligning voxels, within or between subjects, to have the same interpretation. Despite recent attempts to grapple with the complexities of alignment for longitudinal neuroimaging (e.g., 27, 72), little is currently known about how various registration algorithms affect alignment, whether some algorithms are better than others, and when results are fundamentally affected by the choice of registration algorithm.

The notation $\{y_{ij}(v)\ (v \in \mathcal{V}), t_{ij}, \boldsymbol{x}_{ij}\}$, which we introduced in §18.2 for our data structure, implicitly assumes that the ordering of voxels $v \in \mathcal{V}$ is the same for every visit of every subject. But this assumption is often violated, as registration algorithms will attempt to align images such that voxel locations are as close as possible to a theoretically true location. The importance of this point is often dismissed by a statement such as: "for all practical purposes we can assume that the voxels are aligned." While such statements are more acceptable for co-registration (registration of sequences across multiple visits for the same subject, using a subject-specific template), they are highly problematic when images are registered across subjects or to a standard template, such as those produced by the Montreal Neurological Institute (MNI). Disease progression or aging effects may exacerbate the difficulty. Thus a more "honest" notation for responses might be $y_{ij}(v_{ij})$ with $v_{ij} = v + \eta_{ij}(v)$, where the "location error" $\eta_{ij}(v)$ can depend on the true target location, v, as well as on the subject i and visit j. Even the usual mean-zero assumption is debatable for $\eta_{ij}(v)$, as it implies that if one registered an infinite number of longitudinal scans, the average over all visits and/or subjects of the voxel v_{ij} identified as v is actually v. That this is true at every v seems dubious and is impossible to check, as we do not really know what the true location v is. To make any progress we shall assume that $E[\eta_{ij}(v)] = 0$.

If the spatial distortions $\eta_{ij}(v)$ are small, then we have the first-order Taylor approximation $y_{ij}(v_{ij}) \approx y_{ij}(v) + \eta_{ij}(v)y'_{ij}(v)$, where the derivative $y'_{ij}(v)$ is taken along the $v \to v_{ij}$ direction. In practice, this direction will change with the visit and subject in an uncontrollable way. If voxel v is located in tissue where y_{ij} changes very slowly in all directions (e.g., T1-weighted intensity in normal-appearing white matter), then the error could be ignored if spatial distortions are kept small. However, the error cannot be ignored when v is near a relevant tissue boundary (e.g., between white and gray matter), when tissue properties change fast (e.g., fractional anisotropy along the corpus callosum or between corpus callosum and surrounding tissue), or when high-frequency spatial inhomogeneities are observed in the data (e.g., local intensity variations due to preprocessing and/or magnetic field inhomogeneities).

Let $\widetilde{\eta}_{ij}(v) = \eta_{ij}(v)y'_{ij}(v)$. If we treat $\widetilde{\eta}_{ij}(v)$ as another noise variable and assume small spatial deformations, then we have, for example, a modified RIRS model (cf. (18.6)):

$$y_{ij}(v) = [\beta_0(v; \boldsymbol{x}_{ij}) + u_{0i}(v)] + [\beta_1(v; \boldsymbol{x}_{ij}) + u_{1i}(v)]t_{ij} + e_{ij}(v) + \widetilde{\eta}_{ij}(v). \qquad (18.9)$$

The main point here is that $e_{ij}(v) + \widetilde{\eta}_{ij}(v)$ is probably strongly spatially correlated, with a variance and correlation structure that depend on the location v. Regrettably, as yet nothing is known about the structure of the variance and spatial correlation, and whether these components are affected by the visit number or by the specific subject.

Given these seemingly intractable problems, one is forced to decide what simplifying model assumptions to make. The first and easiest option is simply to fit linear mixed-effects models at every voxel while ignoring the effects of $\widetilde{\eta}_{ij}(v)$. Estimated residuals can then be studied to uncover spatial correlations and temporal trends. Second, one could explicitly model residuals as a spatial process plus white noise. Such an approach would be slower, requiring a larger number of voxels (over which the spatial modeling would be applied) and specific spatio-temporal assumptions about the correlation. Third, one could jointly analyze the entire image and specify a low-rank spatial noise process, as in §18.6.2 below. If one is interested in estimating the fixed effects component (e.g., $\beta_0(v; \boldsymbol{x}_{ij}) + \beta_1(v; \boldsymbol{x}_{ij})t_{ij}$ in (18.9)), a practical approach is to estimate it under the independence assumption of residuals both in time and space. Confidence intervals for parameters can then be obtained by bootstrapping subjects, either voxel-wise, by regions of interest (ROI), or at the level of the image (see 19). In general, it may be a good idea to use multiple registration algorithms, and check how heavily the results depend on the particular choice of algorithm.

18.4.6   Borrowing Information "Spatially"

A number of authors have sought to improve upon mass-univariate models by borrowing information across neighboring voxels. The multiscale adaptive regression model of Li et al. (51) incorporates observations from a spherical neighborhood of the given voxel in calculating parameter estimates and test statistics, with the size of the neighborhood chosen adaptively. The marginal models of Skup et al. (81) and Li et al. (50) cited above (§18.4.3.2) extend this spatial smoothing approach to longitudinal data. Other methods for repeated imaging data that borrow information within regions are proposed by Derado et al. (23) and Bernal-Rusiel et al. (10).

Alternatively, instead of a set of separate models, one for each $v \in \mathcal{V}$, one may view the responses at all v as constituting a single functional response (65). Linear models for functional responses are applied to brain imaging data by Zhu et al. (95), Zhu et al. (94), Reiss et al. (68), and others. Relevant references for functional mixed models include Morris et al. (62), Yuan et al. (92) and Zhu et al. (93). Functional-response models that are nonlinear with respect to time are considered by Chiou et al. (16) and Reiss et al. (69); cf. the work of Greven et al. (33), which we discuss below in §18.6.2.

18.5   Nonlinear Models for the Mean

In many applications, the trajectories $\mu_v(\cdot, \boldsymbol{x})$ $(v \in \mathcal{V})$ in (18.1) are expected to be nonlinear with respect to time, especially when a substantial portion of the lifespan is considered. This section concerns (mass-univariate) modeling of nonlinear mean functions. As in §18.4.3.1, we omit covariates $\boldsymbol{x}$ for most of this section to keep the notation straightforward.

18.5.1   Polynomial Models

The simplest and most popular way to model nonlinearity is to posit the quadratic model $\mu_v(t) = \beta_{0v} + \beta_{1v}t + \beta_{2v}t^2$. A more flexible option is to allow μ_v to be either constant, linear, quadratic, or cubic with respect to t. Some authors choose among these models at each location v, by a sequence of hypothesis tests or a model selection criterion such as the Akaike (2) information criterion, and divide the brain into regions exhibiting different polynomial patterns of change (linear, quadratic, etc.). But such polynomial models have a number of limitations. Development or change, in the brain as elsewhere, may proceed in a manner not well described by a polynomial, such as rapid change followed by a plateau. Moreover, with polynomial models, key features of an estimate of μ_v, such as where it attains its peak, can be highly sensitive to the range of ages considered (26).

18.5.2   Nonparametric and Semiparametric Models

The limitations of polynomial models can be overcome by instead allowing each μ_v to be an arbitrary smooth function, modeled by penalized B-splines (e.g., 88). We assume that the mean function can be written as $\mu_v(t) = (t)^T\gamma_v$ for some $\gamma_v \in \mathbb{R}^K$, where $(t) = [b_1(t), \ldots, b_K(t)]^T$ is a set of B-spline basis functions. In the cross-sectional case, γ_v is estimated by penalized least squares:

$$\hat{\gamma}_v = \arg\min_\gamma \sum_{i=1}^I \left[\{y_i - (t_i)^T\gamma\}^2 + \lambda_v \gamma^T \mathbf{P}\gamma \right]. \qquad (18.10)$$

Here λ_v is a positive constant, and $\mathbf{P}$ is often taken to be the $K \times K$ matrix with (i, j) entry $\int b_i''(t) b_j''(t) dt$, which implies that the penalty term $\gamma^T \mathbf{P}\gamma$ in (18.10) is equal to $\int [\mu''(t)]^2 dt$ for $\mu(t) = (t)^T\gamma$. This 2nd-derivative penalty causes smoother fits to be favored, to a degree that is governed by the value of λ_v. When solving (18.10) for each of a large set of regions v, automatic optimal selection of λ_v ($v \in \mathcal{V}$) is essential. Recent work (e.g., 75, 71) has advocated exploiting a correspondence between penalized splines and mixed models to motivate using restricted maximum likelihood (REML) as the criterion for choosing λ_v. Reiss et al. (70) derive a fast technique to optimize the REML criterion for each of a large set of models, which can reduce the computation time for optimal spline-based estimation of μ_v at each of tens of thousands of brain regions from hours to minutes. A similar approach allows for fast region-wise testing of the null hypothesis that μ_v lies in the null space of the penalty in (18.10)—e.g., that μ_v is linear, in the case of a 2nd-derivative penalty— via restricted likelihood ratio tests (RLRT; 18). These "massively parallel" algorithms for region-wise function estimation and hypothesis testing are implemented in the **vows** package (67) for R (63).

We illustrate these procedures using maps of cortical thickness at each of 77,824 vertices, from a longitudinal MRI study conducted at the National Institute of Mental Health by Dr. Jay Giedd, who kindly made the data available to us. The full dataset comprises 1181 cortical thickness maps from 615 individuals belonging to 398 families; most of the individuals were scanned repeatedly, as many as 6 times. We randomly sampled one image per family to obtain a cross-sectional sample with no within-family dependence. We performed restricted likelihood ratio tests of the null hypothesis H_{0v}: μ_v is linear, for each of the 77,824 vertices v. Figure 18.3 displays the 9013 vertices for which significant nonlinearity was detected, in the sense that H_{0v} was rejected with false discovery rate .05 (8). The most prominent region of nonlinearity lies in the temporal poles.

To summarize the collection of region-wise trajectory estimates, Reiss et al. (70) apply a functional k-means algorithm to the estimated first derivatives, yielding clusters of locations v for which the estimate of μ_v is of similar shape. Figure 18.4 displays the 3-cluster solution for our cortical thickness data. The cluster shown in blue, characterized by an increase in thickness followed by decrease, lies mainly in the temporal pole region noted above in Figure 18.3.

While the fast spline-fitting algorithm of Reiss et al. (70) is not directly applicable to longitudinal data, the clustering approach is. Alexander-Bloch et al. (4) estimated vertex-by-vertex mean cortical thickness in each of two longitudinal samples of children: a typically developing group and a group with childhood-onset schizophrenia (COS). They considered a varying-coefficient mixed-effects model (37, 90) of the form (18.7), in which the mean thickness at age t was

$$\mu_v(t) = \begin{cases} f_v(t), & \text{for the controls, and} \\ f_v(t) + h_v(t), & \text{for the COS group.} \end{cases}$$

Upon partitioning the cortex into five "normative development" clusters based on the estimates of f_v, they observed that significant "shape differences" between the two groups (i.e., strong evidence of non-constant h_v) occurred primarily in one of these clusters. Alexander-Bloch et al. (4) used a modified Wald statistic to test the null hypothesis $h_v \equiv 0$; and when

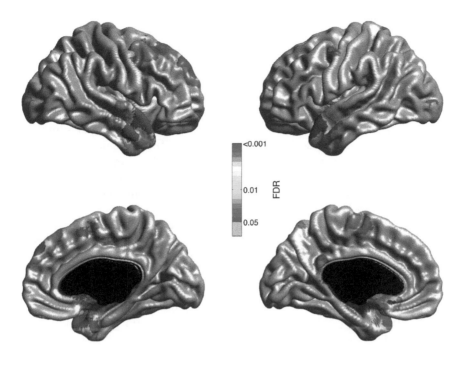

FIGURE 18.3
False discovery rate-corrected p-values for restricted likelihood ratio tests of the null hypothesis H_{0v}: μ_v is linear, for each vertex v.

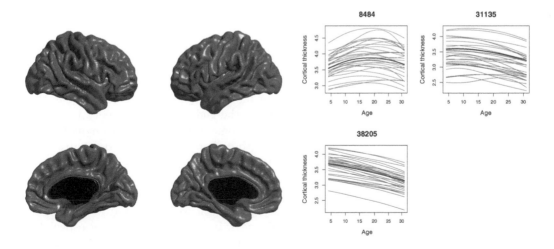

FIGURE 18.4
Left: Division of the brain into 3 clusters of similar cortical thickness trajectories. Right: Examples of mean function estimates for randomly chosen vertices within each cluster, with cluster mean function shown in black and number of vertices in the cluster shown above each subfigure.

this null hypothesis was rejected, they tested the null hypothesis of a constant function $h_v(\cdot)$ against the alternative of an age-varying $h_v(\cdot)$. This modified Wald statistic was developed by Wood (91) for a general class of smooth terms in "extended generalized additive models." Further work is needed to evaluate this approach for the special case of varying-coefficient models.

18.5.3   Analyses with Repeated Cross-Sectional Subsamples

As noted above, Figures 18.3 and 18.4 are derived from a cross-sectional sample of 398 cortical thickness maps randomly subsampled from a longitudinal sample of 1181 maps. In this subsection we perform a nonstandard type of resampling, drawing repeated cross-sectional subsamples of size 398, and apply the fast smoothing method of Reiss et al. (70) to all 77,824 vertices in each of these subsamples—and thereby gain new insights regarding such "mass-produced" trajectory estimates.

First, we conducted a qualitative comparison of vertex-wise cross-sectional and longitudinal function estimates. We drew a longitudinal sample of 398 individuals, one from each family, who had a total of 766 scans. We fitted model (18.7) by penalized splines, i.e., a nonparametric mixed model, for each of the 77,824 vertices; 10 quartic B-spline functions with a 3rd-derivative penalty were used, with knots at equally spaced quantiles of the observed ages. We then drew 10 random cross-sectional subsamples of this longitudinal sample, comprising one randomly selected observation from each subject, and fitted penalized spline regressions to each subsample. For the randomly chosen vertex whose data appear in the left panel of Figure 18.2, the fits based on cross-sectional subsamples (green curves) are quite similar to each other and to the longitudinal estimate (red curve). However, comparing the longitudinal and cross-sectional fits across the brain revealed that the roughness of the function estimate $\hat{\mu}_v$, as gauged by the penalty $\int [\hat{\mu}_v'''(t)]^2 dt$, tended to be lower for the former: the roughness penalty was lower for the longitudinal fit than for any of the 10 cross-sectional fits for 36% of the vertices, and lower than all but one of the 10 for another 15%. An intuitive explanation is suggested by the right panel of Figure 18.2: by separating out the between-subject portion of the variance, the longitudinal analysis avoids incorporating spurious signal that may increase the roughness of cross-sectionally-based function estimates.

Next we used random cross-sectional subsamples to compare the standard 2nd-derivative penalty functional $g \mapsto \int [g''(t)]^2 dt$ with the 3rd-derivative penalty $\int [g'''(t)]^2 dt$. This comparison was motivated by our observation that, when we used cubic B-splines with a 2nd-derivative penalty, the estimated mean function for most vertices was highest at age 3, the lower end of the age range. Since, in Bayesian terms, the 2nd-derivative penalty implies a prior that favors a near-linear fit, we suspected that some of these vertices might in fact attain peak mean thickness at a somewhat later age, but that this was missed because this particular prior caused a downward-sloping line to be favored over a nonlinear curve with an early peak. If so, the problem might be remedied by opting for a 3rd-derivative penalty, corresponding to a prior preference for approximately quadratic, rather than linear, curves. Note that the influential findings of Shaw et al. (79), which we have cited above, relied crucially on estimating the age of peak cortical thickness. In light of this, it seems especially important to understand how the choice of penalty might affect conclusions about when the peak is attained.

For each of 20 cross-sectional subsamples, we estimated μ_v at each vertex using both cubic splines with 2nd-derivative penalties and quartic splines with 3rd-derivative penalties (for brevity we shall refer to the resulting function estimates as "2nd-derivative fits" and "3rd-derivative fits"). We used the generalized cross-validation criterion (GCV; 20) to compare the 2nd- and 3rd-derivative fits. Treating the 20 random subsamples as independent

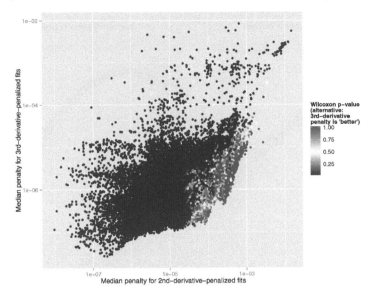

FIGURE 18.5

Median 2nd-derivative penalty (over 20 random cross-sectional subsamples) plotted against median 3rd-derivative penalty, for cortical thickness trajectory estimates based on 2nd- and 3rd-derivative penalties respectively; each point corresponds to one vertex. Wilcoxon signed-rank tests were applied to the differences between the 2nd- and 3rd-derivative fits' GCV scores for the 20 subsamples (see the text). Blue and red dots represent vertices for which the 2nd- and 3rd-derivative penalty, respectively, performed better according to these tests.

samples from an underlying distribution, we applied the Wilcoxon signed-rank test to the differences between the GCV scores attained with the two penalties for each subsample. Roughly speaking, this procedure assesses the null hypothesis that the two penalties yield equally good prediction error. In Figure 18.5, created with the R package `ggplot2` (89), the 2nd-derivative penalty for the 2nd-derivative fits at each vertex is plotted against the 3rd-derivative penalty for the 3rd-derivative fit, with the points color-coded by the signed-rank test p-value. Here we used the one-sided alternative of lower GCV with the 3rd-derivative penalty, so that p-values near 0 (red) imply that the 3rd-derivative fit tends to perform better, whereas p-values near 1 (blue) mean the 2rd-derivative fit is preferred. The 3rd-derivative penalty is seen to yield better performance (lower GCV) for the overwhelming majority of the vertices, but interestingly, the 2nd-derivative fit is preferred for a very specific set of vertices: those for which the 2nd-derivative penalty is relatively high but the 3rd-derivative penalty is not.

For 13.6% (10,554 out of 77,824) of the brain vertices, most of the twenty 2nd-derivative fits attained their highest point at the lower end of the age range, but half or more of the 3rd-derivative fits peaked at a later age. This finding provides one indication of the sensitivity of peak age estimates to the choice of penalty, and a closer look at the repeated cross-sectional fits for two of these 10,554 vertices yields an interesting perspective on the results shown in Figure 18.5. For the vertex whose curve estimates are shown at left in Figure 18.6, the 2nd-derivative fits are essentially downward-sloping lines; evidently the 2nd-derivative penalty's prior predilection towards a linear fit has prevailed for this vertex. But the GCV criterion prefers the 3rd-derivative penalty, suggesting that the true curve is

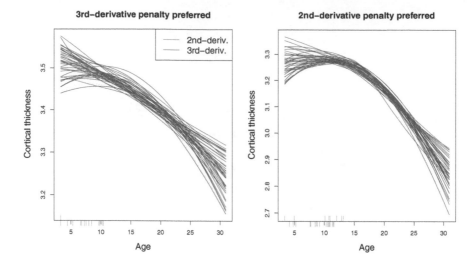

FIGURE 18.6
Two vertices for which 2nd- and 3rd-derivative-penalized fits yield conflicting results regarding age of peak cortical thickness. Specifically, 2nd-derivative fits for most of the 20 random cross-sectional subsamples suggest mean cortical thickness is highest at the lower end of the age range, whereas most of the 3rd-derivative fits attain a peak at a later age (rug plots display the peak ages for each of the curve estimates). GCV scores indicate that the 3rd-derivative penalty yields superior curve estimates for the example at left, while the 2nd-derivative penalty performs better for the example at right.

nonlinear, but the 2nd-derivative fits are "misled" by this prior inclination toward linearity. On the other hand, the right panel refers to one of the vertices represented by blue dots in Figure 18.5, for which the 2nd-derivative fit is relatively highly penalized but the 3rd-derivative fit is not. Here it is the 3rd-derivative fit that seems to be led astray by the prior preference, in this case for an approximately quadratic fit; the GCV criterion favors the 2nd-derivative fits, which are nonlinear but not quadratic. These two examples, while hinting at the limitations of polynomial curves, clearly show that spline fits have their own pitfalls and should be applied with care—and specifically, that different penalties should be considered in practice.

18.6   Beyond Modeling the Mean

This section considers several methods that, instead of focusing on the mean function $\mu_v(\cdot|\boldsymbol{x})$ in (18.1), set the more ambitious goal of understanding individual trajectories of change.

18.6.1   Individual-Specific Curves

For data arising from an accelerated longitudinal design, Harezlak et al. (36) propose the model

$$y_{ijv} = \mu_v(t_{ij}) + g_{iv}(t_{ij}) + \varepsilon_{ijv}, \tag{18.11}$$

which includes not only the group mean function μ_v but subject-specific curves $g_{1v}, \ldots, g_{nv}$ for the n individuals. The authors present a computationally efficient method for fitting this model by penalized splines. This model is highly appealing in that, unlike model (18.1), it offers the potential to capture individual trajectories of change. Most neuroimaging studies, however, do not include enough scans per subject to allow for very precise estimation of subject-specific curves, and this may have posed a barrier to more widespread application of model (18.11).

18.6.2   Modeling Components of Change: Longitudinal Functional Principal Component Analysis

Greven et al. (33) introduce a longitudinal functional principal component analysis (LF-PCA) framework for modeling functional data observed at multiple time points. They apply LFPCA to *curves* derived from a longitudinal diffusion tensor imaging (DTI) study (hence for this subsection the voxel locations will be taken to lie in $[0, 1]$); subsequent work (97, 49) has extended the approach to entire images. Suppose our sample includes I individuals, the ith of whom is observed at time points $t_1, \ldots, t_{J_i}$. At the jth visit, the ith subject contributes a functional observation $y_{ij}(v)$ ($v \in [0, 1]$) which is modeled as

$$y_{ij}(v) = \mu(v, t_{ij}) + x_{i0}(v) + x_{i1}(v)t_{ij} + u_{ij}(v) + e_{ij}(v). \tag{18.12}$$

Here $\mu(\cdot, \cdot)$ is the mean function; $x_{i0}(\cdot)$ and $x_{i1}(\cdot)$ are drawn from a random intercept process and a random slope process, respectively; $u_{ij}(\cdot)$ is drawn from a visit-specific process; and $e_{ij}(\cdot)$ is drawn from a homoscedastic white noise process with variance σ^2 (see 33, for a full specification of the model). Model (18.12) extends the RIRS model (18.6) to responses that are functions rather than scalars. Model fitting relies on the Karhunen–Loève expansions of both the bivariate process generating the random intercept and random slope functions $\boldsymbol{x}_i(\cdot) = [x_{i0}(\cdot), x_{i1}(\cdot)]^T$, and the process generating the visit-specific deviations $u_{ij}(\cdot)$:

$$\boldsymbol{x}_i(v) = \sum_{k=1}^{\infty} \xi_{ik} \boldsymbol{\phi}_k^x(v), \qquad u_{ij}(v) = \sum_{k=1}^{\infty} \zeta_{ijk} \phi_k^u(v).$$

Here $\boldsymbol{\phi}_k^x(\cdot) = [\phi_k^0(\cdot), \phi_k^1(\cdot)]^T$ is the kth eigenfunction of the covariance operator of the bivariate RIRS process; $\phi_k^u(\cdot)$ is the kth eigenfunction of the covariance operator of the visit-specific process; the ξ_{ik}'s are uncorrelated random variables with mean 0 and variance λ_k; and the ζ_{ijk} are uncorrelated with mean 0 and variance ν_k. Truncating these expansions at N_x and N_u components, respectively, and substituting into (18.12) yields

$$y_{ij}(v) \approx \mu(v, t_{ij}) + \sum_{k=1}^{N_x} \xi_{ik} \boldsymbol{\tau}_{ij}^T \boldsymbol{\phi}_k^x(v) + \sum_{k=1}^{N_u} \zeta_{ijk} \phi_k^u(v) + e_{ij}(v) \tag{18.13}$$

where $\boldsymbol{\tau}_{ij} = (1, t_{ij})^T$. Under conditions given by Greven et al. (33), the variation in $y_{ij}(v)$ can be decomposed as

$$\sum_{k=1}^{\infty} \lambda_k + \sum_{k=1}^{\infty} \nu_k + \sigma^2, \tag{18.14}$$

i.e., contributions from the RIRS process, from the visit-specific process, and from random noise.

     As an example we consider a set of curves representing fractional anisotropy (FA) along the corpus callosum, derived by diffusion tensor imaging (DTI) in a sample of patients with multiple sclerosis. The DTI data were collected at Johns Hopkins University and the

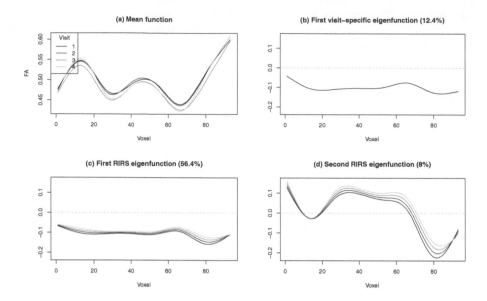

FIGURE 18.7

Longitudinal functional PCA estimates for the corpus callosum FA data. (a) Estimated mean FA curve $\mu(v,t)$, for t set to visits 1, 2, 3, 4. (b) First eigenfunction $\phi_1^u(v)$ of the visit-specific process. (c) First eigenfunction of the RIRS process, represented as $\phi_1^0(v) + t\phi_1^1(v)$ for t set to visits 1, 2, 3, 4. (d) Second eigenfunction of the RIRS process, represented as $\phi_2^0(v) + t\phi_2^1(v)$. Shown in parentheses are the percent of total variation accounted for by each component.

Kennedy-Krieger Institute. FA curves from this sample were analyzed by Greven et al. (33) and in several other papers on methods for longitudinally observed functional data (e.g., 32, 29). The 340 functional responses, obtained from 100 patients with 2–8 visits each, are included in the R package **refund** (41). Each curve consists of FA values at 93 voxels, arranged from anterior to posterior along the corpus callosum.

Using the LFPCA algorithm of Greven et al. (33), as implemented in R code accompanying their paper, we estimated the mean function and eigenfunctions on the right side of (18.13). The estimated mean FA is highest in the most posterior portion, and declines somewhat across the corpus callosum from visit 1 to visit 4 (Figure 18.7(a)). The leading eigenfunction $\phi_1^u(v)$ of the visit-specific process (Figure 18.7(b)) is somewhat akin to an across-the-function location shift. The first two eigenfunctions of the bivariate RIRS process are displayed in Figure 18.7(c)–(d) as $\phi_k^0(v) + t\phi_k^1(v)$ $(k = 1, 2)$ for t representing each of the first four visits. When displayed "dynamically" in this way, these eigenfunctions are seen to represent principal directions of systematic time-dependent deviation from the mean FA curve. It appears from Figure 18.7(c) that the first RIRS eigenfunction has features similar to $\phi_1^u(v)$.

In practice, estimates of the functional principal component scores $\xi_{i1}, \ldots, \xi_{iN_x}$ $(i = 1, \ldots, I)$ may be the most useful product of fitting model (18.13). These scores provide a low-dimensional representation of the change in the entire function for the ith individual, relative to the mean function, over time. This low-dimensional representation can serve as input to descriptive multivariate analyses, such as clustering (49), or can be related to clinical variables of interest.

One limitation of model (18.12) is that it captures only linear change (at each v) with respect to time. To extend the LFPCA framework to modeling of nonlinear change, one might consider the following more general model:

$$y_{ij}(v) = \mu(v, t_{ij}) + \sum_{m=1}^{M} x_{im}(v)\psi_m(t_{ij}) + u_{ij}(v) + e_{ij}(v)$$

(cf. Section 2.2 of 33), which reduces to (18.12) (with a slight shift in subscripts) when $M = 2, \psi_1(t) \equiv 1, \psi_2(t) = t$. The temporal basis functions $\psi_1, \ldots, \psi_M$ can be chosen to capture nonlinear change; natural choices are polynomials or (temporal) functional principal components. In principle, as in the linear case, one can eigendecompose the process generating the M-variate function $[x_{i1}(\cdot), \ldots, x_{iM}(\cdot)]^T$ to obtain a truncated expansion similar to (18.13) and a variance decomposition akin to (18.14).

18.6.3 Modeling the Entire Age-Specific Distribution

The LMS growth chart methodology of Cole and Green (17) fits an age-varying Box-Cox-transformed distribution for a positive parameter of interest, where the mean, standard deviation and transformation parameter change smoothly with age. Chen et al. (15) show how this approach, as modified by Rigby and Stasinopoulos (74), can be used to compare a given individual's brain measure of interest with age-specific norms, via an estimate of the individual's age-specific quantile rank. This methodology can be leveraged to discover neurodevelopmental normalities associated with a disorder, by means of a testing procedure proposed by Chen et al. (15), in which one estimates the age-varying distribution for a control group, and then applies a Kolmogorov-Smirnov-type test to the quantile ranks (based on this estimated distribution) for a set of individuals with a disorder. Chen et al. (15) illustrate their methods with a dataset extracted from the Autism Brain Imaging Data Exchange (ABIDE) sample. In the paper introducing ABIDE, Di Martino et al. (24) considered resting-state functional connectivity, essentially pairwise correlations of the resting fMRI BOLD signals, among the regions shown at left in Figure 18.8. For the edge (pair of regions) indicated, the middle panel of Figure 18.8, in which controls and individuals with autism spectrum disorders (ASD) are represented by blue and red dots respectively, suggests that many of those with ASD fall toward the lower end of the age-specific distribution for the controls—a claim that is supported by the histogram in the right panel. This apparent group difference is not inferred by a conventional regression-based t-test, but is detected by the procedure of Chen et al. (15).

18.6.4 Modeling Local Rates of Change

Recall from §18.3.3 that MLDs collect longitudinal data on subjects with varying baseline ages. Usually the age span of interest (e.g., 20 years) is much longer than the duration on study for each subject (e.g., 3–5 years). Developmental or aging trajectories over long time spans are often nonlinear, especially in the presence of a disease process such as Alzheimer's disease, which greatly accelerates cortical atrophy over normative rates of change (85). However, the duration of subject-level assessments is often so short that only linear models can be fitted to encompass within-subject change.

One possibility is to model local change (i.e., derivatives), assessed via within-subject change in age, as a smoothly varying nonlinear function of between-subject differences in ages. An approach along these lines builds on prior work in estimating differential equations and integral curves from noisy slope fields (1, 46, 64, 55). Reference (1) considers the situation where univariate "local" observations follow an underlying longitudinal model with

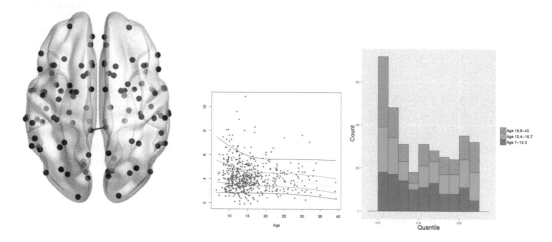

FIGURE 18.8
Left: The line segment indicates an edge connecting two of the regions considered by Di Martino et al. (24). Center: Resting-state functional connectivity for this pair of regions plotted against age (blue dots represent controls; red, ASD), along with estimated 5th, 25th, 50th, 75th, and 95th percentile curves for the control data. Right: Histogram of quantile ranks for the ASD subjects, with respect to the estimated age-varying distribution for the controls. Each bar is divided into three age groups.

a population mean function. Specifically, they consider independent identically distributed pairs $\{y_i(T_i), \dot{y}_i(T_i)\}$, $i = 1, \ldots, n$, where T_i is an unobservable random variable and y_i and $\dot{y}_i$ are the mean and noisily observed first derivative of an underlying dynamical system $f(\cdot)$:

$$y_i(T_i) = f(T_i) \text{ and } \dot{y}_i(T_i) = f'(T_i) + \nu_i,$$

where $E(\nu_i) = 0$ and $\text{Var}(\nu_i) < \infty$. They assume that the system can be characterized by the autonomous ordinary differential equation, $f'(t) = g[f(t)]$, where the unknown smooth function g is estimated nonparametrically and f is estimated by numerically integrating the estimate of g. In our context, the unobserved T_i might be an appropriate way to refer to elapsed time since the unknown onset time of a disorder, and $\dot{y}_i$ might be a slope estimate based on two or more scans.

There are a number of possible elaborations to the approach of (1). For example, local rates of change may depend on the age of the subject, in addition to the mean level (non-autonomous differential equations). It would also be desirable in many situations to include covariates that modify rates of change (e.g., gender or $APOE$ genotype in subjects at risk for Alzheimer's disease). In addition, modeling multivariate outcomes, such as rates of change in multiple regions simultaneously, will be important in some applications.

18.7   Discussion

Modeling change in the brain is a vast topic with numerous important subtopics that we were not able to cover here. For example:

- In this chapter we have generally taken the response y to be a region-wise measure of interest. Recent work has examined the lifespan trajectories of various more complex measures of brain connectivity or network structure, both functional (e.g., 61, 12) and structural (e.g., 3, 22). The distance-based test procedure of Shehzad et al. (80) may prove useful for inferring age-related change in complex brain indices, such as maps of connectivity between a given region and all other brain regions.

- We have largely focused on the dependence of a response y on a time variable t, adjusting for or moderated by covariates $\boldsymbol{x}$. Behavioral scientists often seek to understand more complex relationships among variables, such as mediating effects (57), that may be best captured in a structural equation modeling framework (60). Maxwell and Cole (59) caution against using cross-sectional data to infer longitudinal mediation, and more specifically, Raz and Lindenberger (66) argue that longitudinal data are indispensible for inferring relationships between brain aging and cognitive change. While causal mediation for fMRI time series has received some recent attention (e.g., 53), much more work is needed on related models for the study of change in the brain.

- A number of groups have recently developed measures of effective brain age. These are essentially predicted values of age, derived by applying machine learning algorithms (e.g., support vector regression) with age as response and brain measures as predictors (25, 11). Such latent markers of biological age can be related to behavioral, cognitive, or health-related variables of interest (96)—for example, for predicting conversion from mild cognitive impairment to Alzheimer's disease (28).

We hope and expect that in the coming years, statistical innovations in the areas we have outlined, and others, will play an important role in advancing the science of how change occurs in the human brain.

Acknowledgments

The work of Philip Reiss, Ciprian Crainiceanu and Lan Huo was supported by the National Institutes of Health through grant 1R01MH095836-01A1. Ciprian Crainiceanu was also supported by grants NIH/NINDS R01 NS060910 and NIH/NINDS R01 NS085211. The authors thank Lei Huang, Yin-Hsiu Chen and Thad Tarpey for their contributions to the work described in §18.5.2; Jay Giedd and Aaron Alexander-Bloch, for making available the cortical thickness data analyzed in §18.5.2; Adriana Di Martino and Chao-Gan Yan for providing the ABIDE data referred to in §18.6.3, and Huaihou Chen, for analyses of these data; and Eva Petkova and Xavier Castellanos, for many helpful discussions.

Bibliography

[1] Abramson, I. and H.-G. Müller (1994). Estimating direction fields in autonomous equation models, with an application to system identification from cross-sectional data. *Biometrika 81*(4), 663–672.

[2] Akaike, H. (1973). Information theory and an extension of the maximum likelihood principle. In *Second International Symposium on Information Theory*, pp. 267–281. Akademiai Kiado.

[3] Alexander-Bloch, A., J. N. Giedd, and E. Bullmore (2013). Imaging structural covariance between human brain regions. *Nature Reviews Neuroscience 14*(5), 322–336.

[4] Alexander-Bloch, A. F., P. T. Reiss, J. Rapoport, H. McAdams, J. N. Giedd, E. T. Bullmore, and N. Gogtay (2014). Abnormal cortical growth in schizophrenia targets normative modules of synchronized development. *Biological Psychiatry 76*, 438–446.

[5] Aylward, E. H., N. Minshew, K. Field, B. Sparks, and N. Singh (2002). Effects of age on brain volume and head circumference in autism. *Neurology 59*(2), 175–183.

[6] Bell, R. Q. (1953). Convergence: An accelerated longitudinal approach. *Child Development 24*(2), 145–152.

[7] Bell, R. Q. (1954). An experimental test of the accelerated longitudinal approach. *Child Development 25*(4), 281–286.

[8] Benjamini, Y. and Y. Hochberg (1995). Controlling the false discovery rate: A practical and powerful approach to multiple testing. *Journal of the Royal Statistical Society: Series B 57*(1), 289–300.

[9] Bernal-Rusiel, J. L., D. N. Greve, M. Reuter, B. Fischl, and M. R. Sabuncu (2013a). Statistical analysis of longitudinal neuroimage data with linear mixed effects models. *NeuroImage 66*, 249–260.

[10] Bernal-Rusiel, J. L., M. Reuter, D. N. Greve, B. Fischl, and M. R. Sabuncu (2013b). Spatiotemporal linear mixed effects modeling for the mass-univariate analysis of longitudinal neuroimage data. *NeuroImage 81*, 358–370.

[11] Brown, T. T., J. M. Kuperman, Y. Chung, M. Erhart, C. McCabe, D. J. Hagler, V. K. Venkatraman, N. Akshoomoff, D. G. Amaral, C. S. Bloss, B. J. Casey, L. Chang, T. M. Ernst, J. A. Frazier, J. R. Gruen, W. E. Kaufmann, T. Kenet, D. N. Kennedy, S. S. Murray, E. R. Sowell, T. L. Jernigan, and A. M. Dale (2012). Neuroanatomical assessment of biological maturity. *Current Biology 22*(18), 1693–1698.

[12] Cao, M., J.-H. Wang, Z.-J. Dai, X.-Y. Cao, L.-L. Jiang, F.-M. Fan, X.-W. Song, M.-R. Xia, N. Shu, Q. Dong, M. P. Milham, F. X. Castellanos, X.-N. Zuo, and Y. He (2014). Topological organization of the human brain functional connectome across the lifespan. *Developmental Cognitive Neuroscience 7*, 76–93.

[13] Chan, D., J. C. Janssen, J. L. Whitwell, H. C. Watt, R. Jenkins, C. Frost, M. N. Rossor, and N. C. Fox (2003). Change in rates of cerebral atrophy over time in early-onset Alzheimer's disease: Longitudinal MRI study. *Lancet 362*, 1121–1122.

[14] Chen, G., Z. S. Saad, J. C. Britton, D. S. Pine, and R. W. Cox (2013). Linear mixed-effects modeling approach to fMRI group analysis. *NeuroImage 73*, 176–190.

[15] Chen, H., C. Kelly, F. X. Castellanos, Y. He, X.-N. Zuo, and P. T. Reiss (2015). Quantile rank maps: A new tool for understanding individual brain development. *NeuroImage 111*, 454–463.

[16] Chiou, J. M., H. G. Müller, and J. L. Wang (2003). Functional quasi-likelihood regression models with smooth random effects. *Journal of the Royal Statistical Society: Series B 65*(2), 405–423.

[17] Cole, T. J. and P. J. Green (1992). Smoothing reference centile curves: The LMS method and penalized likelihood. *Statistics in Medicine 11*(10), 1305–1319.

[18] Crainiceanu, C. M. and D. Ruppert (2004). Likelihood ratio tests in linear mixed models with one variance component. *Journal of the Royal Statistical Society: Series B 66*(1), 165–185.

[19] Crainiceanu, C. M., A. M. Staicu, S. Ray, and N. M. Punjabi (2012). Bootstrap-based inference on the difference in the means of two correlated functional processes. *Statistics in Medicine 31*(26), 3223–3240.

[20] Craven, P. and G. Wahba (1979). Smoothing noisy data with spline functions: Estimating the correct degree of smoothing by the method of generalized cross-validation. *Numerische Mathematik 31*(4), 317–403.

[21] Davatzikos, C. and S. M. Resnick (2002). Degenerative age changes in white matter connectivity visualized in vivo using magnetic resonance imaging. *Cerebral Cortex 12*(7), 767–771.

[22] Dennis, E. L., N. Jahanshad, K. L. McMahon, G. I. de Zubicaray, N. G. Martin, I. B. Hickie, A. W. Toga, M. J. Wright, and P. M. Thompson (2013). Development of brain structural connectivity between ages 12 and 30: A 4-Tesla diffusion imaging study in 439 adolescents and adults. *NeuroImage 64*, 671–684.

[23] Derado, G., F. D. Bowman, and C. D. Kilts (2010). Modeling the spatial and temporal dependence in fMRI data. *Biometrics 66*(3), 949–957.

[24] Di Martino, A., C. Yan, Q. Li, E. Denio, F. Castellanos, K. Alaerts, J. Anderson, M. Assaf, S. Bookheimer, M. Dapretto, B. Deen, S. Delmonte, I. Dinstein, D. A. Ertl-Wagner, Band Fair, L. Gallagher, D. P. Kennedy, C. L. Keown, C. Keysers, J. E. Lainhart, C. Lord, B. Luna, V. Menon, N. J. Minshew, C. S. Monk, S. Mueller, R.-A. Müller, M. B. Nebel, J. T. Nigg, K. O'Hearn, K. A. Pelphrey, S. J. Peltier, J. D. Rudie, S. Sunaert, M. Thioux, J. M. Tyszka, L. Q. Uddin, J. S. Verhoeven, N. Wenderoth, J. L. Wiggins, S. H. Mostofsky, and M. P. Milham (2014). The Autism Brain Imaging Data Exchange: Towards a large-scale evaluation of the intrinsic brain architecture in autism. *Molecular Psychiatry 19*, 659–667.

[25] Dosenbach, N. U. F., B. Nardos, A. L. Cohen, D. A. Fair, J. D. Power, J. A. Church, S. M. Nelson, G. S. Wig, A. C. Vogel, C. N. Lessov-Schlaggar, K. A. Barnes, J. W. Dubis, E. Feczko, R. S. Coalson, J. R. Pruett, D. M. Barch, S. E. Petersen, and B. L. Schlaggar (2010). Prediction of individual brain maturity using fMRI. *Science 329*, 1358–1361.

[26] Fjell, A. M., K. M. Walhovd, L. T. Westlye, Y. Østby, C. K. Tamnes, T. L. Jernigan, A. Gamst, and A. M. Dale (2010). When does brain aging accelerate? Dangers of quadratic fits in cross-sectional studies. *NeuroImage 50*(4), 1376–1383.

[27] Fonov, V., A. C. Evans, K. Botteron, C. R. Almli, R. C. McKinstry, and D. L. Collins (2011). Unbiased average age-appropriate atlases for pediatric studies. *Neuroimage 54*(1), 313–327.

[28] Gaser, C., K. Franke, S. Klöppel, N. Koutsouleris, H. Sauer, and the Alzheimer's Disease Neuroimaging Initiative (2013). BrainAGE in mild cognitive impaired patients: Predicting the conversion to Alzheimer's disease. *PLOS ONE 8*(6), e67346.

[29] Gertheiss, J., J. Goldsmith, C. Crainiceanu, and S. Greven (2013). Longitudinal scalar-on-functions regression with application to tractography data. *Biostatistics 14*(3), 447–461.

[30] Giedd, J., F. Lalonde, M. Celano, S. White, G. Wallace, N. Lee, and R. Lenroot (2009). Anatomical brain magnetic resonance imaging of typically developing children and adolescents. *Journal of the American Academy of Child and Adolescent Psychiatry 48*(5), 465–470.

[31] Gogtay, N., J. N. Giedd, L. Lusk, K. M. Hayashi, D. Greenstein, A. C. Vaituzis, T. F. Nugent, D. H. Herman, L. S. Clasen, A. W. Toga, J. L. Rapoport, and P. M. Thompson (2004). Dynamic mapping of human cortical development during childhood through early adulthood. *Proceedings of the National Academy of Sciences 101*(21), 8174–8179.

[32] Goldsmith, J., C. M. Crainiceanu, B. Caffo, and D. Reich (2012). Longitudinal penalized functional regression for cognitive outcomes on neuronal tract measurements. *Applied Statistics 61*(3), 453–469.

[33] Greven, S., C. Crainiceanu, B. Caffo, and D. Reich (2010). Longitudinal functional principal component analysis. *Electronic Journal of Statistics 4*, 1022–1054.

[34] Guillaume, B., X. Hua, P. M. Thompson, L. Waldorp, and T. E. Nichols (2014). Fast and accurate modelling of longitudinal and repeated measures neuroimaging data. *NeuroImage 94*, 287–302.

[35] Hansen, L. P. (1982). Large sample properties of generalized method of moments estimators. *Econometrica 50*, 1029–1054.

[36] Harezlak, J., L. M. Ryan, J. N. Giedd, and N. Lange (2005). Individual and population penalized regression splines for accelerated longitudinal designs. *Biometrics 61*(4), 1037–1048.

[37] Hastie, T. and R. Tibshirani (1993). Varying-coefficient models. *Journal of the Royal Statistical Society: Series B 55*(4), 757–796.

[38] Holland, D., J. B. Brewer, D. J. Hagler, C. Fennema-Notestine, A. M. Dale, and the Alzheimer's Disease Neuroimaging Initiative (2009). Subregional neuroanatomical change as a biomarker for Alzheimer's disease. *Proceedings of the National Academy of Sciences 106*(49), 20954–20959.

[39] Horga, G., T. Kaur, and B. S. Peterson (2014). Annual Research Review: Current limitations and future directions in MRI studies of child-and adult-onset developmental psychopathologies. *Journal of Child Psychology and Psychiatry 55*(6), 659–680.

[40] Hua, X., D. P. Hibar, S. Lee, A. W. Toga, C. R. Jack Jr, M. W. Weiner, P. M. Thompson, and the Alzheimer's Disease Neuroimaging Initiative (2010). Sex and age differences in atrophic rates: An ADNI study with $n = 1368$ MRI scans. *Neurobiology of Aging 31*(8), 1463–1480.

[41] Huang, L., F. Scheipl, J. Goldsmith, J. Gellar, J. Harezlak, M. W. McLean, B. Swihart, L. Xiao, C. Crainiceanu, and P. Reiss (2015). *refund: Regression with Functional Data*. R package version 0.1-13.

[42] Insel, T. R. (2014). Mental disorders in childhood: Shifting the focus from behavioral symptoms to neurodevelopmental trajectories. *Journal of the American Medical Association 311*(17), 1727–1728.

[43] Jones, B. and P. Goos (2012). I-optimal versus D-optimal split-plot response surface designs. *Journal of Quality Technology 44*(2), 85–101.

[44] Jones, B. C., G. Nair, C. D. Shea, C. M. Crainiceanu, I. Cortese, and D. S. Reich (2013). Quantification of multiple-sclerosis-related brain atrophy in two heterogeneous MRI datasets using mixed-effects modeling. *NeuroImage: Clinical 3*, 171–179.

[45] Kent, S., R. Chen, A. Kumar, and C. Holmes (2010). Individual growth curve modeling of specific risk factors and memory in youth with type 1 diabetes: An accelerated longitudinal design. *Child Neuropsychology 16*(2), 169–181.

[46] Koltchinskii, V., L. Sakhanenko, and S. Cai (2007). Integral curves of noisy vector fields and statistical problems in diffusion tensor imaging: Nonparametric kernel estimation and hypotheses testing. *Annals of Statistics 35*(4), 1576–1607.

[47] Kraemer, H. C., J. A. Yesavage, J. L. Taylor, and D. Kupfer (2000). How can we learn about developmental processes from cross-sectional studies, or can we? *American Journal of Psychiatry 157*(2), 163–171.

[48] Laird, N. M. and J. H. Ware (1982). Random-effects models for longitudinal data. *Biometrics 38*(4), 963–974.

[49] Lee, S., V. Zipunnikov, N. Shiee, C. Crainiceanu, B. S. Caffo, and D. L. Pham (2013). Clustering of high dimensional longitudinal imaging data. In *Pattern Recognition in Neuroimaging (PRNI), 2013 International Workshop on*, pp. 33–36. IEEE.

[50] Li, Y., J. H. Gilmore, D. Shen, M. Styner, W. Lin, and H. Zhu (2013). Multiscale adaptive generalized estimating equations for longitudinal neuroimaging data. *NeuroImage 72*, 91–105.

[51] Li, Y., H. Zhu, D. Shen, W. Lin, J. H. Gilmore, and J. G. Ibrahim (2011). Multiscale adaptive regression models for neuroimaging data. *Journal of the Royal Statistical Society: Series B 73*(4), 559–578.

[52] Liang, K.-Y. and S. L. Zeger (1986). Longitudinal data analysis using generalized linear models. *Biometrika 73*(1), 13–22.

[53] Lindquist, M. A. (2012). Functional causal mediation analysis with an application to brain connectivity. *Journal of the American Statistical Association 107*, 1297–1309.

[54] Little, R. J. A. and D. B. Rubin (2002). *Statistical Analysis with Missing Data*. New York: Wiley.

[55] Liu, B. and H.-G. Müller (2009). Estimating derivatives for samples of sparsely observed functions, with application to online auction dynamics. *Journal of the American Statistical Association 104*(486), 704–717.

[56] Loeber, R. and D. P. Farrington (1994). Problems and solutions in longitudinal and experimental treatment studies of child psychopathology and delinquency. *Journal of Consulting and Clinical Psychology 62*(5), 887.

[57] MacKinnon, D. P., A. J. Fairchild, and M. S. Fritz (2007). Mediation analysis. *Annual Review of Psychology 58*, 593–614.

[58] Mathalon, D. H., E. V. Sullivan, K. O. Lim, and A. Pfefferbaum (2001). Progressive brain volume changes and the clinical course of schizophrenia in men: A longitudinal magnetic resonance imaging study. *Archives of General Psychiatry 58*(2), 148–157.

[59] Maxwell, S. E. and D. A. Cole (2007). Bias in cross-sectional analyses of longitudinal mediation. *Psychological Methods 12*(1), 23–44.

[60] McArdle, J. J. (2009). Latent variable modeling of differences and changes with longitudinal data. *Annual Review of Psychology 60*, 577–605.

[61] Menon, V. (2013). Developmental pathways to functional brain networks: Emerging principles. *Trends in Cognitive Sciences 17*(12), 627–640.

[62] Morris, J. S., V. Baladandayuthapani, R. C. Herrick, P. Sanna, and H. B. Gutstein (2011). Automated analysis of quantitative image data using isomorphic functional mixed models, with application to proteomics data. *Annals of Applied Statistics 5*(2A), 894–923.

[63] R Core Team (2013). *R: A Language and Environment for Statistical Computing.* Vienna, Austria: R Foundation for Statistical Computing.

[64] Ramsay, J. O., G. Hooker, D. Campbell, and J. Cao (2007). Parameter estimation for differential equations: A generalized smoothing approach. *Journal of the Royal Statistical Society: Series B 69*(5), 741–796.

[65] Ramsay, J. O. and B. W. Silverman (2005). *Functional Data Analysis* (2nd ed.). New York: Springer.

[66] Raz, N. and U. Lindenberger (2011). Only time will tell: Cross-sectional studies offer no solution to the age–brain–cognition triangle: Comment on Salthouse (2011). *Psychological Bulletin 137*(5), 790–795.

[67] Reiss, P., Y.-H. Chen, L. Huang, L. Huo, R. Tan, and R. Jiao (2015). *vows: Voxelwise Semiparametrics.* R package version 0.4.

[68] Reiss, P., M. Mennes, E. Petkova, L. Huang, M. Hoptman, B. Biswal, S. Colcombe, X. Zuo, and M. Milham (2011). Extracting information from functional connectivity maps via function-on-scalar regression. *NeuroImage 56*, 140–148.

[69] Reiss, P. T., L. Huang, H. Chen, and S. Colcombe (2014). Varying-smoother models for functional responses. arXiv:1412.0778 [stat.ME], available at http://arxiv.org/abs/1412.0778.

[70] Reiss, P. T., L. Huang, Y. H. Chen, L. Huo, T. Tarpey, and M. Mennes (2014). Massively parallel nonparametric regression, with an application to developmental brain mapping. *Journal of Computational and Graphical Statistics 23*(1), 232–248.

[71] Reiss, P. T. and R. T. Ogden (2009). Smoothing parameter selection for a class of semiparametric linear models. *Journal of the Royal Statistical Society: Series B 71*(2), 505–523.

[72] Reuter, M., N. J. Schmansky, H. D. Rosas, and B. Fischl (2012). Within-subject template estimation for unbiased longitudinal image analysis. *NeuroImage 61*(4), 1402–1418.

[73] Ridha, B. H., J. Barnes, J. W. Bartlett, A. Godbolt, T. Pepple, M. N. Rossor, and N. C. Fox (2006). Tracking atrophy progression in familial Alzheimer's disease: A serial MRI study. *Lancet Neurology 5*(10), 828–834.

[74] Rigby, R. A. and D. M. Stasinopoulos (2005). Generalized additive models for location, scale and shape (with Discussion). *Applied Statistics 54*(3), 507–554.

[75] Ruppert, D., M. Wand, and R. Carroll (2003). *Semiparametric Regression.* New York: Cambridge University Press.

[76] Schaie, K. W. (1965). A general model for the study of developmental problems. *Psychological Bulletin 64*(2), 92.

[77] Schaie, K. W. (1986). Beyond calendar definitions of age, time, and cohort: The general developmental model revisited. *Developmental Review 6*(3), 252–277.

[78] Schaie, K. W. and G. I. L. Caskie (2005). Methodological issues in aging research. In D. M. Teti (Ed.), *Handbook of Research Methods in Developmental Science*, pp. 21–39. Malden, MA: Blackwell.

[79] Shaw, P., K. Eckstrand, W. Sharp, J. Blumenthal, J. P. Lerch, D. Greenstein, L. Clasen, A. Evans, J. Giedd, and J. L. Rapoport (2007). Attention-deficit/hyperactivity disorder is characterized by a delay in cortical maturation. *Proceedings of the National Academy of Sciences 104*(49), 19649–19654.

[80] Shehzad, Z., C. Kelly, P. T. Reiss, R. C. Craddock, J. W. Emerson, K. McMahon, D. A. Copland, F. X. Castellanos, and M. P. Milham (2014). A multivariate distance-based analytic framework for connectome-wide association studies. *NeuroImage 93*, 74–94.

[81] Skup, M., H. Zhu, and H. Zhang (2012). Multiscale adaptive marginal analysis of longitudinal neuroimaging data with time-varying covariates. *Biometrics 68*(4), 1083–1092.

[82] Sowell, E. R., P. M. Thompson, C. M. Leonard, S. E. Welcome, E. Kan, and A. W. Toga (2004). Longitudinal mapping of cortical thickness and brain growth in normal children. *Journal of Neuroscience 24*(38), 8223–8231.

[83] Tamnes, C. K., K. B. Walhovd, A. M. Dale, Y. Østby, H. Grydeland, G. Richardson, L. T. Westlye, J. C. Roddey, D. J. Hagler Jr, P. Due-Tønnessen, D. Holland, A. M. Fjell, and the Alzheimer's Disease Neuroimaging Initiative (2013). Brain development and aging: Overlapping and unique patterns of change. *NeuroImage 68*, 63–74.

[84] Thompson, W. K., J. Hallmayer, and R. O'Hara (2011). Design considerations for characterizing psychiatric trajectories across the lifespan: Application to effects of APOE-e4 on cerebral cortical thickness in Alzheimer's disease. *American Journal of Psychiatry 168*(9), 894–903.

[85] Thompson, W. K. and D. Holland (2011). Bias in tensor based morphometry Stat-ROI measures may result in unrealistic power estimates. *NeuroImage 57*(1), 1–4.

[86] Treit, S., C. Lebel, L. Baugh, C. Rasmussen, G. Andrew, and C. Beaulieu (2013). Longitudinal MRI reveals altered trajectory of brain development during childhood and adolescence in fetal alcohol spectrum disorders. *Journal of Neuroscience 33*(24), 10098–10109.

[87] Unger, J. M., G. van Belle, and A. Heyman (1999). Cross-sectional versus longitudinal estimates of cognitive change in nondemented older people: A CERAD study. *Journal of the American Geriatrics Society 47*(5), 559–563.

[88] Wand, M. P. and J. T. Ormerod (2008). On semiparametric regression with O'Sullivan penalized splines. *Australian & New Zealand Journal of Statistics 50*(2), 179–198.

[89] Wickham, H. (2009). *ggplot2: Elegant Graphics for Data Analysis*. New York: Springer.

[90] Wood, S. N. (2011). Fast stable restricted maximum likelihood and marginal likelihood estimation of semiparametric generalized linear models. *Journal of the Royal Statistical Society: Series B 73*(1), 3–36.

[91] Wood, S. N. (2013). On *p*-values for smooth components of an extended generalized additive model. *Biometrika 100*(1), 221–228.

[92] Yuan, Y., J. H. Gilmore, X. Geng, S. Martin, K. Chen, J.-L. Wang, and H. Zhu (2014). FMEM: Functional mixed effects modeling for the analysis of longitudinal white matter tract data. *NeuroImage 84*, 753–764.

[93] Zhu, H., K. Chen, Y. Yuan, and J.-L. Wang (2015). FMPM: Functional mixed processes models for longitudinal functional responses. Under review.

[94] Zhu, H., L. Kong, R. Li, M. Styner, G. Gerig, W. Lin, and J. H. Gilmore (2011). FADTTS: Functional analysis of diffusion tensor tract statistics. *NeuroImage 56*(3), 1412–1425.

[95] Zhu, H., M. Styner, N. Tang, Z. Liu, W. Lin, and J. H. Gilmore (2010). FRATS: Functional regression analysis of DTI tract statistics. *IEEE Transactions on Medical Imaging 29*(4), 1039–1049.

[96] Ziegler, G., R. Dahnke, C. Gaser, and the Alzheimer's Disease Neuroimaging Initiative (2012). Models of the aging brain structure and individual decline. *Frontiers in Neuroinformatics 6*, article 3.

[97] Zipunnikov, V., S. Greven, H. Shou, B. S. Caffo, D. S. Reich, and C. Crainiceanu (2014). Longitudinal high-dimensional principal components analysis with application to diffusion tensor imaging of multiple sclerosis. *Annals of Applied Statistics 8*(4), 2175–2202.

19

Joint fMRI and DTI Models for Brain Connectivity

F. DuBois Bowman

Department of Biostatistics, Mailman School of Public Health, Columbia University

Sean Simpson

Department of Biostatistical Sciences, Wake Forest School of Medicine

Daniel Drake

Department of Biostatistics, Mailman School of Public Health, Columbia University

CONTENTS

19.1	Brain Connectivity		495
	19.1.1	Structural Connectivity	496
	19.1.2	Functional Connectivity	496
19.2	Single Modality Methods		497
	19.2.1	Methods for Functional Connectivity	497
		19.2.1.1 Defining the Spatial Scale for Connectivity Analysis	497
		19.2.1.2 Measures of Association	497
		19.2.1.3 Modeling Approaches	499
		19.2.1.4 Partitioning Methods	500
		19.2.1.5 Network Methods	500
	19.2.2	Methods for Effective Connectivity	502
	19.2.3	Determining Structural Connectivity	502
		19.2.3.1 Diffusion Weighted Imaging and DTI	503
		19.2.3.2 Tractography	504
19.3	Multimodal Approaches		505
	19.3.1	Sequential Procedures	505
	19.3.2	Functional Connectivity with Anatomical Weighting	508
	19.3.3	Modeling Joint Activation and Structural Connectivity	509
		19.3.3.1 Functional Coherence	510
		19.3.3.2 Ascendancy	511
		19.3.3.3 Likelihood Function	511
	19.3.4	Joint ICA	512
	19.3.5	Multimodal Prediction Methods	512
19.4	Conclusion		513
	Bibliography		515

19.1 Brain Connectivity

There has been a rapid thrust in research to unlock mysteries of human brain function spanning several disciplines. Neuroimaging technology provides invaluable tools, which have led to major advancements in our understanding of the brain, with contributions from fields such as neuroscience, psychiatry, neurology, radiology, psychology, statistics, computer science, and biomedical and electrical engineering, among others. The intricate neuroanatomical structure of the brain plays an integral role in supporting brain function, and it independently sheds light on disease and plasticity. While the interdependence between functional and structural properties of the brain is well accepted, current imaging technology that we probe typically yields properties reflecting either structure or function. There is emerging interest and an unmet need to consider joint modeling of data in pursuit of discoveries about normal brain function, disease-related alterations, and the response to treatment for such disorders.

19.1.1 Structural Connectivity

Connectivity research generally seeks to reveal properties of brain circuitry, either functional or structural. Structural connectivity (SC) refers to the underlying white matter axonal connections between different brain regions. The vast number of axonal connections yields an amazingly complex network of interconnected regions.

Neurons are the basic unit of the brain, and humans have roughly 86 billion neurons (Herculano-Houzel 2012), with longstanding estimates as high as 100 billion neurons. Axons are neuron fibers that form millions of white-matter bundles, which serve as lines of transmission in the nervous system. Within this extensive system of white-matter bundles, association bundles join cortical areas within the same hemisphere, commissural bundles link cortical areas in separate hemispheres, and projection fibers connect areas in the cerebral cortex to subcortical structures (Hendelman 2005). This circuitry serves as the anatomical infrastructure enabling electrical and chemical signaling giving rise to complex human thoughts and cognition.

19.1.2 Functional Connectivity

Functional connectivity (FC) studies seek to identify the functional interplay between regions by examining the temporal coherence in neural activity between spatially distinct brain regions, either during the performance of a task or while left to think independently without explicit guidance (called the resting state) (29). In practice, FC studies vary in scope and localization, with the simplest approach examining links between a selected seed brain region and all other regions (nodes) considered. Other approaches seek to dissociate particular brain networks consisting of multiple brain areas that exhibit similar temporal activity profiles or seek to generate complex whole-brain networks (see Figure 19.1). Researchers may compare FC properties between subgroups of subjects (e.g. between patients and healthy controls) and between different scanning sessions (e.g. before and after treatment).

Effective connectivity (EC) analysis targets a stronger directed relationship than what is provided by FC by describing the influence that one brain region exerts on another (29). Conceptually, EC reflects causal relationships. One should distinguish between causal relationships determined from the available data and the underlying physiological causal relationships that may exist at a neuronal level. fMRI data are typically collected on a timescale that is orders of magnitude slower than neuronal activity, and hence the data present inherent limitations to inferring physiological causal drivers at a neuronal level. Nonetheless, progress has been made attempting to establish fMRI-based causal relationships. Also, alternative criteria presented later in this chapter have been established to describe relationships between fMRI-based neural activity in different brain regions, which are slightly less stringent than EC but stronger than FC.

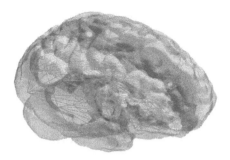

FIGURE 19.1
Functional connectivity map derived from fMRI time series data.

19.2 Single Modality Methods

19.2.1 Methods for Functional Connectivity

The field of statistics has been integral in advancing activation analyses and is making similar strides in connectivity analyses for functional neuroimaging research. The importance of fMRI connectivity analyses in aiding our understanding of normal and abnormal brain function has been well documented (7, 76, 8).

19.2.1.1 Defining the Spatial Scale for Connectivity Analysis

Conceptually, the most refined and comprehensive spatial scale at which one may conduct a connectivity analysis is at the voxel level, where functional connections are established between every pair of voxels. However, functional neuroimaging studies may have upward of $V = 300,000$ voxels included in an analysis. Therefore, a whole-brain comprehensive voxel-level analysis would involve $\binom{V}{2}$ voxel pairs, producing 45 billion connectivity estimates, which poses computational challenges and complexity for interpretations. In practice, one typically considers various strategies to reduce the dimensionality. One approach selects a set of (seed) voxels and assesses the functional association between these regions and every other specified brain area. Simplicity in implementation and interpretation provide the greatest appeal for this approach. However, the limited number of selected seed regions may fail to exploit important information in the data, unless there is a strong neurophysiological basis for selecting the seeds. Even when seed selection is hypothesis driven, usually determined as a brain region, results may be sensitive to the exact voxel locations chosen within the specified brain region(s).

Nodal-based approaches provide a commonly used alternative in which one selects a set of r nodes, with each node corresponding to an area of the brain, and then calculates the associations between all $\binom{r}{2}$ nodal pairs. This process produces an $r \times r$ FC matrix as shown in Figure 19.2, and it is more computationally viable than a comprehensive voxel-level approach. Nonetheless, one must still carefully consider the level of spatial refinement when defining the nodes of interest. Node selection may be hypothesis driven or based on a whole-brain parcellation scheme. Both seed- and nodal-based studies estimate functional connectivity using approaches described in the following subsections, and often these studies aim to identify connectivity differences between groups of interest.

FIGURE 19.2
Functional connectivity matrix derived from fMRI time series data using defined brain regions as nodes.

19.2.1.2   Measures of Association

Functional connectivity methods often estimate measures of association or leverage such association measures for statistical modeling. Association measures applied in neuroimaging analyses consist of both linear approaches, including correlation and coherence, and nonlinear methods such as mutual information and generalized synchronization. Correlation and partial correlation are the simplest and most commonly used measures to quantify functional similarity between brain regions (89, 40). Partial correlation better distinguishes direct from indirect connections. However, when calculating connectivity for a large number of region pairs, partial correlation may underperform, relative to Pearson's correlation, due to the small fraction of indirect connections and the overremoval of signal resulting from the large number of regressors (71). Additionally, the estimation of partial correlation via computation of the inverse of the covariance matrix becomes unstable as the number of regions approaches the number of time points (scans), and the covariance matrix becomes singular when the number of regions surpasses the number of scans.

Coherence is the spectral analogue of correlation (21, 53, 16, 55, 26) and is defined as

$$C_{t_i t_j}(\omega) = \frac{|S_{t_i t_j}(\omega)|^2}{S_{t_i}(\omega) S_{t_j}(\omega)}, \tag{19.1}$$

where $S_{t_i t_j}(\omega)$ is the cross-spectrum at frequency ω of the time series from brain regions i and j, and $S_{t_i}(\omega)$ and $S_{t_j}(\omega)$ are the respective power spectra at frequency ω. Coherence takes a value of 0 in the absence of any linear relationship, and 1 if the time series are perfectly related by a linear magnitude and phase transformation. It is generally estimated either for a single frequency range or for multiple ranges, with results subsequently combined.

Mutual information (MI) captures both linear and nonlinear dependencies between brain region time series (46, 64, 37), and is defined as

$$I(T_i, T_j) = \int_{T_i} \int_{T_j} f\left(t_i, t_j\right) \cdot \log_2 \left(\frac{f\left(t_i, t_j\right)}{f\left(t_i\right) \cdot f\left(t_j\right)} \right) dt_i \, dt_j, \tag{19.2}$$

where $f(t_i, t_j)$ is the joint probability density function of the time series of regions i and j, and $f(t_i)$ and $f(t_j)$ are the respective marginal probability density functions. MI quantifies the shared information between pairs of time series, with $I(T_i, T_j) = 0$ when no information is shared (i.e., the BOLD signals are statistically independent), and monotonically increasing as the amount of information shared increases. Generalized synchronization (or state-space synchrony) also captures nonlinearities by quantifying the interdependence between two signals in state-space reconstructed mappings (23, 77). The practical relevance of these (and other) nonlinear measures for FC remains debatable (40, 58, 54, 85).

19.2.1.3 Modeling Approaches

Modeling approaches may yield estimates of statistical association between brain regions based either on the original time series or on related summaries of the time series. Modeling frameworks may also use basic measures of association, e.g. Pearson correlation, as input and seek to determine other characteristics of FC properties. Several important model-based estimation contributions have been made, and there are opportunities for continued development in this area. As examples, Patel et al. (56) and Chen (18) developed Bayesian modeling approaches, with the former also yielding a strong relationship toward effective connectivity, called ascendancy. For each pair of voxels (or regions) i and j, the approach by Patel et al. (56) quantifies summaries from the original time series defined as $\mathbf{Z} = (Z_1, \ldots, Z_4)'$, where Z_1 is the number of times that both brain regions simultaneously exhibit increased activation, Z_2 is the number of times that one region (say i) is active and the other (say j) is inactive, Z_3 is the number of times that region i is inactive and j is active, and Z_4 is the number of scans for which both are inactive. The model then applies a multinomial likelihood function $Pr(\mathbf{Z} = \mathbf{z}) = \prod_{l=1}^{4} \theta_l^{z_l}$, where $\boldsymbol{\theta} = (\theta_1, \ldots, \theta_4)'$ is the vector of corresponding probabilities of increased activation. The approach uses a conjugate Dirichlet prior for the distribution of $\boldsymbol{\theta}$. Functional connectivity is quantified by the extent to which the estimated probability of joint activation $\hat{\theta}_1$, given the observed data, exceeds the chance-related joint activation expected under statistical independence.

The approach by Chen (18) applies a Bayesian mixture model and has the appealing benefit of unifying voxel and region level analyses for a selected number of seed regions. It is based on a frequently observed distribution of predominantly non-connected voxels and a relatively small number of functionally connected voxels. The method calculates a new measure of connectivity breadth between brain regions, identifies hub regions based on this measure, and identifies particular pairs of voxels that exhibit high probability of functional connectivity.

Varoquaux et al. (84) developed a modeling approach to improve the estimation of an individual participant's connectivity matrix by leveraging information contained in a group of participants' data. Within a framework of Gaussian graphical models, the authors posit the existence of a common conditional independence structure between regions that underlies all subjects' covariance matrices. Using Lasso-based feature selection, they estimate a sparse precision matrix structure for the entire group. Then, armed with this structure, they estimate the covariance matrix for each subject individually.

Varoquaux et al. (83) developed an alternative approach to identifying the sparse structure of the precision matrix. Using a graph theoretical framework, they introduce the notion of a decomposable graphical model in which isolated subsets of densely connected nodes ("cliques") have a relatively few nodes in common ("separating sets"). The corresponding precision matrix has an overlapping block-diagonal structure. The authors devise an efficient algorithm to deduce the decomposable model underlying multiple subjects' rsFC data.

Craddock et al. (20) proposed a multivariate supervised learning paradigm applied to resting-state fMRI data. This approach uses support vector regression to estimate a multivariate connectivity model that predicts the time series associated with one target ROI using the time courses of all voxels not included in the ROI. For a given target ROI, a separate model is fit for each subject; statistics on the resulting model parameters provide spatial indication of the voxels most consistently associated with the ROI in question. Spatial maps of prediction accuracy and model reproducibility were evaluated in the nonparametric prediction accuracy, influence, and reproducibility resampling (NPAIRS) framework.

These, and other, modeling methods tend to inherently allow the identification of group-related connectivity differences. This is in contrast to most association methods in which inference remains a subsequent procedure.

19.2.1.4 Partitioning Methods

Partitioning methods group brain areas together in sets (or clusters) that exhibit more within cluster functional similarity (connectivity) than between cluster similarity. Independent component analysis (ICA) and cluster analysis serve as the two most popular approaches in this category. ICA provides a model-free technique that seeks to dissociate the aggregate set of fMRI time series signals into constituent, spatially independent, subcomponents (15, 4, 5, 35, 34, 25). The resulting subcomponents embody brain regions with coherent patterns of functional activity that can be compared across groups. More formally, let Y be a $T \times V$ matrix containing the spatiotemporal data from T scans and V voxels (or regions) for a participant. ICA decomposes this data into linear combinations of subcomponent source signals as follows:

$$Y = MS + E, \qquad (19.3)$$

where M is a $T \times q$ mixing matrix comprising latent time series for each of the q independent components, S is a $q \times V$ matrix comprising statistically independent spatial signals along the rows and non-Gaussian columns, and E is a $T \times V$ random error matrix representing the variability not accounted for by the independent components. This error matrix is excluded in noise-free ICA (48), yielding $S = M^{-1}Y$, with M^{-1} serving as an unmixing matrix that produces statistically independent signals. ICA has consistently produced brain components that have been attributed to characteristics of resting-state neural processing (3, 22).

Cluster analysis, another data-driven technique, also affords partitioning the brain into functionally similar interconnected units. Many approaches within this class of methods readily apply to fMRI data, including K-means clustering (1), fuzzy clustering (74), hierarchical clustering (9, 10, 78), and other novel clustering procedures (27, 11). When applying both cluster analysis and ICA to data from task fMRI studies, task-related components can easily be evaluated by determining the extent of association with the administered stimuli, while other components are putatively assigned attributions.

Most currently employed FC approaches implicitly assume that the interdependence among the time series of brain regions remains constant throughout an imaging period. However, this "averaging" of the associations among these complex signals may miss important information contained in the dynamic interplay that likely occurs across multiple time scales. Dynamic FC is an emerging area that aims to quantify these changes in connectivity over time. While this field remains nascent, many important developments have already been made (43). Future advancements hope to further uncover how changes in neural relationship patterns govern behavioral and cognitive outcomes.

19.2.1.5 Network Methods

Brain network analyses have rapidly emerged in neuroimaging research over the last decade. The appeal of network methods is that they allow studying the brain as a system, providing insight into how various interacting regions produce complex behaviors (76, 12, 82, 66). Understanding this link between system-level properties and behavioral and health outcomes has profound clinical implications (7, 76, 12, 2).

Network analyses quantify the functional similarity between the time series for all pairs of r nodes or brain regions, creating an interconnected representation of the brain from which system-level properties can then be studied. The resulting $r \times r$ functional connectivity matrix is commonly thresholded to create a binary adjacency matrix that retains stronger connections while removing weaker ones. A schematic exhibiting this binary network generation process is presented in Figure 19.3. Weighted (continuous) network analyses, in which the connectivity matrix is not binarized, have gained traction but still lag behind due to associated computational and methodological challenges (82, 63, 32). Both binary

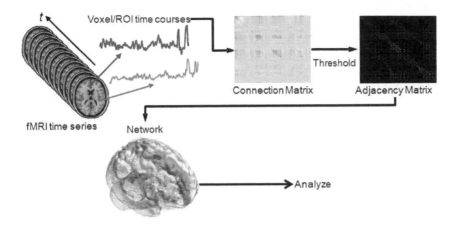

FIGURE 19.3
Schematic for generating brain networks from fMRI time series data. Functional connectivity between brain areas is estimated based on time series pairs to produce a connection matrix. A threshold is commonly applied to the matrix to generate a binary adjacency matrix. From the adjacency matrix, various network analyses can be performed.

and weighted analyses aim to characterize the topological properties of networks and compare these system-level properties across groups and task conditions. Graph measures such as clustering coefficient, path length, efficiency, centrality, and modularity serve as common descriptive topological properties of interest.

Current modeling and comparison approaches for brain networks generally rely on crude comparisons of graph measures or on mass-univariate nodal or edge-based comparisons that ignore the inherent topological properties of the network while also yielding little power to determine significance (90, 31). While useful findings have been gleaned from some univariate approaches, they preclude leveraging the wealth of data present in whole-brain networks to better understand their complex functional organization. This systemic organization confers much of the brain's functional abilities. For example, functional connections may be lost due to an adverse health condition but compensatory connections may develop as a result in order to maintain organizational consistency and functional performance. Consequently, brain network analysis necessitates a multivariate explanatory and predictive (non)linear modeling framework that accounts for the complex dependence structure and allows assessing the effects of multiple variables of interest and topological network features (e.g., demographics, disease status, nodal clustering, nodal centrality, etc.) on the overall network structure. That is, if we have

$$\text{Data} \begin{cases} \boldsymbol{Y}_k : \text{network of participant } k \\ \boldsymbol{X}_k : \text{covariate information (metrics, demographics, etc.),} \end{cases}$$

we seek to model the probability density function of the network given the covariates $P(\boldsymbol{Y}_k|\boldsymbol{X}_k, \boldsymbol{\theta}_k)$, where $\boldsymbol{\theta}_k$ are the parameters that relate the covariates to the network structure.

More recent brain network comparison methods that attempt to better exploit the topological features of network data include the exponential random graph modeling framework (ERGM) (67, 69), the permutation network framework (PNF) (68), and the multivariate distance matrix regression (MDMR) framework (65). While all show promise, they lack the flexibility of the modeling and inferential tools developed for fMRI activation data. The ERGM framework allows efficiently representing complex network data and inherently accounts for higher-order dependence/topological properties, but multiple-participant comparisons can pose problems given that these models were originally developed for the modeling of one network at a time. Moreover, the amount of programming work increases linearly with the number of participants since ERGMs must be fitted and assessed for each participant individually. Incorporating novel metrics (perhaps more rooted in brain biology) may be difficult due to degeneracy issues that may arise (57, 60). While well-suited for substructural assessments, edge-level examinations remain difficult with these models. Additionally, most ERGM developments have been for binary networks; approaches for weighted networks have been proposed but remain in their infancy (45, 24). The PNF approach enables comparing groups of brain networks by assessing the topological consistency of key node sets within and between groups. However, it is a strictly inferential (and not modeling) approach, and thus precludes quantifying and predicting relationships between disease outcomes and network structure, and simulating network structure. Unlike the PNF, the MDMR framework allows controlling for confounding covariates in group comparisons via a "psuedo-F" statistic; however, it too lacks the ability to simulate networks or make predictions. It also fails to account for the dependence in connectivity patterns across voxels. Mixed modeling frameworks show promise for addressing some of these limitations, but have yet to be fully developed for the brain network context (66).

19.2.2   Methods for Effective Connectivity

Effective connectivity analysis seeks to elucidate directional relationships between brain areas. While attempting to infer causality in this manner demands extreme caution given the relatively poor temporal resolution of fMRI data (61, 73, 72), several methods have shown promise. Structural equation models (SEMs) leverage interregional covariances to compute path coefficients containing information on the directional influence of one area on another (49). Dynamic causal modeling (DCM) employs a Bayesian framework to estimate effective connectivity by treating brain signals as part of a deterministic nonlinear dynamic system (30). Granger causality (GC) is a lag-based method that employs multivariate vector autoregressive modeling (MVAR) to infer causal relationships between the activity of two regions (33, 62, 86). Bayesian net modeling approaches search for causal relationships among regions by utilizing directed acyclic graphs (DAGs) to represent brain regions (as sets of random variables) and their conditional dependencies (59). Patel's ascendancy measure, noted earlier, also employs a Bayesian modeling framework to yield hierarchical relationships between functionally connected pairs of regions, e.g. with one region in a functionally connected pair being viewed as a hub node because of increased activity and the other less active region being regarded as a satellite node. Further discussion of these methods and their limitations is provided in Chapters 16 and 20.

19.2.3   Determining Structural Connectivity

Structural connectivity is a measure of association between two brain regions based on the degree to which they are connected via white matter tracts. As these tracts consist of bundles of myelinated axons that carry information between regions, such connections are neurologically relevant—they highlight the structural underpinnings of brain function.

SC analysis proceeds using two distinct steps. First, local measures of diffusivity are obtained through specially designed MRI sequences sensitive to the displacement of water molecules. Regions of highly anisotropic diffusion are associated with white matter tracts: water molecules in fiber bundles are highly constrained in directions perpendicular to the tracts, but diffuse relatively freely along the tract path. Second, via a technique known as tractography, these local measures of diffusivity are linked together to form global estimates of white matter fiber bundles. By quantifying how likely fiber bundles might connect a seed region to a target region, we obtain a measure of SC.

In this section, we first provide an overview of diffusion tensor imaging, the most commonly used MRI-based approach used to measure the directional characteristics of diffusion. We then contrast two approaches to tractography: deterministic and probabilistic. While both approaches give indications of how fiber tracts connect different regions of the brain, the probabilistic approach establishes a framework that naturally leads to a continuous measure of association between two regions, i.e., SC.

19.2.3.1   Diffusion Weighted Imaging and DTI

Structural connectivity analyses leverage information from diffusion weighted imaging, the umbrella term for a category of MRI sequences designed to measure the diffusivity of water molecules. This diffusion, the random displacement of molecules due to thermal agitation, depends on the medium under examination. In an isotropic medium, i.e., one in which the characteristics do not vary with direction, molecules at a given point will diffuse in all directions with equal likelihood. In an anisotropic medium, on the other hand, diffusion may be constrained in some directions, enhanced in others. In the context of the brain, diffusion of cerebral spinal fluid in the ventricles, say, would be isotropic. Diffusion in the fibrous white matter, in contrast, would be anisotropic: diffusion along the direction of the fibers is relatively unimpeded, whereas diffusion perpendicular to the fibers is highly constrained. Diffusion weighted imaging techniques are designed to quantify, in some form or other, the directional characteristics of water diffusion in the brain.

The most common diffusion weighted imaging technique is diffusion tensor imaging (DTI). DTI produces an estimate of a 3x3 symmetric matrix

$$\mathbf{D} = \begin{bmatrix} D_{xx} & D_{xy} & D_{xz} \\ D_{xy} & D_{yy} & D_{yz} \\ D_{xz} & D_{yz} & D_{zz} \end{bmatrix} ,$$

the diffusion tensor, which encodes the spatial properties of diffusivity at a spatial location. $\mathbf{D}$ is defined by six values, so only six non-collinear DWI scans are necessary to determine the diffusion tensor. However, a larger sampling of the direction space is often used in practice to reduce noise. The diffusion tensor is completely characterized by its three mutually orthogonal eigenvectors and corresponding non-negative eigenvalues:

$$\mathbf{D} = \mathbf{\Gamma \Lambda \Gamma}^T ,$$

where $\mathbf{\Lambda}$ is the diagonal matrix composed of the diffusion tensor eigenvalues:

$$\mathbf{\Lambda} = \begin{bmatrix} \lambda_1 & 0 & 0 \\ 0 & \lambda_2 & 0 \\ 0 & 0 & \lambda_3 \end{bmatrix} ,$$

and $\mathbf{\Gamma}$ is the matrix whose columns are the mutually orthogonal diffusion tensor eigenvectors. From this formulation, we can characterize the diffusion tensor as a three-dimensional ellipsoid whose axes are the eigenvectors scaled by their corresponding eigenvalues. The longest axis, often referred to as the longitudinal or axial direction, represents the direction of greatest diffusivity, and its length is related to the magnitude of diffusivity along that direction. In white matter, the axial direction of the diffusion tensor is assumed to align with the fiber bundle orientation.

Various summaries of the eigenvalues of the diffusion tensor are useful to characterize the diffusion at a particular location. One measure, mean diffusivity,

$$\text{MD} = \frac{\lambda_1 + \lambda_2 + \lambda_3}{3}$$

provides an overall estimate of the diffusivity at a particular voxel. It does not, however, provide any indication of the shape of the diffusion tensor ellipsoid. A second measure, fractional anisotropy (FA), quantifies the degree to which the ellipsoid deviates from a spherical shape:

$$\text{FA} = \sqrt{\frac{3}{2}} \frac{\sqrt{(\lambda_1 - \text{MD})^2 + (\lambda_2 - \text{MD})^2 + (\lambda_3 - \text{MD})^2}}{\sqrt{\lambda_1^2 + \lambda_2^2 + \lambda_3^2}}.$$

FA ranges from 0 to 1, where 0 indicates isotropic diffusivity, and 1 signals maximal anisotropy: only one eigenvalue is nonzero and the ellipsoid collapses to a line segment.

19.2.3.2 Tractography

Tractography is the process of stitching together local estimates of diffusion direction at the voxel level to obtain brain-wide maps of white matter tracts. As mentioned, tractography methods are classified as either deterministic or probabilistic.

Deterministic Tractography

Deterministic methods applied to DTI typically rely on the longitudinal direction of the diffusion tensor to map fiber bundles. In its simplest form, deterministic tractography starts at a point within a voxel or region of interest (the seed region), and traces a path in the direction of maximal local diffusivity until it arrives at the boundary of a neighboring voxel. Crossing this boundary, the algorithm turns to follow the direction of maximal diffusivity of the new voxel until, once again, it arrives at a voxel boundary. Proceeding in this fashion for multiple starting points of interest, subject to specified stopping criteria, these techniques generate fiber bundle trajectories, or "streamlines," that map estimates of white matter tracts of the brain. This approach has been dubbed FACT, for Fiber Assessment by Continuous Trajectory(52).

While simple, this approach has a number of drawbacks. For example, the results generated by this approach are sensitive to the (typically arbitrary) orientation of the sampling grid. Simply rotating the coordinates can produce noticeably different streamlines. The sensitivity is exacerbated by the errors inherent in the DTI estimates of the fiber tract orientations due to noise, subject motion, etc. The deterministic tractography algorithm is not very robust to erroneous measurements or missteps in the streamline process. Streamlines typically terminate when they reach areas with low fractional anisotropy, but low FA may reflect the presence of multiple bundle directions (e.g. crossing fibers) rather than the lack of underlying white matter tracts.

Various techniques have been developed to remedy the drawbacks of deterministic tractography. For example, an improvement to FACT that allows transitions across voxel edges and corners (as opposed to just voxel faces), known as FACT-ID (81) has been developed to help mitigate this sensitivity to grid orientation. However, one major limitation persists. Deterministic tractography provides a binary assessment regarding the presence/absence of a stream connecting two locations, but notably there is no uncertainty quantified regarding the existence of the tract. Probabilistic tractography attempts to overcome this shortcoming.

Probabilistic Tractography

Probabilistic tractography methods (6) co-opt similar streamline strategies developed for deterministic tractography, with two fundamental differences. The first major difference lies in how they represent the local diffusivity used to generate streamlines. The deterministic approaches rely on what amounts to the maximum likelihood estimate of the diffusion tensor at each voxel. The probabilistic methods, instead, use the diffusion weighted images to estimate a probability distribution for each of the diffusion tensor-related parameters (at each voxel). As a consequence, in this new framework there is no single direction of maximum diffusion; instead there is a continuum of directions, some more likely than others. The second major difference lies in how the connectivity is evaluated. The deterministic approaches trace one streamline from seed to target to determine if the two are connected. The probabilistic approaches sample the local diffusion directions based on the underlying distributions as they trace the stream, resulting in a different, independent stream each time the process is repeated. This process is replicated many thousands of times; the ratio of the number of times the stream reaches the target to the total number of streams initiated gives an indication of how structurally connected the two regions are. The probabilistic approach alleviates the need for many of the workarounds proposed for the deterministic techniques. For example, typically there is no need to restrict the streamlines to regions with high fractional anisotropy. As different realizations of the stream traverse regions with low fractional anisotropy, they spread out more widely and fewer paths reach the target, naturally leading to an indication of low connectivity. Moving beyond the estimation of the existence of tracts, modeling procedures and network methods, similar to those described above for FC, can be applied to characterize network properties and to examine differences between subgroups of individuals.

19.3 Multimodal Approaches

Ready availability of imaging modalities such as fMRI and DTI enables simultaneous investigations into brain structure and function. Jointly modeling fMRI and DTI data may yield numerous advantages. First, examining solely brain function or only brain structure provides an inherently incomplete picture of neurophysiology. While either of these approaches may yield important insights, a multimodal approach promises a more complete understanding of neurophysiology in healthy individuals and neuropathophysiology of major brain disorders. Moreover, a multimodal approach to connectivity may provide a deeper understanding of the physiological basis for neuroimaging findings, more robust findings stemming from the complementary roles of fMRI and DTI, increased prediction accuracy when targeting clinical or behavioral outcomes, more precise surgical planning, and so forth. However, there is a need for additional development of statistical models to fully leverage such multimodal data.

Note that our use of the term multimodal in this chapter is consistent with the neuroimaging literature, in which the term describes data from two or more imaging techniques that are combined for analysis, in contrast to the conventional use of the term in statistics to describe distributions with multiple modes. The focus here is on multimodal methods targeting FC, in which SC is incorporated as supplemental information. Alternatively, one could primarily target SC, building in FC information, or seek to identify mutual structural–functional relationships.

19.3.1 Sequential Procedures

In this section, we examine methods that quantify the relationship between FC and SC. In particular, we focus on techniques that evaluate functional and structural connectivity separately, and then evaluate similarities and differences.

Koch et al. (44) were among the first to directly examine the relationship between functional and structural connectivity. In a single axial slice, they examined connectivity between crowns of adjacent gyri (inferior frontal to precentral, precentral to postcentral, and postcentral to supramarginal) in both the left and right hemispheres. They manually defined small areas of the selected gyri that included the convex surface of the brain, along with subcortical white matter voxels to serve as seed voxels for tractography.

To quantify FC, they computed temporal correlations between all pairs of voxels in the ROIs. The aggregate FC between regions was defined to be the maximum connectivity between all possible pairs of voxels in the two regions. To quantify SC, the authors used a form of probabilistic tractography; the voxel-wise measure of association was the ratio of the number of paths leaving a white matter source voxel that ended up at a destination voxel relative to the total number of paths simulated. Again, aggregate SC was chosen as the maximum connectivity between source and destination white matter voxel pairs in the corresponding ROIs. Unlike FC, however, two SC values were determined per pair of ROIs, depending on which ROI was considered the source, and which was considered the destination.

No attempt was made to combine measures across subjects, and no statistical analysis relating functional to structural connectivity was performed. Upon plotting SC pairs against corresponding FC values, no clear pattern resulted. However, the authors noted that, while a high degree of FC did not necessarily correspond to high SC, a high degree of SC tended to go along with strong FC.

Little was published in this type of analysis until 2008, when some researchers began with a similar approach, but quantified the similarity of functional and structural using correlation.

Citing Koch et al. (44) as the only directly relevant previous research, Skudlarski et al. (70) attempted a more comprehensive examination of the relationship between functional and structural connectivity. Using T1 anatomical scans, they began by identifying 5000 voxels in a common stereotaxic space that were consistently identified as gray matter across all subjects. Temporal correlations from resting-state scans were then computed for all pairs of the 5000 voxels, resulting in a FC matrix for each subject.

Similarly, using deterministic tractography, SC measures were computed for all pairs of another, distinct set of 5000 voxels consistently identified as white matter across all subjects. From these measures they formed a composite SC matrix that took into account not just direct connections between pairs of voxels, but also paths that took as many as eight hops to join the pairs. Finally, to associate an SC measure to each pair of gray matter voxels, the authors identified the white matter voxels closest to a given gray matter voxel.

To get a general idea of the overall similarity between the two methods, they computed SC vs. FC correlation across all voxel pairs for each subject. After compensating for the effect of decreasing connectivity as a function of geometric distance, the resulting correlations were found significant to $p < 0.0001$ in 39 of 42 subjects.

The authors then examined similarities in spatial connectivity characteristics between the two types of connectivity. First, they computed, for each voxel, the correlation of the corresponding functional and structural seed maps (effectively correlating the corresponding columns of the connectivity matrices), and found virtually all voxels created similar seed maps in both structural and functional connectivity, save for an area in the right middle occipital gyrus. Second, they compared the mean global connectivity maps (each voxel represents the average connectivity with the rest of the 5000 voxels; computed by averaging all columns of the respective connectivity matrices) for functional and structural connectivity. They found that the spatial distribution for the two measures across subjects were similar, save for the insular regions, which displayed strong FC but less strong SC, and the thalamus, with strong SC but no significant FC.

Horn et al. (42) expanded on the voxel-wise approach of Skudlarski et al. (70). Instead of one variant each of FC and SC, they compared the similarities between functional and structural connectivity using two methods for each modality: both full and partial correlations for FC; and global and probabilistic tractography for SC.

They computed the overall correlation between the four combinations of the two functional and two structural connectivity measures, as well as correlations between the two structural measures and the two functional measures. While significant to $p < 0.0001$, the correlation values were relatively low, with R ranging from 0.015 to 0.046 for the full correlation vs. probabilistic tractography. The within-modality agreement was higher, ranging from 0.11 (structural) to 0.15 (functional). To quantify spatial connectivity, they compared the similarities in functional and structural connectivity of mean global connectivity maps for the four combinations of the two functional and two structural connectivity measures. They found that the agreement between functional and structural connectivity was strongest and most robust in the default mode network. While there was little difference between the two FC approaches, they noted that the probabilistic tractography approach failed to identify the medial pre-frontal cortex as a region with high functional–structural connectivity, whereas the global tractography approach did.

Hagmann et al. (36) investigated the topological structure of cortical connectivity networks using diffusion spectrum imaging (DSI) on five subjects. They validated their resulting SC measures in a number of ways, one of which involved correlating their results against FC data derived from resting-state fMRI. Instead of a voxel-wise approach taken by Skudlarski et al. (70), they started with an initial anatomical parcellation of the cortex comprised of 66 elements. They further subdivided the parcellation to obtain a total of 998 subregions, with average surface area of 1.5 cm^2, that cover the entire cortex. Using a form of deterministic tractography on diffusion spectrum imaging data, they computed SC for each pair of subregions. Then, to mitigate the variability in connectivity, they averaged the SC measures across all subregion pairs within the encompassing anatomical regions. Similar steps were taken to obtain corresponding FC measures for these pairs of anatomical regions: first computing temporal correlations between all pairs of subregions, then averaging across subregions within pairs of anatomical regions. To validate their results, they averaged each connectivity measure (functional and structural) across the five subjects and correlated their SC estimates with the corresponding functional ones. They found an agreement to a degree of $R^2 = 0.62$, $p < 10^{-10}$.

Building on the analysis of Hagman et al. (36), Honey et al. (41) examined the ability to infer underlying SC (present or not) between nodes based on FC measures. They compared the distribution of FC for two cases: those pairs for which underlying SC is present, and those for which it is absent. While there was significant overlap between the two, the one associated with existing underlying SC was shifted towards higher values of resting-state FC (rsFC) compared to the zero underlying SC case. From these distributions the authors generated receiver operating curves to quantify the tradeoff between the probability of correctly identifying the presence of underlying SC (true positive) and that of mistaking unconnected subregion pairs for ones that are connected structurally (false positive). They found that inferring underlying SC based on the measured FC was highly inaccurate, to the degree that if an rsFC threshold was chosen to detect 80% of the underlying connections, it would also mis-classify 40% of non-existent connections as actually existing. This high false positive rate was exacerbated by the fact that structurally unconnected pairs outnumbered connected ones by 30 to 1. As a consequence, of the subregion pairs deemed structurally connected based on rsFC values, only 6% would be truly structurally connected.

Hermendstad et al. (38) took a similar approach as Hagmann et al. (36) and Honey et al. (41) by parcelling the brain into a large number of subregions (600 in this case). However, unlike (41), which examined the distribution of FC values conditioned on the presence or absence of underlying SC, Hermendstad et al. restricted their analyses to structural pathways that were consistently connected across 84 subjects, and instead looked at the distribution of functional connectivity values as a function of different categories of structural connection. By thresholding both by the connection density and average fiber length, they obtained five distinct categories of connections: interhemispheric connections that were either short or long; intrahemispheric connections that were dense or long; and the remaining short, sparse, intrahemispheric connections. They computed complementary cumulative distribution functions of FC for each of the five SC categories and used the results to draw qualitative inferences. They found that interhemispheric and dense intra-hemispheric structural connections supported strong FC, but that long intrahemispheric consistently did not. They then reversed the analysis, generating complementary cumulative distribution functions for both structural connection density and average fiber length for low, medium, and high levels of FC. They found that FC increased with connection density; and that FC increased as the intrahemispheric connections grow shorter, but that FC increased with increasing length of interhemispheric connections.

Hinne et al. (39) reworked the purely functional approach of (84), which uses graphical Lasso to estimate the sparse functional connectivity precision matrix. Hinne notes that 1) the coefficient shrinkage associated with the Lasso introduces bias in the estimate of the matrix; 2) the approach only provides a point estimate, which precludes the possibility of drawing inferences about uncertainty of the estimate. To counter these drawbacks, Hinne introduces a Bayesian Functional Connectivity (BFC) analysis which derives the sparseness structure of the precision matrix from the structural connectivity matrix. BFC analysis then amounts to computing a posterior density over sparse precision matrices.

19.3.2   Functional Connectivity with Anatomical Weighting

One approach to multimodal analyses is to use data-driven descriptive methods such as cluster analysis. Bowman et al. present a multimodal approach that combines fMRI with SC information derived from DTI to determine FC via cluster analysis (11). A schematic of the approach is displayed in Figure 19.4. Let f_{ij} represent a measure of functional dissimilarity between regions i and j, e.g. $f_{ij} = h(\rho_{ij}) = 1 - \rho_{ij}$ is used to construct Figure 19.4(c), where ρ_{ij} is the correlation between regions i and j. They defined a distance measure

$$d_{ij} = \left(1 - \frac{\pi_{ij}}{\lambda}\right) f_{ij}, \tag{19.4}$$

where $\pi_{ij} \in [0,1)$ is the probability of SC determined from the DTI data, and $\lambda \in [0,\infty)$ is an attenuation parameter that is optimized empirically, with larger values of λ reducing the impact of anatomical weighting. Cluster analysis proceeds by joining regions with small distances d_{ij}. The anatomical weight function based on $w_{ij} = \left(1 - \pi_{ij}/\lambda\right)$ is shown in Figure 19.4(d). Note that d_{ij} (see Figure 19.4(e)) incorporates SC as a supplemental weight, but still permits clustering of regions (or clusters) based on f_{ij} in the absence of SC, i.e., when $\pi_{ij} \to 0$. The method generally improves network coherence, particularly with increased noise in the fMRI signal.

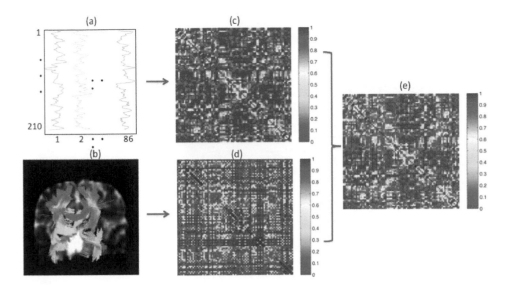

FIGURE 19.4
The anatomically weighted functional connectivity approach uses regional BOLD activity profiles (a) to calculate the functional dissimilarity between each region (c). It also uses DTI tractography results (b) to calculate the probability of no structural white matter connection (d). The anatomical information in (d) is applied as a weight function to the functional information in (c) to produce the weighted results in (e). The extent of weighting is determined empirically from the data and varies across studies as it depends on the degree of separability in the functional data (when the functional signals are easily separable, less anatomical weighting is typically applied).

19.3.3 Modeling Joint Activation and Structural Connectivity

Jointly modeling functional data derived from fMRI and structural data derived from DTI presents a promising pathway toward multimodal analyses and inferences. Such methods define probability models that incorporate functional–structural relationships. One example by Xue et al. (87) extends methodology by Patel et al. (56), which quantifies FC in task-related fMRI data by looking at patterns of joint activation between brain regions.

Let $R_{ikt} = Y_{ikt} - \hat{\mu}_{ik}$ represent the mean-adjusted neural activity for region i, subject k, and scan t. For a fixed constant c, define $A_{ikt} = \mathbf{1}\left(R_{ikt} > c \times \sigma_{ik}\right)$, which serves as an indicator of elevated regional brain activity at time t, where σ_{ik}^2 is the variance of Y_{ikt}. The joint activation between two regions i and j can be characterized by the following variables:

$$
Z_{1n}^* = \sum_{t=1}^{T^*} \mathbf{1}(A_{ikt} = 1, A_{jkt} = 1), \quad Z_{2n}^* = \sum_{t=1}^{T^*} \mathbf{1}(A_{ikt} = 1, A_{jkt} = 0)
$$

$$
Z_{3n}^* = \sum_{t=1}^{T^*} \mathbf{1}(A_{ikt} = 0, A_{jkt} = 1), \quad Z_{4n}^* = \sum_{t=1}^{T^*} \mathbf{1}(A_{ikt} = 0, A_{jkt} = 0).
$$

(19.5)

Z_{1n}^* is the number of times that regions i and j simultaneously exhibit elevated fMRI activity, and Z_{4n}^* is the number of times that both regions remain below an elevated level of activity (referred to here as inactive). It is reasonable to assume that $\mathbf{Z}_k^* = \left(Z_{1k}^*, \cdots, Z_{4k}^*\right)'$

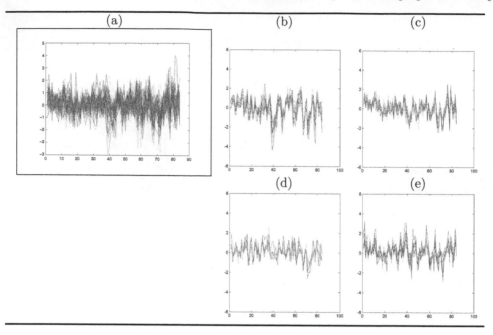

FIGURE 19.5
Clustering results from anatomically weighted functional connectivity approach displaying
(a) time series from 86 regions along with (b)–(e) 4 clusters containing 40 of the regions
collectively. The functional synchrony for regions within the same cluster is evident in (b)–
(e).

follows a multinomial distribution with parameters T^* and $\theta = (\theta_1, \cdots, \theta_4)'$, where

$$\theta_1 = P\left(A_{ikt} = 1, A_{jkt} = 1\right), \quad \theta_2 = P\left(A_{ikt} = 1, A_{jkt} = 0\right)$$
$$\theta_3 = P\left(A_{ikt} = 0, A_{jkt} = 1\right), \quad \theta_4 = P\left(A_{ikt} = 0, A_{jkt} = 0\right). \tag{19.6}$$

To facilitate interpretations, one may standardize Z_{lk}^* ($l = 1, \cdots, 4$) by scaling it with a
specified number of scans T so that the standardized measure, $Z_{lk}(l = 1, 4)$, reflects the
average number of times that i and j are coherent per T scans (e.g., $T = 100$). Therefore,
$\mathbf{Z}$ follows a multinomial distribution with parameters T and θ.

For the anatomical data, S_n^* denotes the DTI-based tractography counts linking regions
i and j out of M^* trials originating from the seed region, say i, in the previously described
FACT procedure. One may assume that S^* follows a binomial distribution with parameters
M^* and π, where π is the probability of SC between regions i and j. Using similar scaling
applied to Z_i^*, scaled counts S can be generated, which follow a binomial distribution with
parameters M and π, e.g. scaling to set $M = 1000$.

19.3.3.1   Functional Coherence

Considering Table 19.1, Xue et al. (87) introduced the following statistic to quantify functional coherence between pairs of brain regions:

$$\kappa = \begin{cases} \frac{\theta_1 + \theta_4 - E}{1 - E} & \text{if } \theta_1\theta_4 > \theta_2\theta_3 \\ 0 & \text{otherwise,} \end{cases} \tag{19.7}$$

TABLE 19.1
Joint activation probabilities for regions i and j.

		Region a		
		Active	Inactive	
Region b	Active	θ_1	θ_3	$\theta_1 + \theta_3$
	Inactive	θ_2	θ_4	$\theta_2 + \theta_4$
		$\theta_1 + \theta_2$	$\theta_3 + \theta_4$	1

where $E = (\theta_1 + \theta_2)(\theta_1 + \theta_3) + (\theta_3 + \theta_4)(\theta_2 + \theta_4)$. κ builds on Cohen's kappa statistic (19) and extends the agreement measure of Patel et al. (2006) by incorporating SC in the likelihood function detailed below and by considering joint coherence. The numerator of κ measures the difference between the probability of coherence and the expected probability of coherence under independence. Nonnegative values of κ are of primary interest, so many applications may restrict attention to $\kappa \in [0, 1]$. κ equals 1 when the probability of being jointly active, θ_1, and jointly inactivate, θ_4, sums to 1, which indicates complete coherence. If there is no agreement between regions i and j, $\kappa = 0$.

19.3.3.2 Ascendancy

Given that i and j are functionally connected, i.e., $\kappa > e_\kappa$ with high probability, a measure of ascendancy quantifies the hierarchical relationship between the regions. The following statistic determines the hierarchical relationship between connected regions based on the relative activity between the regions:

$$\tau_{ij} = \frac{\theta_1 + \theta_2}{\theta_3 + \theta_4} \bigg/ \frac{\theta_1 + \theta_3}{\theta_2 + \theta_4}, \tag{19.8}$$

$\tau_{ij} \in [0, \infty)$. τ_{ij} is interpretable as an odds ratio, specifically as the odds of region i being active, $P(A_i = 1)/(1 - P(A_i = 1))$, relative to the odds of region j being active, $P(A_j = 1)/(1 - P(A_j = 1))$. For functionally connected regions i and j, we say that i is ascendant to j whenever $\tau_{ij} > 1$, indicating that the marginal odds of activation of i are larger than those of j, while $\tau_{ij} < 1$ indicates that j is ascendant to i.

19.3.3.3 Likelihood Function

For any pair of regions i and j, one can base inferences regarding κ and τ_{ij} on the likelihood function:

$$p(\mathbf{Z}, S | \theta, \pi) \propto \prod_{i=1}^{4} \theta_i^{\sum_{k=1}^{n} Z_{ik}} \pi^{\sum_{k=1}^{n} S_k} (1 - \pi)^{n \times M - \sum_{k=1}^{n} S_k} \tag{19.9}$$

(Xue et al. (88)). The likelihood assumes that measures on the same region pair are independent across subjects and over time. The assumption of temporal independence is supported by the standard preprocessing step of pre-whitening to remove temporal correlations.

Using a Bayesian formulation, we express our prior belief about structural connection probabilities π via a beta prior, which takes the form:

$$p(\pi) \propto \pi^{\alpha_0 - 1}(1 - \pi)^{\beta_0 - 1}. \tag{19.10}$$

In the absence of strong *a priori* information, it is reasonable to assume a flat prior for

SC by setting $\alpha_0 = \beta_0 = 1$ for each region pair. A Dirichlet prior is assumed for $\boldsymbol{\theta}$, with parameters $(\alpha(\pi) + \alpha_1, \alpha_2, \alpha_3, \alpha_4)'$, yielding

$$p(\theta|\pi) \propto \frac{\Gamma(\alpha(\pi) + \alpha_1 + \alpha_2 + \alpha_3 + \alpha_4)}{\Gamma(\alpha(\pi) + \alpha_1)} \theta_1^{\alpha(\pi)+\alpha_1-1} \theta_2^{\alpha_2-1} \theta_3^{\alpha_3-1} \theta_4^{\alpha_4-1}. \qquad (19.11)$$

Note that $\alpha(\pi)$ is an increasing function that reflects the assumed relationship between FC and SC. Xue et al. (87) specified $\alpha(\pi) = \{10/\left[(9/\ln(10) - 1)\right]\} \times (10^\pi - 1)$, along with $\alpha_1 = 5$ and $\alpha_2 = \alpha_3 = \alpha_4 = 10$, so that the average value of $\alpha(\pi)$ on $\pi \in [0, 1]$ is 10. They based this prior on previous observations that weak SC corresponds to very few joint functional activations and extremely strong SC is associated with stronger FC, assumed to yield an expected value of θ_1 to be around 0.5 based on their assumptions. Note that the expected value of θ_1 is an increasing function whenever $\alpha(\pi)$ is an increasing function. Sensitivity analyses should be conducted to assess the choice of $\alpha(\pi)$.

19.3.4 Joint ICA

Joint independent component analysis (jICA) provides another tool that leverages information from multiple neuroimaging modalities to produce a more comprehensive understanding of how brain function and structure correlate with behavioral and disease outcomes (14). It enables quantifying covariation across modalities with different temporal and spatial scales and has been employed to combine data from sMRI and fMRI (14), EEG and fMRI (51, 13, 50, 47), MEG and DTI (79), PET and sMRI (75), and fMRI and DTI (28, 80). Joint ICA extends ICA by assuming that two or more modalities share the same mixing matrix. It then proceeds to maximize the independence among joint components by conducting ICA on horizontally concatenated data from each modality. For example, with fMRI and DTI data, the functional time series data are contained in a single vector for each participant (regions/conditions vertically concatenated), and, similarly, the structural FA values are also concatenated into a single spatial vector for each participant. These vectors are then combined across participants to form a participant×time functional matrix $(\boldsymbol{Y}^{\text{fMRI}})$ and participant×voxel structural matrix $(\boldsymbol{Y}^{\text{DTI}})$. These matrices are then scaled, via sums of squares matching, and combined into a single data matrix $\boldsymbol{Y} = \left[\boldsymbol{Y}^{\text{fMRI}} \ \boldsymbol{Y}^{\text{DTI}}\right]$ that can be decomposed with the standard ICA approach. In this case the standard ICA equation $\boldsymbol{Y} = \boldsymbol{MS}$ can also be stated as $\left[\boldsymbol{Y}^{\text{fMRI}} \ \boldsymbol{Y}^{\text{DTI}}\right] = \boldsymbol{M} \left[\boldsymbol{S}^{\text{fMRI}} \ \boldsymbol{S}^{\text{DTI}}\right]$. While jICA provides a model-free approach to fusing and equally weighting data from multiple modalities, its robustness to departures from the common mixing matrix assumption warrants examination. To circumvent this assumption, Sui et al. (80) developed an approach combining multimodal canonical correlation analysis (mCCA) and jICA to allow for multiple mixing matrices.

19.3.5 Multimodal Prediction Methods

Combined characteristics of brain function and structure may serve as important markers for disease, disease progression, and treatment response. There are an increasing number of multimodal neuroimaging analyses targeting prediction and classification, which examine the predictive abilities of brain features such as localized activation, local structural integrity, connectivity measures, and whole brain network properties. This area has great potential for translational impact of neuroimaging. Since many psychiatric and neurological disorders have complex neuropathophysiology, e.g. manifesting in functional and structural alterations of the brain, there is great potential in jointly considering functional and structural markers. The basic idea of generating multimodal neuroimaging predictors or features is conveyed in Figure 19.6. The massive set of brain features can be combined with clinical, genomic, biological, and demographic measures and collectively modeled to determine the predictive ability on an outcome.

We provide an example of a statistical modeling framework that jointly uses FC and SC to predict disease status. Let $\mathbf{Y}_{kij}$ denote FC between brain regions i and j for subject k. Collectively over all gray matter $\mathcal{G}$ in the brain, $\mathbf{Y}_k = \{\mathbf{Y}_{kij}, \forall\, i, j \in \mathcal{G}, i \neq j\}$. Similarly, let $\mathbf{Z}_k$ represent SC over all region pairs falling in white matter $\mathcal{W}$ in the brain. Let $D_k \in \{0, 1\}$ denote the binary disease, progression, or treatment response status, and $\mathbf{W}_k = (W_{k1}, \cdots, W_{kQ})$ be a set of Q covariates.

The objective of our model is to predict the disease status of a subject given imaging data and other covariates. To achieve this goal, we use the posterior samples from the estimation step to calculate the posterior predictive probability. Let θ denote the parameter space, $\mathbf{M}_k = (\mathbf{Y}_k, \mathbf{Z}_k)$ denote the multimodal FC-SC imaging data for subject k, and $\mathbf{A}_k = (\mathbf{M}_k, D_k)$. Suppose we have n training subjects and want to predict the disease status D_{n+1} for a new subject indexed by $n+1$. One may proceed by specifying a probability model $p(\mathbf{A}_k | \mathbf{W}_k, \theta)$ and drawing samples $\theta^{(b)}, b = 1, \ldots, B$, from the joint posterior distribution $P(\theta \mid \{\mathbf{A}_k, \mathbf{W}_k\}_{k=1}^n)$. One approach to incorporating FC–SC relationships would proceed by modeling $p(\mathbf{A}_k | \mathbf{W}_k, \theta) = p(D_k | \mathbf{W}_k, \theta) \times p(\mathbf{Z}_k | D_k, \mathbf{W}_k, \theta) \times p(\mathbf{Y}_k | D_k, \mathbf{Z}_k, \mathbf{W}_k, \theta)$. Predictions can then be based on the posterior predictive probability (suppressing notation reflecting dependence on covariates)

$$\widehat{P}(D_{n+1} = d \mid \mathbf{M}_{n+1}, \{\mathbf{A}_k\}_{k=1}^n) = \frac{P(D_{n+1} = d)\sum_{b=1}^B \omega_d^{(b)}}{\sum_{d'=0,1} P(D_{n+1} = d')\sum_{b=1}^B \omega_{d'}^{(b)}}, \tag{19.12}$$

where $\omega_d^{(b)} = P(\mathbf{M}_{n+1} \mid D_{n+1} = d, \theta^{(b)})$. Then prediction of D_{n+1} is given by

$$\widehat{D}_{n+1} = \arg\max_d \left(P(D_{n+1} = d)\sum_{b=1}^B \omega_d^{(b)} \right). \tag{19.13}$$

19.4 Conclusion

Despite substantial progress for single modality methods, multimodal modeling of brain connectivity remains an important area for involvement by statisticians. For example, little has been done on multimodal network estimation. Researchers have constructed and compared modality-specific networks, but fusing data across modalities remains relatively unexplored. Characterizing multimodal brain networks (DTI and fMRI in particular) will engender a better understanding of brain architectures and their relationships to various outcomes. To our knowledge, the work in (17) provides the only development in this domain. They developed a multi-view spectral clustering technique to derive multimodal networks from fMRI and DTI data.

Attention must be drawn to particular challenges faced when analyzing multimodal data. The dimensionality may become unwieldy. Alignment of information across different temporal scales, say milliseconds to seconds, may cause analytic and interpretative issues. Similarly, integrating information across spatial scales may prove difficult. This issue may include difficulties reconciling different spatial resolution and extent of localization, e.g. stemming from consideration of fMRI and EEG data, as well as challenges merging resting-state FC data from gray matter and SC data rooted in white matter. Also, the emergence of multimodal data raises missing data issues. For instance, some subjects may have scans

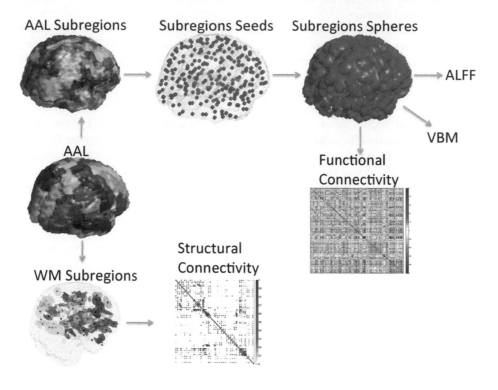

FIGURE 19.6
Multimodal prediction using combined functional and structural connectivity summaries, potentially with other imaging derived markers, clinical, genomic, biological, and demographic measures.

missing for a single modality but may contribute scans for all other modalities. For many analyses, one would be forced to disregard all of the data for such subjects and only use data from subjects for whom there are complete data. The emergence of longitudinal study designs further exacerbates concerns around missing data, since simply missing a single modality at a single measurement occasion may establish grounds for discarding a subject's data for an analysis. Such complete case approaches are naturally inefficient, since they neglect to use a substantial amount of available data, and they may induce bias depending on the reasons for missingness. Some immediate gains may be made by applying existing methods in the statistical literature for handling missing data to neuroimaging, but the enormity and complexity of imaging data prompts the need for additional methodological development. Adding to the potential sources of information, multimodal data may include non-imaging data such as genomic, clinical, demographic, and biologic information, which also exacerbates some of the aforementioned challenges. In summary, multimodal connectivity is an exciting and rapidly expanding area of research, which presents numerous statistical challenges.

Bibliography

[1] D. Balslev, F. A. Nielsen, S. A. Frutiger, J. J. Sidtis, T. B. Christiansen, C. Svarer, S. C. Strother, D. A. Rottenberg, L. K. Hansen, O. B. Paulson, and I. Law. Cluster analysis of activity-time series in motor learning. *Hum Brain Mapp*, 15(3):135–145, Mar 2002.

[2] D. S. Bassett and E. T. Bullmore. Human brain networks in health and disease. *Curr. Opin. Neurol.*, 22(4):340–347, Aug 2009.

[3] C. F. Beckmann, M. DeLuca, J. T. Devlin, and S. M. Smith. Investigations into resting-state connectivity using independent component analysis. *Philos. Trans. R. Soc. Lond., B, Biol. Sci.*, 360(1457):1001–1013, May 2005.

[4] C. F. Beckmann and S. M. Smith. Probabilistic independent component analysis for functional magnetic resonance imaging. *IEEE Trans Med Imaging*, 23(2):137–152, Feb 2004.

[5] C. F. Beckmann and S. M. Smith. Tensorial extensions of independent component analysis for multisubject FMRI analysis. *Neuroimage*, 25(1):294–311, Mar 2005.

[6] T. E. Behrens, M. W. Woolrich, M. Jenkinson, H. Johansen-Berg, R. G. Nunes, S. Clare, P. M. Matthews, J. M. Brady, and S. M. Smith. Characterization and propagation of uncertainty in diffusion-weighted MR imaging. *Magn Reson Med*, 50(5):1077–1088, Nov 2003.

[7] B. B. Biswal, M. Mennes, X. N. Zuo, S. Gohel, C. Kelly, S. M. Smith, C. F. Beckmann, J. S. Adelstein, R. L. Buckner, S. Colcombe, A. M. Dogonowski, M. Ernst, D. Fair, M. Hampson, M. J. Hoptman, J. S. Hyde, V. J. Kiviniemi, R. Kotter, S. J. Li, C. P. Lin, M. J. Lowe, C. Mackay, D. J. Madden, K. H. Madsen, D. S. Margulies, H. S. Mayberg, K. McMahon, C. S. Monk, S. H. Mostofsky, B. J. Nagel, J. J. Pekar, S. J. Peltier, S. E. Petersen, V. Riedl, S. A. Rombouts, B. Rypma, B. L. Schlaggar, S. Schmidt, R. D. Seidler, G. J. Siegle, C. Sorg, G. J. Teng, J. Veijola, A. Villringer, M. Walter, L. Wang, X. C. Weng, S. Whitfield-Gabrieli, P. Williamson, C. Windischberger, Y. F. Zang, H. Y. Zhang, F. X. Castellanos, and M. P. Milham. Toward discovery science of human brain function. *Proc. Natl. Acad. Sci. U.S.A.*, 107(10):4734–4739, Mar 2010.

[8] F. D. Bowman. Brain imaging analysis. *Annual Review of Statistics and Its Application*, 1(1):61–85, 2014.

[9] F. D. Bowman and R. Patel. Identifying spatial relationships in neural processing using a multiple classification approach. *NeuroImage*, 23(1):260–268, 2004.

[10] F. D. Bowman, R. Patel, and C. Lu. Methods for detecting functional classifications in neuroimaging data. *Hum Brain Mapp*, 23(2):109–119, Oct 2004.

[11] F. D. Bowman, L. Zhang, G. Derado, and S. Chen. Determining functional connectivity using fMRI data with diffusion-based anatomical weighting. *Neuroimage*, 62(3):1769–1779, Sep 2012.

[12] E. Bullmore and O. Sporns. Complex brain networks: Graph theoretical analysis of structural and functional systems. *Nat. Rev. Neurosci.*, 10(3):186–198, Mar 2009.

[13] V. Calhoun, L. Wu, K. Kiehl, T. Eichele, and G. Pearlson. Aberrant processing of deviant stimuli in schizophrenia revealed by fusion of fMRI and EEG data. *Acta Neuropsychiatr*, 22(3):127–138, Jun 2010.

[14] V. D. Calhoun, T. Adali, N. R. Giuliani, J. J. Pekar, K. A. Kiehl, and G. D. Pearlson. Method for multimodal analysis of independent source differences in schizophrenia: Combining gray matter structural and auditory oddball functional data. *Hum Brain Mapp*, 27(1):47–62, Jan 2006.

[15] V. D. Calhoun, T. Adali, G. D. Pearlson, and J. J. Pekar. A method for making group inferences from functional MRI data using independent component analysis. *Hum Brain Mapp*, 14(3):140–151, Nov 2001.

[16] C. Chang and G. H. Glover. Time-frequency dynamics of resting-state brain connectivity measured with fMRI. *Neuroimage*, 50(1):81–98, Mar 2010.

[17] H. Chen, K. Li, D. Zhu, T. Zhang, C. Jin, L. Guo, L. Li, and T. Liu. Inferring group-wise consistent multimodal brain networks via multi-view spectral clustering. *Med Image Comput Comput Assist Interv*, 15(Pt 3):297–304, 2012.

[18] S. Chen. New statistical techniques for high-dimensional neuroimaging data. Unpublished, Emory University, 2012.

[19] J. Cohen. A coefficient of agreement for nominal scales. *Educational and Psychological Measurement*, 20:37–46, 1960.

[20] R. C. Craddock, M. P. Milham, and S. M. LaConte. Predicting intrinsic brain activity. *Neuroimage*, 82:127–136, Nov 2013.

[21] C. E. Curtis, F. T. Sun, L. M. Miller, and M. D'Esposito. Coherence between fMRI time-series distinguishes two spatial working memory networks. *Neuroimage*, 26(1):177–183, May 2005.

[22] J. S. Damoiseaux, S. A. Rombouts, F. Barkhof, P. Scheltens, C. J. Stam, S. M. Smith, and C. F. Beckmann. Consistent resting-state networks across healthy subjects. *Proc. Natl. Acad. Sci. U.S.A.*, 103(37):13848–13853, Sep 2006.

[23] J. Dauwels, F. Vialatte, T. Musha, and A. Cichocki. A comparative study of synchrony measures for the early diagnosis of Alzheimer's disease based on EEG. *Neuroimage*, 49(1):668–693, Jan 2010.

[24] B. A. Desmarais and S. J. Cranmer. Statistical inference for valued-edge networks: The generalized exponential random graph model. *PLoS ONE*, 7(1):e30136, 2012.

[25] A. Eloyan, C. M. Crainiceanu, and B. S. Caffo. Likelihood-based population independent component analysis. *Biostatistics*, 14(3):514–527, Jul 2013.

[26] M. Fiecas and H. Ombao. The generalized shrinkage estimator for the analysis of functional connectivity of brain signals. *The Annals of Applied Statistics*, 5:1102–1125, 2011.

[27] P. Filzmoser, R. Baumgartner, and E. Moser. A hierarchical clustering method for analyzing functional MR images. *Magn Reson Imaging*, 17(6):817–826, Jul 1999.

[28] A. R. Franco, J. Ling, A. Caprihan, V. D. Calhoun, R. E. Jung, G. L. Heileman, and A. R. Mayer. Multimodal and multi-tissue measures of connectivity revealed by joint independent component analysis. *IEEE J Sel Top Signal Process*, 2(6):986–997, Dec 2008.

[29] K. J. Friston. Functional and effective connectivity in neuroimaging: A synthesis. *Human Brain Mapping*, 2(1-2):56–78, 1994.

[30] K. J. Friston, L. Harrison, and W. Penny. Dynamic causal modelling. *Neuroimage*, 19(4):1273–1302, Aug 2003.

[31] C. E. Ginestet, A. P. Fournel, and A. Simmons. Statistical network analysis for functional MRI: Summary networks and group comparisons. *Front Comput Neurosci*, 8:51, 2014.

[32] C. E. Ginestet, T. E. Nichols, E. T. Bullmore, and A. Simmons. Brain network analysis: Separating cost from topology using cost-integration. *PLoS ONE*, 6(7):e21570, 2011.

[33] C. W. Granger. Investigating causal relations by econometric models and cross-spectral methods. *Econometrica: Journal of the Econometric Society*, 37(3):424–438, 1969.

[34] Y. Guo. A general probabilistic model for group independent component analysis and its estimation methods. *Biometrics*, 67(4):1532–1542, Dec 2011.

[35] Y. Guo and G. Pagnoni. A unified framework for group independent component analysis for multi-subject fMRI data. *Neuroimage*, 42(3):1078–1093, Sep 2008.

[36] P. Hagmann, L. Cammoun, X. Gigandet, R. Meuli, C. J. Honey, V. J. Wedeen, and O. Sporns. Mapping the structural core of human cerebral cortex. *PLoS Biol.*, 6(7):e159, Jul 2008.

[37] D. Hartman, J. Hlinka, M. Palus, D. Mantini, and M. Corbetta. The role of nonlinearity in computing graph-theoretical properties of resting-state functional magnetic resonance imaging brain networks. *Chaos*, 21(1):013119, Mar 2011.

[38] A. M. Hermundstad, D. S. Bassett, K. S. Brown, E. M. Aminoff, D. Clewett, S. Freeman, A. Frithsen, A. Johnson, C. M. Tipper, M. B. Miller, S. T. Grafton, and J. M. Carlson. Structural foundations of resting-state and task-based functional connectivity in the human brain. *Proc. Natl. Acad. Sci. U.S.A.*, 110(15):6169–6174, Apr 2013.

[39] M. Hinne, L. Ambrogioni, R. J. Janssen, T. Heskes, and M. A. van Gerven. Structurally-informed Bayesian functional connectivity analysis. *Neuroimage*, 86:294–305, Feb 2014.

[40] J. Hlinka, M. Palus, M. Vejmelka, D. Mantini, and M. Corbetta. Functional connectivity in resting-state fMRI: Is linear correlation sufficient? *Neuroimage*, 54(3):2218–2225, Feb 2011.

[41] C. J. Honey, O. Sporns, L. Cammoun, X. Gigandet, J. P. Thiran, R. Meuli, and P. Hagmann. Predicting human resting-state functional connectivity from structural connectivity. *Proc. Natl. Acad. Sci. U.S.A.*, 106(6):2035–2040, Feb 2009.

[42] A. Horn, D. Ostwald, M. Reisert, and F. Blankenburg. The structural-functional connectome and the default mode network of the human brain. *Neuroimage*, 102 Pt 1; Nov 2014.

[43] R. M. Hutchison, T. Womelsdorf, E. A. Allen, P. A. Bandettini, V. D. Calhoun, M. Corbetta, S. Della Penna, J. H. Duyn, G. H. Glover, J. Gonzalez-Castillo, D. A. Handwerker, S. Keilholz, V. Kiviniemi, D. A. Leopold, F. de Pasquale, O. Sporns, M. Walter, and C. Chang. Dynamic functional connectivity: Promise, issues, and interpretations. *Neuroimage*, 80:360–378, Oct 2013.

[44] M. A. Koch, D. G. Norris, and M. Hund-Georgiadis. An investigation of functional and anatomical connectivity using magnetic resonance imaging. *NeuroImage*, 16(1):241–250, May 2002.

[45] P. N. Krivitsky. Exponential-family random graph models for valued networks. *Electron J Stat*, 6:1100–1128, 2012.

[46] S. Ma, V. D. Calhoun, T. Eichele, W. Du, and T. Adali. Modulations of functional connectivity in the healthy and schizophrenia groups during task and rest. *NeuroImage*, 62(3):1694–1704, Sep 2012.

[47] J. Mangalathu-Arumana, S. A. Beardsley, and E. Liebenthal. Within-subject joint independent component analysis of simultaneous fMRI/ERP in an auditory oddball paradigm. *Neuroimage*, 60(4):2247–2257, May 2012.

[48] M. J. McKeown, S. Makeig, G. G. Brown, T. P. Jung, S. S. Kindermann, A. J. Bell, and T. J. Sejnowski. Analysis of fMRI data by blind separation into independent spatial components. *Hum Brain Mapp*, 6(3):160–188, 1998.

[49] A. R. McIntosh and F. Gonzalez-Lima. Structural equation modeling and its application to network analysis in functional brain imaging. *Human Brain Mapping*, 2(1–2):2–22, 1994.

[50] B. Mijović, K. Vanderperren, N. Novitskiy, B. Vanrumste, P. Stiers, B. V. Bergh, L. Lagae, S. Sunaert, J. Wagemans, S. V. Huffel, and M. D. Vos. The "why" and "how" of JointICA: results from a visual detection task. *NeuroImage*, 60(2):1171–1185, Apr 2012.

[51] M. Moosmann, T. Eichele, H. Nordby, K. Hugdahl, and V. D. Calhoun. Joint independent component analysis for simultaneous EEG-fMRI: Principle and simulation. *Int J Psychophysiol*, 67(3):212–221, Mar 2008.

[52] S. Mori, B. J. Crain, V. P. Chacko, and P. C. van Zijl. Three-dimensional tracking of axonal projections in the brain by magnetic resonance imaging. *Ann. Neurol.*, 45(2):265–269, Feb 1999.

[53] K. Muller, G. Lohmann, V. Bosch, and D. Y. von Cramon. On multivariate spectral analysis of fMRI time series. *NeuroImage*, 14(2):347–356, Aug 2001.

[54] Theoden I. Netoff, Thomas L. Carroll, Louis M. Pecora, and Steven J. Schiff. *Detecting Coupling in the Presence of Noise and Nonlinearity*, pages 265–282. Wiley-VCH Verlag GmbH & Co. KGaA, 2006.

[55] H. Ombao and S. Van Bellegem. Evolutionary coherence of nonstationary signals. *IEEE Transactions on Signal Processing*, 56(6):2259–2266, 2008.

[56] R. S. Patel, F. D. Bowman, and J. K. Rilling. A Bayesian approach to determining connectivity of the human brain. *Hum Brain Mapp*, 27(3):267–276, Mar 2006.

[57] P. Pattison, K. Carley, R. Breiger, C.H. Factors, B.H.S. Integration, D.B.S.S. Education, and N.R. Council. *Dynamic Social Network Modeling and Analysis: Workshop Summary and Papers*. National Academies Press, 2003.

[58] E. Pereda, R. Q. Quiroga, and J. Bhattacharya. Nonlinear multivariate analysis of neurophysiological signals. *Prog. Neurobiol.*, 77(1-2):1–37, 2005.

[59] J. D. Ramsey, S. J. Hanson, C. Hanson, Y. O. Halchenko, R. A. Poldrack, and C. Glymour. Six problems for causal inference from fMRI. *Neuroimage*, 49(2):1545–1558, Jan 2010.

[60] A. Rinaldo, S. E. Fienberg, and Y. Zhou. On the geometry of discrete exponential families with application to exponential random graph models. *Electronic Journal of Statistics*, 3:446–484, 2009.

[61] I. Rish, B. Thyreau, B. Thirion, M. Plaze, M. L. Paillere-Martinot, C. Martelli, J. L. Martinot, J. B. Poline, and G. A. Cecchi. Discriminative network models of schizophrenia. In Y. Bengio, D. Schuurmans, J.D. Lafferty, C.K.I. Williams, and A. Culotta, editors, *Advances in Neural Information Processing Systems 22*, pages 252–260. Curran Associates, Inc., 2009.

[62] A. Roebroeck, E. Formisano, and R. Goebel. The identification of interacting networks in the brain using fMRI: Model selection, causality and deconvolution. *Neuroimage*, 58(2):296–302, Sep 2011.

[63] M. Rubinov and O. Sporns. Weight-conserving characterization of complex functional brain networks. *Neuroimage*, 56(4):2068–2079, Jun 2011.

[64] R. Salvador, A. Martinez, E. Pomarol-Clotet, S. Sarro, J. Suckling, and E. Bullmore. Frequency based mutual information measures between clusters of brain regions in functional magnetic resonance imaging. *Neuroimage*, 35(1):83–88, Mar 2007.

[65] Z. Shehzad, C. Kelly, P. T. Reiss, R. Cameron Craddock, J. W. Emerson, K. McMahon, D. A. Copland, F. X. Castellanos, and M. P. Milham. A multivariate distance-based analytic framework for connectome-wide association studies. *Neuroimage*, 93 Pt 1:74–94, Jun 2014.

[66] S. L. Simpson, F. D. Bowman, and P. J. Laurienti. Analyzing complex functional brain networks: Fusing statistics and network science to understand the brain. *Statistics Surveys*, 7:1–36, 2013.

[67] S. L. Simpson, S. Hayasaka, and P. J. Laurienti. Exponential random graph modeling for complex brain networks. *PLoS ONE*, 6(5):e20039, 2011.

[68] S. L. Simpson, R. G. Lyday, S. Hayasaka, A. P. Marsh, and P. J. Laurienti. A permutation testing framework to compare groups of brain networks. *Front Comput Neurosci*, 7:171, 2013.

[69] S. L. Simpson, M. N. Moussa, and P. J. Laurienti. An exponential random graph modeling approach to creating group-based representative whole-brain connectivity networks. *Neuroimage*, 60(2):1117–1126, Apr 2012.

[70] P. Skudlarski, K. Jagannathan, V. D. Calhoun, M. Hampson, B. A. Skudlarska, and G. Pearlson. Measuring brain connectivity: Diffusion tensor imaging validates resting state temporal correlations. *Neuroimage*, 43(3):554–561, Nov 2008.

[71] S. M. Smith, K. L. Miller, G. Salimi-Khorshidi, M. Webster, C. F. Beckmann, T. E. Nichols, J. D. Ramsey, and M. W. Woolrich. Network modelling methods for fMRI. *NeuroImage*, 54(2):875–891, Jan 2011.

[72] S. M. Smith, D. Vidaurre, C. F. Beckmann, M. F. Glasser, M. Jenkinson, K. L. Miller, T. E. Nichols, E. C. Robinson, G. Salimi-Khorshidi, M. W. Woolrich, D. M. Barch, K. Uğurbil, and D. C. Van Essen. Functional connectomics from resting-state fMRI. *Trends Cogn. Sci. (Regul. Ed.)*, 17(12):666–682, Dec 2013.

[73] V. Solo. The dangers of Granger: what they didnt tell you about Granger causality (in fMRI). In *Annu. Meet. Org. Hum. Brain Mapp.* Organization for Human Brain Mapping, June 2011.

[74] R. L. Somorjai and M. Jarmasz. Exploratory analysis of fMRI data by fuzzy clustering: Philosophy, strategy, tactics, implementation. In F. T. Sommer and A. Wichert, editors, *Exploratory Analysis and Data Modeling in Functional Neuroimaging*, pages 17–48. MIT Press, Cambridge, MA, USA, 2003.

[75] K. Specht, R. Zahn, K. Willmes, S. Weis, C. Holtel, B. J. Krause, H. Herzog, and W. Huber. Joint independent component analysis of structural and functional images reveals complex patterns of functional reorganisation in stroke aphasia. *NeuroImage*, 47(4):2057–2063, Oct 2009.

[76] 0. Sporns. *Networks of the Brain.* The MIT Press, 2010.

[77] C. J. Stam and B. W. van Dijk. Synchronization likelihood: An unbiased measure of generalized synchronization in multivariate datasets. *Physica D: Nonlinear Phenomena*, 163(34):236–251, 2002.

[78] L. Stanberry, R. Nandy, and D. Cordes. Cluster analysis of fMRI data using dendrogram sharpening. *Hum Brain Mapp*, 20(4):201–219, Dec 2003.

[79] J. M. Stephen, B. A. Coffman, R. E. Jung, J. R. Bustillo, C. J. Aine, and V. D. Calhoun. Using joint ICA to link function and structure using MEG and DTI in schizophrenia. *Neuroimage*, 83:418–430, Dec 2013.

[80] J. Sui, G. Pearlson, A. Caprihan, T. Adali, K. A. Kiehl, J. Liu, J. Yamamoto, and V. D. Calhoun. Discriminating schizophrenia and bipolar disorder by fusing fMRI and DTI in a multimodal CCA+ joint ICA model. *NeuroImage*, 57(3):839–855, Aug 2011.

[81] P. A. Taylor, K. H. Cho, C. P. Lin, and B. B. Biswal. Improving DTI tractography by including diagonal tract propagation. *PLoS ONE*, 7(9):e43415, 2012.

[82] Q. K. Telesford, S. L. Simpson, J. H. Burdette, S. Hayasaka, and P. J. Laurienti. The brain as a complex system: Using network science as a tool for understanding the brain. *Brain Connect*, 1(4):295–308, 2011.

[83] G. Varoquaux, A. Gramfort, J. B. Poline, and B. Thirion. Markov models for fMRI correlation structure: Is brain functional connectivity small world, or decomposable into networks? *J. Physiol. Paris*, 106(5-6):212–221, 2012.

[84] Gaël Varoquaux, Alexandre Gramfort, Jean-Baptiste Poline, and Bertrand Thirion. Brain covariance selection: better individual functional connectivity models using population prior. In *Advances in Neural Information Processing Systems*, pages 2334–2342, 2010.

[85] Matthias Winterhalder, Björn Schelter, Wolfram Hesse, Karin Schwab, Lutz Leistritz, Daniel Klan, Reinhard Bauer, Jens Timmer, and Herbert Witte. Comparison of linear signal processing techniques to infer directed interactions in multivariate neural systems. *Signal Processing*, 85(11):2137–2160, 2005.

[86] G. R. Wu, W. Liao, S. Stramaglia, H. Chen, and D. Marinazzo. Recovering directed networks in neuroimaging datasets using partially conditioned Granger causality. *Brain Connect*, 3(3):294–301, 2013.

[87] W. Xue, F. D. Bowman, A. V. Pileggi, and A. R. Mayer. A multimodal approach for determining brain networks by jointly modeling functional and structural connectivity. *Front Comput Neurosci*, 9:22, 2015.

[88] W. Xue, F.D. Bowman, and J. Kang. A Bayesian spatial model to predict disease status using imaging data from multiple modalities. *IEEE Transactions on Computational Biology and Bioinformatics*, 2014.

[89] A. Zalesky, A. Fornito, and E. Bullmore. On the use of correlation as a measure of network connectivity. *Neuroimage*, 60(4):2096–2106, May 2012.

[90] A. Zalesky, A. Fornito, and E. T. Bullmore. Network-based statistic: Identifying differences in brain networks. *NeuroImage*, 53(4):1197–1207, Dec 2010.

20

Statistical Analysis of Electroencephalograms

Yuxiao Wang, Lechuan Hu, and Hernando Ombao

Department of Statistics, University of California, Irvine

CONTENTS

20.1	Introduction	524
20.2	Spectral Analysis of a Single-Channel EEG	524
	20.2.1 Brief Description of the Data	525
	20.2.2 Fourier-Domain Approach	525
	20.2.2.1 The Fourier Regression Model and Variance Decomposition	525
	20.2.2.2 The Spectrum of a Single-Channel Time Series	528
	20.2.2.3 Estimating the Spectrum via Periodograms	528
	20.2.2.4 Other Periodogram-Based Estimation Methods	531
	20.2.2.5 Examples of Smoothing Periodograms	531
	20.2.2.6 Multitaper Method (MTM)	531
	20.2.3 Time-Domain Approach	532
	20.2.3.1 Moving Average (MA) Model	532
	20.2.3.2 Autoregressive (AR) Model	533
	20.2.3.3 Autoregressive Moving Average (ARMA) Model	534
	20.2.3.4 The Spectra of MA, AR and ARMA Processes	535
	20.2.3.5 Second-Order Autoregressive [AR(2)] Model	537
	20.2.3.6 Estimating the Spectrum	538
	20.2.4 Estimating the Spectrum Using Multiple EEG Traces	539
	20.2.4.1 Other Averaged Estimators	539
	20.2.4.2 Estimating Power in Specific Frequency Bands	540
	20.2.4.3 Detecting Outliers	540
	20.2.5 Confidence Intervals	540
20.3	Spectral Analysis of Multichannel EEG	541
	20.3.1 Fourier-Domain Approach	542
	20.3.1.1 The Fourier–Cramér Representation	542
	20.3.1.2 The Spectral Matrix of an EEG	542
	20.3.1.3 Non-Parametric Estimator of the Spectral Matrix	544
	20.3.2 Time-Domain Approach	544
	20.3.3 Estimating Partial Coherence	545
	20.3.4 Estimating the Spectral Matrix Using Multiple EEG Traces	548
	20.3.4.1 Estimating the Spectral Matrix in Specific Frequency Bands	549
	20.3.5 Modeling and Inference on Connectivity	549
	20.3.5.1 Granger Causality	550
	20.3.5.2 Partial Directed Coherence (PDC)	550
	20.3.5.3 Summary of Metrics for Connectivity	551

20.4   Spectral Analysis for High-Dimensional Data 551
        20.4.1   Methods for Fitting VAR Model on Multivariate Time Series 552
                 20.4.1.1   Least Squares Estimation 552
                 20.4.1.2   LASSO 552
                 20.4.1.3   LASSLS 553
        20.4.2   EEG Data Analysis via LASSLS Methods 554
                 20.4.2.1   VAR Modeling on High-Dimensional Multichannel EEG .. 554
                 20.4.2.2   Inference on Directed Connectivity 555
20.5   Source Localization and Estimation 555
        20.5.1   Overview of Source Models for EEG Data 555
                 20.5.1.1   Dipole Source Model 556
                 20.5.1.2   Independent Source Model 558
                 20.5.1.3   A Generalized Model of EEG Signals 559
        20.5.2   Inverse Source Reconstruction 559
                 20.5.2.1   Parametric Methods 559
                 20.5.2.2   Imaging Methods 560
                 20.5.2.3   Summary 562
        Bibliography 562

20.1   Introduction

Electroencephalogram (EEG) is a technique for recording brain activity which has been widely used in cognitive studies and in clinical applications. It is appealing for a number of reasons: it is non-invasive, relatively inexpensive to collect, highly portable and hence can be easily integrated in the clinical setting, and most of all because of its ability to capture the temporal dynamics of brain activity. Examples of the applications of EEG in studying the brain and detecting neurological disorders are described in (30), (41), (30) and (24) among many others.

This chapter broadly covers statistical modeling of electroencephalograms. In Section 20.2, we discuss the basic concepts of spectral analysis and time-domain modeling for single-channel recordings. In Section 20.3, we discuss the approaches to analyzing multichannel EEGs. Measures of dependence such as coherence, partial coherence and partial directed coherence are developed to study inter-relationships between the different EEG channels. Estimation and inference methods are discussed. In Section 20.4, we discuss current approaches to dealing with high dimensionality of EEG data. In particular, we shall cover penalized regression. Finally, in Section 20.5, methods for the estimation and localization of the underlying brain sources (including blind source separation and physical modeling techniques) are developed.

20.2   Spectral Analysis of a Single-Channel EEG

The common approach to analyzing electroencephalograms (EEGs) is to use all channels to investigate neuronal dynamics locally (within region or channel) and globally (between regions or between channels). In this section, we introduce basic ideas of time series modeling, both in the time domain and frequency domain. The first part of this chapter shall cover techniques for analyzing neuronal activity recorded at a single channel. The latter part will focus on modeling brain connectivity using all channels.

Let $X(t), t = 1, \ldots, T$ be an EEG recording for one epoch at a single channel. For simplicity, we assume that the duration of an epoch is one second and the sampling rate is 1000 Hertz. Thus, there are $T = 1000$ time points in a single epoch. The primary goal here is to determine the oscillatory content of the EEG. A formal way to study this is via the spectrum, which is the distribution of the total variance (energy) in the EEG across all frequencies on the range $(0, 500)$ Hertz. We shall consider both the Fourier-domain (Section 20.2.2) and time-domain (Section 20.2.3) approaches to characterize and estimate the spectrum.

In this chapter, we shall assume that the EEG signal in this epoch is weakly stationary, which means that the following have to be satisfied: (1) The mean of the EEG, denoted $\mu(t) = \mathbb{E}X(t)$, is constant over t during the entire epoch. This indicates that there are no deterministic trends in the signal. (2) The variance of the EEG, denoted $\mathbb{V}\mathrm{ar}\,X(t)$, is also constant over the epoch. (3) The covariance between EEG values at times t and s, denoted $\gamma(h) = \mathbb{C}\mathrm{ov}[X(t), X(s)]$, depends only on the lag $h = |t - s|$. To illustrate the idea of weak stationarity, consider this particular example: the covariance between EEG values at times $s = 1$ and $t = 2$ is the same as that between $s = 100$ and $t = 101$ and so on. To estimate the mean, covariance and correlation when the signal is weakly stationary, we can pool information across all time points within the epoch. In particular, to estimate the lag-one auto-covariance $\gamma(1)$, we can use all pairs of $[X(1), X(2)]$, $[X(2), X(3)]$, $\ldots$, $[X(T - 1), X(T)]$.

20.2.1 Brief Description of the Data

To illustrate the methods, we will analyze the EEG of one participant in the EEG study approved by the Institutional Review Board of the University of California, Irvine. The overall aim of this study is to identify biomarkers from resting-state EEG that could predict a subject's ability to learn a new motor skill. In addition to examining global activity (across the entire brain network), we are interested in localized activity, especially in the motor-related regions. During the awake resting-state phase, subjects sat in front of a computer monitor and were instructed to fix their gaze on a cross displayed at the center of a computer monitor for approximately three minutes. While the original EEGs were recorded over 256 channels, only $N = 194$ were used in this analysis due to serious artifacts in 62 channels that could not be corrected.

The following preprocessing steps were performed: low-pass filtered at 100 Hertz, segmented into non-overlapping 1-second epochs, and detrended by fitting a quadratic polynomial curve to each channel at each epoch. The epochs were visually inspected for contamination by overt muscle activity, such as from neck or cheek movements, and removed from further analysis. Next, EEG data underwent an Infomax ICA decomposition. Components attributed to muscle artifact were automatically rejected, and the remaining components were inspected to identify eye blinks, eye movements, and heart rhythms. The final dataset consists of 160 epochs, and each epoch was recorded over 1 second, and $T = 1000$ time points for each epoch. Figure 20.1 shows the location of the EEG channels projected on a 2D space, Figure 20.2 shows the EEG activities for one epoch, and Figure 20.3 shows the EEG signal for one channel in the supplementary motor area (SMA) for all 160 epochs.

20.2.2 Fourier-Domain Approach

20.2.2.1 The Fourier Regression Model and Variance Decomposition

As noted, an EEG can be seen as a super-position of waveforms with varying amplitudes. Here we shall use the family of Fourier waveforms to represent an EEG signal. A particular Fourier waveform is denoted by $\psi_k(t) = \frac{1}{\sqrt{T}} \exp(i2\pi \frac{k}{T}t)$, $t = 1, \ldots, T$ where k represents

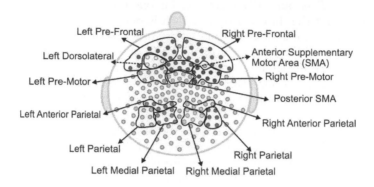

FIGURE 20.1
EEG topography.

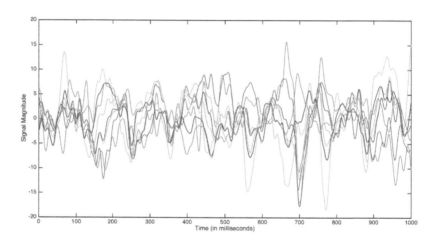

FIGURE 20.2
Plot of EEG signals for 8 channels (1 epoch).

the number of cycles over one epoch. Here, for a signal of length T, the number of cycles k have values in the set $\{-(\frac{T}{2}-1),\ldots,\frac{T}{2}\}$. Without loss of generality, we have assumed T to be even. The frequency index $k = \frac{T}{2}$ corresponds to the waveform with the Nyquist frequency (half of the sampling rate). Thus, when the EEG is sampled at the rate of 1000 Hertz, then the index $k = \frac{T}{2}$ corresponds to 500 Hertz (or 500 cycles over the one-second epoch).

While EEGs are real-valued signals, the building blocks (i.e., the Fourier waveforms) are complex-valued. The real and imaginary parts of $\psi_k(t)$ are, respectively, $\frac{1}{\sqrt{(T)}}\cos(2\pi\frac{k}{T}t)$ and $\frac{1}{\sqrt{(T)}}\sin(2\pi\frac{k}{T}t)$. Representations of real-valued signals that use complex-valued waveforms have the advantage that it is easy to align the observed signals with the Fourier waveforms by adjusting the phases of the Fourier waveforms. Moreover, to ensure that the representation guarantees that the signal is real-valued, we impose that the coefficients for the negative-valued indices ($k = -1, \ldots, -(\frac{T}{2}-1)$) are complex conjugates of the coefficients for the positive-valued coefficients.

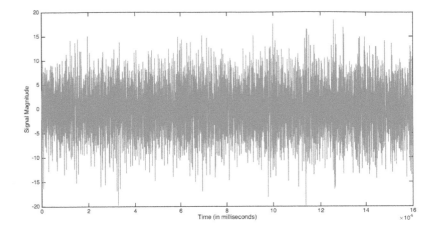

FIGURE 20.3
Plot of EEG signals for one channel (in the SMA region) for 160 epochs.

One can model the EEG using regression by using these Fourier waveforms as the basis functions or as the "explanatory variables." However, unlike a standard regression model, the coefficients (or the amplitudes) associated with each Fourier waveform is random. The discrete Fourier regression model of $X(t)$ (also the discrete Cramér representation) is given by

$$X(t) = \sum_{k=-\left(\frac{T}{2}-1\right)}^{\frac{T}{2}} A(\omega_k)\psi_k(t) \tag{20.1}$$

where $A(\omega_k)$ are uncorrelated random amplitudes with mean 0 and variance denoted by $\mathbb{V}\mathrm{ar}\, A(\omega_k) = f(\omega_k)$. Thus, one could use the above model as a data-generating mechanism. The data $X(t)$ is obtained after the amplitudes are randomly drawn from their respective distributions.

Due to the fact that the random coefficients $A(\omega_k)$ are uncorrelated, the total variance of the EEG $X(t)$ is derived to be

$$\mathbb{V}\mathrm{ar}\, X(t) = \sum_{k=-\left(\frac{T}{2}-1\right)}^{\frac{T}{2}} \mathbb{V}\mathrm{ar}\, A(\omega_k)\,|\psi_k(t)|^2 = \sum_{k=-\left(\frac{T}{2}-1\right)}^{\frac{T}{2}} f(\omega_k)\frac{1}{T}.$$

The key insight here is that the variance of the EEG at any time t is the average of the variance of the coefficients across all the frequencies k. A waveform whose amplitude has a very small variance (relative to the others) will have negligible contribution to the total variance of the EEG. In fact it will be difficult to detect the presence of such oscillation in the EEG because these amplitudes have a distribution that is centered at 0. When the variance of the distribution is small, the amplitude will also be small with high probability.

More formally, when $X(t)$ is zero mean and stationary, then it admits the Cramér representation

$$X(t) = \int_{-0.5}^{0.5} A(\omega)\exp(i2\pi\omega t)dZ(\omega), \tag{20.2}$$

where $A(\omega)$ is the transfer function and the increment $dZ(\omega)$ is a random process over frequencies ω that satisfies $\mathbb{E}dZ(\omega) = 0$ for all ω; $\mathbb{V}\mathrm{ar}\, dZ(\omega) = d\omega$; and for $\omega \neq \lambda$,

$\mathbb{C}\text{ov}[dZ(\omega), dZ(\lambda)] = 0$. Based on the Cramér representation, we now derive the variance of $X(t)$. In the following derivation, we denote B^* to be the complex conjugate of B,

$$
\begin{aligned}
\mathbb{V}\text{ar}\, X(t) &= \mathbb{E}\left[X(t)X^*(t)\right] \\
&= \mathbb{E}\left[\int A(\omega)\exp(i2\pi\omega t)dZ(\omega)\int A^*(\lambda)\exp(-i2\pi\lambda t)dZ^*(\lambda)\right] \\
&= \int\int A(\omega)A^*(\lambda)\exp[i2\pi(\omega-\lambda)t]\,\mathbb{C}\text{ov}[dZ(\omega), dZ(\lambda)] \\
&= \int |A(\omega)|^2 d\omega,
\end{aligned}
$$

which is consistent with the results above for the discretized Cramér representation.

20.2.2.2   The Spectrum of a Single-Channel Time Series

Suppose that the single-channel EEG $X(t)$ is a weakly stationary time series with mean $\mathbb{E}X(t) = 0$, variance $\mathbb{V}\text{ar}\, X(t) = \sigma_{XX}(0)$, and auto-covariance $\gamma_{XX}(h) = \mathbb{C}\text{ov}[X(t+h), X(t)]$. The theoretical derivations in time series (e.g., need to show existence of the spectrum and consistency of estimators of the spectrum) all need to make the assumption that the EEGs decorrelate "sufficiently quickly," which is formally stated as $\sum_{h=-\infty}^{\infty} |\gamma_{XX}(h)| < \infty$. The above condition implies that there exists some finite lag H so that the covariance between observations lagged at least H time units (e.g., milliseconds) apart are negligible. The spectrum and auto-covariance function of $X(t)$ are defined via the Fourier (and inverse Fourier) transforms as follows

$$
f_{XX}(\omega) = \sum_{h=-\infty}^{\infty} \gamma_{XX}(h)\exp(-i2\pi\omega h), \quad \text{where } \omega \in (-0.5, 0.5) \tag{20.3}
$$

$$
\gamma_{XX}(h) = \int_{-0.5}^{0.5} f_{XX}(\omega)\exp(i2\pi\omega h)d\omega \quad \text{where } h = 0, \pm 1, \ldots. \tag{20.4}
$$

When we set the lag $h = 0$ in Equation 20.4, the lag 0 auto-covariance $\gamma_{XX}(0)$ becomes the variance and thus we arrive at the decomposition

$$
\mathbb{V}\text{ar}\, X(t) = \int_{-0.5}^{0.5} f_{XX}(\omega)d\omega. \tag{20.5}
$$

From the above results, the spectrum of a weakly stationary time series with representation given in Equation 20.22 is $f_{XX}(\omega) = |A(\omega)|^2$. The spectrum tells us the contribution of each oscillation ψ_k, denoted $f_{XX}(\omega)d\omega$ to the variance of $X(t)$.

20.2.2.3   Estimating the Spectrum via Periodograms

Suppose that we observe EEG $X(t)$ over one epoch where $t = 1, \ldots, T$. We will assume that the sample mean $\overline{X} = \frac{1}{T}\sum_{t=1}^{T} X(t)$ is zero. If $\overline{X}$ is not zero, then it must be removed before proceeding. The first step is to compute the Fourier coefficients

$$
d_X(\omega_k) = \sum_{t=1}^{T} X(t)\exp(-i2\pi\omega_k t) \quad \text{where } \omega_k = \frac{k}{T}, \ k = -\left(\frac{T}{2}-1\right), \ldots, \frac{T}{2}. \tag{20.6}
$$

Note that each Fourier coefficient is the inner product (or the cross-covariance) between the observed EEG $\mathbf{X} = [X(1), \ldots, X(T)]'$ and each of the Fourier waveforms

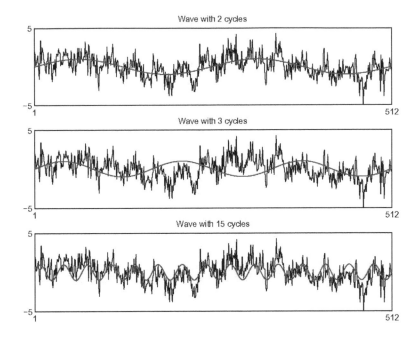

FIGURE 20.4

The Fourier coefficients are interpreted as the inner product (cross-covariance) between the EEG **X** and each of the Fourier waveforms ψ_k.

$\psi_k = [\psi_k(1), \ldots, \psi_k(T)]'$ (See Figure 20.4). The Fourier coefficients are computed using the fast Fourier transform (FFT), which is implemented in both the R and MATLAB® software.

The second step is to compute the Fourier periodograms

$$I_{XX}(\omega_k) = \frac{1}{T} d_{XX}(\omega_k) d^*_{XX}(\omega_k) = \frac{1}{T} |d_{XX}(\omega_k)|^2.$$

When T is large, then the periodograms are approximately unbiased, i.e., $\mathbb{E} I_{XX}(\omega) \approx f_{XX}(\omega)$. This follows from the result in (3), which we state below. Under mild assumptions on the signal $X(t)$,

$$I_{XX}(\omega_k) \quad \dot{\sim} \quad f_{XX}(\omega_k) \xi(\omega_k), \tag{20.7}$$

where $\xi(\omega_k)$ are approximately uncorrelated for $k = 0, 1, \ldots, \frac{T}{2}$. Moreover, for $\omega_k = 0, \frac{1}{2}$, $\xi(\omega_k) \dot{\sim} \chi^2(1)$; and $\xi(\omega_k) \dot{\sim} \frac{\chi^2(2)}{2}$ for $\omega_k = \frac{1}{T}, \ldots, \frac{\frac{T}{2}-1}{T}$. However, the above result indicates that the periodograms are not mean-squared consistent because $\mathbb{V}\text{ar}\, I_{XX}(\omega) \approx [f_{XX}(\omega)]^2$, which does not go to 0 even as the time series length T increases. To reduce the variance, one can take local averages of the periodograms over small neighborhoods of frequency. For example, the local average of the periodograms around ω_k (taken over a neighborhood with $2m + 1$ frequencies) is

$$\widehat{f}_{XX}(\omega_k) = \frac{1}{2m+1} \sum_{\ell=-m}^{m} I_{XX}(\omega_{k+\ell}). \tag{20.8}$$

Remarks.

1. The variance of the estimator of the spectrum decreases as the size of the frequency window $(2m + 1)$ increases. Denote the average of the *true* spectrum to be $\overline{f}_{XX}(\omega_k) = \frac{1}{2m+1} \sum_{\ell=-m}^{m} f_{XX}(\omega_{k+\ell})$ and average of the squares of the spectrum to be $\overline{f^2}_{XX}(\omega_k) = \frac{1}{2m+1} \sum_{\ell=-m}^{m} f_{XX}^2(\omega_{k+\ell})$. Then the variance of the spectrum estimator is

$$\mathrm{Var}\, \widehat{f}_{XX}(\omega_k) \approx \left[\frac{1}{2m+1}\right]^2 \sum_{\ell=-m}^{m} \left[f_{XX}(\omega_{k+\ell})\right]^2 \approx \frac{1}{2m+1} \overline{f^2}_{XX}(\omega_k).$$

On the one hand, it is desirable that we let m grow in order to reduce the variance (produce smoother estimates). However, one must be careful because, on the other hand, increasing m could lead to increased bias. That is, $\mathbb{E}\widehat{f}_{XX}(\omega_k) \approx \overline{f}_{XX}(\omega_k)$ where $\overline{f}_{XX}(\omega_k) \approx f_{XX}(\omega_k)$ when m is small. This bias–variance tradeoff is well known in non-parametric regression.

2. For spectral estimation, there are data-driven approaches to finding the optimal bandwidth. The gamma-deviance generalized cross-validation procedure in (25) selects the $\widehat{m}$ that minimizes the deviance, which is a more general concept of the least squares error between the estimator and the true value, which is based on the log likelihood. Under Gaussianity, the deviance is equivalent to the squared error between the estimator and the true value. Here, the periodograms are distributed as gamma random variables rather than Gaussian. For each candidate smoothing parameter $m \in \mathcal{M}$, we compute the smoothed periodogram denoted $\widehat{f}_{XX,m}(\omega_k)$. The Gamma generalized cross-validation criterion is

$$\mathrm{GCV}(m) = \frac{1}{L+1} \sum_{k=0}^{L} q_k$$

$$\left\{ \frac{-\log[I_{XX}(\omega_k)/\widehat{f}_{XX,m}(\omega_k)] + [I_{XX}(\omega_k) - \widehat{f}_{XX,m}(\omega_k)]/\widehat{f}_{XX,m}(\omega_k)}{\left(\frac{2m}{2m+1}\right)^2} \right\}. \qquad (20.9)$$

The optimal smoothing parameter is $\widetilde{m} = \mathrm{argmin}_{\mathcal{M}}\mathrm{GCV}(m)$.

3. The estimator $\widehat{f}_{XX}(\omega_k)$ in Equation (20.9) is obtained by putting equal weight of $W_\ell = \frac{1}{2m+1}$ on every periodogram $I_{XX}(\omega_{k+\ell})$ in the neighborhood. Thus, an estimator with more general weights is

$$\widehat{f}_{XX}(\omega_k) = \sum_{\ell=-m}^{m} W_{\ell,m} I_{XX}(\omega_{k+\ell}),$$

where the weights satisfy the following: $0 \leq W_{\ell,m} \leq 1$ and $\sum_\ell W_{\ell,m} = 1$. It is often desirable to put heavier weight at the center of the neighborhood so that $I_{XX}(\omega_k)$ gets the most weight in the estimator. One such weight is based on the Gaussian kernel. Define $w_{\ell,k} = \exp(-\frac{(\omega_{\ell+k}-\omega_k)^2}{2\sigma^2})$, then $W_{\ell,k} = \frac{w_{\ell,k}}{\sum_{\ell=-m}^{m} w_{\ell,k}}$.

4. At this point it is instructive to note that the sum of the squares of the observed EEG is equal to the sum of the periodograms, i.e.,

$$\sum_{t=1}^{T} \left[X(t)\right]^2 = \sum_{k=-\left(\frac{T}{2}-1\right)}^{\frac{T}{2}} I_{XX}(\omega_k).$$

In engineering language, the total energy of the EEG signal is preserved in the frequency domain. This is because the Fourier waveforms ψ_k are orthogonal with each having norm T or, equivalently, the normalized Fourier waveforms $\frac{1}{\sqrt{T}}\psi_k$ are orthonormal. However, the smoothed periodograms no longer sum to the total energy of the time series.

20.2.2.4 Other Periodogram-Based Estimation Methods

The spectrum (or log spectrum) can be estimated non-parametrically using smoothing splines as demonstrated by (38) by treating the bias-corrected log periodogram as a decomposition of the log spectrum (to be estimated by smoothing splines) and a stochastic term which has mean zero and homogenous variance across the frequency domain. One of the criticisms of smoothing splines estimates is that they do not capture localized features such as abrupt changes, peaks, and troughs. To address these limitations, wavelets-based estimation methods were developed by (22), (12) and (39). Here, the functional of interest, the log spectrum, is assumed to be approximated via an expansion of a wavelet basis. The coefficients of these wavelet functions are estimated by thresholding. A classical approach, given by Welch in (40), entails breaking up the time series into short (possibly overlapping) blocks, computing the periodograms at each of these blocks, and then averaging these periodograms. This approach is implemented in many softwares (e.g., MATLAB®) though its limitation is that the frequency resolution is potentially reduced due to having fewer observations at each of the smaller blocks.

20.2.2.5 Examples of Smoothing Periodograms

In order to illustrate the spectral estimate for univariate time series, we use a single-channel EEG data series obtained from a channel at the SMA region, as plotted in Figure 20.5. The spectrum estimate via the raw periodogram and the smoothed periodogram using a moving average with half-window size of 10 and 50 time points are shown in Figure 20.6. As the window size increases, the spectrum estimate is getting smoother. One way to determine the "optimal" bandwidth is using Gamma GCV.

20.2.2.6 Multitaper Method (MTM)

The multitaper method (MTM) uses the average of a set of independent estimates of the power spectrum, by multiplying the original signals by orthogonal taper function, then averaging the periodogram computed for each tapered signal (35). The taper functions are pre-constructed as optimal functions that are (1) normalized, (2) orthogonal, and (3) minimize the spectral leakage due to finite length of observation. The optimal taper functions belong to the function family named discrete prolate spheroidal sequences (DPSS). Let $h_1(t), \ldots, h_K(t)$ be the K taper functions; the periodogram computed from the kth taper function, denoted by $I_{XX}^{(k)}(\omega)$, can be computed as

$$I_{XX}^{(k)}(\omega) = \frac{1}{T}|\sum_{t=1}^{T} h_k(t)X(t)\exp(-i2\pi\omega t)|^2. \qquad (20.10)$$

The multitaper method averages the obtained K periodograms, that is,

$$\widehat{f}_{XX}(\omega) = \frac{\sum_{k=1}^{K}\lambda_k I_{XX}^{(k)}(\omega)}{\sum_{k=1}^{K}\lambda_k}. \qquad (20.11)$$

As an example of spectrum estimation, we use the EEG signal at a channel in the SMA region of length $T = 10000$, as plotted in Figure 20.5. The sampling frequency of the data

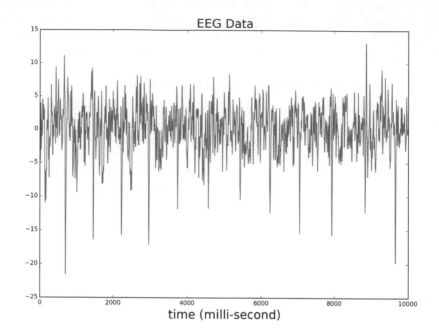

FIGURE 20.5
Plot of EEG signal at channel 23 (in SMA region).

is 1000, which means the temporal resolution of the signal is 1 millisecond. Figure 20.6 shows the power spectrum density (PSD) estimate for the EEG signal via various methods including the periodogram, smoothed periodogram using a rectangular window, ARMA model, and multitaper method. The taper functions are plotted in Figure 20.7, which are the five leading Slepian functions with length $T = 10000$ and half-bandwidth $= 2.5$.

20.2.3 Time-Domain Approach

One way to estimate the spectrum of an EEG in an epoch is by fitting a parametric time domain model to the EEG traces. We consider three general models for weakly stationary time series, namely, moving average (MA), autoregressive (AR), and autoregressive moving average (ARMA). Both the AR and MA models are special cases of the ARMA model. Here, we will discuss each of these models and derive the parametric form of their spectra.

20.2.3.1 Moving Average (MA) Model

The moving average model of lag or order Q is defined as follows. Before doing so, consider the white noise sequence $W(t)$ with mean $\mathbb{E}W(t) = 0$ and autocovariance sequence

$$\mathbb{C}\text{ov}[W(s), W(t)] = \begin{cases} \sigma_W^2, & \text{if } s = t \\ 0, & \text{if } s \neq t. \end{cases} \quad (20.12)$$

The EEG recording in an epoch $X(t)$ (with zero mean) is a moving average of order Q, written MA(Q), if it can be represented as

$$X(t) = W(t) + \theta_1 W(t-1) + \ldots + \theta_Q W(t-Q). \quad (20.13)$$

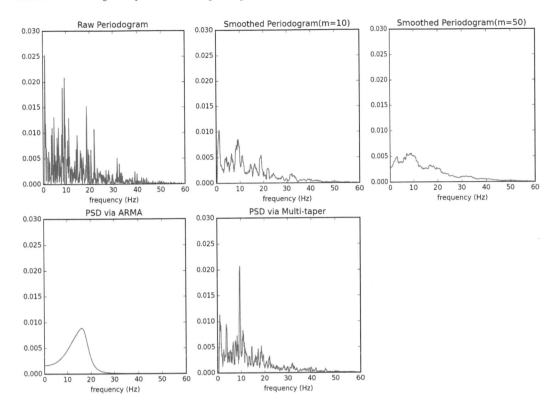

FIGURE 20.6
Power spectrum density estimate via periodogram, smoothed periodogram, multitaper method, and ARMA method.

Thus, the moving average process is a linear combination of the current and past white noise values. It is one-sided since the current value $X(t)$ does not depend on the future white noise values. At this point, it is helpful to introduce some notation. First, for a positive integer d, the backshift operator B^d is defined as follows: $B^d Y(t) = Y(t - d)$. When the backshift operator B^d is applied to $Y(t)$, the output is $Y(t - d)$. Next, we define the MA polynomial function $\Theta(B)$ to be

$$\Theta(B) = 1 + \theta_1 B + \theta_2 B^2 + \ldots + \theta_Q B^Q. \qquad (20.14)$$

Then the EEG $X(t)$ follows an MA(Q) representation if it can be written as $X(t) = \Theta(B)W(t)$.

20.2.3.2 Autoregressive (AR) Model

One of the most common models applied to EEGs is the autoregressive (AR) model. The EEG $X(t)$ is autoregressive of order P, denoted AR(P), if it has the representation

$$X(t) = \phi_1 X(t - 1) + \ldots + \phi_P X(t - P) + W(t). \qquad (20.15)$$

Next, the AR polynomial function of the backshift operator is

$$\Phi(B) = 1 - \phi_1 B - \phi_2 B^2 - \ldots - \phi_P B^P. \qquad (20.16)$$

Then the EEG $X(t)$ follows an AR(P) representation if it can be written as $\Phi(B)X(t) = W(t)$.

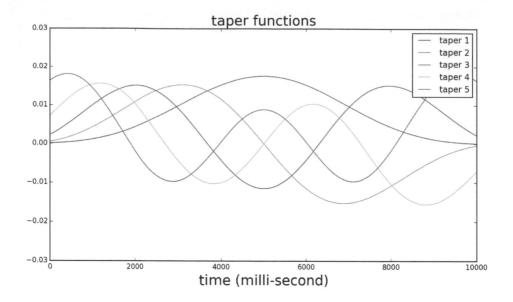

FIGURE 20.7
The five leading Slepian sequences (taper functions) for T = 10000 and half-bandwidth = 2.5.

We now discuss the concept of causality in time series. (This is different from Granger causality, which will be discussed in Section 20.3.5.1). The EEG $X(t)$ is represented as a causal linear process if it does not depend on future white noise values. In other words, $X(t)$ is causal if

$$X(t) = \sum_{j=0}^{\infty} c_j W(t-j),$$

where the filter coefficients $\{c_j\}$ are absolutely summable, i.e., $\sum_{j=0}^{\infty} |c_j| < \infty$. For example, a first-order AR(1) time series $X(t) = \phi_1 X(t-1) + W(t)$ where $\phi_1 \in (-1, 1)$ is causal because it can be represented by

$$X(t) = \sum_{j=0}^{\infty} c_j W(t-j) = \sum_{j=0}^{\infty} \phi_1^j W(t-j).$$

An AR(P) time series $X(t)$ is causal if the roots $z_1, \ldots, z_P$ of the AR function

$$\Phi(z) = 1 - \phi_1 z - \ldots - \phi_P z^P$$

all have magnitudes greater than 1. Note that some of these roots might be complex-valued. If a root, say z_1, is complex-valued, then it is necessary that there exists another root, say z_2, that is a complex conjugate of z_1, i.e., $z_2 = z_1^*$.

20.2.3.3　Autoregressive Moving Average (ARMA) Model

A time series $X(t)$ is an autoregressive moving average model of order (P, Q), denoted ARMA(P, Q), if it can be expressed as

$$
\begin{aligned}
X(t) = {} & \phi_1 X(t-1) + \ldots + \phi_P X(t-P) \\
& + W(t) + \theta_1 W(t-1) + \ldots + \theta_Q W(t-Q),
\end{aligned}
\tag{20.17}
$$

which can be written in terms of the AR and MA polynomial operators $\Phi(B)$ and $\Theta(B)$ as follows: $\Phi(B)X(t) = \Theta(B)W(t)$.

A key requirement here for the process to be a valid ARMA is that the polynomial functions $\Phi(z)$ and $\Theta(z)$ do not share a common factor. Without this requirement, the models might not be identifiable. This is nicely motivated in (31). Suppose that the time series $X(t)$ is white noise, i.e., $X(t) = W(t)$. Then it follows that $0.9X(t-1) = 0.9W(t-1)$ and hence

$$X(t) - 0.9X(t-1) = W(t) - 0.9W(t-1)$$
$$\text{or equivalently } (1 - 0.9B)X(t) = (1 - 0.9B)W(t),$$

which suggests that $X(t)$, which is white noise, is also ARMA(1,1). To avoid this identifiability problem, we impose that $\Phi(z)$ and $\Theta(z)$ do not share a common factor. This example is not a valid ARMA(1,1) because the AR and MA polynomial functions contain the common factor $(1 - 0.9z)$.

20.2.3.4 The Spectra of MA, AR and ARMA Processes

We shall derive the spectrum of the ARMA(P, Q) process, which includes the AR and MA as special cases. Define $Y(t)$ to be a time series that is a linearly filtered version of another time series $Z(t)$ with zero mean, i.e.,

$$Y(t) = \sum_{j=-\infty}^{\infty} c_j Z(t-j) \text{ where } \sum_{j=-\infty}^{\infty} |c_j| < \infty.$$

Denote the auto-covariance of $Y(t)$ and $Z(t)$ to be, respectively, $\gamma_{YY}(h)$ and $\gamma_{ZZ}(h)$; and the corresponding spectra to be $f_{XX}(\omega)$ and $f_{YY}(\omega)$. Define the transfer function of the filter coefficients $\{c_j, j = -\infty, \ldots, \infty\}$ to be

$$C(\omega) = \sum_{j=-\infty}^{\infty} c_j \exp(-i2\pi\omega j).$$

The auto-covariance of $Y(t)$ can be expressed as follows

$$\gamma_{YY}(h) = \mathbb{C}\text{ov}\left[\sum_{j=-\infty}^{\infty} c_j Z(t+h-j), \sum_{r=-\infty}^{\infty} c_r Z(t-r)\right]$$

$$= \sum_{j=-\infty}^{\infty} \sum_{r=-\infty}^{\infty} c_j c_r \gamma_{ZZ}(h-j+r)$$

$$= \sum_{j=-\infty}^{\infty} \sum_{r=-\infty}^{\infty} c_j c_r \int f_{ZZ}(\omega) \exp(i2\pi\omega(h-j+r)) d\omega$$

$$= \int \sum_{j=-\infty}^{\infty} c_j \exp(-i2\pi\omega j) \sum_{r=-\infty}^{\infty} c_r \exp(i2\pi\omega r) f_{ZZ}(\omega) \exp(i2\pi\omega h) d\omega$$

$$= \int |C(\omega)|^2 f_{ZZ}(\omega) \exp(i2\pi\omega h) d\omega.$$

Since $\gamma_{YY}(h) = \int f_{YY}(\omega) \exp(i2\pi\omega h) d\omega$, by the uniqueness of the spectrum, it follows that

$$f_{YY}(\omega) = |C(\omega)|^2 f_{ZZ}(\omega).$$

This result will be utilized to derive the spectrum of MA, AR, and ARMA processes.

Remark 1. The result above tells us that filtering alters the spectrum of the original time series $Z(t)$ in a particular way. The spectrum of the resulting time series is equal to the original time series $f_{ZZ}(\omega)$ modulated by the square of the gain function $|C(\omega)|^2$.

Remark 2. The generic domain of the spectrum is $\omega \in (0, 1/2)$. We often rescale this so that it accurately reflects the sampling rate and the Nyquist frequency. So in our example where the sampling rate is 1000 Hertz, the domain of the spectrum is $(0, 500)$ Hertz. In practice, we may not plot the spectrum for the entire domain if the EEG has been low-pass filtered.

Spectrum of the white noise. Recall that the spectrum of the white noise is $\gamma_{WW}(h)$ is equal to σ_W^2 when $h = 0$; and 0 when $h \neq 0$. The spectrum of the white noise time series is

$$f_{WW}(\omega) = \sum_{h=-\infty}^{\infty} \gamma_{WW}(h) \exp(-i2\pi\omega h) = \sigma_W^2,$$

which is constant over all frequencies. Such a signal is called "white" because white light contains waveforms of different frequencies—all with the same intensity.

The spectrum of the MA process. Suppose that the EEG $X(t)$ follows the MA(Q) model. Recall that an MA(Q) time series is a filtered version of past white noise values, i.e.,

$$X(t) = \sum_{j=-\infty}^{\infty} c_j W(t-j) \quad \text{where} \quad c_j = \theta_j, \; j = 1, \ldots, Q.$$

$c_0 = 1$ and $c_j = 0$ for all $j \leq -1$ and $j \geq Q+1$. The transfer function is $C(\omega) = 1 + \sum_{j=1}^{Q} \theta_j \exp(-i2\pi\omega j)$. The spectrum of the EEG $X(t)$ is

$$f_{XX}(\omega) = |C(\omega)|^2 f_{WW}(\omega) = |1 + \sum_{j=1}^{Q} \theta_j \exp(-i2\pi\omega j)|^2 \sigma_W^2.$$

The spectrum of the AR process. Suppose that the EEG $X(t)$ follows the AR(P) model. The AR(P) representation is $\Phi(B)X(t) = W(t)$ where $\Phi(B) = 1 - \phi_1 B - \ldots - \phi_P B^P$. Define $\Phi(\omega) = 1 - \phi_1 \exp(-i2\pi\omega) - \ldots - \phi_P \exp(-i2\pi\omega P)$. By equating the spectra on both sides of the representation, we have

$$|\Phi(\omega)|^2 f_{XX}(\omega) = f_{WW}(\omega)$$
$$f_{XX}(\omega) = |1 - \phi_1 \exp(-i2\pi\omega) - \ldots - \phi_P \exp(-i2\pi\omega P)|^{-2} \sigma_W^2.$$

The spectrum of the ARMA process. Suppose that the EEG $X(t)$ has an ARMA(P,Q) model $\Phi(B)X(t) = \Theta(B)W(t)$ where

$$\Phi(B) = 1 - \sum_{p=1}^{P} \phi_p B^p \qquad \Theta(B) = 1 + \sum_{q=1}^{Q} \theta_q B^q.$$

The transfer functions of the AR and MA filters are, respectively,

$$\Phi(\omega) = 1 - \sum_{p=1}^{P} \phi_p \exp(-i2\pi\omega p)$$

$$\Theta(\omega) = 1 + \sum_{q=1}^{Q} \theta_q \exp(-i2\pi\omega q).$$

The spectrum of the EEG is

$$f_{XX}(\omega) = |\Phi(\omega)|^{-2}|\Theta(\omega)|^2\sigma_W^2. \tag{20.18}$$

20.2.3.5 Second-Order Autoregressive [AR(2)] Model

Second-order auto-regressive AR(2) models are interesting because they can accurately represent oscillations. Thus, AR(2) models can be useful for analyzing EEGs. The oscillatory properties of AR(2) processes are determined by the roots of the AR polynomial equation, which we now explain in detail. Recall that the AR(2) process $X(t)$ can be written as $\Phi(B)X(t) = W(t)$ where the polynomial operator is $\Phi(B) = 1 - \phi_1 B - \phi_2 B^2$. Denote the polynomial equation to be $\Phi(z) = 1 - \phi_1 z - \phi_2 z^2$ and let z_1 and z_2 be the roots of the AR(2) polynomial equation so that $\phi(z_1) = \phi(z_2) = 0$.

In order for the $AR(2)$ to be a causal process, the roots must have magnitudes greater than 1, i.e., $|z_1| > 1$ and $|z_2| > 1$. The roots z_1 and z_2 are either (a) both real-valued and identical; (b) both real-valued but unique; (c) both complex-valued and hence complex conjugates of each other. Under (c), time series realizations of the AR(2) process have distinct oscillatory features that are determined by the roots. The Cartesian representation of the roots z_1 and z_2 is

$$z_1 = z_1^R + iz_1^I \quad z_2 = z_1^R - iz_1^I$$

where i is the imaginary number, z_1^R and z_1^I are real-valued numbers. Clearly we have constrained z_2 to be the complex conjugate of z_1, i.e., $z_2 = z_1^*$. These complex numbers can be represented via polar coordinates. Let M be the magnitude of the roots and hence $M = \sqrt{[z_1^R]^2 + [z_1^I]^2}$. Again, for causal processes, M must be greater than 1. Let ξ be the phase angle of the roots and hence $\xi = \frac{1}{2\pi}\tan^{-1}(z_1^I/z_1^R)$. In the polar representation, M is the distance of z_1 and z_2 from the origin (i.e., the complex number $0 + i0$). Thus, the polar representation of the roots z_1 and z_2 is

$$z_1 = M\exp(i2\pi\xi) \quad \text{and} \quad z_2 = M\exp(-i2\pi\xi).$$

The spectrum of the AR(2) process with roots given above will have power that is concentrated around ξ. Moreover, as $M \to 1^+$, the power becomes increasingly more concentrated around ξ; and the power becomes more diffused (less concentrated) when M becomes larger. To illustrate this point, we plot in Figure 20.8 the spectra (left) for different values of ξ, $M = 1.1$, and sampling frequency $\Omega_s = 100$ Hz; and the simulated time series (right). The spectra of the AR(2) processes are concentrated at the typical frequency bands: delta, theta, alpha, beta, and gamma, respectively.

Remark. One could decompose an EEG $X(t)$ into latent (unobserved) AR(2) components. Define $Y_1(t)$ be an AR(2) process with spectra concentrated at the delta band; $Y_2(t)$ for theta; $Y_3(t)$ for alpha; $Y_4(t)$ for beta; and $Y_5(t)$ for gamma. Denote the corresponding spectra of these latent AR(2) processes to be $f_\ell(\omega)$. Moreover, the latent processes $Y_\ell(t)$, $\ell = 1, \ldots, 5$

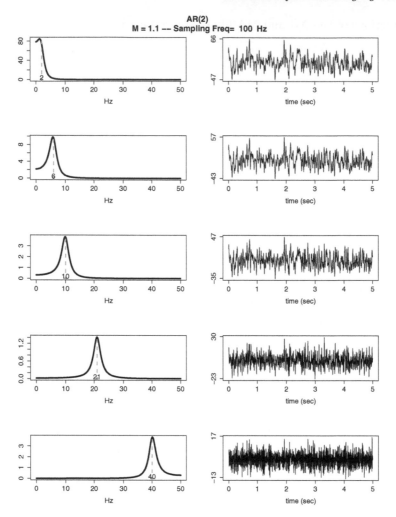

FIGURE 20.8

Left: Spectra for the AR(2) process with different peak frequency; $\Omega_s \xi = 2, 6, 10, 21, 40$ which correspond to delta, theta, alpha, beta and gamma frequency bands. Right: Simulated signals from the corresponding AR(2) process.

will be assumed to be independent. The EEG $X(t)$ and its spectrum can be decomposed as, respectively,

$$X(t) = \sum_{\ell=1}^{5} B_\ell Y_\ell(t) \quad \text{and} \quad f_{XX}(\omega) = \sum_{\ell=1}^{5} B_\ell^2 f_\ell(\omega).$$

20.2.3.6 Estimating the Spectrum

To estimate the spectrum of the EEG process via the time-domain approach, the first step is to determine the optimal ARMA model based on the observed data $\{X(t), t = 1, \ldots, T\}$. Here, we select the best model using the Bayesian Schwarz information criterion (BIC).

We now discuss the elements of the criterion. First, we fix the set of all ARMA(p,q) models from the best model selected. The AR order p ranges within the set $\mathcal{P} = \{0, \ldots, P_{\max}\}$ and the MA order q ranges within the set $\mathcal{Q} = \{0, \ldots, Q_{\max}\}$. The maximum values $P_{\max}$

and Q_{max} are pre-specified and it is recommended that $P_{max} + Q_{max}$ not exceed $0.15T$ where T is the total number of time points in an epoch. Note that the optimal ARMA is actually an AR model if the optimal MA order $\widehat{Q}$ is 0.

The optimal orders $\widehat{P}$ and $\widehat{Q}$ satisfy the criterion

$$[\widehat{P}, \widehat{Q}] = \underset{\mathcal{P} \times \mathcal{Q}}{\arg\min} \left[\log(\widehat{\sigma}_W^2) + \frac{(P + Q + 1)\log(T)}{T} \right],$$

where $\widehat{\sigma}_W^2$ is the MLE estimation of the noise variance $\mathbb{V}\mathrm{ar}\, W(t)$. The BIC criterion penalizes the complexity model via $\frac{(P+Q+1)\log(T)}{T}$. Without the penalty, the criterion will very often choose the model with the largest order since the squared error either decreases or stays constant (i.e., it never increases) as more parameters are included in the model.

Suppose that the ARMA model that is optimal for the EEG signal $X(t)$ is ARMA$(\widehat{P}, \widehat{Q})$. Denote the estimators of the ARMA parameters as $\widehat{\phi}_p$ and $\widehat{\theta}_q$ where $p = 1, \ldots, \widehat{P}$ and $q = 1, \ldots, \widehat{Q}$. The parametric estimator for the spectrum is

$$
\begin{aligned}
\widehat{f}_{XX}(\omega) &= |\widehat{\Phi}(\omega)|^{-2}|\widehat{\Theta}(\omega)|^2 \widehat{\sigma}_W^2 \quad \text{where} \quad &(20.19)\\
\widehat{\Phi}(\omega) &= 1 - \widehat{\phi}_1 \exp(-i2\pi\omega) - \ldots - \widehat{\phi}_{\widehat{P}} \exp(-i2\pi\omega\widehat{P})\\
\widehat{\Theta}(\omega) &= 1 + \widehat{\theta}_1 \exp(-i2\pi\omega) + \ldots + \widehat{\theta}_{\widehat{Q}} \exp(-i2\pi\omega\widehat{Q}).
\end{aligned}
$$

The spectrum estimated via ARMA model for a single-channel EEG (Figure 20.5) is shown in Figure 20.6. Compared to the results from the periodogram and multitaper methods, the power spectrum density (PSD) estimate via ARMA model is smoother.

20.2.4 Estimating the Spectrum Using Multiple EEG Traces

In a designed experiment, the EEG is segmented into R non-overlapping epochs where an epoch is typically a 1-second recording. If the entire recording across R epochs is stationary, which roughly means that the brain process did not evolve over the entire recording, then it is sensible to combine information across the different epochs in a meaningful way.

One approach is to simply take the average of the periodograms computed from all epochs. More formally, let $\{X^r(t)\}$ be the EEG recorded during epoch r and denote the corresponding periodogram to be $I_{XX}^r(\omega_k)$. The spectral estimate is

$$\widehat{f}_{XX}(\omega) = \frac{1}{R}\sum_{r=1}^{R} I_{XX}^r(\omega).$$

Remark 1. Note that as the number of epochs R increases, the variance of the estimator $\widehat{f}_{XX}(\omega)$ decreases. This suggests that it would not be necessary to smooth the periodograms at each epoch r. In fact, when R is large, one should avoid smoothing because smoothing degrades frequency resolution.

Remark 2. If the number of epochs is small, say $R < 30$, it might be necessary to do mild smoothing of the periodograms in each epoch. We recommend trying out different smoothing bandwidths that give sufficiently smooth overall estimates without degrading the frequency too much.

20.2.4.1 Other Averaged Estimators

The principle of averaging epoch-specific estimates can be extended to other types of estimators. For example, one could compute the average of Welch periodograms or multitaper periodograms. A data-adaptive approach for estimating spectrum via aggregating information from multiple signals is described in (4). Moreover, one could obtain spectral estimates for each epoch using time-domain methods and then average these across all epochs. More formally, let $\widehat{f}_{XX}^{r}(\omega)$ be the estimator for epoch r. The estimator could be any of the ones already discussed (e.g., raw periodograms, mildly smoothed periodograms, Welch estimator, ARMA-based estimator). The final estimator would then be

$$\widehat{f}_{XX}(\omega) = \frac{1}{R}\sum_{r=1}^{R}\widehat{f}_{XX}^{r}(\omega).$$

20.2.4.2   Estimating Power in Specific Frequency Bands

We are often interested in spectral power at a frequency band as opposed to a singleton frequency. For example, the spectrum at the alpha band and beta band which cover 8–12 Hertz and 12–32 Hertz are, respectively,

$$f_{XX}(\text{alpha}) = \int_{\frac{8}{\Omega_s}}^{\frac{12}{\Omega_s}} f_{XX}(\omega)d\omega \quad \text{and} \quad f_{XX}(\text{beta}) = \int_{\frac{12}{\Omega_s}}^{\frac{32}{\Omega_s}} f_{XX}(\omega)d\omega.$$

20.2.4.3   Detecting Outliers

Methods for detecting potential outliers in EEG recordings have been discussed and demonstrated in (23) and its references.

20.2.5   Confidence Intervals

We will make the following assumptions: (1) the EEGs were independently generated across several epochs (i.e., the brain response in one epoch is independent of the brain responses in other epochs) and (2) the EEGs at all the epochs were produced from the same underlying process (there is stationarity across the entire period of consideration such as the resting state). For the situation where the underlying brain process might have evolved, a method that models the non-stationarity properties both within a trial and across trials has been developed in (7).

Let $I_{XX}^{r}(\omega_k)$ be the periodogram at frequency ω_k computed for epoch r and let $L = \frac{T}{2}$. Under stationarity (and mild assumptions on the covariance structure), the periodograms $I_{XX}^{r}(\omega_0), \ldots, I_{XX}^{r}(\omega_L)$ are all approximately independent.

$$I_{XX}^{r}(\omega_k) \quad \dot{\sim} \quad \chi^2(1)f_{XX}(\omega_k) \ \text{ for } \ k = 0, L$$

$$I_{XX}^{r}(\omega_k) \quad \dot{\sim} \quad \frac{\chi^2(2)}{2}f_{XX}(\omega_k) \ \text{ for} k = 1, \ldots, L-1.$$

We have the following approximate distribution for the periodogram

$$I_{XX}^{r}(\omega_k)\dot{\sim}\frac{\chi^2(2)}{2}f_{XX}(\omega_k)$$

for $k = 1, \ldots, L-1$. Therefore, for large R, we have the following asymptotic distribution for the averaged periodogram across independent trials

$$\widehat{f}_{XX}(\omega_k) = \frac{1}{R}\sum_{r=1}^{R}I_{XX}^{r}(\omega_k)\dot{\sim}\frac{\chi_{2R}^2}{2R}f_{XX}(\omega_k). \qquad (20.20)$$

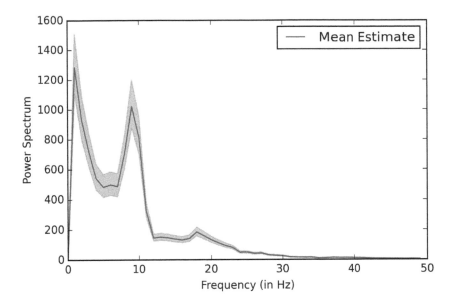

FIGURE 20.9

95% confidence interval for the power spectrum of signals at channel 23, computed using 160 epochs.

An approximate $1 - \alpha$ confidence interval for $f_{XX}(\omega_k)$ can be constructed as

$$\left(\frac{2R\widehat{f}_{XX}(\omega_k)}{\chi^2_{2R}(1 - \alpha/2)}, \ \frac{2R\widehat{f}_{XX}(\omega_k)}{\chi^2_{2R}(\alpha/2)}\right). \tag{20.21}$$

As an example, we compute the spectrum estimate for univariate time series with $R = 160$ trials (the concatenated signal is plotted in Figure 20.3). The estimate of the power spectrum density (PSD) and its corresponding 95% confidence interval is shown in Figure 20.9.

20.3 Spectral Analysis of Multichannel EEG

Let $\mathbf{X}(t) = [X_1(t), \dots, X_N(t)]'$ where $t = 1, \dots, T$ is an EEG recording for one epoch at N channels. As in the previous section, the duration of an epoch was one second and the sampling rate was 1000 Hertz. Thus, $T = 1000$. The primary goal here is to study dependence between different channels. There are many ways to characterize dependence. We shall focus on cross-coherence, partial coherence, and partial directed coherence as the primary measures of dependence between a pair of channels.

Coherence is a frequency-specific (or frequency-band-specific) measure that has an intuitive interpretation. Consider decomposing the EEGs into different frequency oscillations (see Figure 20.10). Coherence between two EEGs at, say, the alpha band, is the squared cross-correlation between the alpha components of the two EEGs. Like correlation versus partial correlation, coherence does not distinguish between direct and indirect dependence between a pair of channels. For example, the dependence between the two EEGs $X_1(t)$ and

$X_2(t)$ could be driven by a third channel $X_3(t)$. To differentiate between direct and indirect dependence, we remove the linear effect of all the channels on each of $X_1(t)$ and $X_2(t)$ and then take the squared cross-correlation between the residuals. This is exactly the essence of the other measure, partial coherence, but this will be computed in a more mathematically elegant way (by inverting the spectral matrix) rather than that suggested by the intuitive explanation.

Both spectral measures of dependence, coherence, and partial coherence, can be estimated non-parametrically, using periodograms or parametrically using time domain models such as the vector autoregressive moving average (VARMA). For high-dimensional EEGs, we propose to estimate partial coherence by combining the parametric and non-parametric methods. We first estimate the spectral matrix by taking a weighted average of the periodogram-based estimator (non-parametric) and a well-conditioned matrix (e.g., diagonal matrix or a parametric estimator derived from fitting a VARMA or VAR model to the multichannel EEG). The weighted average is often well conditioned (i.e., the ratio of the largest to the smallest eigenvalues is of small order), thus making it possible to invert the matrix. The off-diagonal elements of the inverse estimated spectral matrix are then standardized to form partial coherence. One limitation of coherence and partial coherence is that they both measure undirected associations between a pair of channels. Here, we use partial directed coherence to measure the strength of information flow from one channel into another. This measure is derived from fitting a VAR model to the multichannel EEG.

As in the single-channel case, we shall assume the N-channel EEG signal $\mathbf{X}(t)$ in an epoch to be weakly stationary. That is, (1) the mean of each EEG channel, denoted $\mu_n(t) = \mathbb{E}X_n(t)$, $n = 1, \ldots, N$ is constant over t during the entire epoch; (2) the variance of the EEG at each channel, denoted $\mathbb{V}\text{ar}\, X_n(t)$, is also constant over the epoch; (3) the covariance between EEG $X_n(t)$ and $X_m(s)$ is expressed as follows: $\mathbb{C}\text{ov}[X_n(t), X_m(s)] = \gamma_{nm}(t - s)$ and $\mathbb{C}\text{ov}[X_m(t), X_n(s)] = \gamma_{mn}(t - s)$. Note here that $\gamma_{nm}(h)$ is not necessarily equal to $\gamma_{mn}(h)$. The conditions on weak stationarity can also be equivalently expressed in terms of the spectrum and cross-spectrum. These are stated as follows: (1) the auto-spectrum of each EEG, denoted $f_{nn}(\omega)$, depends only on frequency ω and does not change over the entire epoch t; and (2) the cross-spectrum between a pair of EEGs (denoted $f_{nm}(\omega)$ also does not change over t. The assumptions on weak stationarity mean that we can pool information across all time points within the epoch to model and estimate the oscillatory properties of the EEG.

20.3.1   Fourier-Domain Approach

20.3.1.1   The Fourier–Cramér Representation

Let $\mathbf{X}(t)$ be a zero mean and stationary time series. Then it admits the Cramér representation

$$\mathbf{X}(t) = \int_{-0.5}^{0.5} \mathbf{A}(\omega) \exp(i 2\pi \omega t) d\mathbf{Z}(\omega) \tag{20.22}$$

where $\mathbf{A}(\omega)$ is a $N \times N$ transfer function matrix and the increment $d\mathbf{Z}(\omega)$ is a random process over frequencies ω that satisfies $\mathbb{E}d\mathbf{Z}(\omega) = \underline{0}$ for all ω; $\mathbb{C}\text{ov}[d\mathbf{Z}(\omega), d\mathbf{Z}(\omega)] = \mathbf{I}_N d\omega$; and for $\omega \neq \lambda$, $\mathbb{C}\text{ov}[d\mathbf{Z}(\omega), d\mathbf{Z}(\lambda)] = \mathbf{0}$.

20.3.1.2   The Spectral Matrix of an EEG

- The spectral density matrix: First we define the lagged variance–covariance matrix $\Gamma_\ell = \mathbb{C}\text{ov}[\mathbf{X}(t), \mathbf{X}(t - \ell)]$, which consists of auto- and cross-covariances. The power spectral

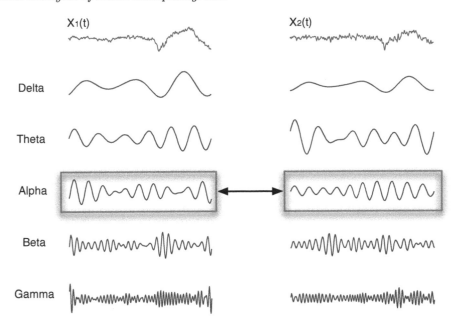

FIGURE 20.10
Decomposition of EEG signals into oscillations of different frequency bands.

density (PSD) matrix is defined as the Fourier transform of Γ_ℓ, which is

$$\mathbf{f}_{XX}(\omega) = \sum_{\ell=-\infty}^{\infty} \Gamma_\ell \exp\left(-i2\pi\omega\ell\right). \tag{20.23}$$

Moreover, based on the Cramér representation, the power spectral density (PSD) matrix $\mathbf{f}_{XX}(\omega)$ can also be represented using

$$\mathbf{f}_{XX}(\omega) = \mathbf{A}(\omega)^* \mathbf{A}(\omega).$$

- The auto-spectrum and cross-spectrum: The auto- spectrum of $X_i(t)$ at frequency ω is given by $\mathbf{f}_{ii}(\omega)$, and the cross-spectrum between $X_i(t)$ and $X_j(t)$ at frequency ω is given by $\mathbf{f}_{ij}(\omega)$.

- The coherence and partial coherence: The squared coherence between X_i and X_j is defined by $\rho_{ij}^2 = \frac{|\mathbf{f}_{ij}(\omega)|^2}{|\mathbf{f}_{ii}(\omega)||\mathbf{f}_{ii}(\omega)|}$.
The partial coherence between EEG at channels n and m at frequency ω is the squared cross-correlation between the ω-oscillations between these two channels after removing the ω-oscillations at the different channels. Define the inverse of the spectral matrix $\mathbf{f}_{XX}(\omega)$ to be $\Gamma(\omega) = \mathbf{f}^{-1}(\omega)$ and denote these elements as $\Gamma_{nm}(\omega)$. Partial coherence between the n and m channels is defined to be

$$\gamma_{nm}(\omega) = \frac{|\Gamma_{nm}(\omega)|^2}{\Gamma_{nn}(\omega)\Gamma_{mm}(\omega)}.$$

20.3.1.3 Non-Parametric Estimator of the Spectral Matrix

Suppose that we observe EEG $\mathbf{X}(t)$ over one epoch where $t = 1, \ldots, T$. Here, we will assume that the sample mean $\overline{\mathbf{X}} = \frac{1}{T}\sum_{t=1}^{T}\mathbf{X}(t) = \mathbf{0}$. If $\overline{\mathbf{X}}$ is not zero, then it must be removed before proceeding with computing the Fourier transform. Here, we illustrate the non-parametric estimation method for the spectral matrix via periodograms.

The first step is to compute the Fourier coefficients

$$\mathbf{d_X}(\omega_k) = \sum_{t=1}^{T} \mathbf{X}(t)\exp(-i2\pi\omega_k t) \tag{20.24}$$

at the fundamental (Fourier) frequencies $\omega_k = \frac{k}{T}$ where $k = -\left(\frac{T}{2}-1\right), \ldots, \frac{T}{2}$. Note that $\mathbf{d_X}(\omega_k)$ is a $N \times 1$ vector for each frequency ω_k and the n-th element is the Fourier coefficient of the component $X_n(t)$. The second step is to compute the $N \times N$ Fourier periodograms matrix

$$\mathbf{I}(\omega_k) = \frac{1}{T}\mathbf{d_X}(\omega_k)\mathbf{d_X^*}(\omega_k).$$

The diagonal elements $I_{nn}(\omega_k)$ are the auto-periodograms and the off-diagonal elements $I_{nm}(\omega_k)$ are the cross-periodograms between the channels $X_n(t)$ and $X_n(t)$. An estimate of the spectral matrix is obtained by smoothing the periodogram matrices across frequencies

$$\widehat{\mathbf{f}}(\omega_k) = \frac{1}{2m+1} \sum_{\ell=-m}^{m} \mathbf{I}(\omega_{k+\ell}). \tag{20.25}$$

The estimator for the spectrum of $X_i(t)$ is given by $\widehat{\mathbf{f}}_{ii}(\omega)$, and the estimator for the cross-spectrum between $X_i(t)$ and $X_j(t)$ at frequency ω is given by $\widehat{\mathbf{f}}_{ij}(\omega)$. The estimator of the squared coherence between X_i and X_j is $\widehat{\rho}_{ij}^2 = \frac{|\widehat{\mathbf{f}}_{ij}(\omega)|^2}{|\widehat{\mathbf{f}}_{ii}(\omega)||\widehat{\mathbf{f}}_{ii}(\omega)|}$

Illustration. The cross-spectrum of signals for 5 EEG channels (Figure 20.11) is estimated by periodogram using data from one epoch. The logarithms of the absolute value of the cross-spectrum estimate for each frequency band are shown in Figures 20.12–20.16.

20.3.2 Time-Domain Approach

- VARMA and its corresponding spectral matrix: Let $\mathbf{X}(t) \in \mathbb{R}^N$ be multivariate time series with dimension N. Suppose that $\mathbf{X}(t)$ follows a VARMA(P, Q) model $\Phi(B)\mathbf{X}(t) = \Theta(B)\mathbf{W}(t)$ where $W(t)$ has mean $\mathbf{0}$ and covariance matrix Σ_W, and

$$\Phi(B) = 1 - \sum_{p=1}^{P} \phi_p B^p \quad \Theta(B) = 1 + \sum_{q=1}^{Q} \theta_q B^q.$$

The spectral matrix of $\mathbf{X}$ can be determined by

$$\mathbf{f}(\omega) = \Phi(\omega)^{-1}\Theta(\omega)\Sigma_W\Theta(\omega)^*(\Phi(\omega)^*)^{-1}$$

where $\Phi(\omega) = \mathbf{I} - \sum_{p=1}^{P}\phi_p\exp(-i2\pi\omega p)$ and $\Theta(\omega) = \mathbf{I} + \sum_{q=1}^{Q}\theta_q\exp(-i2\pi\omega q)$

- VAR and VMA and their corresponding spectral matrices: As a special case of the VARMA model, the spectral matrix of a VAR(P) process $\mathbf{X}(t) = \sum_{p=1}^{P}\phi_p\mathbf{X}(t-p) + \mathbf{W}(t)$ (i.e., $\Phi(B)\mathbf{X}(t) = \mathbf{W}(t)$) can be computed as

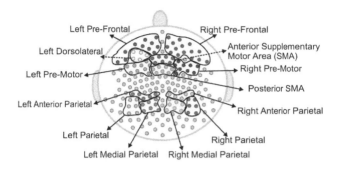

(a) Chanel Location

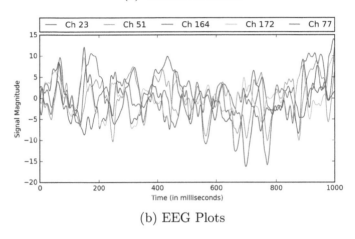

(b) EEG Plots

FIGURE 20.11

EEG plot for multiple channels at one epoch, channels include 23 (SMA), 51 (left M1), 164 (right M1), 172 (right antPr), and 77 (left antPr).

$$\mathbf{f}(\omega) = \Phi(\omega)^{-1}\Sigma_W(\Phi(\omega)^*)^{-1}.$$

Similarly, the spectral matrix of a MA(Q) process $\mathbf{X}(t) = \mathbf{W}(t) + \sum_{q=1}^{Q} \theta_q \mathbf{W}(t-q)$ (i.e., $\mathbf{X}(t) = \Theta(B)\mathbf{W}(t)$) can be computed as

$$\mathbf{f}(\omega) = \Theta(\omega)\Sigma_W\Theta(\omega)^*.$$

20.3.3 Estimating Partial Coherence

Partial coherence between EEG at channels n and m at frequency ω is the squared cross-correlation between the ω-oscillations between these two channels after removing the ω-oscillations at the different channels. Define the inverse of the $P \times P$ spectral matrix to be $\Gamma(\omega) = \mathbf{f}^{-1}(\omega)$ and denote these elements as $\Gamma_{nm}(\omega)$. Partial coherence between the n and m channels is defined as

$$\gamma_{nm}(\omega) = \frac{|\Gamma_{nm}(\omega)|^2}{\Gamma_{nn}(\omega)\Gamma_{mm}(\omega)}.$$

To estimate partial coherence, we must obtain the inverse of the estimate of the spectral

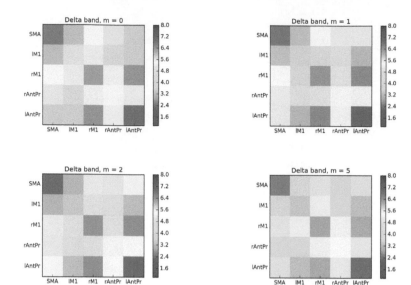

FIGURE 20.12
Periodogram (log scale) and smoothed periodogram (m = 1, 2, 5) of EEG channels 23 (SMA), 51 (left M1), 164 (right M1), 172 (right antPr), and 77 (left antPr), averaged over the delta band.

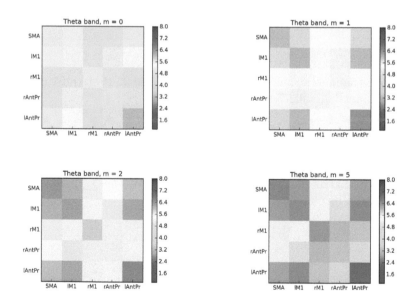

FIGURE 20.13
Periodogram (log scale) and smoothed periodogram (m = 1, 2, 5) of EEG channels 23 (SMA), 51 (left M1), 164 (right M1), 172 (right antPr), and 77 (left antPr), averaged over the theta band.

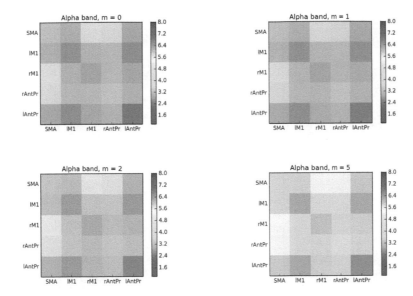

FIGURE 20.14
Periodogram (log scale) and smoothed periodogram (m = 1, 2, 5) of EEG channels 23 (SMA), 51 (left M1), 164 (right M1), 172 (right antPr), and 77 (left antPr), averaged over the alpha band.

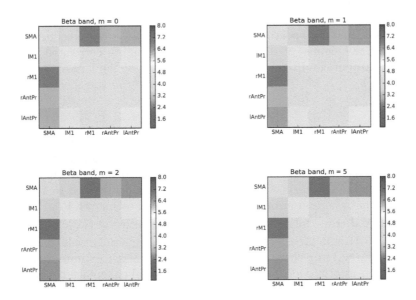

FIGURE 20.15
Periodogram (log scale) and smoothed periodogram (m = 1, 2, 5) of EEG channels 23 (SMA), 51 (left M1), 164 (right M1), 172 (right antPr), and 77 (left antPr), averaged over the beta band.

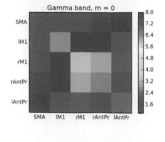
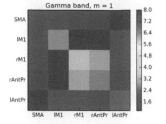
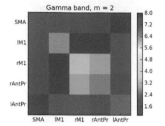
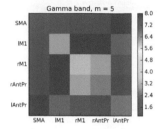

FIGURE 20.16
Periodogram (log scale) and smoothed periodogram (m = 1, 2, 5) of EEG channels 23 (SMA), 51 (left M1), 164 (right M1), 172 (right antPr), and 77 (left antPr), averaged over the gamma band.

matrix. This requires that the spectral matrix estimate have a good condition number. However, when the EEG signals at different channels are highly correlated, the spectral matrix estimate has a poor condition number (i.e., the ratio of the largest to the smallest eigenvalue is large). Fiecas et al. (6) (8) proposed an estimator of the spectral matrix that takes the form

$$\widehat{\mathbf{f}}(\omega) = \lambda(\omega)\widetilde{\mathbf{f}}(\omega) + [1 - \lambda(\omega)]\mathbf{V}(\omega).$$

where $\widetilde{\mathbf{f}}(\omega)$ is the non-parametric estimator, which is a mildly smoothed periodogram matrix and $\mathbf{V}(\omega)$ is a well-structured and well-conditioned matrix. Examples of $\mathbf{V}(\omega)$ include spectral matrix estimators derived from fitting a VAR model to the multichannel EEG, or simply a diagonal matrix. The term $\lambda(\omega) \in (0,1)$ is the weight attached to $\widetilde{\mathbf{f}}(\omega)$, which is proportional to the mean-squared error of the well-structured estimator $\mathbf{V}(\omega)$. When $\mathbf{V}(\omega)$ has a high mean-squared error, relative to the mean-squared error of $\widetilde{\mathbf{f}}(\omega)$, then the relative weight $W(\omega)$ given to $\widetilde{\mathbf{f}}(\omega)$ is large. The estimator $\widehat{\mathbf{f}}(\omega)$ is generally well-conditioned and has a lower mean-squared error than the classical non-parametric estimator, which is the smoothed periodogram matrix. Denote $\widehat{\Gamma}(\omega) = \widehat{\mathbf{f}}^{-1}(\omega)$. Partial coherence between channels n and m is given by

$$\widehat{\gamma}_{nm}(\omega) = \frac{|\widehat{\Gamma}_{nm}(\omega)|^2}{\widehat{\Gamma}_{nn}(\omega)\widehat{\Gamma}_{mm}(\omega)}.$$

20.3.4 Estimating the Spectral Matrix Using Multiple EEG Traces

In a designed experiment, the EEG is segmented into R non-overlapping epochs where an epoch is typically a 1-second recording. If the entire recording across R epochs is stationary, which roughly means that the brain process did not evolve over the entire recording, then it is sensible to combine information across the different epochs in a meaningful way.

One approach is to simply take the average of the periodograms computed from all epochs. More formally, let $\{\mathbf{X}^r(t)\}$ be the EEG recorded during epoch r and denote the corresponding periodogram to be $\mathbf{I}_{XX}^r(\omega)$. The spectral estimate is

$$\widehat{\mathbf{f}}_{XX}(\omega) = \frac{1}{R}\sum_{r=1}^{R}\mathbf{I}_{XX}^r(\omega).$$

Remark 1. Note that as the number of epochs R increases, the variance of the estimator $\widehat{\mathbf{f}}_{XX}(\omega)$ decreases. This suggests that it would not be necessary to smooth the periodograms at each epoch r. In fact, when R is large, one should avoid smoothing because smoothing degrades frequency resolution.

Remark 2. If the number of epochs is small, say $R < 30$, it might be necessary to do mild smoothing of the periodograms in each epoch. We would recommend trying out different smoothing bandwidths that give sufficiently smooth overall estimates without degrading the frequency too much.

The principle of averaging epoch-specific estimates can be extended to other types of estimators. For example, one could compute the average of Welch periodograms or multitaper periodograms. Moreover, one could obtain spectral estimates for each epoch using time-domain methods and then average these across all epochs. More formally, let $\widehat{\mathbf{f}}_{XX}^r(\omega)$ be the estimator for epoch r. The estimator could be any of the ones already discussed (e.g., raw periodograms, mildly smoothed periodograms, Welch estimator, ARMA-based estimator). The final estimator would then be

$$\widehat{\mathbf{f}}_{XX}(\omega) = \frac{1}{R}\sum_{r=1}^{R}\widehat{\mathbf{f}}_{XX}^r(\omega).$$

20.3.4.1 Estimating the Spectral Matrix in Specific Frequency Bands

We are often interested in spectral power at a frequency band as opposed to a singleton frequency. For example, the spectrum at the alpha band and beta band which cover 8–12 Hertz and 12–32 Hertz are, respectively,

$$\mathbf{f}_{XX}(\text{alpha}) = \int_{\frac{8}{\Omega_s}}^{\frac{12}{\Omega_s}} \mathbf{f}_{XX}(\omega)d\omega \quad \text{and} \quad \mathbf{f}_{XX}(\text{beta}) = \int_{\frac{12}{\Omega_s}}^{\frac{32}{\Omega_s}} \mathbf{f}_{XX}(\omega)d\omega.$$

20.3.5 Modeling and Inference on Connectivity

In multivariate time series analysis in neuroimaging signals such as EEG and fMRI, the inter-channel connectivity is of particular interest. The term connectivity may refer to structural, functional, and effective connectivity according to (10):

- Structural connectivity refers to the anatomical structure of the brain, which can be studied via analysis of fMRI.

- Functional connectivity refers to undirected temporal correlations between neurophysiological events. Functional connectivity can be studied based on coherence or correlation.

- Effective connectivity refers to the directed effects of one neural activity over another. Approaches to characterizing effective connectivity include Dynamic Causal Modeling (DCM)(11), Granger Causality Modeling (GCM) (16), and Transfer Entropy (29).

20.3.5.1   Granger Causality

The concept of Granger causality (GC) was first proposed in (14). The basic idea of GC states that for random variables $X(t)$ and $Y(t)$ with $t = \{1, \ldots, T\}$, we say X "Granger-causes" (GC) Y if Y can be better predicted using the past value of both X and Y than it can using the past observation of Y alone. According to (15) the causality relationship is based on two principles: (1) the cause must happen prior to its effect and (2) the cause must contain unique information for predicting future values of the effect.

The limitations of Granger causality include the following (1) It does not necessarily mean physiological causality, (2) GC can only account for linear relationship, and (3) GC relies on the stationarity of the analyzed the signals. Nevertheless, we shall explore its utility in this study given that it is widely used in neuroscience.

To illustrate the idea of Granger causality, consider the following two auto-regressive models

$$X(t) = \sum_{p=1}^{P} \phi_p X(t-p) + \epsilon(t) \tag{20.26}$$

$$X(t) = \sum_{p=1}^{P} \phi_p' X(t-p) + \sum_{p=1}^{P} \psi_p' Y(t-p) + \epsilon'(t). \tag{20.27}$$

If model 20.27 is significantly better than 20.26, that is

$$\mathbf{E}\left[\sum_{t=P+1}^{T} (X(t) - \sum_{p=P}^{p} \phi_p X(t-p))^2\right] \; > \; \mathbf{E}\left[\sum_{t=P+1}^{T} (X(t) - \sum_{p=1}^{P} \phi_p' X(t-p) \right.$$
$$\left. - \sum_{p=1}^{P} \psi_p' Y(t-p))^2\right],$$

then we can conclude that Y Granger causes (GC) X.

20.3.5.2   Partial Directed Coherence (PDC)

Partial directed coherence (PDC) has been introduced in making inference on frequency specific connectivity between signals (1). It is a directed measurement of connectivity and is based on the vector autoregressive model of the signals. Partial directed coherence can be treated as a frequency description of Granger causality.

Let $\mathbf{X}(t) = [X_1(t), \ldots, X_M(t)]$ be a stationary N-dimensional time series with mean zero. A vector autoregressive (VAR) model with order P for $\mathbf{X}(t)$ is given by

$$\mathbf{X}(t) = \sum_{p=1}^{P} \phi_k \mathbf{X}(t-k) + \mathbf{W}(t), \tag{20.28}$$

where $\phi_p \in \mathbb{R}^{N \times N}$ is the coefficient matrix of the VAR model at lag p and $\mathbf{W}(t) \in \mathbb{R}^N$ is a Gaussian white noise process with zero mean and covariance matrix Σ_W. The causality of $\mathbf{X}(t)$ requires the coefficient matrices satisfy $\det(I - \sum_{p=1}^{P} \phi_p z^p) \neq 0$ for all $z \in \mathbb{C}$ and $|z| \leq 1$.

Let $\Phi(\omega) = I - \sum_{p=1}^{P} \phi_p \exp(-i2\pi\omega p)$ be the Fourier transform of ϕ_p at frequency ω. Then

TABLE 20.1

Summary of metrics for connectivity

Approach	Model Based	Data Driven	Functional Connectivity	Effective Connectivity
Dynamic Causal Modeling (DCM)	✓			✓
Fourier Coherence		✓	✓	
Wavelet Coherence		✓	✓	
Granger Causality		✓	✓	✓
Partial Directed Coherence (PDC)		✓	✓	✓
Directed Transfer Function (DTF)		✓	✓	✓

the partial directed coherence (PDC) for $\mathbf{X}$ can be represented as

$$\pi_{ij}^2(\omega) = \frac{|\Phi_{ij}(\omega)|^2}{\sum\limits_{k=1}^{N} \Phi_{kj}(\omega)\Phi_{kj}^*(\omega)}. \tag{20.29}$$

$\pi_{ij}^2(\omega)$ provides a measure of the linear influence of X_j on X_i at frequency ω. The partial directed coherence $\pi_{ij}^2(\omega)$ takes values from interval $[0, 1]$ and satisfies $\sum_i \pi_{ij}^2(\omega) = 1$, which means each time series has unit outflow.

20.3.5.3 Summary of Metrics for Connectivity

Table 20.1 summarizes the popular methods for characterizing the connectivity between channels in multivariate time series, and their ability of measuring functional and effective connectivity.

20.4 Spectral Analysis for High-Dimensional Data

The vector autoregressive (VAR) model is often used in modeling temporal dependency between multichannel time series signals. The parameters of a VAR model can be used for subsequent analysis for connectivity in both the time domain (e.g. Granger-causality, correlation, partial correlation) and frequency domain (e.g. coherence, partial coherence, partial directed coherence and directed transfer function). For a multivariate time series $X(t)$ of dimension N that follows a VAR model, the dynamics of the time series can be fully characterized by the AR coefficients and covariance matrix for the noise. However, the dimension of the parameter is a quadratic function of the dimension N, which restricts the capability of the VAR model when the dimension of $X(t)$ is high (i.e., N is large), due to lack of power in model estimation. In this section, we will introduce popular methods that handle VAR model fitting for time series of high dimension. The methods include classic Least Square Estimation (LSE), LASSO[1] regression, and a two-step approach called LASSLS[2] (19).

---

[1] Least absolute shrinkage and selection operator.
[2] LASSLS is a two-step approach combining LASSO and LSE.

20.4.1   Methods for Fitting VAR Model on Multivariate Time Series

Let $\mathbf{X}(t) \in \mathbb{R}^N$ be a multivariate random process that follows the following vector autoregressive (VAR) model

$$\mathbf{X}(t) = \sum_{p=1}^{P} \Phi_p \mathbf{X}(t-p) + \varepsilon(t) \tag{20.30}$$

where $\varepsilon(t)$ is a multivariate Gaussian random variable with mean 0 and variance matrix Σ_E. In this section, we will discuss methods that are used to estimate the unknown parameters $\Phi_1, \ldots, \Phi_P$ using observations $\mathbf{X}(1), \ldots, \mathbf{X}(T)$.

20.4.1.1   Least Squares Estimation

A simple and quick algorithm is worth mentioning since it reformulates the problem to a linear regression equation. Specifically, the VAR(P) model described in Equation 20.30 can be rearranged to a multivariate regression model (Equation 20.31):

$$\underbrace{\begin{bmatrix} \mathbf{X}(T)' \\ \vdots \\ \mathbf{X}(P+1)' \end{bmatrix}}_{Y} = \underbrace{\begin{bmatrix} \mathbf{X}(T-1)' & \cdots & \mathbf{X}(T-P)' \\ \vdots & \ddots & \vdots \\ \mathbf{X}(P)' & \cdots & \mathbf{X}(1)' \end{bmatrix}}_{X} \underbrace{\begin{bmatrix} \Phi_1' \\ \vdots \\ \Phi_P' \end{bmatrix}}_{B} + \underbrace{\begin{bmatrix} \varepsilon(T)' \\ \vdots \\ \varepsilon(P+1)' \end{bmatrix}}_{E} \tag{20.31}$$

Denote $Y = [y_1, y_2, ..., y_N]$, $B = [b_1, b_2, ..., b_N]$, $E = [e_1, e_2, ..., e_N]$, where y_k, b_k, e_k are the kth column vector of Y, B, and E, respectively, for $(k = 1, 2, \ldots, N)$; then we have

$$\underbrace{y_k}_{m \times 1} = \mathbb{X} \underbrace{b_k}_{q \times 1} + \underbrace{e_k}_{m \times 1}, \quad e_k \overset{indep}{\sim} N_m(\vec{0}, \sigma_{kk} I_m) \tag{20.32}$$

where $m = T - P$, $q = N \times P$, and σ_{kk} is the kth diagonal element of Σ_E. Note that (20.31) is decomposed to sub-linear regression problems of estimating $\{b_k\}_{k=1}^N$ in a parallel manner and all the entries of connectivity matrices are included in $\{b_k\}_{k=1}^N$.

The most common approach to fit a linear regression model is via least squares (LS), which provides an unbiased estimate for the coefficients B. The least squares (LS) method fits the model via minimizing the sum of squared errors (Equation 20.33):

$$\widehat{b}_k = \underset{b_k \in \Re^q}{\mathrm{argmin}} \|y_k - \mathbb{X} b_k\|^2, \tag{20.33}$$

which is able to give an unbiased estimate $(\mathbb{X}'\mathbb{X})^{-1}\mathbb{X}'y_k$ for b_k.

20.4.1.2   LASSO

In many regression problems, the estimators can be improved in terms of mean squared error by incorporating the effects of shrinkage, especially when the original problem is ill-conditioned. The shrinkage of the parameter estimation can be achieved by adding a penalty term to the original objective function being minimized. The penalty terms include the L_1 norm of the parameters (LASSO), Tikhonov regularization of the parameters (Ridge regression) or the combination of the two (Elastic net regression). In this section, we will introduce LASSO regression, which can achieve both parameter shrinkage and variable selection. The LASSO method estimates the parameter via minimizing the sum of SSE and L_1 norm of the parameters, as described in Equation 20.34

$$\tilde{b}_k = \underset{b_k \in \mathfrak{R}^q}{\text{argmin}} \|y_k - \mathbb{X}b_k\|^2 + \lambda \|b_k\|_1. \tag{20.34}$$

The tuning parameter λ determines the strength of the penalty, which balances (1) fitting a linear model with least squares and (2) shrinking the coefficients b_k. The L_1 penalty causes some coefficients to be zeros exactly, thus it leads to a sparse estimate $\tilde{b}_k$. In general, the tuning parameter λ controls the sparsity of the solution by controlling the strength of the shrinkage. During the process to implement algorithms for LASSO, advantage was taken of the recent demonstration by (9) that estimation of generalized linear models with convex penalties can be handled by cyclical coordinate descent and computed along a regularization path.

20.4.1.3   LASSLS

A two-step procedure for estimating connectivity matrices named LASSLS was developed in (19). LASSLS has inherited low bias for non-zero estimates and high specificity for zero-estimates from LSE and the LASSO separately. The approach consists of the following two steps:

- Step 1: Apply LASSO to identify entries in ϕ_k whose estimates should be set to 0.

- Step 2: Fit LSE with the constriction that zero entries estimates from Step 1 are set to 0.

Algorithm 1 LASSLS algorithm

---

procedure TWO-STEP ESTIMATION

   Step 1:

      Generate a sequence of (P, λ) and randomly divide data to K folds

      For a possible choice of (P, λ), leave one fold as test data at each time

      Train 20.32 with LASSO method on other folds and compute $\{\tilde{\Phi}_k\}_{k=1}^P$ for $\{\Phi_k\}_{k=1}^P$

      Based on $\{\tilde{\Phi}_k\}_{k=1}^P$, calculate prediction error on test set and finally take average

      Select (P, λ) with the lowest average prediction error

      Obtain estimate $\{\tilde{b}_k\}_{k=1}^N$ for $\{b_k\}_{k=1}^N$ in 20.32 of lag P using LASSO method with λ

   Step 2:

     if $\tilde{b}_k^j = 0$ **then**

        Set $b_k^j = 0$.

     end if

     if $\tilde{b}_k^j \neq 0$ **then**

        Keep b_k^j.

     end if

     Obtain estimate $\{\hat{b}_k\}_{k=1}^N$ for $\{b_k\}_{k=1}^N$ in 20.32 with LSE under above constriction

     Obtain estimate $\{\hat{\Phi}_k\}_{k=1}^P$ for $\{\Phi_k\}_{k=1}^P$ by arranging $\{\hat{b}_k\}_{k=1}^N$

end procedure

The LASSLS algorithm can be implemented as follows: To obtain the optimal VAR order P and penalty parameter λ, a K-fold cross-validation test in Step 1 is employed. A sequence of candidates of $\{P_j, \lambda_k\}$ will be pre-specified and the optimal values are selected such that the average of prediction error on test data is minimized.

Under a high sparsity condition, $\{\tilde{b}_k\}_{k=1}^{N}$ have sign consistency assured by the LASSO estimator (32), which means that for a sufficiently large sample size

$$\Pr(\text{sgn}(\tilde{b}_k) = \text{sgn}(b_k)) \to 1, \tag{20.35}$$

where $\text{sgn}(b_k)$ is the sign function with value of 1, 0, or -1 corresponding to $b_k > 0$, $b_k = 0$, or $b_k < 0$ respectively. Therefore, after Step 1 the true non-connectivity relationship can be identified. Then in Step 2, the non-zero entries will be re-estimated. Since constraints have been applied for LSE in Step 2, the computing is simplified compared with the original LSE approach. Simulations have been performed and the performance of LSE, LASSO and LASSLS have been evaluated in (19). The simulation results indicate that in the VAR process with sparse coefficients, LSE lacks the ability to give specificity for true zero coefficients, since its parameter estimates do not contain exact zeros, while the estimate from LSE has general lower bias across all the entries. LASSO is able to identify most true zero entries of the parameters, while the methods lead to an estimate with higher bias compared to that of LSE and LASSLS. The LASSLS method has inherited the specificity of true zero values from LASSO, and consequently can capture true zero values as well as LASSO. The comparison of the mean squared error (MSE) indicates that LASSLS outperforms both LSE and LASSO when the true VAR process has sparse coefficients.

20.4.2 EEG Data Analysis via LASSLS Methods

In this section, we will illustrate analysis of brain connectivity using EEG data via the LASSLS method.

20.4.2.1 VAR Modeling on High-Dimensional Multichannel EEG

The EEG channels can be grouped as in Table 20.2, according to the brain regions they belong to. The connectivity at the region level can be inferred via the blocks of connectivity matrices (VAR coefficients matrices). We will illustrate the application of the VAR model on the analysis of connectivity between EEG channels at the resting state using EEG recordings with 10 trials. Totally, there are 5 connectivity matrices (VAR coefficient matrices) of 158

TABLE 20.2

EEG channel grouping.

Brain Region	Pre-frontal	Dorsolateral Pre-frontal		Pre-motor
Channel #	1,2,...,30	31,...,40		41,...,58
Brain Region	Anterior SMA	Supplementary motor		Posterior SMA
Channel #	69,...,75	59,...,68		76,...,82
Brain Region	Primary motor	Parietal	Lateral Parietal	Anterior parietal
Channel #	83,...,96	97,...,132	133,...,148	149,...,158

dimensions to estimate, each of which indicates the connectivity with time lag from 1 to 5. Figure 20.17 visualizes the estimation for the VAR coefficients using matrix imaging, where small squares represent the entries of the connectivity matrix (coefficient matrix), and blanks are assigned to entries whose value is zero, therefore non-connectivity is easy to identify. Red color is assigned to a positive entry and blue to a negative entry, and the strength of connectivity is implied by the color key. It is observable that most entries of $\hat{\Phi}_2$ and $\hat{\Phi}_3$ are blank and limited non-zero estimates are close to zero, which implies that the functional connectivity at lags 2 and 3 is almost nonexistent.

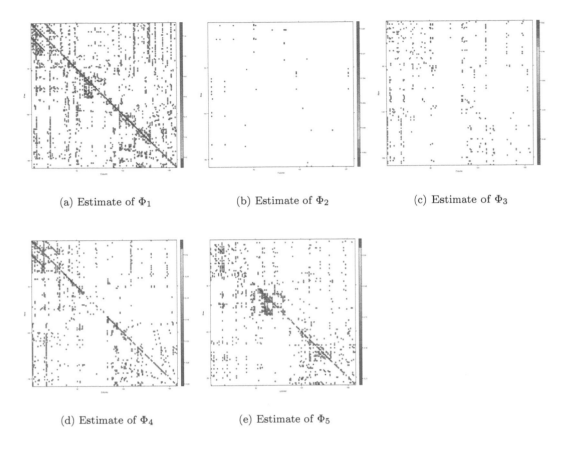

(a) Estimate of Φ_1 (b) Estimate of Φ_2 (c) Estimate of Φ_3

(d) Estimate of Φ_4 (e) Estimate of Φ_5

FIGURE 20.17
Estimate of VAR coefficients.

20.4.2.2   Inference on Directed Connectivity

With the estimation of the VAR model of the EEG signals, the directed connectivity measured by partial directed coherence (PDC) can be estimated subsequently, for both the inter-channel level and inter-region level. PDC estimates at different frequency bands are visualized in Figure 20.18 (theta band), Figure 20.19 (beta band), and Figure 20.20 (gamma band).

20.5   Source Localization and Estimation

20.5.1   Overview of Source Models for EEG Data

It is also of interest to study the properties of the unobserved neuronal activities from EEG data. EEG signals are not direct measurements of neuronal activity. Therefore, the first step to studying brain activity is to reconstruct sources that generate the observed EEG signals. Two classes of source models have been developed. One class depends on the physical head-forward model, which assumes the EEG sources are dipole currents that are located within

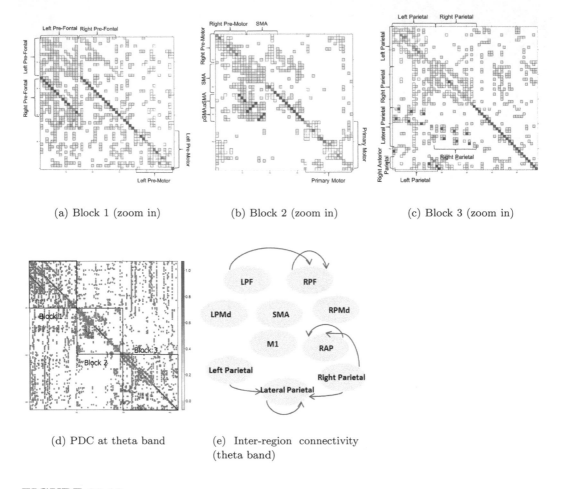

(a) Block 1 (zoom in)   (b) Block 2 (zoom in)   (c) Block 3 (zoom in)

(d) PDC at theta band   (e) Inter-region connectivity
                               (theta band)

FIGURE 20.18
Connectivity at the theta band.

the brain region. The other class of model assumes that the sources represent independent brain networks, which might be either localized or distributed.

20.5.1.1   Dipole Source Model

According to (33), the potential at location r in the brain or on the scalp surface can be expressed as the integral of the contributions of sources

$$X(r,t) = \int_B G_H(r,r')P(r',t)dV(r') \tag{20.36}$$

where $P(r',t)$ represents the dipole moment per unit volume at location r', where r' ranges over the whole brain region B; $G_H(r,r')$ is Green's function, which depends on the properties of the volume conductor and the locations of source r' and measurement location r.

In order to place the inverse problem in the EEG in a fundamental context, according to (24), the entire brain volume can be parceled into N voxels, each of volume ΔV, each having a strength $p_n(r_n,t) = P(r_n,t)\Delta V$. The potential given by Equation 20.36 can be

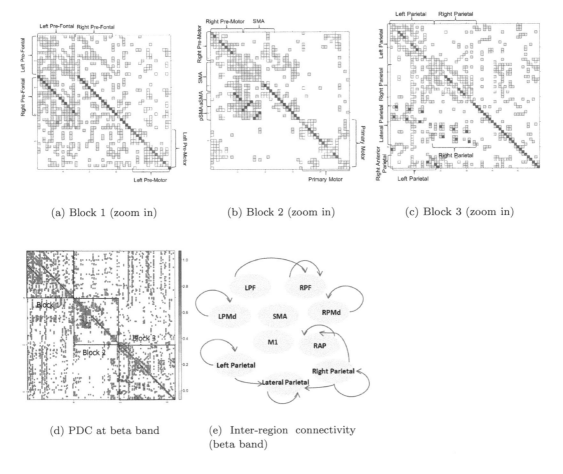

(a) Block 1 (zoom in) (b) Block 2 (zoom in) (c) Block 3 (zoom in)

(d) PDC at beta band (e) Inter-region connectivity (beta band)

FIGURE 20.19
Connectivity at the beta band.

written as a finite sum over contributions from N voxels:

$$X(r_k, t) = \sum_{n=1}^{N} G_n(r_k, r_n) p_n(r_n, t) \ . \tag{20.37}$$

According to (2), the algebraic formulation of EEG measurement at location r can be written as

$$\mathbf{X}(r) = \sum_i a(r, r_{q_i}, \Theta_{q_i}) q_i, \tag{20.38}$$

where $a(r, r_q, \Theta_q)$ is formed as the solution to the electric-forward problem for a dipole with unit amplitude and orientation Θ. For the simultaneous activation of N dipoles located at r_{q_n}, the EEG recording for M locations can be written as

$$\mathbf{X}(t) = \mathbb{A}\mathbf{Z}(t) \tag{20.39}$$

where $\mathbf{X}(t) = [X(r_1, t), ..., X(r_M, t)]'$ is the EEG measurement for M locations at time t, $A_{mn} = a(r_m, r_{q_n} \Theta_{q_n})$ is the gain matrix relating the dipoles to the sensors, and $\mathbf{Z}(t) = [q_1(t), ..., q_N(t)]'$ is the source amplitudes. Each column of $\mathbb{A}$ is the forward field

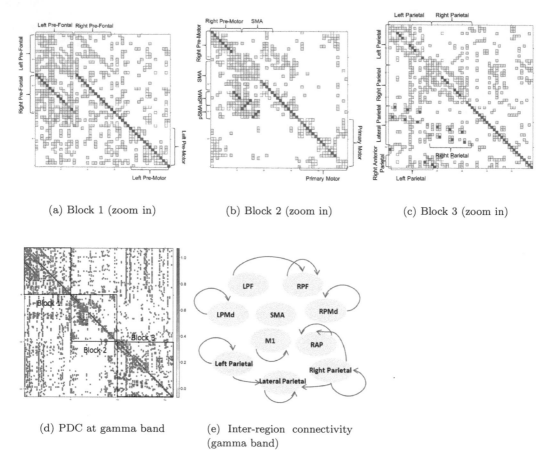

(a) Block 1 (zoom in)   (b) Block 2 (zoom in)   (c) Block 3 (zoom in)

(d) PDC at gamma band   (e) Inter-region connectivity
                               (gamma band)

FIGURE 20.20
Connectivity at the gamma band.

of the current dipole, sampled as M discrete locations of the EEG sensors.

20.5.1.2   Independent Source Model

Section 20.5.1.1 models the EEG signals as the potential generated by dipole currents within brain regions due to neuronal activities, where the sources might be correlated. An alternative way of modeling EEG signals is to treat it as the output of a number of statistically independent potential-generating systems. The systems are spatially fixed but can be either spatially restricted or widely distributed, according to (20). The independent systems can be interpreted as independent brain activity networks. Figure 20.21 shows a graphical illustration of the mixing structure for the signals of two independent sources (Z_1 and Z_2) and four electrodes ($X_1, \ldots, X_4$). The element A_{mn} in mixing matrix $\mathbb{A}$ can be interpreted as the loading of the n-th source signal Z_n on the EEG signal at the m-th electrode (X_m). If $A_{mn} = 0$, for some m and n, we conclude that $Z_n(t)$ has no contribution to the EEG signal $X_m(t)$, since the loading for source $Z_n(t)$ is zero.

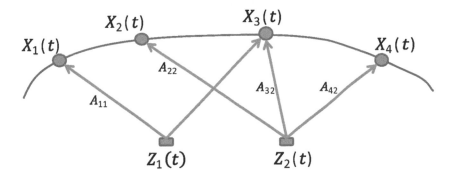

FIGURE 20.21

Graphical representation of the latent source model. The directions of the arrows represent a dependence relationship.

20.5.1.3 A Generalized Model of EEG Signals

According to Section 20.5.1.1 and 20.5.1.2, a general model of EEG data can be written as

$$\mathbf{X}(t) = \mathbb{A}\mathbf{Z}(t) + \mathbf{N}(t) \tag{20.40}$$

where $\mathbf{X}(t) \in \mathbb{R}^M$ is the recorded EEG signal for M electrodes at time t; $\mathbf{Z}(t) \in \mathbb{R}^N$ represents the activity of N source signals at time t due to brain activities and is not directly observed; and $\mathbb{A} \in \mathbb{R}^{M \times N}$ is a matrix representing instantaneous source mixing. The noise term $\mathbf{N}(t) \in \mathbb{R}^M$ represents either measurement error or machine error at time t.

Let $\mathbf{X} = [\mathbf{X}(1), \ldots, \mathbf{X}(T)]$, $\mathbf{Z} = [\mathbf{Z}(1), \ldots, \mathbf{Z}(T)]$, and $\mathbf{N} = [\mathbf{N}(1), \ldots, \mathbf{N}(T)]$, then we have $\mathbf{X} \in \mathbb{R}^{M \times T}$, $\mathbf{Z} \in \mathbb{R}^{N \times T}$ and $\mathbf{N} \in \mathbb{R}^{N \times T}$. The model in Equation 20.40 can be written in the following equivalent form:

$$\mathbf{X} = \mathbb{A}\mathbf{Z} + \mathbf{N}. \tag{20.41}$$

20.5.2 Inverse Source Reconstruction

According to (2), the lead field matrix $\mathbb{A}$ and the source activity $\mathbf{Z}(t)$ can be estimated based on the physical modeling of the head. The lead field matrix $\mathbb{A}$ can be estimated from a spherical head model or a realistic head model. The solution for spherical head model can be computed analytically (28), however the solution for a realistic head model can only be computed numerically. The most popular approaches to estimating the lead field matrix $\mathbb{A}$ are BEM (boundary element method (18)) and FEM (finite element method (42)). The only thing unknown in inverse source reconstruction is the dipole activity q_n, $n = 1, \ldots N$. In general, there are two approaches to estimation of EEG sources: parametric and imaging methods.

20.5.2.1 Parametric Methods

Parametric methods typically assume the sources can be represented by a few equivalent current dipoles of unknown location. The dipole moments are estimated using non-linear numerical methods including simplex search, genetic algorithm, and simulated annealing

(36). In the presence of noise, the forward model at time can be written as

$$\mathbf{X}(t) = \mathbb{A}\mathbf{Z}(t) + \mathbf{N}(t),\tag{20.42}$$

where $\mathbf{X}(t) = [X_1(t), \ldots, X_M(t)]'$ is the EEG measurement at M locations $\{r_1, \ldots, r_M\}$; $A_{mn} = a(r_m, r_{q_n}\Theta_m)$ is the gain matrix relating the dipoles to the sensors, $\mathbf{Z}(t) = [q_1(t), \ldots, q_N(t)]'$ is the source amplitudes, and $\mathbf{N}(t) \in \mathbb{R}^M$ is the measurement and machine noise.

Let $\mathbf{X} = [\mathbf{X}(1), \ldots, \mathbf{X}(T)]$, $\mathbf{Z} = [\mathbf{Z}(1), \ldots, \mathbf{Z}(T)]$, and $\mathbf{N} = [\mathbf{N}(1), \ldots, \mathbf{N}(T)]$, the model in Equation 20.42 can be written in the following equivalent form:

$$\mathbf{X} = \mathbb{A}\mathbf{Z} + \mathbf{N}.\tag{20.43}$$

The goal of dipole fits is to determine the best set $\{r_i, \Theta_i\}$ of number N sources and model parameters, and sources $\mathbf{Z}$ that best describe the data. Least square estimation (LSE) is one of these approaches. The LSE approach minimizes the cost function in the form

$$J_{LS}(r_{q_i}, \Theta_i, \mathbf{Z}) = \|\mathbf{X} - A(\{r_{q_i}, \Theta_i\})\mathbf{Z}\|_F^2 \tag{20.44}$$

where the subscript F indicates the Frobenius norm, that is

$$\|\mathbf{X} - A(\{r_{q_i}, \Theta_i\})\mathbf{Z}\|_F^2 = \mathrm{Tr}[(\mathbf{X} - A(\{r_{q_i}, \Theta_i\})\mathbf{Z})(\mathbf{X} - A(\{r_{q_i}, \Theta_i\})\mathbf{Z})^T].$$

The cost function 20.44 can be minimized by exhaustive scanning through the whole solution space. However, it is too computationally expensive, even for small values of N. Some non-linear optimizations based on directed search algorithms have been developed, which are more efficient than the exhaustive searching algorithm. The non-linear optimization algorithms also suffer from the problem of easily being trapped in local optima.

One disadvantage of the dipolar fits model is that the correct number of dipole sources N is unknown a priori and it has to be estimated. Approaches to estimating the number of sources include stepwise selection, multiple signal classification (MUSIC), principal component analysis (PCA), independent component analysis (ICA), and second-order blind identification (SOBI).

20.5.2.2   Imaging Methods

Imaging methods assume that the primary sources are intracellular currents in dendritic trunks of the pyramidal neuron in cerebral cortex (34, 2). It performs inference on dipole moments at a fixed set of locations within the brain and, therefore, the inverse problem in this case is linear.

Distributed source models reconstruct the electric activity at each point of the grid of the solution space (3D for brain activity, 2D for cortical activity). It makes an assumption on the number of dipoles in the brain by pre-defining a cortex grid and treating each grid point as an equivalent source. It overcomes the problem with the dipolar model described in Section 20.5.2.1 that the exact number of dipoles could not be determined a priori. However, in the distributed source models, the number of source points N is much greater than the number of EEG electrodes M, which makes the source inversion problem ill-conditioned. Typically the number of sensors on the scalp can be 32–256, while the number of equivalent dipole sources on the cortex depends on the spacing of the meshgrid on the cortex area. For example, a meshgrid on the cortex area with 3-mm spacing, the number of grid points (voxels) can be greater than 10^4. In order to solve the ill-conditioned inversion problem, regularization must be applied; for example, constraints that are based on biophysical knowledge or other imaging techniques.

The basic assumptions for distributed source models are as follows:

(1) Each point can be treated as a dipole with fixed location; the only thing that varies is magnitude and orientation.

(2) The observed EEG signals are linearly related to the source magnitude through a mixing matrix A.

In general, the approach for source imaging minimizes a cost function that combines both the estimation error and a penalty term

$$\widehat{\mathbf{Z}} = \arg\min_{\mathbf{Z}} \|\mathbf{X} - \mathbb{A}\mathbf{Z}\|_Q^2 + f(\mathbf{Z}), \tag{20.45}$$

where the Q-norm in Equation (20.45) is defined as follows

$$\| \mathbf{X} - \mathbb{A}\mathbf{Z} \|_Q^2 = \mathrm{Tr}[(\mathbf{X} - \mathbb{A}\mathbf{Z})^T Q^{-1} (\mathbf{X} - \mathbb{A}\mathbf{Z})],$$

with $Q = \mathbb{E}[\mathbf{N}(t)\mathbf{N}(t)^T]$ representing the variance–covariance matrix of the noise term $\mathbf{N}(t)$. The penalty term $f(\mathbf{Z})$ in Equation (20.45) represents the a priori assumptions.

Note that if $f(\mathbf{Z})$ can be written as $f(\mathbf{Z}) = \sum_{t=1}^{T} f_t(\mathbf{Z}(t))$, then Equation (20.45) can be reformulated as

$$\widehat{\mathbf{Z}} = \arg\min_{\mathbf{Z}} \sum_{t=1}^{T} \Big[(\mathbf{X}(t) - \mathbb{A}\mathbf{Z}(t))^T Q^{-1} (\mathbf{X}(t) - \mathbb{A}\mathbf{Z}(t)) + f_t(\mathbf{Z}(t)) \Big]. \tag{20.46}$$

Therefore, for any given time t, the estimator of $\mathbf{Z}(t)$ can be constructed using

$$\widehat{\mathbf{Z}}(t) = \arg\min_{\mathbf{Z}(t)} \Big[(\mathbf{X}(t) - \mathbb{A}\mathbf{Z}(t))^T Q^{-1} (\mathbf{X}(t) - \mathbb{A}\mathbf{Z}(t)) + f_t(\mathbf{Z}(t)) \Big]. \tag{20.47}$$

A few approaches have been developed for cortical source imaging based on different assumptions on the regularization term $f(\mathbf{Z})$. The minimum norm estimate (MNE) method sets the penalty term to be $f(\mathbf{Z}) = \lambda \|\mathbf{Z}\|_R^2$, which means there's no a priori information used, except that it assumes the current distribution has small overall intensity (L_2 penalty). The minimum norm estimator (MNE) favors weak and localized activation patterns. That is, superficial solution points in a 3D brain is favored since there is less activation needed to have similar impact on observations, compared to some "deeper" sources. In order to overcome the disadvantage of MN that it favors superficial sources, weighted minimum norm (WMN) (13) approaches have been developed where different weighting strategies have been applied. The WMN estimator uses the regularization function $f(\mathbf{Z}) = \lambda \|W\mathbf{Z}\|_F^2$, where W is determined in multiple ways, for example, $W \in \mathbb{R}^{N \times N}$ can be chosen as a diagonal matrix, where the diagonal elements are proportional to the norm of the columns of the mixing matrix $\mathbb{A}$. That is, $W_{nn} = \sqrt{\sum_{m=1}^{M} (\mathbb{A}_{mn})^2}$. Other popular source imaging methods that use L_2 regularization include Laplacian weighted minimum norm (LORETA)(27) and standardized low-resolution brain electromagnetic tomography (sLORETA)(26), both of which achieve spatially smoothed source reconstruction.

The LORETA method is one of the most used source imaging methods. It sets the penalty term in Equation (20.45) to be $f(\mathbf{Z}) = \|DW\mathbf{Z}\|_F^2$, where $D \in \mathbb{R}^N$ is used to implement the discrete spatial Laplacian operator and W is a diagonal matrix for the column normalization of $\mathbb{A}$. The Laplacian matrix D can be written as

$$D_{ij} = \frac{1}{h^2} \begin{cases} -6 & \text{if} \|\mathbf{r}_i - \mathbf{r}_j\| = 0 \\ 1 & \text{if} \|\mathbf{r}_i - \mathbf{r}_j\| = h \\ 0 & \text{otherwise} \end{cases} \tag{20.48}$$

TABLE 20.3
Summary of methods for source reconstruction.

Methods	Parameter and Signals	Dimensions	Objective function to be minimized	Remarks
Dipole fits	$r_n, \phi_n, \mathbf{Z}_n$	$N < M$	$\|\mathbf{X} - \sum_{n=1}^{N} a(r_n, \phi_n)\mathbf{Z}_n\|_F^2$	estimate location and strength
LORETA	$\mathbf{Z}$	$N > M$	$\|\mathbf{X} - \mathbb{A}\mathbf{Z}\|_F^2 + \lambda\|DW\mathbf{Z}\|_F^2$	smooth solution
MNE	$\mathbf{Z}$	$N > M$	$\|\mathbf{X} - \mathbb{A}\mathbf{Z}\|_Q^2 + \lambda\|\mathbf{Z}\|_R^2$	prefer surface/weak source
WMN	$\mathbf{Z}$	$N > M$	$\|\mathbf{X} - \mathbb{A}\mathbf{Z}\|_Q^2 + \lambda\|W\mathbf{Z}\|_F^2$	weighted according to depth
MC	$\mathbf{Z}(t)$	$N > M$	$\|\mathbf{X}(t) - \mathbb{A}\mathbf{Z}(t)\|_Q^2 + \lambda\|W\mathbf{Z}(t)\|_1$	sparse solution
LAURA	$\mathbf{Z}$	$N > M$	$\|\mathbf{X} - \mathbb{A}\mathbf{Z}\|_F^2 + \lambda\|WB\mathbf{Z}\|_F^2$	follow Maxwells' law

where h is the minimum inter-grid distance. The LORETA estimator for $\mathbf{Z}$ can be obtained as

$$\widehat{\mathbf{Z}} = \arg\min_{\mathbf{Z}} \|X - A\mathbf{Z}\|_F^2 + \lambda\|DW\mathbf{Z}\|_F^2. \tag{20.49}$$

LORETA favors the solution with smooth spatial distribution, provided that it puts a penalty on spatial roughness, which is measured by the Laplacian of the weighted sources. The assumption is related to the physiological in the sense that it assumes neighboring sources are correlated. It also needs to be pointed out that solutions obtained using LORETA are over-smoothed in some situations.

20.5.2.3 Summary

Table 20.3 summarizes a few popular methods for reconstructing sources from EEG recordings, include dipole fits, LORETA(27), Minimum Norm (MNE) (17), Weighted Minimum Norm (WMN) (13), Minimum Current (MC) (21) (37), and Local Autoregressive Average (LAURA) (5).

Bibliography

[1] Baccalá, L. A. and Sameshima, K. (2001). Partial directed coherence: A new concept in neural structure determination. *Biological Cybernetics*, 84(6):463–474.

[2] Baillet, S., Mosher, J. C., and Leahy, R. M. (2001). Electromagnetic brain mapping. *Signal Processing Magazine, IEEE*, 18(6):14–30.

[3] Brockwell, P. J. and Davis, R. A. (2006). *Introduction to Time Series and Forecasting*. Springer Science & Business Media.

[4] Bunea, F., Ombao, H., and Auguste, A. (2006). Minimax adaptive spectral estimation from an ensemble of signals. *Signal Processing, IEEE Transactions on*, 54(8):2865–2873.

[5] de Peralta Menendez, R. G., Andino, S. G., Lantz, G., Michel, C. M., and Landis, T. (2001). Noninvasive localization of electromagnetic epileptic activity. I. Method descriptions and simulations. *Brain Topography*, 14(2):131–137.

[6] Fiecas, M. and Ombao, H. (2011). The generalized shrinkage estimator for the analysis

of functional connectivity of brain signals. *The Annals of Applied Statistics*, pages 1102–1125.

[7] Fiecas, M. and Ombao, H. (2015). Modeling the evolution of dynamic brain processes during an associative learning experiment. Technical report, University of California, Irvine.

[8] Fiecas, M., Ombao, H., Linkletter, C., Thompson, W., and Sanes, J. (2010). Functional connectivity: Shrinkage estimation and randomization test. *Neuroimage*, 49(4):3005–3014.

[9] Friedman, J., Hastie, T., and Tibshirani, R. (2010). Regularization paths for generalized linear models via coordinate descent. *Journal of Statistical Software*, 33(1):1.

[10] Friston, K. J. (1994). Functional and effective connectivity in neuroimaging: A synthesis. *Human Brain Mapping*, 2(1-2):56–78.

[11] Friston, K. J., Harrison, L., and Penny, W. (2003). Dynamic causal modelling. *Neuroimage*, 19(4):1273–1302.

[12] Gao, H.-Y. (1997). Choice of thresholds for wavelet shrinkage estimate of the spectrum. *Journal of Time Series Analysis*, 18(3):231–251.

[13] Gorodnitsky, I. F., George, J. S., and Rao, B. D. (1995). Neuromagnetic source imaging with FOCUSS: A recursive weighted minimum norm algorithm. *Electroencephalography and clinical Neurophysiology*, 95(4):231–251.

[14] Granger, C. W. (1969). Investigating causal relations by econometric models and cross-spectral methods. *Econometrica: Journal of the Econometric Society*, pages 424–438.

[15] Granger, C. W. (1980). Testing for causality: a personal viewpoint. *Journal of Economic Dynamics and control*, 2:329–352.

[16] Granger, C. W. (1988). Some recent development in a concept of causality. *Journal of econometrics*, 39(1):199–211.

[17] Hämäläinen, M. S. and Ilmoniemi, R. (1994). Interpreting magnetic fields of the brain: Minimum norm estimates. *Medical & Biological Engineering & Computing*, 32(1):35–42.

[18] Hamalainen, M. S. and Sarvas, J. (1989). Realistic conductivity geometry model of the human head for interpretation of neuromagnetic data. *Biomedical Engineering, IEEE Transactions on*, 36(2):165–171.

[19] Hu, L. and Ombao, H. (2015). Modeling high-dimensional multichannel electroencephalograms. Technical report, University of California, Irvine.

[20] Makeig, S., Bell, A. J., Jung, T.-P., Sejnowski, T. J., et al. (1996). Independent component analysis of electroencephalographic data. *Advances in Neural Information Processing Systems*, pages 145–151.

[21] Matsuura, K. and Okabe, Y. (1995). Selective minimum-norm solution of the biomagnetic inverse problem. *Biomedical Engineering, IEEE Transactions on*, 42(6):608–615.

[22] Moulin, P. (1994). Wavelet thresholding techniques for power spectrum estimation. *Signal Processing, IEEE Transactions on*, 42(11):3126–3136.

[23] Ngo, D., Sun, Y., Genton, M. G., Wu, J., Srinivasan, R., Cramer, S. C., and Ombao, H. (2015). An exploratory data analysis of electroencephalograms using the functional boxplots approach. *Frontiers in neuroscience*, 9.

[24] Nunez, P. L. and Srinivasan, R. (2006). *Electric Fields of the Brain: The Neurophysics of EEG*. Oxford University Press.

[25] Ombao, H. C., Raz, J. A., Strawderman, R. L., and Von Sachs, R. (2001). A simple generalised cross-validation method of span selection for periodogram smoothing. *Biometrika*, 88(4):1186–1192.

[26] Pascual-Marqui, R. D. et al. (2002). Standardized low-resolution brain electromagnetic tomography (sloreta): Technical details. *Methods Find Exp Clin Pharmacol*, 24(Suppl D):5–12.

[27] Pascual-Marqui, R. D., Lehmann, D., Koenig, T., Kochi, K., Merlo, M. C., Hell, D., and Koukkou, M. (1999). Low resolution brain electromagnetic tomography (LORETA) functional imaging in acute, neuroleptic-naive, first-episode, productive schizophrenia. *Psychiatry Research: Neuroimaging*, 90(3):169–179.

[28] Rush, S. and Driscoll, D. A. (1969). EEG electrode sensitivity: An application of reciprocity. *Biomedical Engineering, IEEE Transactions on*, (1):15–22.

[29] Schreiber, T. (2000). Measuring information transfer. *Physical Review Letters*, 85(2):461.

[30] Schwartz, R. S., Brown, E. N., Lydic, R., and Schiff, N. D. (2010). General anesthesia, sleep, and coma. *New England Journal of Medicine*, 363(27):2638–2650.

[31] Shumway, R. H. and Stoffer, D. S. (2010). *Time Series Analysis and Its Applications: With R Examples*. Springer Science & Business Media.

[32] Song, S. and Bickel, P. J. (2011). Large vector auto regressions. arXiv preprint arXiv:1106.3915.

[33] Srinivasan, R. and Deng, S. (2012). *Multivariate Spectral Analysis of Electroencephalogram*, pages 1–24. CRC Press.

[34] Teplan, M. (2002). Fundamentals of EEG measurement. *Measurement Science Review*, 2(2):1–11.

[35] Thomson, D. J. (1982). Spectrum estimation and harmonic analysis. *Proceedings of the IEEE*, 70(9):1055–1096.

[36] Uutela, K., Hamalainen, M., and Salmelin, R. (1998). Global optimization in the localization of neuromagnetic sources. *Biomedical Engineering, IEEE Transactions on*, 45(6):716–723.

[37] Uutela, K., Hämäläinen, M., and Somersalo, E. (1999). Visualization of magnetoencephalographic data using minimum current estimates. *NeuroImage*, 10(2):173–180.

[38] Wahba, G. (1980). Automatic smoothing of the log periodogram. *Journal of the American Statistical Association*, 75(369):122–132.

[39] Walden, A. T., Percival, D. B., and McCoy, E. J. (1998). Spectrum estimation by wavelet thresholding of multitaper estimators. *Signal Processing, IEEE Transactions on*, 46(12):3153–3165.

[40] Welch, P. D. (1967). The use of fast Fourier transform for the estimation of power spectra: A method based on time averaging over short, modified periodograms. *IEEE Transactions on Audio and Electroacoustics*, 15(2):70–73.

[41] Wu, J., Srinivasan, R., Kaur, A., and Cramer, S. C. (2014). Resting-state cortical connectivity predicts motor skill acquisition. *NeuroImage*, 91:84–90.

[42] Yan, Y., Nunez, P., and Hart, R. (1991). Finite-element model of the human head: scalp potentials due to dipole sources. *Medical and Biological Engineering and Computing*, 29(5):475–481.

21

Advanced Topics for Modeling Electroencephalograms

Hernando Ombao

Department of Statistics, University of California, Irvine

Anna Louise Schröder

Department of Statistics, London School of Economics

Carolina Euán

Departmento de Estadistica, Centro de Investigaciones en Matematicas

Chee-Ming Ting and Balqis Samdin

Center for Biomedical Engineering, Universiti Teknologi Malaysia

CONTENTS

21.1	Introduction	568
21.2	Clustering of EEGs	572
	21.2.1 Proposal: The Spectral Merger Clustering Method	572
	21.2.1.1 Total Variation Distance	572
	21.2.1.2 Hierarchical Spectral Merger Algorithm	573
	21.2.2 Analysis of Epileptic Seizure EEG Data	574
21.3	Change-Point Detection	580
	21.3.1 Existing Methods and Challenges	580
	21.3.2 The FreSpeD Method	582
	21.3.2.1 Comparison to the Other Approaches	584
	21.3.3 Analysis of the Multichannel Seizure EEG Data	585
	21.3.3.1 Seizure Localization	585
	21.3.3.2 Seizure Onset Estimation and Potential Precursors	585
21.4	Modeling Time-Varying Connectivity Using Switching Vector Autoregressive Models	588
	21.4.1 Background on Vector Autoregressive (VAR) Models	589
	21.4.1.1 Stationary VAR Model	589
	21.4.1.2 Time-Varying VAR Model	590
	21.4.1.3 Switching VAR (SVAR) Model	591
	21.4.2 Parameter Estimation	592
	21.4.3 Estimating Dynamic Connectivity States in Epileptic EEG	593
21.5	Best Signal Representation for Non-Stationary EEGs	599
	21.5.1 Overview of Signal Representations	599
	21.5.2 Overview of SLEX Analysis	600
	21.5.3 Selecting the Best SLEX Signal Representation	603
	21.5.4 SLEX Analysis of Multichannel Seizure EEG	606

21.6   Dual-Frequency Coherence Analysis 608
         21.6.1   Overview and Historical Development 610
         21.6.2   The Local Dual-Frequency Cross-Periodogram 612
         21.6.3   Formalizing the Concept of Evolutionary Dual-Frequency Spectra ... 612
                  21.6.3.1   Harmonizable Process: Discretized Frequencies 612
                  21.6.3.2   A New Model: The Time-Dependent Harmonizable Process   613
                  21.6.3.3   Dual-Frequency Coherence between Bands 613
         21.6.4   Inference on Local Dual Frequency Coherence 614
         21.6.5   Local Dual Frequency Coherence Analysis of EEG Data 615
                  21.6.5.1   Description of the Data and Experiment 615
                  21.6.5.2   Implementation Details 615
                  21.6.5.3   Results and Discussion 616
         21.6.6   Conclusion 617
21.7   Summary 618
         Bibliography 621

The chapter will cover some of the advanced topics for modeling electroencephalograms with complex structures. We focus in particular on the non-stationarity of epileptic seizure EEGs, which can be manifested in a number of ways such as evolution of the spectra and coherence; and changes in the parameters of time-domain models. Here, we illustrate a procedure for clustering EEG channels based on the similarity of their spectra. Since seizure EEGs are non-stationary, we show that the channel groupings also evolve across the seizure process. Second, we present the FreSpeD method for detecting the points in time when there are changes in either the spectrum or coherence. It is very interesting that the method detected changes in the coherence at precise frequency bands immediately preceding seizure onset. These changes are very subtle. They are not detectable by mere visual inspection and were previously undetected by other methods. Third, we present a different approach to spectral analysis by modeling the EEG as a process that is represented by different states which change over time. Each state is uniquely characterized by a vector autoregressive process. A change in the state implies that there is a change in the EEG process. The fourth topic treats the problem of modeling non-stationary time series from the point of view of finding some optimal model that captures the non-stationarity in the signal. The SLEX method essentially first builds a collection of models where each has a unique SLEX basis and then selects the best model using a complexity-penalized Kullback–Leibler criterion. The fifth topic introduces the new concept of time-varying dual-frequency coherence (e.g., coherence between the alpha oscillation in one channel and the beta oscillation in another channel). This approach is used to analyze the cross-dependence structure between channels during a visual-motor task experiment.

21.1   Introduction

In this chapter, we consider a number of approaches for modeling and analyzing multi-channel electroencephalograms when they exhibit "non-stationary" behavior. An example are the traces in Figure 21.1 which were recorded around an epileptic seizure episode. One observes an increase in the variance or wave amplitudes at seizure onset. This epileptic seizure recording captures brain activity of a subject who suffered a spontaneous epileptic seizure while being connected to the EEG. The recording is digitized at 100 Hz and about

EEG at the T3 (left temporal) channel

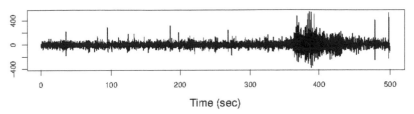

Time (sec)

T4 (right temporal) channel

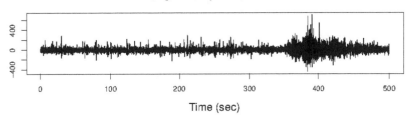

Time (sec)

FIGURE 21.1

Top: EEG recordings at the left temporal channel (T3). Bottom: right temporal channel (T4). Sampling rate 100 Hertz. Duration of recording is 500 seconds.

500 seconds long, providing us with a time series of length $T = 50000$. Data is collected at 21 channels, 19 bipolar scalp electrodes placed according to the 10-20 system (see Figure 21.2), and two sphenoidal electrodes placed intracranially at the base of the temporal lobe. The data has been previously analyzed, for example in (57), where the primary focus was on obtaining the optimal representation from a family of models defined by the smooth complex exponential (SLEX) library (see Section 21.5). Seizure EEGs are realizations of a rapidly changing electrophysiological process resulting from abnormal firing behavior of a network of neuronal subpopulations. These aberrations in neuronal electrical activity are expressed by fluctuating amplitudes of waveforms, changing spectral decompositions and evolving cross-dependence between channels. A complete analysis of this type of a brain signal requires a variety of advanced tools which the authors have developed. We discuss recently developed methods that uncover many characteristics of this signal that could help us better understand not only the evolution of the seizure process but also potential changes in neurophysiology that precede seizure onset.

Another example concerns EEG traces during the execution of a hand motor task in Figure 21.3. In this experiment, the participant was instructed to move the joystick to either the left or to the right. The cognitive process in each trial is as follows: the participant receives the external input, processes the information, and then finally executes the task. Throughout the process, different brain regions are engaged at different levels and time points. A visual inspection of the EEG traces does not reveal obvious evidence of non-stationarity (such as changes in variance). However, it can be detected in the analysis of coherence, frequency-specific linear dependence between channels. More specifically, in Section 21.6, we present a model that rigorously characterizes how dependence between different oscillatory components from a pair of channels (e.g., between beta oscillation in one signal and alpha oscillation in another) evolves during a trial. Note too that a non-zero coherence between different frequency bands at different channels already implies non-stationarity.

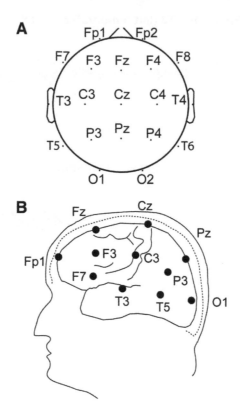

FIGURE 21.2
EEG scalp topography.

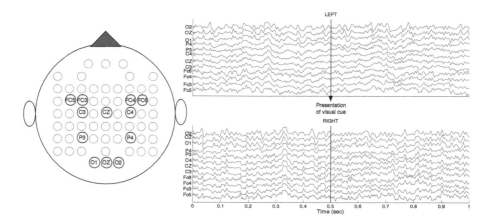

FIGURE 21.3
Left: EEG scalp topography. Right: traces of the EEG signals for one trial for leftward and rightward movements.

As a review, a multichannel time series with P channels, denoted $\mathbf{X}(t) = [X_1(t), \ldots, X_P(t)]'$, is stationary (or strictly stationary) when the joint distributions $\{\mathbf{X}(s), \ldots, \mathbf{X}(t)\}$ and $\{\mathbf{X}(s+h), \ldots, \mathbf{X}(t+h)\}$ are identical for any $s <= t$ and shift $h \neq 0$. In other words, under stationarity, shifting the time indices does not change the distribution. This implies the conditions of weak stationarity: (1) $\mathbf{EX}(t) = \mu$ which does not change with time; and (2) $\mathbb{C}\text{ov}[\mathbf{X}(t), \mathbf{X}(t+h)]$ may depend on h but not on t. From condition (2) it also follows that the variance–covariance matrix of $\mathbf{X}(t)$ does not change over time. From the spectral point of view, condition (2) is equivalent to (2'): the auto-spectrum (decomposition of power at each channel) and coherence between all pairs of channels remain constant over time.

Departures from non-stationarity can be expressed in many ways. The most obvious ones to spot are changes in the mean or changes in the variance. In this chapter, we will consider a range of such departures. First, under stationarity, any form of grouping or cluster structure of the different EEG channels should be time-invariant. In Section 21.2, we introduce a method to cluster channels according to their auto-spectral characteristics. This particular notion of clustering depicts the topographical distribution of the channels based on the similarity of their spectra. A change in clustering signifies that at least one channel undergoes a change in its spectral content. This channel changes its membership from one cluster to another. The method of (22) for clustering EEG channels uses the total variation distance as the distance metric. With this method we demonstrate that the clustering of EEGs evolves over the duration of the epileptic seizure.

Another departure from non-stationarity is change in the dependence or connectivity between channels as measured by cross-coherence. This type of change cannot be detected by visual inspection. In Section 21.3, we describe a method for frequency-specific change-point detection, FreSpeD (66). The method can be used to data-adaptively identify seizure onset and seizure focal point, but it also provides new insights into the preictal brain electrophysiology. In Section 21.2, we demonstrate that, without any prior information or model assumption, the FreSpeD discovered subtle changes in the coherence, which occurred within few seconds before the seizure onset. These preliminary changes do not necessarily have to be concentrated only on the lesion or affected area (e.g., on the left temporal region for this particular patient). In fact, these changes can be spatially generalized, engaging both neighboring and inter-hemispheric regions. They get more pronounced and spatially concentrated leading to the seizure.

While most of methods in this chapter examine changes in the spectral properties of the multichannel EEGs, the focus of our discussion in Section 21.4 is to capture changes expressed in the time-domain properties. We present a model for seizure EEGs where we assume that there is a finite number of underlying electrophysiological states; with each state being characterized by a unique vector autoregressive (VAR) process. The underlying models for each state are identifiable. Moreover, at each time point, only one state is active and over time, states are allowed switch abruptly. A change in the state implies a change in the VAR parameters and hence a change in the process.

In Section 21.5, we present a different approach to non-stationarity. In place of identifying points in time when the EEG process undergoes a change in its stochastic properties, the SLEX method finds an optimal representation for the non-stationary EEG. The method starts with constructing the SLEX library, a collection of bases, each of which consists of time-localized Fourier waveforms). For each SLEX basis in the library, a corresponding SLEX model is constructed. The method then finds the model that optimizes the penalized Kullback–Leibler criterion, which essentially balances fit and model complexity. The model is defined by a unique basis which in turn gives a unique (dyadic) segmentation of time from which we indirectly identify estimates of change point locations (but restricted on dyadic points).

21.2 Clustering of EEGs

Clustering methods are important to localize and characterize spatial patterns of EEG signals. There are three different types of time series clustering algorithms (47): methods based on comparison of raw data, feature-based methods, and methods based on models fitted to the data. The main method in this section is feature-based, as we will consider the spectrum of the EEG to be the central feature for classification purposes.

Examples of the clustering approaches that have been applied in brain signals are (45) for fMRI data and (37) for EEG data. In the first one, the area between the variations of the signals around their means was used as a similarity measure in the Growing Neural Gas (GNG) clustering algorithm. In the second one, the shortest distance on a torus between time series of phase differences was used with a modified K-means cluster algorithm for detection of phase synchronization on analog signals of the cerebral cortex.

In addition, clustering algorithms based on the spectrum have been used for classification purposes on earthquake and explosion data. Reference (40) proposes clustering procedures based on the Kullback–Leibler and Chernoff information between the spectral densities. These measures feed a hierarchical clustering algorithm and a k-means algorithm. However, as the data is non-stationary, (68) proposes the use of the time-varying spectrum and a local version of the Kullback–Leibler measure to address the problem of non-stationarity. Also, (36) considered the non-stationary case using the time-varying spectrum. They develop a discriminant scheme based on the Kullback–Leibler divergence between the SLEX spectra of the different classes.

21.2.1 Proposal: The Spectral Merger Clustering Method

Clustering methods seek to find groups in data so that members of the same group are alike, while members of different groups are as dissimilar as possible. A clustering procedure is determined by two components: (1.) a dissimilarity measure and (2.) a clustering algorithm. The dissimilarity measure depends on the question of interest. For example, if we want to get clusters with the channels that are more correlated, then coherence can be used as dissimilarity measure. The clustering algorithm is going to determine the mechanism to build the clusters. Most authors agree that there are two broad families of clustering algorithms, namely, partitioning and hierarchical clustering.

The spectral merger clustering (SMC) method uses the total variation distance as the dissimilarity measure and the hierarchical spectral merger algorithm. This method was developed in (22). The proposed SMC method is able to identify brain regions that are synchronized in a sense that these regions share similar oscillations or waveforms. The resulting clusters of EEGs serve as a proxy for segmenting the brain cortical surface. The following is a brief description of the elements of this method.

21.2.1.1 Total Variation Distance

The total variation distance (TVD) between two densities, f and g, is defined as

$$d_{TV}(f,g) = 1 - \int \min\{f(\omega), g(\omega)\} d\omega. \tag{21.1}$$

This equation allows a simple interpretation of the total variation distance in graphical terms. This distance is equal to 1 minus the intersection of the areas below the densities f and g. Thus, densities with a bigger common area are considered to be less dissimilar. Figure 21.4 illustrates the situation for two density functions, the blue (pink) area represents the non-common area and it is the value of the TVD.

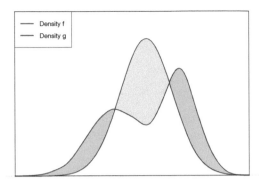

FIGURE 21.4
The TV distance measures the similarity between the two densities. The blue (pink) area is the value of the TV distance.

Since the spectra are not probability densities, they have to be normalized by dividing the estimated spectrum by the sample variance $\widehat{\gamma}(0)$. We are going to denote the normalized estimated spectral density as $\widehat{f}(\omega) = \widehat{g}(\omega)/\widehat{\gamma}(0)$.

21.2.1.2 Hierarchical Spectral Merger Algorithm

For our zero-mean P-channel EEG signal $\mathbf{X}(t) = [X_1(t), \ldots, X_P(t)]'$, the hierarchical spectral merger algorithm starts with P clusters, one for each individual channel.

1. Estimate the spectral density for each cluster using the smoothed periodogram and compute the TVD between their spectra.

2. Find the two clusters that have the lowest TVD and save this value as a characteristic.

3. Merge the signals in the two closest clusters and replace the two clusters by this new one.

4. Repeat steps 1–3 until there is only one cluster left. After each iteration, the estimated spectral density for each cluster will be the average overall spectrum in the same cluster.

The characteristic saved in step 2 represents the marginal cost of using $k-1$ instead of k clusters and one could base a decision on the number of clusters on these costs. If a large value is observed, then it may be preferable to keep k clusters instead of restricting oneself to $k-1$ clusters. When two clusters merge, the signals in each one of them should be considered as coming from the same spectral density, hence it is reasonable to take the average spectra over all the channels in the same cluster in order to get a better estimation of the original spectral density, by reducing the estimation variability. **Remark.** This algorithm can be slightly modified at step 4; instead of taking the average spectra, we can concatenate the time series in a cluster to get a new estimated spectral density. This procedure also produces a better estimation of the spectral density and in some applications, as the study of ocean waves, can be reasonable. We are going to refer to this second version as a hierarchical merger algorithm. Both algorithms compute the distance between the new cluster and the old clusters based on an updated estimated spectra.

To illustrate our method, consider two different AR(2) processes. Both their spectra have energy concentrated at the alpha band (at 10 Hz), but one also has an energy concentration

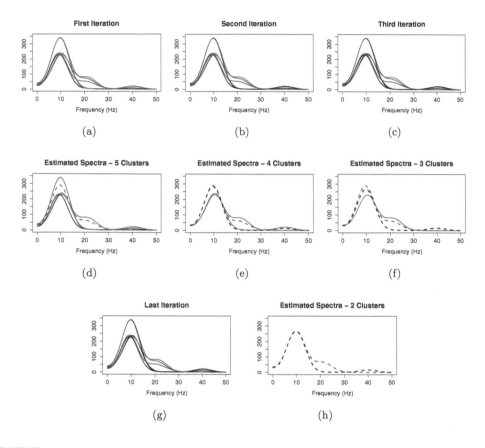

FIGURE 21.5
Dynamic of the hierarchical spectral merger algorithm. (a), (b), (c), and (g) show the clustering process for the spectra. (d), (e), (f) and (h) show the evolution of the estimated spectra, which improves when we merge the series on the same cluster.

at the beta band (at 21 Hz) and the other at the gamma band (at 40 Hz). We sample three time series for each process, 10 seconds of each one with a sampling frequency of 100 Hz ($t = 1, \ldots, 1000$). The dynamic of the clustering method is shown in Figure 21.5. We start with six clusters; at the first iteration we find the closest spectra, represented in Figure 21.5(a) with the same color (red). After the first iteration we merge these series and get 5 estimated spectra, one per cluster; Figure 21.5(d) shows the estimated spectra where the new one is represented by the dashed red curve. We can follow the dynamic in Figures 21.5(b), 21.5(e), 21.5(c) and 21.5(f). By the end of the SMC method, Figure 21.5(g), the obtained clusters are the same as the simulation setting.

21.2.2 Analysis of Epileptic Seizure EEG Data

We applied the proposed clustering method to the epileptic seizure EEG data described in the introduction of this chapter. Our goal is to analyze the changes on the clustering of the EEG signals before, during and after the epileptic seizure. Figure 21.6 shows the EEG signal at channel T3, the gray region (330.01–340 sec, $t = 33001 - 34000$) corresponds to a period before the epileptic seizure, the red region (342.01–352 sec, $t = 34201 - 35200$) contains

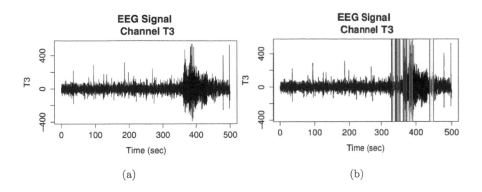

FIGURE 21.6

EEG signal from channel T3. (a) Complete signal. (b) Segments involved in the analysis to compare EEG signals before (gray), during (red) and after (blue, green and yellow) the epileptic seizure.

the initial point of the epileptic seizure, and finally the blue, green and yellow regions (362.01–372 sec, $t = 36201 - 37200$, 380.01–390 sec, $t = 38001 - 39000$, and 440.01–450 sec $t = 44001 - 45000$) are segments after the epileptic seizure.

HMClust is an R package developed to analyze EEG data using clustering based on the TVD and the hierarchical spectral merger clustering. It is available at http://ucispacetime.wix.com/spacetime!project-a/cxl2. To estimate the spectral densities we used the smoothed periodogram with a parzen window (bandwidth value = 100, chosen using the Gamma deviance generalized cross-validation criterion, c.f. (55)).

We apply clustering using the TVD and the hierarchical spectral merger algorithm. The following R code is an example of how to use the HMClust package for one of the subsets.

```
library(HMClust)
X[[1]]<-scale(Data[33001:34000,],scale=FALSE)
Clust1<-HMC(X[[1]],freq=100,Merger=2)
```

Clust1 has three elements, Clust1$Diss.Matrix, Clust1$min.value and Clust1$Groups. Clust1$Diss.Matrix corresponds to the initial dissimilarity matrix, the TVD between each EEG signal for all channels. Figure 21.7 shows these matrices for the five subsets. We observe that before the epileptic seizure (Figure 21.7(a)) the signals have influence from different frequencies and they do not share the same frequencies. After the epileptic seizure (Figures 21.7(c), 21.7(d) and 21.7(e)) a clustering appears. However, these are not the final clustering results.

Figure 21.8 shows the trajectories of the minimal TVD obtained when we use the hierarchical spectral merger algorithm. We can conclude that there is no significant clustering before the epileptic seizure by looking at the TVD trajectory for the first segment. It is almost constant for any number of clusters and bigger than 0.2. However, after the epileptic seizure (blue, green and dashed black curves) there is a significant decrease of the distance from 1 to 6.

In almost all the clustering algorithms, choosing the number of clusters is a difficult problem. Based on several simulation studies, for the SMC method an empirical criterion is proposed. The main idea is to identify the smallest number of clusters where the TVD value is reasonably small, as this way we can hope for better interpretability. Consider the

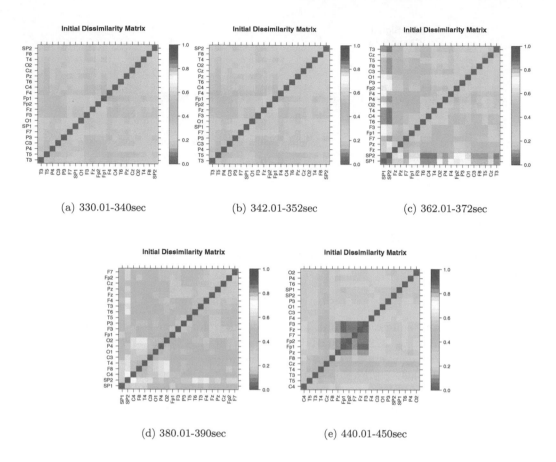

(a) 330.01-340sec   (b) 342.01-352sec   (c) 362.01-372sec

(d) 380.01-390sec   (e) 440.01-450sec

FIGURE 21.7
Initial values of the TVD between each channel for time segments (a) 33001–34000, (b) 34201–35200, (c) 36201–37200, (d) 38001–39000 and (e) 44001–45000.

trajectories plotted in Figure 21.8. The criterion considers the smaller number of clusters where the speed of decreasing for all these trajectories is slowed, i.e., the point where we observed an elbow appears. To obtain the final clustering results based on the empirical criterion we are going to fix the same number of clusters for all subsets at 6. We used the function *cutk* to extract from the Clust1$Groups the members of each one of the 6 clusters.

Figure 21.9 shows the location of the clusters at the cortical surface for subsets before (21.9(a)), during (21.9(b)) and after (21.9(c), 21.9(d), 21.9(e)) the epileptic seizure. Channels with same color belong to the same cluster. In the preictal period we cannot observe a clustering: almost all channels belong to the sky blue cluster. The spectra show energy concentration in lower frequencies and only channels located in the anterior frontal region indicate activity in the beta band (see Figure 21.10(a)). This can be related to basal and visual activity. During the seizure, a more spatial pattern on the clustering can be observed. A strong clustering of channels from the left parietal and left temporal regions appears, due to a higher influence of the delta band (see Figure 21.10(b)). However, there is still a dominated cluster (blue sky) with channels in the frontal region and right temporal. In both segments, before and during (starting) the seizure, clusters have channels with low oscillations (spectra with low peak frequency) that are just slightly different. This behavior

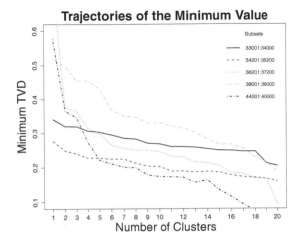

FIGURE 21.8
Trajectories of the minimum value of the TVD resulting using the SMC method.

was not unexpected, due to the initial dissimilarity between channels (see Figures 21.7(a) and 21.7(b)).

After the epileptic seizure, energy concentrates in the gamma band, in several channels, resulting in some clusters that are well defined and with bigger differences compared to the others. During the period 362.01–372 sec ($t = 36201 - 37200$), channels from the left parietal and left temporal regions are synchronized in beta and gamma bands (see Figure 21.10(c)). Notice that Fz, Pz and F4 channels are not clustered with others because these channels are the only ones that present less activity on the higher frequencies, beta and gamma band. During the period 380.01–390 sec ($t = 38001 - 39000$), activity on the gamma band is common for all clusters and the main differences between clusters is driven by the influence of other frequency bands. Channels on the frontal region get synchronized with the channels in the left parietal and left temporal regions. However in this subset, the cluster (red) is dominated by activity on the theta and gamma bands with almost null influence of the beta band. Comparing this cluster (red one, after seizure) with the blue sky cluster before the seizure, this cluster has almost all channels but the channels are synchronized in high and lower frequencies.

After 90 sec of the epileptic seizure, in the last subinterval, almost all the channels are grouped into one big cluster (again) dominated by delta band activity. Showing that all activity on higher frequencies disappears. The clustering structure is similar to before-seizure clustering, and is driven just by small differences in lower frequencies.

Notice that the SP1 and SP2 channels are synchronized in all subsets, which is not surprising since both channels are located inside the brain and very close to each other.

Consider the channels T3 (left temporal) and T4 (right temporal). They are spectrally synchronized only before the epileptic seizure. Later, when the seizure starts, T4 shows activity in the beta band while T3 continues with activity in lower frequencies. Post-seizure, T3 is synchronized with the left parietal and left temporal regions and T4 is synchronized with the respective regions in the right hemisphere. However, this spectral synchronicity disappears after 90 sec.

Comparing SMC results with other approaches. Using the SMC method we observe that only lower frequencies are mostly involved in both the preictal period and the

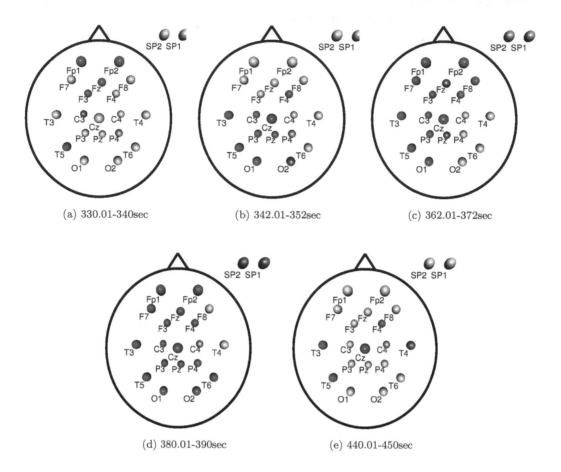

(a) 330.01-340sec   (b) 342.01-352sec   (c) 362.01-372sec

(d) 380.01-390sec   (e) 440.01-450sec

FIGURE 21.9

Location of each cluster at the cortical surface for subsets before, during and after the epileptic seizure. Time segments (a) 33001–34000, (b) 34201–35200, (c) 36201–37200, (d) 38001–39000 and (e) 44001–45000. Notice that $SP1$ and $SP2$ are located inside the brain, as a visualization tool, we plotted these channels on the right upper corner.

last subinterval (approximately 90 seconds post-seizure). In contrast, immediately following seizure onset, the higher frequency bands dictated the clustering distribution of the channels. Moreover, immediately following the seizure onset but before the last subinterval, the channels were clustered similarly but the clustering was heavily influenced by the beta and gamma frequency bands. These results were consistent with the dynamics obtained using the two-state Markov-switching VAR models (see Section 21.4 where the one of the states was active between 350–440 seconds. The SMC results are clear evidence of non-stationarity in our dataset, and that the main changes on the signals were after 340 seconds, which is in line with the results using the FreSpeD method (Section 21.3).

The SMC method, as others, has some limitations: 1) A larger number of channels will require a larger time of computing, due to the hierarchical structure. 2) How to choose the number of clusters is still a subjective criterion and we cannot measure the uncertainty. 3) Comparing clusters between segments/epochs will be more difficult when the number of segments/epochs is big, and a model for these comparison needs further research.

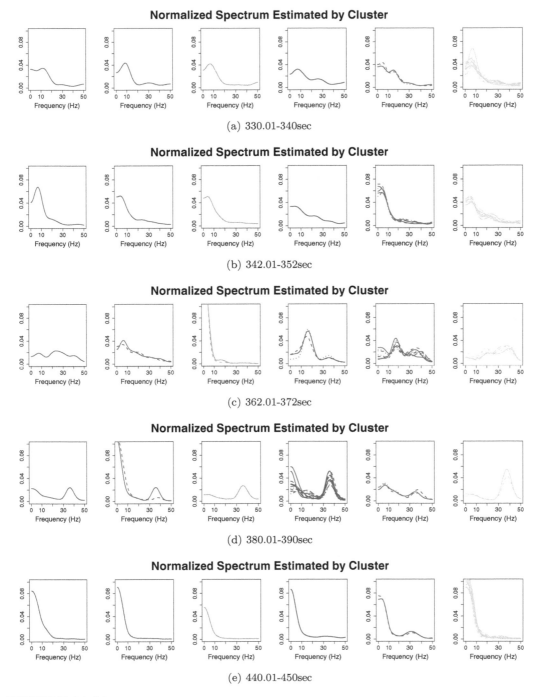

FIGURE 21.10
Spectrum estimated for each channel by cluster for subsets before, during and after the epileptic seizure. Time segments (a) 33001–34000, (b) 34201–35200, (c) 36201–37200, (d) 38001–39000 and (e) 44001–45000.

21.3 Change-Point Detection

This section discusses change-point detection for non-stationary EEG data. Analysing the evolution of dynamic brain processes is among the most valuable uses for EEG recordings. In this context, the segmentation of the non-stationary process into short, approximately stationary segments is a direct output and we will use both terms here interchangably whenever possible. Estimates of the number of change points, their locations and the nature of the change between adjacent segments can help improve understanding of complex brain processes, especially those whose stochastic properties, such as the spectrum, evolve over time. Examples are the evaluation of reaction times for a control group versus a subject group suffering a clinical condition, the detection of patterns in EEG traces for the use of Brain Computer Interfaces, the diagnosis of sleep disorders or the analysis of EEG traces recorded before and during epileptic seizures, such as the data described in the introduction. A change-point analysis of this seizure recording is presented in Section 21.3.3, where we discuss the questions of seizure onset estimation in time and localization of the seizure focal point in the brain topography. We also present findings on the identification of potential electrophysiological markers that are precursors of seizure onset.

21.3.1 Existing Methods and Challenges

We restrict our attention in this section to the offline detection of change points that reflect an event-related temporal evolution. However, we point out the existence of a strand of literature related to the segmentation of EEG data for the analysis of the static brain condition such as resting state or seizure state. These approaches typically decompose the time series into short blocks and analyse spectral patterns on these blocks (25). In the context of seizure analysis, This is typically done in the framework of classification or clustering (see e.g. 1, 74, 44). There exists a wide range of parametric and nonparametric approaches to change-point detection in EEG data. All share the assumption that the brain process can be described as approximately piecewise-stationary with sudden transitions between these pieces, which are defined by a set of change points $\mathbb{N} = \{\eta_k, k = 1, 2, \ldots, N\}$. The minimum distance between adjacent change points is bounded below by $\delta \leq \min_k(\eta_k - \eta_{k-1})$. By convention, $\eta_0 = 1$ and $\eta_{N+1} = T$. The goal is to discover the number and location of change points, where each change point represents the transition from one stationary process to an in some sense different stationary process. In this framework, the time series can consist of up to $N + 1$ different stationary processes, unlike the approach of regime-switching processes (see Section 21.4). To be detectable, the change size has to be bounded away from zero, and the theoretical requirements on this bound depend typically on the number of observations T. Change-point detection in EEG data has a long history and we focus on some more recent advances.

The earliest works in this field are parametric and the vast majority of early works focuses on single-channel data; for an overview see Pardey et al. (59), Chen and Gupta (11). Biscay et al. (5) introduce a Bayesian approach to EEG segmentation and discuss this using the example of a piecewise AR(p) process. Their change-point detection method is an example of global optimization, which is, in large datasets, substantially more computationally demanding than sequential change-point detection using binary segmentation (BS). Within a binary segmentation algorithm, at each step we test for the existence of one change point on an interval, which equals the full time interval $\{1, \ldots, T\}$ in the first step. If there is a change detected on a given interval, the data is split at the location of the estimated change point and the test procedure is repeated on the two resulting subintervals, the one on the left of the change point and the one on its right. This procedure terminates when no more changes are detected on any subinterval. Subject to boundary constraints, changes can be detected anywhere on the interval, which makes this procedure substantially more flexible than standard dyadic segmentation.

There also exists extensive literature on nonparametric methods for change-point detection in EEG data. Nonparametric methods can be preferable over their parametric counterparts because they require very few assumptions on the true process, apart from some form of (piecewise-)stationarity. Ombao et al. (56) discuss segmentation using a library of smooth localized complex exponentials (SLEX), which are orthogonal and time-localized versions of the Fourier complex exponentials. Their segmentation method uses a criterion based on log spectral energy with the goal of characterizing the data through a unique, optimal basis, which implies approximate change-point locations. The method is computationally fast and allows for valuable insights when applied to seizure EEG, but change-point locations are restricted to be dyadic. The SLEX method is discussed in detail in Section 21.5. Brodsky et al. (7) introduce a nondyadic, and in this sense more flexible, method using a cumulative-sum (CUSUM) test statistic that is applied sequentially using BS. The core method is easy to implement and fast, but requires additional two-stage post-processing and restricts the EEG analysis to alpha-band activity. Kaplan et al. (41) extend this to detect change points in the spectral energy at various frequency bands. The authors do not explicitly address the resulting multiple-testing problem. Saab and Gotman (62) use wavelet decomposition and feature extraction for segmentation in the context of automatic seizure detection and seizure onset warning. This approach detects change points at different channels and frequency bands, but, as the authors point out, it requires subject-specific parameter tuning. Furthermore, similar to the other previously mentioned works, this method ignores possible changes in the coherence between channels, which are inevitable during an event-related activation of a particular area in the brain.

In the recent past, the complex connectivity between different brain areas has moved into the focus of neuroscientists. With improved computational and storage technologies, methods have been devised that simultaneously aim at the detection of change points at different EEG channels and in the coherence between channels. These methods are computationally fast and can thus be applied to longer time series, as those resulting from high-resolution EEG recordings. For instance, Ombao et al. (57) generalize the Auto-SLEX in a multichannel setting and provide segmentwise characterizations of energy spectra and coherence. Davis et al. (18) and Kirch et al. (42) introduce parametric change-point detection methods for multivariate EEG data. In Davis et al. (18) an automatic piecewise autoregressive modeling procedure is presented which estimates an optimal partitioning of the time series and segment-specific AR-lag estimates using the minimum-description-length principle. Kirch et al. (42) devise a parametric change-point detection method based on the assumption that the underlying process follows (approximately) a sparsified VAR(p) model. The method is able to detect either a single change point or two change points in an epidemic (two-regime) setting, which is of limited use in applications on seizure data. However, this approach can be generalized using BS as shown in Schröder and Ombao (66). Terrien et al. (71) discuss the segmentation of bivariate data for the estimation of synchronization measures between channels. Their method is based on a modification of the best basis algorithm (13).

In Schröder and Ombao (66), we introduce a frequency-specific method for change-point detection using a thresholded multivariate CUSUM statistic with binary segmentation. The advantage of frequency-specific change-point detection is an additional dimension of interpretability. As discussed in previous work (e.g. 5, 41), the assignment of a change to one or several frequency bands can greatly improve physicians' and neuroscientists' understanding the nature of this change, because it can be associated with typical causes of band-related energy shifts.

There exist many more approaches to change-point detection in a multivariate time series. While the theoretical properties of these may be well documented and the numerical results convincing, as for e.g. Preuß et al. (60), the computational difficulties arising for large time series are pronounced and render them inopportune for the analysis of EEG traces with several minutes length.

21.3.2   The FreSpeD Method

We discuss now the nonparametric frequency-specific change-point detection method FreSpeD, which was introduced in Schröder and Ombao (66). We focus on this method since to the best of our knowledge, at the time of writing, there is no other multichannel EEG change-point detection method that simultaneously provides directly interpretable results for the distribution of change points in time, space, and frequency domain. The method is computationally fast and provides flexible (non-dyadic and non-parametric) segmentation.

The FreSpeD method is built on the premise that the multichannel EEG signal is represented as a blockwise quasi-stationary process. Formally, let $\mathbf{X}(t) = [X_1(t), \ldots, X_P(t)]'$ be a zero-mean P-channel EEG signal that is stationary on each of the $N+1$ segments defined by the set of change-point locations $\mathbb{N} = \{\eta_k, k = 1, 2, \ldots, N\}$,

$$\mathbf{X}(t) = \sum_{k=1}^{N+1} \mathbb{I}_t^k \int_{-0.5}^{0.5} \exp(i2\pi\omega t) d\mathbf{Z}^k(\omega), \tag{21.2}$$

where the indicator function is defined $\mathbb{I}_t^k = 1$ if $t \in \{\eta_{k-1} + 1, \ldots, \eta_k\}$ and $d\mathbf{Z}^k(\omega) = [dZ_1^k(\omega), \ldots dZ_P^k(\omega)]'$ is a zero-mean random increment process with covariance

$$\mathbb{C}\mathrm{ov}(d\mathbf{Z}^k(\omega)) = \mathbf{f}^k(\omega)d\omega = \begin{bmatrix} f_{1,1}^k(\omega) & \cdots & f_{1,P}^k(\omega) \\ \cdots & \cdots & \cdots \\ f_{P,1}^k(\omega) & \cdots & f_{P,P}^k(\omega) \end{bmatrix} d\omega. \tag{21.3}$$

This is a member of the family of locally stationary processes as described in Dahlhaus (17).

The FreSpeD method estimates change points in the time-varying autospectra and cross-coherence,

$$f_{p,p}(t,\omega) = \sum_{k=1}^{K} \mathbb{I}_t^k f_{p,p}^k(\omega) \tag{21.4}$$

$$\rho_{p,q}(t,\omega) = \sum_{k=1}^{K} \mathbb{I}_t^k \rho_{p,q}^k(\omega), \tag{21.5}$$

where the coherence on segment k is defined as $\rho_{p,q}^k(\omega) = |f_{p,q}^k(\omega)|^2/|f_{p,p}^k(\omega)|\| f_{q,q}^k(\omega)|$. The underlying spectra and coherences, $f_{p,p}(t,\omega)$ and $\rho_{p,q}(t,\omega)$, evolve in a piecewise constant manner with change points being restricted to the set $\mathbb{N}$. At any change point η_k, one, several, or all components $f_{p,p}(t,\omega)$ and/or $\rho_{p,q}(t,\omega)$ change. The change-point locations η_k are estimated using a series of locally computed spectral estimates. Consider the averaged local periodogram (a generalization of 76)

$$\widehat{\boldsymbol{f}}(t_v,\omega_l) = \frac{1}{M} \sum_{m=1}^{M} \boldsymbol{d}^m(t_v,\omega_l)\boldsymbol{d}^{m,*}(t_v,\omega_l). \tag{21.6}$$

Here, $d^m(t_v, \omega_l)$ denotes the discrete Fourier transform of $\mathbf{X}(t)$ of the mth segment of the interval indicated by t_v and the asterisk denotes the complex conjugate. This is result of a partitioning of $[1, T]$ into intervals of length ν, which are defined by the boundaries $\{t_v, v = \nu, 2\nu, \ldots, T_\nu\}$ where $T_\nu = T/\nu$. For variance reduction within each part of length ν the Fourier transforms are computed on each of M equally-long, non-overlapping blocks, and then averaged.

Remarks. It should be noted that within the framework of local stationarity, in probability, as T grows and ν remains constant, the process is stationary on each of the short intervals between the points t_v. Note also that we are not looking pointwise at frequencies ω anymore, but in practice are considering averages over frequency bands, which we denote ω_l, like in Sections 21.2 and 21.6. These bands can be mapped to the standard frequency bands in EEG analysis, such as alpha and beta bands. The number of frequency bands L depends on the window length ν and the number of subsegments M, $L = \nu/(2M)$ for even ν/M. This discretization is necessary in any data application. It can be shown that the true spectral energy and coherence are well-approximated by the estimates $\widehat{f}_{p,p}(t_v, \omega_l)$ and $\widehat{\rho}_{p,q}(t_v, \omega_l) = |\widehat{f}_{p,q}(t_v, \omega_l)|^2 / (\widehat{f}_{p,p}(t_v, \omega_l) \widehat{f}_{q,q}(t_v, \omega_l))$.

FreSpeD estimates change points simultaneously over all frequency bands ω_l for each channel's autospectrum and the channel pair cross-coherence. Consider the thresholded sum of the CUSUM test statistic, defined on the interval $[s, e] : t_1 \le s < e \le T_\nu$ with threshold τ,

$$\mathfrak{C}_{s,b,e}(z) = \sum_{l=1}^{L} \mathcal{C}_{s,b,e}(z_l) \mathbb{I}(\mathcal{C}_{s,b,e}(z_l) > \tau) \tag{21.7}$$

$$\mathcal{C}_{s,b,e}(z_l) = \left| \sqrt{\frac{e-b}{n(b-s+1)}} \sum_{t_v=s}^{b} z_l(t_v) - \sqrt{\frac{b-s+1}{n(e-b)}} \sum_{t_v=b+1}^{e} z_l(t_v) \right| \Big/ \sigma_{s,e} \tag{21.8}$$

Here, $\sigma_{s,e}$ denotes the standard deviation of z_l on $[s, e]$. For any panel of processes that can be described as (approximately) piecewise-constant functions plus noise, z_l, the statistic $\mathfrak{C}_{s,b,e}(z)$, is maximized at one of their change points, and even for large L this statistic can consistently detect change points that occur only in a small subset of the group while being robust to false positives (12).

We now briefly describe the intuition behind this statistic: For a single time series z_l, $\mathcal{C}_{s,b,e}(z_l)$ is maximized where the contrast between $[z_l(s), \ldots, z_l(b)]$ and $[z_l(b+1), \ldots, z_l(e)]$ is maximized. If this contrast, scaled by the process variability, exceeds a threshold τ, then we consider this a potential change-point candidate. We provide theoretical and practical guidance for the choice of the threshold in Schröder and Ombao (66).

Within FreSpeD, all P autospectra and $P(P-1)/2$ different coherence pairs of the time series $\mathbf{X}(t)$ can be analysed separately. This is a fast approach to analysing EEG data with potentially many channels, because the method can be parallelized. We now describe the algorithm in pseudo code.

function FRESPED(z, s, e, τ)
    if $e - s \le 1$ **then**
        STOP
    else
        $b_0 := \arg\max_b \mathfrak{C}_{s,b,e}(z)$
        if $\mathfrak{C}_{s,b,e}(z) > \tau$ **then**
            add b_0 to the set of estimated change-points $\mathbb{N}$
            FRESPED(z, s, b_0, τ)
            FRESPED($z, b_0 + 1, e, \tau$)

```
        else
            STOP
        end if
    end if
end function
```
This function is called for $z_l = \widehat{f}_{p,p}(t_v, \omega_l)$ and $z_l = \widehat{\rho}_{p,q}(t_v, \omega_l)$, $\forall p, q \neq p$.

A fast implementation of FreSpeD is available in the R package FreSpeD. This implementation also allows the user to specify several optionalities, such as the use of the median absolute deviation estimator in place of the sample standard deviation, which may be more robust in particular on short intervals $[s, e]$. Furthermore, FreSpeD has an optional step to diminish the risk of detecting a spurious change point: if at any stage of the algorithm we have a zero thresholded sum statistic in the immediate neighbourhood of a change-point candidate, i.e. $\min_{t=-\delta}^{\delta} \mathfrak{C}_{s,b_0+t,e}(z) = 0$, we ignore this point and set b_0 to be the argument of the next-largest thresholded sum statistic. This is continued until either $\min_{t=-\delta}^{\delta} \mathfrak{C}_{s,b_0+t,e}(z) \neq 0$ or no candidate points are left. Finally, if the user expects changes to occur simultaneously at all or a subset of autospectra and cross-coherences, a postprocessing step may be considered to merge the change-point estimates. In simulation studies, a simple merging approach taking the average change-point location of a cluster shows convincing results. Here a cluster is defined as a group of estimated change points that are less than the minimum change-point distance δ apart.

21.3.2.1   Comparison to the Other Approaches

The approach discussed here shares a number of characteristics with the other ones presented in this chapter. Generally, we consider a process that suddenly changes, as the switching model discussed in Section 21.4. Both can be used to detect seizure onset in multichannel EEG, but the results of our method are, in this application, more precise (compared to the physician's estimate). Unlike the switching model, our approach uses information from the spectral features of the data. The clustering method of Section 21.2 also uses these spectral data features, but they are extracted on pre-determined segments to evaluate the brain process prior to, during and post-seizure.

Just like the FreSpeD method, the SLEX method discussed in Section 21.5 can be used for the purpose of change-point detection in multichannel EEG data. It is computationally fast and able to handle long time series because like the FreSpeD method it does not require global optimization to segment the data. However, the SLEX method, has been designed to describe data through a unique, optimal basis, and therefore change-point estimates are a by-product and only approximate due to the dyadic segmentation approach.

The method of Gorrostieta et al. (30) (see Section 21.6) takes advantage of replicated trials and therefore does not require the smoothing applied by the FreSpeD method. It focuses fully on connectivity between brain regions and only uses information from autospectra indirectly. However, events such as seizure can hardly be recorded within multiple (i.i.d.) trials, which renders the method inappropriate for the analysis of spontaneous brain excitability.

In the application on seizure data, our method is different from all other methods discussed here in the following ways. First, we are able to identify two types of potential seizure precursors, namely (A) an instance directly prior to seizure onset, where changes are detected in the coherence between a range of EEG channels and (B) an instance several minutes prior to the seizure, where changes in the autospectra at channels close to the seizure focal point are detected. Both events were not detected by any of the methods presented here, nor by previous work on this data, to the best of our knowledge. Second, we are able to detect the seizure focal point as direct output of our method. The only other method able to localize seizure is the SLEX method using principal components. Third, the change points detected by FreSpeD are directly attributable to channels (for changes in the autospectra) or channel pairs (for coherence changes) and to frequencies, unlike parametric (e.g. VAR) methods and unlike methods using a dimension-reduction step (e.g. PCA).

21.3.3   Analysis of the Multichannel Seizure EEG Data

Our goal is to provide new insights in the dynamics of the preictal and ictal brain process using the FreSpeD method. We analyse a dataset from a live seizure recording as described in the chapter's introduction and focus on results related to the characterization of the seizure and the identification of potential precursors. We will analyse the data with respect to change points in the multivariate second-order structure, i.e. we consider both changes in the channel-specific autospectra as well as in channel-pair coherences. Regarding the spatial dimension of this analysis, it should be noted that the 10–20 system is to date commonly used in clinical settings, while neuroscientists frequently experiment with much finer spatial resolution such as high-density 256-channel EEG sensor arrays. The FreSpeD method is devised to handle the resulting curse of dimensionality, with the computational time being bounded by $O(P^2)$ and no memory-intense requirements of high-dimensional matrix inversion.

21.3.3.1   Seizure Localization

The identification of the seizure centre in time, space and frequency is of high importance to physicians considering measures such as lobe resection or lesionectomy (see e.g. 51). We start this analysis by discussing the spatial dimension. FreSpeD identifies 413 change points overall, out of which 105 are in the channel-specific autospectra and the remaining in pairwise coherences. The estimated change-point distribution can be used directly to identify the seizure focal point spatially: over the full recording, most change points are located in the autospectra of channels T3 and T5, respectively, each of which contains eight change points. Figure 21.11 illustrates where these changes are located in time and in how many frequency bands they can be identified. For a given change point, the interpretation of thresholded CUSUM statistics is not straightforward, as it depends on the number of frequency bands exceeding the threshold and the extent of change at each band. The number of frequency bands in which the energy level changes at a specific change point is a more robust and easily interpretable statistic, illustrating the frequency-width or spread of a change. Both channels show changes at and shortly after the seizure onset, but overall there are more frequency bands with energy changes in the autospectrum of channel T3, which is where the physician identified the seizure focal point. This data-adaptive seizure localization by FreSpeD using information from the time and frequency domains is potentially of great value to neuroscientists: an automated mechanism to support a physician's judgement diminishes the risk of human error in the visual inspection of EEG traces and is more time efficient.

21.3.3.2   Seizure Onset Estimation and Potential Precursors

The evolution of brain processes leading up to a seizure is of high interest in the context of seizure warning systems. Figure 21.12 shows the cumulative sum of detected change points over time, respectively, for autospectra, coherences, and total. We see that there are barely any changes detected in the preictal period, with the total number of change points rising sharply around 340 sec ($t = 34000$). This sudden increase corresponds with

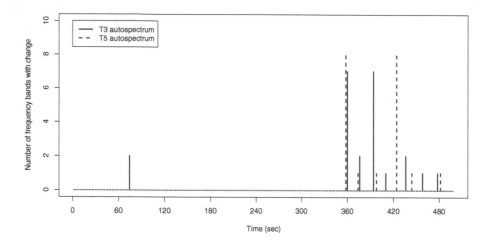

FIGURE 21.11
Change points in the autospectra of channels T3 and T5. The figure shows time (x-axis) versus the number of frequency bands, where energy changes are detected (y-axis).

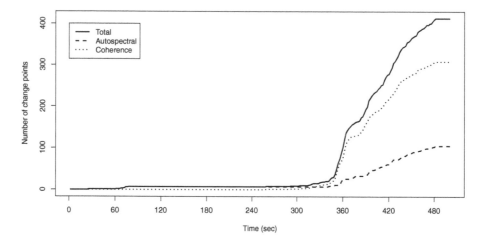

FIGURE 21.12
Cumulative sum of detected change points over all channels and/or channel pairs.

what the physician identified as seizure onset. This illustrates that FreSpeD can be used to data-adaptively identify not only the seizure focal point, but also the seizure onset.

With the aim of an improved understanding of changes in the spectral features of the brain process, we now focus on highlighting a number of findings regarding preictal EEG. However, it should be noted that seizure precursors can vary between different seizures of one patient and between patients. In the recording at hand, we find that in the 78 sec immediately before seizure onset ($t = 26000$–34000) FreSpeD detects 13 changes in the cross-coherence of several channel pairs and none in channel-specific autospectra. The coherence changes can be described as slowly building up, as changes are more clustered towards the end of this time interval and change magnitudes tend to increase from a fairly low level. Furthermore, the changes are evenly distributed over the brain with 15 channels being involved, 2 centrally located, 7 on the left and 6 on the right hemisphere. The changes are mostly visible in the theta and gamma bands. This is a newly detected pattern in a dataset that has received much attention in earlier works (e.g. 57, 18). It shows that even before the seizure onset, dependencies between brain regions change, both within the seizure area and with more distant regions. This would not be visually identifiable: physicians determine seizure onset by visual inspection of EEG traces and hence can only identify changes at single-channel variance and spectral energy. Thus, the detection of these types of change require a powerful and easily interpretable statistical procedure. The observation made here could potentially change the paradigm for seizure characterization that it is beyond abnormal local changes. It is worthwhile to point out that this observation is an immediate output of our method, as the detected changes have a direct interpretation with respect to frequency band and topographical location.

The FreSpeD method detects another group of preictal change points at the very start of the recording. Between these two clusters there are 182 sec in the recording where no change point is detected. Opposed to the second cluster of change points detected immediately before seizure onset, this early cluster consists of seven channel-specific changes in the autospectrum and contains no cross-coherence change. The detected changes all occur between 60 sec ($t = 6000$) and 78 sec ($t = 7800$) into our recording, apart from the very first change happening within the first 30 sec of recording. In light of the instability of the CUSUM statistic at the very borders of the interval and the small magnitude of the detected change, it can be debated if this first change point is spurious. However, just as each of the following six change points, it is detected in the lower end of the beta band, a frequency range where we detect overall relatively few change points (see Table 21.1). Within this cluster of six autospectra that quickly change sequentially, all channels are located in the back half of the brain and none in the frontal lobe. The largest change in terms of magnitude is detected in channel T5, whereas channel T3 is the only channel with a change detected also in the mid-range of the beta band. While the temporal distance to the seizure onset makes a direct link unlikely, the fact that the FreSpeD method independently picks up changes at six channels simultaneously, early in the second minute of the recording, reflects event-related changes in global brain activity that may warrant further investigation.

Looking at the occurrence of change points over frequencies (Table 21.1), most often we can identify changes in the energy and coherence at low frequency ranges, in particular in the theta band. While we identify comparably few changes at higher frequencies, the cumulative absolute magnitude is larger. This is in line with visual inspection and literature on seizure data analysis (77, 39): during normal brain states, energy concentration at high frequencies is low, but we observe a pronounced sudden increase at seizure onset. However, the observation of more, but smaller (and therefore possibly not visually detectable) change points at lower frequencies has not received much attention to date and is a new contribution of the FreSpeD method. It suggests that during seizure, low-frequency energy and coherence vary more often on a generally short range.

TABLE 21.1
Frequency-specific proportion of change points and change magnitude, in percent. Change magnitude is measured as the sum over thresholded CUSUM statistics, over time, frequency and EEG channels and channel pairs.

Frequency band (approx.)	Min. Hz	Max. Hz	Prop. of total change points		Prop. of change magnitude	
			Autospectra	Coherences	Autospectra	Coherences
Theta	5	10	58.1	32.2	54.8	51.3
Alpha	10	15	33.3	24.0	23.1	32.4
Beta	15	20	24.8	23.5	15.0	30.6
Beta	20	25	24.8	18.6	28.3	28.6
Beta	25	30	28.6	16.7	41.2	31.2
Gamma	30	35	30.5	18.4	50.5	36.9
Gamma	35	40	38.1	23.2	73.5	50.3
Gamma	40	45	41.0	24.0	80.6	55.6
Gamma	45	50	35.2	23.7	83.9	55.3

21.4 Modeling Time-Varying Connectivity Using Switching Vector Autoregressive Models

Effective connectivity in brain network analysis refers to the influence that one neuronal region exerts over another (27). This connectivity is a directed causal influence, which can be interpreted as the information flow between regions. In recent fMRI studies (see e.g. 10, 2, 46, 52, 43), it is reported that brain functional connectivity, instead of the traditionally assumed stationarity across time, can be highly dynamic even in the resting state. To estimate dynamic functional connectivity, these studies use the sliding window approach. However, one limitation of the sliding-window methods is that they are ineffective when there are abrupt changes. For these situations, Lindquist et al. (49) suggests that time-varying volatility models are successful in capturing these instantaneous changes in the undirected functional connectivity. In addition, recent evidence from fMRI suggests the time-varying functional connectivity tends to be clustered in recurring quasi-stable temporal regimes with relatively long periods, which reveals state-related dynamic behavior of the brain connectivity (2, 14, 4). Most of the above-mentioned studies only examined un-directed dynamic functional connectivity. Methods for studying dynamic effective connectivity are still limited. In this work, we consider the problem of quantifying the dynamic changes, both instantaneous and state related, in the effective connectivity.

In this section, we summarize the framework for estimating effective connectivity dynamics using non-stationary vector autoregressive (VAR) processes developed in Samdin et al. (64). This method was originally applied to fMRI data, but the results for EEG recordings of an epileptic seizure patient are similarly convincing. In this framework, we consider two types of VAR processes to model the latent directed connectivity dynamics in the highly non-stationary multichannel EEG signals: (1) the VAR model with time-varying parameters or time-varying VAR (TV-VAR), and (2) the VAR model with Markov-regime switching or switching VAR (SVAR). The TV-VAR and the SVAR models capture, respectively, the dynamic connectivity with instantaneous change at each time point, and that with regime shift between quasi-stationary states. In addition, the dynamic connectivity features are typically embedded in various physiological and instrumental noise with a very low signal-to-noise ratio (SNR). To address this problem, we employ the state-space representation for both VAR models, which allows for modeling of the underlying time-evolving

VAR connectivity structures as a latent process in the state equation, which is indirectly observed in the observation equation that accounts for the various noise sources in EEGs. Moreover, the state-space formulation enables efficient parameter estimation, smoothing and inference of the latent connectivity dynamics from the EEGs. Given the noisy EEGs, the hidden processes of TV-VAR coefficients and the dynamic regimes can be estimated recursively in time using the Kalman filter (KF), which is optimal in the mean-squares sense, and further refined by the Rauch–Tung–Striebel smoother (also called the Kalman smoother (KS)) based on the future observations. The maximum likelihood estimation of the state-space models can be accomplished by using the iterative expectation-maximization (EM) algorithm. The TV-VAR model with EM estimation and KF has been used in our recent work (63) for estimating time-varying directed coherence from EEG data. This differs from previous studies using TV-VAR for connectivity analysis of fMRI (34) and EEG (58), which depend on some sub-optimal heuristic methods for parameter selection.

The estimation framework consists of three stages. First, the TV-VAR coefficients are extracted by Kalman filter to capture the time-varying directed connectivity between EEGs. Second, the EEG time courses are partitioned into distinct temporal segments according a finite number of discrete, quasi-stationary states, based on the K-means clustering of the TV-VAR coefficients as features. Finally, a Markov-switching VAR model is used to estimate the transitions between these connectivity regimes. In this context, the data can be modelled as switching between different states, each state being characterized by a unique VAR process. The switching model is initialized using least-squares estimates of a stationary VAR for each regime, based on the K-means segmentation. The change points of the connectivity states are refined by the Kalman smoother, and the more accurate estimates of the VAR parameters for each state are obtained by the EM algorithm. Allen et al. (2) also proposed a framework for estimating a state-related dynamics pattern of functional connectivity with the transition states by using only sliding windows and K-means clustering, which gives sub-optimal connectivity estimates. A switching linear dynamic system (SLDS) was also used in Smith et al. (70) to estimate effective connectivity in fMRI. The limitation of this study is that the initial parameters need to be pre-specified manually. This is time consuming and does not guarantee optimal estimates, as the EM algorithm is known to be sensitive to initial parameter estimates and tends to be trapped in a local maximum. Our framework based on the K-means clustering of the TV-VAR features automatically and data-adaptively select the better initial parameters for the switching VAR model. This framework is important for a number of applications in neuroscience data, such as identifying the states that correspond to cognitive regime vs. resting state or differentiating seizure from non-seizure states.

21.4.1 Background on Vector Autoregressive (VAR) Models

21.4.1.1 Stationary VAR Model

Assume the multichannel EEG data of dimension P and length T can be modeled as a vector autoregressive process of order L (VAR(L)), $\mathbf{X}_t = [X_{1,t}, \ldots, X_{P,t}]', t = 1, \ldots, T$ with the representation

$$\mathbf{X}_t = \mathbf{\Phi}_1 \mathbf{X}_{t-1} + \ldots + \mathbf{\Phi}_L \mathbf{X}_{t-L} + \mathbf{v}_t, \qquad (21.9)$$

where $\mathbf{v}_t$ is a $P \times 1$ white noise with $\mathbf{x}(\mathbf{v}_t) = \mathbf{0}$ and $\mathbb{C}\mathrm{ov}(\mathbf{v}_t) = \mathbf{R}$ for all t. The VAR coefficient matrix $\mathbf{\Phi}_\ell = [\phi_{\ell ij}], 1 \leq i, j \leq P$ quantifies the effective connectivity at lag ℓ between different brain regions as measured by the EEGs. There exists a directed influence in the Granger-causality sense with direction from region j to region i for any connection strength $|\phi_{ij}| > 0$. This model has been used to infer the assumed time-invariant effective connectivity in fMRI data, by fitting on the entire time course (33, 75, 31, 72).

The model parameters will be estimated by formulating the model (21.9) as a multivariate linear model, $\mathbf{X} = \mathbf{U}\boldsymbol{\beta} + \mathbf{E}$, where $\mathbf{X} = [\mathbf{X}_1', \ldots, \mathbf{X}_T']'$ is the EEG time series with the dimension $T \times P$; $\mathbf{U}$ is the matrix consisting of previous observed data with the dimension $(T_L) \times PL$ matrix

$$
\mathbf{U} = \begin{pmatrix}
\mathbf{X}_L' & \mathbf{X}_{L-1}' & \cdots & \mathbf{X}_1' \\
\mathbf{X}_{L+1}' & \mathbf{X}_L' & \cdots & \mathbf{X}_2' \\
\vdots & \vdots & \cdots & \vdots \\
\mathbf{X}_{T-1}' & \mathbf{X}_{T-2}' & \cdots & \mathbf{X}_{T-L}'
\end{pmatrix},
$$

$\boldsymbol{\beta} = [\boldsymbol{\Phi}_1', \ldots, \boldsymbol{\Phi}_L']'$ is $PL \times P$ and contains the coefficients of all the VAR matrices and $\mathbf{E} = [\mathbf{v}_1', \ldots, \mathbf{v}_T']'$ is the collection of noise terms at all channels and time points. The conditional least squares (LS) estimators of $\boldsymbol{\beta}$ and $\mathbf{R}$ given time points $t = L+1, \ldots, T$ are defined respectively as $\hat{\boldsymbol{\beta}} = (\mathbf{U}'\mathbf{U})^{-1}\mathbf{U}'\mathbf{X}$ and $\hat{\mathbf{R}} = (1/(T-L))(\mathbf{X} - \mathbf{U}\hat{\boldsymbol{\beta}})'(\mathbf{X} - \mathbf{U}\hat{\boldsymbol{\beta}})$.

We can compute the directed coherence (DC), the linear dependence between signals at specific frequency, ω, from the estimated VAR parameters. Let us denote the cross-spectral density matrix of $\mathbf{X}_t$ by $\mathbf{f}(\omega) = [f_{ij}(\omega)], 1 \leq i, j \leq P$, for $\omega \in [0, 2\pi]$, whose ij-th element is the cross-spectrum between the component i and j of $\mathbf{X}_t$. Then, under a VAR process, the matrix $\mathbf{f}(\omega)$ admits a unique factorization $\mathbf{f}(\omega) = \mathbf{H}(\omega)\mathbf{R}\mathbf{H}(\omega)^*$ where $(.)^*$ denotes the conjugate transpose, and $\mathbf{H}(\omega)$ is the transfer function defined by the inverse of the Fourier transform of the coefficient matrices

$$
\mathbf{H}(\omega) = \left(\mathbf{I} - \sum_{\ell=1}^{L} \boldsymbol{\Phi}_\ell \mathrm{e}^{-i\ell\omega} \right)^{-1}.
$$

Then, the DC is defined as the normalized cross-spectrum

$$
\mathrm{DC}_{ij}(\omega) = \frac{\left| f_{ij}(\omega) \right|^2}{f_{ii}(\omega) f_{jj}(\omega)} \tag{21.10}
$$

with $\mathrm{DC}_{ij}(\omega) \in [0, 1]$. A value close to one indicates strong directed influence from the brain region j to i, at frequency ω.

21.4.1.2 Time-Varying VAR Model

We extend the static VAR model from the previous section into a time-varying VAR (TV-VAR) model where the VAR coefficient matrices are allowed to change with time in order to capture the dynamic structure of effective connectivity. The TV-VAR model is formulated as

$$
\mathbf{X}_t = \sum_{\ell=1}^{L} \boldsymbol{\Phi}_{\ell t} \mathbf{X}_{t-\ell} + \mathbf{v}_t \tag{21.11}
$$

where $\{\boldsymbol{\Phi}_{\ell t}, \ell = 1, \ldots, L\}$ are $P \times P$ matrices of TV-VAR coefficients at lag ℓ and time t and $\mathbf{v}_t$ is a $P \times 1$ white Gaussian observational noise with zero mean and covariance matrix $\mathbf{R}$. By defining $\mathbf{a}_t = vec([\boldsymbol{\Phi}_{1t}, \ldots, \boldsymbol{\Phi}_{Lt}])$, the $PL \times 1$ state vector of TV-VAR coefficients at time point t, re-arranged from the matrices $\{\boldsymbol{\Phi}_{\ell t}\}_{\ell=1}^{L}$, the TV-VAR model can be written in state-space form, analogously to the VAR model, as in (3)

$$
\mathbf{a}_t = \mathbf{a}_{t-1} + \mathbf{w}_t \tag{21.12}
$$
$$
\mathbf{X}_t = \mathbf{C}_t \mathbf{a}_t + \mathbf{v}_t \tag{21.13}
$$

with $\mathbf{C}_t = \mathbf{I}_P \otimes \mathbf{Y}_t'$ where $\mathbf{I}_P$ is a $P \times P$ identity matrix, $\mathbf{Y}_t = [\mathbf{X}_{t-1}', \ldots, \mathbf{X}_{t-L}']'$, and $\otimes$ denotes the Kronecker product. The TV-VAR process (21.11) is rewritten in a compact form (21.13) as the observation equation, with a linear mapping $\mathbf{C}_t$ consisting of the past observations. In the state equation (21.12), the hidden state $\mathbf{a}_t$ is assumed to follow a first-order Gauss–Markov process and $\mathbf{w}_t$ is a Gaussian state noise with mean zero and $LP^2 \times LP^2$ covariance matrix $\mathbf{Q}$. Both $\mathbf{R}$ and $\mathbf{Q}$ are assumed to be time-invariant. We denote by $\boldsymbol{\theta} = (\mathbf{R}, \mathbf{Q})$ the model parameters of the TV-VAR state-space model.

21.4.1.3 Switching VAR (SVAR) Model

The Markov-switching VAR (SVAR) is a quasi-stationary model consisting of S independent underlying processes, where each is characterized completely by a unique VAR model, subject to Markovian regime shift.

$$\mathbf{X}_t = \sum_{\ell=1}^{L} \boldsymbol{\Phi}_{[S_t],\ell} \mathbf{X}_{t-\ell} + \mathbf{v}_t \tag{21.14}$$

where $S_t \in \{1, \ldots, S\}$ is the discrete latent state variable as a model indicator. The regimes of distinct VAR structures change over time according to the switch variables, $S_t, t = 1, \ldots, T$, which follow a hidden Markov chain with transition matrix $Z = [z_{ij}], 1 \le i, j \le S$ where

$$z_{ij} = P(S_t = j | S_{t-1} = i) \tag{21.15}$$

denotes the probability of transition from state i to j. At each time point t, only one latent process (and hence only one VAR process) is "active" (or turned on). The remaining are turned off. One goal here is to determine the regimes where each state is active.

The SVAR model of order L, SVAR(L) in (21.14) can be written in a switching linear Gaussian state-space form as follows

$$\mathbf{X}_t = \mathbf{A}_{[S_t]} \mathbf{X}_{t-1} + \mathbf{w}_t \tag{21.16}$$
$$\mathbf{Y}_t = \mathbf{H} \mathbf{X}_t + \mathbf{v}_t \tag{21.17}$$

where the SVAR(L) signal $\mathbf{X}_t$ is re-written as an underlying process in the form of VAR(1) in the state equation, where $\mathbf{A}_{[S_t]}$ is a $PL \times PL$ lag-one coefficient matrix indexed by the S_t and with the following structure

$$\mathbf{A} = \begin{pmatrix} \boldsymbol{\Phi}_1 & \boldsymbol{\Phi}_2 & \cdots & \boldsymbol{\Phi}_{L-1} & \boldsymbol{\Phi}_L \\ \mathbf{I}_P & \mathbf{0} & \cdots & \mathbf{0} & \mathbf{0} \\ \mathbf{0} & \mathbf{I}_P & \cdots & \mathbf{0} & \mathbf{0} \\ \vdots & & \ddots & & \vdots \\ \mathbf{0} & \mathbf{0} & \cdots & \mathbf{I}_P & \mathbf{0} \end{pmatrix}$$

and $\mathbf{w}_t = [\boldsymbol{\eta}_t', \mathbf{0}, \ldots, \mathbf{0}]'$ is the $PL \times 1$ state noise. The latent process $\mathbf{X}_t$ is indirectly observed in various noise sources modeled by $\mathbf{v}_t$ as noisy observations $\mathbf{Y}_t$ in the observation equation with $\mathbf{H} = [\mathbf{I}_P, \mathbf{0}, \ldots, \mathbf{0}]$

Both observation noise and the state noise, $\mathbf{v}_t$ and $\mathbf{w}_t$, are assumed as i.i.d. Gaussian processes, i.e. $\mathbf{v}_t \sim N(\mathbf{0}, \mathbf{R}_{[S_t]})$ and $\mathbf{w}_t \sim N(\mathbf{0}, \mathbf{Q}_{[S_t]})$ respectively, with the observation and state noise covariance matrices, $\mathbf{R}_{[S_t]}$ and $\mathbf{Q}_{[S_t]}$, switching with the state variables S_t. Since the Gaussian distribution varies between the S regimes, the full model is the approximation of a mixture of Gaussian distributions. Here, the regime dynamics are piecewise linear. The matrices $\mathbf{A}_j$, $j = 1, \ldots, S$ describe effective connectivity that varies across states. Instead of hard alignment of each time point t to the corresponding state, we can evaluate the probability of activation for each state, $P(S_t = j | \mathbf{Y}_{1:T})$, which is termed "soft-switching." In this model, we set the matrix $\mathbf{H}$ to be the identity matrix. We denote all model parameters as $\boldsymbol{\Theta} = \{(\mathbf{A}_j, \mathbf{Q}_j, \mathbf{R}_j), j = 1, \ldots, S\}$.

21.4.2 Parameter Estimation

In this section, we describe the parameter estimation of the developed TV-VAR and SVAR state-space models, which involve sequential estimation of the hidden state parameters as well as the model parameters. Given the observations $\mathbf{X}_t, t = 1, \ldots, T$, the latent state vectors of the TV-VAR coefficients $\mathbf{a}_t, t = 1, \ldots, T$ in the TV-VAR model (21.12) and (21.13) can be estimated sequentially in time using the Kalman filter. The maximum likelihood estimation of $\boldsymbol{\theta}$ can be performed using the iterative expectation-maximization (EM) algorithm (see e.g. 63, 67, 28).

Similarly for the SVAR state-space model, the aims are to estimate the underlying denoised EEG signals $\mathbf{X}_t$ and the latent switch variable S_t. This involves estimating, sequentially in time, the filtered densities or probabilities $p(\mathbf{X}_t|\mathbf{Y}_{1:t})$ and $P(S_t|\mathbf{Y}_{1:t})$ given the noisy EEG observations up to time t, $\mathbf{Y}_{1:t} = \{\mathbf{Y}_1, \ldots, \mathbf{Y}_t\}$, and the more accurate smoothed densities $p(\mathbf{X}_t|\mathbf{Y}_{1:T})$ and $P(S_t|\mathbf{Y}_{1:T})$ given the available future observations $\mathbf{Y}_{1:T} = \{\mathbf{Y}_1, \ldots, \mathbf{Y}_T\}$. We estimate the filtered and smoothed densities of $\mathbf{X}_t$ given state j at time t, by the Kalman filter and Kalman smoother, respectively,

$$\mathbf{X}_{t|t}^j = \mathbf{x}[\mathbf{X}_t|\mathbf{Y}_{1:t}, S_t = j] \tag{21.18}$$

$$V_{t|t}^j = \mathbb{C}\text{ov}[\mathbf{X}_t|\mathbf{Y}_{1:t}, S_t = j] \tag{21.19}$$

$$\mathbf{X}_{t|T}^j = \mathbf{x}[\mathbf{X}_t|\mathbf{Y}_{1:T}, S_t = j] \tag{21.20}$$

$$V_{t|T}^j = \mathbb{C}\text{ov}[\mathbf{X}_t|\mathbf{Y}_{1:T}, S_t = j] \tag{21.21}$$

$$V_{t,t-1|T}^j = \mathbb{C}\text{ov}[\mathbf{X}_t, \mathbf{X}_{t-1}|\mathbf{Y}_{1:T}, S_t = j] \tag{21.22}$$

where $\mathbf{X}_{t|t}^j$ and $V_{t|t}^j$ are the mean and covariance of the filtered density $p(\mathbf{X}_t|\mathbf{Y}_{1:t}, S_t = j)$; $\mathbf{X}_{t|T}^j$ and $V_{t|T}^j$ are the mean and covariance of the smoothed density $p(\mathbf{X}_t|\mathbf{Y}_{1:T}, S_t = j)$ given state j at time t. $V_{t,t-1|T}^j$ is the cross-variance of joint density $p(\mathbf{X}_t, \mathbf{X}_{t-1}|\mathbf{Y}_{1:T}, S_t = j)$ The estimates of filtered and smoothed state occupancy probability of being state j at time t are also computed

$$\mathbf{M}_{t|t}^j = P(S_t = j|\mathbf{Y}_{1:t}) \tag{21.23}$$

$$\mathbf{M}_{t|T}^j = P(S_t = j|\mathbf{Y}_{1:T}). \tag{21.24}$$

We use a extension of the EM algorithm suggested by Murphy (53) to estimate the model parameters $\boldsymbol{\Theta}$ by maximizing the model likelihood $L(\mathbf{Y}_{1:T}|\boldsymbol{\Theta}) = p(\mathbf{X}_{1:T}|\mathbf{Y}_{1:T}, \boldsymbol{\Theta})[\log p(\mathbf{X}_{1:T}|\mathbf{Y}_{1:T}, \boldsymbol{\Theta})]$. In the expectation step (E-step), the sufficient statistics are obtained from the smoothed estimates

$$P_t = \hat{E}[\mathbf{X}_t\mathbf{X}_t'] = V_{t|T} + \mathbf{X}_{t|T}\mathbf{X}_{t|T}' \tag{21.25}$$

$$P_{t,t-1} = \hat{E}[\mathbf{X}_t\mathbf{X}_{t-1}'] = V_{t,t-1|T} + \mathbf{X}_{t|T}\mathbf{X}_{t-1|T}' \tag{21.26}$$

where $\mathbf{X}_{t|T}$, $V_{t|T}$ and $V_{t,t-1|T}$ are quantities of the smoothed densities $p(\mathbf{X}_t|\mathbf{Y}_{1:T})$ and $p(\mathbf{X}_t, \mathbf{X}_{t-1}|\mathbf{Y}_{1:T})$, corresponding to 21.20 to 21.22 by marginalizing out the state variable j of the $p(\mathbf{X}_t|\mathbf{Y}_{1:T}, S_t = j)$ and $p(\mathbf{X}_t, \mathbf{X}_{t-1}|\mathbf{Y}_{1:T}, S_t = j)$ using Gaussian approximation. We refer the KF/FS approach for estimating the state parameters of the switching model respectively as switching KF (SKF) and switching KS (SKS), as in Murphy (53).

In the maximization step (M-step), the model parameters for regime j are updated as follows

$$\hat{\mathbf{A}}_j \quad = \quad \left(\sum_{t=2}^{T} W_t^j P_{t,t-1} \right) \left(\sum_{t=2}^{T} W_t^j P_{t-1} \right)^{-1} \tag{21.27}$$

$$\hat{\mathbf{Q}}_j \quad = \quad \left(\frac{1}{\sum_{t=2}^{T} W_t^j} \right) \left(\sum_{t=2}^{T} W_t^j P_t - \hat{\mathbf{A}}_j \sum_{t=2}^{T} W_t^j P'_{t,t-1} \right) \tag{21.28}$$

$$\hat{z}_{ij} \quad = \quad \frac{\sum_{t=2}^{T} P(S_{t-1} = j, S_t = i | \mathbf{Y}_{1:T})}{\sum_{t=1}^{T-1} W_t^j} \tag{21.29}$$

where the weights $W_t^j = \mathbf{M}_{t|T}^j$ are computed from the smoothing step. Here, we assume the number of states is known, and the observation noise covariance $\mathbf{R}_j$ is initialized from a distinct stationary VAR model for each regime and not updated by the EM algorithm.

21.4.3   Estimating Dynamic Connectivity States in Epileptic EEG

We apply our non-stationary VAR model framework to analyzing state-related changes in effective brain connectivity that can be associated with epileptic seizure recorded in EEG data. The onset and the duration of the seizure is unknown a priori. Our data-driven approach simultaneously allows for an automatic partitioning of the epileptic event in the EEG signals at different channels into distinct dynamic regimes, as well as estimation of a set of directed dependencies between the channels for each partition. The overview of the proposed framework is shown in Figure 21.13. First, we use Kalman filtering to extract adaptively a sequence of TV-VAR coefficients in (21.12)–(21.13) from the EEG channels as features to capture the directed brain connectivity over time. Second, the EEG time series are partitioned into time intervals corresponding to seizure-free and seizure states. This is based on K-means clustering of the time-varying VAR connectivity matrices, with $K = 2$ assumed known. In this analysis, we assume that there are only two states: seizure and non-seizure. We understand that there are possible micro-states even within both seizure and non-seizure states. The determination of the number of states in an epileptic event will be further investigated in future works. Finally, we employ a two-state Markov-switching VAR model (21.16)–(21.17) that captures the dynamics of connectivity structures for the normal and seizure brain states, with distinct state-dependent VAR models switching between the two states according to a hidden Markov chain.

The autoregressive coefficient matrices of the SVAR model for each state are initialized by the LS estimates $\widehat{\boldsymbol{\Phi}}_{[j]}^{\text{LS-KM}}, j = 1, 2$ of two stationary VAR models fitted separately on the seizure and non-seizure EEG segments resulting from the K-means partitioning. Given the initialized model, the regime partitions are refined using the switching KF which computes the latent state sequence $\widehat{S}_t^{\text{SKF}}$, indicating which brain connectivity states are most likely to occur at each time point. The time state alignment is then further adjusted using the switching KS, which estimates $\widehat{S}_t^{\text{SKS}}$ based on the observations of the entire time course. The SVAR parameters are re-estimated from the EEG signals iteratively using the EM algorithm (21.25)–(21.29). We use the EM updates in (21.27) for the VAR coefficient matrices, denoted by $\widehat{\boldsymbol{\Phi}}_{[j]}^{\text{EM}}$, to infer the directed brain connectivity for each state. The estimation framework is summarized in Algorithm 1. Simulation results with a VAR process with 2 regimes in Samdin et al. (64) showed the superior performance of EM-estimated SVAR model-based estimators in producing the most accurate state assignments and lowest errors in the VAR parameter estimates for each regime, compared to that based on the K-means clustering only.

Algorithm 1: An Estimation Framework for State-Dependent Effective Connectivity Changes

---

Input: P-channel EEG signals $\mathbf{Y}_t, t = 1, \ldots, T$ and number of states S.

Step 1: <u>Extraction of TV-VAR Features</u>

- Given model $\boldsymbol{\theta}$ (21.12)–(21.13), compute the TV-VAR coefficient $\widehat{\mathbf{a}}_t = \mathbf{X}[\mathbf{a}_t|\mathbf{Y}_{1:t}]$ by KF.

Step 2: <u>Initial Estimation of Connectivity Regimes by Clustering the TV-VAR Coefficients</u>

- Partition $\{\mathbf{Y}_1, \ldots, \mathbf{Y}_T\}$ into S distinct clusters $\mathbf{C} \in \{C_1, \ldots, C_S\}$ using K-means algorithm

$$\arg\min_{\mathbf{C}} \sum_{j=1}^{S} \sum_{\mathbf{Y}_t \in C_j} \left\| \widehat{\mathbf{a}}_t - \boldsymbol{\mu}_j \right\|^2, \text{ where } \boldsymbol{\mu}_j \text{ is the centroid of } C_j.$$

- Generate state sequence $\widehat{S}_t^{\mathrm{KM}} = \arg\min_{j \in \{1, \ldots, S\}} \left\| \widehat{\mathbf{a}}_t - \boldsymbol{\mu}_j \right\|^2, t = 1, \ldots, T.$

- Compute $\widehat{\boldsymbol{\Phi}}_{[j]}^{\mathrm{LS\text{-}KM}}$ by LS-fitting a VAR to time segments $C_j = \{\mathbf{Y}_t : \widehat{S}_t^{\mathrm{KM}} = j\}$ for state $j = 1, \ldots, S.$

Step 3: <u>Estimation of SVAR by EM Algorithm</u>

- At iteration $r = 0$, initialize $\widehat{\boldsymbol{\Theta}}^{(0)}$ based on estimates from Step 2.

- **EM Iteration**, for $r \geq 1$,

  - E-Step: Given $\widehat{\boldsymbol{\Theta}}^{(r)}$,
    * Compute filtered densities $p(\mathbf{X}_t|\mathbf{Y}_{1:t})$ and $P(S_t|\mathbf{Y}_{1:t})$ by SKF.
    * Compute smoothed densities $p(\mathbf{X}_t|\mathbf{Y}_{1:T})$ and $P(S_t|\mathbf{Y}_{1:T})$ by SKS.
    * Compute the expectations by (21.25)–(21.26).

  - M-Step: Re-estimate $\widehat{\boldsymbol{\Theta}}^{(r+1)}$ using (21.27)–(21.29).

- **Until** convergence, return $\widehat{\boldsymbol{\Theta}}^* = \widehat{\boldsymbol{\Theta}}^{(r)}$.

- **Output:**

  - Refined state sequence given $\widehat{\boldsymbol{\Theta}}^*$

    $$\widehat{S}_t^{\mathrm{SKF}} = \arg\max_{j \in \{1, \ldots, S\}} P(S_t|\mathbf{Y}_{1:t}), \text{ and } \widehat{S}_t^{\mathrm{SKS}} = \arg\max_{j \in \{1, \ldots, S\}} P(S_t|\mathbf{Y}_{1:T}),$$
    $$t = 1, \ldots, T.$$

  - Improved connectivity estimates $\widehat{\boldsymbol{\Phi}}_{[j]}^{\mathrm{EM}}$ from the EM update $\widehat{\mathbf{A}}_j^*$, for $j = 1, \ldots, S.$

We evaluate the framework for estimating dynamic directed connectivity between the data recorded at the left parietal channel P3 and the temporal channels T3 and T4 before and during an epileptic seizure. We used SVAR models of order three, as selected using the Schwarz's Bayesian criterion for this data. Figure 21.14(a) shows the TV-VAR coefficient matrices (vectorized) at different time lags, estimated by the Kalman filter from the three EEG channels. The results suggest that the directed connectivity between different brain

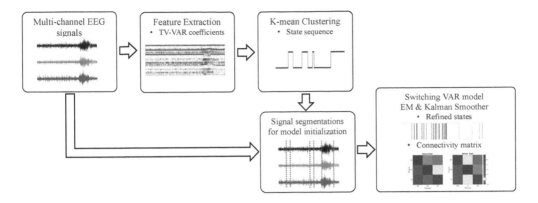

FIGURE 21.13

A framework for estimating quasi-stationary dynamic effective connectivity states in multi-channel EEG signals, based on TV-VAR and SVAR processes.

regions exhibits changes over time, instead of frequently assumed stationarity (33, 75). The non-stationarity of coherence between these channels has also been revealed using the SLEX method (Figure 5.5). However, the TV-VAR features provide additional insights on the direction of information flows between the brain regions. The estimates also indicate strongest connections at lag 1, followed by lags 2 and 3. Despite the presence of changes in the finer time scale (here with the resolution of 0.01 seconds), the directed connectivity tends to be clustered to quasi-stationary discrete states (normal and seizure) for long periods (e.g. the seizure event has a time span of about one minute as in Figure 21.14(c)), with relatively slow connectivity dynamics within each state. This is indicated by the slower variations in the estimated TV-VAR coefficients within each regime (normal: 0–350 sec and 450–500 sec; seizure 350–450 sec), compared to the rapid transition between these regimes at 350 sec. Besides, strengthened and more abrupt changes in connectivity can also be clearly seen during the seizure event with an onset around 350 sec, compared to the normal brain activity. This implies a regime shift in the directional dependence structure in transition between the normal and seizure state.

Figure 21.14(b) shows the inferred brain states at each point in time using different approaches within the discussed framework of TV-VAR feature extraction, K-means clustering and the SVAR model. The data is fitted assuming two states which can be interpreted as "normal" (blue) and "seizure" (red) state. Both switching KF and switching KS under the EM-updated model provide better tracking of the connectivity state dynamics, in terms of more accurate state assignments. In contrast, using only K-means clustering based on the TV-VAR coefficients yields substantial misclassification, particularly in the non-seizure intervals, because of a low signal-to-noise ratio in the time series of VAR coefficients. Moreover, the switching models are able to more precisely localize the change points in connectivity at the seizure onset, compared to the K-means clustering with a imprecise delayed detection of the onset. The improved fit can be due to the more accurate model parameter estimates, updated iteratively by the EM algorithm, and the refined temporal boundaries of the states resulting from the Kalman smoothing based on both past and future observations. Among the switching model estimates, we can see that using the SKS improves the state estimates by smoothing of some spurious estimates by the SKF. Figure 21.14(c) shows the EEG data overlaid by the estimated states by the SKS, which demonstrates the reasonably well detected temporal change points between states. This suggests that the seizure onset can be characterized not only by changes in channel-specific auto-spectrum and un-directed coherence (as detected by the FreSpeD method in Figure 3.1 and 3.2), but also in the directionality of the connections between channels. The figure also reveals detected change points in directed connectivity even before the seizure onset, as also reported in the coherence using the FreSpeD.

Figure 21.15 shows the inferred directed connectivity matrices at three lags between the three channels for the normal and seizure states. Note that the connectivity structures are substantially different between the two states, and across time lags. At lag 1, we found strong directed influences from both P3 and T4 to T3 in the seizure state, which are absent in the normal state. Inversely, during the normal state, the directed connections detected are from T3 to both P3 and T4, which are absent during the seizure. This implies a sharp contrast between the normal and seizure brain activity based on the directionality of the connections, with information inflows to T3 during the seizure, compared to the outflows from T3 in the normal brain. Besides, many connections exhibit change of signs, from positive to negative correlations and vice verse, comparing the two states. The largest number of bi-directional connections are observed for lag 3, which decreases at the shorter lags (i.e. less in lag 2 followed by lag 1).

To further analyze the directed connectivity in specific frequencies, the directed coherence for each brain state were computed respectively from the VAR coefficient matrices. The results are shown in Figure 21.16. The directed information flows between the channels occurred mostly within the alpha band (8–12 Hz) and beta frequency band (12–30 Hz), and slightly at the delta frequency band (1–4 Hz), a finding consistent with the undirected coherence analysis using the SLEX model on the same data (see Section 21.5.3). Most of the directed interactions are concentrated at the alpha band during the epileptic seizure state, and at the beta band during the non-ictal (normal) brain state. The decrease in directed coherence at the beta band during seizure are clearly apparent compared to the normal state. Besides, we clearly observed, at the seizure state, an enhanced uni-directed coherence at the alpha band, from both T3 and P3 (channels at the left hemisphere of the brain) to T4 (channel at the right hemisphere) (T3→T4; P3→T4), compared to the opposite directions from T4 to T3 and P3 (T4→T3; T4→P3). The directed influence from T3 to P3 (T3→P3) is also stronger than that from P3 to T3 (P3→T3). This suggests the propagation of seizure activity across the brain hemispheres at alpha oscillations from the left temporal to the right temporal lobe, with T3 being the source of influences to both T4 and P3.

We have presented a unified framework based on the SVAR model and clustering of TV-VAR features, for identifying regime changes in effective connectivity structure, induced by a finite number of latent switching brain states, which is potentially important to neuroscience studies. The method is capable of detecting simultaneously both the abrupt switches between connectivity regimes and the slow variations within the regime, and provides measures of dependence with directionality. Application to multi-channel epileptic EEGs show the ability of the procedure to segment the ictal and non-ictal periods, and reveal abrupt pre-seizure activities. However, there are few limitations of the framework that we will address in future work. First, the number of brain states is assumed known up to a coarse approximation (possibly via neurologists' assessment or restrictive experimental settings). However, it is often undetermined a priori, particularly in situations where the mental activities are unconstrained such as during the resting state, which probably needs to be estimated from the data. Second, the VAR orders are supposed fixed across distinct regimes. As an extension, one could allow for varying orders for different regimes, and derive a procedure to perform order selection for each regime.

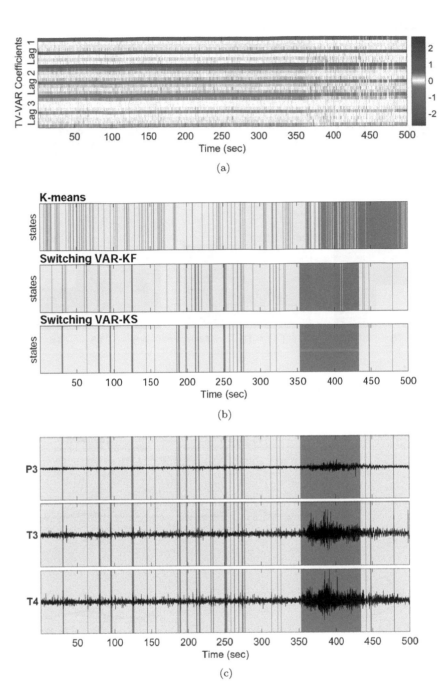

FIGURE 21.14
Estimation of state-related dynamics of directed connectivity between the channels P3, T3 and T4. (a) Estimated TV-VAR coefficient matrices (each vectorized with dimension $P^2 \times L = 3^2 \times 3 = 27$) using Kalman filtering at different time lags $\ell = 1, 2, 3$. (b) Inferred states (blue:normal; red:seizure) at each time point, for different stages of our general framework combining K-means clustering of the TV-VAR coefficients, $\widehat{S}_t^{KM}$ (top), Switching KF, $\widehat{S}_t^{SKF}$ (middle) and switching KS, $\widehat{S}_t^{SKS}$ (bottom) based on a EM-estimated two-state SVAR(3) model. (c) EEG data overlaid by final estimated states.

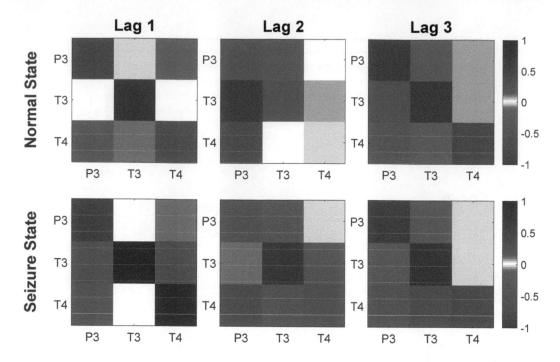

FIGURE 21.15
Estimated VAR connectivity matrices (at three lags) between the three epileptic EEG channels, $\widehat{\boldsymbol{\Phi}}_{[j],\ell}^{\mathrm{EM}}, \ell = 1, 2, 3$, for the non-ictal and the ictal brain state, $j = 1, 2$.

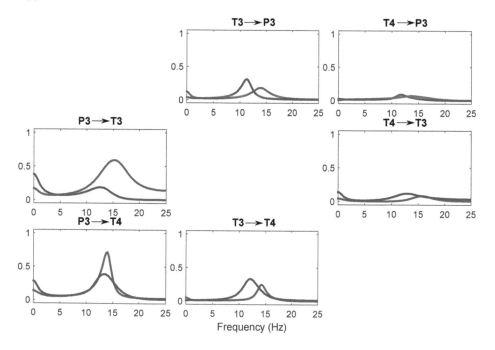

FIGURE 21.16
Estimates of directed coherence between EEGs for the normal (blue) and the epileptic (red) brain state, computed from the estimated VAR parameters in Figure 21.15.

21.5 Best Signal Representation for Non-Stationary EEGs

Epileptic seizure EEG data displays non-stationary behavior (see Figure 21.1). The variance of the EEG (or wave amplitude) evolves over time. Moreover, when one zooms into localized time segments, it becomes apparent that the decomposition of oscillatory waveforms also changes. Classical Fourier analysis is not adequate for studying such signals. In this section, we shall analyze these brain signals using the SLEX (smooth localized complex exponential) library. The SLEX library consists of several bases; each basis is composed of orthonormal localized Fourier waveforms. To estimate the evolutionary spectra at each channel and the evolutionary coherence between a pair of channels, our approach is to select a basis from the SLEX library that gives the best representation for the non-stationary EEG signal.

21.5.1 Overview of Signal Representations

To motivate this approach, we consider the classical representations of time series. Let $\mathbf{X}(t) = [X_1(t), \ldots, X_P(t)]'$ be a zero-mean P-channel EEG signal. Under stationarity, $\mathbf{X}(t)$ can be expressed as a linear combination of sine and cosine waveforms with random coefficients (amplitudes and phases) via the Cramér (spectral) representation

$$\mathbf{X}(t) = \int_{-1/2}^{1/2} \mathbf{A}(\omega) \, \exp(\, i2\pi\omega t) d\mathbf{Z}(\omega) \qquad (21.30)$$

where $\mathbf{A}(\omega)$ is the transfer function matrix of dimension $P \times P$ and $d\mathbf{Z}(\omega)$ is a P-dimensional zero mean orthonormal increment random process, i.e., $\mathbb{Cov}[d\mathbf{Z}(\omega), d\mathbf{Z}(\lambda)] = \mathbf{0}$ when $\omega \neq \lambda$ and $\mathbb{Var}[d\mathbf{Z}(\omega)] = \mathrm{Id}\, d\omega$ where Id is the $P \times P$ identity matrix. The spectral matrix of $\mathbf{X}(t)$ is $\mathbf{f}(\omega) = \mathbf{A}(\omega)\mathbf{A}^*(\omega)$ which is a $P \times P$ complex-valued Hermitian. The Cramér representation uses the Fourier waveforms as building blocks. The transfer function does not vary across time and therefore the spectral matrix and all spectral quantities change over frequency but remain constant in time.

The classical representation is not adequate for processes whose spectra evolve with time. Dahlhaus (16) developed a generalized representation that also uses the Fourier waveforms as building blocks but allows the transfer function to evolve over time. To simplify ideas, we use its approximate representation

$$\mathbf{X}_T(t) \approx \int_{-1/2}^{1/2} \mathbf{A}(t/T, \omega) \, \exp(\, i2\pi\omega t)d\mathbf{Z}(\omega) \qquad (21.31)$$

where $\mathbf{A}(t/T, \omega)$ is the transfer function matrix that is defined on rescaled time t/T. A special case of the above Dahlhaus representation is the change-point model (piecewise quasi-stationary) in Equation (21.2). The spectral matrix, defined on rescaled time $u \in [0, 1]$ and frequency $\omega \in (-0.5, 0.5)$, is $\mathbf{f}(u, \omega) = \mathbf{A}(u, \omega)\mathbf{A}^*(u, \omega)$. The Dahlhaus model provides an asymptotic framework under which one can establish consistency of the estimator for the time-varying spectral matrix.

This section shall introduce a complementary approach that utilizes the library of *localized* Fourier waveforms. The remainder of this section is organized as follows. The basic ideas on the SLEX waveforms and transform is first discussed, followed by the method for fitting the SLEX model and estimating the time-varying spectral properties.

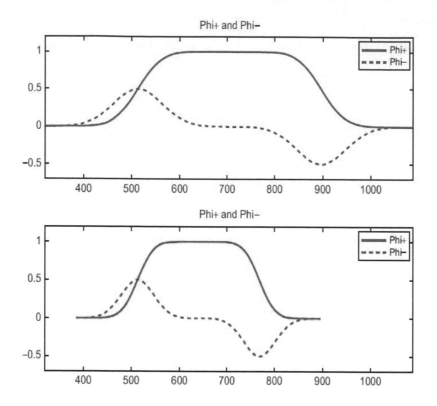

FIGURE 21.17

Smooth window pairs $\Psi_{+,B}(u)$ and $\Psi_{-,\omega,B}(u)$. These windows can be stretched or compressed. In the top picture, B is approximately the rescaled interval $(\frac{500}{1000}, \frac{900}{1000})$; in the bottom picture, B is approximately $(\frac{500}{1000}, \frac{750}{1000})$.

21.5.2   Overview of SLEX Analysis

The SLEX waveforms. While many localized and orthonormal basis functions can be used to analyze non-stationary time series (e.g., wavelets and wavelet packets), there is a strong rationale for using time-localized generalizations of the *Fourier* waveforms. To analyze non-stationary EEG, we propose to use the SLEX (**S**mooth **L**ocalized Complex **EX**ponential) waveforms $\phi_\omega(u) = \Psi_+(u)\exp(i2\pi\omega u) + \Psi_-(u)\exp(-i2\pi\omega u)$ where $\omega \in (-1/2, 1/2)$; and $u \in \mathcal{I} = [-\eta, 1+\eta]$, $0 < \eta < 0.5$. The windows, plotted in Figure 21.17, come in pairs so that when Ψ_+ is specified, the other window Ψ_- is determined. Moreover, these windows can be compressed or dilated depending on how long the EEG remains stationary in a block. Plots of the SLEX waveforms are given in Figure 21.18. The SLEX waveforms, being complex-valued, are potentially useful for representing non-stationary EEG because they naturally capture the time-lag structure between components of a multichannel EEG via the phases of the time-varying cross spectra. Note that if the two EEG components X_p and X_q are related by $X_p(t) = X_q(t-\ell)$, then the phase of the cross-spectrum between these two depends linearly on the lag ℓ. However, for non-stationary signals, this lead–lag relationship could vary over time, which can be easily accommodated by the SLEX library. Analysis based on the SLEX gives results that are easy to interpret because they are time-dependent generalizations of Fourier analysis of stationary time series.

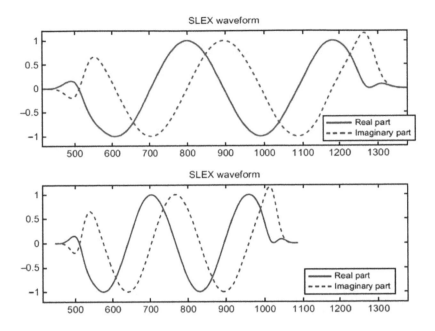

FIGURE 21.18
Examples of the SLEX waveforms at different scales and locations. The SLEX waveforms can be dilated or compressed as well as shifted.

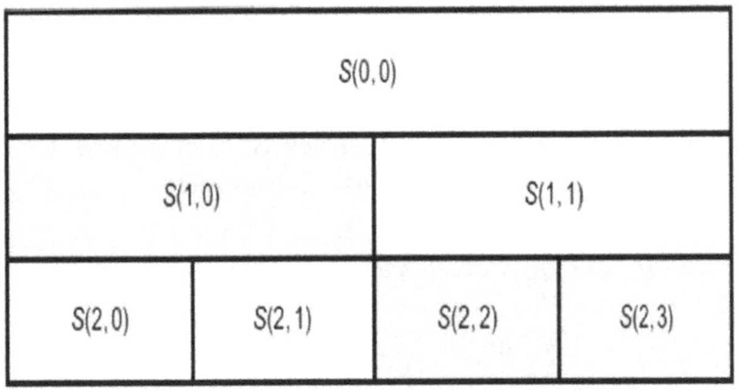

$S(0,0)$			
$S(1,0)$		$S(1,1)$	
$S(2,0)$	$S(2,1)$	$S(2,2)$	$S(2,3)$

FIGURE 21.19
A SLEX library with level $J = 2$. The shaded blocks represent one basis from the SLEX library.

The SLEX library. The SLEX library is a collection of bases, and each basis is composed of the SLEX waveforms which are localized. Thus they are able to capture the local spectral features of the time series. Moreover, the SLEX library allows a flexible and rich representation of the EEG. To illustrate these ideas, we construct a SLEX library in Figure 21.19 with level $J = 2$. There are 7 dyadic time blocks in this library. These are denoted as $S(0,0)$, which covers the entire time series; $S(1,0)$ and $S(1,1)$, which are the two half blocks and $S(2,b), b = 0, 1, 2, 3$, which are the four quarter blocks. Note that in general, for each resolution level $j = 0, 1, \ldots, J$, there are 2^j time blocks each having length $T/2^j$. We will adopt the notation $S(j, b)$ to denote the block b on level j where $b = 0, 1, \ldots, 2^j - 1$. The time blocks $S(j, b)$ correspond to the rescaled blocks $B(j, b)$ in the following manner: $S(0, 0)$ corresponds to $[0, 1]$; $S(1, 0)$ corresponds to $(0, 1/2)$; $S(2, 3)$ corresponds to $[3/4, 1]$. From this specific library with maximal level $J = 2$ there are five possible bases and one particular basis is composed of blocks $S(1, 0), S(2, 2), S(2, 3)$, which correspond to the shaded blocks in Figure 21.19. We point out that each basis is allowed to have multi-resolution scales, i.e., a basis can have time blocks with different lengths. This is ideal for processes whose intervals of stationarity have lengths that also vary with time. For example, there are longer stationary segments during non-seizure as opposed to the seizure state.

To choose the finest time scale (or deepest level) J, it would be helpful to seek advice regarding an appropriate time resolution of EEGs. In general, the blocks should be sufficiently small to ensure that the EEG is quasi-stationary in these blocks. However, to control the variance of the spectral estimator the blocks should not be unnecessarily small.

Computing the SLEX transform. The SLEX transform is a collection of coefficients corresponding to all the SLEX waveforms in the entire SLEX library. Thus, it is a redundant transform: a time series of length T will have JT total number of coefficients. Consider the SLEX vector (discretized waveform) with support on the discretized time block $S(j, b)$ consisting of time points $\{\alpha_0, \ldots, \alpha_1 - 1\}$. Furthermore, define $|S| = \alpha_1 - \alpha_0$ and the overlap $\epsilon = [\eta |S|]$ where [.] denotes the greatest integer function. The SLEX waveforms provide smooth transitions across blocks and hence its support is actually an "expanded" block $\{\alpha_0 - \epsilon, \ldots, \alpha_1 - 1 + \epsilon\}$. The SLEX vector with support S oscillating at frequency ω_k is

$$\phi_{S,\omega_k}(t) = \Psi_+ \left(\frac{t - \alpha_0}{|S|} \right) \exp\left(i 2\pi \omega_k (t - \alpha_0) \right)$$

$$+ \Psi_- \left(\frac{t - \alpha_0}{|S|} \right) \exp\left(-i 2\pi \omega_k (t - \alpha_0) \right),$$

where $\omega_k = k/|S|, k = -\frac{|S|}{2} - 1, \ldots, \frac{|S|}{2}$.

The SLEX coefficients can actually be computed using the fast Fourier transform (FFT). Let $X_p(t)$ be one component of a P-channel time series $\mathbf{X}(t)$ of length T. The SLEX coefficients (corresponding to $X_p(t)$) on block $S(j, b)$ are defined as

$$
\begin{aligned}
d_{j,b}^p(\omega_k) &= (M_j)^{-1/2} \sum_t X_p(t) \overline{\phi_{j,b,\omega_k}(t)} \\
&= (M_j)^{-1/2} \sum_t \Psi_+ (\frac{t - \alpha_0}{|S|}) X_p(t) \; \exp[-i 2\pi \omega_k (t - \alpha_0)] \\
&\quad + (M_j)^{-1/2} \sum_t \Psi_- (\frac{t - \alpha_0}{|S|}) X_p(t) \; \exp[i 2\pi \omega_k (t - \alpha_0)]
\end{aligned}
$$

where $M_j = |S(j, b)| = T/2^j$. In the implementation, the "edge" blocks in each level j, namely $S(j, 0)$ and $S(j, 2^j - 1)$, are padded with zeros when we compute the SLEX transform. Finally, by using the FFT, the number of operations needed to compute the SLEX transform has order of magnitude $O(PT(\log_2 T)^2])$

Computing the SLEX periodogram matrix. The $P \times 1$ vector of SLEX coefficients at block $S(j, b)$ and frequency ω_k, $\mathbf{d}_{j,b}(\omega_k) = [d_{j,b}^1(\omega_k), \ldots, d_{j,b}^P(\omega_k)]'$. The SLEX periodogram matrix is $\mathbf{I}_{j,b}(\omega_k) = \mathbf{d}_{j,b}(\omega_k)\mathbf{d}_{j,b}^*(\omega_k)$ where $\mathbf{d}^*$ is the complex conjugate transpose of $\mathbf{d}$. The diagonal elements of $\mathbf{I}_{j,b}$ are the SLEX auto-periodograms $I_{j,b}^{p,p}(\omega_k) = \left| d_{j,b}^p(\omega_k) \right|^2$ while off-diagonal elements are the SLEX cross-periodograms $I_{j,b}^{p,q}(\omega_k) = d_{j,b}^p(\omega_k) d_{j,b}^{q*}(\omega_k)$. Analogous to the Fourier periodogram matrices, we also smooth the SLEX periodogram matrices across frequency $\widetilde{\mathbf{I}}_{j,b}(\omega_k) = \frac{1}{2M+1} \sum_{r=-L}^{L} \mathbf{I}_{j,b}(\omega_{k+r})$ where $2M+1$ is the size of the smoothing window (in frequency) to produce a mean-squared consistent estimator.

21.5.3 Selecting the Best SLEX Signal Representation

We now select the model from a family of SLEX models that best represents the multichannel EEG data. The first step is to build a family of SLEX models where each model has a spectral representation in terms of a unique SLEX basis. The second step is to select the model that best represents the time series data using some optimality criterion. The SLEX method in (57) uses the penalized Kullback–Leibler (log energy) criterion, which balances fit (as measured by the likelihood) and complexity to prevent selecting the model with unnecessarily many blocks. The model selection step is equivalent to choosing the optimal dyadic segmentation of the multichannel EEG. After the best model (or the best segmentation) is selected, estimates for the time-varying spectral matrix, coherence and partial coherence are extracted at the blocks that define the best segmentation. As part of the model selection step, we address the problem of high dimensionality and multi-collinearity in the multichannel EEG. The approach of Ombao et al. (57) is to systematically extract a set of non-redundant spectral information that will be used in model selection and further analysis. This is accomplished by computing the time-varying eigenvalue–eigenvector decomposition of the SLEX spectral matrix, a generlization of the frequency domain principal components analysis (PCA, c.f. 6). The output of this step is the zero-coherency (uncorrelated) SLEX principal components with time-varying spectra which will be utilized in the model selection step.

Building a family of SLEX models. The family of SLEX models consists of signal representations, each of which uses a unique basis from the SLEX library. Thus, each model corresponds to a unique dyadic segmentation of the time series. Let $\mathcal{B}$ be a collection of rescaled time blocks in $[0,1]$ for one particular segmentation and let $\{\phi_{B,\omega}(t), \ B \in \mathcal{B}\}$ be one particular basis. Note that the blocks B are rescaled analogues (i.e., they are subsets of the unit interval $(0,1)$) with corresponding blocks of S defined on integer t. Define $\Theta_B(\omega)$ as the transfer function defined on block B and $d\mathbf{Z}_B(\omega)$ to be a P-dimensional zero mean orthonormal increment random process. The SLEX model that corresponds to the segmentation $\mathcal{B}$ is

$$\mathbf{X}(t) = \sum_{B \in \mathcal{B}} \int_{-0.5}^{0.5} \Theta_B(\omega)\phi_{B,\omega}(t)d\mathbf{Z}_B(\omega). \tag{21.32}$$

The spectral quantities. The $P \times P$ Hermitian SLEX spectral matrix at the rescaled time-frequency point (u,ω) (where u belongs a block B) is defined to be $\mathbf{f}(u,\omega) = \Theta_B(\omega)\Theta_B^*(\omega)$. The autospectrum of the p-th component $X_p(t)$ on (u,ω) is the p-th element on the diagonal denoted by $f_{p,p}(u,\omega)$. The cross-spectrum between the p-th and q-th components is the (p,q) element $f_{p,q}(u,\omega)$. Finally, the cross-coherence between the p-th and q-th components is defined to be $\rho_{p,q}(u,\omega) = \frac{|f_{p,q}(u,\omega)|^2}{f_{p,p}(u,\omega)f_{q,q}(u,\omega)}$. Note that these same quantities were also used in determining change points in Section 21.3.

The complexity-penalized Kullback–Leibler criterion. The Kullback–Leibler (KL) criterion derived in Ombao et al. (57) has two components: (1) the KL (or log energy) part, which measures divergence (or difference) between the candidate SLEX model and the unknown process that generated the observed EEG; and (2) the complexity penalty part that prevents the unnecessary splitting of a stationary mother block into children blocks. This criterion explicitly takes into account both the auto- and cross-correlation information from all components of the multichannel EEG.

Consider a candidate model $\mathcal{M}_\mathcal{B}$ where $\mathcal{B}$ corresponds to a particular segmentation of the EEG data. Let B be one block in the basis $\mathcal{B}$ and denote $\mathcal{C}(B)$ to be its corresponding KL value. The total KL for the candidate model $\mathcal{M}_\mathcal{B}$ is added over all blocks $\mathcal{C}(\mathcal{B}) = \sum_{B \in \mathcal{B}} C(B)$. We state the complexity-penalty KL criterion. In discretized block $S_{j,b}$ there is a total of $M_j = T/2^j$ time points and thus also a total number of M_j discrete frequency values. The KL value on block $B(j, b)$ is

$$C(j, b) = \sum_{k=-M_j/2+1}^{M_j/2} \log \det \widetilde{\mathbf{I}}_{j,b}(\omega_k) + \beta_{j,b}(P)\sqrt{M_j} , \qquad (21.33)$$

where $\widetilde{\mathbf{I}}_{j,b}$ is the smoothed periodogram matrix on $S(j, b)$ and $\beta_{j,b}(p)$ is the data-driven complexity penalty for block $S_{j,b}$. Let $h_{j,b}$ be the bandwidth used in smoothing the SLEX periodogram matrix. A simple version of the complexity parameter as used in (57) is $\beta_{j,b}(P) = P \beta_{j,b}$ where $\beta_{j,b}$ takes the form $\beta_{j,b} = \beta_{j,b}(h_{j,b}) = \log_{10}(\mathrm{e})/\sqrt{h_{j,b}} \sqrt{2 \log M_j}$. Finally, the complexity-penalized KL value for the model $\mathcal{M}_\mathcal{B}$ is $\mathcal{C}(\mathcal{B}) = \sum_{B(j,b) \in \mathcal{B}} C(j, b)$. As an illustration, the cost for the model defined by the shaded blocks in Figure 21.19 is the sum of the cost at each of these blocks: $C(1, 0) + C(2, 2) + C(2, 3)$.

The Algorithm for Selecting the Best Model. The best segmentation $\mathcal{B}^*$ or equivalently the best model $\mathcal{M}_{\mathcal{B}^*}$ for the data is the one that minimizes the complexity-penalized KL criterion, i.e.,

$$\mathcal{B}^* = \mathrm{argmin}_\mathcal{B} \mathcal{C}(\mathcal{B}).$$

In the actual implementation, we will utilize the best basis algorithm, which is a bottom-up algorithm whose essential idea is to compare the cost at a parent block $(B(j, b))$ and the children blocks $(C(j+1, 2b-1)$ and $C(j+1, 2b))$. If $C(j, b) < C(j+1, 2b-1) + C(j+1, 2b)$ then we choose the parent block $S(j, b)$. Otherwise, we choose the children blocks.

Remarks on Model Selection

(1.) *On approximating the true process by a SLEX model.* The procedure selects, within the family of Gaussian SLEX processes, the minimizer of the Kullback–Leibler divergence between the candidate models and the true underlying process that generated the data. This is equivalent to finding the best Kullback–Leibler approximation of a piecewise-stationary covariance SLEX process to the data.

(2.) *On extracting non-redundant information.* The components of many brain signals are usually highly collinear. High multi-collinearity in the EEG data suggests that one should perform dimension reduction. One way to accomplish this is via SLEX principal components analysis (see below). In the computation of the complexity-penalized KL criterion, we replace the multivariate time series by the SLEX principal components. Thus, the model selection procedure is conducted by taking into account the full information on the multivariate spectra.

(3.) *On the complexity penalty parameter $\beta_{j,b}$.* This will be determined from the data. The search algorithm for the best segmentation (or block partitions) can be considered as a variant of the Dyadic CART algorithm in (21).

SLEX Principal Components Analysis (PCA). For stationary multivariate time series, frequency domain PCA is motivated in (6) as follows. Let $\mathbf{X}(t)$ be a P-variate zero mean EEG with spectral density matrix $\mathbf{f}(\omega)$. Suppose now that we want to approximate $\mathbf{X}(t)$ by a Q-variate process $(Q \leq P)$ $\mathbf{U}(t)$ with uncorrelated (zero coherency) components defined to be $\mathbf{U}(t) = \sum_{\ell=-\infty}^{\infty} \mathbf{c}'_{t-\ell}\mathbf{X}(\ell)$, where $\{\mathbf{c}_r\}$ is a $P \times Q$ filter matrix and each component is absolutely summable. The filter coefficients $\{\mathbf{c}_r\}$ are derived via the following reconstruction criterion. Suppose that, from the reduced time series $\mathbf{U}(t)$, we want to be able to reconstruct the original time series $\mathbf{X}(t)$ by $\widehat{\mathbf{X}}(t) = \sum_{\ell=-\infty}^{\infty} \mathbf{b}_{t-\ell}\mathbf{U}(\ell)$ where the filter $\mathbf{b}_r$ is a $P \times Q$ matrix that satisfies $\sum_{r=-\infty}^{\infty} |\mathbf{b}_r| < \infty$. We want $\widehat{\mathbf{X}}(t)$ to be the minimizer of the mean square approximation error criterion $E\left[\left(\mathbf{X}(t) - \widehat{\mathbf{X}}(t)\right)^* \left(\mathbf{X}(t) - \widehat{\mathbf{X}}(t)\right)\right]$ where $\mathbf{A}^*$ denotes the complex conjugate transpose of $\mathbf{A}^*$. In this discussion, we suppose that the eigenvalues of $\mathbf{f}(\omega)$ are unique and we let $v^1(\omega) > v^2(\omega) > \ldots, > v^Q(\omega)$ be the eigenvalues with corresponding eigenvectors $V^1(\omega), V^2(\omega), \ldots, V^Q(\omega)$. The solution is to choose $\mathbf{c}_\ell = \int_{-1/2}^{1/2} \mathbf{c}(\omega)\exp(i2\pi\ell\omega)d\omega$, where $\mathbf{c}(\omega)$ is the matrix consisting of eigenvectors $V^1(\omega), \ldots V^Q(\omega)$. The spectrum of the m-th principal component $U_m(t)$ at frequency ω is the m-th largest eigenvalue $v^m(\omega)$. We refer the reader to the excellent discussion in Shumway and Stoffer (69) of the applications of frequency domain PCA in stationary time series.

These ideas are extended to the non-stationary case filter coefficients $\{\mathbf{c}_r\}$ to vary over time. Here, multivariate non-stationary time series is decomposed into the SLEX principal components, which are non-stationary components that have zero coherency. The time-varying filter and SLEX PC spectra defined on rescaled blocks $B(j,b)$ are obtained by performing an eigenvalue–eigenvector decomposition of the estimated spectral density matrix $\tilde{\mathbf{I}}_{j,b}(\omega_k)$ for each ω_k on the time block $S(j,b)$. The best model (or best segmentation) is obtained by applying the penalized log energy criterion on the SLEX PC. The spectra of the SLEX PCs are simply the eigenvalues of the spectral density matrix. Denote $v_{j,b}^1(\omega_k), \ldots, v_{j,b}^P(\omega_k)$ to be the P eigenvalues arranged in decreasing magnitude. If the reduced dimension $Q \leq P$ is known, then the penalized log energy criterion (21.33) at block $S(j,b)$ can be defined in terms of the Q SLEX PCs to be

$$\mathcal{C}(j,b) = \sum_{k=-M_j/2+1}^{M_j/2} \sum_{q=1}^{Q} \log(v_{j,b}^q(\omega_k)) + Q\,\beta_{j,b}\,\sqrt{M_j} \qquad (21.34)$$

where $2M_j = T/2^j$ is the total number of observations in every block level j. In practice, however, Q is rarely known and there is no consensus on the best approach to selecting Q even in the stationary situation. We propose a data adaptive approach that does not require the user to specify Q. The basic idea is to assign a weight to each SLEX component that is proportional to its variance (spectrum). Essentially, SLEX PCs with larger eigenvalues are given more weight and those with smaller eigenvalues are given smaller weights. The weight $w_{j,b}^p(\omega_k)$ of the SLEX PC with the p-th largest eigenvalue is defined to be

$$w_{j,b}^p(\omega_k) = v_{j,b}^p(\omega_k)/\sum_{p=1}^{P} v_{j,b}^p(\omega_k). \qquad (21.35)$$

The cost at block $S(j,b)$ is

$$\mathcal{C}(j,b) = \sum_{k=-M_j/2+1}^{M_j/2} \sum_{p=1}^{P} w_{j,b}^p(\omega_k)\,\log v_{j,b}^p(\omega_k) + \beta_{j,b}\,\sqrt{M_j}, \qquad (21.36)$$

where, as before, $\log(v_{j,b}^p(\omega_k))$ is the logarithm of the spectrum of the p-th principal component at frequency ω_k in block $S(j,b)$ having applied PCA to the optimally smoothed periodogram matrix. Note that this cost is based on the "weighted" eigenvalues. Hence, we do not need the factor Q in the complexity penalty term.

One advantage of the approach of weighting the eigenvalues is that the "optimal" number Q need not be explicitly specified as it implicitly renders irrelevant those components that do not contribute much to the variance. From a numerical point of view, it also avoids computational problems since the term $w^p \log(v^p)$ is assigned the value "zero" when v^p and w^p are both close to 0, i.e., when the absolute and relative contribution to variance are, respectively, small.

Obtaining the Spectral Estimates. Let $\mathcal{B}^*$ be the basis that corresponds to the best model. To estimate the time-varying spectral matrix $\mathbf{f}(u,\omega)$ at rescaled time u and frequency ω, suppose that $B(j,b)$ is the time block in the basis $\mathcal{B}^*$ that corresponds to the rescaled time u. The estimate of the SLEX spectral density matrix at (u,ω) is defined to be $\widehat{\mathbf{f}}(u,\omega) = \widetilde{\mathbf{I}}_{j,b}(\omega)$, which is the kernel-smoothed periodogram matrix. The auto-spectral estimate of the p-th component $X_p(t)$ defined on (u,ω) is $\widehat{f}_{pp}(u,\omega)$; the cross-spectral estimate between the p-th and q-th components is $\widehat{f}_{pq}(u,\omega)$; and the cross-coherence estimate between the p-th and q-th components is $\widehat{\rho}_{pq}(u,\omega) = \frac{|\widehat{f}_{pq}(u,\omega)|^2}{\widehat{f}_{pp}(u,\omega)\widehat{f}_{qq}(u,\omega)}$.

To compute the confidence intervals for the SLEX auto-spectra, we state the asymptotic results in (56). For $\omega \in (0,1/2)$, $\widehat{f}_{p,p}(u,\omega)/f_{p,p}(u,\omega) \overset{\cdot}{\sim} \chi^2_{2M_j h_{j,b}}/(2M_j h_{j,b})$ where $h_{j,b}$ is the smoothing bandwidth and $M_j h_{j,b}$ is the number of frequency indices in the smoothing span. To obtain the confidence intervals for the SLEX coherence, define $\widehat{r}_{p,q}(u,\omega) = \tanh^{-1}[\widehat{\rho}_{p,q}(u,\omega)]$. Then $\widehat{r}_{p,q}(u,\omega)$ is asymptotically normal with mean and variance approximately equal to $r_{p,q}(u,\omega)$ and $1/[2(2M_j h_{j,b} - P)]$, respectively. This follows readily from (29) and (6).

21.5.4 SLEX Analysis of Multichannel Seizure EEG

We report the results in Ombao et al. (57), which used only a subset of the EEG data centered around seizure onset. The time index for the segment of the data is $t = 29605$ to $t = 37796$, with a length of $T = 8192$. This segment of the data corresponds to the physical time $[296, 378]$ seconds. The first step in the SLEX method is to build the family of SLEX models and select the one that is best according to our penalized log energy criterion. As suggested by the neurologist, levels $J = 6$ or $J = 7$ were used and both resulted in identical best models. Prior to model selection, the SLEX PCs were obtained via the time-varying eigenvalue–eigenvector decomposition of the SLEX matrix. The best model has segmentation defined by change points that occur at approximately 296 seconds + $\{20, 30, 36, 38, 40, 61, 72, 73, 74, 77\}$ seconds. According to the neurologist, the physical manifestations of seizure became evident at around $296 + 40$ sec from the start of recording. However, prior to this, the SLEX analysis revealed that changes in the electrical activity of the brain were already beginning to take place even before the physical symptoms were observed. In this context it can be noted that, unlike the FreSpeD method, discussed in Section 21.3, the SLEX method does not directly identify which channels (or pairs of channels) and frequencies (or frequency bands) change.

The SLEX method systematically filters non-redundant information. This is noteworthy because, in the absence of any additional information about the patient or this seizure episode, one would have to examine all $\frac{18!}{2!16!} = 153$ pairwise cross-correlations (and cross-coherences), which can be overwhelming. The SLEX method guides the user to focus on the most interesting channels. The first and second SLEX PCs altogether account for approximately 70% of the variance in the EEG data. We focus our attention only to the first two SLEX PCs.

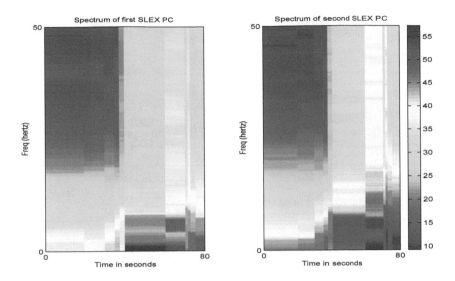

FIGURE 21.20
Time-varying spectra of the first and second SLEX principal components.

The time-varying spectra of the first SLEX PC (left side in Figure 21.20) account primarily for the increase in power in the lower frequencies after the onset of seizure. The second SLEX PC (right side in Figure 21.20), on the other hand, accounts for the spread of power from the delta band (0–4 Hertz) to the alpha band (8–12 Hertz). Even without using any patient information, the SLEX method identified the $T3$ channel (left temporal lobe) as one of the important channels, which is the location of a brain lesion that is believed to be implicated in seizure for this particular patient. We observe that, for the first SLEX PC, most of the weights are concentrated on the $T3$ channel and $T4$ channel (right temporal lobe). For the second SLEX PC, the weights are quite diffused at the temporal and frontal lobe areas.

The estimates of the SLEX time-varying spectra conveys the information that the distribution of power over frequency indeed evolves during the seizure process. We see that power at the lower frequencies is increased and that power is spread to middle and higher frequencies during seizure. It is important to note that features of this multichannel seizure EEGs were captured by the first and second SLEX PCs.

Dependence or connectivity between brain areas was also investigated. The first eigenvector suggested to focus further analysis on two networks, namely, (i) coherence between $T3$ and the other channels and (ii) coherence between $T4$ and the other channels. In Figures 21.21 and 21.22, the connectivity between brain areas changes throughout the duration of the epileptic seizure. In Figure 21.21, it is interesting that coherence between $T3$ and those at the left side of the brain, namely, left parietal ($P3$), left frontal ($F3$) and left central ($C3$), are more similar to each other than coherence between $T3$ and the counterparts on the right side ($P4$, $F4$ and $C4$ respectively) are quite different suggesting differences in patterns of evolution of seizure between those on the side of the affected region (near $T3$) and the opposite side. Moreover, one observes in Figure 21.22 the bilaterality (symmetry) of the coherence, i.e., the coherence pattern between $T4$ (right temporal lobe) and the channels on the right side are more similar to each other compared to the channels

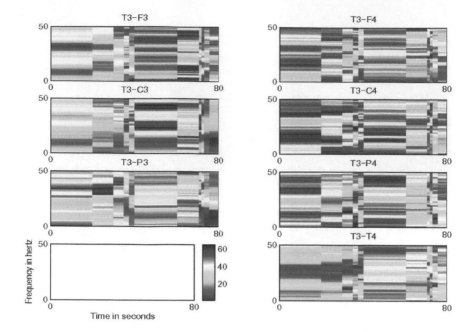

FIGURE 21.21
Left: SLEX coherence estimates between $T3$ and channels on the left side of the brain namely $F3, C3, P3$. Right: SLEX coherence estimates between $T3$ and the channels on the right side of the brain namely $F4, C4, P4, T4$. The color index here is actually coherence $\times$ 100 so that the values range from 0 to 100.

on the left side. This observation of similar synchrony patterns between channels on the same side of the brain around seizure is interesting though not completely understood. One plausible explanation is that the channels on the same side are likely to be projected from similar populations of neurons and thus are expected to behave similarly.

21.6   Dual-Frequency Coherence Analysis

In this chapter, we have considered studying connectivity from different points of view. In Section 21.3, coherence (which is a frequency-specific measure of connectivity) is used to detect and estimate change points in non-stationary time series. Our focus here is to consider more complex types of dependence that are present in EEGs. On modeling connectivity between different brain regions the key questions are (a) what are meaningful measures of dependence and (b) how do we estimate and test for the statistical significance of dependence? Motivated by these current challenges, (30) develops a new approach for investigating dependence structures in general non-stationary signals that are neglected by standard spectral methods. Here, we will apply the recently developed local dual-frequency coherence model in the seizure EEG analysis. To briefly describe the context of the problem, consider the decomposition of EEG recordings at the right frontal-central channel (FC4)

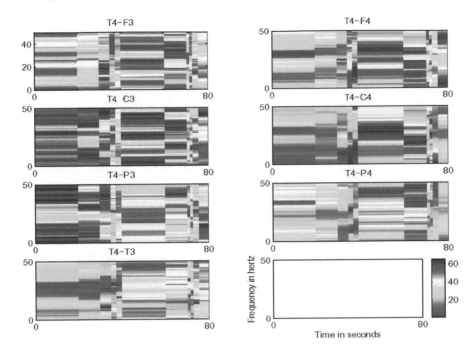

FIGURE 21.22
Left: SLEX coherence estimates between $T4$ and channels on the left side of the brain namely $F3, C3, P3, T3$. Right: SLEX coherence estimates between $T4$ and the channels on the right side of the brain namely $F4, C4, P4$. The color index here is actually coherence $\times$ 100 so that the values range from 0 to 100.

and the left parietal channel (P3) (see Figure 21.23). The EEG at each channel is decomposed into a range of oscillations from low frequency (delta band) to high frequency (gamma band). One of the scientific goals here is to model the interaction (connectivity) between the alpha oscillations of channel FC4 and the beta oscillations of channel P3. In particular, we will attempt to answer whether or not increased alpha activity in the cortical area is associated with excitation of beta oscillations in another area.

The standard approach to measuring dependence between oscillations is via cross-coherence (or simply coherence). In the stationary setting, this is discussed in Chapter 20. In the non-stationary setting this is discussed in Section 21.3 in the context of change-point detection, where cross-coherence is a feature, and in Section 21.5 in the context of obtaining the optimal signal representation which takes into account the time-varying cross-coherence between channels. Under the Cramér representation, dependence between components of a multichannel EEG $\mathbf{X}_t$ is directly captured by the covariance of the random increments $d\mathbf{Z}(\omega)$. We now describe a more intuitive interpretation of cross-coherence developed in Ombao and van Bellegem (54) which is described as follows. First, apply a filter to both channels so that the resulting filtered signals contain power only at a specific band. Second, compute the cross-correlation of these filtered signals. Ombao and van Bellegem (54) demonstrate that the expectation of the squared cross-correlation of filtered signals is asymptotically equivalent to cross-coherence.

A major limitation of standard cross-coherence analysis is that it examines between-channel interactions only at the *same frequency* (or frequency band), i.e., dependence be-

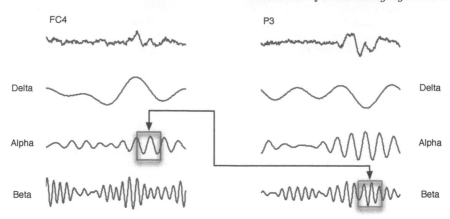

FIGURE 21.23

Decomposition of the electroencephalograms at the right frontal channel (FC4) and the left parietal channel (P3) into the delta (0–4 Hertz), alpha (8–12 Hertz) and beta (12–30 Hertz) oscillations. The goal is to estimate dependence between the alpha band oscillations at the FC4 channel with the beta band oscillations at the P3 channel.

tween low-frequency oscillations in one channel and low-frequency oscillations in another. This limitation should be addressed especially in neuroscience where several studies (e.g. 38, 8, 20) have suggested local neuronal populations firing at some rate "coactivate" with other neuronal populations firing at a different rate. Thus, there is a need for more sophisticated statistical tools to investigate these complex interactions.

21.6.1   Overview and Historical Development

The concept of dependence between oscillations at different frequencies has already been established in the context of harmonizable processes (see 50, 65, 48, 35, 32). Formally, a time series $\mathbf{X}_t$ belongs to the class of harmonizable processes if it admits the representation

$$\mathbf{X}(t) = \int_{-0.5}^{0.5} \exp(i2\pi\omega t)d\mathbf{Z}(\omega), \tag{21.37}$$

where now the zero-mean random increments $d\mathbf{Z}(\omega)$ may be *correlated* across different frequencies so that

$$\mathbb{C}\mathrm{ov}[d\mathbf{Z}(\omega_k), d\mathbf{Z}(\omega_l)] = \boldsymbol{f}(\omega_k, \omega_l)d\omega_k d\omega_l \ . \tag{21.38}$$

Here, $\boldsymbol{f}(\omega_k, \omega_l)$ is a complex-valued $P \times P$ matrix called the generalized spectral matrix or the Loève spectral matrix. For harmonizable processes, the dual-frequency coherence between a pair of time series X_p, X_q, at a pair of frequencies (ω_k, ω_l) is defined to be

$$\rho_{p,q}(\omega_k, \omega_l) = \big|\mathrm{Cor}[dZ_p(\omega_k), dZ_q(\omega_l)]\big|^2 = \frac{\big|f_{p,q}(\omega_1, \omega_2)\big|^2}{f_{p,p}(\omega_k, \omega_k)f_{q,q}(\omega_l, \omega_l)}$$

where $f_{p,q}(\omega_k, \omega_l)$ is the (p, q)-th element of $\boldsymbol{f}(\omega_k, \omega_l)$. The concept of dual-frequency coherence only partially addresses the problem of capturing of cross-frequency dependence. It is not adequate because it captures only the *globally averaged* dependence between oscillatory components.

In this chapter, we present a new concept and inference for a *time-varying* coherence analysis across *different* frequencies (and frequency bands) under the setting where there is data obtained from replicated trials. So far this chapter has illustrated different methods on the epileptic seizure EEG data which is observational (not induced) and is non-replicated. To illustrate estimation and inference for dual-frequency coherence, we shall use EEG data that is collected over many replicated trials. Under this setting, the estimator of dual-frequency is obtained averaging cross-periodograms across trials. Thus, the resulting estimator is a more precise estimator because it requires very little (or none at all) trial-specific smoothing across frequency.

We now briefly describe the estimation procedure developed in (30). First, around each time point we form a local window; compute the time-localized dual-frequency cross-periodograms; and finally average across replicated trials. These cross-periodograms are computed at each pointwise dual-frequency pair (ω_k, ω_l). In addition, a parallel approach for estimating coherence between frequency *bands* was also developed by averaging the periodograms across frequencies at bands. Unlike most approaches for pointwise-in-frequency analysis for data obtained only from a single-trial, it is not necessary to apply smoothing of the cross-periodograms over frequency since our estimator's variance is reduced by averaging across replicated trials instead. By avoiding smoothing in the frequency plane, the proposed approach maintains the resolution in the frequency domain and preserves localized peaks.

Similar to the change-point detection method discussed in Section 21.3, the local dual-frequency periodogram approach in (30) is non-parametric and therefore does not suffer from the risk of model misspecification, unlike parametric spectral estimation methods. Moreover, the estimation approach takes advantage of the fast Fourier transform (FFT) and thus can be readily implemented to analyze large time series datasets. It is related to the classical windowed Fourier analysis, which is widely used in signal processing. Here, the sliding time window idea is applied for a different purpose, i.e., for estimating dependence between a pair of frequencies rather than a single frequency.

We now review some periodogram-based methods for spectral estimation for stationary and locally stationary processes.

We first briefly point out the different estimands (spectral quantities) under different settings (stationary, locally stationary, harmonizable). Define the general support of the spectral matrix in the frequency domain to be the square $D = [-\frac{1}{2}, \frac{1}{2}] \times [-\frac{1}{2}, \frac{1}{2}]$. For stationary processes, the random increments are uncorrelated across frequencies. Hence, the support is confined only to the diagonal of this square (i.e., the line that contains points where the frequencies are identical). When a process is "locally stationary," as in the sense of Priestley (61) and Dahlhaus (17) or the SLEX model in (55, 56) and Section 21.5, the spectral matrix still lives on the diagonal of D but it is allowed to evolve over time. For harmonizable processes, the support of the Loève spectrum is the entire frequency square D. However, estimation theory imposes constraints on the structure of the Loève spectrum, e.g., requiring its support to live on lines (48). Moreover, when there are no replicates, there are constraints on both the support and smoothness along these spectral lines that are necessary in order to deal with the insufficient number of observations to estimate the Loève spectrum on the entire square D.

Throughout this section, let $\boldsymbol{X}^r(t)$ be the signal recorded during the r-th replicate where $r = 1, \ldots, R$. We assume the EEG signal to be i.i.d. across replicates r. These time series are all aligned according to the experimental conditions so that the starting time point for each replicate is the time when the common stimulus was presented at each of the replicates. In the following, both the length of the time series per replicate T and the local time window size N within each replicate are assumed to be even.

21.6.2   The Local Dual-Frequency Cross-Periodogram

The concept of a Loève spectrum is generalized in (30) to the case where the dual-frequency spectrum $\boldsymbol{f}_t(\omega_k, \omega_l)$ and coherence $\rho_{p,q,t}(\omega_k, \omega_l)$ evolve over time. As in the above, we fix the local time window to have size N so that our predefined discrete Fourier frequencies take the form $\omega_k = \frac{k}{N}$ where $k = -(\frac{N}{2} - 1) \ldots \frac{N}{2}$. The time window size of N must be chosen carefully to avoid bias arising from non-stationarity and aliasing. When the period of quasi-stationarity is brief, then estimates derived from a large N would be corrupted due to non-stationarity of the large window. However, a small N could introduce aliasing because high-frequency oscillations would appear aliased as low-frequency oscillations in this small window.

To obtain some data-analogue measure of the underlying time-localized cross-oscillatory interactions between component X_p at frequency ω_k and component X_q at frequency ω_l, we build on the windowing procedure above by first computing the dual-frequency periodogram for trial r at a local time block around t. In the following discussion, we shall assume that we have replicated signals (from repeated trials) and we shall assume that the underlying brain process across the different trials are identical. Define the P-dimensional vector of Fourier coefficients at frequency ω (at all channels) around time point t computed for trial r to be

$$\boldsymbol{d}^r(\omega_k) = \frac{1}{\sqrt{N}} \sum_{s=t-(N/2-1)}^{t+N/2} \boldsymbol{X}_s^r \exp(-i2\pi\omega_k s)$$

where $\omega_k = k/N$, $k = -\left(\frac{N}{2} - 1\right), \ldots, \frac{N}{2}$. Next, denote $\boldsymbol{d}^{r,*}(\omega_k)$ to be the complex-conjugate transpose of the Fourier vector $\boldsymbol{d}^r(\omega_k)$. The local dual-frequency periodogram matrix from trial r and the averaged local dual-frequency periodogram are, respectively,

$$\boldsymbol{I}_t^r(\omega_k, \omega_l) = \boldsymbol{d}_t^r(\omega_k)\boldsymbol{d}_t^{r,*}(\omega_l) \quad \text{and} \quad \widehat{\boldsymbol{f}}_t(\omega_k, \omega_l) = \frac{1}{R} \sum_{r=1}^{R} \boldsymbol{I}_t^r(\omega_k, \omega_l). \tag{21.39}$$

To estimate the strength of the time-localized dependence between the ω_k-oscillations at component X_p and the ω_l-oscillations at component X_q, we have

$$\widehat{\rho}_{p,q,t}(\omega_k, \omega_l) = \frac{|\widehat{f}_{p,q,t}(\omega_k, \omega_l)|^2}{\widehat{f}_{p,p,t}(\omega_k, \omega_k)\widehat{f}_{q,q,t}(\omega_l, \omega_l)}. \tag{21.40}$$

21.6.3   Formalizing the Concept of Evolutionary Dual-Frequency Spectra

21.6.3.1   Harmonizable Process: Discretized Frequencies

We formally define the time-constant Loève spectrum and develop a model based on discrete Fourier frequencies. One has to be very careful in determining the frequency resolution when this discrete frequency model is generalized to the time-dependent setting. For now, we shall set M to be a fixed integer that determines frequency resolution and let $\omega_k = \frac{k}{M}$. The spectral representation for a P-variate harmonizable time series defined on the discretized Fourier frequency and recorded during the r-th trial is

$$\boldsymbol{X}^r(t) = \frac{1}{\sqrt{M}} \sum_{k=-(\frac{M}{2}-1)}^{\frac{M}{2}} \exp(2\pi i\omega_k t) \, \boldsymbol{z}_k^r \,, \quad t = 1, \ldots, T \,, r = 1, \ldots, R \,, \tag{21.41}$$

where the P-dimensional complex random vectors $\mathbf{z}_k^r = (z_{1k}^r, \ldots, z_{Pk}^r)$, with $\mathbf{z}_{-k}^r = \mathbf{z}_k^{r,*}$, are defined to be $d\mathbf{Z}^r(\omega_k)$, the zero mean increment process in the dual-frequency domain.

Remarks. Note that within this discrete-frequency model, which is inspired by the SLEX (Smooth Localized Complex Exponential) model of the non-stationary random process in (56), our target spectrum is $\boldsymbol{f}(\omega_k, \omega_l) = \mathbb{C}\mathrm{ov}(d\mathbf{Z}^r(\omega_k), d\mathbf{Z}^r(\omega_l))$, defined on a fixed equidistant grid of Fourier frequencies in the frequency square D.

21.6.3.2   A New Model: The Time-Dependent Harmonizable Process

The model in Equation (21.41) is generalized to the time-dependent case under which the new concept of an evolutionary dual-frequency spectrum (EDS) is rigorously defined. We shall impose a smooth time-variation on $\boldsymbol{f}(\omega_k, \omega_l)$ in exactly the same way as it is done in the literature of estimation of evolutionary (single-frequency) spectra using infill asymptotics for $T \to \infty$. Therefore, let us define the target now to be the time-varying Loève spectrum $\boldsymbol{f}(u, \omega_k, \omega_l)$, $u \in (0, 1)$, which changes smoothly in (rescaled) time u. To parallel the most general specification in (15), we will assume that there exists a universal positive constant Q such that for each T, each $t = 1, \ldots, T$, $|\boldsymbol{f}_t(\omega_k, \omega_l) - \boldsymbol{f}(t/T, \omega_k, \omega_l)| \leq \frac{Q}{T}$. The time-dependent harmonizable process with evolutionary dual-frequency spectrum $\boldsymbol{f}(u, \omega_k, \omega_l)$, $u \in (0, 1)$ is formally defined as follows.

Definition. For fixed T and N, a P-variate time series $\mathbf{X}_T^r(t)$ recorded during the r-th replicate is said to be an evolutionary discrete-frequencies harmonizable process if it admits the representation

$$\mathbf{X}_{t,T}^r = \frac{1}{\sqrt{N}} \sum_{k=-\frac{N}{2}+1}^{\frac{N}{2}} \exp(2\pi i \omega_k t)\, d\mathbf{Z}_{t,T}^r(\omega_k)\ ,\ t = 1, \ldots, T\ , \qquad (21.42)$$

where $\boldsymbol{f}_{t,T}(\omega_k, \omega_l) = \mathbb{C}\mathrm{ov}(d\mathbf{Z}_{t,T}^r(\omega_k), d\mathbf{Z}_{t,T}^r(\omega_l))$ is the time-varying dual-frequency spectrum (common across all replicates r) and the random increments $d\mathbf{Z}_T^r(t)(\omega_k)$ inherit all the properties of the z_k^r defined in Equation (21.41).

21.6.3.3   Dual-Frequency Coherence between Bands

The general methodology derived for pointwise frequencies can be easily tailored to coherence analysis of frequency bands. As noted, in many practical applications frequency bands are used more often than pointwise frequencies. For EEG analysis, statistical inference on the cross-oscillatory bandwise dependence will be based on only five frequency bands (delta, theta, alpha, beta and gamma). Thus, for multiple testing correction, it will be sufficient to use existing methods such as Bonferroni and the false discovery rate (FDR). We implement the latter in our analyses.

To conduct inference on cross-dependence between frequency bands, we first define the average cross-spectral power at bands (Ω_1, Ω_2) between a pair of channels (p, q) to be

$$f_{p,q}(u, \Omega_1, \Omega_2) = \frac{1}{KL} \sum_{\omega_k \in \Omega_1} \sum_{\omega_l \in \Omega_2} f_{p,q}(u, \omega_k, \omega_l)$$

where K and L are the total number of frequencies in bands Ω_1 and Ω_2. Under this definition the band-averaged cross-spectral matrix between bands Ω_1 and Ω_2 at rescaled time $u \in [0, 1]$ (where $t = [uT]$) is

$$\boldsymbol{f}(u, \Omega_1, \Omega_2) = \frac{1}{KL} \sum_{\omega_k \in \Omega_1} \sum_{\omega_l \in \Omega_2} \boldsymbol{f}(u, \omega_k, \omega_l). \qquad (21.43)$$

The corresponding estimator of $f_{p,q}(u, \Omega_1, \Omega_2)$ based on R replicates is

$$\widehat{f}_{p,q,[uT]} = \frac{1}{R}\frac{1}{KL} \sum_{\omega_k \in \Omega_1} \sum_{\omega_l \in \Omega_2} I^r_{p,q,t}(\omega_k, \omega_l).$$

Given the fixed set of number of frequencies in each frequency band and the unbiased property in each pair of single frequencies established in (30), the unbiased property for the frequency band estimators immediately follows. Moreover, the asymptotic normality property for frequency band estimators follows, c.f. (30) for assumptions and lemmata. For R sufficiently large, with $t = [uT]$ for a given fixed $u \in (0,1)$, the vector

$$\left[\widehat{f}_{p,p,[uT]}(\Omega_1, \Omega_1), \ \widehat{f}_{q,q,[uT]}(\Omega_2, \Omega_2), \ \Re\widehat{f}_{p,q,[uT]}(\Omega_1, \Omega_2), \ \Im\widehat{f}_{p,q,[uT]}(\Omega_1, \Omega_2) \right]'$$

is approximately normal with asymptotic mean

$$\left[f_{p,p}(u, \Omega_1, \Omega_1), \ f_{q,q}(u, \Omega_2, \Omega_2), \ \Re f_{p,q}(u, \Omega_1, \Omega_2), \ \Im f_{p,q}(u, \Omega_1, \Omega_2) \right]'$$

and variance $\frac{1}{R}\mathbf{\Phi}(u, \Omega_1, \Omega_2)$. The precise form of the variance is complicated and long; we refer the reader to the Appendix section of (30).

21.6.4 Inference on Local Dual Frequency Coherence

We now present a formal inference procedure for testing the presence of significant association between the predefined pair of Fourier frequencies (ω_k, ω_l), around a specific rescaled time u, i.e. as before we again adopt the notation $t = [uT]$ in the sequel. The results presented here are general and would hold to the case where we consider coherence between any non-overlapping bands Ω_1 and Ω_2. The main interests are to (1) derive confidence intervals for the evolutionary dual-frequency coherence; (2) test the null hypothesis $H_0 : \rho_{pq}(u, \omega_k, \omega_l) = 0$ for a specific time u and pair of frequencies (ω_k, ω_l) (or frequency bands); and (3) to test for differences in coherence across factors such as patient groups and experimental conditions.

Inference using results on the exact beta distribution. Denote the estimand to be the evolutionary dual-frequency coherence $\rho := \rho_{p,q}(u, \omega_k, \omega_l)$ and the estimator, derived from R trials, to be $\widehat{\rho} := \widehat{\rho}_{p,q,[uT]}(\omega_k, \omega_l)$. The exact probability density function of the square root of the evolutionary dual coherence estimator, denoted $\widehat{\rho}$, is given by

$$f(\widehat{\rho} \mid \rho, R) = \frac{(R-1)(1-\rho)^R(1-\widehat{\rho})^{R-2}}{(1-\rho\widehat{\rho})^{2R-1}} \ {}_2F_1(1-R, 1-R; 1; \rho) \tag{21.44}$$

where ${}_2F_1$ is the Gaussian hypergeometric function. This was first developed in (26) and then later developed in (9) and (6). Under the null hypothesis $H_0 : \rho = 0$, the exact probability density function from Equation (21.44) reduces to

$$f(\widehat{\rho} \mid \rho = 0, R) = (R-1)(1-\widehat{\rho})^{R-2} \sim \text{Beta}(1, R-1), \tag{21.45}$$

which is a beta-random variable with the shape parameters $(1, R-1)$. This result will be utilized to identify the statistically significant (i.e., strictly positive) evolutionary dual-frequency coherence.

21.6.5 Local Dual Frequency Coherence Analysis of EEG Data

21.6.5.1 Description of the Data and Experiment

The EEG recording from the visual-motor experiment is described in the Introduction. These same EEG recordings were analyzed in Fiecas et al. (24) and Fiecas and Ombao (23), but the limitation of the earlier analyses is that they only study interactions between oscillations at the same frequency (i.e., the single-frequency coherence). The novelty of the current analysis is that it actually examines interactions between oscillations at different frequencies using the dual-frequency coherence method. Thus, this new analysis identifies other complex dependence structure in the signals that are missing in the standard coherence analysis.

The EEG signals are recorded from an experiment where a visual cue is presented, at each trial, to a right-handed participant who is instructed to move a hand-held joystick from a central position to either the right or to the left side. Our analysis focuses on a subset of $P = 12$ channels from a standardized EEG topography consisting of 64 scalp electrodes. These 12 channels are believed to be most involved in visual-motor actions. The 12 channels are located in the fronto-central (FC3, FC5, FC4 and FC6), central (C3, CZ, and C4), parietal (P3 and P4), and the occipital (O1, OZ and O2) regions. An approximate topography of the location of these channels is given in Figure 21.3. The EEG signals are digitized at a sampling rate of 512 Hertz. Each trial consists of a one-second recording ($T = 512$). The visual cue is presented at $t = 256$ (equivalent to 500 milliseconds) and the recording continues for another 500 milliseconds post stimulus presentation. The order of presentation of the visual cue (left vs. right) is random. There are a total of $R = 118$ replicated identical trials for each direction. In Figure 21.3 we show the EEG trace of one trial for each direction at the indicated channels. Prior to statistical analysis, these single-trial EEGs are bandpass filtered at $[0.2, 100]$ Hertz and standardized to have zero mean and unit variance. In this analysis, we are primarily interested in the cross-frequency interactions between the alpha band (8–12 Hertz) and the beta band (16–30 Hertz), which are both implicated in many cognitive processing including these visual-motor tasks. Here, we apply our proposed approach to investigate these alpha–beta interactions ($\alpha \leftrightarrow \beta$) and how these interactions may evolve within a trial.

21.6.5.2 Implementation Details

On computing the local dual-frequency band coherence. The time-dependent dual-frequency coherence between the alpha band ($\Omega(\alpha)$) and beta band ($\Omega(\beta)$), denoted $\rho_{p,q,[uT]}(\Omega(\alpha), \Omega(\beta))$, was estimated for each pair of the 12 channels (p, q) according to Equations (21.39) and (21.40). For each trial r and time point t, we formed a window of size $N = 100$ and then we computed the dual-frequency periodogram $I_t^r(\omega_k, \omega_l)$. The local dual-frequency periodogram at the alpha and beta *bands* were obtained by averaging across all discrete frequencies contained in the bands:

$$\widehat{f}_{[uT]}^r(\Omega(\alpha), \Omega(\beta)) = \frac{1}{KL} \sum_{k=1}^{K} \sum_{l=1}^{L} I_t^r(\omega_k, \omega_l) \tag{21.46}$$

where ω_k is a Fourier frequency in $\Omega(\alpha)$; ω_l is a Fourier frequency in $\Omega(\beta)$, and $K = 5$ and $L = 15$ are the total number of frequencies in the $\Omega(\alpha)$ and $\Omega(\beta)$ bands, respectively. We then computed the average of the local dual-frequency periodograms across the R replicates to obtain

$$\widehat{f}_{[uT]}(\Omega(\alpha), \Omega(\beta)) = \frac{1}{R} \sum_{r=1}^{R} \widehat{f}_{[uT]}^r(\Omega(\alpha), \Omega(\beta)),$$

The dependence between the alpha band activity at the frontal-central channels and the beta band activity at the other channels are shown in Figure 21.24; alpha activity in the central and beta activity in all other channels are shown in Figure 21.25; alpha activity in the parietal and beta in all other channels are in shown in Figure 21.26. The alpha band was selected in the analysis because the study in Del Percio et al. (19) reported that it plays a role in visuo-motor performance of athletes. Moreover, (73) demonstrates that the beta band is highly implicated during motor planning.

On testing for the significance of the local dual-frequency coherence. To test the null hypothesis $H_0 : \rho_{pq}(u, \Omega(\alpha), \Omega(\beta)) = 0$ at each time point $u = t/T$ where u is in $\mathcal{I} = (\frac{N}{2T}, 1 - \frac{N}{2T})$ (to avoid boundary effects), we used the exact null distribution of magnitude squared coherence in Equation (21.45). To account for multiple testing for each pair of frequency bands across all time points we applied the false discovery rate procedure (FDR) at level 0.05. Non-significant dual frequency coherences are indicated by the white color. Significant values are represented according to the color scale provided.

On testing for differences in dual-frequency coherence between the right vs. left movements. Differences between right and left visual cues were tested by applying the Fisher-z transformation on the coherence values and then using the normal approximation derived in Gorrostieta et al. (30). We test, again using FDR at 0.05 level, the hypothesis $H_0 :$ $\rho_{p,q,\text{left}}(u, \Omega(\alpha), \Omega(\beta)) - \rho_{(p,q),\text{right}}(u, \Omega(\alpha), \Omega(\beta)) = 0$ for all time points $u \in \mathcal{I}$. Significant differences are illustrated by black points in both left and right plots on Figures 21.24, 21.25 and 21.26.

21.6.5.3   Results and Discussion

The highly significant findings from Figures 21.24, 21.25 and 21.26 relate to the coherence between the alpha band in the frontal-central (FC) channels and the beta band at the central (C) channels. The alpha and beta bands were of primary interest because each of these oscillations plays a significant individual role during movement based on studies that demonstrate links between beta band activity motor behavior, which is generally attenuated during active movements. Alpha band activity is believed to reflect neural activity related to stages of motor response during a continuous monitoring task. Using the proposed dual-frequency coherence method, (30) obtains significant interactions between the alpha and beta bands. Another important finding from this analysis is that these significant alpha–beta interactions are not static but evolve within a trial. This indicates the highly dynamic nature of brain responses while processing these visual cues.

The local dual-frequency coherence method also identified differences in the brain responses for the left vs. rightward movements. For the leftward movement condition, (30) noted that at around 600 milliseconds (or 100 milliseconds post visual cue presentation), alpha band activity at the left frontal-central channels (FC5 and FC3) is significantly related to the beta activity on central channels and the left parietal channel (P3) (see Figure 21.24). For a rightward movement condition, alpha activity from left frontal-central channels (FC5 and FC3) is also significantly correlated with beta band activity in central channels. However, unlike the leftward condition, this correlation occurs almost immediately after the presentation of the cue, before the 600-millisecond time mark. One plausible explanation for this phenomenon is the right-handedness of the participants in the study. Significant differences founded in the time interval (600–640 milliseconds) are summarized in Figure 21.27. These novel findings could potentially open up new lines of hypothesis on the dynamic nature of cross-frequency neuronal interactions.

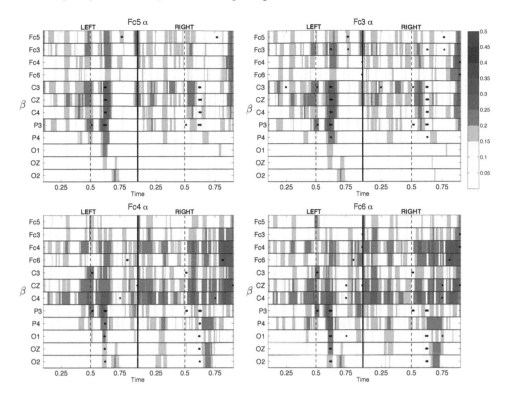

FIGURE 21.24
Local evolutionary dual coherence estimate between alpha activity at the frontal-central channels and beta activity in the rest of the channels. Each plot has estimates for both the left and right conditions. Vertical dashed lines at time 0.5 sec for the left and right conditions represent time when the visual cue was presented. The color indicates the magnitude of the coherence estimates. White indicates insignificant coherence at FDR level 0.05. The black dots denote statistically significant differences between coherence values in the left vs. right conditions.

21.6.6 Conclusion

In this section, we presented the evolutionary dual coherence (EDC), which is a novel measure of frequency-domain coherence for non-stationary multichannel EEGs. The proposed measure is more general than the classical coherence concept because (a) it measures the interactions between oscillations at different frequencies; and (b) it allows this dependence measure to change over time. This approach, developed in (30), is a solution to the long-standing problem of an evolving dual-frequency coherence. To the best of our knowledge, this is the only model that currently rigorously defines the unknown quantity of evolving dual-frequency coherence. Even in the less complicated situation where the dual coherence does not change over time (i.e., classical harmonizable process), the only existing model with a rigorous estimation is given in (48). However, that approach imposes the severe constraint that dual-frequency coherence between components must occur *only* at frequencies that lie parallel to the main diagonal of the Loève spectral matrix. The implication here is that such processes are almost periodically correlated (i.e., processes whose correlation structure repeats in time over some period). Our proposed model allows a more general dual-coherence structure and hence is more flexible for physical modeling of brain signals. To perform sta-

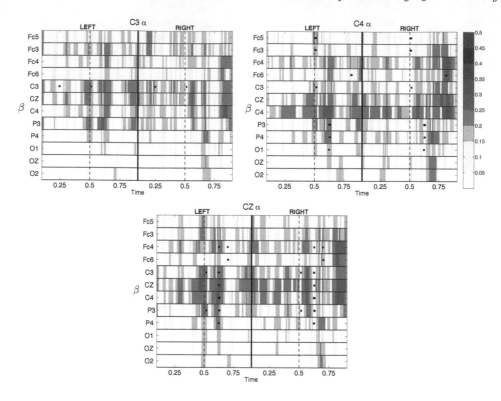

FIGURE 21.25
Local evolutionary dual coherence estimate between alpha activity at the central channels and beta activity in the rest of the channels. Each plot has estimates for both the left and right conditions. Vertical dashed lines at time 0.5 sec for the left and right conditions represent time when the visual cue was presented. The color indicates the magnitude of the coherence estimates. White indicates insignificant coherence at FDR level 0.05. The black dots denote statistically significant differences between coherence values in left vs. right conditions.

tistical inference on the evolutionary dual-frequency coherence (EDC), a non-parametric estimator was developed in the context of an experimental setting with replicated data. The approach—both the estimator and the modeling framework—takes full advantage of the replicated time series available in experimental settings.

21.7   Summary

We cover in this chapter a number of rigorously developed approaches for modeling and analyzing multichannel EEG recordings as realizations of complex and highly dynamic brain processes. The complexity of these signals is manifested in the richness of their attributes, which the methods presented here evaluate in the time and frequency domains and with different but related goals, all of which form part of an improved understanding of brain processes recorded by EEG. We use two datasets to illustrate our methods: In Sections 21.2, 21.3, 21.4 and 21.5, we analyse EEGs recorded during an epileptic seizure. Seizure EEGs are

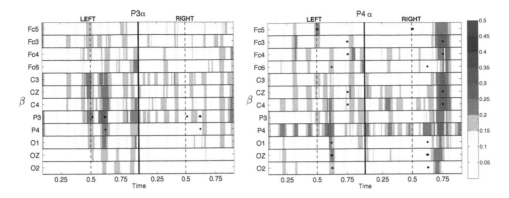

FIGURE 21.26

Local evolutionary dual coherence estimate between alpha activity at the parietal channels and beta activity in the rest of the channels. Each plot has estimates for both the left and right conditions. Vertical dashed lines at time 0.5 sec for the left and right conditions represent the time when the visual cue was presented. The color indicates the magnitude of the coherence estimates. White indicates insignificant coherence at FDR level 0.05. The black dots denote statistically significant differences between coherence values in left vs. right conditions.

realizations of a rapidly changing electrophysiological process resulting from abnormal firing behavior of network of neuronal subpopulations. These aberrations in neuronal electrical activity are expressed by fluctuating amplitudes of waveforms, changing spectral decompositions and evolving cross-dependence between channels. A complete analysis of this type of a brain signal requires a variety of advanced tools, which the authors have developed.

In Section 21.6, we analyze data from repeated trials of a visual-motor task. The strong statistical framework that can be built under the assumption of trials that are all realizations of the same stochastic process allows us to make inference in a higher level of complexity, namely the analysis of coherence between different channels and different frequency bands.

In Section 21.2, we use the SMC method for two main purposes: to cluster cortical regions according to their spectral behavior and to investigate whether or not the cluster formation varies between time intervals pre-, during and post-seizure. The results show the dynamic behavior of the seizure EEGs as the cluster memberships change over time. The SMC method also identifies the frequency bands that are primarily responsible for the cluster formation. It is interesting to note that the dominant frequencies bands (i.e., those that dictate how the channels are clustered together) also evolve over time from low frequencies in the pre-ictal state to the beta-gamma bands immediately following seizure onset.

While the SMC method demonstrates evolution in the cluster formations between predetermined time periods before, during and after the seizure, the FreSpeD method presented in Section 21.3 directly identifies change points in frequency bands of the autospectra and cross-coherences in the multi-channel EEG recording. The method directly provides the user with estimates of change points in time, each of which can be attributed to one or several EEG channels and frequency bands.

It is remarkable that the FreSpeD method is sensitive even to subtle changes in the cross-coherence immediately prior to the seizure onset. This is an example of a feature that the FreSpeD method discovers, which has not been detected in previous analyses (e.g. 57, 18), and may be considered a potential spectral precursor marker of seizure.

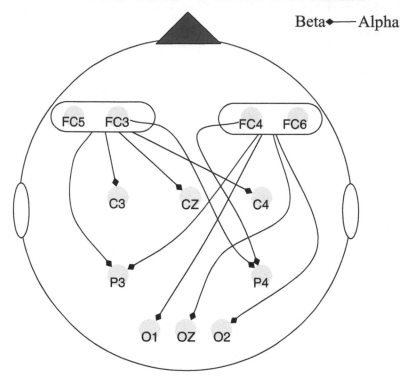

Significant differences in the interval (600−640) milliseconds from fronto-central channels to the rest of the channels

FIGURE 21.27

Significant differences between the left vs. right directions were observed over the time interval $(600, 640)$ milliseconds which is also equivalent to 100–140 milliseconds post visual cue presentation. This figure refers to differences in coherence of alpha activity in *right frontal-central channels* with beta activity in the rest of the channels and differences in coherence of alpha activity in left frontal-central channels with beta activity in the rest of the channels.

Both the SMC and FreSpeD methods use the spectral properties of EEG signals to identify clusters and change points. The spectral matrix provides an intuitive interpretation for signals such as decompositions of waveforms and correlations between oscillatory components. To estimate the spectral matrix using non-parametric methods (i.e., based on periodograms), a stretch or time blocks of data is required. Parametric methods for time-domain models are able to handle EEG data with rapidly changing stochastic properties since they do not require blocks of data. This flexibility comes at the cost of risking model misspecification, which is less of a problem with nonparametric approaches.

We present here one such parametric method, which analyses dynamic EEG signals in the time domain via the switching VAR (SVAR) model. Under this model, we assume two states, each of which can be described by a VAR process. As expected, the SVAR approach gives a clear segmentation of the data into seizure and non-seizure states. Moreover, the SVAR method suggests that seizure onset can also be characterized in the directionality of the connections between channels, and not only by changes in channel-specific auto-spectrum and coherence, as detected by the FreSpeD method. The SVAR method detects changes in directed connectivity even before seizure onset, which is consistent with the results reported in the coherence using the FreSpeD method. However, the results from applying the FreSpeD method are more easily interpretable for clinicians who view EEG recordings from a frequency-focused perspective.

The SLEX method presented in Section 21.5 is a different approach to analyzing non-stationary signals. Its main goal is the identification of the model (or localized Fourier-type basis that optimally represents the EEG signals). The model selection uses the SLEX principal components, which are dimension-reduced signal summaries. Applied to the seizure EEG data, the method selected a basis that detects the approximate seizure onset, consistent with the findings by the FreSpeD and SVAR methods. It automatically gives spectral and coherence estimates, just like FreSpeD, which can be interpreted as time-dependent analogues of Fourier measures for stationary signals. One feature of the SLEX method is that it seeks a representation in terms of an orthonormal basis, which is mathematically elegant and computationally attractive. However, it suffers from the drawback that the waveforms must have localized support with edges at dyadic points. Thus the SLEX method only identifies the approximate change-point location, unlike the SVAR and FreSpeD methods, which can more accurately identify the change points. One major advantage of the SLEX method, especially when compared to the iterative EM algorithm applied in the SVAR method, is the use of computationally efficient algorithms such as the fast Fourier transform and the best basis algorithm (see 13), which allows for an application to massive and high-dimensional signals.

The first four approaches discussed in this chapter are complementary in the sense that they interrogate different aspects of the seizure EEG data. As we have illustrated here, a single model or method cannot capture the diverse features of these signals. Rather, an assembly of methods is needed in order to uncover the different facets of the signals. The results from these methods, taken together, provide a fuller characterization of the rich and complex brain processes.

Yet we note that these four methods combined can be further improved to give a more complete picture of the brain electrophysiology. The SMC, FreSpeD, SVAR and SLEX methods do not investigate dependence-different oscillatory components of a pair of channels (e.g., alpha oscillations in one channel and beta oscillations in another). This is significant especially now that the evidence for interactions between different oscillations is accumulating. However, the spectral representations in FreSpeD (piecewise quasi-stationary) and SLEX directly assume that the random increments are uncorrelated across frequencies. The uncorrelated increments assumption is also implied in the SMC and SVAR approaches. To address this limitation, Gorrostieta et al. (30) developed the evolving dual-frequency model that is useful for studying how interactions between oscillations at different bands can change across time. The four methods can be further generalized by taking dual-frequency coherence into account when modeling and estimating change points. One could use dual frequency as a feature for determining clusters, identifying change points and selecting the best signal representation.

Bibliography

[1] H Adeli, S Ghosh-Dastidar, and N Dadmehr. A wavelet-chaos methodology for analysis of EEGs and EEG subbands to detect seizure and epilepsy. *Biomedical Engineering, IEEE Transactions on*, 54(2):205–211, 2007.

[2] EA Allen, E Damaraju, SM Plis, EB Erhardt, T Eichele, and VD Calhoun. Tracking

whole-brain connectivity dynamics in the resting state. *Cerebral Cortex*, 24:663–676, 2012.

[3] M Arnold, XHR Milner, H Witte, R Bauer, and C Braun. Adaptive AR modeling of nonstationary time series by means of Kalman filtering. *Biomedical Engineering, IEEE Transactions on*, 45(5):553–562, 1998.

[4] A Baker, M Brookes, I Rezek, S Smith, T Behrens, P Smith, and M Woolrich. Fast transient networks in spontaneous human brain activity. *eLife*, 3(3):1–18, 2014.

[5] R Biscay, M Lavielle, A González, I Clark, and P Valdés. Maximum a posteriori estimation of change points in the EEG. *International Journal of Bio-Medical Computing*, 38(2):189–196, 1995.

[6] DR Brillinger. *Time Series: Data Analysis and Theory*, volume 36. Siam, 1981.

[7] BE Brodsky, BS Darkhovsky, AY Kaplan, and SL Shishkin. A nonparametric method for the segmentation of the EEG. *Computer Methods and Programs in Biomedicine*, 60(2):93–106, 1999.

[8] RT Canolty, E Edwards, SS Dalal, M Soltani, SS Nagarajan, HE Kirsch, MS Berger, NM Barbaro, and RT Knight. High gamma power is phase-locked to theta oscillations in human neocortex. *Science*, 313(5793):1626–1628, 2006.

[9] G Carter and C Knapp. Estimation of the magnitude-squared coherence function via overlapped fast Fourier transform processing. *Audio and Electroacoustics*, 21(4): 337–344, 1973.

[10] C Chang and GH Glover. Time–frequency dynamics of resting-state brain connectivity measured with fMRI. *Neuroimage*, 50(1):81–98, 2010.

[11] J Chen and AK Gupta. *Parametric Statistical Change Point Analysis: With Applications to Genetics, Medicine, and Finance*. Springer Science & Business Media, 2011.

[12] H Cho and P Fryzlewicz. Multiple-change-point detection for high dimensional time series via sparsified binary segmentation. *Journal of the Royal Statistical Society: Series B*, 7(2):475–507, 2014.

[13] RR Coifman and MV Wickerhauser. Entropy-based algorithms for best basis selection. *Information Theory, IEEE Transactions on*, 38(2):713–718, 1992.

[14] I Cribben, R Haraldsdottir, L Atlas, T Wager, and M Lindquist. Dynamic connectivity regression: Determining state-related changes in brain connectivity. *NeuroImage*, 61 (4):907–20, 2012.

[15] R Dahlhaus. Fitting time series models to nonstationary processes. *Annals of Statistics*, 25(1):1–37, 1997.

[16] R Dahlhaus. A likelihood approximation for locally stationary processes. *Annals of Statistics*, 28(6):1762–1794, 2000.

[17] R Dahlhaus. Locally stationary processes. *Handbook of Statistics, Time Series Analysis: Methods and Applications*, pages 351–408, 2012.

[18] RA Davis, TCM Lee, and GA Rodriguez-Yam. Structural break estimation for nonstationary time series models. *Journal of the American Statistical Association*, 101(473): 223–239, 2006.

[19] C Del Percio, C Babiloni, M Bertollo, N Marzano, M Iacoboni, F Infarinato, R Lizio, M Stocchi, C Claudio Robazza, G Cibelli, S Comani, and F Eusebi. Visuo-attentional and sensorimotor alpha rhythms are related to visuo-motor performance in athletes. *Human Brain Mapping*, 30(11):3527–3540, 2009.

[20] T Delmiralp, Z Bayraktaroglu, D Lenz, S Junge, N Busch, B Maess, M Ergen, and C Herrmann. Gamma amplitudes are coupled to theta phase in human EEG during visual perception. *International Journal of Psychophysiology*, 64(1):24–30, 2007.

[21] L Donoho. CART and best-ortho-basis: A connection. *The Annals of Statistics*, 25(5): 1870–1911, 1997.

[22] C Euán, H Ombao, and J Ortega. Detection of spectral synchronicity in brain signals by a clustering approach. Submitted, 2015.

[23] M Fiecas and H Ombao. The generalized shrinkage estimator for the analysis of functional connectivity of brain signals. *Annals of Applied Statistics*, 5(2):1102–1125, 2011.

[24] M Fiecas, H Ombao, C Linkletter, W Thompson, and J Sanes. Functional connectivity: Shrinkage estimation and randomization test. *Neuroimage*, 49(4):3005–3014, 2010.

[25] AA Fingelkurts, AA Fingelkurts, and AY Kaplan. The regularities of the discrete nature of multi-variability of EEG spectral patterns. *International Journal of Psychophysiology*, 47(1):23–41, 2003.

[26] RA Fisher. The general sampling distribution of the multiple correlation coefficient. *Proceedings of the Royal Society A: Mathematical, Physical and Engineering Sciences*, 121(788):654–673, 1928.

[27] KJ Friston. Functional and effective connectivity: A review. *Brain Connectivity*, 1(1): 13–36, 2011.

[28] Z Ghahramani and GE Hinton. Parameter estimation for linear dynamical systems. Technical report, CRG-TR-96-2, University of Toronto, Department of Computer Science, 1996.

[29] NR Goodman. Statistical analysis based on a certain multivariate complex Gaussian distribution. *Annals of the Mathematical Statistics*, 34(1):152–177, 1963.

[30] C Gorrostieta, H Ombao, and R von Sachs. Time-dependent dual-frequency coherence analysis. Submitted, 2015.

[31] Cristina Gorrostieta, Mark Fiecas, Hernando Ombao, Erin Burke, and Steven Cramer. Hierarchical vector auto-regressive models and their applications to multi-subject effective connectivity. *Frontiers in Computational Neuroscience*, 7, 2013.

[32] A Hanssen, Y Larsen, and LL Scharf. Complex time-frequency and dual frequency spectra of harmonizable processes. In *Proceedings of EUSIPCO*, pages 1577–1580, 2010.

[33] L Harrison, William D Penny, and Karl Friston. Multivariate autoregressive modeling of fMRI time series. *NeuroImage*, 19(4):1477–1491, 2003.

[34] Martin Havlicek, Jiri Jan, Milan Brazdil, and Vince D Calhoun. Dynamic Granger causality based on Kalman filter for evaluation of functional network connectivity in fMRI data. *Neuroimage*, 53(1):65–77, 2010.

[35] H Hindberg and A Hanssen. Generalized spectral coherences for complex-valued harmonizable processes. *IEEE Transactions on Signal Processing*, 55(6):2407–2413, 2007.

[36] H-Y Huang, H Ombao, and DS Stoffer. Discrimination and classification of nonstationary time series using the SLEX model. *Journal of the American Statistical Association*, 99(467):763–774, 2004.

[37] A Hutt and MHJ Munk. Detection of phase synchronization in multivariate single brain signal by a clustering approach, In *Coordinated Activity in the Brain*, pages 149–164. Springer Series in Computational Neuroscience. Springer, New York, 2009.

[38] O Jensen and L Colgin. Cross-frequency coupling between neuronal oscillations. *Trends in Cognitive Sciences*, 11(7):267–269, 2007.

[39] P Jiruska, J Csicsvari, AD Powell, JE Fox, W-C Chang, M Vreugdenhil, X Li, M Palus, AF Bujan, and RW Dearden. High-frequency network activity, global increase in neuronal activity, and synchrony expansion precede epileptic seizures in vitro. *Journal of Neuroscience*, 30(16):5690–5701, 2010.

[40] Y Kakizawa, RH Shumway, and M Taniguchi. Discrimination and clustering for multivariate time series. *Journal of the American Statistical Association*, 93(441):328–340, 1998.

[41] A Kaplan, J Röschke, B Darkhovsky, and J Fell. Macrostructural EEG characterization based on nonparametric change point segmentation: Application to sleep analysis. *Journal of Neuroscience Methods*, 106(1):81–90, 2001.

[42] C Kirch, B Muhsal, and H Ombao. Detection of changes in multivariate time series with application to EEG data. *Journal of the American Statistical Association, in press*, 2014.

[43] A Kucyi and K Davis. Dynamic functional connectivity of the default mode network tracks daydreaming. *NeuroImage*, 100:471–80, 2014.

[44] TS Kumar, V Kanhangad, and RB Pachori. Classification of seizure and seizure-free EEG signals using multi-level local patterns. In *Digital Signal Processing, 2014 19th International Conference on*, pages 646–650. IEEE, 2014.

[45] N Lachiche, J Hommet, J Korczak, and A Braud. Neuronal clustering of brain fMRI images. *Pattern Recognition and Machine Intelligence: Lecture Notes in Computer Science*, 3776:300–305, 2005.

[46] N Leonardi, J Richiardi, M Gschwind, S Simioni, J-M Annoni, M Schluep, P Vuilleumier, and D Van De Ville. Principal components of functional connectivity: A new approach to study dynamic brain connectivity during rest. *Neuroimage*, 83:937–950, 2013.

[47] TW Liao. Clustering of time series data: A survey. *Pattern Recognition*, 38:1857–1874, 2005.

[48] KS Lii and M Rosenblatt. Spectral analysis for harmonizable processes. *Annals of Statistics*, 30(1):258–297, 2002.

[49] MA Lindquist, Y Xu, MB Nebel, and BS Caffo. Evaluating dynamic bivariate correlations in resting-state fMRI: A comparison study and a new approach. *Neuroimage*, 101:531–546, 2014.

[50] M Loève. *Probability Theory.* The University Series in Higher Mathematics, 1955.

[51] A Matysiak, PJ Durka, E Martinez Montes, M Barwiñski, P Zwoliñski, M Roszkowski, and KJ Blinowska. Time-frequency-space localization of epileptic EEG oscillations. *Acta Neurobiologiae Experimentalis*, 65(4):435, 2005.

[52] R Monti, P Hellyer, D Sharp, R Leech, C Anagnostopoulos, and G Montana. Estimating time-varying brain connectivity networks from functional MRI time series. *NeuroImage*, 103:427–43, 2014.

[53] KP Murphy. Switching Kalman filter models. Compaq Cambridge Res. Lab., Cambridge, MA, Tech. Rep. 98–100, 1998.

[54] H Ombao and S van Bellegem. Evolutionary coherence of nonstationary signals. *Signal Processing, IEEE Transactions on*, 56(6):2259–2266, 2008.

[55] H Ombao, JA Raz, RL Strawderman, and R von Sachs. A simple generalised cross-validation method of span selection for periodogram smoothing. *Biometrika*, 88(4): 1186–1192, 2001.

[56] H Ombao, JA Raz, R von Sachs, and W Guo. The SLEX model of a non-stationary random process. *Annals of the Institute of Statistical Mathematics*, 54(1):171–200, 2002.

[57] H Ombao, R von Sachs, and W Guo. SLEX analysis of multivariate nonstationary time series. *Journal of the American Statistical Association*, 100(470):519–531, 2005.

[58] Amir Omidvarnia, Ghasem Azemi, Boualem Boashash, John M. Otoole, Paul B. Colditz, and Sampsa Vanhatalo. Measuring time-varying information flow in scalp EEG signals: Orthogonalized partial directed coherence. *IEEE Transactions on Biomedical Engineering*, 61(3):680–693, 2014.

[59] J Pardey, S Roberts, and L Tarassenko. A review of parametric modelling techniques for EEG analysis. *Medical Engineering & Physics*, 18(1):2–11, 1996.

[60] P Preuß, R Puchstein, and H Dette. Detection of multiple structural breaks in multivariate time series. *Journal of the American Statistical Association, In Press*, 2014.

[61] MB Priestley. Evolutionary spectra and non-stationary processes. *Journal of the Royal Statistical Society Series B*, 27(2):204–237, 1965.

[62] ME Saab and J Gotman. A system to detect the onset of epileptic seizures in scalp EEG. *Clinical Neurophysiology*, 116(2):427–442, 2005.

[63] B Samdin, C-M Ting, S-H Salleh, M Hamedi, and AB Mohd Noor. Estimating dynamic cortical connectivity from motor imagery EEG using Kalman smoother & EM algorithm. In *Statistical Signal Processing, 2014 IEEE Workshop on*, pages 181–184. IEEE, 2014.

[64] B Samdin, C-M Ting, H Ombao, and S-H Salleh. A unified estimation framework for state-related changes in effective brain connectivity. IEEE Trans. on Biomedical Engineering, 2016, In Press.

[65] LL Scharf, B Friedlander, and DJ Thomson. Covariant estimators of time-frequency descriptors for nonstationary random processes. In *Conference Record of Thirty-Second Asilomar Conference on Signals, Systems and Computers*, pages 808–811. IEEE, 1998.

[66] AL Schröder and H Ombao. FreSpeD: Frequency-specific change-point detection in multichannel EEG seizure recordings. Revised, 2016.

[67] RH Shumway. Discriminant analysis for time series. *Handbook of Statistics*, 2:1–46, 1982.

[68] RH Shumway. Time-frequency clustering and discriminant analysis. *Statistics & Probability Letters*, 63(3):307–314, 2003. ISSN 0167-7152.

[69] RH Shumway and DS Stoffer. *Time Series Analysis and Its Applications: With R Examples*. Springer Science & Business Media, 2010.

[70] JF Smith, A Pillai, K Chen, and B Horwitz. Identification and validation of effective connectivity networks in functional magnetic resonance imaging using switching linear dynamic systems. *Neuroimage*, 52(3):1027–1040, 2010.

[71] J Terrien, G Germain, C Marque, and B Karlsson. Bivariate piecewise stationary segmentation: Improved pre-treatment for synchronization measures used on non-stationary biological signals. *Medical Engineering & Physics*, 35(8):1188–1196, 2013.

[72] CM Ting, A Seghouane, SH Salleh, and AM Noor. Estimating effective connectivity from fMRI data using factor-based subspace autoregressive models. *IEEE Sig. Process. Lett.*, 22:757–761, 2014.

[73] C Tzagarakis, N Ince, A Leuthold, and G Pellizzer. Beta-band activity during motor planning reflects response uncertainty. *Journal of Neuroscience*, 30(34):11270–11277, 2010.

[74] AT Tzallas, MG Tsipouras, and D Fotiadis. Epileptic seizure detection in EEGs using time–frequency analysis. *Information Technology in Biomedicine, IEEE Transactions on*, 13(5):703–710, 2009.

[75] P A Valdés-Sosa, J Sánchez-Bornot, A Lage-Castellanos, M Vega-Hernández, J Bosch-Bayard, L Melie-García, and E Canales-Rodríguez. Estimating brain functional connectivity with sparse multivariate autoregression. *Philosophical Transactions of the Royal Society B: Biological Sciences*, 360(1457):969–981, 2005.

[76] P Welch. The use of fast Fourier transform for the estimation of power spectra: A method based on time averaging over short, modified periodograms. *IEEE Transactions on Audio and Electroacoustics*, pages 70–73, 1967.

[77] GA Worrell, L Parish, SD Cranstoun, R Jonas, G Baltuch, and B Litt. High-frequency oscillations and seizure generation in neocortical epilepsy. *Brain*, 127(7):1496–1506, 2004.

Index

A

AAL, *see* Automated Anatomical Labeling atlas
ABIDE, *see* Autism Brain Imaging Data Exchange
acquisition
    BOLD fMRI, 152–153
    fMRI image, and reconstruction, 214–220
    multisequence sMRI, 115–116
    signal generation, fMRI, 146–148
activation, 9, 229–230
ADC, *see* Apparent diffusion coefficient-based model
ADHD200 prediction competition, 402
ADJUST algorithm, 187
ADNI, *see* Alzheimer's Disease Neuroimaging Initiative
Advanced Normalization Tools (ANTS), 116–117, 124–126, 130–131, 234
affine registration, 122, 239
AFNI, *see* Analysis for functional neuroimaging
age-specific distribution, 485
Akaike's information criterion (AIC), 26, 423, 477
algorithms, *see also* specific algorithm
    ADJUST, 187
    automated segmentation, 53
    CT reconstruction, 7
    deformable surface algorithms, 236
    Dijkstra's, 448
    Dyadic CART, 604
    fuzzy clustering, 53
    GNG clustering algorithm, 572
    HARDI reconstruction, 80
    IsoData algorithm, 44
    LASSLS, 553–554
    LASSO, 26–27, 90, 551, 552–553
    Levenberg-Marquardt algorithm, 24
    "massively parallel," 478
    measures, surface-based, 48
    morphometric measures, 49–50
    Nelder-Mead simplex algorithm, 24
    non-local means, 81
    non-local threshold algorithms, 44
    N41TK, 119
    Otsu's method, 43–44
    PET reconstruction, 7
    PHYCAA+, 286
    registration, surface-based, 41
    segmentation, 43–44
    simulated annealing, 85
    statistical parametric maps, 51
    SuBLIME, 133
    surface-based methods, 41, 48
    tractography, 83–86
    unbiased NLM, 81
    watershed segmentation, 44
    weighted least square, 79
Allen Human Brain Atlas, 46
Alzheimer's Disease Neuroimaging Initiative (ADNI), 468, 470
A Mathematical Model for Understanding the Statistical (AMMUST) framework, 220, 222, 227–228
AMMUST, *see* A Mathematical Model for Understanding the Statistical framework
amplitude spectrum, 188
analysis for functional neuroimaging (AFNI), 37, 116, 167, 267, 286, 350
anatomical connectivity, 420
anatomical references, functional imaging, 52
anatomical segmentation using proximities (ASP), 234
anisotropic Wiener filtering, 81
ANOVA (analysis of variance)
    ICC, 274
    preprocessing pipeline testing, 278
    random effects t-test, 343
    statistical parametric networks, 457
antipodal symmetry, 88
apodization, 224

apparent diffusion coefficient (ADC)-based model, 76
AR, *see* Autoregressive model
AR2, *see* Second-order autoregressive model
Archimedean spiral, 145
arcs, networks, 444
area, cortical surface features, 242
ARMA, *see* Autoregressive moving average model
arterial spin labeling (ASL), 148, 152
artifacts
    acquisition and reconstruction, 116
    BOLD fMRI, 152–153
    removal from EEG, 184–187
    structural MRI, 39–40
ascendancy measure, 502
ASL, *see* Arterial spin labeling
ASP, *see* Anatomical segmentation using proximities
association measures, 498
asymptotic theory, 86
atlases
    AAL, 371, 381, 421, 444
    Allen Human Brain Atlas, 46
    brain, 53
    Brodman atlas, 45, 371
    Colin27 atlas, 45
    LONI Probabilistic Brain Atlas, 421
    multi-subject MRI, 45–46
    population-based atlases, 45–46
    single-subject atlas, 45
    Talairach atlas, 45, 421
Autism Brain Imaging Data Exchange (ABIDE), 485
autocorrelation function, 188–189
Automated Anatomical Labeling (AAL) atlas
    effective connectivity, 421
    FC of rs-fMRI, 381
    functional connectivity, 371
    vertex set, 444
automated segmentation algorithms, 53
autoregressive (AR) model
    change-point direction, 580
    localization, 161
    single-channel EEG, 533–537
autoregressive moving average (ARMA) model
    localization, 161
    multichannel EEG, 549
    noise and nuisance signal, 156

single-channel EEG, 532, 534–539
Auto-SLEX method, 581

B
ball and stick model, 71
Balloon model
    BOLD, 153–154
    DCM, 428–429
    nonlinear models, 324–325
    Volterra series model, 325
bands, coherence between, 613–614
basis function methods
    nonparametric/semiparametric models, 478
    PET, 25–26
    region-wise linear models, 475
    registration, 90
Bayesian models and techniques
    ARMA model, 538–539
    background, 7
    change-point direction, 580
    DCM, 429
    dynamic connectivity states estimation, 594
    EEG neurophysics, 179
    false discovery rate, 362
    functional connectivity, 370, 499, 502
    HRF, time domain, 316
    joint activation/structural connectivity, 511
    mixed-effects/fixed-effects analyses, 342
    multimodal approaches, 508
    PCA, 402
    PET, 28
    prediction, 282
    reconstruction, 80
    repeated cross-sectional subsamples, 480
    sampling mechanisms, 88
    sparse Bayesian Learning approach, 28
    spatial-adaptive estimation methods, 81
    structural equation models, 423, 426
    surface-based registration, 42
    tractography algorithms, 85
    volumetric registration, 41
$B0$ correction, 228
BEM, *see* Boundary element method
Benjamini-Hochberg procedure, 363
Bernoulli random variables, 451, 454–455
Bessel function, 75, 77

BET, *see* Brain extraction tool
beta distribution, exact, 614
between-subject, MRI brain registration, 277–278
bias field correct, 40
bi-exponential nonlinear model, 326
"Big Data" problems, 359
bilateral filtering, 81
bi-modal distribution, 43
binary labels, 44
binary segmentation (BS), 580–581
biophysical models, 450
blocked designs, fMRI, 157
blood oxygenation level dependent (BOLD) fMRI
    acquisition artifacts, 152–153
    background, 6–7, 11
    Balloon model, 324–325
    brain variability, 338
    combined prediction *vs.* spatial reproducibility metrics, 277
    concepts, 148–150
    DCM, 428–429
    effective connectivity, 422, 435
    frequency domain, 322
    functional connectivity, 498
    general linear model, 310
    Granger causality, 431
    group-subject model, 341
    HRF, 313, 322
    image reconstruction, fMRI, 205
    LF-PCA, 485
    localization, 161
    modeling signal and noise, 153–155
    pipeline optimization, 290
    prediction, 283
    preprocessing, 270, 278–279, 290
    rs-fMRI, 165–166, 278
    simulated dataset, 268
    spatial limitations, 150–151
    structural equation models, 423, 425–427
    structural MRI, 39
    temporal limitations, 151–152
    time domain, 313
Bonferroni procedures
    dual-frequency coherence between bands, 613
    false discovery rate, 360
    family-wise error rate, 356–358, 360

group studies, functional neuroimaging, 337
    multiplicity corrections, 355–356
    spatial dependence, 363
    statistical parametric maps, 51
    thresholding networks, 446
bootstrapping, 86–87, 477
borrowing information "spatially," 477
boundary element method (BEM), 178, 559
Box-Cox transformed distribution, 485
brain
    activation, 139, 165
    atlas, 53
    cancers, 17, 19
    overview, 36–37
    shape and function variability, 337–338
    tissue segmentation, 44–45
brain, multivariate decompositions
    background, 10
    factor analysis, 410
    high-dimensional models, 404–405, 407–408
    homotopic group ICA, 407
    independent component analysis, 405–408
    introduction, 399–400
    non-negative matrix decomposition, 409–410
    partial least squares regression, 409
    population value decomposition, 410
    principal component analysis, 400–402
    projection pursuit, 410
    singular value decomposition, 400–402
    sparce PCS framework, 409
    structured PCA models, 403–405
    tensor (multi-way) statistical analysis, 410–411
brain, surface analysis
    affine registration, 239
    area and curvatures, 242
    background, 8
    cortical surface features, 241–243
    data smoothing, 243–248
    diffeomorphic registration, 241
    diffusion smoothing, 243–245
    gray matter volume, 242–243
    heat kernel smoothing, 246–248
    introduction, 233–235
    iterated kernel smoothing, 245–246
    linear models, 248–249
    local parameterization, 236

longitudinal models, 251–252
multivariate linear models, 249–250
quadratic polynomial, 236
random field theory, 252–254
small-n large-p problems, 250
spherical harmonic representation,
    237–238
statistical inference, 248–254
surface flattening, 236
surface parameterization, 235–238
surface registration, 239–241
thickness, 241
brain connectivity
basic concepts, 495–496
functional connectivity, 496–502,
    508–509
statistical methods, 9–10
structural connectivity, 496, 502–505,
    509–512
brain connectivity, joint fMRI and DTI
    models
anatomical weighting, 508–509
ascendancy, 511
background, 10
deterministic tractography, 504
diffusion weighting imaging, 503–504
effective connectivity methods, 502
functional coherence, 510–511
joint activation modeling, 509–512
joint ICA, 512
likelihood function, 511–512
measures of association, 498
modeling approaches, 499
multimodal approaches, 505–513
network methods, 500–502
partitioning methods, 500
prediction methods, multimodal,
    512–513
probabilistic tractography, 505
sequential procedures, 506–508
single modality methods, 497–505
spatial scale, 497
structural connectivity determination,
    502–505
tractography, 504–505
brain extraction tool (BET), 119
Brainmap database, 167
Brodman atlas, 45, 371
BS, *see* Binary segmentation
Butterworth filter, 381

C
C computer language, 113
canonical aspects
HRF, 315, 320–322, 326
ICA, 407
SPM99 HRF, 326
variates analysis (CVA), 283, 289–293,
    402
"capping," 42
CAR, *see* Conditional autoregressive model
Carp's arguments, 280
Cartesian operations
characteristic path length, 447
DSI, 73–74
HYDI, 74
image formation, 145
image processing implications, 227
QBI, 75
second-order autoregressive model, 537
signal equation and k-space coverage,
    215
CAT, *see* Computed axial tomography
causal inference, *see* Effective connectivity
    and causal inference
CCoh, *see* Cross-coherence
CCor, *see* Cross-correlation
CD, *see* Critical-difference interval
cerebrospinal fluid (CSF)
brain surface analysis, 234
defined, 36
EEG neurophysics, 177
image contrast, 143
intensity normalization, 127
multi-subject structural MRI, 40
rs-fMRI, 166
segmentation, 43
spatial registration, 123
change, modeling components of
age-specific distribution, 485
background, 9
borrowing information "spatially," 477
cross-sectional data, 473
cross-sectional designs, 469–470
discussion, 486–487
individual-specific curves, 482–483
introduction, 468
linear models, 472–477
local rates of change, 485–486
longitudinal data, 473–474
longitudinal functional PCA, 483–485
marginal models, 474

mean trajectory, 472–477

misalignment complications, 475–477

mixed-effects models, 473–474

multi-cohort longitudinal designs, 471–472

nonlinear models, 477–482

nonparametric models, 477–480

notation, 468–469

polynomial models, 477

region-wise linear models, 472–477

relative efficiency, mean function estimation, 474–475

repeated cross-sectional subsamples, 480–482

semiparametric models, 477–480

single-cohort longitudinal designs, 470–471

change-point detection

existing methods/challenges, 580–582

FreSpeD method, 582–585

multichannel seizure EEG data, 585–588

overview, 580

potential precursors, 585–588

seizure localization, 585

seizure onset estimation, 585–588

changes, 8–9

characteristic path length, 447–448, 458

CHARMED model, 71

chemical-shift artifact, 39

Chernoff technique, 572

Chi distribution, 80

Children's Hospital Boston, 129

circle integration, 75

classical statistics, 336

class labels, 44

"cliques," 499

cluster-mass testing, 345

clusters and clustering

analysis, 500

coefficient, 448–449

epileptic seizure data, 574–579

hierarchical spectral merger algorithm, 573–574

overlap, pipeline evaluation metrics, 271–272

overview, 572

size significance, 347–350

spectral merger clustering method, 572–574

total variation distance, 572–573

Cohen kappa statistic, 511

coherence

between bands, 613–614

edge set, 446

functional connectivity, 498

multichannel EEG, 541–543

stationary data analysis, 190–193

Colin27 atlas, 45

combined prediction *vs.* spatial reproducibility metrics, 276–277

comparisons, network analysis, 455–459

compartmental approaches, 20–24, 28–29

complexity-penalized Kullback-Leibler criterion, 603–604

complexity theory, 442

complex-valued fMRI activation, 229–230

computed axial tomography (CAT/CT), 5

computer tomography (CT), 41

conditional autoregressive (CAR) model, 363–364

conditional Granger causality, 432

conductivity of skull, 178

confidence intervals, single-channel EEG, 540–541

connected triples, 448

connection density, edge set, 446

connectivity

data analysis, 162–164

directed, high-dimensional data, 555

multichannel EEG, 549–551

overview, 420

contrast, 143, 340

Cook's distance, 79

co-registration, 52, 159

corrections for multiplicity

background, 9

dimensions for consideration, 364–365

false discovery rate control, 360–362

family-wise error rate, 356–360

introduction, 355–356

spatial dependence, accounting for, 362–364

correlations, 122, 275, *see also* Cross-correlation

cortical folding, 49

cortical surface features, 8, 241–243

cotan estimation, 244

Cramer spectral representation, 12, 527–528, 599

critical-difference (CD) interval, 288

cross-coherence (CCoh), 373–374, 378, 384

cross-correlation (CCor), 371–372, 378, 384

cross-periodograms, 612, *see also* Periodogram methods

cross-sectional and longitudinal designs

age-specific distribution, 485

background, 9

borrowing information "spatially," 477

cross-sectional data and designs, 469–470, 473

discussion, 486–487

individual-specific curves, 482–483

introduction, 468

linear models, 472–477

local rates of change, 485–486

longitudinal data, 473–474

longitudinal functional PCA, 483–485

marginal models, 474

mean trajectory, 472–477

misalignment complications, 475–477

mixed-effects models, 473–474

multi-cohort longitudinal designs, 471–472

nonlinear models, 477–482

nonparametric models, 477–480

notation, 468–469

polynomial models, 477

region-wise linear models, 472–477

relative efficiency, mean function estimation, 474–475

repeated cross-sectional subsamples, 480–482

semiparametric models, 477–480

single-cohort longitudinal designs, 470–471

cross-sectional data and designs, 469–470, 473

cross-sectional intensity analysis, 131–133

cross spectrum, 190–191

cross-validation bandwidth selection method, 82

CSF, *see* Cerebrospinal fluid

CT, *see* Computed axial tomography; Computer tomography

cumulative-sum (CUSUM) test statistic

change-point direction, 581

FreSpeD method, 583

seizure localization, 585, 587

curvature, cortical surface features, 242

CUSUM, *see* Cumulative-sum test statistic

"cutting the potato" example, 129

CVA, *see* Canonical variates analysis

D

DAGs, *see* Directed acrylic graphs

Dahlhaus representation, 599

data

connectivity, 162–164

format, fMRI, 166–168

localization, 160–162

prediction, 164–165

resting-state, fMRI data, 381

signal generation, 141–144

single-channel EEG, 525

smoothing, brain surfaces, 243–248

structure, multisequence sMRI, 114–115

databases, 166–169

datasets, 93–94, 268, 285–286

DBM, *see* Deformation-based morphometry

DBVPM, *see* Dynamic Bayesian variable partition model

DCE-MRI, *see* Dynamic contrast enhanced MRI

DCM, *see* Dynamic causal models

DCR, *see* Dynamic connectivity regression

dead salmon example, 359

decompositions, *see* Multivariate decompositions

deconvolution process, 29

"default mode network" (DMN), 381

deformable surface algorithms, 236

deformation-based morphometry (DBM), 47–48, 51

degenerate tensors, 86

degree distributions, 449–450, 453

delta functions, 76, 90, 313

density-integrated topology, 457–458

density of a graph, networks, 443–444

dependent clusters, 364–365

desynchronization, 180, 194

DET, *see* Temporal detrending

deterministic tractography, 83–84, 506

Dice coefficient, 272

DICOM (Digital Imaging and Communications in Medicine) format, 167

diffeomorphic registration, 241

differential networks, 457

diffusion gradient sequence, 66

diffusion magnetic resonance imaging (dMRI)

acronyms, 95

ADC-based model, 72–73

background, 6, 9–10
basic concepts, 66–69
datasets, 93–94
diffusion gradient sequence, 66
diffusion orientation transform, 77
diffusion propagator imaging, 77–78
diffusion spectrum imaging, 73–74
diffusion tensor imaging, 69–70
diffusion weighted imaging, 66–69
eigenvectors of diffusion tensor, 69–70
exact Q-Ball imaging, 76
free diffusion, 67
generalized DTI, 71–72
glossary, 95
group analysis, 90–93
high angular resolution diffusion
    imaging, 70–78
high-order tensor model, 72–73
hybrid diffusion imaging, 74–75
introduction, 66
mixture of tensor model, 70–71
noise components, 79–80
public resources, 93–95
Q-Ball imaging, 75–76
reconstruction, 79–83
registration, 89–90
restricted diffusion, 67–69
sampling mechanisms, 87–89
scalar indices, 69–70
simple harmonic oscillator
    reconstruction and estimation, 78
software, 94–95
spatial-adaptive estimation methods,
    80–83
spherical deconvolution, 77
spherical polar Fourier imaging, 78
tractography algorithms, 83–86
uncertainty, estimated diffusion
    quantities, 86–87
voxelwise estimation methods, 79–80
diffusion orientation transform (DOT), 77,
    80
diffusion propagator imaging (DPI), 77–78
diffusion smoothing, 243–245
diffusion spectrum imaging (DSI), 73–74,
    507
diffusion tensor, 67
diffusion tensor imaging (DTI)
    anatomical connectivity, 420
    connectivity, 162
    diffusion gradient sequence, 69–70

diffusion weighting imaging, 503–504
functional connectivity, 370
generalization, 70–73
joint ICA, 512
LF-PCA, 483–484
MRI scanners, 142
MRI to fMRI, 147
structural connectivity, 389, 503
structured PCA, 403
vertex set, 445
diffusion tensor imaging (DTI), joint fMRI
    models
    anatomical weighting, 508–509
    ascendancy, 511
    deterministic tractography, 504
    diffusion weighting imaging, 503–504
    effective connectivity methods, 502
    functional coherence, 510–511
    joint activation modeling, 509–512
    joint ICA, 512
    likelihood function, 511–512
    measures of association, 498
    modeling approaches, 499
    multimodal approaches, 505–513
    network methods, 500–502
    partitioning methods, 500
    prediction methods, multimodal,
        512–513
    probabilistic tractography, 505
    sequential procedures, 506–508
    single modality methods, 497–505
    spatial scale, 497
    structural connectivity determination,
        502–505
    tractography, 504–505
diffusion weighted imaging (DWI), 66–69,
    142, 503
Dijkstra's algorithm, 448
dimension reduction methods, *see* Effective
    connectivity and causal inference
dipole moment, EEG neurophysics, 177
Dirac-delta function, 245
directed acrylic graphs (DAGs), 502
directed graphs, 444, 450
direct effects, effective connectivity and
    causation, 434–435
Dirichlet distribution, 499
disconnected graphs, 447, 458
discrete prolate spheroidal sequences
    (DPSS), 531
discretized frequencies, 612–613

DISTATIS, 275, 289
DMN, *see* "Default mode network"
dMRI, *see* Diffusion magnetic resonance
        imaging
DOT, *see* Diffusion orientation transform
DPI, *see* Diffusion propagator imaging
DPSS, *see* Discrete prolate spheroidal
        sequences
drift
        acquisition artifacts, 152
        HRF, time domain, 313, 318
        noise and nuisance signal, 155–156
        structural equation models, 426
DSI, *see* Diffusion spectrum imaging
DTI, *see* Diffusion tensor imaging
dual-frequency coherence analysis
        coherence between bands, 613–614
        cross-periodograms, 612
        discretized frequencies, 612–613
        EEG data, 615–618
        Evolutionary Dual Frequency Spectra,
            612–618
        exact beta distribution, inference on,
            614
        historical developments, 610–611
        implementation details, 615–618
        inference, local dual frequency
            coherence, 614
        introduction, 608–610
        overview, 610–611
        time-dependent harmonizable process,
            613
DWI, *see* Diffusion weighted imaging
Dyadic CART algorithm, 604
dynamic Bayesian variable partition model
        (DBVPM), 376
dynamic causal models (DCM)
        BOLD, 154
        effective connectivity, 421
        effective connectivity and causal
            inference, 427–429
        functional connectivity, 502
        Granger causality, 431
        multichannel EEG, 549
        structural equation models, 426, 427
dynamic connectivity regression (DCR), 376
dynamic connectivity state estimation,
        593–598
dynamic contrast enhanced MRI
        (DCE-MRI), 112

E
EAP, *see* Ensemble average propagator
EC, *see* Effective connectivity; Euler
        characteristic
echo-planar imaging (EPI), 145
echo time (TEs), 39, 53
ECoG, *see* Electrocorticogram
EDC, *see* Evolutionary dual-frequency
        coherence analysis
"edge contrast," 457
edges, networks, 443
edge set, 446
EDS, *see* Evolutionary Dual Frequency
        Spectra (EDS)
EDSS, *see* Expanded Disability Status Scale
EEG, *see* Electroencephalograms
EEG-Lab, 267
effective connectivity (EC), 162, 163, 496
effective connectivity (EC), and causal
        inference
        causation, 432–435
        concepts, 420–422
        dynamic causal models, 427–429
        Granger causality, 429–432
        introduction, 419–420
        models, 422–432
        structural equation models, 422–427
eigenvectors of diffusion tensor, 69–70
Einstein summation convention, 71
EKG, *see* Electrocardiography
electrocardiography (EKG), 185
electrocorticogram (ECoG)
        edge set, 446
        network analysis, 442
        synchronization, 182
        vertex set, 445
electroencephalograms and
            electroencephalography (EEG)
        analysis, statistical methods, 10
        artifact removal, 184–187
        background, 4, 6
        DCM, 427–428
        edge set, 446
        experimental design, 11–12
        functional connectivity, 370
        Granger causality, 431
        ICA, 406
        introduction, 175–177
        joint ICA, 512
        MRI to fMRI, 148
        multichannel, 548–549

network analysis, 442
neurophysics, 177–179
nonstationary data analysis, 194–197
preprocessing, 183–184
recording, 182–183
single channel, 539–540
stationary data analysis, 187–193
structural equation models, 423
synchronization, 180–182
vertex set, 445
electroencephalograms and
electroencephalography (EEG),
modeling
background, 6, 10
change-point detection, 580–588
clustering of EEGs, 572–579
coherence between bands, 613–614
concept formalization, 612–614
discretized frequencies, 612–613
dynamic connectivity state estimation,
593–598
EEG data analysis, 615
epileptic EEG, 593–598
evolutionary dual-frequency coherence
analysis, 608–618
exact beta distribution, 614
FreSpeD method, 582–585
harmonizable process, 612–613
historical developments, 610–611
implementation details, 615–618
inference procedure, 614
introduction, 568–571
local dual-frequency
cross-periodogram, 612
multichannel seizure EEG data,
585–588
non-stationary EEGs, best signal
representation, 599–608
parameter estimation, 592–593
signal representations, 599
SLEX, 600–606
spectral merger clustering method,
572–574
stationary VAR model, 589–590
switching VAR models, 588–598
time-dependent harmonizable process,
613
time-varying VAR model, 590–591
electroencephalograms and
electroencephalography (EEG),
statistical analyses

background, 10
connectivity, model and inference,
549–551
data description, 525
dipole source model, 556–557
Fourier-domain approach, 525–532,
542–544
generalized model, EEG signals, 559
Granger causality, 550
high-dimensional data, 551–555
independent source model, 558
introduction, 524–525
inverse source reconstruction, 559–562
multichannel EEG, 541–551
multiple EEG traces, 548–549
partial directed coherence, 550–551
single-channel EEG, 524–541
source localization and estimation,
555–562
source models, 555–556
specific frequency band spectral matrix
estimation, 549
spectral matrix estimation, 548–549
time-domain approach, 532–539,
544–545
VAR modeling, multichannel EEG, 554
electromyography (EMG), 185
electrooculographic (EOG) signals, 185
EM, *see* Expectation-maximization
algorithm
embedded SEMs, 425
EMG, *see* Electromyography
ensemble average propagator (EAP)
DOT, 77
DPI, 78
DSI, 74
estimated diffusion quantities,
uncertainty, 86
generalized DTI, 72
HYDI, 75
reconstruction, 79
registration, 89–90
restricted diffusion, 68
sampling mechanisms, 87–88
SD methods, 77
spatial-adaptive estimation methods,
83
SPFI, 78
entropy, 122–123
EOG, *see* Electrooculographic signals
EP, *see* Evoked potentials

EPI, *see* Echo-planar imaging
epileptic EEG, 593–598
epileptic seizure data, 574–579
Erdős-Rényi random graphs, 451–452
ERGM, *see* Exponential random graph
    models
ERP, *see* Event related potentials
error control, type 1, 346
E-step, *see* Expectation step
estimation
    diffusion quantities, uncertainty, 86–87
    dipole source model, 556–557
    group analysis, functional
        neuroimaging, 341–342
    network analysis, 455–459
    parameters, switching VAR models,
        592–593
    PET, 24–27
    resting-state, fMRI data, 382
Euclidian characteristics
    diffusion smoothing, 245
    individual subject's curves, 285
    surface area and curvatures, 242
    surface-based registration, 41
    surface data smoothing, 243
    volumetric registration, 41
Euler characteristic (EC), 8, 359
Euler's approximation, 85
Euler scheme, 245
euSEM, *see* "Extended unified SEM"
event-related designs, fMRI, 157–158
event related potentials (ERP), 12, 194
evoked potentials (EP), 194
evolutionary dual-frequency coherence
    (EDC) analysis
    coherence between bands, 613–614
    concept formalization, 612–614
    discretized frequencies, 612–613
    EEG data analysis, 615
    exact beta distribution, 614
    harmonizable process, 612–613
    historical developments, 610–611
    implementation details, 615–618
    inference procedure, 614
    introduction, 608–610
    local dual-frequency
        cross-periodogram, 612
    overview, 610–611
    time-dependent harmonizable process,
        613

Evolutionary Dual Frequency Spectra
    (EDS), 612–616
exact beta distribution, 614
exact Q-Ball imaging, 75–77
Expanded Disability Status Scale (EDSS),
    403
expectation-maximization (EM) algorithm
    background, 5
    dynamic connectivity states estimation,
        593
    HDICA, 408
    ICA, 406–407
    mixed-effects/fixed-effects analyses,
        342–343
    reconstruction, 79
    similarity metric ranking approaches,
        276
    SVAR models, 589
expectation step (E-step), 592
experimental design, 11–12, 157–158
exponential random graph models (ERGM)
    functional connectivity, 501–502
    future directions, 385
    network analysis, 453–454
"extended unified SEM" (euSEM), 426
extraction, non-redundant information, 604

F
FA, *see* Fractional anisotropy
F statistic, 343
FACT, *see* Fiber Assessment by Continuous
    Trajectory
factor analysis, 410
FADTTS, *see* Functional analysis of
    diffusion tensor tract statistics
    pipeline
false detection rate, 51
false discovery rate (FDR)
    control, 364
    dual-frequency coherence between
        bands, 613
    functional neuroimaging data, 360–362
    local dual-frequency coherence, 616
    localization, 162
    random field theory, 252
    spatial context, statistical inference,
        344
    spatial dependence, 363
    spatial pattern reproducibility, 275
    statistical parametric maps, 51
    statistical parametric networks, 456

thresholding networks, 446
false positive rates (FPR), 268
family-wise error rate (FWER)
 control, 364
 false discovery rate, 360
 functional neuroimaging data, 356–360
 localization, 162
 permutation tests, 346
 spatial context, statistical inference,
  344
 statistical parametric maps, 51
fast Fourier transform, *see also* Fourier
  transform and methods
 DSI, 73–74
 dual-frequency coherence analysis, 611
 image formation, 145
 preprocessing EEG, 184
 single-channel EEG, 529
 stationary data analysis, 187–188
fast ICA, 405
FC, *see* Functional connectivity
FDR, *see* False discovery rate
FEM, *see* Finite element method
[^{18}F]FDG, *see* Fluorodeoxyglucose
FGMM, *see* Finite Gaussian mixture model
Fiber Assessment by Continuous Trajectory
  (FACT), 504, 510
FiberCup phantom dataset, 86
fiber-tracking methods, 83–86, 87, 89–91
fiducial marker, 122
fiducial registration error (FRE), 122
field of view (FOV), 53, 147–148, 347
finite element method (FEM), 178,
  244–245, 559
finite Gaussian mixture model (FGMM), 44
finite impulse response (FIR), 155, 317
"finite strain" (FS), 90
FIR, *see* Finite impulse response
first-level variance, 339
Fisher-z transformation, 371, 373, 616
fixed effects analyses, 339
fixed pipelines, 289–292
FLAIR, *see* Fluid attenuated inversion
  recovery
fluid attenuated inversion recovery
  (FLAIR), 110–113, 115–116, 132
fluorodeoxyglucose ([^{18}F]FDG), 18, 19–20,
  27
FMEM, *see* Functional mixed-effects model
fMRI, *see* Functional magnetic resonance
  imaging

FMRIB Software Library, 37, 117, 124, 167
fMRI Data Center (fMRIDC), 167
fMRISTAT, 248
forced-choice recognition test (REC),
  285–286
foreground from background segmentation,
  43–44
Fourier transform and methods
 background, 5
 cross-coherence/partial
  cross-coherence, 373
 DSI, 73–74
 dual-frequency coherence analysis,
  611–612
 exact QBI, 76
 fMRI, 206–214, 230
 FreSpeD method, 583
 Granger causality, 431–432
 heat kernel smoothing, 247
 HRF, frequency domain, 321–322
 image formation, 144–145
 image processing, 220, 223
 image reconstruction, 206
 local dual-frequency coherence,
  614–615
 multichannel EEG, 542, 544
 nonstationary data analysis, 194
 non-stationary EEG signal
  representation, 599–600
 Nyquist ghost k-space correction, 218,
  220
 preprocessing EEG, 184
 reconstruction isomorphism
  representation, 220–222
 restricted diffusion, 68
 single-channel EEG, 525–532
 SLEX method, 600
 spherical harmonic representation, 238
 stationary data analysis, 187–189
 structural MRI, 38–39, 43, 51
 VAR models, 590
FOV, *see* Field of view
FPR, *see* False positive rates
fractional anisotropy (FA)
 DTI, 69–70
 group analysis, 92
 LF-PCA, 483–484
"fraction of pipelines," 290
FRE, *see* Fiducial registration error
Fréchet mean, 246
free diffusion, 67

FreeSurfer
    gray matter volume, 243
    preprocessing pipeline optimization,
        267
    small-*n* large-*p* problems, 250
    spatial registration, 122
    structural MRI, 37, 46, 48
"frequency dragging," 196
frequency-specific change-point detection
        (FreSpeD) method
    change-point detection, 582–585
    dynamic connectivity states estimation,
        596
    multichannel seizure EEG data, 585
    SLEX method, 584–585, 606
    summary, 619–620
frequentist perspective, 364
Friedman multiple-treatment test, 288
Frobenius norm, 401
FRT, *see* Funk-Radon transform
FS, *see* "Finite strain"
FSL software, 119, 124, 130, 167, 267, 350,
        375, 381
F-test, 51, 336
full width, half maximum (FWHM)
    image processing implications, 224
    individual subject's curves, 284
    preprocessing, 160
    random field theory, 254
    spatial regularization, 344
    surface data smoothing, 243
functional analysis of diffusion tensor tract
        statistics (FADTTS), 91–92
functional connectivity (FC)
    background, 8–9
    basic concepts, 420, 496–497
    cross-coherence, 373–374
    cross-correlation, 371–372
    future directions, 384–389
    independent component analysis,
        374–375
    introduction, 369–370
    methods and measures, 371–376
    mutual information, 374
    overview, 162–163
    partial cross-coherence, 373–374
    partial cross-correlation, 371–372
    preprocessing, 381
    principal component analysis, 374–375
    problems, 384–389
    resting-state, 381–384

simulation study, 376–381
    stability selection, 372–373
    stationary data analysis, 191
    time-varying connectivity, 375–376
Functional Connectomes Project, 166, 167,
        168
functional data analysis approach, 29
functional magnetic resonance imaging
        (fMRI)
    acquisition, 146–148, 152–153, 217–220
    artifacts, 152–153
    background, 4, 6–7
    BOLD fMRI, 148–155
    connectivity, 162–164
    co-registration, 159
    data and data analysis, 141–148,
        160–168
    databases, 166–169
    DCM, 427–428
    edge set, 446
    effective connectivity, 422
    experimental design, 11, 157–158
    family-wise error rate, 357
    fMRI resting-state analysis, 278–280
    functional connectivity, 370
    future developments, 168–169
    Granger causality, 431
    high temporal resolution multiband
        data, 168
    image contrast, 143–144
    image formation, 144–146
    introduction, 139–141
    joint ICA, 512
    k-space correction, 217–220
    large-scale data bases, 168–169
    localization, 160–162
    longitudinal imaging studies, 168
    modeling signal and noise, 153–156
    MRI scanners, 142
    MR physics, 142–143
    multi-modal analysis, 169
    network analysis, 442
    noise and nuisance signal, 155–156
    normalization, 159–160
    prediction, 164–165
    preprocessing, 158–160
    realignment, 159
    reconstruction, 217–220
    resting state, 165–166
    scanner, 141–142
    signal generation, 141–144

slice-time correction, 159
smoothing, 160
software, 166–168
spatial limitations, 150–151
task-based analysis, 280
task preprocessing optimization,
    280–293
temporal limitations, 151–152
within-subject model, 340
functional magnetic resonance imaging
    (fMRI), image reconstruction
acquisition and reconstruction, 214–220
background, 7
complex-valued fMRI activation,
    229–230
Fourier transform, 206–214
image processing, 220–228
introduction, 205–206
k-space, 214–220
Nyquist ghost correction, 217–220
reconstruction isomorphism
    representation, 220–228
signal equation, 214–217
space coverage, 214–217
functional magnetic resonance imaging
    (fMRI), joint DTI models
anatomical weighting, 508–509
ascendancy, 511
deterministic tractography, 504
diffusion weighting imaging, 503–504
effective connectivity methods, 502
functional coherence, 510–511
joint activation modeling, 509–512
joint ICA, 512
likelihood function, 511–512
measures of association, 498
modeling approaches, 499
multimodal approaches, 505–513
network methods, 500–502
partitioning methods, 500
prediction methods, multimodal,
    512–513
probabilistic tractography, 505
sequential procedures, 506–508
single modality methods, 497–505
spatial scale, 497
structural connectivity determination,
    502–505
tractography, 504–505
functional magnetic resonance imaging

(fMRI), linear and nonlinear time
    series analysis
background, 9
Balloon model, 324–325
bi-exponential nonlinear model, 326
frequency domain, 321–322
future directions, 327
HRF, 312–322
introduction, 309–310
methods comparison, 320–321
multi-subject analysis, 322–323
nonlinear models, 323–327
nonparametric models, 317–320
parametric models, 313–316
semi-parametric approaches, 323
single-level analysis, 310–312
time domain, 312–321
Volterra series models, 325, 326–327
functional mixed-effects model (FMEM), 92
functional neuroimaging data, multiplicity
    corrections
background, 9
dimensions for consideration, 364–365
false discovery rate control, 360–362
family-wise error rate, 356–360
introduction, 355–356
spatial dependence, accounting for,
    362–364
functional neuroimaging group studies
background, 9
brain shape and function variability,
    337–338
cluster-mass testing, 345
estimation, 341–342
fixed effects analyses, 339
group analysis, functional
    neuroimaging, 339–343
group-subject model, 340–341
inference strategy examples, 347–350
introduction, 335–337
mass univariate framework, 344
mixed effects analyses, 339
mixed effects model, 341
notations, 340
permutation testing, 346
random effects t-test, 343
randomized parcellation-based
    inference, 345
region size, 344–345
spatial context, statistical inference,
    344–345

spatial regularization, 344
standard statistical tests, 354
statistical inference, 342–345
threshold-free cluster enhancement, 345
type 1 error control, 346
within-subject model, 340
Funk-Radon transform (FRT), 75–76
future developments
    FC analysis, fMRI data, 384–389
    functional regression methods, 92
    high-dimensional risk prediction
        methods, 93
    high temporal resolution multiband
        data, 168
    large-scale data bases, 168–169
    linear and nonlinear models, 327
    longitudinal imaging studies, 168
    multi-modal analysis, 169
fuzzy clustering algorithm, 53
FWER, *see* Family-wise error rate
FWHM, *see* Full width, half maximum

G
GA, *see* Geodesic anisotropy
gadolinium delayed-enhanced MRI, 402
Gamma deviance generalized
        cross-validation criterion, 575
Gamma functions
    BOLD, 154–155
    HRF, 314, 315, 320–321
Gamma generalized cross-validation
        criterion (Gamma-GCV), 530–531
Gaussian Nave Bayes (GNB) model,
        289–290, 292–293
Gaussian techniques
    artifact removal, 185
    automated segmentation, 234
    brain surfaces analysis, 234, 235
    brain tissue segmentation, 44, 45
    diffusion smoothing, 243–244, 245
    DPI, 78
    DTI, 69, 71
    dynamic causal models, 429
    error control with permutation testing,
        346
    family-wise error rate, 359
    functional connectivity, 374, 381
    functional neuroimaging, 340, 341
    GDTI, 72, 73
    general linear models, 248
    heat kernel smoothing, 247

high-dimensional ICA, 408
ICA, 405, 406, 407
image formation, 146
image processing, 223–224
individual subject's curves, 284
inhomogeneity correction, 118, 119
iterated kernel smoothing, 245
joint fMRI and DTI models, 499
local dual frequency coherence, 614
longitudinal models, 251
multivariate general linear models, 250
noise components and voxelwise
        estimation methods, 9
non-linear least squares method, 24
nonparametric models, 317
optimal preprocessing pipelines, 286,
        290
parameter estimation, 592
partial directed coherence, 550
prediction, 283
projection pursuit, 410
random effect t-test, 343
reconstruction, 79
segmentation, 43, 44
skull stripping, 120
SLEX model, 604
smoothing, 160
spatial pattern reproducibility, 275
spectrum estimation via periodograms,
        530
SPFI, 78
spherical harmonic representation, 237
structural equation models, 426
structural MRI, 51
surface-based registration, 42
surface data smoothing, 243
SVAR model, 591
synchronization and EEG, 180
time-varying VAR model, 590, 591
tractography algorithms, 85
VAR model, multivariate time series,
        552
volumetric registration, 41
Gauss-Newton method, 342
GC, *see* Granger causality
GCV, *see* Generalized cross-validation
        criteria
GDTI, *see* Generalized DTI
GEE, *see* Generalized estimating equations
generalized autocalibrating partial parallel
        acquisition (GRAPPA), 228

generalized cross-validation criteria (GCV), 480–482

generalized DTI (GDTI), 71–72, 80

generalized estimating equations (GEE), 474

generalized fractional anisotropy (GFA), 76

generalized linear model (GLM), *see also* Linear and nonlinear models
 denoising, 294
 future directions, 385
 localization, 160–161
 prediction, 283
 preprocessing, rs-fMRI, 278
 small-*n* large-*p* problems, 250
 spatial pattern reproducibility, 275
 structural MRI, 51
 surfaces statistical inference, 248
 voxel-based morphometry, 47
 within-subject model, 340

generalized method of moments (GMM) approach, 474

genetics, imaging, 12

genome-wide association study (GWAS), 53

geodesic anisotropy (GA), 70

geodesic distance, 243, 246

GFA, *see* Generalized fractional anisotropy

ghosting artifact, 39

Gibbs characteristics, 45, 223, 238

Giedd, Jay, 478

GLM (generalized linear model), *see* Linear and nonlinear models

global efficiency, 448

global tractography algorithms, 83, 85–86

GM, *see* Gray matter

GMM, *see* Generalized method of moments approach

GNB, *see* Gaussian Nave Bayes model

Goltz, Friedrich, 369

gradient echo-echo planar imaging (GRE-EPI), 214, 217

gradient sampling indices (GSI), 88–89

Gram-Charlier A series, 72

Granger causality (GC)
 autoregressive model, 534
 edge set, 446
 effective connectivity, 421, 422
 effective connectivity and causal inference, 429–432
 functional connectivity, 502
 multichannel EEG, 549–550
 VAR models, 589

graphical lasso feature selection
 functional connectivity, 378
 rs-fMRI, 383
 stability selection, 372

graphical models, PET, 27–28

graphs, networks, 443

GRAPPA, *see* Generalized autocalibrating partial parallel acquisition

gray matter (GM)
 brain surface analysis, 234
 defined, 36, 54
 image contrast, 143
 inhomogeneity correction, 118
 intensity normalization, 127
 MRI to fMRI, 148
 multimodal predication methods, 513
 multi-subject structural MRI, 40
 subcortical volumes, 46
 surface-based registration, 41
 volume, cortical surface features, 242–243

Green's function, 177, 556

GRE-EPI, *see* Gradient echo-echo planar imaging

group analysis, 90–93, 340–343

group-level analysis, 161

group-subject model, 340–341

Growing Neural Gas (GNG) clustering algorithm, 572

GSI, *see* Gradient sampling indices

GWAS, *see* Genome-wide association study

H

Hammersley-Clifford theorem, 45

Hanning window, 74

Hann window, 377

HARDI, *see* High angular resolution diffusion imaging

harmonizable process, 612–613

HCP, *see* Human Connectome Project (HCP)

HDICA, *see* High-dimensional ICA

HDLSS, *see* High-dimensional low sample size

"healthy survivor" effect, 470

heartbeat, noise and nuisance signal, 156

heat kernel smoothing, 246–248

heavy-tailed degree distributions, 450

hemodynamic response function (HRF)
 BOLD, 153–155
 effective connectivity, 422

fMRI, 157–158
    general linear models, 310
    localization, 161
    structural equation models, 426–427
Hermite polynomial, 72
Hermitian properties
    image processing implications, 227
    non-stationary EEG signal
        representation, 599
    one-dimensional Fourier transform, 207
    SLEX method, 603
    two-dimensional Fourier transform, 211
hierarchical spectral merger algorithm,
        573–574
high angular resolution diffusion imaging
        (HARDI)
    diffusion orientation transform, 77
    diffusion propagator imaging, 77–78
    diffusion spectrum imaging, 73–74
    estimated diffusion quantities,
        uncertainty, 87
    generalized DTI, 70–73
    hybrid diffusion imaging, 74–75
    Q-Ball imaging, 75–76
    reconstruction, 80
    reconstruction challenge phantom
        dataset, 86
    sampling mechanisms, 89
    simple harmonic oscillator
        reconstruction and estimation, 78
    spherical deconvolution, 77
    spherical polar Fourier imaging, 78
high-dimensional data
    directed connectivity, inference, 555
    LASSLS, 553–555
    LASSO, 552–553
    least squares estimation, 552
    VAR model, multivariate time series,
        552–554
high-dimensional ICA (HDICA), 407–408
high-dimensional low sample size (HDLSS),
        402
high-dimensional structured PCA, 404–405
higher-order singular value decomposition
        (HOSVD), 376
high-order tensor (HOT) model, 72–73, 76,
        80
high temporal resolution multiband data,
        168
Hilbert space and transform, 188, 237
homodyne interpolation, 227

homogeneous polynomial basis, 72
homotopic group ICA, 407
HOSVD, *see* Higher-order singular value
        decomposition
HOT, *see* High-order tensor model
Hotelling T^2-statistic, 48, 249
HRF, *see* Hemodynamic response function
Human Brain Mapping, 384
Human Connectome Project (HCP), 84,
        167, 169, 370, 442
hybrid diffusion imaging (HYDI), 74–75
hysteresis thresholding, 44

I

ICA, *see* Independent component analysis
ICC, *see* Intra-class correlation coefficient
I-efficiency, 475
image reconstruction, fMRI
    acquisition and reconstruction, 214–220
    background, 7
    complex-valued fMRI activation,
        229–230
    Fourier transform, 206–214
    image processing, 220–228
    introduction, 205–206
    k-space correction, 217–220
    k-space coverage, 214–217
    Nyquist ghost k-space correction,
        217–220
    reconstruction isomorphism
        representation, 220–228
    signal equation, 214–217
    space coverage, 214–217
image registration, *see* Registration
images and imaging
    acquisition, sMRI, 37–40
    contrast, signal generation, 143–144
    fMRI image reconstruction, 220–228
    formation, signal generation, 144–146
    genetics, sMRI, 53
    identification, multisequence sMRI,
        112–113
    inverse source reconstruction, 560–562
    "normalization," 115
imaging, modalities
    diffusion magnetic resonance imaging,
        65–95
    electroencephalography, 175–197
    functional magnetic resonance imaging,
        139–169

multisequence clinical structural brain MRI, 109–133
PET, 17–30
structural magnetic resonance imaging, 35–54
imaging, multivariate decompositions
  background, 10
  factor analysis, 410
  high-dimensional models, 404–405, 407–408
  homotopic group ICA, 407
  independent component analysis, 405–408
  introduction, 399–400
  non-negative matrix decomposition, 409–410
  partial least squares regression, 409
  population value decomposition, 410
  principal component analysis, 400–402
  projection pursuit, 410
  singular value decomposition, 400–402
  sparce PCS framework, 409
  structured PCA models, 403–405
  tensor (multi-way) statistical analysis, 410–411
implementation, EDC, 615–618
impulse response function (IRF), 22, 24, 178
independent component analysis (ICA)
  artifact removal, 185–187
  background, 10
  connectivity, 420
  FC analysis, fMRI data, 374–375
  inverse source reconstruction, 560
  preprocessing, rs-fMRI, 279
  time-varying connectivity, 376
independent voxels, 364–365
indifference-zone ranking, 278
individual pipelines, 289–292
individual-specific curves, 482–483
inference, *see also* Effective connectivity (EC), and causal inference
  brain, surface analysis, 248–254
  EEG connectivity, 549–551
  Evolutionary Dual Frequency Spectra, 614
  functional neuroimaging, 342–345, 347–350
Infomax ICA decomposition, 525
information, spatial borrowing, 477
inhomogeneity correction, 114–115, 117–119
inner cortical surface, 241

Institutional Review Board (UC-Irvine), 525
intensity normalization, 126–129
interhemispheric connections, 508
International Consortium for Brain Mapping (ICBM152), 46
interpolation, multisequence sMRI, 119–121
inter-stimulus intervals (ISI), 312–313, 314
inter-subject registration, 160
intra-class correlation coefficient (ICC)
  fMRI task-based analysis, 280
  pipeline evaluation metrics, 272–274
  preprocessing, 279–280
  reproducibility, 282
  rs-fMRI, 279
  structured PCA, 403
intrahemispheric connections, 508
intuitive description, data structure problems, 114–115
inverse logit model, 322
inverse source reconstruction, 559–562
inverse transformation sampling scheme, 40–41
Ioannidis, John, 280
IRF, *see* Impulse response function; Iterative residual fitting
ISI, *see* Inter-stimulus intervals
IsoData algorithm, 44
isotrophy, 88
iterated kernel smoothing, 245–246
iterative residual fitting (IRF), 238

J
Jaccard properties, 272, 292
Jacobian properties, 48, 242
JADE, 410
Java Image Science Toolkit (JIST), 117, 119, 129
jittering, 152
Johns Hopkins University, 483
joint entropy, 122–123
joint fMRI and DTI models, brain connectivity
  anatomical weighting, 508–509
  ascendancy, 511
  background, 10
  deterministic tractography, 504
  diffusion weighting imaging, 503–504
  effective connectivity methods, 502
  functional coherence, 510–511
  joint activation modeling, 509–512

joint ICA, 512
likelihood function, 511–512
measures of association, 498
modeling approaches, 499
multimodal approaches, 505–513
network methods, 500–502
partitioning methods, 500
prediction methods, multimodal,
      512–513
probabilistic tractography, 505
sequential procedures, 506–508
single modality methods, 497–505
spatial scale, 497
structural connectivity determination,
      502–505
tractography, 504–505
joint posterior distribution, 513
joint probability density function, 498

K
Kalman filter (KF), 589, 592–594
Kalman smoothing (KS), 589, 592, 595
kappa statistic, 272
Karhunen-Loève expansions, 483
Kennedy-Kriger Institute, 484
kinetics, *see* Tracer kinetic modeling
Kolmogorov-Smirnov-type test, 485
Kronecker product
      HRF, time domain, 318
      image processing implications, 223
      reconstruction isomorphism
            representation, 221
      TV-VAR model, 590
k-space, *see* Fourier transform and methods
Kullback-Leibler criterion
      clustering of EEGs, 572
      combined prediction *vs.* spatial
            reproducibility metrics, 277
      SLEX method, 568, 571, 603–604

L
Lagrangian multiplier and evolution, 241,
      426
Laguerre functions, 78
landmark-based registration, 122
Laplace-Beltrami operator
      diffusion smoothing, 244
      exact QBI, 76
      quadratic polynomial, 236
      reconstruction, 80
Laplace equation, 77

Laplacian methods, 193, 244
large deformation diffeomorphic metric
      mapping (LDDMM), 241
large-scale data bases, 168–169
LASSLS algorithm, 551, 553–555
lasso feature selection
      functional connectivity, 378, 499
      rs-fMRI, 383
      stability feature, 373
LASSO regression
      high-dimensional data, 551, 552–554
      PET, 26–27
      reconstruction, 80
LDDMM, *see* Large deformation
      diffeomorphic metric mapping
least squares (LS), 72, 478, 552
least squares estimation (LSE)
      high-dimensional data, 552, 554
      HRF, time domain, 316
      inverse source reconstruction, 560
      small-*n* large-*p* problems, 250
Lebesgue measure, 238
Legendre polynomials, 76, 237, 282
lesions, 129–131
LesionTOADs software, 129
Levenberg-Marquardt algorithm, 24, 316
LF-PCA, *see* Longitudinal functional
      principal component analysis
likelihood ratio statistic, 343
likelihood ratio test, 423
linear and nonlinear models, *see also*
      Generalized linear model (GLM)
      background, 9
      Balloon model, 324–325
      bi-exponential nonlinear model, 326
      brain surface statistical inference,
            248–249
      frequency domain, 321–322
      future directions, 327
      HRF, 312–322
      introduction, 309–310
      mean trajectory, 472–482
      methods comparison, 320–321
      modeling components of change,
            477–482
      multi-subject analysis, 322–323
      nonlinear models, 323–327
      nonparametric models, 317–320
      overview, 323–324
      parametric models, 313–316
      semi-parametric approaches, 323

single-level analysis, 310–312
structural equation models, 426
time domain, 312–321
Volterra series models, 325, 326–327
linear discriminant analysis, 283
linear minimum mean square error
(LMMSE) estimator, 81
linear structural equation models (LSEMs),
423, 425
LMMSE, *see* Linear minimum mean square
error estimator
local dual-frequency cross-periodogram, 612
local efficiency, clustering coefficient, 449
localization, data analysis, 160–162
localizing areas of activation, 9
local parameterization, 236
local rates of change, 485–486
local tractography algorithms, 83–84
Loève spectrum, 610, 611, 612
Logan plot, 27–28
log-Euclidean metric, 82, 246
logit functions, 316
log-likelihood function, 81, 82, 372
log-log regression, 450
longitudinal models, *see also* Cross-sectional
and longitudinal designs
brain surface statistical inference,
251–252
future developments, 168
multisequence sMRI, 131–133
longitudinal properties
imaging studies, 168
LF-PCA, 483–485
relaxation, 143
LONI Pipeline Processing Environment, 117
LONI Probabilistic Brain Atlas, 421
LORETA, 561–562
low reproducibility, 337
LSE, *see* Least squares estimation
LSEM, *see* Linear structural equation
models
LTI system and theory, 154, 322

M
MA, *see* Moving average model
magnetic resonance imaging (MRI)
background, 4, 5–6
between-subject, brain registration,
277–278
scanner hardware, 141–142
magnetic resonance (MR) physics, 142–143

magnetic susceptibility artifacts, 39–40
magnetoencephalography (MEG)
background, 4, 6, 7
DCM, 427–428
edge set, 446
functional connectivity, 370
Granger causality, 431
iterated kernel smoothing, 245
joint ICA, 512
MRI to fMRI, 148
network analysis, 442
structural equation models, 423
vertex set, 445
magnitude-only (MO) model, 229
MANCOVA, 48
marginal models, 474
Markov chain Monte Carlo (MCMC)
approach, 28, 277, *see also* Monte
Carlo methods
Markov chain Monte Carlo maximum
likelihood estimation (MCMC
MLE), 454
Markov random fields (MRFs)
brain tissue segmentation, 45
spatial-adaptive estimation methods,
81, 82
spatial regularization, 344
Markov-switching VAR models, 578, 591,
593, *see also* Switching VAR
(SVAR) models
MARM, *see* Multiscale adaptive regression
modeling
MASM, *see* Multiscale adaptive smoothing
model
"massively parallel" algorithms, 478
mass-univariate approach, 335, 344, 455
mathematical models, network analysis,
450, *see also* Algorithms
MATLAB, 113, 117, 167, 529, 531
maximization step (M-step), 592–593
maximum a posteriori (MAP) estimates,
41–42
maximum entropy, 122–123
maximum likelihood estimation (MLE)
ARMA model, 539
ICA, 405
rs-fMRI, 382
maximum likelihood function, 341–342
maximum test statistic, 360
mCCA, *see* Multimodal canonical
correlation analysis

MCMC, *see* Markov chain Monte Carlo approach

MCMC MLE, *see* Markov chain Monte Carlo maximum likelihood estimation

MD, *see* Mean diffusivity

MDMR, *see* Multivariate distance matrix regression

MDS, *see* Multi-dimensional scaling

mean diffusivity (MD), 69–70

mean networks, 457

mean squared displacement (MSD), 75

mean-squared error (MSE), 323, 554

mean-squared function, 529

mean trajectory, 472–482

measures of association, 498

medial model-based approach, 91

medical image processing, analysis, and visualization (MIPAV), 113, 117, 119, 124

MEG, *see* Magnetoencephalography

meta-analysis, 11, 445

meta-data storage, 166–167

"method backwards" approach, 114

MGLM, *see* Multivariate general linear models

mHARDI, *see* Multiple shell data

MI, *see* Mutual information

millisecond-scale modulations, 177

minimal deformation target template, 46

minimum norm estimate (MNE), 561

Minkowski functional, 252

MIPAV, *see* Medical image processing, analysis, and visualization

misalignment complications, 475–477

missing data, 471

mistakes, *see* Problems and pitfalls

mixed effects analyses, 339

mixed-effects model, 336, 341, 473–474

mixture of tensor model, 70–71

MLDs, *see* Multi-cohort longitudinal designs

MLE, *see* Maximum likelihood estimation

MNE, *see* Minimum norm estimate

MO, *see* Magnitude-only model

modalities overview, 4–7

models and modeling
    ball and stick model, 71
    BOLD fMRI, 153–155
    CHARMED model, 71
    dipole source model, 556–557

    dynamic causal models, 427–429
    effective connectivity and causal inference, 422–432
    exponential random graph, 453–454
    fMRI, 155–156
    generalized model, EEG signals, 559
    Granger causality, 429–432
    high-order tensor model, 72–73
    independent source model, 558
    mathematical, network analysis, 450
    mixed effects model, 341
    network analysis, 450–455
    PET, 26–30
    source localization and estimation, 555–562
    source models, 555–556
    stochastic block, 454–455
    structural equation models, 422–427
    time-domain approach, 532–539
    within-subject model, 340

models and modeling, components of change
    age-specific distribution, 485
    background, 9
    borrowing information "spatially," 477
    cross-sectional data and designs, 469–470, 473
    discussion, 486–487
    individual-specific curves, 482–483
    introduction, 468
    linear models, 472–477
    local rates of change, 485–486
    longitudinal data, 473–474
    longitudinal functional PCA, 483–485
    marginal models, 474
    mean trajectory, 472–477
    misalignment complications, 475–477
    mixed-effects models, 473–474
    multi-cohort longitudinal designs, 471–472
    nonlinear models, 477–482
    nonparametric models, 477–480
    notation, 468–469
    polynomial models, 477
    region-wise linear models, 472–477
    relative efficiency, mean function estimation, 474–475
    repeated cross-sectional subsamples, 480–482
    semiparametric models, 477–480

single-cohort longitudinal designs, 470–471
models and modeling, electroencephalograms (EEG)
background, 6, 10
change-point detection, 580–588
clustering of EEGs, 572–579
coherence between bands, 613–614
concept formalization, 612–614
discretized frequencies, 612–613
dynamic connectivity state estimation, 593–598
EEG data analysis, 615
epileptic EEG, 593–598
evolutionary dual-frequency coherence analysis, 608–618
exact beta distribution, 614
FreSpeD method, 582–585
harmonizable process, 612–613
historical developments, 610–611
implementation details, 615–618
inference procedure, 614
introduction, 568–571
local dual-frequency cross-periodogram, 612
multichannel seizure EEG data, 585–588
non-stationary EEGs, best signal representation, 599–608
parameter estimation, 592–593
signal representations, 599
SLEX, 600–606
spectral merger clustering method, 572–574
stationary VAR model, 589–590
switching VAR models, 588–598
time-dependent harmonizable process, 613
time-varying VAR model, 590–591
models and modeling, linear and nonlinear
background, 9
Balloon model, 324–325
bi-exponential nonlinear model, 326
frequency domain, 321–322
future directions, 327
HRF, 312–322
introduction, 309–310
methods comparison, 320–321
multi-subject analysis, 322–323
nonlinear models, 323–327
nonparametric models, 317–320
parametric models, 313–316
semi-parametric approaches, 323
single-level analysis, 310–312
time domain, 312–321
Volterra series models, 325, 326–327
mono-exponential decay assumption, 72
monotonic transformations, 458
Monte Carlo methods, *see also* Markov chain Monte Carlo approach
estimated diffusion quantities, uncertainty, 86–87
family-wise error rate, 359
tractography algorithms, 84
Montreal Neurological Institute (MNI)
brain surface analysis, 234
FC of rs-fMRI, 381
fMRI data, 167
group studies, functional neuroimaging, 336
homotopic group ICA, 407
misalignment, complications from, 476
preprocessing, 160
spatial registration, 124
templates and atlases, 46
Moore-Penrose inverse, 401
Morlet wavelet transform, 194
motion and motion correction, 39, 116, 156, 287, 311
moving average (MA) model, 532–533, 535–537
MRF, *see* Markov random fields
MRtrix, 84
MSD, *see* Mean squared displacement
MSE, *see* Mean-squared error
M-step, *see* Maximization step
multi-atlas label fusion techniques, 119
multi-center studies, 52–53
multichannel EEG
connectivity, model and inference, 549–551
Fourier-Cramér representation, 542
Fourier-domain approach, 542–544
Granger causality, 550
multiple EEG traces, 548–549
non-parametric estimator, 544
overview, 541–542
partial directed coherence, 550–551
specific frequency band spectral matrix estimation, 549
spectral matrix, 542–544, 548–549
time-domain approach, 544–545

multichannel seizure EEG
  change-point detection, 585–588
  SLEX method, 606–608
multi-cohort longitudinal designs (MLDs),
    471–472
multi-coil data, 150–151
multi-dimensional scaling (MDS), 275
multi-modal analysis, 10, 169
multimodal canonical correlation analysis
    (mCCA), 512
multinomial likelihood function, 499
multiple comparisons problem, 252
multiple EEG traces
  multichannel EEG, 548–549
  single-channel EEG, 539–540
multiple shell data, 70, 76, 77, 78
multiple signal classification (MUSIC), 560
multiplicity corrections, functional
      neuroimaging data
  background, 9
  dimensions for consideration, 364–365
  false discovery rate control, 360–362
  family-wise error rate, 356–360
  introduction, 355–356
  spatial dependence, accounting for,
      362–364
multiscale adaptive regression modeling
    (MARM), 82–83
multiscale adaptive smoothing model
    (MASM), 321–322
multisequence structural MRI (sMRI)
  acquisition, 115–116
  background, 8
  conclusions, 133
  cross-sectional intensity analysis,
      131–133
  data structure, 114–115
  handling and storage, 113
  images identification, 112–113
  inhomegeneity correction, 117–119
  intensity normalization, 126–129
  interpolation, 119–121
  introduction, 110–112
  intuitive description, data structure
      problems, 114–115
  lesion maping, 130–131
  longitudinal intensity analysis, 131–133
  pitfalls and mistakes, 114
  preprocessing, 116–129
  reconstruction, 115–116
  skull stripping, 119

software, 118–119, 121, 123–129
spatial registration, 121–126
multi-subject data and analysis, 322–323,
    326–327
multi-subject MRI
  brain tissue segmentation, 44–45
  deformation-based morphometry, 47–48
  foreground from background
      segmentation, 43–44
  introduction, 40
  morphometry, 46–50
  other measures, 49–50
  registration, 40–43
  segmentation, 43–45
  statistical analyses, 50–51
  statistical parametric maps, 51
  subcortical volumes, 46–47
  surface-based measures, 48–49
  surface-based registration, 41–43
  templates and atlases, 45–46
  tensor-based morphometry, 47–48
  volumetric registration, 41
  voxel-based morphometry, 47
multi-subject MRI morphometry
  deformation-based type, 47–48
  other measures, 49–50
  subcortical volumes, 46–47
  surface-based measures, 48–49
  tensor-based type, 47–48
  voxel-based type, 47
multitaper method, 531–532, 539
multivariate decompositions
  background, 10
  factor analysis, 410
  high-dimensional models, 404–405,
      407–408
  homotopic group ICA, 407
  independent component analysis,
      405–408
  introduction, 399–400
  non-negative matrix decomposition,
      409–410
  partial least squares regression, 409
  population value decomposition, 410
  principal component analysis, 400–402
  projection pursuit, 410
  singular value decomposition, 400–402
  sparce PCS framework, 409
  structured PCA models, 403–405
  tensor (multi-way) statistical analysis,
      410–411

multivariate distance matrix regression (MDMR), 501–502
multivariate general linear models (MGLMs), 249–250
multivariate normal (MVN) data, 376–377
multivariate pattern analysis (MVPA), 276
multivariate vector autoregressive modeling (MVAR), 502
multi-voxel pattern analysis (MVPA), 164–165
MUSIC, *see* Multiple signal classification
mutual information (MI)
    edge set, 446
        FC analysis, fMRI data, 374
        functional connectivity, 498
MVAR, *see* Multivariate vector autoregressive modeling
MVPA, *see* Multivariate pattern analysis; Multi-voxel pattern analysis
MW, *see* White matter

N

N3, *see* Nonparametric nonuniform intensity normalization method
Nadaraya-Watson estimator, 246
Nadaya-Watson kernel regression, 246
naïve ordinary least squares (N-OLS), 473, 474
National Institute of Mental Health, 478
National Institutes of Health (NIH), 132, 370
NAWM, *see* Normal-appealing white matter
near-infrared spectroscopy (NIRS), 175
Nelder-Mead simplex algorithm, 24
networks
    background, 10
    biophysical models, 450
    characteristic path length, 447–448
    clustering coefficient, 448–449
    connectivity, 163–164
    construction, 443–447
    degree distribution, 449–450
    density-integrated topology, 457–458
    edge set, 446
    Erdős-Rényi random graphs, 451–452
    estimation and comparison, 455–459
    exponential random graph models, 453–454
    introduction, 441–443
    mathematical models, 450
    network models, 450–455
    notation, 443–444
    preferential attachment, 453
    small-world networks, 452
    statistical models, 450
    statistical parametric networks, 456–457
    stochastic block models, 454–455
    thresholding networks, 446–447
    topology, descriptive measures, 447–450
    vertex set, 444–445
    weighted networks comparison, 458–460
NeuroImage, 384
neuroimage preprocessing
    background, 8
    between-subject, MRI brain registration, 277–278
    case study, 280–293
    changes, quantification, 268–269
    choices, 369–370
    cluster overlap, 271–272
    correlations, spatial pattern reproducibility, 275
    evaluation metrics, 270–277
    fMRI resting-state analysis, 278–280
    fMRI task-based analysis, 280
    fMRI task preprocessing optimization, 280–293
    individual subject's curves, 284–285
    intra-class correlation coefficient, 272–274
    introduction, 264–266
    literature review, 277–280
    optimization, 280–293
    pipelines, 267–280, 285–292
    prediction metrics, 276–277, 280–281, 282–284
    preprocessing, 267–270, 277–280, 285–292, 369–370
    problems and pitfalls, 293–295
    pseudo-ROC curves, 271
    reproducibility metrics, 280–282
    similarity metric ranking approaches, 275–276
    simulated dataset utility, 268
    spatial pattern reproducibility, 275
    spatial reproducibility, 276–277
neuroimaging, 4, 11
Neuroimaging Informatics Technology Initiative (NIFTI), 167
Neuroimaging Informatics Tools and

Resources Clearinghouse (NITRC), 269, 381
neurophysics, electroencephalography, 177–179
Neurosynth database, 167
Newton-Raphson optimization, 408, 454
NIFTI, *see* Neuroimaging Informatics Technology Initiative
NIRS, *see* Near-infrared spectroscopy
NITRC, *see* Neuroimaging Informatics Tools and Resources Clearinghouse
NLLS, *see* Non-linear least squares
NMR, *see* Nuclear magnetic resonance
NNMD, *see* Non-negative matrix decomposition
nodal-based approaches, 497
node degrees, 450
nodes, networks, 443
noise components, 79–80, *see also* Signal-to-noise ratio (SNR)
N-OLS, *see* Naïve ordinary least squares
non-binary label assignments, 44
nondegenerate tensors, 86
non-linear least squares (NLLS), 24–25, 26
nonlinear models, *see also* Linear and nonlinear models
    Balloon model, 324–325
    bi-exponential nonlinear model, 326
    mean trajectory, 477–482
    modeling components of change, 477–482
    overview, 323–324
    Volterra series models, 325
non-local means (NLM) algorithm, 81, 82
non-local threshold algorithms, 44
non-negative matrix decomposition (NNMD), 409–410
non-parametric estimator, 544
nonparametric models
    hemodynamic response function, time domain, 317–320
    modeling components of change, 477–480
    PET, 28–29
nonparametric nonuniform intensity normalization (N3) method, 119, 233
nonparametric prediction accuracy, influence, and reproducibility resampling, *see* NPAIRS
non-redundant information extraction, 604

nonstationary data analysis, 194–197
non-stationary EEGs, best signal representation
    complexity-penalized Kullback-Leibler criterion, 603–604
    Kullback-Leibler criterion, 603–604
    multichannel seizure EEG, 606–608
    overview, 599
    PCA, 604–606
    signal representations, 599
    SLEX method, 600–606
    spectral quantities, 603
normal-appealing white matter (NAWM), 128
normality hypothesis, 336
normalization, preprocessing, 159–160
NPAIRS (nonparametric prediction accuracy, influence, and reproducibility resampling)
    combined prediction *vs.* spatial reproducibility metrics, 276–277
    fMRI task optimization, 281
    joint fMRI and DTI models, 499
    prediction, 283
    preprocessing pipeline optimization, 289
N4ITK algorithm, 119
nuclear magnetic resonance (NMR), 37
Nyquist frequency and methods
    noise and nuisance signal, 156
    preprocessing EEG, 183–184
    single-channel EEG, 526, 536
    temporal limitations, 151
Nyquist ghost correction
    fMRI acquisition and reconstruction, 217–220, 230
    image processing implications, 223, 227

O

ODF, *see* Orientation distribution function
one-dimensional Fourier transform, 206–210
OpenMRI Project, 167
open statistical problems, 384–385
optimization, preprocessing pipelines, 268
ordered subsets expectation maximization (OSEM), 7
order of a graph, networks, 443
ordinary least squares (OLS), 311, 473
Organisation for Human Brain Mapping, 294
orientation distribution function (ODF)

DPI, 78
DSI, 74
estimated diffusion quantities,
    uncertainty, 86
exact QBI, 76
HYDI, 75
QBI, 75–76
reconstruction, 79
registration, 89–90
sampling mechanisms, 87–88
SD methods, 77
spatial-adaptive estimation methods,
    80, 83
tractography algorithms, 83
OSEM, *see* Ordered subsets expectation
    maximization
Otsu's method, 43–44
outer cortical surface, 241
outliers, 79, 287, 540
^{15}O-Water PET, 19

P

Papadakis scheme, 89
parallel imaging, 150–151
parameterization, 235–238
parametric maps, 51
parametric methods, 559–560
parametric models, 313–316
parcellation
    functional connectivity, 497
    randomized parcellation-based
        inference, 345
    vertex set, 445
Parseval's theorem, 188
partial correlation, 498
partial cross-coherence (PCCoh), 373–374,
    381–382
partial cross-correlation (PCCor), 371–372,
    378, 381–382
partial directed coherence (PDC), 550–551
partial least squares (PLS) regression, 409
partial volume correction, 29
partial volume effects, 40
patch-based techniques, 119
Patel's ascendancy measure, 502
patient motion, *see* Motion and motion
    correction
Patlak plot, 27–28
PCA, *see* Principal component analysis
PCCoh, *see* Partial cross-coherence
PCCor, *see* Partial cross-correlation

PCR, *see* Principal component regression
PD, *see* Proton density-weighted image
PDC, *see* Partial directed coherence
PDF, *see* Probability density function
Pearson correlation
    combined prediction *vs.* spatial
        reproducibility metrics, 277
    functional connectivity, 498, 499
    reproducibility, 282
    spatial pattern reproducibility, 275
penalized sum of squared errors (PSSE),
    319–320
penalty function, 81
periodogram methods
    ARMA model, 539
    cross-coherence/partial
        cross-coherence, 373–374
    FreSpeD method, 582
    functional connectivity, 377
    multichannel EEG, 544, 549
    multiple EEG traces, 539–540
    single channel EEG, 528–531, 530–532
    SLEX method, 602, 604
permutation network framework (PNF),
    501–502
permutation tests
    group studies, functional neuroimaging,
        336
    preprocessing pipeline testing, 278
    statistical parametric maps, 51
    type 1 error control, 346
Perona-Malik-like smoothing, 81, 82
perturbation theory, 86
PET, *see* Positron emission tomography
PGSE, *see* Pulsed gradient spin-echo
    sequence
phantom datasets, 86
phase-locked change, 194
phase-only coherence, 191
phase-only (PO) activation, 229–230
PHYCAA+ algorithm, 286
physics, 38, 142–143
physiological correction, 287
physiological noise, 156
pipeline evaluation metrics
    cluster overlap, 271–272
    correlations, spatial pattern
        reproducibility, 275
    intra-class correlation coefficient,
        272–274
    overview, 270–271

prediction metrics, 276–277
pseudo-ROC curves, 271
similarity metric ranking approaches,
 275–276
spatial pattern reproducibility, 275
spatial reproducibility, 276–277
pitfalls, *see* Problems and pitfalls
pivotality hypothesis, 346
plasma input functions models, 20–23
PLS, *see* Partial least squares regression
PNF, *see* Permutation network framework
PO, *see* Phase-only activation
point-spread function, 150
Poisson model, 314, 316
polynomial models, 477
polynomial parameterization, 236
population-based atlases, 45–46
population value decomposition (PVD), 410
positive regression dependence, 362
positron emission tomography (PET)
 background, 4, 5, 6, 7, 19–20
 basis function methods, 25–26
 Bayesian models, 28
 compartmental approaches, 20–24
 estimation and statistical methods,
  24–27
 functional connectivity, 370
 graphical models, 27–28
 introduction, 17–18
 joint ICA, 512
 modeling considerations, 29–30
 model selection, 26–27
 MRI to fMRI, 148
 non-linear least squares, 24–25
 non-parametric models, 28–29
 PCA, 402
 plasma input functions models, 20–23
 principal component analysis, 374
 reference tissue models, 23–24
 structural equation models, 423
 tracer kinetic modeling, 20–24
posterior cingulate cortex (PCC), 381
post-gadolinium injection, 112
potential precursors, 585–588
power, lack of, 337
power laws, 450, 453
power spectrum density (PSD)
 ARMA model, 539
 confidence intervals, 541
 multichannel EEG, 542–543
 single-channel EEG, 532

power spectrum estimation, 188
PPD, *see* "Preservation of principle
  direction"
PPI, *see* Psychophysiological interaction
  analysis
prediction, 164–165, 276–277
preferential attachment, 453
preprocessing
 changes quantification, 268–269
 choices, 369–370
 compartmental analysis, 28–29
 EEG signals, 183–184
 overview, 8
 pipeline, 264–266
 resting-state, fMRI data, 381
preprocessing, multisequence sMRI
 basic concepts, 116–117
 cross-sectional intensity analysis,
  131–133
 inhomegeneity correction, 117–119
 intensity normalization, 126–129
 interpolation, 119–121
 lesion maping, 130–131
 longitudinal intensity analysis, 131–133
 skull stripping, 119
 software, 118–119, 121, 123–129
 spatial registration, 121–126
preprocessing, neuroimaging
 background, 8
 between-subject, MRI brain
  registration, 277–278
 case study, 280–293
 changes, quantification, 268–269
 choices, 369–370
 cluster overlap, 271–272
 correlations, spatial pattern
  reproducibility, 275
 evaluation metrics, 270–277
 fMRI resting-state analysis, 278–280
 fMRI task-based analysis, 280
 fMRI task preprocessing optimization,
  280–293
 individual subject's curves, 284–285
 intra-class correlation coefficient,
  272–274
 introduction, 264–266
 literature review, 277–280
 optimization, 280–293
 pipelines, 267–280, 285–292
 prediction metrics, 276–277, 280–281,
  282–284

preprocessing, 267–270, 277–280,
    285–292, 369–370
problems and pitfalls, 293–295
pseudo-ROC curves, 271
reproducibility metrics, 280–282
similarity metric ranking approaches,
    275–276
simulated dataset utility, 268
spatial pattern reproducibility, 275
spatial reproducibility, 276–277
preprocessing pipelines
    datasets, 285–286
    fixed pipelines, 289–292
    independent test results, 292–293
    individual pipelines, 289–292
    steps selection, 286–289
    study and optimization, 267–270
    testing, 277–280
"preservation of principle direction" (PPD),
    90
principal component analysis (PCA)
    background, 10
    connectivity, 163, 420
    FC analysis, fMRI data, 374–375
    functional connectivity, 500
    global signal regression, 287
    inverse source reconstruction, 560
    modeling components of change,
        483–485
    SLEX method, 604–606
    SVD, 400–402
principal component regression (PCR), 409
principal curvatures, 242
probabilistic tractography algorithms, 83–84
probability density function (PDF), 276,
    614
"problem forward" approach, 114
problems and pitfalls
    FC analysis, fMRI data, 384–389
    multisequence sMRI, 114
    neuroimage preprocessing, 293–295
product density ICA, 406
projection pursuit, 410
projection-slice theorem, 76
propagation-separation (PS) methods, 81,
    82, 322
proton density-weighted (PD) image, 54,
    110
PS, *see* Propagation-separation methods
PSD, *see* Power spectrum density
"pseudo-F" statistic, 502

pseudo-ROC curves, 271
PSSE, *see* Penalized sum of squared errors
psychophysiological interaction (PPI)
    analysis, 375, 420–421
public resources, dMRI, 93–95
PubMed, 140
pulsed gradient spin-echo (PGSE) sequence,
    66–67
PVD, *see* Population value decomposition
PyMVPA, 284
Python language, 113

Q
Q-Ball imaging
    DOT, 77
    DPI, 77–78
    HYDI, 74–75
    overview, 75–76
    reconstruction, 80
    spatial-adaptive estimation methods,
        83
q-space
    Cartesian sampling lattice, 73–74
    sampling mechanisms, 88
    spatial-adaptive estimation methods,
        81
quadratic polynomial, 236
quantification, 86

R
R software, *see also* Software
    ANTsR, 117
    FreSpeD, 584
    functions, 168
    ggplot 2 package, 584
    HMClust, 575
    multisequence sMRI, 110–113
    single-channel EEG, 529
    vows package, 478
radiotracers, PET, 29
ramp sampling, 227
random effects t-test, 343
random field theory (RFT)
    background, 7–8
    brain surface statistical inference,
        252–254
    family-wise error rate, 360
    statistical parametric maps, 51
random graph models, 453–454
random graphs, 451–452

random intercept model, 473

random intercept/random slope (RIRS) model, 474, 476, 483–484

randomized parcellation-based inference, 345

rank preservation, structured PCA, 405

rates of change, local, 485–486

Rauch-Tung-Striebel smoother, 589

RAVENS maps, 404

Rayleigh distribution, 43, 146

reachability, 447

realignment, 159

REC, *see* Forced-choice recognition test

receiver operating characteristic (ROC) curves, 268, 271, 294

recognition (REC), 285–286, 289–290, 292–293

reconstruction
    basic concepts, 79
    isomorphism representation, 220–228
    multisequence sMRI, 115–116
    noise components, 79–80
    sampling mechanisms, 88
    spatial-adaptive estimation methods, 80–83
    voxelwise estimation methods, 79–80

recording EEG signals, 182–183

red-blue-green (RGB) map, 70

Reeb graph, 247

"reference electrodes," 182

reference tissue models, 23–24

region size, 344–345

regions of interest (ROI)
    anatomical references, 52
    defined, 54
    effective connectivity, 421
    functional connectivity, 371, 378, 499
    group analysis, 90–91
    misalignment, complications from, 477
    multimodal approaches, 506
    network construction, 443
    region-wise linear models, 472
    rs-fMRI, 381–382
    tractography algorithms, 84
    vertex set, 445

region-wise linear models, 472–477

registration
    between-subject, MRI brain registration, 277–278
    brain surface statistical inference, 239–241

defined, 54
dMRI, 89–90
misalignment, complications from, 475–476
multi-subject MRI, 40–43
surface-based, 41–43
volumetric, 41

Reich, Daniel, 132

relative efficiency, mean function estimation, 474–475

relative phase, 191

relevance vector regression, 28

reliability
    automated segmentation algorithms, 53
    ICC, 272–273
    registration, 90
    tractography algorithms, 86

REML, *see* Restricted maximum likelihood

re-orientation, 90

repeated cross-sectional subsamples, 480–482

repetition bootstrapping methods, 87

reproducibility, group studies, 337

resolution, MRI to fMRI, 147–148

respiration, noise and nuisance signal, 156

resting-state analysis
    fMRI, 165–166, 381–384, 403, 453, 499
    functional connectivity, 507
    preprocessing pipelines, testing, 278–280

restricted diffusion, 67–69

restricted likelihood ratio test (RLRT), 478

restricted maximum likelihood (REML), 251, 323, 478

return-to-origin (RTO) probability, 75

RFT, *see* Random field theory

"rich club" of nodes, 450

"rich get richer" concept, 453

Rician noise and distribution
    estimated diffusion quantities, uncertainty, 86
    reconstruction, 79
    spatial-adaptive estimation methods, 81
    structural MRI, 43, 51

Riemannian properties
    diffusion smoothing, 244
    surface area and curvatures, 242
    surface parameterization, 236

rigid registration, 121

RIRS, *see* Random intercept/random slope
    model
RLRT, *see* Restricted likelihood ratio test
RMSE, *see* Root mean square error
ROC, *see* Receiver operating characteristic
    curves
ROI, *see* Regions of interest
root mean square error (RMSE), 122
RPBI, 347–349
rsFC, *see* Resting-state functional
    connectivity
rs-fMRI, *see* Resting-state fMRI
RTO, *see* Return-to-origin probability

S
salmon example, 359
sampling mechanisms, 87–89
Satterthwaite method, 311
SBM, *see* Stochastic block models
SC, *see* Structural connectivity
scalar indices, 69–70
scale-free networks, 450, 453
scanner signal generation, 141–142
Schwartz's Bayesian models and techniques,
    538–539, 594
second-level variance, 339
second-order autoregressive (AR2) model,
    537–538
second-order blind identification (SOBI),
    560
segmentation, 36, 40
seizure localization, 585
seizure onset estimation, 585–588
SEM, *see* Structural equation models
semi-parametric approaches and models,
    323, 477–480
SENSE, *see* Sensitivity encoding
sensitivity encoding (SENSE), 227–228
separable least squares, 25
sequences, sMRI, 38–39
SFIR, *see* Smooth finite impulse response
SH, *see* Spherical harmonic basis and
    representation
Shannon-Nyquist sampling criteria, 206, 210
sHARDI methods, 74, 76, *see also* Single
    shell data
Shepp and Logan's CT reconstruction
    algorithm, 7
Shepp and Vardi's PET reconstruction
    algorithm, 7

SHORE, *see* Simple harmonic oscillator
    reconstruction and estimation
shrinkage
    cross-coherence/partial
        cross-coherence, 374
    LASSO regression, 552
    rs-fMRI, 383
Sigma filter, 81
signal attenuation, 67
signal equation, 214–217
signal generation
    acquisition of data, 146–148
    image contrast, 143–144
    image formation, 144–146
    MR physics, 142–143
    scanner, 141–142
signal processing, *see*
        Electroencephalography
signal representations, non-stationary
        EEGs, 599
signal-to-noise ratio (SNR)
    brain surface analysis, 233
    DOT, 77
    image formation, 145
    multi-subject analysis, 323
    pipeline evaluation metrics, 271
    preprocessing choices, 270
    pseudo-ROC curves, 271
    QBI, 75
    reconstruction, 80
    sampling mechanisms, 88
    structural MRI, 51
    SVAR models, 588
    temporal limitations, 151
similarity metric ranking approaches,
        275–276
simple harmonic oscillator reconstruction
        and estimation (SHORE), 78
simulated annealing algorithm, 85
simulated dataset utility, 268
simulation study, 376–381
simultaneous multi-slice (SMS), 152, 228
single-channel EEG
    autoregressive model, 533–537
    autoregressive moving average model,
        534–537
    confidence level, 540–541
    data description, 525
    Fourier-domain approach, 525–532
    Fourier regression model, 525–528
    introduction, 524–525

moving average model, 532–533, 535–537
multiple EEG traces, 539–540
multitaper method, 531–532
outlier detection, 540
periodogram-based estimation methods, 528–531
second-order autoregressive model, 537–538
single-channel time series spectrum, 528
specific frequency band power estimation, 540
spectrum estimation, 528–531, 538–540
time-domain approach, 532–539
variance decomposition, 525–528
single-channel time series spectrum, 528
single-cohort longitudinal designs (SLDs), 470–471
single-level analysis, 310–312
single shell data, 70, 75, 77
single-subject analysis, 161
single-subject atlas, 45
singular value decomposition (SVD)
HRF, time domain, 315
PCA and, 400–402
PCA/ICA, 375
skull conductivity, 178
skull stripping, 119
SLDS, *see* Switching linear dynamic system
SLDs, *see* Single-cohort longitudinal designs
SLEX method, *see* Smooth localized complex exponential library
slice-timing correction, 159, 287
sLORETA, 561–562
small-*n* large-*p* problems, 250
small-world networks, 452
SMC, *see* Spectral merger clustering method
smooth finite impulse response (SFIR), 317, 320–321
smoothing, 80–83, 160
smooth localized complex exponential (SLEX) library
change-point direction, 580
clustering of EEGs, 572
complexity-penalized Kullback-Leibler criterion, 603–604
dynamic connectivity states estimation, 595–596
evolutionary dual-frequency coherence analysis, 613
FreSpeD comparison, 584–585
Kullback-Leibler criterion, 603–604
multichannel seizure EEG, 606–608
non-stationary EEG signal representation, 599, 600–602
overview, 568, 569, 571
PCA, 604–606
spectral quantities, 603
SMS, *see* Simultaneous multi-slice
SNR, *see* Signal-to-noise ratio
SOBI, *see* Second-order blind identification
"soft-switching," 591
software, *see also* R software
AFNI, 37, 116, 167, 267, 286, 350
ANTs, 116–117, 124–126, 130–131
dMRI, 94–95
fMRI, 166–168
FMRIB Software Library, 37, 117, 124, 167
FSL, 119, 124, 130, 167, 267, 350, 375, 381
inhomegeneity correction, 118–119
interpolation, 121
JIST, 117, 119, 129
LesionTOADs, 129
multisequence sMRI, 127–129
open source, 167–168
preprocessing MRI data, 37
skull stripping, 119
solutions offered by, 350
spatial registration, 123–126
statistical parametric mapping, 37, 167, 267
SurfStat, 251
source localization and estimation
dipole source model, 556–557
generalized model, EEG signals, 559
independent source model, 558
inverse source reconstruction, 559–562
parametric methods, 559–560
source models, 555–556
space coverage, 214–217
sparse Bayesian Learning approach, 28
sparse PCA, 409
sparsified VAR model, 581
spatial-adaptive estimation methods, 80–83
spatial considerations
alignment, registration, 89–90
borrowing of information, 477

context, statistical inference, 343–345

dependence, 362–364

inhomogeneity corrections, 114–115

limitations, 150–151

pattern reproducibility, 275

registration, 121–126

regularization, 344

reproducibility, 276–277

resolution, 147–148

specific frequency band power estimation, 540

specific frequency band spectral matrix estimation, 549

spectral matrix, 542–544, 548–549

spectral merger clustering (SMC) method, 572–574, 619–620

spectral quantities, SLEX method, 603

spectrum estimation, 528–531, 538–540

SPFI, *see* Spherical polar Fourier imaging

spherical deconvolution (SD), 77

spherical harmonic (SH/SPHARM) basis and representation

brain surfaces analysis, 237–238

correspondence, 239–240

HOT model, 73

QBI, 75

SD methods, 77

surface registration, 239

spherical polar Fourier imaging (SPFI), 78

SPIN, *see* Statistical principles of image normalization

SPM, *see* Statistical parametric maps

SPN, *see* Statistical parametric networks

SSD, *see* Sum of shared differences

SSE, *see* Sum of squared errors

SS-OLS, *see* Summary statistics ordinary least squares

stability, 284, 372–373

standard statistical tests, 341, 354

STAPLE, 275

stationary data analysis, 187–193

stationary VAR model, 589–590

statistical analysis

affine registration, 239

area and curvatures, 242

background, 8

cortical surface features, 241–243

data smoothing, 243–248

diffeomorphic registration, 241

diffusion smoothing, 243–245

gray matter volume, 242–243

heat kernel smoothing, 246–248

introduction, 233–235

iterated kernel smoothing, 245–246

linear models, 248–249

local parameterization, 236

longitudinal models, 251–252

multi-subject MRI, 50–51

multivariate linear models, 249–250

quadratic polynomial, 236

random field theory, 252–254

small-n large-p problems, 250

spherical harmonic representation, 237–238

statistical inference, 248–254

surface flattening, 236

surface parameterization, 235–238

surface registration, 239–241

thickness, 241

statistical analysis, electroencephalograms (EEG)

background, 10

connectivity, model and inference, 549–551

data description, 525

dipole source model, 556–557

Fourier-domain approach, 525–532, 542–544

generalized model, EEG signals, 559

Granger causality, 550

high-dimensional data, 551–555

independent source model, 558

introduction, 524–525

inverse source reconstruction, 559–562

multichannel EEG, 541–551

multiple EEG traces, 548–549

partial directed coherence, 550–551

single-channel EEG, 524–541

source localization and estimation, 555–562

source models, 555–556

specific frequency band spectral matrix estimation, 549

spectral matrix estimation, 548–549

time-domain approach, 532–539, 544–545

VAR modeling, multichannel EEG, 554

statistical inference

group analysis, functional neuroimaging, 342–343

group studies, functional neuroimaging, 344–345

spatial context, 343–345
statistical methods and models
    background, 7–10
    brain surfaces analysis, 233–254
    cross-sectional/longitudinal data,
        467–487
    effective connectivity and causal
        inference, 419–436
    electroencephalograms, 523–562,
        567–621
    functional connectivity, 369–389
    functional neuroimaging group studies,
        335–350
    image reconstruction, fMRI, 205–230
    joint fMRI and DTI models, 495–514
    linear and nonlinear models, 309–327
    multiplicity corrections, 355–365
    multivariate decompositions, 399–411
    network analysis, 441–460
    neuroimage preprocessing, 263–295
statistical parametric maps (SPMs), *see
        also* Neuroimage preprocessing
    brain surface analysis, 234
    defined, 54
    multi-subject MRI, 51
    preprocessing, 117
    software, 37
    spatial registration, 124
statistical parametric networks (SPNs),
        455, 456–457
statistical principles of image normalization
        (SPIN), 127
steady-state evoked potentials (SSEP), 196
Stejskal-Tanner equation, 69
stereostatic coordinate system, 45
stochastic block models (SBMs), 454–455
storage, 113, 166–167
structural connectivity (SC), 389, 496
structural equation models (SEMs)
    effective connectivity, 421
    effective connectivity and causal
        inference, 422–427
    functional connectivity, 502
    Granger causality, 431
structural imaging modalities, 5
structural magnetic resonance imaging
        (sMRI)
    anatomical references, functional
        imaging, 52
    artifacts, 39–40
    background, 5–6, 8

brain tissue segmentation, 44–45
deformation-based morphometry, 47–48
edge set, 446
foreground from background
    segmentation, 43–44
glossary, 53–54
image acquisition, 37–40
image reconstruction, 38
imaging genetics, 53
introduction, 36–37
joint ICA, 512
morphometry, 46–50
MRI scanners, 142
multi-center studies, 52–53
multi-subject MRI, 40–51
network analysis, 442
other measures, 49–50
partial volume corrections, 29
physics, 38
registration, 40–43
segmentation, 43–45
sequences, 38–39
statistical analyses, 50–51
statistical parametric maps, 51
subcortical volumes, 46–47
surface-based measures, 48–49
surface-based registration, 41–43
templates and atlases, 45–46
tensor-based morphometry, 47–48
tumor detection, 51–52
volumetric registration, 41
voxel-based morphometry, 47
structural magnetic resonance imaging
        (sMRI), multisequence
acquisition, 115–116
background, 8
conclusions, 133
cross-sectional intensity analysis,
    131–133
data structure, 114–115
handling and storage, 113
images identification, 112–113
inhomegeneity correction, 117–119
intensity normalization, 126–129
interpolation, 119–121
introduction, 110–112
intuitive description, data structure
    problems, 114–115
lesion maping, 130–131
longitudinal intensity analysis, 131–133
pitfalls and mistakes, 114

preprocessing, 116–129
reconstruction, 115–116
skull stripping, 119
software, 118–119, 121, 123–129
spatial registration, 121–126
structural neuroimaging, 8–9
structured MLDs, 471–472
structure preservation, structured PCA, 405
Student's *t*-test and distribution, 311, 336
subcortical volumes, 46–47
subject attrition, 471
subject-specific networks, 456
SuBLIME algorithm, 133
summary network, 456
summary statistics ordinary least squares
(SS-OLS), 473, 474
sum of shared differences (SSD), 122
sum of squared errors (SSE)
high-dimensional data, 552, 553
HRF, the time domain, 315, 316
support vector machines (SVM), 283–284
surface area expansion factor, 49
surface-based measures, 48–49
surface-based registration, 41–43, 122
surface data smoothing, 243–248
surface flattening, 236
surface parameterization, 235–238
surface registration, 239–241
SurfStat package, 251
susceptibility artifacts, 152
SVD, *see* Singular value decomposition
SVM, *see* Support vector machines
switching linear dynamic system (SLDS),
589
switching VAR (SVAR) models
background, 588, 589–591
dynamic connectivity state estimation,
593–598
epileptic EEG, 593–598
overview, 588–589, 591
parameter estimation, 592–593
SMC method comparison, 578
stationary VAR model, 589–590
summary, 620–621
time-varying VAR model, 590–591
synchronization, 180–182
synchronization likelihood, 446

T
Talairach atlas and coordinates, 45, 421

T_1- and T_2-weighted images, 54, 146, *see
also* Structural MRI (sMRI)
target registration error (TRE), 122
task-based analysis, 280
task preprocessing optimization
datasets, 285–286
fixed pipelines, 289–292
independent test results, 292–293
individual pipelines, 289–292
individual subject's curves, 284–285
prediction metrics, 280–281, 282–284
preprocessing pipeline optimization,
285–292
reproducibility metrics, 280–282
steps selection, 286–289
Taylor techniques
DCM, 428
HRF, time domain, 315, 316
misalignment, complications from, 476
multi-subject analysis, 323
TBM, *see* Tensor-based morphometry
TE, *see* Transfer entropy
templates, multi-subject MRI, 45–46
temporal detrending (DET), 287
temporal limitations, 151–152
temporal resolution, 147–148
tensor-based morphometry (TBM), 47–48
tensor statistical analysis, 410–411
TEs, *see* Echo time (TEs)
test results, 292–293
Tetraortho scheme, 89
TFCE statistic, 347–350
thickness, cortical surface features, 241
3-D signals, 78
threshold-free cluster enhancement, 345
thresholding
analysis, 446–447
challenge of, 365
construction, 446–447
function, density-integrated topology,
458
Tikhonov regularization, 178, 318–319, 552
time-dependent harmonizable process, 613
time-domain approach
multichannel EEG, 544–545
single-channel EEG, 532–539
time series analysis, linear and nonlinear
models
background, 9
Balloon model, 324–325
bi-exponential nonlinear model, 326

frequency domain, 321–322
future directions, 327
hemodynamic response function, time
    domain, 312–321
HRF, 309–310, 321–322
methods comparison, 320–321
multi-subject analysis, 322–323
nonlinear models, 323–327
nonparametric models, 317–320
parametric models, 313–316
semi-parametric approaches, 323
single-level analysis, 310–312
time domain, 312–321
Volterra series models, 325, 326–327
time-varying connectivity, 375–376, 427
time-varying VAR (TV-VAR) model, 588,
    590–592
    dynamic connectivity states estimation,
        593–596
TMT, *see* Trail-Making Test
tomography, 5, 7
topology, descriptive measures, 447–450
total variation distance (TVD), 572–573
TPR, *see* True positive rates
TR, *see* Repetition time
trace metric, 82
tracer kinetic modeling, 20–24
tractography algorithms, 83–86
Trail-Making Test (TMT), 286, 289–290,
    292–293
transfer entropy (TE), 549
transitivity, clustering coefficient, 449
transverse relaxation, 143, 215
TRE, *see* Target registration error
"treatment causes death" statement, 433
true positive rates (TPR), 268
t-test, 25, 51, 343
Tucker decomposition, 411
Tukey filter, 223
tumor detection, 51–52
tutorial, *see* Multisequence structural MRI
TVD, *see* Total variation distance
TV-VAR, *see* Time-varying VAR model
two-dimensional Fourier transform, 210–214
type 1 error control, 346

U
unbiased NLM algorithm, 81
UNC-Chapel Hill, 91
uncertainty, 86–87, 89
undirected graphs, 372–373

University of North Carolina, 129
unstructured MLDs, 471
unweighted networks, 444

V
validation system, registration, 90
variability preservation, 405
variance-covariance matrix, 341–342
variance decomposition, 525–528
variation distance, 572–573
VARMA, *see* Vector autoregressive moving
    average
VAR model, *see* Vector autoregressive
    (VAR) model
VBM, *see* Voxel-based morphometry
vector autoregressive moving average
    (VARMA), 542, 544
vector autoregressive (VAR) model, *see also*
    Switching VAR models
    high-dimensional data, 551
    multichannel EEG, 544–545, 548, 550
    multivariate time series, 552–554
vertex set, 444–445
vertices, networks, 443
VMA models, 544–545
Volterra series model, 325, 326–327
volume conduction, 192
volumetric registration, 41, 43
voxel-based analysis, 90–91
voxel-based morphometry (VBM), 47
voxel-similarity-based registration, 122
voxelwise estimation methods, 79–80
v-space, 81

W
Wald statistic, 342, 478, 480
watershed segmentation, 44
wavelets, partial volume corrections, 29
weakly stationary time series, 194, 528
weighted graphs
    characteristic path length, 447–448
    comparisons, 459
    density-integrated topology, 458
weighted least square algorithm, 79–80
weighted minimum norm (WMN), 561
weighted networks, 444, 458–460
Welch periodograms
    cross-coherence/partial
        cross-coherence, 373
    multichannel EEG, 549
    multiple EEG traces, 539–540

"well separation," 425
Westfall-Young algorithms, 51, 346
white matter (WM)
    defined, 36, 54
    DWI and DTI, 503
    group analysis, 92
    image contrast, 143
    inhomogeneity correction, 118
    intensity normalization, 128
    longitudinal and cross sectional
        intensity analysis, 132
    MRI to fMRI, 148
    multi-subject structural MRI, 40
    rs-fMRI, 166
    sMRI, 39
    surface-based registration, 41
white noise spectrum, 536
Wiener filtering, 81

Wilcoxon signed-rank test, 481
wild bootstrapping methods, 87
Williams index, 275
"windowed sinc," 120, 121
within-subject model, 340
WMN, *see* Weighted minimum norm
World Wide Web, preferential attachment,
    453
Worsley random field approaches, 233, 362

X
X-rays, 4

Y
"yoking" images, 470

Z
zero filling, 223, 224, 226